Smart Polymers

Smart Polymers

Basics and Applications

Edited by
Asit Baran Samui

CRC Press
Taylor & Francis Group
Boca Raton London New York

CRC Press is an imprint of the
Taylor & Francis Group, an **informa** business

First edition published [2022]
by CRC Press
6000 Broken Sound Parkway NW, Suite 300, Boca Raton, FL 33487-2742

and by CRC Press
4 Park Square, Milton Park, Abingdon, Oxon, OX14 4RN

CRC Press is an imprint of Taylor & Francis Group, LLC

© 2022 Taylor & Francis Group, LLC

Library of Congress Cataloging-in-Publication Data
Names: Samui, Asit Baran, editor.
Title: Smart polymers : basics and applications / edited by Asit Baran Samui.
Other titles: Smart polymers (CRC Press)
Description: Seventh edition. | Boca Raton, FL : CRC Press, 2022. |
Includes bibliographical references and index.
Identifiers: LCCN 2021048743 | ISBN 9780367480776 (hbk) |
ISBN 9781032228877 (pbk) | ISBN 9781003037880 (ebk)
Subjects: LCSH: Smart materials. | Polymers.
Classification: LCC TA455.P58 S59 2022 |
DDC 620.1/92–dc23/eng/20220114
LC record available at https://lccn.loc.gov/2021048743

ISBN: 9780367480776 (hbk)
ISBN: 9781032228877 (pbk)
ISBN: 9781003037880 (ebk)

DOI: 10.1201/9781003037880

Typeset in Times
by Newgen Publishing UK

About the Editor

Asit Baran Samui completed his PhD from the University of Mumbai, India, in 1986. He has served as a materials scientist in the Naval Materials Research laboratory, DRDO, India. For the past few years, he has been engaged in teaching at various universities, including his current one, the Institute of Chemical Technology, India. He was engaged in multidisciplinary research activities in DRDO, resulting in the development of a range of products. The work included speciality polymers, polymers for thermal management and sub-PPT sensing, extensive studies on conducting polymers, nanoceramics, and smart polymers. He led a group of scientists in DRDO engaged in the development of smart materials. His current teaching areas are polymer science, nanotechnology, and smart polymers. At present, he is teaching on phase change materials, molecular imprinting polymers, polymer composites, and catalyst-based pyrolysis of waste polymer.

Dr Samui has written 120 publications in peer-reviewed journals, including several review papers and 9 Indian patents. There are about a dozen book chapters to his credit, including a handbook, and an encyclopedia, across multiple publishers. He has successfully guided a dozen PhD students and several MSc and MTech students. Regarding editing, he has been a guest editor for one issue of *Defence Science Journal*, India. He is also a member of the editorial advisory boards of several journals.

Contributors

Bikash Chandra Chakraborty
Former Outstanding Scientist
Naval Materials Research Laboratory (DRDO)
Ambernath, Maharashtra, India

Swaroop Gharde
Department of Metallurgical & Materials Engineering
Defence Institute of Advanced Technology
Pune, India

Pankaj E. Hande
Department of Chemistry
Indian Institute of Technology
Bombay, Mumbai, India

Ramanand Jagtap
Department of Polymer and Surface Engineering
Institute of Chemical Technology
Mumbai, India

Sarang Jamdade
Indian Coast Guard
Ministry of Defence
Government of India, India

Ravindra Kale
Institute of Chemical Technology
Mumbai, India

Vinayak Kamble
Department of Polymer and Surface Engineering
Institute of Chemical Technology
Mumbai, India

Balasubramanian Kandasubramanian
Defence Institute of Advanced Technology
Pune, India

Faisal Kholiya
CSIR-Central Salt & Marine Chemicals Research Institute
Bhavnagar, Gujarat, India

Prakash Mahanwar
Department of Polymer and Surface Engineering
Institute of Chemical Technology
Mumbai, India

Ramavatar Meena
Academy of Scientific and Innovative Research (AcSIR)
Ghaziabad, Uttar Pradesh, India
CSIR-Central Salt & Marine Chemicals Research Institute
Bhavnagar, Gujarat, India

Junaid Parkar
Department of Polymer and Surface Engineering
Institute of Chemical Technology
Mumbai, India

Manoranjan Patri
Naval Materials Research Laboratory
Ambernath, Maharashtra, India

Sushil S. Pawar
Department of Paint Technology
Naval Materials Research Laboratory
Ambernath, Maharashtra, India

Swati S. Rao
Naval Materials Research Laboratory
Ambernath, Maharashtra, India

Asit Baran Samui
Department of Polymer and Surface Engineering
Institute of Chemical Technology
Mumbai, India

A.K. Sidharth
Department of Polymer and Surface Engineering
Institute of Chemical Technology
Mumbai, India

Praveen Srinivasan
Naval Materials Research Laboratory
Ambernath, Maharashtra, India

Alips Srivastava
Naval Materials Research Laboratory
Ambernath, Maharashtra, India

Swati Sundararajan
Zuckerberg Institute of Water Research, Jacob Blaustein Institutes
    for Desert Research
Ben-Gurion University of the Negev
Beersheba, Israel

Kirti Thakur
Department of Metallurgical and Materials Engineering
Defence Institute of Advanced Technology
Pune, India

Preface

The search for newer and newer materials is prompted by the need to add extra attributes to material properties. While doing so, a new class of materials have evolved, which are called "smart materials." Research on smart polymers became popular due to their adaptability and ease of processing. Smart polymers are macromolecules capable of undergoing a rapid, reversible change from one phase to another, triggered by small shifts in the local environment, which may be a variation in temperature, pH, light (power or frequency), magnetic force, biological molecules, etc. Smart polymers comprise a large number of polymers with various potential applications with unique characteristics. These polymers are able to respond to stimuli in several ways by altering shape, solubility, wettability, color, conductivity, light-transmitting abilities, and surface characteristics. The most important factor is that these materials undergo fast changes on the application of stress and revert back to the initial state on removal of the stress and can function for a large number of cycles.

Temperature-responsive polymers undergo a fast change in color and/or dimension, such as in the case of hydrogels. Temperature-triggered drug delivery, diagnostics, tissue engineering, bio-separation, thermally switchable devices, sensing applications, etc. have enhanced functionalities due to the thermo-responsive nature of polymers. Photoactive polymers and polymers for optical recording properties respond to a specific wavelength of light, while they change dimension and phase respectively. Their photoactive nature offers the fast response required for photo-actuation, photo-switching, photo-cross-linking, controlled release, surface patterning, optical writing, and others. Shape memory polymers are able to hold the shape created under stress and revert back to their original form at certain temperature or certain wavelengths of light. Polymers with shape memory properties have been developed up to an advanced stage due to their immense potential in biotechnology and other smart applications. Self-healing polymers have been conceptualized to enhance the service life of the system by repairing the damage automatically without any external influence. Various strategies allow the fractured surface to heal and return to its original condition immediately after damage. Polymer hydrogels and polymers for drug delivery exhibit smart behavior by changing shape under the influence of specific pH or temperature. Both these stimuli have a variety of commercial applications, including drug encapsulation and drug delivery. Polymeric light emitting diodes have immense potential in color variation and mixing under an applied voltage. Molecularly imprinting polymers provide a new molecular cage based on size, shape, and secondary interaction to detect

liquids and gases at a very low level. Ion imprinting polymers, having similar characteristics, can detect and remove toxic metal ions, even from biological fluids. Optical sensing polymers also provide systems with a detection ability at trace level. As an example, polymers with a high level of conjugation can be used for sensing as the target molecules change the conjugation state, which is expressed in a definite change of fluorescence. Solid polymer electrolytes for fuel cells transport the hydronium ion through its membrane by hopping. Having low density, this has been studied thoroughly and has been developed via nanotechnology to improve the ion transporting properties, while maintaining stability in the cell environment. Hydrophobic and super-hydrophobic coatings maintain a contact angle higher than 90° and 150° respectively and protect the surface from the ingress of water. These surfaces are important to industry due to various industrial requirements and many products have been commercially exploited and many more are on their way. In the area of textiles and paints, smart behavior adds to both decorative and functional attributes. Vibration damping polymers are required to protect machinery from damage. Also, polymers protect workers from a noisy environment that causes psychological damage. The polymers with a specific glass transition temperature dampen the vibrations with a specific frequency. Thermal storage polymers, storing latent heat, are required as an alternative mode of energy storage as they can partially cater to the growing energy demand. New polymeric materials are being chemically synthesized or formulated in such a way that they sense specific environmental changes. Approaching the material from chemistry and an engineering perspective, the smart polymers presented comprise the basic understanding of the materials and their properties, together with the most reliable and current data available on their preparation, fabrication, and application methods, respectively.

This book will be of interest to a wide range of professionals, such as researchers working in the fields of materials sciences, thermal and photoresponsive polymers, drug delivery, optical data storage, trace level sensing, smart surfaces, self-healing polymers, sensing, paints, textiles, soft matter, and polymer science/engineering. Chemists, chemical engineers, mechanical engineers, and others involved in the polymer industry will also be interested in polymers which have vibration damping and thermal energy storage properties. As future technologies will depend on the smart behavior of materials and more, it is prudent to make the basic knowledge and vital information on important aspects of smart polymers available to students, researchers, and industry. The compilation of specific areas makes the book of interest to a wide range of readers.

1

Introduction to Smart Polymers

Asit Baran Samui
Institute of Chemical Technology, Mumbai, India

Abbreviations

UV	ultra-violet
LCP	liquid crystal polymer
CA	cinnamic acid
LCST	lower critical solution temperature
PNIPAM	poly(N-isopropylacrylamide)
SMP	shape memory polymer
SPH	super-porous hydrogel
DCPD	dicyclopentadiene
HDI	hexamethylene diisocyanate
GMA	glycidyl methacrylate
MGP	methyl-α-D-glucopyranoside
DBTDL	dibutyltin dilaurate
SHM	structural health monitoring
WVTR	water vapor transmission rate
TRPG	thermal-responsive polymeric hydrogel
PCM	phase change material
WORM	write once and read many times
CD-R	recordable CD
LED	light emitting diode
PLED	polymeric LED
NIR	near-infrared
OLED	organic light emitting diode
SWCNT	single-walled carbon nanotube
MIP	molecular-imprinted polymer
IIP	ion-imprinted polymer
PMP	pinacolyl methyl phosphonate
PEMFC	polymer electrolyte membrane fuel cell
SAC	silicone-acrylic copolymer
PANI	polyaniline
T_g	glass transition temperature
FRC	fiber-reinforced composites
CFRP	carbon fiber-reinforced polymer composites
MWCNT	multi-walled carbon nanotubes
CNT	carbon nanotube

The first report of smart materials was published in the 1800s, which was based on their magnetostrictive characteristics.[1] During the development of smart materials research, the term "smart" was disseminated widely, expanding from its base, and finally acquired the meaning of "the ability to change physico-chemical appearance, properties and structure under certain stress in an anticipated manner."

Chapter 2 deals with photoresponsive polymers. Photochromic spiropyrans (SP) can be tethered to the surfaces of single-walled carbon nanotubes through molecular self-assembly,[2] which are conjugated with conductive polymers,[3] to derive photosensitive devices. Both ultra-violet (UV) light and visible-light irradiation can make the device conductance swing between two distinct states either by reversible conformation-induced doping or proton transfer.[4]

Density functional theory calculations indicate that, in the closed form, the electric dipole moment of the molecule is 6.4 D, while in the open form, it is more than double, i.e., 13.9 D.[4] Substantial variations in the electrostatic environment are caused by such changes in the dipole moment. It is expected that photochromic SP, when employed as one component of the gate dielectric, can alter the dielectric capacitance of the latter during illumination. In effect, the device performance is modulated. The methodology and the resultant performance form the basis for new types of ultrasensitive devices for chemical and environmental sensing in a non-invasive manner. The results are expected to offer new opportunities for fabricating more rational designs of multifunctional molecular sensors and devices.

Triggering the release of ingredients from microcapsules by light irradiation is an application that has great interest in the agricultural and cosmetic fields. UV and visible light-sensitive microcapsules are thus desirable for such purposes. As a general strategy, microcapsules are made sensitive to light by incorporating light-sensitive polymers, functional dyes, and metal nanoparticles.[5,6] Nano-engineered polymeric capsules, modified by incorporating Bacteriorhodopsin into their shell, can be used for controlled release.[7] The molecules act as a light-driven proton pump, which in turn controls the pH change at which the pores are opened and thus the shell permeability is revised.

Liquid crystals, when combined with polymers, form attractive materials with optical properties (liquid crystal polymers, LCPs) and the responsiveness of liquid crystals, along with their mechanical properties and processing, is similar to that of polymers. These materials find important applications in displays,[8] in smart

DOI: 10.1201/9781003037880-1

1

devices, such as artificial muscles which can respond to electric fields,[9] heat,[10] and light.[11] Light, being a clean energy, can be manipulated conveniently and controlled precisely, and is considered most interesting and convenient among various stimulating sources. In order to realize photoresponsive smart polymers, the LCPs either have to be blended with or functionalized by photosensitive molecules such as azobenzenes, cinnamic acid (CA), cinnamylidene, bisbenzylidene, etc. Azobenzene is the most popular and conveniently used photosensitive molecule because of its fast response on exposure to an appropriate wavelength of light.[12] In fact, considerable research on using trans-cis isomerization as the photoactive trigger has been done, using azobenzene compounds.[13] Photoactivity adds interesting attributes to the photoactive liquid crystalline elastomer. Polymer networks can be aligned or randomized at will by varying the irradiation characteristics and similarly the phase type during polymerization.

Temperature-responsive or, in other words, thermo-responsive polymers are polymers which exhibit a drastic and discontinuous change of their physical properties with a change in temperature and thus fall in the category of stimuli-responsive materials.[14]

Among all stimuli-responsive polymers, thermo-responsive polymers probably constitute the largest class. Their unique and characteristic property of a lower critical solution temperature (LCST) has been intensely studied in recent decades.[15,16] The LCST is a fascinating phenomenon found for various polymer solutions and has significance for practical applications. Temperature-responding polymers, having a fine hydrophobic-hydrophilic balance in their structure, make the chains collapse or expand in response to the modified hydrophobic and hydrophilic interactions between the polymeric chains and the aqueous media during small temperature changes around the critical temperature.[17] The function of an enzymatic receptor can be modulated when this type of polymer is conjugated close to its active place.[18] The transition between the extended and coiled form of the molecule is able to make the receptor switch on and switch off.[19]

The combination of ionizable and hydrophobic functional groups in a copolymer will be sensitive to both temperature and pH.[20] Thermo-sensitive polymers are either combined with polyelectrolytes in the form of an interpenetrating polymer network[21] or completely new monomers are developed that respond simultaneously to both the stimuli.[22] Hydrophobic modified poly(N-isopropylacrylamide) (PNIPAM) polymer, that consists of short azobenzene segments repeatedly inserted within the main chain, forms flower micelles in cold water, which can be collapsed by changing the temperature according to the hydration/dehydration transition of the temperature-responsive PNIPAM.[23]

Shape memory materials are considered stimuli-responsive materials, as they can change their shape under the influence of an external stimulus. The subject is dealt with in Chapter 10. A common stimulus for changing shape is a change in temperature, which is called a thermally-induced shape memory effect. Shape memory behavior can be observed for a range of polymers with substantially different chemical compositions. The material, that is deformed into a temporary shape, returns to its original shape due to external environmental stimuli, such as chemicals, temperature, or pH, etc. A combination of the polymer structure,

polymer morphology, applied processing, and programming technology contributes to designing a shape memory polymer (SMP). SMPs possess two material phases, such as the glassy and the rubbery phases, which play a definite role in SMP functioning as, in the glassy phase, the material is rigid and cannot easily be deformed, whereas, in the rubbery phase, it has high deformability. The thermal transition for the fixation of the temporary shape can also be selected as the melting point. Strain-induced crystallization of the switching segment can be introduced by cooling the material which is strained above the temperature of transition.[24]

For minimally invasive surgery, degradable implants can be inserted into the human body in a compressed shape through a small incision, placed in the correct position, which regain their required shape after warming up to body temperature. After a preordained period, the implant is degraded, is absorbed in the body and, thus, additional surgery is not required to remove the implant.[24] Polymer gels can also undergo swelling or shrinkage in response to changes in the external conditions, which may be either pH, ionic strength, electric fields, or light. Hydrogels with hydrophobic, crystallizable side chains and cross-linked poly(vinyl alcohols) exhibit a one-way shape memory effect under the thermal trigger.[25] Shape memory polymers (SMPs) can be used in almost every area of daily life, such as self-repair of auto bodies, kitchen utensils, switches, sensors, intelligent packaging, etc. Industrial applications cover a wide range, such as sensors and actuators, in the electronic industry, in the textile industry, coating applications, sporting goods, biomedical devices and systems, etc.[26,27] Specifically, applications are observed in smart fabrics,[28] heat-shrinkable tubes for electronics or films for packaging,[29] self-deployable sun sails in spacecraft,[30] self-disassembling mobile phones,[31] intelligent medical devices,[32] or implants for minimally invasive surgery.[33] A family of hybrid inorganic-organic SMPs, comprising inorganic nanoparticles as net points and hundreds of polymer chains making the net point a junction, leads to significant increases in the elastic modulus without the typical loss of sharpness in the transition temperature and shape memory behavior.[34] This type of new design opens the way for the synthesis of multifunctional SMPs with tunable physical properties and transition temperatures. SMPs are likely to find more and more applications in the coming years because they are lighter, cheaper, and more flexible than metal-based alloys.

Hydrogels, a cross-linked water-swollen network, with a high degree of flexibility very similar to natural tissue, have exceptional promise due to the possibility of a wide range of applications.[35] This is discussed in Chapter 5. Hydrogels can be designed with controllable responses in the form of shrinkage or expansion during changes in the surrounding environmental conditions. The physical and chemical stimuli include temperature, electric or magnetic field, light, pressure, sound, pH, solvent composition, ionic strength, and molecular species.[36] In multilayer hydrogel systems, different proteins are encapsulated into each layer and control of the protein release mode and the rate are made possible by independently regulating the cross-linking density of each layer.[37] Considering the ample ways of influencing hydrogels, a large number of systems have been developed for applications in the area of hygienic products,[37] agriculture,[38] drug delivery systems,[39], sealing,[37] coal dewatering,[40] food additives,[41] pharmaceuticals,[42] biomedical applications,[43] tissue engineering and regenerative medicines,[44] diagnostics,[45] wound dressing,[46] separation of biomolecules or cells,[47] barrier materials to regulate biological adhesions,[48] and biosensors.[49]

Super-porous hydrogels (SPHs), a different category of water-absorbent polymer systems, have high mechanical strength and elastic properties. Further, SPH hybrids have many superior properties, such as, in the stretched condition, they are unbreakable. Their elastic and rubbery properties make SPH hybrids suit various applications requiring resilient gels. Together with the resilience, the elastic water-swollen SPH hybrids can resist various types of stresses, including tension, compression, bending, and twisting.[36]

If materials' properties can be fashioned to extend their lifetime, the environmental footprint can be reduced, which plays a crucial role in today's development planning of new technologies. A detailed discussion is incorporated in Chapter 6. The ability to self-repair has thus become an extremely important property of materials.[50,51] Self-healing materials have the built-in ability to repair damage without the need for any external involvement. Generally, cracks and other types of damage at a microscopic level affect the thermal, electrical, and acoustical properties of materials, and the cracks' growth can lead to the eventual failure of the material. In general, cracks are hard to detect at an early stage, which are normally detected by manual inspections, and then the repairs are carried out. On the other hand, self-healing materials prevent degradation through the repair mechanism of the micro-damage.[52] The majority of self-healing materials are applied to make "self-healing" coatings, including microencapsulation,[53, 54] the introduction of reversible physical bonds such as hydrogen bonding,[55] ionomers,[56] and chemical bonds.[57] Microencapsulation remains the commonest method, which uses, for example, microencapsulated dicyclopentadiene (DCPD) monomer and Grubbs' catalyst to self-heal epoxy polymer.[54,58] For corrosion protection, encapsulation of a number of materials, for example, hexamethylene diisocyanate (HDI),[59] glycidyl methacrylate (GMA),[60] epoxy resin,[61] linseed oil,[62] etc. has been performed. Studies have established that microencapsulation effectively protects the metal against corrosion and extends the service life of a coating.

Self-healing composites can be divided mainly into three groups: (1) capsule-based; (2) vascular; and (3) intrinsic self-healing materials.[63] A thermoplastic/thermosetting semi-interpenetrating network is known to possess a repeatable self-healing ability. A soluble linear polymer, such as thermoplastic poly (bisphenol-A-co-epichlorohydrin), can be combined with thermosetting epoxy resin to make a self-healing composite in which the thermoplastic takes an active part during healing.[64] Reversible breaking and reforming of dynamic covalent,[54,65,66] or non-covalent bonds[67] induced by temperature, electromagnetic radiation, and/or chemical surroundings (pH, ionic strength, concentrations, and redox reactions) can be of great use to material chemistry and a welcome addition to self-healing applications. Multiphase supramolecular thermoplastic elastomers that combine high modulus and toughness with spontaneous healing capability could be designed in the form of hydrogen-bonding brush polymers, which self-assemble into a hard-soft microphase-separated

system with a self-healing capability.[68] The self-healing can be realized in the presence of atmospheric carbon dioxide (CO_2) and water (H_2O).[69] Methyl-α-D-glucopyranoside (MGP), having four -OH groups, can be cross-linked by reactions with HDI and PEG in the presence of a dibutyltin dilaurate (DBTDL) catalyst to form polyurethane networks. When subjected to controllable mechanical damage, such networks exhibit self-repair ability in air under ambient conditions.

Self-health monitoring is another smart technology required to prevent the sudden failure of the structure. Structural health monitoring (SHM) is a damage identification strategy for aerospace, civil, and mechanical systems, which prevents catastrophic failure. Thus, the strategy makes the structure smart. Three different techniques for health monitoring and damage detection of composite structures are usually considered: (1) piezoelectric sensors and actuators like PVDF films bonded to composite laminate; (2) magnetostrictive material such as Terfenol-D in particulate form embedded in one of the layers of composite laminate; and (3) experimental modal analysis of composite laminate by a laser doppler scanning vibrometer to record the vibration signatures of the structure.[70] P3HT-based multilayer thin films can be used for photocurrent-based strain sensing, which, unlike traditional strain sensors, does not require an external power supply for continuous operation as it requires light by which it generates photocurrent.[71]

The response of the polymer to pH or salt level makes the hydrogel "smart." One important application of this property is drug delivery. The subject is introduced in Chapter 7. The hydrogel is called a carrier when loaded with medicine. As the swelling of the hydrogel increases, the chains of the network move further apart. As the chains move away from each other, the drug can diffuse more quickly through the hydrogel to the skin or other target areas. Drug delivery technologies are required to optimize the drug release profile, absorption, distribution, etc. to improve the product efficacy, safety, as well as the patient's comfort. Generally, the administration of the drug is done through the mouth, skin, nasal, vaginal, ocular, rectal, and inhalation routes. The current trend is for drug delivery to the targeted organ, drug release via external influence, etc. In fact, a drug delivery system should have the ability to respond to physiological conditions, sense the changes, and alter the drug-release profile accordingly.[72] Stimuli-responsive polymeric drug delivery is another field of controlled drug delivery. These systems closely follow the normal physiological process of the disease state where the amount of drug released is controlled, according to the physiological need.[73] Thermo-responsive hydrogels can be classified into two groups based on the origin of thermo-sensitivity in aqueous swelling: the first is based on polymer-water interactions and the second is the polymer-polymer interaction along with the polymer-water interaction.[74,75] Hyperthermia treatment of cancer patients is usually performed at 42°C, which indicates that the temperature-responsive delivery systems are required to have their LCST above body temperature.[76] As the electro-responsive hydrogels generally shrink or swell in an electric field, they can be applied in drug delivery systems, artificial muscle, or biomimetic actuators.[77] Magnetic nanoparticles may be embedded in hydrogels which can carry therapeutic agents that are released upon heating.[78] A magnetic field can be used to modulate the permeability of the polyelectrolyte microcapsules prepared by layer-by-layer self-assembly.[79] On application of external alternating magnetic fields, the embedded Co@Au nanoparticles undergo rotation that disturbs and distorts the capsule wall and drastically increases its permeability to macromolecules. One of the pioneering approaches of exploiting ultrasound in drug delivery involved directing the ultrasonic waves directly at the polymers or the hydrogel matrix.[80] Ultrasonically-regulated drug delivery can afford the repeated modulation ability of release rates of substances from an external position.[81] Light as stimulus can be beamed instantaneously in specific amounts with precision. This makes light-responsive hydrogels very attractive. UV or visible light-sensitive polymers, are equally viable, however, the visible-light-responsive polymers and hydrogels are preferred due to their safe, inexpensive, ready availability, and clean and easily manipulatable nature.[77] The pH-responsive polymers containing pendant acidic or basic groups undergo conformational shifts with changes in the environmental pH, leading to changes in the swelling behavior of the hydrogels. Changes in ionic strength affect the size of the ionizable groups containing polymeric micelles, polymer solubility, and the fluorescence-quenching kinetics of the chromophores bound to electrolytes respectively.[82] Glucose-responsive hydrogel systems can maintain a self-regulated insulin release in response to the concentration of glucose in the blood.[83] For the application of polymers in medicine, biotechnology, and nanotechnology, the protein is modified with polymers.[84] This strategy improves their stability, solubility, and biocompatibility. Further, this type of modification is also likely to modify the catalytic properties of the enzymes to make active and selective biocatalysts.[85] Biodegradable biopolymers can be electrospun into mats with a predetermined fiber arrangement and structural features, which are functionalized through secondary processing, to display specific biochemical characteristics.[86] Bacterial cellulose, produced from inexpensive sources, has attracted huge interest for various biotechnology applications in the environment, electrochemistry, food, and health industries.[87] Among various uses, the important ones may be defined as the removal of metal ions from wastewater, use as dietary fiber and packing material and as skin-burn, bone, and articular repair therapy, and as an artificial blood vessel in the area of health.

Continuous efforts are being made to develop degradable polymers for a large number of applications, covering biomedical, sensors and actuators, packaging, food science, photolithography, etc. A large number of degradable materials have been designed to exhibit tuneable degradation properties, which are controlled by an external stimulus, such as temperature, pH, enzymes, ultrasound, light, etc.[88] The advance of diverse research areas such as nonlinear optics, photolithography, photoprotecting chemical synthesis, as well as tissue engineering and biopatterning technology has inspired researchers to construct novel photolabile macromolecular assemblies, which encourage the development of photodegradable polymers.[89]

Smart textiles can be made by incorporating smart polymers or chemically integrating smart polymers into textiles to derive a unique addition to the inherent property of the latter. Smart textiles are elaborated in Chapter 8. The addition of smart functions can result in the development of luminescent textiles,[90] textile displays,[91] self-cleaning textiles,[92] temperature-regulated textiles,[93] and self-moving textiles.[94] Another class of textiles is where computing, digital components, and electronics are implanted to carry out functions such as determining the location of a person via GPS, monitoring of biometric data, such as heart rate, assessing the environmental factors, such as temperature and the presence of toxic gases, together with real-time feedback in the form of electrical stimuli, changing color. These are some of the textile-based technologies now available.[95,96] Along with sensors, energy-harvesting systems are also being developed to create a new form of wearable technology. Smart breathable cotton fabrics can be made by using a temperature-sensitive copolymer, such as poly (N-tert-butylacrylamide-ran-acrylamide: 27: 73) as a coating.[97] The coated fabrics show a temperature-responsive water vapor transmission rate (WVTR). The surface roughness of thermal-responsive polymeric hydrogels (TRPG) treated substrate can increase the hydrophobicity and hydrophilicity of the treated surfaces to a great extent.[98] At 40°C, the contact angle is 149:3±2:5°, whereas, it is 0° at 25°C. Photochromic materials are generally used in jacquard fabrics, embroideries, and prints in different garments for decoration. Microencapsulation of liquids and solids is an innovative micro packaging technology, which can make the textile smart by adding extra functionality to it.[99] Microcapsules may contain color-changing polymers, which have photochromic or thermochromic properties. The microcapsules may also contain phase change materials (PCMs), which can be locked into fibers by incorporating them in resin and coating the fabric with it. A flexible energy harvester made from polyvinylidene fluoride film attached to a curved substrate in a shell shape is used to harvest energy from human motion.[100] Mechanical energy is converted into electrical energy during the fast state transition of the shell structure. The fabric with embedded piezoelectric shell structures is integrated as an energy harvester in a wearable textile, which, when worn on the elbow joints and fingers, generates a high output power of 0.21 mW even during slow and irregular motion. The smart application of polymers in fabricating solar cells is realized by using lightweight woven polymer fibers in nanogenerators to create a smart fabric. This hybrid power textile, sized 4 cm x 5 cm, can charge a 2 mF commercial capacitor up to 2 V in 1 minute under ambient sunlight together with mechanical excitation, such as human motion and the wind blowing.[101] Continuous charging of an electronic watch, directly charging a cell phone, and driving water-splitting reactions can be performed by the textile. With the rapid development in the field of electronics, such as computer, multimedia, and network applications, the volume of data has increased enormously, which demands a comparable development in optics so that storage and transfer of mass information can achieve very high density and transfer rates.[102]

The polymeric materials for holographic data recording are discussed in Chapter 9. Data recording and reading are performed by optical means and are rigged to the extent that they are not vulnerable to corruption from magnetic and semi-conductor memory. Together with extraordinary high memory, the recording medium can be used for write once and read many times (WORM) and also for the write-erase-write protocol. The basic principle of writing is that the monochromatic laser beam, concentrated on a microspot of the recording medium, changes the physical or chemical properties of the spot. The change occurred allows the spot to be distinguished from the surrounding medium. Irreversible change of the incorporated dye in the recording layer occurs during laser irradiation when making a recordable CD (CD-R).[103] In a phase change optical disc, the laser beam-irradiated track portion undergoes crystalline melting, which quenches to an amorphous state after the laser is removed.[104] To erase the data, it is irradiated with an erasing laser so that the temperature rises above the crystallization temperature. In this condition, the recorded portion disappears to restores the crystalline phase. After the introduction of photochromism production by the two-photon excitation during the 1970s, it was demonstrated during the 1980s that it could lead to three-dimensional data storage.[105] Physical phenomena for data reading and recording have been developed, which were followed by the development of large numbers of chemical systems for the medium. In order to increase the storage capacity, it is possible to incorporate two or even more of these data layers, but the laser interacts with every layer that it passes through when reaching the target layer. Thus, noise is created in another layer. The 3D optical data storage methods overcome this problem by involving nonlinear data reading and writing methods. Practically, a laser is brought to a focus at a particular depth in the recording medium corresponding to a particular information layer. To write the next piece of information, the depth of the focus is changed.

Improvements in material sensitivity, resolution, diffraction efficiency, and material stability are observed via (1) two-photon absorption techniques to store and retrieve information in the volume of the material layer;[106] (2) incorporation of nanoparticles to reduce shrinkage of the recording medium due to polymerization;[107] (3) introduction of a thermally curable recording medium suitable for mass production;[108] (4) incorporation of chain transfer agents in the formulation to improve the high spatial frequency performance of the materials;[109] (5) incorporation of quantum dots in the medium to improve the refractive index modulation and fluorescence;[106,110] and (6) development of PMMA-based photopolymer materials to increase the thermal stability.[111]

A light-emitting diode (LED) is a two-lead semi-conductor light source, which functions by applying a suitable voltage to the leads. Electrons recombine with holes within the device, releasing energy in the form of photons. In a polymeric LED (PLED), polymer is incorporated between the leads. The band gap of the emissive polymer incorporated between the leads determines the emission color. In 1990, electroluminescence from semi-conducting polymer was exhibited by Burroughs and co-workers at the University of Cambridge.[112] Immediately after the discovery, the PLEDs' development progressed rapidly to gain in efficiency,[113] to be able to emit all colors from infrared to ultra-violet,[114] together with the efficient generation of white light,[115] and the display of transparent devices.[116] PLED, being flexible, can be made in the form of thin films or sheets,

which can cater to a large range of applications, such as television or computer screens that can be rolled up and carried. It can be used for military warfare as a camouflage material, as the application of a small voltage can change the color of military vehicle or the garment, making those indistinguishable from the surroundings. The poly (2,7-carbazole) derivatives reflect green light in their neutral state and in the oxidized state they exhibit transmissive brown.[117] These two colors are known to be compatible with adaptive camouflage for military vehicles and personnel. Products made by using PLED displays, are ultra-light, ultra-thin displays, with low power consumption and excellent readability. Further, spillage and environmental pollution are very low for this material. PLEDs are considered an appropriate material for developing products such as virtual reality headsets, a wide range of thin, lightweight, full-color, portable computing, communications and information-related products, and flexible displays. In fact, some materials like this are already on the market.

Multiple-layered LED is designed to achieve optimal charge injection, transport, and light emission.[118] Incorporation of nanoparticles into one efficient polymer, poly (2-methoxy 5-(2-ethylhexoxy)-1-4-phenylene-vinylene) (MEH-PPV) can improve the photovoltaic efficiency. In such cases, CdSe,[119] C60,[120] and TiO_2[121] nanoparticles are used to make the polymer/nanoparticle composite films, which exhibit an increase in current and luminance output by an order.. Efforts are being made to develop multicolor and white PLEDs for future technology. The performance of a multi-primary color display based on PLED/OLED (organic light emitting diode) can be analyzed by using the fitting curves of the characteristics of devices (i.e., current density, voltage, luminance).[122] Six primary colors together with the white color format are designed to form a seven-primary colors format that contributes to energy saving and an increase in power efficiency. There are also attempts to blend or design copolymers to make an emission white in color. OLED is being developed at a brisk pace in parallel.

The delocalization of electronic structures of conjugated polymers is associated with distinct optical and electrochemical properties.[123] The large extinction coefficients and high emission characteristics are conducive to sensing applications. Further, the binding or unbinding events are associated with substantial variation of the optical or electrochemical response. Compared to small molecules, a large change in signal occurs due to the intra- and inter-chain energy transfer from a large number of conjugated units.

Conjugated cationic fluorescent polymer thin films are found to be instantaneously quenched by volatile amines in the gas phase at low ppm concentrations.[124] A mini-array with various counter-anions could differentiate amines. It has been established that conjugated polymers can be used to detect explosives or various diseases like Alzheimer's, etc. and also actuators can be made, which can sense a change in the surrounding medium.

Unique electronic and optical properties of single-walled carbon nanotubes (SWCNTs) include a band-gap producing near-infrared (NIR) fluorescence emission via exciton recombination, which is highly sensitive to its local environment. Further, the penetration depth of the NIR emission of SWCNTs is very high for biological tissue.[125] Large Stokes shift, the absence of photobleaching and blinking of SWCNTs[126] led to the development of molecular sensors for biological systems.[127] As unmodified SWCNTs are insoluble in water, fluorescent sensors are designed by functionalizing SWCNTs non-covalently with amphiphilic heteropolymers, phospholipids, and polynucleic acids to produce highly stable suspensions of individual SWCNTs.[128]

Sensing of the metal ion or organic molecules at an ultra-low concentration is very necessary, considering their importance in a living system and also in products. Molecular imprinting/ion imprinting is an artificial method of creating selective recognition sites in a cross-linked polymer matrix and the imprinted polymer is called a molecular-imprinted polymer (MIP) or an ion-imprinted polymer (IIP). This subject is detailed in Chapter 12. In simple terms, the imprinting is akin to welding of various parts of a lock with the key inside during welding. After welding, the key is removed and the keyhole will allow only this particular key to enter and open the lock. Among many available methods for recognition/removal of toxic compounds/metal ions, MIP/IIP offers excellent selectivity toward the target molecules, in the practical relevant condition. Another important attribute is that the recognition is possible in the presence of similar molecules/metal ions due to their high selectivity. As an example, to avoid using the costly conventional drug desferrioxamine B, Fe(III)-imprinted poly(hydroxyethyl methacrylate-N-methacryloyl-(L)-glutamic acid (HEMA-MAGA) beads are employed for selective removal of iron from human plasma.[129] Surface ion-imprinted amine-functionalized silica gel sorbent is employed for the selective removal of arsenic (V) from aqueous solution.[130] Determination of zinc in natural water samples, rice, wheat, red beans, kidney beans, watermelon seeds, and chickpeas, etc. is done by using nanopore size Zn(II)-IIP.[131] The double template imprinting methodology is the combination of two efficient imprinting methods at different scales, such as molecular imprinting and micelle templating synthesis, which can be precisely controlled not only at adsorption sites but also at the pore structures.[132] By using acrylamide as a monomer to make DNT- and TNT-imprinted polymer material, it is possible to detect both by using a quartz crystal microbalance, while minimum sensitivity remains at 15.65 mM of TNT.[133] Pinacolyl methyl phosphonate (PMP), a chemical warfare agent, can be detected at low concentrations by using MIP.[134] An MIPs-based fiber-optic-based system can be woven into fabric or wallpaper, which can be placed at the entrance of buildings to detect threats being transported through those areas. Greater selectivity of m-nitrophenol over its isomer p-nitrophenol, and other similar compounds, phenol, and 3-chlorophenol ensures the wide range of applicability of MIPs for molecular recognition.[135]

As a clean alternative to non-renewable fossil fuel, fuel cells represent systems, which can produce clean and silent power and the product of cell reaction is water. Polymer electrolyte membrane fuel cells (PEMFCs) are well suited for vehicle application due to their low weight compared to other fuel cells. Solid polymer electrolytes are discussed in Chapter 14. The basic requirements of fuel cell membranes are high proton conductivity, low methanol/water permeability, good mechanical and thermal stability, and cost-effectiveness.[136] Perfluorosulfonic acid ionomer membranes (Nafion®) are the most effective polymer electrolyte membranes, as their unique ionic properties result from the incorporation of perfluorovinyl ether groups terminated with sulfonate groups onto a tetrafluoroethylene (Teflon) backbone, and they also have excellent thermal and mechanical stability.[137] Radiation-induced grafting to fabricate the membranes is a cost-effective method due to the use of inexpensive commercial materials and the methodology is based on established industrial processes.[138] The poly (arylene ether), having fluorine atoms in the backbone, exhibits improved swelling behavior and enhanced hydrophilic/hydrophobic phase separation.[139]

Among the proton-conducting fillers used in Nafion composite membranes, layered zirconium phosphate (α-ZrP), a Bronsted acid with the ability to donate protons, is the most investigated compound.[140,141] The water retention property of Nafion at elevated temperatures is enhanced due to the functioning of α-ZrP particles as nano tanks.

Various approaches have been followed to enhance the hydrolytic stability of sulfonated polyimides, such as the use of monomers with flexible linkages,[42] diamine monomers with high nucleophilicity, diamines with the sulfonic acid group in a side-chain,[143] trifunctional monomers with cross-linkable groups to initiate cross-linking,[144] forming semi-interpenetrating polymer networks with appropriate materials, etc.[145]

Mesoporous acid-free silica xerogels with an average pore size of 3.7 nm can be used to prepare 0.5 μm thin supported films, whose proton conductivity reaches 2.0×10^{-2} S cm^{-1} and is highly sensitive to RH.[146] The copolymers of poly(2,5-benzimidazole) (ABPBI) and poly[2,2'-(p-phenylene)-5,5'-bibenzimidazole] (pPBI) (AB-pPBI) perform the dual role of enhancing the stability of ABPBI and improve the processability of rigid pPBI. The acid-doped copolymer membranes have higher dimensional stability and mechanical properties than the acid-doped ABPBI membrane.[147] The addition of silicate-based nanomaterials to the polymer membrane is used to improve the mechanical, thermal, and barrier properties, together with a water uptake of the composite membranes, which results in superior performance at a higher temperature compared to that of the virgin membrane.[148] With increasing demand for current and future portable electronic items, the requirement of small, flexible lightweight power sources is constantly increasing. A highly flexible, ultra-light and thin PEMFC is fabricated by a combination of a thin flow-field plate with a laser-machined metal current collector that can be bent and designed in multiple shapes.[149] The 0.992-mm-thick air-breathing fuel cell demonstrated a total power of 508 mW and performance degradation of <10% after 200 repeated bending tests.

Hydrophic and super-hydrophobic polymer coatings are described in Chapter 11. Hydrophobic coating, having a contact angle of more than 90° and excellent water repellency, can be designed as smart, functional coatings that combine hydrophobicity with additional functionality for specific industrial, consumer, medical, or military applications. The smart coatings can be self-cleaning coatings with the ability to break down into less problematic byproducts, responsive polymer coatings which can change their surface topography in response to light, to vary wettability or surface friction,[150] and a smart corrosion detector capsule, which detects a corrosion signal and releases a healing agent to repair the damage. Many examples of super-hydrophobic behavior are observed in nature, such as a lotus leaf, which is designed with hierarchical surface projections (epicuticular wax crystals) that do not allow water droplets to spread on the surface.[151] They can be applied as transparent super-hydrophobic coatings for various uses, including architecture, vehicle windows, eyeglasses, and optical windows for electronics.[152]

Super-hydrophobic surfaces have the capacity to reduce drag in water as the result of a significant repulsion feature that limits the interaction of water at the interface due to the presence of a thin layer of air.[153] Drag reduction in super-hydrophobic materials has undergone more in-depth study in recent years.[151] In addition, super-hydrophobic coatings offer a means for the control of fluids as demonstrated in an experiment in which a water droplet is electrostatically manipulated on a super-hydrophobic surface.[154] Super-hydrophobic surfaces (water contact angle $\theta > 150°$) with a low contact angle hysteresis are considered for a number of industrial applications, such as reduction of ice and snow adhesion, self-cleaning of surfaces, textiles with stain resistance, surfaces having corrosion prevention, and foul release surfaces.[155] Super-liquid-repellent surfaces are fabricated via the one-step sol-gel coating technology, which contains colloidal silica nanoparticles and a fluoroalkyl silane coupling agent.[156] The films exhibiting contact angles of 150° and 120° for water and dodecane, respectively, have great application potential.

Wireless communicating smart textiles are fabricated by integrating polymer-glass-metal fiber composites of sub-1 mm dia, indistinguishable from the textile host, using conventional weaving processes.[96] Due to their water-repellent behavior, the minimal frequency shift is observed with the coating, while maintaining a high signal level.

A nonfluorinated super-hydrophobic–super-oleophilic polythiophene coating with reversibility to the super-hydrophilic-and-oleophobic surface has been developed by using electrochemical polymerization on a two-dimensional layered colloidal particle template.[157] Further, simultaneous reversible electrochromic and extreme wettability properties are witnessed by changing the applied voltage. It is expected that the coating will be used for dual applications in self-cleaning coatings, channeling of flow properties, controlled membrane separation, and regenerable surfaces, together with electrochromic behavior by varying the applied potential. Maintaining super-hydrophilicity and super-hydrophobicity on the same surface in two-dimensional micropatterns opens possibilities for a wide range of applications from the cell, droplet, and hydrogel microarrays for screening to surface tension confined microchannels for separation and diagnostic devices.[158] In the liquid-floating microgyroscope, the super-hydrophobic coating is applied to the rotor and stator's sliding surface to reduce resistance during its rotation.[159] The PTFE-PPS super-hydrophobic composite coating is prepared by spraying to achieve a maximum surface contact angle of 158° with the mass ratio of PTFE and PPS at 1:1.[160] The addition of silica nanoparticles further enhances the contact

angle. Paint composition containing an ethanolic suspension of perfluorosilane-coated titanium dioxide nanoparticles can be sprayed, dipped, or extruded onto both hard and soft materials to make a self-cleaning surface, even when dipped in oil.[161] The paint formulations are found to be suitable for use on clothes, paper, glass, and steel for a myriad of self-cleaning applications.

Fluorocarbon-free super-hydrophobic surfaces with water contact angles of ~155° can be created from low-cost and easily synthesized aluminum oxide nanoparticles functionalized carboxylic acids, having highly branched hydrocarbon (HC) chains.[162] These functionalized nanoparticles generate hydrophobic surfaces when applied to a variety of substrates. Sprayable fluorine-free waterborne super-hydrophobic paint (water contact angle: 158°) is prepared by combining waterborne silicone-acrylic copolymer (SAC) and silica sol.[163] The final stable cross-linked network structure in the composite exhibits outstanding mechanical durability, chemical and thermal stability, together with excellent self-cleaning ability in the air as well as in an oily environment, and high efficiency in oil-water separation. Multi-walled carbon nanotubes are incorporated into the photocurable resins to increase the surface roughness and mechanical strength.[164] The results exhibit that the behavior of hydrophilic material can macroscopically change to hydrophobic if the surface has the proper microstructured features. The adhesive force can be controlled in the range of 23–55 μN by varying the number of eggbeater arms and the application of the technology can extend to water droplet manipulation, a three-dimensional cell culture, microreactor, oil spill clean-up, and oil-water separation.

The smart way of protecting a metal surface from corrosion is to use a conducting polymer, which passivates the surface as well as protects it from corrosion. Corrosion protection of metal surfaces has been analyzed in Chapter 3. The protective activity of polyaniline, originating from the passivation of the metal surface, was stressed by DeBerry[165] and Wessling.[166,167] The total action of polyaniline (PANI) against corrosion was presented by Schauer et al.[168] Their study concluded that an oxide layer is formed as a barrier coat at the first stage. If there is any defect site in the coating, there may be an onset of corrosion at that point. The conductivity of PANI acts to prevent corrosion by separating the cathodic and anodic partial reactions. Finally, when PANI is dedoped to an emeraldine base, it acts to improve the barrier properties. The smartness is also imparted by tuning the coating's action toward releasing anti-corrosion agents on demand. If the surrounding area is corroding in nature and there is an onset of corrosion, the anti-corrosion agent is released. Thus, the active corrosion protection aims to release the active and repairing material within a short time after the coating matrix is broken and corrosion of the substrate is initiated. A new generation of anti-corrosion coatings that possess passive matrix functionality, as well as the ability to respond actively during changes in the local environment, can be developed.[169] The inhibition of inorganic ions can also be performed by the incorporation of exchangeable ions associated with cation- and anion-exchange solids.[170]

The onset of coordinated chain motion is observed around the glass transition temperature (T_g). If the vibration energy is applied to a polymer at the glass transition temperature, the chain motion facilitates the absorption of mechanical energy that is converted into heat. Analytical accounts of vibration damping by polymers is given in Chapter 13. The corresponding frequency to T_g also allows the vibrational energy to be absorbed similarly. Vibration damping materials are required to exhibit large viscous losses in response to deformation due to vibration. These losses are expressed as loss modulus. Damping of a structure can be accomplished by using either passive or active modes. In passive modes, the inherent ability of certain materials to absorb the vibration energy is used, whereas, in the active mode, sensors are used to sense and measure the vibration and activate the damping actions. The sensors can be designed by using piezoelectric devices.

For high areal density applications, the surface quality of the storage medium can adversely affect the accuracy of the reading device, the ability to store data, and the replication qualities of the substrate. Thus, vibration damping performance can be improved by integrating either high modulus or high damping monolithic vibration damping materials, comprising a polymer, such as a polycarbonate.[171] The vibration damping ability in concrete is required to minimize the hazard in bridges, buildings, and other civil infrastructure systems. In fiber-reinforced composites (FRC), the major role played by the fibers is to bridge across the cracked matrix, which increases the toughness of the composite by maintaining the energy absorption mechanisms, and the impact energy will be consumed by the debonding and pull-out processes of the fibers.[172] Significant changes in the vibration response have been observed with macro- and microscopic inclusions. If thin microstructures and channels are carved within these materials and are filled with a high-viscosity fluid, the response changes drastically. Constrained and unconstrained vibration-damping experiments accomplished on a replicated array of oil-filled microchannels, designed within a block made up of polydimethylsiloxane, exhibit a vibration suppression ability.[173] Increase in the fundamental frequency occurs due to change in the stiffness of the block and an increase in the damping ratio and the loss factor, originating from the development of a slip boundary condition between the oil and the microchannel walls, causing frictional dissipation of energy.

The study of vibration damping characteristic of nanocomposites and carbon fiber-reinforced polymer composites (CFRPs) containing multi-walled carbon nanotubes (MWCNTs) indicates the enhancement of the damping ratio of the hybrid composites with the addition of carbon nanotubes (CNTs).[174] Increasing dynamic loss modulus and the loss factor of the nanocomposites and the corresponding CFRPs is the result of sliding at the CNT–matrix interfaces. From a comparison of glass and glass hybrid composites in an active vibration control study, it is observed that the hybrid composite is comparatively more efficient than the glass reinforced composite.[175]

This chapter summarizes the contents of this book, while efforts are made to provide a wide overview of *smart polymers* and their interesting applications. The book aims to be a useful technical resource for chemists, physicists, chemical engineers,

mechanical engineers, and experts associated with the polymer industry dealing with various types of polymers for medical, automotive, aerospace, marine engineering, or defense requirements. Further, the students and researchers in polymer science will find it an interesting learning and application resource.

REFERENCES

1. K.A. Bogdanowicz. "Liquid crystalline polymers for smart applications." PhD dissertation. Tarragona, Spain, Universitat Rovira I Virgili (2015).
2. X. Guo, L. Huang, S. O'Brien, P. Kim, and C. Nuckolls. "Directing and sensing changes in molecular conformation on individual carbon nanotube field effect transistors." *J. Am. Chem. Soc.* 127(43) (2005): 15045–15047.
3. X. Guo, D. Zhang, G. Yu, M. Wan, J. Li, Y. Liu, and D. Zhu. "Reversible photoregulation of the electrical conductivity of spiropyran-doped polyaniline for information recording and nondestructive processing." *Adv. Mater.* 16(7) (2004): 636–640.
4. R. Schroeder, L.A. Majewski and M. Grell. "All-organic permanent memory transistor using an amorphous, spin-cast ferroelectric-like gate insulator." *Adv. Mater.* 22(30) (2010): 3282–3287.
5. M. Bedard, B.D.U. Geest, A.G. Skirtach, H. Mohwald, and G.B. Sukhorukov. "Polymeric microcapsules with light-responsive properties for encapsulation and release." *Adv. Colloid Interface Sci.* 158(1–2) (2010): 2–14.
6. A.P. Esser-Kahn, S.A. Odom, N.R. Sottos, S.R. White, and J.S. Moore. "Triggered release from polymer capsules." *Macromolecules* 44(14) (2011): 5539–5553.
7. S. Erokhina, L. Benassi, P. Bianchini, A. Diaspro, V. Erokhin, and M.P. Fontana. "Light-driven release from polymeric microcapsules functionalized with bacteriorhodopsin." *J. Am. Chem. Soc.* 131(28) (2009): 9800–9804.
8. R.A.M. Hikmet. "Electrically induced light scattering from anisotropic gels." *J. Appl. Phys.* 68(9) (1990): 4406–4412.
9. K. Urayama, S. Honda, and T. Takigawa. "Deformation coupled to director rotation in swollen nematic elastomers under electric fields." *Macromolecules* 39(5) (2006): 1943–1949.
10. Y. Yusuf, Y. Ono, Y. Sumisaki, P.E. Cladis, H.R. Brand, H. Finkelmann, and S. Kai. "Swelling dynamics of liquid crystal elastomers swollen with low molecular liquid crystals." *Phys. Rev. E* 69(2) (2004): 021710.
11. Y. Yu, M. Nakano, and T. Ikeda. "Photomechanics: Directed bending of a polymer film by light." *Nature* 425 (2003): 145–146.
12. F. Weigert. "Ein photochemisches Modell der Retina." *Pflügers Archiv* 190 (1921): 177–197.
13. H. Rau. *Photochemistry and Photophysics*, vol. II. Ed. J.F. Rabek, Boca Raton, FL: CRC Press, 1990.
14. A.S. Hoffman. "'Intelligent' polymers in medicine and biotechnology." *Artif. Organs.* 19(5) (1995): 458–467.
15. E.S. Gil and S.M. Hudson. "Stimuli-responsive polymers and their bioconjugates." *Prog. Polym. Sci.* 29(12) (2004): 1173–1222.
16. I. Dimitrov, B. Trzebicka, A.H.E. Muller, A. Dworak, and C.B. Tsvetanov. "Thermosensitive water-soluble copolymers with doubly responsive reversibly interacting entities." *Prog. Polym. Sci.* 32(11) (2007): 1275–1343.

17. M.R. Aguilar, C. Elvira, A. Gallardo, B. Vázquez, and J.S. Roman. "Smart polymers and their applications as biomaterials." In N. Ashammakhi, R. Reis, and E. Chiellini (Eds.), *Topics in Tissue Engineering*, vol. 3. Madrid: III Biomaterials, 2007.

18. Z. Ding, G. Chen and A.S. Hoffman. "Properties of polyNIPAAm-trypsin conjugates: Unusual properties of thermally sensitive oligomer–enzyme conjugates of poly(N-isopropylacrylamide)–trypsin." *J Biomed Mater Res. Part A* 39(3) (1998): 498–505.

19. P.S. Stayton, T. Shimoboji, C. Long, A. Chilkoti, G. Ghen, J.M. Harris, and A.S. Hoffman "Control of protein-ligand recognition using a stimuli-responsive polymer." *Nature* 378(6556) (1995): 472–474.

20. D. Kuckling, H-JP. Adler, K.F. Arndt, L. Ling, and W.D. Habicher. "Temperature and pH-dependent solubility of novel poly(N-isopropylacrylamide) copolymers." *Macromol. Chem. Phys.* 201(2) (2000): 273–280.

21. L. Verestiuc, C. Ivanov, E. Barbu, and J. Tsibouklis. "Dual-stimuli-responsive hydrogels based on poly(N-isopropyl acrylamide)/chitosan semi-interpenetrating networks." *Int. J. Pharm.* 269(1) (2004): 185–194.

22. N. Gonzalez, C. Elvira, and J. San Román. "Novel dual-stimuli-responsive polymers derived from ethyl pyrrolidine." *Macromolecules* 38(22) (2005): 9298–9303.

23. O. Boissiere, D. Han, L. Tremblay, and Y. Zhao. "Flower micelles of poly (N-isopropylacrylamide) with azobenzene moieties regularly inserted into the main chain." *Soft Matter* 7 (2011): 9410–9415.

24. A. Lendlein and S. Kelch. "Shape-memory polymers." *Angew. Chem. Int. Ed.* 41(12) (2002): 2034–2057.

25. M. Uchida, M. Kurosawa, and Y. Osada. "Swelling process and order-disorder transition of hydrogel containing hydrophobic ionizable groups." *Macromolecules* 28(13) (1995): 4583–4586.

26. Y. Zhu, J. Hu, L-Y. Yeung, Y. Liu, F. Ji, and K-W. Yeung. "Development of shape memory polyurethane fiber with complete shape recoverability." *Smart. Mater. Struct.* 15(5) (2006): 1385–1394.

27. U.S. D'hollander, G. Van Assche, B. Van Mele, and F.D. Prez Ugent. "Novel synthetic strategy toward shape memory polyurethanes with a well-defined switching temperature." *Polymer* 50(19) (2009): 4447–4454.

28. S. Mondal and J.L. Hu. "Temperature stimulating shape memory polyurethane for smart clothing." *Indian J. Fibre Textile Res.* 31(1) (2006): 66–71.

29. A. Charlesby. *Atomic Radiation and Polymers*. New York: Pergamon Press, 1960.

30. D. Campbell, M. Lake, M. Scherbarth, E. Nelson, and R. Six. "Elastic memory composite material: an enabling technology for future space structures." Paper presented at Structures, Structural Dynamics, and Materials Conference, Austin, Texas, 18–21 April 2005.

31. H. Hussein and D. Harrison. "Investigation into the use of engineering polymers as actuators to produce 'automatic disassembly' of electronic products." In T. Bhamra and B. Hon (Eds.), *Design and Manufacture for Sustainable Development 2004*. Weinheim: Wiley-VCH, 2004.

32. H.M. Wache, D.J. Tartakowska, A. Hentrich, and M.H. Wagner. "Development of a polymer stent with shape memory effect as a drug delivery system." *Journa. Mater. Sci.: Mater. Med.* 14(2) (2003): 109–112.

33. A. Lendlein and R. Langer. "Biodegradable, elastic shape-memory polymers for potential biomedical applications." *Science* 296(5573) (2002): 1673–1676.

34. P. Agarwal, M. Chopra, and L.A. Archer. "Nanoparticle netpoints for shape-memory polymers." *Angew. Chem.* 123(37) (2011): 8829–8832.

35. Y. Li, G. Huang, X. Zhang, B. Li, Y. Chen, T. Lu, T.J. Lu, and F. Xu. "Magnetic hydrogels and their potential biomedical applications." *Adv. Funct. Mater.* 23(6) (2013): 660–672.

36. M.A. Enas. "Hydrogel: Preparation, characterization, and applications: A review." *J. Adv. Res.* 6(2) (2015):105–121.

37. A. Singh, P.K. Sharma, V.K. Garg, and G. Garg. "Hydrogels: A review." *Int. J. Pharm. Sci. Rev. Res.* 4(2) (2010): 97–105.

38. A.K. Saxena. "Synthetic biodegradable hydrogel (Pleura Seal) sealant for sealing of lung tissue after thoracoscopic resection." *J. Thoracic Cardiovasc Surg.* 139(2) (2010): 496–497.

39. H. Mehrdad, A. Amir, and R. Pedram. "Hydrogel nanoparticles in drug delivery." *Adv. Drug Deliv. Rev.* 60(15) (2009): 1638–1649.

40. X. Sun, G. Zhang, Q. Shi, B. Tang, and Z.J. Wu. "Preparation and characterization of water-swellable natural rubbers." *J. Appl. Polym. Sci.* 86(13) (2002): 3212–3217.

41. X. Chen, B.D. Martin, T.K. Neubauer, R.J. Linhardt, J.S. Dordick, and D.G. Rethwisch. "Enzymatic and chemoenzymatic approaches to synthesis of sugar-based polymer and hydrogels." *Carbohydr. Polym.* 28(1) (1995): 15–21.

42. N. Kashyap, N. Kumar, and M.N.V. Ravikumar. "Hydrogels for pharmaceutical and biomedical applications." *Crit. Rev. Ther. Drug Carr. Syst.* 22(2) (2005): 107–149.

43. D.F. Stamatialis, B.J. Papenburg, M. Girone, S. Saiful, S.N.M. Bettahalli, S. Schmitmeier, and M. Wessling. "Medical applications of membranes: Drug delivery, artificial organs, and tissue engineering." *J. Membr. Sci.* 308(1–2) (2008): 1–34.

44. L. Zhang, K. Li, W. Xiao, L. Zheng, Y. Xiao, H. Fan, and X. Zhang. "Preparation of collagen-chondroitin sulfate–hyaluronic acid hybrid hydrogel scaffolds and cell compatibility in vitro." *Carbohydr. Polym.* 84(1) (2011):118–125.

45. L.H. Van der Linden, S. Herber, W. Olthuis. P. Bergveld. "Patterned dual pH-responsive core-shell hydrogels with controllable swelling kinetics and volume." *Analyst* 128(4) (2003): 325–331.

46. P. Sikareepaisan, U. Ruktanonchai, and P. Supaphol. "Preparation and characterization of asiaticoside-loaded alginate films and their potential for use as effectual wound dressings." *Carbohydr. Polym.* 83(4) (2011): 1457–1469.

47. F. Wang, Z. Li, M. Khan, K. Tamama, P. Kuppusamy, W.R. Wagner, C.K. Sen, and J. Guan. "Injectable, rapid gelling and highly flexible hydrogel composites as growth factor and cell carriers." *Acta Biomater.* 6(6) (2010): 1978–1991.

48. D. Roy, J.N. Cambre, and S.S. Brent. "Future perspectives and recent advances in stimuli-responsive materials." *Prog. Polym. Sci.* 35(12) (2010): 278–301.

49. P. Krsko, T.E. McCann., T-T. Thach, T.L. Laabs, H.M. Geller, and M.R. Libera. "Length-scale mediated adhesion and directed growth of neural cells by surface-patterned poly(ethylene glycol) hydrogels." *Biomaterials* 30(5) (2009): 721–729.

50. R.P. Wool. "Self-healing materials: A review." *Soft Matter* 4(3) (2008): 400–418.

51. Y. Yang and M.W. Urban. "Self-healing polymeric materials." *Chem. Soc. Rev* 42(17) (2013): 7446–7467.

52. S.K. Ghosh. *Self-Healing Materials: Fundamentals, Design Strategies, and Applications.* 1st edn. Weinheim: Wiley-VCH, 2008.

53. B. Aissa, D. Therriault, E. Haddad, and W. Jamroz. "Self-healing materials systems: Overview of major approaches and recent developed technologies." *Adv. Mater. Sci. Engg.* 2012 (2012): 1–17.

54. S.R.White, N. Sottos, P. Geubelle, J. Moore, M.R. Kessler, S. Sriram, E. Brown, and S. Viswanathan. "Autonomic healing of polymer composites." *Nature* 409 (2001): 794–797.

55. Y. Chen and Z. Guan. "Multivalent hydrogen bonding block copolymers self-assemble into strong and tough self-healing materials." *Chem. Commun.* 50(74) (2014): 10868–10870.

56. W.H. Binder. *Self-Healing Polymers: From Principles to Applications.* 1st edn. Weinheim: Wiley-VCH Verlag GmbH (2013), pp. 315–334.

57. Y.L. Liu and T.W. Chuo. "Self-healing polymers based on thermally reversible Diels–Alder chemistry." *Polym. Chem.* 4(7) (2013): 2194–2205.

58. M.R. Kessler, N.R. Sottos, and S.R. White. "Self-healing structural composite materials." *Composites Part A: Applied Science and Manufacturing* 34(8) (2003): 743–753.

59. M. Huang and J. Yang. "Facile microencapsulation of HDI for self-healing anticorrosion coatings." *J. Mater. Chem.* 21(30) (2011): 11123–11130.

60. L.M. Meng, Y.C. Yuan, M.Z. Rong, and M.Q. Zhang. "A dual mechanism single-component self-healing strategy for polymers." *J. Mater. Chem.* 20(29) (2010): 5969–6196.

61. H.H. Jin, C.L. Mangun, D.S. Stradley, J.S. Moore, N.R. Sottos, and S.R. White. "Self-healing thermoset using encapsulated epoxy-amine healing chemistry." *Polymer* 53(2) (2012): 581–587.

62. C. Suryanarayana, K.C. Rao, and D. Kumar. "Preparation and characterization of microcapsules containing linseed oil and its use in self-healing coatings." *Prog. Org. Coat.* 63(1) (2008): 72–78.

63. B.J. Blaiszik, S.L.B. Kramer, S.C. Olugebefola, J.S. Moore, N.R. Sottos, and S.R. White. "Self-healing polymers and composites." *Annu. Rev. Mater. Res.* 40 (2010): 179–211.

64. S.A. Hayes, F.R. Jones, K. Marshiya, and W. Zhang. "A self-healing thermosetting composite material." *Compos. Part A: Appl. Sci. Manuf.* 38(4) (2007): 1116–1120.

65. K. Imato, M. Nishihara, T. Kanehara, Y. Yamamoto, A. Takahara, and H. Otsuka. "Self-healing of chemical gels cross-linked by diarylbibenzofuranone-based trigger-free dynamic covalent bonds at room temperature." *Angew. Chem. Int. Ed.* 51(5) (2012): 1138–1142.

66. H. Ying, Y. Zhang, and J. Cheng. "Dynamic urea bond for the design of reversible and self-healing polymers." *Nat. Commun.* 5 (2014): 3218.

67. R.P. Sijbesma, F.H. Beijer, L. Brunsveld, B.J. Folmer, J.K. Hirschberg, R.F. Lange, J.K. Lowe, and E. Meijer. "Reversible polymers formed from self-complementary monomers using quadruple hydrogen bonding." *Science* 278 (1997): 1601–1604.

68. Y. Chen, A.M. Kushner, G.A. Williams, and Z. Guan. "Multiphase design of autonomic self-healing thermoplastic elastomers." *Nat. Chem.* 4 (2012): 467–472.

69. Y. Ying and W.U. Marek. "Self-repairable polyurethane networks by atmospheric carbon dioxide and water." *Angew. Chem. Int. Ed.* 53(45) (2014): 12142–12147.

70. H.D. Sarode and R.R. Wayakole. "Structural health monitoring of composites." *Int. Engg. Res. J. Spl. Issue* (2016): 423–427.

71. D. Ryu and K.J. Loh. "Strain sensing using photocurrent generated by photoactive P3HT-based nanocomposites." *Smart Mater. Struct.* 21(6) (2012): 065016.

72. P. Gupta, K. Vermani, and S. Garg. "Hydrogels: From controlled release to pH-responsive drug delivery." *Drug Discov. Today* 7(11) (2002): 569–579.

73. J. Kost and R. Langer. "Responsive polymeric delivery systems." *Adv. Drug Deliv. Rev.* 46(1–3) (2001): 125–148.

74. Y.H. Bae, T. Okano, and S.W. Kim. "Insulin permeation through sensitive hydrogels." *J. Control Rel.* 9(3) (1989): 271–279.

75. Y. Hirokawa and T. Tanaka. "Volume phase transition in a nonionic gel." *J. Chem. Phys.* 81(12) (1984): 6379–6380.

76. S. Chatterjee, K.I. Kwon, and K. Park. "Smart polymeric gels: Redefining the limits of biomedical devices." *Prog. Polym. Sci.* 32(8–9) (2007): 1083–1122.

77. Y. Qiu and K. Park. "Environment-sensitive hydrogels for drug delivery." *Adv. Drug Deliv. Rev.* 53 (2001): 321–339.

78. L.L. Lao and R.V. Ramanujan. "Magnetic and hydrogel composite materials for hyperthermia applications." *J. Mater. Sci. Mater. in Med.* 15(10) (2004): 1061–1064.

79. Z. Lu, M.D. Prouty, Z. Guo, V.O. Golub, C.S.S.R. Kumar, and Y.M. Lvov. "Magnetic switch of permeability for polyelectrolyte microcapsules embedded with Co@Au nanoparticles." *Langmuir* 21(5) (2005): 2042–2050.

80. S. Sershen and J. West. "Implantable polymeric systems for modulated drug delivery." *Adv. Drug. Deliv. Rev.* 54(9) (2002): 1225–1235.

81. K. Kost, K. Leong, and R. Langer. "Ultrasonically controlled polymeric drug delivery." *Macromol. Chem. Macromol. Symp.* 19(1) (1988): 275–285.

82. K. Szczubialka, M. Jankowska, and M. Nowakowska. "'Smart' polymeric nanospheres as new materials for possible biomedical applications." *J. Mater. Sci., Mater. Med.* 14(8) (2003): 699–703.

83. E.S. Gil and S.M. Hudson. "Stimuli-responsive polymers and their bioconjugates." *Prog. Polym. Sci.* 29(12) (2004): 1173–1222.

84. M.A. Gauthier and H.A. Klok. "Polymer-protein conjugates: An enzymatic activity perspective." *Polym. Chem.* 1(9) (2010): 1352–1373.

85. A. Díaz-Rodríguez and B.G. Davis. "Chemical modification in the creation of novel biocatalysts." *Curr. Opin. Chem. Biol.* 15(2) (2011): 211–219.

86. J. Venugopal and S. Ramakrishna. "Applications of polymer nanofibers in biomedicine and biotechnology." *Appl. Biochem. Biotechnol.* 125(3) (2005): 147–158.

87. M. Tiboni, A. Grzybowski, and J.D. Fontana. "Bacterial cellulose: A peculiar polymer for biotechnological applications." In J.M. Fontana, M.A. Tiboni, and M.A. Grzybowski (Eds.), *Cellulose and Other Naturally Occurring Polymers.* Kerala: Research Signpost, 2014, pp. 1–10.

88. L.S. Nair and C.T. Laurencin. "Biodegradable polymers as biomaterials" *J. Biomater. Sci. Polym. Ed.* 32(8–9) (2007): 762–798.

89. G. Pasparakis, T. Manouras, P. Argitis, and M. Vamvakaki. "Photodegradable polymers for biotechnological applications." *Macromol. Rapid Commun.* 33(3) (2012): 183–198.

90. S. Agrawal, L. Cincotta, E.D. Kingsley, and N. Lane. "Chromic luminescent compositions and textiles." US Patent no. US 20140103258 A1 (2014).

91. I. Sayed, J. Berzowska, and M. Skorobogatiy. "Jacquard-woven photonic bandgap fiber displays." *Res. J. Textile Apparel* 14(4) (2010): 97–105.

92. K. Qi, X. Chen, and Y. Liu. "Facile preparation of anatase/SiO_2 spherical nanocomposites and their application in self-cleaning textiles." *J. Mater. Chem.* 17 (2007): 3504–3508.

93. S. Mondal. "Phase change materials for smart textiles—an overview." *Appl. Therm. Eng.* 28(11–12) (2008): 1536–1550.

94. J. Hu, Y. Zhu, J. Lu, L.Y. Yeung, and K.W. Yeung. "Uniqueness of shape memory fibers in comparison with existing man-made fibers." Paper presented at 9th Asian Textile Conference, Federation of Asian Professional Textile Associations (Taiwan) (2007).

95. N.G. Zheng, Z.H. Wu, and M. Lin. "Infrastructure and reliability analysis of electric networks for E-textiles." *IEEE Trans. Syst. Man Cybernetics* 40(1) (2010): 36–51.

96. J.F. Gu, S. Gorgutsa, and M. Skorobogatiy. "Soft capacitor fibers for electronic textiles." *Appl. Phys. Lett.* 97(13) (2010): 133305.

97. N.S. Save, M. Jassal, and A.K. Agrawal. "Smart breathable fabric." *JIT*. 34(3) (2005): 139–155.

98. J. Hu, H. Meng, G. Li, and S.I. Ibekwe. "A review of stimuli-responsive polymers for smart textile applications." *Smart Mater. Struct.* 21(5) (2012): 053001.

99. P. Monllor, M.A. Bonet, and F. Cases. "Characterization of the behavior of flavor microcapsules in cotton fabrics." *Eur. Polym. J.* 43(6) (2007): 2481–2490.

100. B. Yang and K-S. Yun. "Piezoelectric shell structures as wearable energy harvesters for effective power generation at low-frequency movement." *Sens. Actuators A Phys.* 188(1) (2012): 427–433.

101. J. Chen, Y. Huang, N. Zhang, H. Zou, R. Liu, C. Tao, X. Fan, and Z.L. Wang. "Micro-cable structured textile for simultaneously harvesting solar and mechanical energy." *Nat. Energy* 1 (2016): 16138.

102. Y. Wang, Y. Wu, H. Wang, M. Huang, and Y. Wang. "Optical data storage." In G. Campardo, F. Tiziani, and M. Iaculo (Eds.), *Memory Mass Storage*. Berlin: Springer (2011).

103. H. Mustroph, M. Stollenwerk, and V. Bressau. "Current development in optical data storage with organic dye." *Angew. Chem. Int. Ed.* 45(13) (2006): 2016–2035.

104. T. Ohta. "Phase-change optical memory promotes the DVD optical disk." *J. Optoelectron. Adv. Mater.* 3(3) (2001): 609–626.

105. D.A. Parthenopoulos and P.M. Rentzepis. "Three-dimensional optical storage memory." *Science* 245(4920) (1989): 843–845.

106. X. Li, C. Bullen, J.W.M. Chon, R.A. Evans, and M. Gu. "Two-photon-induced three-dimensional optical data storage in CdS quantum-dot-doped photopolymer," *Appl. Phys. Lett.* 90(16) (2007): 161116.

107. N. Suzuki, Y. Tomita, and T. Kojima. "Holographic recording in TiO_2 nanoparticle-dispersed methacrylate photopolymer films." *Appl. Phys. Lett.* 81(22) (2002): 4121–4123.

108. M.R. Gleeson, J.T. Sheridan, F-K. Bruder, T. Rölle, H. Berneth, M-S. Weiser, and T. Fäcke. "Comparison of a new self-developing photopolymer with AA/PVA based photopolymer utilizing the NPDD model." *Opt. Expr.* 19(27) (2011): 26325–26342.

109. J. Guo, M.R. Gleeson, S. Liu, and. J.T. Sheridan. "Non-local spatial frequency response of photopolymer materials containing chain transfer agents: part I. Theoretical Modeling." *J. Optics* 13(9) (2011): 095601

110. X. Liu, Y. Tomita, J. Oshima, K. Chikama, K. Matsubara, T. Nakashima, and T. Kawailess. "Holographic assembly of semiconductor CdSe quantum dots in polymer for volume Bragg grating structures with diffraction efficiency near 100%." *Appl. Phys. Lett.* 95(26) (2009): 261109.

111. L.P. Krul, V. Matusevich, D. Hoff, R. Kowarschik, Y.I. Matusevich, G.V. Butovskaya, and E.A. Murashko. "Modified polymethylmethacrylate as a base for thermo-stable optical recording media." *Opt. Expr.* 15(14) (2007): 8543–8549.

112. J.H. Burroughs, D.C.C. Bradley, A.R. Brown, R.N. Marks, K. Mackay, R.H. Friend, P.L. Burns, and A.B. Holmes. "Light-emitting diodes based on conjugated polymers." *Nature* 347 (1990): 539–541.

113. H. Becker, H. Spreitzer, W. Kreuder, E. Kluge, H. Schenk, I. Parker, and Y. Cao. "Soluble PPVs with enhanced performance—a mechanistic approach." *Adv. Mater.* 12(1) (2000): 42–48.

114. H. Suzuki. "Temperature dependence of the electro-luminescent characteristics of light-emitting diodes made from poly(methylphenylsilane)." *Adv. Mater.* 8(8) (1996): 657–659.

115. B.W. D'Andrade, M.E. Thompson, and S.R. Forrest. "Controlling exciton diffusion in multilayer white phosphorescent organic light emitting devices." *Adv. Mater.* 14(2) (2002): 147–151.

116. V. Bulovic, G. Gu, P.E. Burrows P.E, S.R. Forrest, and M.E. Thompson. "Transparent light-emitting devices." *Nature* 380 (1996): 29.

117. S. Beaupre, A-C. Breton, J. Dumas, and M. Leclerc. "Multicolored electrochromic cells based on poly (2,7-carbazole) derivatives for adaptive camouflage." *Chem. Mater.* 21(8) (2009): 1504–1513.

118. D.Y. Kim, S.K. Lee, J.L Kim, J.K. Kim, H. Lee, H. N. Cho, S.I. Hong, and C.Y. Kim. "Improved efficiency of polymer LEDs using electron transporting layer." *Synth. Met.* 121(1–3) (2001): 1707–1708.

119. J.H. Park, S-I. Park, T-H. Kim, and O.O. Park. "Enhanced electroluminescence in emissive polymer/CdSe double-layer films." *Thin Solid Films* 515(5) (2007): 3085–3089.

120. K.W. Lee, K.H. Mo, J.W. Jang, N.K. Kim, W. Lee, I-M. Kim, and C.E. Lee. "Charge transport in low-concentration MEH-PPV conjugated polymer/fullerene composites." *Curr. Appl. Phys.* 9(6) (2009): 1315–1317.

121. S.A. Carter, J.C. Scott, and P.J. Brock. "Enhanced luminance in polymer composite light emitting devices." *Appl. Phys. Lett.* 71(9) (1997): 1145–1147.

122. Y. Xiong, F. Deng, S. Xu, and S. Gao. "Performance analysis of multi-primary color display based on OLEDs/PLEDs." *Opt. Commun.* 398 (2017): 49–55.

123. R. Sebastien and T.M. Swager. "Conjugated amplifying polymers for optical sensing applications." *ACS Appl. Mater. Interfac.* 5(11) (2013): 4488–4502.

124. S. Rochat and T.M. Swager. "Fluorescence sensing of amine vapors using a cationic conjugated polymer combined with various anions." *Angew. Chem. Int. Ed.* 53(37) (2014): 9792–9796.

125. M.J. O'Connell, S.M. Bachilo, C.B. Huffman, V.C. Moore, M.S. Strano, E.H. Haroz, K.L. Rialon, P.J. Boul, W.H. Noon, C. Kittrell, … and R.E. Smalley. "Band gap fluorescence from individual single-walled carbon nanotubes." *Science* 297(5581) (2002): 593–596.

126. D.A. Heller, S. Baik, T.E. Eurell, and M.S. Strano. "Single-walled carbon nanotube spectroscopy in live cells: Towards long-term labels and optical sensors." *Adv. Mat.* 17(23) (2005): 2793–2799.

127. S. Kruss, A.J. Hilmer, J. Zhang, N.F. Reuel, B. Mu, and M.S. Strano. "Carbon nanotubes as optical biomedical sensors." *Adv. Drug Del. Rev.* 65(15) (2013.): 1933–1950.

128. S.K. Samanta, M. Fritsch, U. Scherf, W. Gomulya, S.Z. Bisri, and M.A. Loi. "Conjugated polymer-assisted dispersion of single-wall carbon nanotubes: The power of polymer wrapping." *Acc. Chem. Res.* 47(8) (2014): 2446–2456.

129. H. Yavuz, R. Say, and A. Denizli. "Iron removal from human plasma based on molecular recognition using imprinted beads." *Mater. Sci. Eng: C* 25(4) (2005): 521–528.

130. H-T. Fan, X. Fan, J. Li, M. Guo, D. Zhang, F. Yan, and T. Sun. "Selective removal of arsenic(V) from aqueous solution using a surface-ion-imprinted amine-functionalized silica gel sorbent." *Ind. Eng. Chem. Res.* 51(14) (2012): 5216–5223.

131. F. Shakerian, S. Dadfarnia, A. Mohammad, and H. Shabani. "Synthesis and application of nanopore size ion-imprinted polymer for solid phase extraction and determination of zinc in different matrices." *Food Chem.* 134(1) (2012): 488–493.

132. S. Dai, M.C. Burleigh, Y.H. Ju, H.J. Gao, J.S. Lin, S.J. Pennycook, C.E. Barnes, and Z.L. Xue. "Hierarchically imprinted sorbents for the separation of metal ions." *J. Am. Chem. Soc.* 122(5) (2000): 992–993.

133. C. Tancharoen, W. Sukjee, C. Sangma, and T. Wangchareansak. "Molecularly imprinted polymer for explosive detection." Paper presented at Asian Conference on Defence Technology (ACDT 2015), Phetchaburi, Thailand, 23–25 April 2015.

134. J.W. Boyd, G.P. Cobb, G.E. Southard, and G.M. Murray. "Development of molecularly imprinted polymer sensors for chemical warfare agents." *J. H. Apl. Tech. Digest* 25(1) (2004): 44–49.

135. T. Kanai, C. Sanskriti, P. Vislawath, and A.B. Samui. "Acrylamide based molecularly imprinted polymer for detection of m-nitrophenol." *J. Nanosci. Nanotechnol.* 13(4) (2013): 3054–3061.

136. A. Smitha, S. Sridhar, and A.A. Khan. "Solid polymer electrolyte membranes for fuel cell applications—a review." *J. Membr. Sci.* 259(1–2) (2005): 10–26.

137. K.A. Mauritz and R.B. Moore. "State of understanding of nafion." *Chemical Reviews* 104(10) (2004): 4535–4585.

138. J. Park, T. Takayama, M. Asano, Y. Maekawa, and K. Kudo. "Graft-type polymer electrolyte membranes for fuel cells prepared through radiation-induced graft polymerization into alicyclic polybenzimidazoles." *Polymer* 54(17) (2013): 4570–4577.

139. D.S. Kim, G.P. Robertson, Y.S. Kim, and M.D. Guiver. "Copoly(arylene ether)s containing pendant sulfonic acid groups as proton exchange membranes." *Macromolecules* 42(4) (2009): 957–963.

140. B. Mecheri, A. D'Epifanio, E. Traversa, and S. Licoccia. "Effect of an ormosil-based filler on the physico-chemical and electrochemical properties of nafion membranes." *J. Power Sources* 169(2) (2007): 247–252.

141. M.A. Navarra, C. Abbat, F. Croce, and B. Scrosati. "Temperature-dependent performances of a fuel cell using a superacid zirconia-doped Nafion polymer electrolyte." *Fuel Cells* 9(3) (2009): 222–225.

142. Z.X. Hu, T. Ogou, M. Yoshino, O. Yamada, H. Kita, and K.I. Okamoto. "Direct methanol fuel cell performance of sulfonated polyimide membranes." *J. Power Sources* 194(2) (2009): 674–682.

143. X. Chen, P. Chen, and K. Okamoto. "Synthesis and properties of novel side-chain-type sulfonated polyimides." *Polym. Bull.* 63(1) (2009): 1–14.

144. E. Higuchi, N. Asano, K. Miyatake, H. Uchida, and M. Watanabe. "Distribution profile of water and suppression of methanol crossover in sulfonated polyimide electrolyte membrane for direct methanol fuel cells." *Electrochim. Acta* 52(16) (2007): 5272–5280.

145. C.H. Lee and Y.Z.J. Wang. "Synthesis and characterization of epoxy-based semi-interpenetrating polymer networks sulfonated polyimides proton-exchange membranes for direct methanol fuel cell applications." *Polym. Sci., Part A: Polym. Chem,* 46(6) (2008): 2262–2276.

146. M.T. Colomer, F. Rubio, and J.R. Jurado. "Transport properties of fast proton conducting mesoporous silica xerogels." *J. Power Sources* 167(1) (2007): 53–57.

147. S.K. Kim, T.H. Kim, J.W. Jung, and J.C. Lee. "Copolymers of poly(2,5-benzimidazole) and poly[2,2'-(p-phenylene)-5,5'-bibenzimidazole] for high-temperature fuel cell applications." *Macromol. Mater. Eng.* 293(11) (2008): 914–921.

148. A.K. Mishra, S. Bose, T. Kuila, N.H. Kim, and J.H. Lee. "Silicate-based polymer-nanocomposite membranes for polymer electrolyte membrane fuel cells." *Prog. Polym. Sci.* 37(6) (2012): 842–869.

149. T. Park, Y.S. Kang, S. Jang, S.W. Cha, M. Choi, and S.J. Yoo. "A rollable ultra-light polymer electrolyte membrane fuel cell." *NPG Asia Materials* 9(5) (2017): e384.

150. J. Stumpel, D. Broer, and A. Schenning. "Stimuli-responsive photonic polymer coatings." *Chem. Commun* 50(100) (2014): 15839–15848.

151. G.D. Bixler and B. Bhushan. "Bioinspired rice leaf and butterfly wing surface structures combining shark skin and lotus effects." *Soft Matter* 8(44) (2012): 11271–11284.

152. X. Zhang, F. Shi, J. Nia, Y. Jiang, and Z. Wang. "Superhydrophobic surfaces: from structural control to functional applications." *J. Mater. Chem.* 18(6) (2008): 621–633.

153. R. Truesdell, A. Mammoli, P. Vorobieff, F.V. Swol, and C.J. Brinker. "Drag reduction on a patterned superhydrophobic surface." *Phys. Rev. Lett.* 97(4) (2006): 044504.

154. M. Washizu. "Electrostatic actuation of liquid droplets for microreactor applications." *IEEE Trans. Ind. Appl.* 34(4) (1998): 732–737.

155. Z. Guo, W. Liu, and B-L. Su. "Superhydrophobic surfaces: From natural to biomimetic to functional." *J. Colloid Interface Sci.* 353(2) (2010): 335–355.

156. M. Hikita, K. Tanaka, T. Nakamura, T. Kajiyama, and A. Takahara. "Super-liquid-repellent surfaces prepared by colloidal silica nanoparticles covered with fluoroalkyl groups." *Langmuir* 21(16) (2005): 7299–7302.

157. R.B. Pernites, R.R. Ponnapati, and R.C. Advincula. "Superhydrophobic–superoleophilic polythiophene films with tunable wetting and electrochromism." *Adv. Mater.* 23(28) (2011): 3207–3213.

158. E. Ueda and P.A. Levkin. "Emerging applications of superhydrophilic-superhydrophobic micropatterns." *Adv. Mater.* 25(9) (2013): 1234–1247.

159. M.A. Samaha, H.V. Tafreshi, and M. Gad-el-Hak. "Superhydrophobic surfaces: From the lotus leaf to the submarine." *Comptes Rendus Mécanique* 340(1–2) (2012): 18–34.
160. R. Weng, H. Zhang, and X. Liu. "Spray-coating process in preparing PTFE-PPS composite super-hydrophobic coating." *AIP Adv.* 4(3) (2014): 031327.
161. Y. Lu, S. Sathasivam, J. Song, C.R. Crick, C.J. Carmalt, and I.P. Parkin. "Robust self-cleaning surfaces that function when exposed to either air or oil." *Science* 347(6226) (2015): 1132–1135.
162. S. Alexander, J. Eastoe, A.M. Lord, F. Guittard, and A.R. Barron. "Branched hydrocarbon low surface energy materials for superhydrophobic nanoparticle derived surfaces." *ACS Appl. Mater. Interfaces* 8(1) (2016): 660–666.
163. H. Ye, L. Zhu, W. Li, H. Liu, and H. Chen. "Simple spray deposition of the water-based superhydrophobic coatings with high stability for flexible applications." *J. Mater. Chem. A* 5 (20) (2017): 9882–9890.
164. Y. Yang, X. Li, X. Zheng, Z. Chen, Q. Zhou, and Y. Chen. "3D-printed biomimetic super-hydrophobic structure for microdroplet manipulation and oil/water separation." *Adv. Mater.* 29 (48) (2017): 1704912.
165. D.W. DeBerry. "Modification of the electrochemical and corrosion behavior of stainless steels with an electroactive coating." *J. Electrochem. Soc.* 132(5) (1985): 1022–1026.
166. B. Wessling. "Passivation of metals by coating with polyaniline: Corrosion potential shift and morphological changes." *Adv. Mater.* 6(3) (1994): 226–228.
167. B. Wessling. "Corrosion prevention with an organic metal (polyaniline): Surface ennobling, passivation, corrosion test results." *Mater. Corros.* 47(8) (1996): 439–445.
168. T. Schauer, A. Joos, L. Dulog, and C.D. Eisenbach. "Protection of iron against corrosion with polyaniline primers." *Prog. Org. Coat.* 33(1) (1998): 20–27.
169. D.V. Andreeva and D.G. Shchukin. "Smart self-repairing protective coatings." *Materials Today* 11(10) (2008): 24–30.
170. R.B. Leggat, W. Zhang, R.G. Buchheit, and S.R. Taylor. "Performance of hydrotalcite conversion treatments on AA2024-T3 when used in a coating system." *Corrosion* 58(4) (2002): 322–328.
171. R. Hariharan, G.C. Davis, and S. Subramanian. "Vibration damping monolithic." US Patent no. US 6,441,123 B1 (2002).
172. A. Bentur and S. Mindess. *Fibre Reinforced Cementitious Composites.* London: Taylor & Francis, 2007.
173. R.K. Singh, R. Kant, S.S. Pandey, M. Asfer, B. Bhattacharya, P.K. Panigrahi, and S. Bhattacharya. "Passive vibration damping using polymer pads with microchannel arrays." *J. Microelectromech. Syst.* 22(3) (2013): 695–707.
174. S.U. Khan, C.Y. Li, N.A. Siddiqui, and J-K. Kim. "Vibration damping characteristics of carbon fiber-reinforced composites containing multi-walled carbon nanotubes." *Compos. Sci. Technol.* 71(12) (2011): 1486–1494.
175. P.S.S. Kumar, K. Karthik, and T. Raja. "Vibration damping characteristics of hybrid polymer matrix composite." *IJMME-IJENS* 15(1) (2015): 42–47.

2

Photoresponsive Polymers

Asit Baran Samui
Institute of Chemical Technology, Mumbai, India

CONTENTS

DOI: 10.1201/9781003037880-2

Abbreviations

LC	liquid crystalline
PEG	polyethylene glycol
ATRP	atom transfer radical polymerization
NMR	nuclear magnetic resonance
HECPE	hyperbranched liquid crystalline polyester epoxy
PLCP	photoactive liquid crystalline polymer
LECPE	linear liquid crystalline polyester epoxy
LCD	liquid crystal display
PDLCs	polymer-dispersed liquid crystals
PVA	polyvinyl alcohol
LDPE	low-density polyethylene
CLCPs	cross-linked LC polymers
LCBCs	liquid crystalline block copolymers
LCPs	liquid crystalline polymers
BCs	block copolymers
LCNs	liquid crystalline networks
IPNs	interpenetrating polymer networks
PDMS	polydimethyl siloxane
OBA	ortho nitrosobenzaldehyde
IOLs	intraocular lenses
SLCPs	supramolecular liquid crystalline polymers
PSS	poly(styrene sulfonate)
PAH	poly(allylamine hydrochloride)
LCPMs	liquid crystalline polymer microparticles
PS-P4VP	polystyrene-block-poly (4-vinyl pyridine)
5CB	4′-pentyl-4-cyanobiphenyl
SCNPs	polymer single-chain nanoparticles
PG	poly(l-glutamic acid)

PFO	perfluorooctane
SPNs	semiconducting polymer nanoparticles
PLCBC	photoresponsive liquid crystalline block copolymers
PEO	polyethylene oxide
Azo	azobenzene
CNT	carbon nanotube
FLC	ferroelectric liquid crystal
LCEs	liquid crystalline elastomers
T_g	glass transition temperature
VA	vertical alignment
PSVA	polymer-sustained vertical alignment
PBMA	polybutyl methacrylate
Paz	azo polymer
LbL	layer-by-layer
PEM	polymer electrolyte membrane
HOMO	higher occupied molecular orbital
LUMO	lower unoccupied molecular orbital
PE	polyethylene
SHM	structural health monitoring
CAD	computer aided design

2.1 Introduction

Photoresponsive polymers are polymers which respond to light irradiation via a change in chemical structure or physical properties. The responses vary with the wavelength, strength, and mode, etc. of light irradiation. Bending of polymer film, changes in the viscosity of polymeric solutions, changes in the electrical conductivity, or even color changes can be created by exposing the polymer to a suitable wavelength of light. Nanoporous liquid crystalline (LC) polymers, nanochannels, photorearrangement for solar energy harvesting, structural health monitoring, photoactuation, etc. are also realized by using photoresponsive polymers. The permeability of films to gases can also be effected by using photosmart material. Drug release is also aided by photoresponsive segments in polymer molecules. Various design possibilities of these polymers promote further research, which leads to more innovative application possibilities. Another interesting area developed over the years is photocleavable polymers. The polymer, with a photoresponsive segment chemically incorporated in the polymer, cleaves during exposure to light. Self-healing materials, photo-controlled adhesion, photoactuation, and energy harvesting have also been developed with these types of polymers.[1]

Apart from photoresponsive polymers, the application of irradiation is widespread in polymer synthesis and processing. Photoresponsive polymers can be designed directly from photoresponsive monomers or a photoresponsive catalyst. For clean and controlled application-oriented processing of polymers, photo-cross-linking is practiced. Therefore, photopolymerization will be discussed first, which will be followed by photo-cross-linking and the details of photoresponsive behavior.

2.1.1 Photopolymerization and Photo-Cross-Linking

Photopolymerization involves the synthesis of polymers and the cross-linking of polymers by irradiation at a suitable wavelength. The rapid cure rate, room temperature reaction, and solvent-free formulations of photopolymerization have clear advantages compared to thermal polymerization. By using these techniques, ultraviolet (UV) curing inks, adhesives, etc. can be formulated. Other areas of application of photopolymerization are processing of materials for electronics, optics, membranes, coatings, and modification of surfaces. The technique has been extended to the areas of preparation of biomaterials for tissue engineering,[2] biosensors,[3] dental restorations,[4] cell adhesion controlling surface modifications,[5] and drug delivery.[6]

Photoresponsive polymers are synthesized via a large number of synthetic routes. The most common is the copolymerization of photoresponsive monomers and other non-photoresponsive monomers or liquid crystalline monomers or grafting of the photoresponsive moiety onto reactive polymers. Following this, a large number of new methods have been developed with better properties or facile synthetic procedures. Photoresponsive polymers can be synthesized by using click chemistry-based reaction of a diyne derivative of a siloxane with an azido derivative of substituted aniline.[7] The resulting polymer exhibits trans-cis photoisomerization at a reasonable rate.

The direct preparation of photoresponsive film can be done by reacting azo acrylate as the monomer and azobenzene diacrylate as the cross-linker. Thermal polymerization is conducted in a polyamide-coated glass cell. The polyamide layer acts as an aligning layer for azo mesogens. The mixture of monomer, cross-linker, and a thermal initiator is heated and poured into the glass cell. After polymerization at an elevated temperature, the free-standing film is obtained.[8]

Polyethylene glycol (PEG) can be conjugated with the o-nitrobenzyl group, and then dichlorophenoxyacetic acid can be grafted in the self-assembly of photoresponsive micelles with a core-shell design.[9] Photoresponsive micelles with herbicide show controlled release behavior under radiation. The spiropyran-containing polymethacrylate end capped with silicone oligomers is synthesized via atom transfer radical polymerization (ATRP).[10] The polymer aggregates can be formed by reversible self-assembly because of physical cross-linking via stacking of charged merocyanines. Furthermore, the silicone moiety ensures the stability of the aggregates. It finds a use as a nanocarrier, especially for encapsulating and releasing polar substances in nonpolar solvents.

An ATRP solution, comprising a nitrobenzyl methacrylate and a pentamethyl diethyl tetramine and cuprous bromide mixture, when brought into contact with a silicon wafer coated with a silane initiator monolayer, forms a silicone surface-grafted photosensitive polymeric brush.[11] Under the influence of UV light, elimination of the o-nitrobenzyl units from the polymer side chains occurs, which imparts enhanced hydrophilicity to the surface that improves the wettability.

There are various techniques by which photo-cross-linking takes place. The two main techniques are:

1. photo-cross-linking via formation of cyclobutane rings;
2. cross-linking by dimerization of nitrenes.

2.1.1.1 Photo-Cross-Linking via Formation of Cyclobutane Rings

A large number of photo-cross-linkable polymers used for photoimaging respond to radiation by dimerization via $2\pi + 2\pi$ type cycloaddition (Figure 2.1). The double bond (π bond) can be part of an aliphatic chain, ring structures, such as poly(naphthyl vinyl acrylate), maleimide, coumarin, stilbene, cinnamic acid, chalcone, etc.[12]

2.1.1.2 Cross-Linking by Dimerization of Nitrenes

Nitrenes are formed via the decomposition of azide groups under UV irradiation. The dimerization of nitrenes with two unpaired electrons results in the formation of azo moiety, which is used as the technique for photo-cross-linking (Figure 2.2).

1,2,3-thiadiazoles are known to decompose under irradiation to diradicals.[13] The diradicals can dimerize by following various pathways. One of the pathways is furnished in (Figure 2.3).[14]

Aromatic dimethacrylates find application in various areas, from dentistry to microelectronics.[15] Having good photorefractive, optical, and charge-transporting properties, the carbazole-based

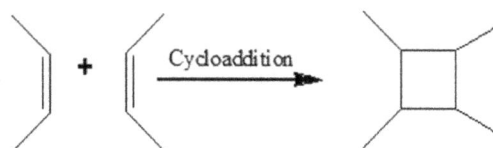

FIGURE 2.1 Schematics for dimerization via $2\pi + 2\pi$ type cycloaddition.

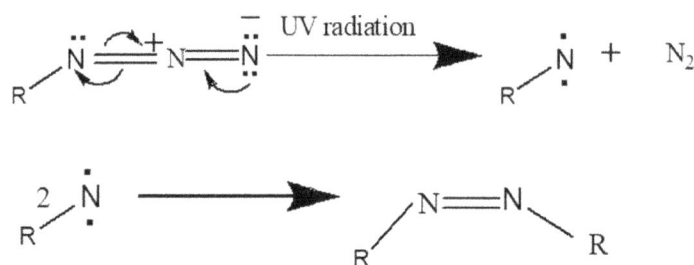

FIGURE 2.2   Schematics for decomposition of azide group and subsequent dimerization under UV radiation.

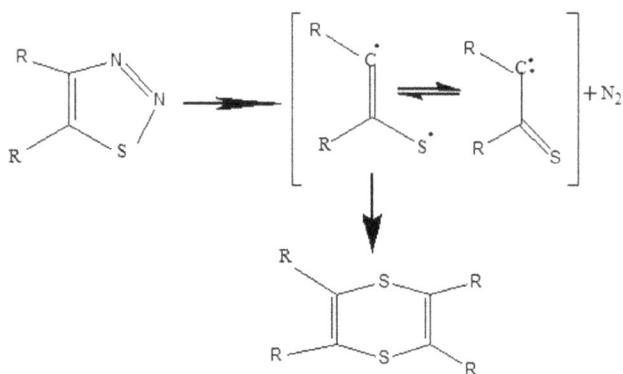

FIGURE 2.3   Schematics for decomposition of 1,2,3-thiadiazoles followed by dimerization under irradiation.

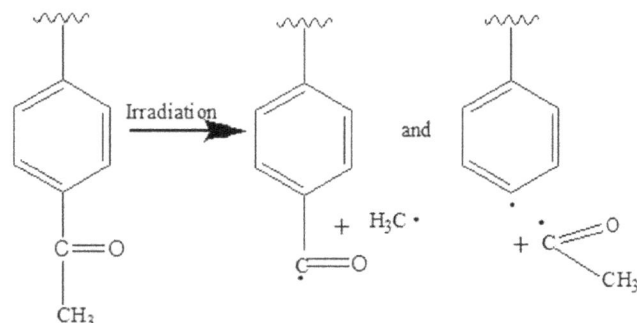

FIGURE 2.4   Schematics for cleavage of Poly(p-vinylacetophenone) under irradiation.

polymer can be used by cross-linking it by exposing it to light. Pendant ketone groups of a polymer can be cleaved by suitable irradiation via Norrish reaction. Poly(p-vinylacetophenone) can be cleaved by irradiating with UV radiation (>300 nm) (Figure 2.4). The radicals generated initiate cross-linking.

Polymers having a photo-cross-linkable alkyne group can be cross-linked photochemically with $W(CO)_6$ as a catalyst.[16] The tungsten pentacarbonyl complex is formed via the decomposition of $W(CO)_6$ under irradiation in the presence of alkyne, which rearranges to a vinylidene derivative that acts as an intermediate for the polymerization of alkynes. Metallopolymers, such as methacrylate side groups containing polyferrocenylsilane, undergo cross-linking with suitable radiation. A biological UV curable adhesive design is used to suit the requirement, such as fast curing in contact with living tissues, controlling the heat generation from polymerization and controlling the properties of the resultant material.[17] Photo-cross-linking has certain interesting features such as photoreversibility. Usually, the cross-linking is done at a wavelength above 300 nm and uncross-linked at a wavelength below 290 nm.

2.1.2   Change of Properties Due to Photo-Cross-Linking

Various changes occur due to photo-cross-linking. The UV-VIS spectroscopy can be employed to detect the rupture of the olefinic double bond followed by $2\pi + 2\pi$ cycloaddition and cross-linking. As an example, the increase of irradiation time for hyperbranched benzylidene polyester epoxies leads to the decrease of absorption intensity together with a decrease in the wavelength[18] (Figure 2.5).

The change in inherent viscosity with the irradiation of the polymer solution can be followed to distinguish between intermolecular and intramolecular cycloaddition. The intermolecular cycloaddition is confirmed by the increase in the inherent viscosity of hyperbranched polyester epoxies with irradiation[19] (Figure 2.6). The presence of substituents slows down the photocycloaddition reaction.

The ^{1}H NMR is also used to analyze the irradiated polymer solutions to detect the intermolecular photocycloaddition.[20] New peaks are detected around 2.7 (hyperbranched liquid crystalline polyester epoxy; HECPE) and 2.9 ppm in the ^{1}H NMR spectrum of benzylidene polyester (terminal epoxy and pendant hydroxyl groups) and the intensity increases with the duration of exposure (Figure 2.7).

The steric factor also affects the refractive index of the polymer film undergoing cross-linking via cyclobutane ring formation. Figure 2.8 shows the rapid decrease of the refractive index with time for hyperbranched polyethers with benzylidene units.[21] This decrease originates from a large change in the molecular polarizability that leads to substantial change in contribution from the bond refraction to the total molar refraction.[22]

2.1.3   Photoactive Additives for Cross-Linking Polymer Films

Polymer thin films are important due to their widespread use, particularly for the fabrication of sensors and micro-electronic, opto-electronic, and biomedical instruments and also for their

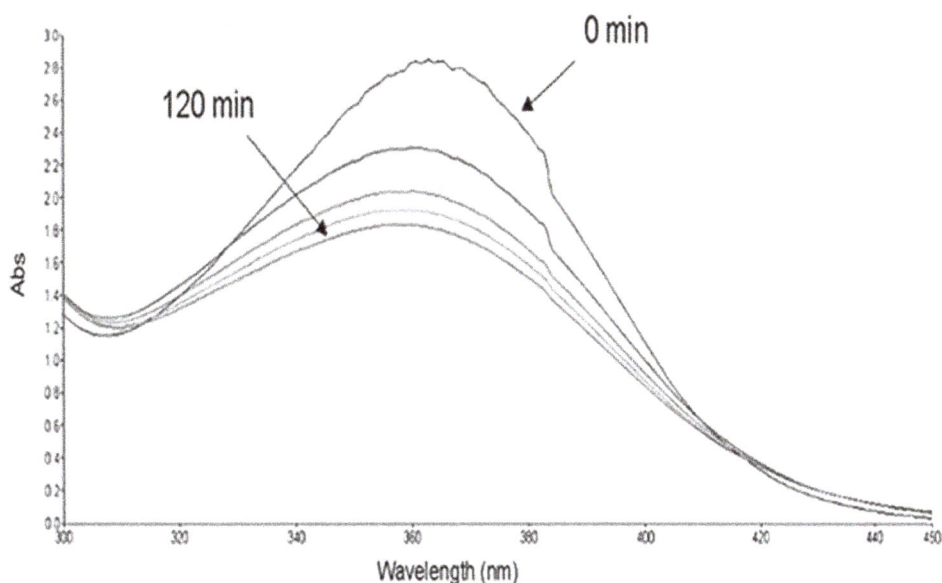

FIGURE 2.5 UV–vis spectra of HPE–T 32 film irradiated for different time intervals.

Source: [18]. Reprinted with permission from John Wiley & Sons.

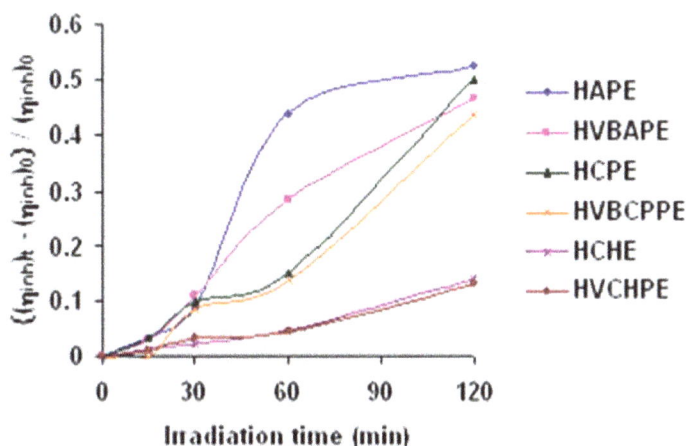

FIGURE 2.6 Effect of substituents, cyclic and acyclic units, and ring size on photoviscosity of polymer solutions.

Source: [19]. Reprinted with permission from John Wiley & Sons.

surface and mechanical properties.[23] In thin film technology, the challenge is the stabilization of a polymer film on a surface so that the continuous wetting of the polymer on the substrate is ensured. By adding various functional materials, the dewetting can be minimized. The direct and universal method employs cross-linking the polymers by incorporating a photoactive additive such as a bis-benzophenone derivative.[24] The strategy can be effectively applied for photo-induced formation of gels and grafting polymers to polymer surfaces.

2.1.4 Stimuli-Responsive Polymers

The design of stimuli-responsive molecules with other functional properties is one of the important goals for both researchers as well as industry. The structure of such switchable materials is chosen in such a way that they can be controlled by various stimuli. Stimuli-responsive polymers are the polymers whose behavior can be manipulated by varying the nature of their surrounding media.[25] By exposing the polymer to stimuli, such as temperature, pH, redox, ionic environment, radiation, etc., their physical/chemical properties can be varied. Light, as a stimulant, can be localized in time and space and can be manipulated from a remote location. Generally, the photochemical processes are self-sufficient without the need for additional reagents, and the by-products are mostly limited.[26] Photoresponsive polymers can be used both in solution as well as in the solid phase. A drug encapsulated into a photoresponsive hydrophobic micellar core of a block copolymer can be delivered by beaming light to a precise location.[27] The hydrophobic/hydrophilic balance is disturbed by light, which triggers the disturbance in the micelles, resulting in the release of the drug.

Surface properties can be modulated by incorporating photoresponsive polymers on the surface, as the polymers are able to shift their properties as required, which can be used to pattern the surface and govern the wettability of surfaces[28] and also for non-stick surfaces.[29] Thin film formation by photopolymers has also very important applications in the area of surface patterning and design of porous membranes. Photoresponsive groups are used to make smart polymer gels and hydrogels. Photoresponsive block copolymers can be designed that are able to form various nanostructures via irradiation.

Other important classes of photoresponsive materials can be designed by either using photochromic molecules for doping liquid crystals (LCs) or by chemically incorporating photoresponsive moiety into LC polymers. The self-organized superstructures formed are efficiently controlled by light. The light-driven information generated in photoresponsive LCs can be transmitted, and this property is used in various photonic technologies.[30]

2.1.5 Commonly Used Chromophores in Photoactive Liquid Crystalline Polymers (PLCP)

When irradiated with light of an appropriate wavelength, a photochromic compound is able to switch from one isomeric form to another with a different absorption spectrum and different physicochemical properties, including redox potential, dielectric constant, and so on. These molecular property changes can be employed in various photonic devices, such as data storage and optical switch, 2D/3D actuators. Some chromophores are given in Figure 2.9.

The photoactive molecules can be incorporated into polymer architecture for various functions. Among the various approaches, the first involves physical mixing of photochromic dopant with liquid crystal polymer as guest and host respectively. In this case, the properties of the guest component vary with light without any changes in the matrix itself. The second depends on the linking of both mesogenic and photoactive moieties via covalent bonds in a polymer. In the third category, the non-covalent interactions, such as hydrogen or ionic bonding, are used to link the photochromic moiety to the LC polymer chain.

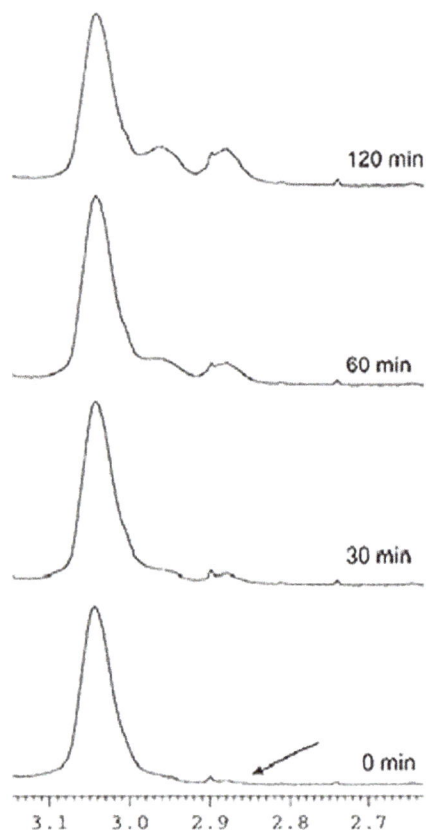

FIGURE 2.7 1H NMR spectrum of LECPE in DMSO-d6 during photolysis (0–120 min).

Source: [20]. Reprinted with permission from John Wiley & Sons.

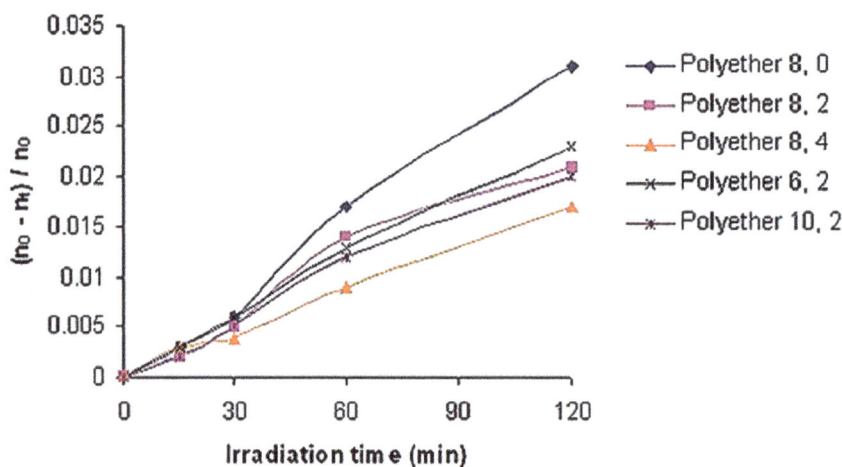

FIGURE 2.8 Effect of substitution and spacer length on the refractive index of polymer films.

Source: [21]. Reprinted with permission from John Wiley & Sons.

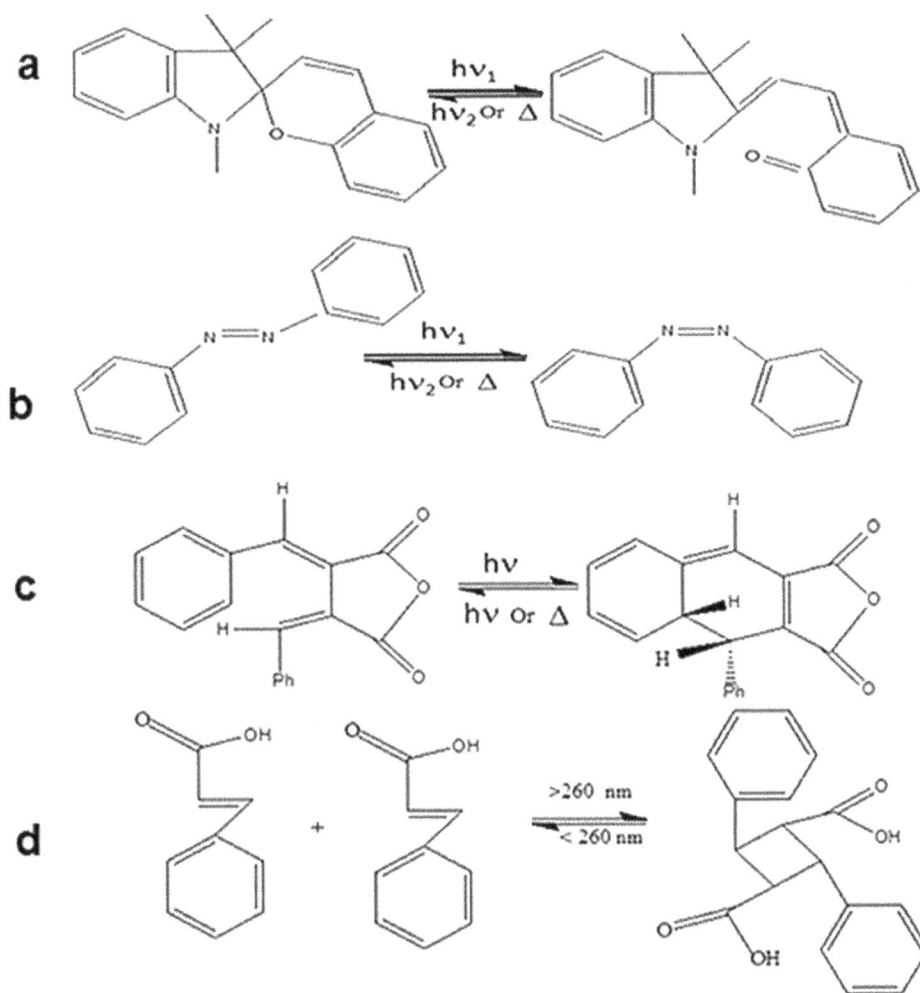

FIGURE 2.9 Common chromophores and their photoactivity: (a) photoisomerization of fulgides via ring closure and back; (b) photoisomerization of azobenzene from trans to cis and back; (c) photoactivity of spiropyrans via ring opening and back; (d) photodimerization of cinnamic acid derivative via cyclobutane ring formation.

2.2 Photoactive Liquid Crystalline Polymers

These polymers are known to have film/fiber/coating-forming properties together with their ability to be tailor-made, which make them a very important material in various inorganic and organic photoresponsive compounds. The orientation and morphology of the polymers control their optical and electrical properties.[31]

2.2.1 LCP/Chromophore Blend

Azobenzene-containing polymers receive the most attention of the chromophores due to their easy synthesis and exceptional optical responses. In the guest-host system, rod-like trans-azobenzene molecules fit well in the LC orientation. With irradiation by a linearly polarized light of an appropriate wavelength, the azobenzene units can be converted to the *cis* form that results in the destabilization of the LC orientation. In fact, the LCD is designed with the guest-host system that depends on the alignment of the LC and the reorientation of the dye guest on the application of electric voltage.[32]

2.2.2 Photoactive Liquid Crystalline Polymer Blends and Composites

Polymer-dispersed liquid crystals (PDLCs), having many commercial applications, are a dispersion of liquid crystal droplets in a solid polymer matrix. These tiny micron-sized droplets exhibit various exceptional behaviors. Under normal conditions, there is random orientation of a dispersed LC director in various LC droplets. However, under the influence of a strong external electric field, the PDLC film turns transparent because of the alignment of the LC director along the electric field. This behavior has been used to make electro-optical instruments.

2.2.3 Photoactive Polymer Dye/LC Blend

The azobenzene polymer networks stabilize liquid crystals as well as alter their orientation through irradiation with a suitable wavelength. It is implied that any azobenzene orientation change due to irradiation can change the director fields of the surrounding liquid crystal molecules, resulting in their changed orientation. To regulate the average liquid crystal orientation over macroscopic-length scales, this approach is of considerable importance.[33]

2.2.4 Photoactive LC Polymers (PLCP)

2.2.4.1 Molecular Design of Photoactive LC Polymers

One of the specific features of LC polymers is that both flexible and rigid (mesogenic) fragments are arranged in such a way that rod and disc-like structures are realized. The mesogens are placed in various ways, such as along the main chain, connected as a side chain, and as a structural unit in hyperbranched architectures. The important feature of PLCP is that the structure does not change during the transition from the LC state to the isotropic state. The LC structure is thus frozen in the glassy state, which is distinctly advantageous over the small LC molecules.

2.2.4.2 Main Chain Photoactive LC Polymers

Main chain photoactive polymers can be designed with various architectures; the most common is the attachment of chromophore and LC sideway via a covalent bond (Figure 2.10 (a)). Generally, the spacers are chemically connected between the LC and the chromophore so that alignment is facilitated. Another arrangement may be the LC in the main chain and the chromophore in the side chain.

2.2.4.3 Side Chain Photoactive LC Polymers

Both the LC and the chromophore can be linked as a side group of a polymer chain (Figure 2.10 (b)). This way, the mobility of the LC and the chromophore is isolated from the main chain. The presence of a spacer makes it easier for decoupled pendant groups to undergo orientation and reorientation under the action of light. This is a convenient arrangement to synthesize the acrylate monomer having pendant chromophore and LC groups respectively. The polymer is then formed by copolymerization of both the monomers using radical polymerization.

2.2.4.4 Hyperbranched LC Polymers

The presence of branch points or multiple numbers of end groups constitutes the branched polymers that have properties in between linear polymers and polymer networks (Figure 2.10 (c)).

The hyperbranched architecture is one of the branch polymers which can be designed to be amorphous as well as crystalline in nature. Multiple numbers of end groups with different natures impart varied characteristics to the polymers. The end groups can be modified to make polymers with the desired properties. These polymers have additional processing advantages, such as low viscosity. The free radical, condensation, ring opening polymerization and others can be used to synthesize such polymers.

2.2.5 Effect of Trans-Cis Isomerization on the Matrix

If the combined matrix of LC and the most studied azo is irradiated with an appropriate radiation, there is isomerization in the azo, which converts from a linear rod to a bent cis form. This process destabilizes the LC orientation and the system goes from LC to an isotropic state. The system goes back to its original form slowly after withdrawing the radiation or when irradiated with another wavelength light or heated. The simplest mechanism is presented in Figure 2.11.

This is the basis of various applications in which mechanical motion is required. Molecular mobility can be converted to micro to macroscopic mobility.

The elaborate photo-induced bending principle can be shown schematically in Figure 2.12.

Under UV irradiation the rod-like trans-azobenzene moieties, fitting within the LC polymer arrangement, undergo isomerization

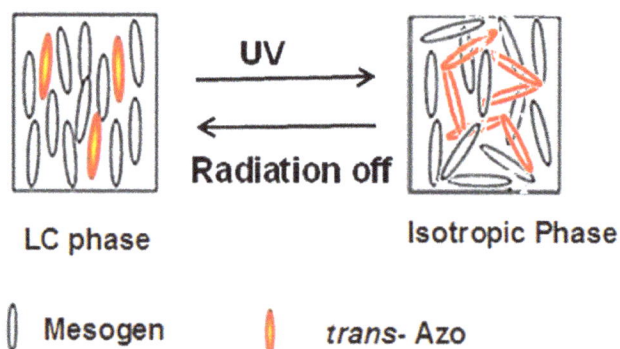

FIGURE 2.11 Pictorial representation of trans-cis isomerization under irradiation and resultant LC-Iso transition and back process.

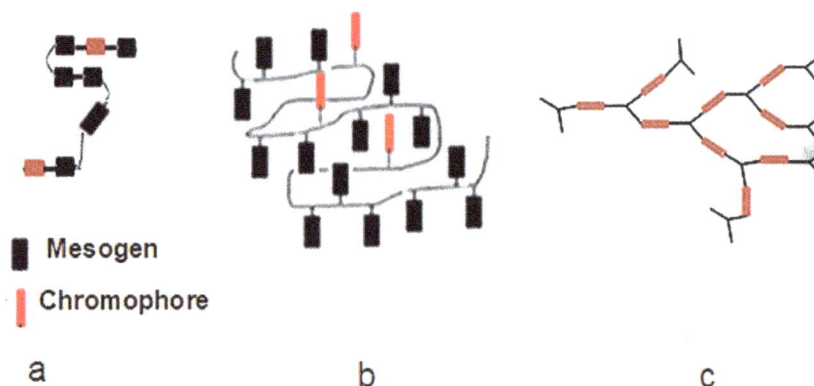

FIGURE 2.10 Photoactive liquid crystalline polymers (Representative architectures): (a) main chain; (b) side chain; (c) hyperbranched.

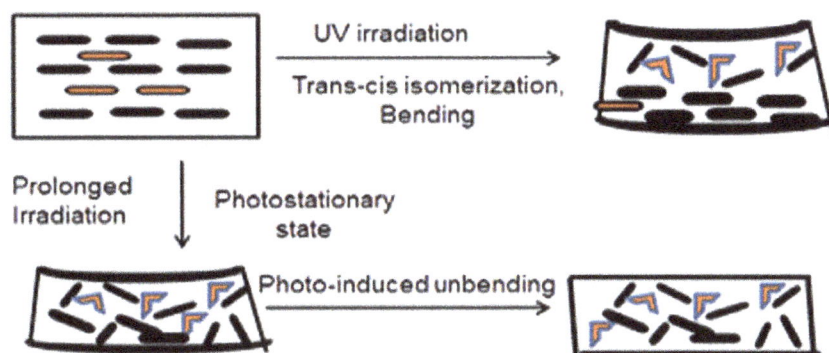

FIGURE 2.12 Schematic representation of the photo-induced bending and subsequent unbending of homogeneously aligned azobenzene-containing cross-linked LC polymers.

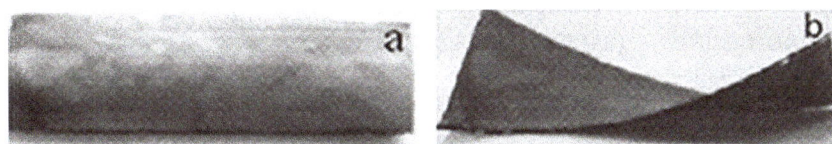

FIGURE 2.13 Photomechanical bending of azo-siloxane polymer under irradiation with UV light (a) before irradiation; (b) after irradiation.

Source: [7]. Reproduced with permission from John Wiley and Sons.

to the cis-form. This causes disruption of the homogeneous arrangement and strains the system. The incident light is absorbed by only a thin surface layer with high azobenzene concentration, and the resultant strain is asymmetric as only the surface layer creates strain. This type of strain forces the sample to bend toward the light source for in-plane-aligned mesogens. Continuous irradiation results in photo-induced unbending because the loss of the asymmetric strain after the photostationary state in the cis-azobenzene concentration is reached throughout the film.[34] Visible light can also result in unbending via back-isomerization of cis- to trans form. A single nanosecond pulse is enough to bend the sample, with corresponding response below one millisecond duration.

It can also be stated that the photo-induced bending direction is not dependent only on the initial chromophore alignment. The bending is also dictated by the polarization direction of the excitation beam.[35] Also, the bending direction of the polymer film is affected by the internal composition variation in the LC polymer matrix,[36] as well as the nature of bonding between the azobenzene and the polymer.[37] The photo-induced bending, resulting from the gradient in the LC alignment order caused by isomerization, is reversible. The out-of-plane twisting and helical motions of photoactuators are relatively recent and important developments. The development of complex motion, such as twisting and helical motion, has been accomplished by following complex systems produced by nature.[38] Many insects are found to produce various light-generating motions like bending, twisting, and sweeping.[39] Photoisomerization is associated with the generation of mechanical stress. This stress is the primary requirement for developing various molecular machines. The stress generated by small molecules being very weak, the azo moiety is incorporated into the polymer network.[40] By incorporating azobenzene into the liquid crystalline

elastomer, large mechanical deformations are possible.[41] However, only 2D motions are realized by this strategy. The strategy of using a bilayer film comprising photoactive LCE as one layer and non-photoactive polymer as the second layer function improves the photoresponse as well as the deformation characteristics. Siloxane polymers containing azo moieties making one layer and PVA the other layer show significant bending during irradiation at 365 nm with an irradiation intensity of 5.5–6.5 mW/cm[27] (Figure 2.13).

The photoactive layer can be designed also by direct casting azobenzene molecules on non-photoactive film, for example, unstretched LDPE.[42] This bilayer film has unique photoresponse characteristics, as it shows photo-induced bending to the photoactive side, irrespective of the radiation direction. This type of bilayer design with small photoactive molecules maintains the ability to convert the amphiphilic molecules to more hydrophilic forms and thus control is achieved from a remote location. A high photoresponse, quick recovery, and high mechanical properties can all be achieved.

2.2.6 Photoresponse of Cross-Linked Liquid Crystalline Polymers

Photomobile polymer materials have the ability to convert light energy directly into mechanical work without using any electrical or mechanical systems. However, the deformation in amorphous polymer materials occurring in an isotropic way produces too small an effect to be practically utilized. The deformation can be increased by using cross-linked LC polymers (CLCPs).[43] The initial alignment of mesogens, the cross-linking density and polarization direction of actinic light, act in controlling the deformation.[44] On irradiation with UV light, photochromic and mesogenic groups containing diarylethene CLCP films exhibit bending toward the irradiation source.[45] The bent films come back to the initial flat position by irradiating with visible light. The photocyclization/photocycloreversal of the diarylethene derivative occurs during exposure to UV/visible light.

2.2.7 Liquid Crystalline Block Copolymers

Liquid crystalline block copolymers (LCBCs) have generated interest among researchers as well as in industry due to their great potential for future applications. The self-organizing liquid crystalline polymers (LCPs) and the block copolymers (BCs) with the inherent property of microphase separation can be combined into one organic system of LCBCs to design a smart system that has the ability to be influenced by more than one driving force to form and control the self-organized nanostructure.[46] The microphase-separated nanostructures of LCBCs exhibit a pronounced effect on the performance of the LCs. Also, the diverse nanostructures formation is affected by the LC ordering behavior. The microphase separated amorphous BCs-derived nanostructures find applications as nanotemplating and nanofabrication materials.[47] The control of PLCBCs' nature by irradiation allows for the manipulation of supramolecularly self-assembled nanostructures in polymer films with designed characteristics. In azobenzene-containing LCPs, the azobenzene chromophore possesses the dual characteristics of the photoresponsive moiety and the mesogenic group. The refractive index can be modulated by regulating the polymer films by photo-induced phase transformation, molecular alignment, and cooperative molecular motion. Together with photoresponsive properties, it also influences the PLCBCs' bulk films toward microphase separation that attributes the assembled nanostructures with light-controllable performance.[48] Due to the immiscibility of one PLCP and the other non-LCP blocks in PLCBCs, the PLCP segment self-assembles into nanostructures. The microphase separation results in the confinement of mesogenic PLCBCs blocks into submicron or nanoscale domains at a much lower volume ratio than the non-LCP blocks. Thus, the confinement largely influences the LC orientational or positional order in such a narrow nanospace. Due to the nanoscale microphase separation, the scattering of visible light is avoided and the optical performance of block-copolymer films is improved, which is responsible for their application in optical devices and actuators.

2.2.8 Photoresponsive Soft Ionic Crystals

Ion pair formation between the counter-ions is responsible for the formation of various states such as crystals, LCs, and liquids, through numerous interactions.[49] Different functionalities can be attached to oppositely charged species so that the states and types of assembly of the ion pairs can be controlled to achieve the desired properties. By incorporating stimuli-responsive moieties in the ion pairs, the system can be controlled by applying an external stimulus. By incorporating azobenzene, various ion pairing assemblies can be realized due to different geometries and ratios of trans and cis forms. Photo-induced phase transitions via the isomerization of the azobenzene derivatives in the bulk phase are constrained because of the restriction of free volumes.[50] Ion

pair assemblies are formed due to non-covalent interactions, and the external conditions can be manipulated to achieve the required arrangement of the components. Azobenzene carboxylates as photo-responsive charged species can generate high crystalline assemblies and exhibit phase transitions.[51] Azobenzene anions are coupled with bulky cations (tetrabutylammonium ion) to form ion pairs so that free volumes are available near the azobenzene moieties. Because of the formation of loosely connected ion pairs by non-covalent interactions, the azobenzene unit can easily undergo isomerization.

2.2.9 Photoresponsive Liquid Crystalline Epoxy Networks with Dynamic Ester Bonds

The covalently cross-linked structures in commonly used siloxane and acrylate-based LC networks (LCNs) do not allow these to be reprocessed to make a new shape or repair the material. Continued attempts have been made to make use of the dynamic covalent chemistries to design polymers as this allows the covalent bonds to break and reform during the application of external stimuli.[52] As an example, the epoxy and LC epoxy networks can be modified by incorporating dynamic ester bonds, which improves the processability of the material.[53] Three different functionalities such as azobenzene chromophores, liquid crystals, and dynamic ester bonds, can be combined in a material which will be able to respond to external stimuli at the molecular level.[54] Excellent processability along with dual stimuli-induced shape memory and self-healing properties are possible without any further processing and tuning. Further advances in the processes and technology can be made by incorporating other smart ingredients, such as magnetic nanoparticles or single-walled nanotubes to harness additional smart characteristics. The stoichiometry can be varied to tailor the properties.

2.2.10 Interpenetrating Polymer Networks (IPNs)

By combining soft and tough polymers with an azo polymer network with a large photo-induced force, their photo mobile nature can be greatly improved. Interpenetrating polymer networks (IPNs), with various processing and tailor-made advantages, can be designed to combine photoactive azo polymers and polydimethylsiloxane (PDMS).[55] The introduction of PDMS into an azopolymer network does not influence the alignment of the mesogens. The phase-separated IPN film exhibits superior bending speed compared to the pure azo polymer film (Figure 2.14).

2.2.11 Polymer Brushes

Polymer brushes can be designed by covalently attaching two different types of polymer with one end hooked to a solid support while maintaining high grafting density.[56] Two different polymers are grafted either separately or with a diblock strategy.[57,58] The nanophase separation in the film can occur as the surface topography and energy undergo changes due to variation in the surrounding environment in the form of solvent quality, pH value, and temperature.[59] Further, the azobenzene containing polymer topography can be changed in the glassy state of the

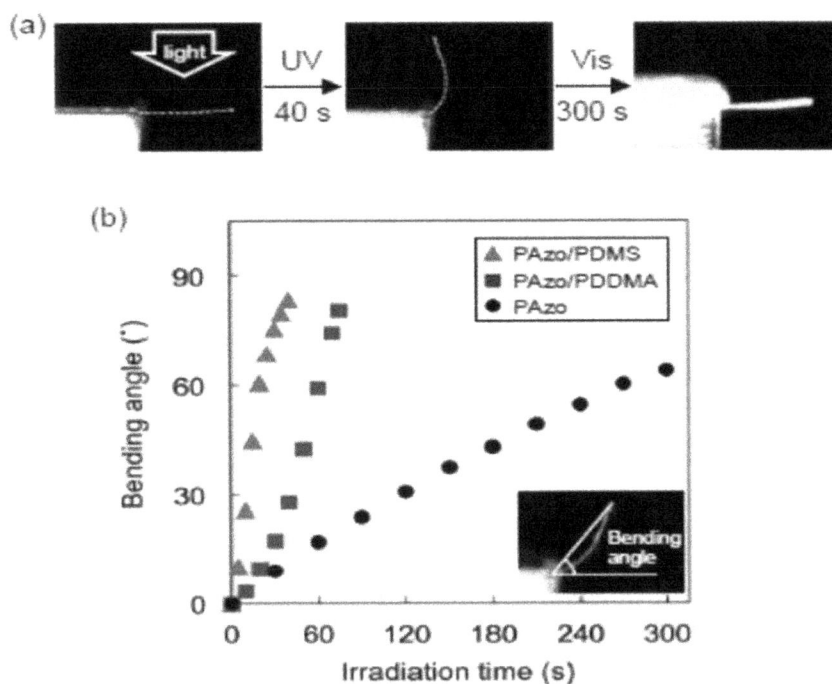

FIGURE 2.14 (a) Photo-induced bending behavior of Pazo/PDMS film upon irradiation with UV (365 nm, 10 mWcm⁻¹) and visible light (>540 nm, 40 mWcm⁻¹) at 30°C; (b) bending angle as a function of irradiation time of UV light.

Note: Size of the films: 3 mm.

polymer film by using a photoisomerization reaction. The strong optomechanical stresses can cause local scission of the covalently bound chains of azobenzene-functionalized polymer brushes that result in an irreversible nanopatterning of brushes.[60] The polarization of two interfering laser beams can be controlled during irradiation that changes the local distribution of the electrical field vector and can be used for reversible switching of polymer topography.[61]

2.2.12 Polymers that Shed Small Molecules, Ions, or Free Radicals upon Irradiation

There are some polymers which release molecules, ionic species, or free radicals under suitable irradiation. As an example, poly [p-(formyloxy) styrene], synthesized from poly (p-hydroxy styrene), undergoes decarboxylation smoothly under ultraviolet light irradiation[62] (Figure 2.15).

This reaction takes place in solution as well as in a solid state. Such rearrangement in a thin polymer film can be used to develop an image as the exposed and unexposed area will have dissimilar properties. Sharp images are formed from this type of material when it is used in photolithographic processes.

2.2.13 Photorefractive Polymers

When exposed to light, the refractive index of an organic photorefractive material undergoes a temporary change. Materials having both an electro-optic and a photo-conducting nature exhibit this effect. By beaming non-uniform light intensity on such

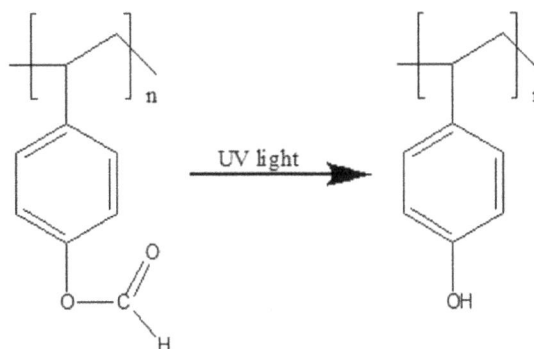

FIGURE 2.15 Decarboxylation of poly[p-(formyloxy)styrene] under UV light.

material, the charge is generated in the illuminated regions, which migrates to the dark region and finally get trapped. An internal electric field is generated due to charge redistribution, which leads to the refractive index variation via the electro-optic effect. Thus, a reversible refractive index grating is formed that can be erased by irradiation with a uniform intensity of light. A spatial phase shift between the illumination pattern and the refractive index grating is also observed.[63] Two types of photorefractive polymers are observed as the main groups, such as the functionalized one with all the individual components linked to the polymer backbone, and the other one is the composite with low molecular weight dopants. The nonlinear optical responses of typical nematic liquid crystals are enhanced by using methyl-red dye or an azobenzene liquid crystal. The main contributing factors are the orientational photorefractivity with photo-induced space charge fields, dopant-nematic molecular torque, and order parameter modification as a result of trans-cis isomerization of the azo-dopants.[64]

If the liquid crystals are stabilized by polymers and a low concentration of polymeric electron acceptor is maintained, strong photorefractive properties are realized. The generation of charge in nematic liquid crystals and its transport properties can be modified by selective doping. As an example, a 1,4,5,8-naphthalenediimide electron acceptor, with attached acrylate functionality, can be polymerized in an aligned nematic LC to form a composite.[65] The photopolymerization results in the formation of an anisotropic gel-like medium so that the LC can freely reorient in the presence of a space charge field while the polymerized regions can hold the charge-trapping sites. The trapping sites can thus hold the charge for a longer duration and facilitate a holographic grating with higher resolution PSLC as compared to nematic LC alone.

The organic/inorganic hybridized nanocomposites are also able to exhibit a photorefractive effect. It is reported that the bifunctional [p-nitrophenylazo]carbazolyl functionalized polymer can be used as a matrix for polymeric charge-transport and second-order nonlinear optical phenomena, and incorporation of CdS nanoparticles as photosensitizers results in an enhanced photorefractive effect.[66] The HgS or PbS nanocrystals can be used to make photorefractive polymeric nanocomposites that can be useful for the communication wavelength of 1.3 μm. The PbS nanocrystals-based nanocomposite exhibits a net gain in excess of the associated absorption loss.[67] Photorefractive organic materials can be used for a rewritable holographic 3D display, intensification of lasers, signal restoration, data storage, and so on.

2.2.14 Photoreversibility and Biocompatibility

Because of its transparent nature, PDMS is used for making intraocular lenses (IOLs). As an example, a photoreversible PDMS-coumarin network is such a material, whose shape and other properties can be adjusted after operation without resorting to any invasive action.[68] Usually, these polymers are cured via coumarin dimerization to form a cyclobutane ring by irradiation with UV light. The uncross-linking of the cured polymer occurs via breaking of cyclobutane ring by exposing it to another wavelength of light which is usually below 300 nm.

2.2.15 Photocleavable Polymers

These types of polymers contain metastable photochromic groups in their structures, which are stable under normal condition. However, under irradiation, these photochromic groups crack from the polymer, resulting in its degradation.[69] Both main chain and side chain photocleavable polymers are possible. It is also possible to construct both reversible and irreversible photocleavable polymers. The former can be reconstructed after cleaving as required and the latter can be used only once. Various chromophore linkers are designed for various applications. For example, o-nitro-benzyl alcohol derivative transforms into

o-nitrosobenzaldehyde (OBA) species upon exposure to UV light.[70] Cinnamic acid dimer via cyclobutane ring formation is known as truxillic acid, which can be broken into the original monomers (cinnamic acid) via irradiation at a wavelength below 260 nm.[71] The polymer chains can be connected at the flanks of the two carboxyl groups so that they can be broken from the middle of the chain. Azo moiety is known for its trans-cis photoisomerization. Azo and α-cyclodextrin form a light-responsive supramolecular complex through host-guest interaction, which is reversible in nature, that makes it a special photocleavable functionality.[72] The photocleavable groups can easily be modified, which allows the tailoring of the polymer structures by controlling the location of the light-responsive groups. The polymer structure can be synthesized with the required characteristics while the location of the photocleavable groups and their number can be varied as desired. A single photocleavable species can be designed in such a way that the polymer chains initiate from two sides, resulting in its placement in the middle. Block copolymers can have photocleavable moiety at the junction points of different polymers blocks, and that is expected to have very advanced applications. The self-assembling nature of block copolymers, leading to various nanostructures, makes smart stimuli-responsive polymer assemblies.[73] The nano-objects' self-assembly disruption or disassembly can also be done photochemically. For controlled drug delivery applications, single point disruption mostly is not sufficient to effect the desired release of hydrophobic drugs incorporated in the micelle core or vesicle membrane, due to the retention of the integrity of the hydrophobic polymer chains after getting separated from the hydrophilic moiety. In order to degrade the polymer into smaller fragments by irradiation, the number of photocleavable moieties has to be increased, which results in enhanced drug release. Another way of enhancing the drug release characteristics is to make a polymer with a photocleavable group in each monomer unit. With increasing scopes of application, a range of possible applications beckons.

2.3 Photoresponsive Microparticles

Photoresponsive microparticles are attractive as microscale objects are interesting due to the combination of the microparticle properties and their photoresponsive properties, such as change of refractive index, morphologies, and others.[74] Supramolecular LCPs (SLCPs), formed via hydrogen bonding, have the potential to be used as advanced materials.[75] The polymer microparticles interfere with the supramolecular self-assembly. Microparticles can be made by incorporating photosensitive chromophores to realize interesting properties, such as wrinkled morphologies.[76] The microparticles with a wrinkled surface afford a large specific surface area, which allows easy observation of deformation during irradiation. As an example, porphyrinoids can be encapsulated in hollow microcapsules by following layer-by-layer assembly of poly(styrene sulfonate) and poly(allylamine hydrochloride), which are pH-sensitive and are used to differentiate the fluorophore interactions at the microcapsule interface or in the microcapsule shell.[77]

2.3.1 Microparticle-Enhanced Photoresponse of PDLC-Like Films

By using photoactive LC polymer microparticles as dopants and a host material without any photoresponsive properties, hybrid LC films can be prepared. Mechanical stretching can be used to produce topological shape change, and mesogenic alignment in the LC polymer microparticles (LCPMs). This alignment enables the films to bend toward the source of UV irradiation, which is a behavior exactly similar to PDLC.[78] The hydrophobic LC microparticles with rough morphologies have superior interactions with hydrophilic polymer substrates. By dispersing wrinkled LCPMs in polyvinyl alcohol, a hybrid film can be obtained which has the ability to convert light energy to a mechanical one. The LCPMs have the ability to combine the nanoscopic changes in the molecular structure of photoactive mesogens into the macroscopic motion of the hybrid films. Even an LCP content of less than 5% is sufficient to produce a photomechanical transition in the hybrid LC films. Furthermore, by controlling the irradiation location, photo-induced mobility is precisely manipulated.

2.3.2 Photoresponsive Microcapsules

Microcapsules are formed by encapsulating an active agent or core material covered by a shell, as the latter protects the former from the adverse effect of the environment and also protects it from interference from the matrix chemistry during processing. The release of core materials from polymeric capsules in a controlled dose has potential applications in the area of self-healing, fragrance release, and drug delivery, etc. The active component is required to be triggered on demand. External stimuli such as pH, temperature, light, magnetic force, electric field, mechanical force, enzymes, etc. are used as the trigger. Photoresponsive release systems use the trigger from a remote location. When designing a photo-responsive release system, one of the most convenient and popular materials used is an azobenzene-derived polymer. Photochromic compounds such as azobenzene and stilbene derivatives are mostly used because of their E-Z isomerization on photoirradiation and they are used to photo-modulate the membrane properties, such as permeability and wettability. Stilbene-based photoresponsive microcapsules such as poly(α-methylstilbenesebacoate-co-α-methylstilbene isophthalate) with various core materials release the core during UV irradiation. The release occurs due to a change in surface roughness taking place as a result of UV-induced isomerization.[79]

2.4 Photoresponse in the Nano-Dimension

2.4.1 Two-Dimensional Nano-Dot Array Engineering of the Block Copolymer

It has been observed that block copolymer systems have the potential to produce low-cost periodic nanostructures by self-assembling, which can be applied to design high-density data storage systems, optical storage media, advanced nanostructured membranes, nanopatterned surfaces, and so on.[80] The Langmuir Blodget method is one of the finest methods for the construction of a nanostructured surface of block copolymers with a smoothness level at the nanometer scale. An amphiphilic block copolymer solution in a highly volatile solvent can be spread on water, which leads to the aggregation of hydrophobic polystyrene blocks to form irregular surface micelles. On the other hand, LC-induced spreading can result in the formation of a highly ordered dot array.[81] For example, polystyrene-block-poly (4-vinyl pyridine) (PS-P4VP) can be spread along with 4'-pentyl-4-cyanobiphenyl (5CB) to produce a highly ordered hexagonal dot array of PS. This happens due to the initial adsorption of the 5CB monolayer on the water surface acting like a 2D lubricant solvent to help the system reach an equilibrium state. The P4VP block length and the PS together control the dot-to-dot distance and the dot size, respectively.[82] The removal of 5CB molecules from the nano-films of the surface micelle is quite easy. The polymer arrays are used in the area of lithography and as scaffolds for deposition of various functional materials.

2.4.2 Photoactive Polymer Single-Chain Nanoparticles

Polymer single-chain nanoparticles (SCNPs) have attracted attention due to their simple preparation methods and sizes below 20 nm, with potential for use in drug delivery, as a nano-reactor, and coatings.[83] LC order is retained and the photoisomerization of azobenzene occurs in tiny size SNCPs under severe confinement constraints. More interesting features of LC-SCNPs, such as significant fluorescence emission, are observed for dispersion in a not-so-good solvent. Photo-deformation from a spherical to a stretched shape takes place by exposing it to linearly polarized irradiation. The dependence of the photo-deformation of LC-SCNPs on the excitation wavelength is evident from the observation that the stretching of nanoparticles is perpendicular to the polarization at 365 nm wavelength, while it shows parallel placement parallel to the polarization below 400/500 nm irradiation. Thus, ultra-small LCs forming multifunctional SCLCPs have application potential in the area of bio-imaging and LC technology.

The electrostatic assembly of the cationic conjugated polymer (perfluorooctane; PFO) and anionic poly(l-glutamic acid) with the anti-cancer drug doxorubicin (PFO/PG-Dox) forms polymer nanoparticles that exhibit simultaneous imaging and disease therapeutics properties.[84] The quenching of the fluorescence of PFO in PFO/PG-Dox nanoparticles by Dox occurs via an electron transfer mechanism. After the PFO/PG-Dox nanoparticles come into contact with the cancer cells, the PFO fluorescence is activated due to the release of Dox from the hydrolysis of the poly(l-glutamic acid). Thus, this nanoparticle system does the dual role of delivering Dox to the targeted cancer cells as well as controlling the Dox release, which is evident from the fluorescence signal by the PFO. The fluorescence, in turn, images the cancer cells.

The storage behavior of photon energy by semi-conducting polymer nanoparticles (SPNs) of size < 40 nm is observed via chemical defects, which emit a long-NIR afterglow at 780 nm with a long half-life.[85] When used *in vivo*, the intensity of the afterglow is found to be 100 times brighter than commonly used inorganic afterglow agents. This has the ability to image high-contrast lymph nodes and tumors in living mice with more than 100 times signal-to-background ratio, compared to that of NIR fluorescence imaging.

Multifunctional polymer SCNPs, prepared from a side-chain LCP (an azo-based coumarin derivative of acrylate copolymer) have a higher fraction of photoisomerizable azobenzene components while photodimerizable coumarin is introduced as a minor component.[86] Thus, the polymer possesses both photo-active mesogen and photo-cross-linking ability. The SCNPs of a size below 20 nm remain non-fluorescent in THF dispersion, while significant fluorescence is possible in chloroform dispersion due to agglomeration. Photo-induced deformation of the azobenzene LCSCNPs depends on the linearly polarized excitation wavelength. All-polymer transparent nano-composites can be designed by combining LCSCNPs and poly(methyl methacrylate) (PMMA) in which the azobenzene mesogens can be oriented by mechanical stretching.

2.4.3 Nanotemplating

A PEO-based photoresponsive LC block copolymers (PLCBC) film can be selectively doped with Ag+ and then treated with vacuum ultraviolet (VUV) to reduce the Ag+ ions together with the removal of template film that results in an ordered array of silver (Ag) nanoparticles.[46] The nanotemplates of the PLCBC films are able to maintain regularity in the arrangement of the high-dense Ag nanoparticles. This unique kind of PLCBC photolithography does not have the size limitation as observed with conventional lithography. The nanopatterns with ordered periodicity can be effectively employed in the field of photonics, plasmonics, and metal connectivity in molecular electronics.[87] The modified shape of various kinds of functional nanomaterials, such as RuO_2 having electrical conductivity, Fe with magnetic properties, or organic conducting polymers' supramolecular self-assembly, can be obtained from self-assembling phenomena of PLCBC nanotemplates.[88]

2.4.4 Nanochannels

The oriented PEO cylinders in the PLCBC film, functioning as nanochannels, allow the etching agent to pass through the film to reach the substrate surface.[89] PLCBC lithography can thus be used as a wet nanopatterning technique for substrates such as silicon wafers, metals, semi-conductors, glasses, and polymers as a low cost alternative. The azobenzenes, chalcone, or stilbene act as photoresponsive mesogens in preparing PLCBCs.[90] Because of their stability enhancement by the photo-cross-linking reaction, they are well suited for nanochannel application. These PLCBC-based membranes have potential applications for separation technologies, as electrolytes for fuel cells, and ion channels.

2.4.5 Nanoporous Liquid Crystalline Polymers

Nanoporous LCPs, having a large surface area, with pores in the nano-dimension along with high permeability, maintain a high adsorption ability.[91] In order to make these materials versatile,

their behavior from capture to release of species needs to be controlled by a distant trigger.[92] Isomerization of the most studied photoactive compounds based on azobenzene results in a change in acid strength,[93] which leads to a variation of the binding ability. The pore size can be changed by hooking on to a host molecule, which can drive the selectivity toward cations of different sizes.[94] Lyotropic LC assemblies with azobenzene components having pores in the nano dimension that can be switched on demand, are also possible.[95]

Gas-permeable LC membranes can be designed, which will have the ability to switch from an impermeable state to a permeable state, that is from an ordered to a disordered state.[96] Anisotropic as well as photo-triggered ion conduction is possible with smectic LCs with azobenzene moiety.[97] A nanoporous, hydrogen bonded, thermotropic, smectic LC polymer network of nano dimensions with a pore interior of anionic character and the ability to selectively adsorb cationic dye, is realized via self-organization.[98] The hydrogen-bonded, smectic, liquid crystalline polymer network can be cross-linked with azobenzene to make nanoporous LC materials with controlled binding sites.[91] The porous channels of a nano dimension can be produced in the polymer film by exposure to radiation.

2.4.6 Nanostructured Photoresponsive Fillers

For practical applications, a material structure assembly can be used in the conversion of radiation-induced nanoscale effects for macroscopic deformation. Thus, the photoactuator design depends on methods associated with the conversion of energy as well as the assembly of structures. A combination of a nanostructured filler of a photoresponsive nature, such as azobenzene LC (azo-LC), and a non-photoresponsive polymer, such as poly(vinyl alcohol), makes a composite anisotropic film, whose photo-induced nanoscale deformation is amplified to generate its macroscopic motion.[99]

2.4.7 Motion of Adsorbed Nanoparticles

For the introduction of motion of nanoparticles adsorbed on top of soft polymer films, polymer surfaces are designed in such a way that during irradiation with external stimuli their shape and chemical composition are altered.[100]

By applying different combinations of interference patterns, dynamic topographical changes in the polymer films are observed (Figure 2.16).[101] The adsorbed particles on top of such a dynamically fluctuating surface change their positions during irradiation. On irradiation, the majority of the particles get aligned along the topographical minima through either translational or rotational motion. Not all the big clusters are able to change their place, while others undergo rotation.

2.5 Types of Photoresponse

Various important types of responses have been observed for various photoresponsive polymers, which can produce multifunctional materials.

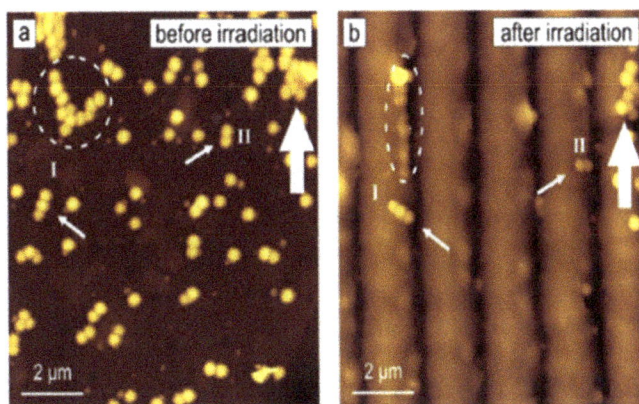

FIGURE 2.16 AFM micrographs of the polymer film covered with silica particles of 300 nm in diameter (a) before; and (b) after successive irradiation with the interference pattern of ±45° polarization combination.

2.5.1 Photoionization

Photoionization has the ability to induce photoactuation.[102] Incorporation of leuco derivatives of triphenylmethane in polyacrylamide gels leads to its large reversible deformation upon UV irradiation. This is the result of the photoionization of triphenylmethane into colored triphenylmethyl cation. There is also a large increase in volume because of swelling.

2.5.2 Photothermal Actuation

Photothermal actuators function by incorporating various photosensitive materials in the matrix, which can convert light into heat. Usually, carbon materials such as CNT, noble metal nanoparticles, metal complexes, or photothermal polymers, are used to manifest photothermal behavior that can lead to actuation. In the bilayer photothermal actuator, the mismatch between the thermal expansion coefficient induces bending under irradiation. The photothermal-conversion and thermal–mechanical components can be combined so that the photoactuators can be operated by the photothermal effect.[103]

The performance of a photoactuator is decided by the coupling of a photothermal with a thermal–mechanical conversion. Thus, the perfection of assembling the two components becomes important. As a dispersed structure allows local and internal heating, a large heat-exchange area and a high photo-mechanical energy conversion efficiency are provided by homogeneous mixing of two components.[104]

2.5.3 Enhanced Photocontrol of Actuation

The simultaneous application of photochemistry and photo-induced thermal influence to control the photo-mechanical actuation in a single device is a major challenge when making a sophisticated smart system. Polymer-grafted gold nanorods can be doped into azobenzene liquid-crystalline dynamic networks to design hybrid azobenzene liquid crystalline dynamic networks that are able to respond to near infra-red (NIR) and UV–Vis light.[105] The trans-cis photoisomerization of azobenzene and the

photothermal effect originating from the surface plasmon resonance of the Au nanorods are used to make a combination of two mechanisms to have greater control on motion and directions. By inscribing molecular alignments in different regions of the bilayer, the actuators' molecular alignment can be inscribed to make plastic "athletes," which are able to execute planned tasks. Light-driven caterpillar-inspired walkers with the ability to crawl forward, a polymer "crane," capable of performing macroscopic movements to do work, etc., can be designed using this methodology.

2.5.4 Visible Light Actuation

Actuator functioning under the influence of visible light makes a great addition to the inventory of smart materials that have a wide range of applicability. Graphene oxide makes an important nanocomponent film by adding to a polymer matrix, which under irradiation with visible light exhibits reversible photomechanical behavior that is quite stable and rapid.[106]

2.5.5 Infra-Red (IR) Light Actuation

During exposure to IR radiation, a polymer-CNT composite exhibits photomechanical actuation by expanding at a small pre-strain.[107] Also, it contracts at a larger pre-strain when exposed to IR radiation. This photoactuation behavior mainly depends on the kink instabilities generated by the nanotubes upon photon absorption, and the polymer matrix-induced orientation of the CNTs within the homogeneous polymer matrix.

2.5.6 Direct Light Propulsion of Bulk Graphene Material

The electron-hole pairs in semi-conductors can be excited by irradiation. Graphene with exceptional structure and properties and the bulk three-dimensional linked graphene material facilitate radiation absorption at various wavelengths, and the emission of energetic electrons propels the bulk material efficiently. The opportunity for bulk-scale light manipulation has potential application in the areas of solar sails and sunlight-induced space transportation. This propulsion force is quite large compared to the radiation pressure.[108]

2.5.7 Photo-Modulations of Surface Wrinkles Formed on Elastomer Sheets

Surface wrinkles, observed abundantly in nature, can offer tools with multipurpose ability for microscale surface fabrication.[109] Lateral compression or shrinkage beyond the buckling stress results in mechanical instability, leading to surface wrinkles. Usually, the fabrication involves laminated bilayer films comprising a combination of a hard skin layer and an elastomeric substrate underneath. Photo-alignable radical initiator-initiated polymerization shrinkage in the uniaxially aligned LC media and light-initiated modulations of the wrinkle characteristics are the two common modes of photomodulation.[110] Thus, the application of light and mechanical stimuli can fabricate an active surface with new functions.

2.5.8 Photo-Modulation of Ferroelectric Liquid Crystals

Spontaneous polarization behavior of ferroelectric LC (FLC) phases facilitates their fast response to the electric fields applied.[30] This qualifies the FLCs to be used in microdisplays as active switching constituents. FLCs are able to exhibit bistable switching (the electric field reversal reverses the spontaneous polarization direction) behavior. The photoresponsive FLC, derived by doping it with a photochromic compound having a cis-trans isomerization ability, exhibits photochemical switching of the spontaneous polarization of FLC.[111] The switching potential of the FLC host is altered due to the isomerization of the photochromic compound. By using chiral azobenzene dopant, much faster response times can be achieved. High contrast images can be produced with these FLCs and the reversing of contrast can be done by changing the polarity of the applied field.[112] Further, the temperature-independent FLC devices can be designed using FLCs with the ability to keep the signs of spontaneous polarization unchanged with temperature. High contrast images are recorded by optical recording in a film of photochromic FLC doped with an azobenzene-based dopant, whose contrast can be inverted via the polarity change of the applied field. Under bistable boundary conditions, the optically recorded image remains stored even after the electric field is withdrawn.

2.5.9 Photoisomerization in Biological Systems

In photoregulated biological processes, geometrical isomerization of photoisomerizable molecules embedded in a protein matrix triggers changes in the conformation of the proteins or the state of assembly in the membrane that result in excitation in the nerve or the regulation of enzymatic activity.[113] The spirobenzopyrans, attached as a side group to poly(methyl methacrylate), isomerize to merocyanines. The corresponding changes observed are extreme coloration and enhancement of the dipole moment.

2.5.10 Rewritable, Reprogrammable, Dual Light-Responsive Polymer Actuators

Photo-induced asymmetric bending can be realized by using liquid crystalline networks (LCNs) using multiple photoactive dyes, which generate artificial cilia in response to the lights of various wavelengths.[114] Also, a dual-layer-designed LCN actuator can be bent or twisted by the frequency modulation of light.[115] In both methods, only the simple design of patterning is possible, whereas complex patterns are difficult. Fabrication of reusable and reprogrammable photoresponsive polymer actuators is possible by adopting simple methods, such as the incorporation of a photochromic azo dye with pH sensitivity in a liquid crystalline matrix, and this can be adopted to fabricate polymer actuators with a reusable and reprogrammable nature.[116] By employing an acid treatment, a variety of polymer films can be obtained with various decorations. The acid-induced patterns on the polymer films being reversible, erasing old patterns and writing new patterns are possible, and thus soft actuators can easily be derived. The active molecules with two stable states can be used by following this approach to fabricate rewritable actuators. It has been demonstrated that the combination of two photoresponsive dyes possessing pH-dependent stable states can generate various complex patterns (Figure 2.17).

2.5.11 Light-Induced Refractive Index Modulation

One very interesting property of photoactive liquid crystalline elastomers (LCEs) (azo-functionality) is that they exhibit a good opto-optical response. The combined response of the LCE matrix to the trans-cis photoisomerization process generates large light-induced adjustments of the optical refractive index, which is related to this response.[117] This collective response to isomerization can be used for the optical patterning of LCEs and is found in the simple fabrication ability of tunable optical diffraction structures, which can easily be controlled by employing external strain and/or temperature variations.[118] The azobenzene moiety attached as a pendant group, and with good compatibility with the nematically arranged matrix, affords largely localized modification of optical birefringence of the medium. It undergoes significant refractive index variation under UV irradiation. However, with cross-linked azo-derivatives, a small variation occurs due to more delocalized illumination-induced perturbations.

2.5.12 Humidity and Photo-Induced Dual Actuation

Responses of smart polymers to both humidity and light give the materials additional features for practical applications. However, the humidity response is usually associated with a lack of reproducibility. This can be tackled by using hydrophobic cross-linked LC polymers (CLCP) film, comprising a photoactive azo moiety performing both as mesogen and chromophore together with another mesogen.[119] The film exhibits actuation by variation of either humidity or UV light. The combination of hydration of the oxygen-containing groups and the alignment of LCs induces fast and reversible actuation of CLCP film in the perpendicular direction to that of the LCs' alignment by varying the humidity. The CLCP film absorbs moisture from the humid air because of the

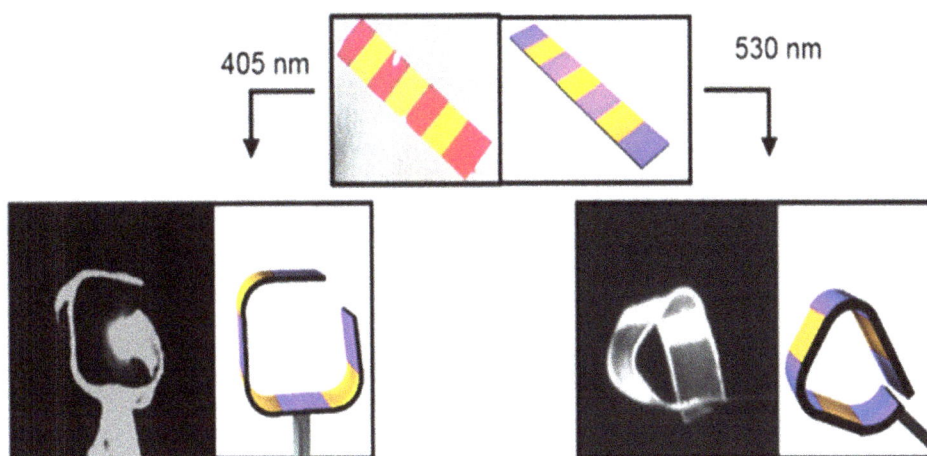

FIGURE 2.17 Specific bending of a patterned film. The same film is exposed with 405 nm (left) and 530 nm (right) light. The film specifically bends at the yellow region when exposed to 405 nm and at the magenta region when exposed with 530 nm. When switching off the light, the film unbends to the flat state. At the right of each image, a schematic of the patterned film and its bending behavior is shown.

formation of hydrogen bonds between it and the H_2O molecules. A "tree" structure designed by using a CLCP film undergoes bending toward the non-exposed side during the approach of a human finger without direct contact. Also inchworm walk and tumbling locomotions are possible. The CLCP film also exhibits bending under UV light along the direction of LC alignment. The dual-responsive CLCP films and technique can be employed for possible applications in electronics and sensors.

2.5.13 Switchable Glass Transition Temperature

Because of the ability of light to control the mechanical properties of photoresponsive polymer materials, the development of self-healing materials, photo-controlled adhesion, and energy harvesting has become possible. The aggregation state in a photoresponsive polymer powder can be changed from solid to liquid by irradiation of UV light as the cis isomer behaves like a plasticizer that reduces the T_g below room temperature.

Polymers with variable glass transition temperatures (T_g) are useful to heal cracks in high-T_g polymers and also as an additive during the processing of hard polymers at room temperature without using any plasticizer. Reversible photoswitching of the azobenzene groups from trans to cis suppresses the glass transition temperature (T_g).[1] Azo polymers in trans-form are solids as the T_g is above room temperature, whereas this is liquid in the cis-form when the T_g is below room temperature. This solid-to-liquid transition in these polymers can result in the reduction of the surface roughness of azopolymer films. It also repeatedly heals the cracks in azopolymers and maintains the adhesion of azopolymers in transfer printing.

2.5.14 Light-Powered Molecular Devices

Micro-machines, capable of generating constant mechanical work during continuous irradiation, can be designed by adopting the actuation procedure via photo-induced expansion/contraction or bending. The actuator parts, functioning by a reversible molecular shape, change with irradiation, and can be used to design sunlight-powered motors. By careful selection it is possible to amplify these molecular forces into macroscopic actuating "artificial muscles" or "molecular devices."[120] After photo-irradiation, a cross-linked liquid crystalline polymer film (CLCP) can be designed to undergo rotational motion.[121] A continuous ring made from a laminated film of azobenzene with a thin polyethylene sheet can be fixed onto a pulley system as a "belt," which exhibits counter-clockwise rotational motion when exposed simultaneously to both UV and visible light at +45 and -45 degrees to each of the pulleys.

Considering the harmful nature of UV radiation, it is always necessary to develop photoactuators responsive to visible light for biotechnological applications. Photo-responsive azotolane (azo-LCN) moieties can be used to develop micro-robots responsive to visible light, able to lift and move an object with much higher weight as compared to that of a robotic arm.[122–124] In fact, the azo-LCN films on PE substrates can be used in such a way that these films are connected by joints to function as a wrist, a hand, and so on, and by addressing individual photoactive sections, the robotic arm can be commanded to perform complex movements.

2.5.15 Making Waves in a Photoactive Polymer Film

Photoisomerization of azo moiety is followed by the slow relaxation of the cis form to the trans form. This can be avoided by incorporating azobenzene derivatives which have the ability to quickly relax thermally from cis-to-trans form.[125] The thermal relaxation time of an azobenzene can be reduced by using strategies, such as the addition of a push-pull group or by introducing a tautomerizable azo-hydrazone.[126] By mixing photoactive polymer films, based on copolymers of hydrogen-bonded azopyridine or the azobenzene of the ortho-hydroxyl group, with a mixture of LC acrylate and LC diacrylate, continuous, directional, macroscopic mechanical waves are created by being illuminated continuously with light. A device made by attaching LC networks to a plastic frame moves while the wave travels through the sample. The orientation of the film determines the travel direction. Small objects attached to the frame can be carried over long distances (centimeters). Thus, the photoactive films are also considered to have the potential for application in light-driven system in the area of self-cleaning, energy harvesting, and transport.

2.5.16 Photochromic Effect on Surface Contact Angle

Azo polyelectrolytes exhibit a decreased contact angle of water on the surface with UV irradiation, together with exhibiting the photochromic effect. The type of azo chromophores and the degree of functionalization determine the variation of contact angle. As an example, a large increase in the contact angle and wettability of the material is observed during irradiation of a copolymer of butyl methacrylate with ultraviolet light, which goes back to its original state once returned to the dark.[127]

2.5.17 Changes in Viscosity and Solubility of Polymeric Solutions

Photo-induced changes in the conformation of macromolecules may lead to a change in their viscosity in solutions.[128] For example, the viscosity in water of a carboxyl containing polymer and disulfonic acid-based polymer solutions can be reduced. Similarly, there is a decrease of the specific viscosity of a maleic anhydride-based copolymer with an azobenzene-based side chain to the extent of 24–30% in 1,4-dioxane and 1–8% in tetrahydrofuran. The decrease in viscosity is reversible.

2.5.18 Photoactive Gate Dielectrics

By using photoactive gate dielectrics, the performance of an organic transistor can be significantly enhanced by reversibly adjusting with lights of various wavelengths. The design can be used for both types of organic transistors on a transparent substrate with the additional attribute of flexibility. Photochromic spiropyran (SP), linked to nanocarbon substrate surfaces, can be combined with conductive polymers to make photosensitive devices.[129] The device conductance can be switched from one to another state by the application of UV and visible light. The reason behind the phenomenon is the reversible conformation-induced doping or proton transfer. The density functional theory calculations have established that the closed form has the electric dipole moment of 6.4 D while that of the open form is 13.9 D.[130] The large change in dipole moment has the ability to change the electrostatic environment of the devices to a considerable extent. Thus, by using the photochromic SP as one component of the gate dielectric, the dielectric capacitance of the gate dielectric is changed due to illumination, which leads to modulation of the device performance. New types of ultrasensitive devices can be designed with this principle for non-invasive sensing in chemical and environmental parameters.

2.6 Alignment of Liquid Crystal and Liquid Crystalline Polymers

In liquid crystal displays (LCDs), the LC alignment layer plays an important role as it influences the optical and electrical properties of LCDs.[131] The surface and interface boundaries determine the LC orientation.[132] The mechanical rubbing of a polymer surface is a well-known technique for the alignment of LCs during the formation of liquid crystal display devices, using LCs of low molecular weight.[133] The essential part of LCDs is the LC alignment layer. Vertical alignment (VA) displays are a type of LCDs, which have the liquid crystals arranged vertically to the glass substrates. The liquid crystals remain in a perpendicular position to substrate at nil position without any applied voltage that creates a black display between crossed polarizers. The LC changes to a tilted position by applying voltage that allows light to pass through and create a gray-scale display. The tilt produced by voltage determines the characteristics of the display. The advantages of the VA mode of liquid crystals are the contrast ratio, the wide viewing angle, the simultaneous applicability of reflective and transmissive mode, and so on.[134] As a result, the VA mode has attracted huge research attention and hastened its adoption in industry to produce various types of LCDs, such as cell phones, televisions, etc.[135]

To control the pretilt angle in order to achieve perfect vertical alignment, methods such as rubbing vertical alignment,[136] polymer-sustained vertical alignment (PSVA),[137] or photo-induced vertical alignment are used.[138] The PSVA technology based manufacturing process, conducted in an LC cell using LC and UV-curable monomers, produces a fast response, high transmittance. The UV radiation is used together with the voltage higher than the Fréedericksz transition voltage.[139] The pretilt angle is decided by the particular network formed by the UV radiation. Photo-alignment methods for LCs have been considered to be very advantageous compared to rubbing mode as they are performed in the non-contact mode at high resolution.[140] The surface photo-alignment of nematic LCs on a polymer substrate is adopted to produce an LC display panel.[141] The orientation control of LC materials in both three dimensions and in-plane has become important when designing LC-based display panels.

The vertical alignment of mesogens, normally seen with LC films, is not suitable for effective photoreactions and the desired orientation of photoactive mesogens should be horizontal. Generally, rod-like azobenzene units are chemically linked at the para-position in respect of the azo group with the polymer backbone via a spacer. For use as a command surface to control the nematic LCs in plane, a para connecting-type architecture produces some difficulties, as homeotropically oriented mesogens need large photon doses for the induction of in-plane anisotropy and the non-retention of alignment for a long time. When connected in the ortho-position, the efficiency of the azobenzene mesogens in prompting the in-plane orientation can be increased to a considerable extent together with the reduction of the light dose.[142]

Highly stable in hot, extreme environmental conditions, the main chain LC polymers with favorable pretilt angles can be used for nematic LCs and other uses requiring particular alignment conditions.[143] However, such robustness, causing a restricted motion function against their alignment, requires large amounts of light exposure.

Block copolymers, forming nanometer-level structures after phase separation, have attracted attention for their potential commercial application in next-generation nano-lithography.[144] The interfaces between blocks of side chain LC block polymers are usually found parallel to the oriented mesogens.[145] The alignment of microphase-separated structures of LC block copolymers is thus controlled by the LC block arrangement characteristics.[146] In-plane photo-alignment can be carried out with linearly polarized light for azobenzene-containing block copolymer films with microphase-separated cylinder morphologies.

By annealing polymer systems such as polybutyl methacrylate (PBMA)-azo polymer (PAz), the mesogens are directed parallel to the substrate. As a general rule, the PBMA blocks segregate to the air interface due to its lower surface tension and high flexibility, thus helping it to arrive at the planar molecular orientation. After achieving the planar orientation, the linearly polarized light can be used to carry out photo-alignment.[147] It is quite interesting to note that the skin layer can induce photo-alignment of non-photoresponsive LC polymers. Side chain LC polymer films of phenyl benzoate can be covered with the above polymer as a skin layer.[148] The suitability of various types of substrates, including silica plates and flexible polymer films, advocate the appropriateness of this method for device fabrication.

Multidomain to monodomain orientation is another photo-induced process occurring during photo-alignment of azobenzene LCs. Microphase-separated nanostructures in LC block copolymers' (LCBCs) thin film can be regulated by light. The continuous LC phase can easily change the orientation of the phase-separated nanocylinders from in-phase to out-of-plane during its photo-triggered phase transition.[139] The optimization of the molecular weight of the polymer, film thickness, light intensity, and relative humidity is required for operating at room temperature. Nanotechnology requires precise and rapid control of functional nanostructures at room temperature, and this method will be useful for the fast control of nanostructures in nanofabrication and nanoengineering.[149]

2.7 Applications of Photoresponsive Polymers

2.7.1 Smart Micro Packaging Via Layer-by-Layer Assembly

Smart micro packaging can be designed by layer-by-layer (LbL) arrangements of polymers in the form of films or shells. A suitable stimulus is employed to decompose the film so that the drug is automatically released. Drugs can be filled in pre-fabricated tiny containers that can be introduced in some systems. Layer-by-layer assembly can be designed to release an overall drug content up to 5 wt% during a planned hour span, which is being used as coatings on implants or patches on skin.[150]

To conduct the release of material from selected microchambers, focused laser radiation can be used with precision.[151] Incorporation of inorganic particles in the multilayer shells adds additional properties, such as clay which imparts strength and rigidity, or metal and metal oxide nanoparticles which impart responsiveness to light and a magnetic field.[152] By incorporating metal nanoparticles in a polymer electrolyte membrane (PEM),

the plasmon absorption band of the nanoparticles occurs during irradiation. The absorbed light energy is dissipated as heat, which raises the local temperature that destroys the surrounding PEM film and allows the entrapped material to be released in the surrounding medium.

2.7.2 Conjugated Polymer-Based Solar Cells

Conjugated polymers with sufficient absorption capability of sunlight have several important advantages over silicon-based solar cells: (1) they can reduce manufacturing costs because of their processability in solution by ink-jet printing, roll-to-roll processing, and so on; (2) they have adjustable physical properties; and (3) the feasibility of the application on bent and warped surfaces.[153] For use in an organic photovoltaic solar cell, the conjugated polymer is required to possess a large absorption coefficient, a low band gap, a high charge mobility, environmental stability, solubility, and so on.[154] Typically, a conjugated polymer with an energy gap varying from 2 to 1.2 eV can absorb from 25–80% of the solar energy, while absorbing up to 1000 nm and increasing the cell efficiency by a factor of two or three. However, it reduces absorption at shorter wavelengths, which can be addressed by using a solar cell in which blending a large band gap as well as a narrow band gap polymers is performed.[155]

The photoactive layer is designed with the desired morphology as it has great control over the device performance. For enhanced charge separation and diffusion in the polymer layer, the blend network is made by controlling the solvent evaporation kinetics and the thermal annealing parameters. The HOMO/LUMO level of the polymer needs careful tuning to maintain an efficient electron transfer, the air stability of polymer and an open circuit voltage. Device engineering is required to provide the extrinsic stability of the conjugated polymer in air. Polymers based on diketopyrrolopyrrole, fluorene, thiophene, isothianaphthene/ thienopyrazine, benzothiadiazole/aza-benzothiadiazole/Se-benzothiadiazole, silacyclopentadienes, phenylene, etc. are the materials established for better suitability in the solar cell.

2.7.3 Photorearrangement of Polymers for Harvesting Solar Energy

Harvesting light energy is considered possible by polymeric materials that have the ability to mimic photosynthesis. Among other properties, the choice, location, and tacticity of chromophores are very important for this type of energy harvesting by the relevant polymers.[156] Among the chromophores hooked to polymeric chains, fused ring structures, including heteroatoms in the ring, form very stable excimers, while increasing the rings is not efficient.[157] Vast numbers of polymers have been added over time.

Harvesting light energy can be done by a different approach, such as the pendant groups absorb light energy, rearrange, and then the rearrangement back to the original structure releases the energy. As an example, the photo rearrangements of norbornadiene to quadricyclane and back are of considerable interest because of the storing of the photo energy as strain energy in the latter that is recovered by reverse process[158] (Figure 2.18).

This photoisomerization reaction, also known as valence isomerization, is associated with electron reshuffling and formation or breaking of new π and σ bonds. Polymers prepared with the norbornandiene moieties absorb in the visible region and their isomerization is very rapid while showing efficient fatigue resistance.[159]

The polypeptide modified by spiropyran can be used to prepare the molecular-scale machine which converts light into mechanical energy.[159] During exposure to light, the reversible coil structure converts to the α-helix form. The ring-opening of the spiropyran group leading to the formation of charged species is effected by irradiation with UV radiation. The open ring reverses back to spiropyran ring structure in sunlight (Figure 2.19).

Energy is harvested in a photosynthesis unit by antenna pigments and transferred to a reaction center for redox reactions.[160] Around the reaction center, the antenna chromophores are arranged according to the order of energy gradient.[161] The arrangement of this type facilitates the sequential transfer of energy and it transfers efficiently to the reaction centers over small distances.

2.7.4 Actuator

Laminated films comprising unstretched low-density polyethylene (PE) films and cross-linked azobenzene LCP (homogeneous) film, prepared by thermal compression, exhibit

bending from both ends when exposed to UV radiation and the initial state is regained by irradiating with visible light.[162] On the other hand, laminate with the homeotropic arrangement of LCP bends from the center of the film on exposure to UV irradiation and reverts back to its original state on exposure to visible light. The stress generated for the former is at 100–210 kPa, whereas it is lower at 100–60 kPa for the homeotropic arrangement. A light-driven LCE micro-actuator device, with a gripping facility and capability to mimic the capturing property of a specific plant, can be fabricated on an optical fiber to generate power as well as a sensor for detecting environmental variation.[163] The grip is closed during the passing of an object across the field of view.

Continuous, directional, and macroscopic mechanical waves can be generated on azobenzene-functionalized polymer film by irradiating it continuously with a feedback loop.[116] The motion is caused by the contraction of one side, and the expansion of other side due to reaction by irradiation. The wave formation characteristics open up wide possibilities for applications in light-driven locomotion, self-cleaning, energy harvesting, and so on.

Catapult motion on a polymer film, capable of transporting an object at reasonable velocity, can be induced by moderate-intensity blue-light irradiation.[164] The red light, having the ability to remove thermal effects completely and to penetrate deep into live tissue, can be used to control soft actuator working on the principle of triplet-triplet annihilation-based luminescence.[165] Photoresponsive LCP can be used to make tubular micro-actuators for controlling the fluid slugs.[114] The capillary forces for liquid propulsion can also be generated by irradiation-induced asymmetric deformation of an actuator. Using a gripper based on the same materials, a light-controlled micro swimming "robot" can be made to perform functions like swimming to transport.[166] A light-controlled (532 nm laser) LCE, containing artificial muscles-based walking micro-robot, is able to walk with the help of four conical legs, and the alternate contraction and expansion of the body.[167]

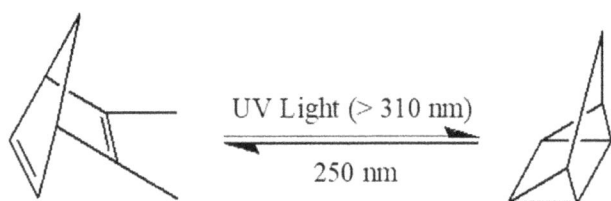

FIGURE 2.18 Photorearrangement of norbornadiene under UV radiation.

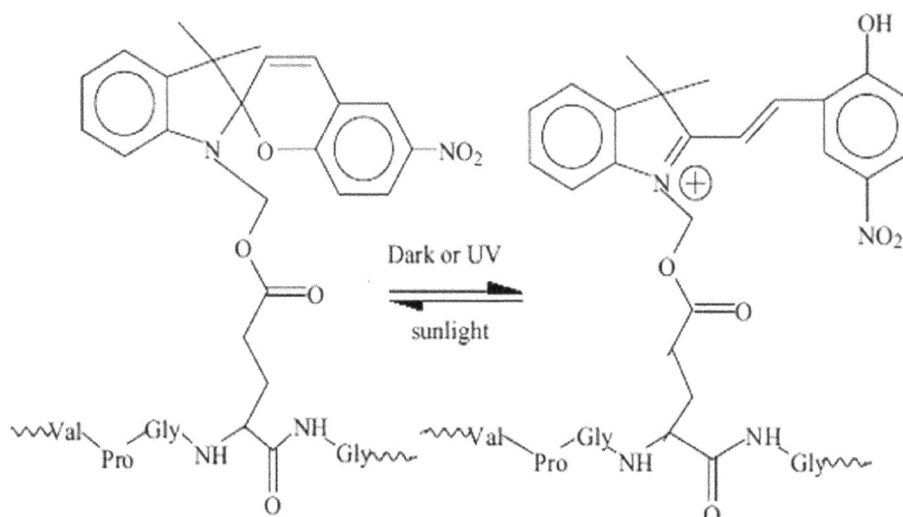

FIGURE 2.19 Light-triggered reversible ring opening of spiropyrans attached polymer.

2.7.5 Strain Sensing by Photoactive P3HT-Based Nanocomposites

Structural health monitoring (SHM) sensors suffer many deficiencies in the form of high energy demand, heavy weight, and indirect mode of damage detection. The flaws in the system can be eliminated by using thin composite film made by combining polythiophenes with multi-walled carbon nanotubes (MWCNTs).[168] With irradiation of light, the nanocomposite is able to generate a photocurrent, whose magnitude is dependent on the applied strain. Under a tensile cyclic load, the photocurrents generated in thin films correlate well with strain, exhibiting linear strain sensitivities. The general phenomenon of increased photocurrent generation originates from the alignment of polymer and reduction of hopping distances.

2.7.6 4D Printing

Using 3D printing, 3D objects can be fabricated by using software and computer-aided design (CAD). The field of rapid prototyping was enriched by the development of 3D printing because of its lower cost and easy personification of CAD data.[169] The combination of smart material and 3D printing results in a 3D object, which functions like a smart polymer responding to external stimuli. These 3D printed objects with the ability to respond to external stimuli are called "4D printed objects," considering time acting as the fourth dimension.[170] 4D printed objects with smart materials undergo various changes, including self-folding.[171] 4D printed actuators help in making micro/nano-actuators and other smart devices and have also been applied in biomedical devices,[172] security,[173] precisely designed surfaces for use in optics and electronic devices,[174] etc. Although 4D printing has gained momentum, the photoresponse is yet to be realized.

2.7.7 Soft Robot with Thermo- and Photo-Reversible Bonds

Capabilities such as camouflage, perception, and cognition have been demonstrated by soft material-based robots. Although different from a metal-based conventional robot, these soft robot functions depend on their design.[175] To design a standard robot, the structural components need to be combined with actuator components. Avoiding cumbersome steps to combine two opposing functions, a single component can be used to design a polymeric robot by using a crystalline shape memory polymer network with the incorporation of thermo- and photo-reversible covalent bonds. For example, a crystalline polyurethane network with plenty of esters and cinnamic acid derivatives in the structure exhibits thermally reversible bond exchanges in both urethane and ester linkages and photoreversible cross-linking via the dimerization of cinnamates.[176] The photo-reversible cinnamate chemistry has the advantages originating from its stability in ambient light so that the device remains stable and its reversibility is with another wavelength radiation. Its easy fabrication has raised the expectation of combining with various properties, such as surface control, drug elution, biodegradability, etc.[177]

REFERENCES

1. H. Zhou, C. Xue, P. Weis, Y. Suzuki, S. Huang, K. Koynov, G.K. Auernhammer, R. Berger, H.J. Butt, and S. Wu. "Photoswitching of glass transition temperatures of azobenzene-containing polymers induces reversible solid-to-liquid transitions." *Nat. Chem.* 9(2) (2017): 145–151.
2. K.T. Nguyen and J.L. West. "Photopolymerizable hydrogels for tissue engineering applications." *Biomaterials* 23(22) (2002): 4307–4314.
3. P. Alves, J.F.J. Coelho, J. Haack, A. Rota, A. Bruinink, and M.H. Gil. "Surface modification and characterization of thermoplastic polyurethane." *Eur. Polym. J.* 45(5) (2009): 1412–1419.
4. A. Gatti, A.N.S. Rastelli, S.J.L. Ribeiro, Y. Messaddeq, and V.S. Bagnato. "Polymerization of photocurable commercial dental methacrylate-based composites." *J. Therm. Anal. Calorim.* 3(3) (2007): 631–634.
5. P. Alves, S. Pinto, J-P. Kaiser, A. Bruinink, H.C. Sousa, and M.H. Gil. "Surface grafting of a thermoplastic polyurethane with methacrylic acid by previous plasma surface activation and by ultraviolet irradiation to reduce cell adhesion." *Colloids Surf. B: Biointerfaces* 82(4) (2011): 371–377.
6. A.E. Rydhholm, S.K. Reddy, K.S. Anseth, and C.N. Bowman. "Development and characterization of degradable thiol-allyl ether photopolymers." *Polymer* 48(15) (2007): 4589–4600.
7. S. Pandey, B. Kolli, S.P. Mishra, and. A.B. Samui. "Siloxane polymers containing azo moieties synthesized by click chemistry for photo-responsive and liquid crystalline applications." *J. Polym. Sci. Part A: Polym. Chem.* 50(7) (2012): 1205–1215.
8. C. Li, C.W. Lo, D. Zhu, C. Li, Y. Liu, and H. Jiang. "Synthesis of a photoresponsive liquid crystalline polymer containing azobenzene." *Macromol. Rapid Commun.* 30(22) (2009): 1928–1935.
9. K. Ding, L. Shi, L. Zhang, T. Zeng, Y. Yin, and Y. Yi. "Synthesis of photoresponsive polymeric pro-pesticide micelles based on PEG for the controlled release of a herbicide." *Polym. Chem.* 7(4) (2016): 899–904.
10. L. Ma, J. Li, D. Han, H. Geng, G. Chen, and Q. Li. "Synthesis of photoresponsive spiropyran-based hybrid polymers and controllable light-triggered self-assembly study in toluene." *Macromol. Chem. Phys.* 214(6) (2013): 716–725.
11. A.A. Brown, O. Azzaroni, and W.T.S. Huck. "Photoresponsive polymer brushes for hydrophilic patterning." *Langmuir* 25(3) (2009): 1744–1749.
12. A. Ravve. "Photoresponsive polymers." In *Light-Associated Reactions of Synthetic Polymers*. New York: Springer, 2006, pp. 246–315.
13. V.A. Bakulev and W. Dehaen. "Chemical properties of 1,2,3-thiadiazoles." In E.C. Taylor, P. Wipf, and A. Weissberger (Eds.), *Chemistry of Heterocyclic Compounds: A Series of Monographs*. Hoboken, NJ: John Wiley & Sons Inc., 2004, pp. 113–154.
14. G.A. Delzenne and U. Laridon. "Photoable polymers." In *Proceedings of S.P.E. Conference on Photopolymers*. Ellenville, NY: Springer (1967), pp. 6–7.
15. H.C. Ng and J.E. Guillet. "Photochemistry of ketone polymers. 18. Effects of solvent, ketone content, and ketone structure on the photolysis of styrene-vinyl aromatic ketone copolymers." *Macromolecules* 18(11) (1985): 2294–2299.

16. L. Abdellah, B. Boutevin, G. Caporiccio, and F. Guida-Pietrasanta. "Study of photocrosslinkable polysiloxanes bearing gem di-styrenyl groups: Synthesis and thermal properties." *Eur. Polym. J.* 39(1) (2003): 49–56.

17. R.S. Benson. "Use of radiation in biomaterials science." *Nucl Instrum Methods Phys Res B: Beam Interactions with Materials and Atoms* 191(1–4) (2002): 752–757.

18. V.S. Rao and A.B. Samui. "Photoactive liquid crystalline polymers: A comprehensive study of linear and hyperbranched polymers synthesized by a_2b_2, a_2b_3, a_3b_2, and a_3b_3 approaches." *J. Polym. Sci. Part A: Polym. Chem.* 49(6) (2011): 1319–1330.

19. V.S. Rao and A.B. Samui. "Molecular engineering of photoactive liquid crystalline polyester epoxies containing benzylidene moiety." *J. Polym. Sci. Part A: Polym. Chem.* 46(23) (2008): 7637–7655.

20. M. Murali, V.S. Rao, and A.B. Samui. "Liquid crystalline photoactive hyperbranched and linear benzylidene polyester with terminal epoxy and pendant hydroxyl groups," *J. Polym. Sci. Part A: Polym. Chem.* 45(14) (2007): 3116–3123.

21. V.S. Rao and A.B. Samui. "Structure-property relationship of photoactive liquid crystalline polyethers containing benzylidene moiety." *J. Polym. Sci. Part A: Polym. Chem.* 47(8) (2009): 2143–2155.

22. G. Langer, T. Kavc, W. Kern, G. Kranzelbinder, and E. Toussaere. "Refractive index changes in polymers induced by deep UV irradiation and subsequent gas phase modification." *Macromol Chem Phys*, 202(18) (2001): 3459–3467.

23. S. Granick, S.K. Kumar, E.J. Amis, M. Antonietti, A.C. Balazs, A.K. Chakraborty, G.S. Grest, C. Hawker, P. Janmey, ... and C.R. Safinya. "Macromolecules at surfaces: Research challenges and opportunities from tribology to biology." *J. Polym. Sci., Part B: Polym. Phys.* 41(22) (2003): 2755–2793.

24. G.T. Carroll, M.E. Sojka, X. Lei, N.J. Turro, and J.T. Koberstein. "Photoactive additives for cross-linking polymer films: Inhibition of dewetting in thin polymer films." *Langmuir* 22(18) (2006): 7748–7754.

25. A.P. Esser-Kahn, S.A. Odom, N.R. Sottos, S.R. White, and J.S. Moore. "Triggered release from polymer capsules." *Macromolecules* 44(14) (2011): 5539–5553.

26. O. Bertrand and J-F. Gohy. "Photo-responsive polymers: Synthesis and applications." *Polym. Chem.* 8(1) (2017): 52–73.

27. N. Fomina, J. Sankaranarayanan, and A. Almutairi. "Photochemical mechanisms of light-triggered release from nanocarriers." *Adv. Drug Deliv. Rev.* 64(11) (2012): 1005–1020.

28. M. Dübner, N.D. Spencer, and C. Padeste. "Light-responsive polymer surfaces via postpolymerization modification of grafted polymer-brush structures." *Langmuir* 30(49) (2014): 14971–14981.

29. J. Deng, X. Liu, W. Shi, C. Cheng, C. He, and C. Zhao. "Light-triggered switching of reversible and alterable biofunctionality via β-cyclodextrin/azobenzene-based host-guest interaction," *ACS Macro Lett.* 3(11) (2014): 1130–1133.

30. H.K. Bisoyi and Q. Li. "Light-driven liquid crystalline materials: From photo-induced phase transitions and property modulations to applications." *Chem. Rev.* 116(24) (2016): 15089–15166.

31. A. Bobrovsky, V. Shibaev, G. Elyashevich, E. Rosova, A. Shimkin, V. Shirinyan, and K-L. Cheng. "Photochromic composites based on porous stretched polyethylene filled by nematic liquid crystal mixtures." *Polym. Adv. Technol.* 21(2) (2010): 100–112.

32. I. Grabcheva, I. Moneva, E. Wolarzb, and D. Bauman. "Fluorescent 3-oxy benzanthrone dyes in liquid crystalline media." *Dyes and Pigments* 58(1) (2003): 1–6.

33. L. Corvazier and Y. Zhao. "Induction of liquid crystal orientation through azobenzene-containing polymer networks." *Macromolecules* 32(10) (1999): 3195–3200.

34. A. Shimamura, A. Priimagi, J-I. Mamiya, M. Kinoshita, T. Ikeda, and A. Shishido. "Photoinduced bending upon pulsed irradiation in azobenzene-containing crosslinked liquid-crystalline polymers." *J. Nonlinear Opt. Phys. Mater.* 20(4) (2011): 405–413.

35. N. Tabiryan, S. Serak, X.M. Dai, and T. Bunning. "Polymer film with optically controlled form and actuation." *Opt. Express* 13(19) (2005): 7442–7448.

36. C.L. Van Oosten, D. Corbett, D. Davies, M. Warner, C.W.M. Bastiaansen, and D.J. Broer. "Bending dynamics and directionality reversal in liquid crystal network photoactuators." *Macromolecules* 41(22) (2008): 8592–8596.

37. A. Priimagi, A. Shimamura, M. Kondo, T. Hiraoka, S. Kubo, J. Mamiya, M. Kinoshita, T. Ikeda, and A. Shishido. "Location of the azobenzene moieties within the cross-linked liquid-crystalline polymers can dictate the direction of photoinduced bending." *ACS Macro Lett.* 1(1) (2012): 96–99.

38. S.J. Gerbode, J.R. Puzey, A.G. McCormick, and L. Mahadevan. "How the cucumber tendril coils and overwinds." *Science* 337(6098) (2012): 1087–1091.

39. T.L. Hendrick, B. Cheng, and X. Deng. "Wingbeat time and the scaling of passive rotational damping in flapping flight." *Science* 324(5924) (2009): 252–255.

40. S. Tsoi, J. Zhou, C. Spillmann, J. Naciri, T. Ikeda, and B. Ratna. "Liquid-crystalline nano-optomechanical actuator." *Macromol. Chem. Phys.* 214(6) (2013): 734–741.

41. M. Camacho-Lopez, H. Finkelmann, P. Palffy-Muhoray, and M. Shelley. "Fast liquid-crystal elastomer swims into the dark." *Nat. Mater.* 3(2004): 307–310

42. Z. Liu, R. Tang, D. Xu, J. Liu, and H. Yu. "Precise actuation of bilayer photomechanical films coated with molecular azobenzene chromophores." *Macromol. Rapid Commun.* 36(12) (2015): 1171–1176.

43. E-K. Fleischmann and R. Zentel. "Liquid-crystalline ordering as a concept in materials science: from semiconductors to stimuli-responsive devices." *Angew. Chem. Int. Ed.* 52(34) (2013): 8810–8827.

44. Y. Yu, M. Nakano, and T. Ikeda. "Photomechanics: Directed bending of a polymer film by light." *Nature* 425(6954) (2003): 145.

45. J-I. Mamiya, A. Kuriyama, N. Yokota, M. Yamada, and T. Ikeda. "Photomobile polymer materials: Photoresponsive behavior of cross-linked liquid-crystalline polymers with mesomorphic diarylethenes." *Chem. Eur. J.* 21(8) (2015): 1–5.

46. H. Yu. "Photoresponsive liquid crystalline block copolymers: From photonics to nanotechnology." *Prog. Polym. Sci.* 39(4) (2014): 781–815.

47. S.B. Darling. "Directing the self-assembly of block copolymers." *Prog. Polym. Sci.* 32(10) (2007): 1152–1204.

48. H.F. Yu, T. Iyoda, and T. Ikeda. "Photoinduced alignment of nanocylinders by supramolecular cooperative motions." *J. Am. Chem. Soc.* 128(34) (2006): 11010–11011.

49. Y. Haketa and H. Maeda. "Dimension-controlled ion-pairing assemblies based on π-electronic charged species." *Chem. Commun.* 53(20) (2017): 2894–2909.

50. K. Ichimura. "Photo-isomerization properties in downsized crystals." *Chem. Commun.* 12 (2009): 1496–1498.

51. R. Yamakado, M. Hara, S. Nagano, T. Seki, and H. Maeda. "Photo-responsive soft ionic crystals: Ion-pairing assemblies of azobenzene carboxylates." *Chem. Eur. J.* 23(39) (2017): 9244–9248.

52. R.J. Wojtecki, M.A. Meador, and S.J. Rowan. "Using the dynamic bond to access macroscopically responsive structurally dynamic polymers." *Nat. Mater.* 10 (2011): 14–27.

53. Z. Pei, Y. Yang, Q. Chen, E.M. Terentjev, and Y. Wei. "Mouldable liquid-crystalline elastomer actuators with exchangeable covalent bonds." *Nat. Mater.* 13 (2014): 36–41.

54. Y. Li, O. Rios, J.K. Keum, J. Chen, and M.R. Kessler. "Photoresponsive liquid crystalline epoxy networks with shape memory behavior and dynamic ester bonds." *ACS Appl. Mater. Interfaces* 8(24) (2016): 15750–15757.

55. T. Ube, K. Minagawaa, and T. Ikeda. "Interpenetrating polymer network of liquid-crystalline azobenzene polymer and poly(dimethylsiloxane) as photo mobile materials." *Soft Matter* 13(35) (2017): 5820–5823.

56. C.J. Galvin and J. Genzer. "Applications of surface-grafted macromolecules derived from post-polymerization modification reactions." *Prog. Polym. Sci.* 37(7) (2012): 871–906.

57. S. Santer, A. Kopyshev, J. Donges, J. Rühe, X. Jiang, B. Zhao, and M. Müller. "Memory of surface patterns in mixed polymer brushes: Simulation and experiment." *Langmuir* 23(1) (2007): 279–285.

58. M.R. Tomlinson and J. Genzer. "Evolution of surface morphologies in multivariant assemblies of surface-tethered diblock copolymers after selective solvent treatment." *Langmuir* 21(25) (2005): 11552–11555.

59. M. Müller. "Phase diagram of a mixed polymer brush." *Phys. Rev. E.* 65(3) (2002): 30802.

60. A. Kopyshev, C.J. Galvin, J. Genzer, N. Lomadze, and S. Santer. "Polymer brushes modified by photosensitive azobenzene containing polyamines." *Polymer* 98 (2016): 421–428.

61. N.S. Yadavalli, S. Loebner, T. Papke, E. Sava, N. Hurduc, and S. Santer. "A comparative study of photoinduced deformation in azobenzene-containing polymer films." *Soft Matter* 12(9) (2016): 2593–2603.

62. J.M.J. Frechet, T.G. Tessier, C.G. Willson, and H. Ito. "Poly[p-(formyloxy)styrene]: Synthesis and radiation-induced decarbonylation." *Macromolecules* 18(3) (1985): 317–321.

63. G.G. Malliaras, V.V. Krasnikov, H.J. Bolink, and G. Hadziioannou. "Photorefractivity in poly(N-vinyl carbazole)-based polymer composites." *Pure Appl. Opt.* 5(5) (1996): 631–643.

64. I.C. Khoo, M.Y. Shih, A. Shishido, P.H. Chen, and M.V. Wood. "Liquid crystal photorefractivity – towards supra-optical nonlinearity." *Optical Materials* 18(1) (2001): 85–90.

65. G.P. Wiederrecht and M.R. Wasielewski. "Photorefractivity in polymer-stabilized nematic liquid crystals." In Proceedings of the SPIE 3471: *Xerographic Photoreceptors and Organic Photorefractive Materials IV*, (1998); doi: 10.1117/12.328153.

66. L. Ding, L. Huang, and Y.J. Zhong. "Photorefractivity in a bi-functional polymer nanocomposite sensitized by CdS nanoparticle." *Wuhan Univ. Technol.-Mat. Sci. Edit.* 25(5) (2010): 550–554.

67. J.G. Winiarz, L. Zhang, J. Park, and P.N. Prasad. "Inorganic: Organic hybrid nanocomposites for photorefractivity at communication wavelengths." *J. Phys. Chem. B.* 106(5) (2002): 967–970.

68. R. Jellali, V. Bertrand, M. Alexandre, N. Rosière, M. Grauwels, M-C.D. Pauw-Gillet, and B. Jerome. "Photoreversibility and biocompatibility of polydimethylsiloxane-coumarin as adjustable intraocular lens material." *Macromol Biosci.* 17(7) (2017): 1600495.

69. Q. Yan, D. Han, and Y. Zhao. "Main-chain photoresponsive polymers with controlled location of light-cleavable units: From synthetic strategies to structural engineering." *Polymer Chemistry* 4(19) (2013): 5026–5037.

70. A. Patchornik, B. Amit, and R.B. Woodward. "Photosensitive protecting groups." *J. Am. Chem. Soc.* 92(21) (1970): 6333–6335.

71. A. Lendlein, H. Jiang, A. Junger, and R. Langer. "Light-induced shape-memory polymers." *Nature* 434(7035) (2005): 879–882.

72. A. Harada. "Cyclodextrin-based molecular machines." *Acc. Chem. Res.* 34(6) (2001): 456–464.

73. C. Tonhauser, A.A. Golriz, C. Moers, R. Klein, H.J. Butt, and H. Frey. "Stimuli-responsive y-shaped polymer brushes based on junction-point-reactive block copolymers." *Adv. Mater.* 24(41) (2012): 5559–5563.

74. Y. Li, Y. He, X. Tong, and X.J. Wang. "Photoinduced deformation of amphiphilic azo polymer colloidal spheres." *J. Am. Chem. Soc.* 127(8) (2005): 2402–2403.

75. T. Kato, N. Mizoshita, and K. Kishimoto. "Functional liquid-crystalline assemblies: Self-organized soft materials." *Angew. Chem., Int. Ed.* 45(1) (2006): 38–68.

76. D. Han, X. Tong, Y. Zhao, and Y. Zhao. "Block copolymers comprising π-conjugated and liquid crystalline subunits: Induction of macroscopic nanodomain orientation." *Angew. Chem. Int. Ed.* 49(48) (2010): 9162–9165.

77. R. Teixeira, V. Vaz Serra, P.M.R. Paulo, S.M. Andradea, and S.M.B. Costa. "Encapsulation of photoactive porphyrinoids in polyelectrolyte hollow microcapsules viewed by fluorescence lifetime imaging microscopy (FLIM)." *RSC Adv.* 5(96) (2015): 79050–79060.

78. H. Yu, C. Dong, W. Zhou, T. Kobayashi, and H. Yang. "Wrinkled liquid-crystalline microparticle-enhanced photoresponse of PDLC-like films by coupling with mechanical stretching." *Small* 7(21) (2011): 3039–3045.

79. K.A. Bogdanowicz, B. Tylkowski, and M. Giamberini. "Preparation and characterization of light-sensitive microcapsules based on a liquid crystalline polyester." *Langmuir* 29 (2013): 1601–1608.

80. C.H. Arnaud. "Molecular-scale lithography." *Chem. Eng. News* 86(42) (2008): 57–59.

81. S. Nagano, Y. Matsushita, Y. Ohnuma, S. Shinma, and T. Seki. "Formation of a highly ordered dot array of surface micelles of a block copolymer via liquid crystal-hybridized self-assembly." *Langmuir* 22(12) (2006): 5233–5236.

82. S. Nagano, Y. Matsushita, S. Shinma, and T. Seki. "Two-dimensional nano-dot array engineering of block copolymer surface micelles on water surface." *Thin Solid Films* 518(2) (2009): 724–728.

83. S. Mavila, O. Eivgi, I. Berkovich, and N.G. Lemcoff. "Intramolecular cross-linking methodologies for the synthesis of polymer nanoparticles." *Chem. Rev.* 116(3) (2016): 878–961.

84. X. Feng, F. Lv, L. Liu, H. Tang, C. Xing, Q. Yang, and S. Wang. "Conjugated polymer nanoparticles for drug delivery and imaging." *ACS Appl. Mater. Interfaces* 2(8) (2010): 2429–2435.

85. Q. Miao, C. Xie, X. Zhen, Y. Lyu, H. Duan, X. Liu, J.V. Jokerst, and K. Pu. "Molecular afterglow imaging with

bright, biodegradable polymer nanoparticles." *Nature Biotechnol.* 35(11) (2017): 1102–1110.

86. W. Fan, X. Tong, G. Li, and Y. Zhao. "Photoresponsive liquid crystalline polymer single-chain nanoparticles." *Polym. Chem.* 8(22) (2017): 3523–3529.

87. J. Li, K. Kamata, S. Watanabe, and T. Iyoda T. "Template- and vacuum-ultraviolet-assisted fabrication of an Ag-nanoparticle array on flexible and rigid substrates." *Adv. Mater.* 19(9) (2007): 1267–1271.

88. S. Suzuki, K. Kamata, H. Yamauchi, and T. Iyoda. "Selective doping of lead ions into normally aligned PEO cylindrical nanodomains in amphiphilic block copolymer thin films." *Chemistry Letters* 36(8) (2007): 978–979.

89. R. Watanabe, K. Kamata, and T. Iyoda. "Smart block copolymer masks with molecule-transport channels for total wet nanopatterning." *J. Mater. Chem.* 18(45) (2008): 5482–5491.

90. T. Yamamoto, K. Kimura, M. Komura, Y. Suzuki, T. Iyoda, S. Asaoka, and H, Nakanishi. "Block copolymer permeable membrane with visualized high-density straight channels of poly(ethylene oxide)." *Adv. Funct. Mater.* 21(5) (2011): 918–926.

91. H.P.C. van Kuringen, Z.J.W.A. Leijten, A.H. Gelebart, D.J. Mulder, G. Portale, D.J. Broer, and A.P.H.J. Schenning. "Photoresponsive nanoporous smectic liquid crystalline polymer networks: Changing the number of binding sites and pore dimensions in polymer adsorbents by light." *Macromolecules* 48(12) (2015): 4073–4080.

92. F.D. Jochum and P.D. Theato. "Temperature- and light-responsive smart polymer materials." *Chem. Soc. Rev.* 42(17) (2013): 7468–7483.

93. T. Fukaminato, E. Tateyama, and N. Tamaoki. "Fluorescence photoswitching based on a photochromic pKa change in an aqueous solution." *Chem. Commun.* 48(88) (2012): 10874–10876.

94. M. Natali and S. Giordani. "Molecular switches as photocontrollable 'smart' receptors." *Chem. Soc. Rev.* 41(10) (2012): 4010–4029.

95. C.S. Pecinovsky, E. S. Hatakeyama, and D.L. Gin. "Polymerizable photochromic macrocyclic metallomesogens: Design of supramolecular polymers with responsive nanopores." *Adv. Mater.* 20(1) (2008): 174–178.

96. J. Liu, M. Wang, M. Dong, L. Gao, and J. Tian. "Reversible photoinduced switching of permeability in a cast non-porous film comprising azobenzene liquid crystalline polymer." *Macromol. Rapid Commun.* 32(19) (2011): 1557–1562.

97. C.G. Nardele, V.M. Dhavale, K. Sreekumar, and S.K. Asha. "Ionic conductivity probed in main chain liquid crystalline azobenzene polyesters." *J. Polym. Sci., Polym. Chem.* 53(5) (2015): 629–641.

98. H.P.C. van Kuringen, G.M. Eikelboom, I.K. Shishmanova, D.J. Broer, and A.P.H.J. Schenning. "Responsive nanoporous smectic liquid crystal polymer networks as efficient and selective adsorbents." *Adv. Funct. Mater.* 24(32) (2014): 5045–5051.

99. H.F. Yu, C. Dong, W.M. Zhou, T. Kobayashi, and H.A. Yang. "Photomechanical behaviors of hybrid films of liquid-crystalline polymer microparticles." *Small* 7(21) (2011): 3039–3045.

100. N.S. Yadavalli, T. König, and S. Santer. "Selective mass transport of azobenzene-containing photosensitive films towards or away from the light intensity." *J. Soc. Inf. Disp.* 23(4) (2015): 154–162.

101. S. Loebner, J. Jelken, N.S. Yadavalli, E. Sava, N. Hurduc, and S. Santer. "Motion of adsorbed nano-particles on azobenzene containing polymer films." *Molecules* 21(12) (2016): 1663.

102. M. Irie and D. Kungwatchakun. "Photoresponsive polymers: Mechanochemistry of polyacrylamide gels having triphenylmethane leuco derivatives." *Makromol. Chem., Rapid Commun.* 5(12) (1984): 829–832.

103. Y. Hu, Z. Li, T. Lan, and W. Chen. "Photoactuators for direct optical-to-mechanical energy conversion: From nanocomponent assembly to macroscopic deformation." *Adv. Mater.* 28(47) (2016): 10548–10556.

104. Y. Yamamoto, K. Kanao, T. Arie, S. Akita, and K. Takei. "Air ambient-operated PNIPAM-based flexible actuators stimulated by human body temperature and sunlight." *ACS Appl. Mater. Interfaces* 7(20) (2015): 11002–11006.

105. X. Lu, H. Zhang, G. Fei, B. Yu, X. Tong, H. Xia, and Y. Zhao. "Liquid-crystalline dynamic networks doped with gold nanorods showing enhanced photocontrol of actuation." *Adv. Mater.* 30(14) (2018):1706597.

106. L. Yu and H.F. Yu. "Light-powered tumbler movement of graphene oxide/polymer nanocomposites." *ACS Appl. Mater. Interfaces* 7(6) (2015): 3834–3839.

107. S.V. Ahir and E.M. Terentjev. "Photomechanical actuation in polymer-nanotube composites." *Nat. Mater.* 4(6) (2005): 491–495.

108. T.F. Zhang, H.C. Chang, Y.P. Wu, P.S. Xiao, N.B. Yi, Y.H. Lu, Y.F. Ma, Y. Huang, K. Zhao, …, and Y.S. Chen. "Macroscopic and direct light propulsion of bulk graphene material." *Nat. Photonics* 9(7) (2015): 471–476.

109. J. Rodríguez-Hernández. "Wrinkled interfaces: Taking advantage of surface instabilities to pattern polymer surfaces." *Prog. Polym. Sci.* 42 (2015): 1–41.

110. T. Seki, D. Yamaoka, T. Takeshima, Y. Nagashima, M. Hara, and S. Nagano. "Photo-modulations of surface wrinkles formed on elastomer sheets." *Mol. Cryst. Liq. Cryst.* 644(1) (2017):52–60.

111. T. Ikeda, T. Sasaki, and K. Ichimura. "Photochemical switching of polarization in ferroelectric liquid-crystal films." *Nature* 361(1993): 428–430.

112. L. Komitov, O. Tsutsumi, C. Ruslim, T. Ikeda, K. Ichimura, and K. Yoshino. "Optical recording using a photochromic ferroelectric liquid crystal." *J. Appl. Phys.* 89(12) (2001): 7745–7749.

113. M. Irie, A. Menju, and K. Hayashi. "Photoresponsive polymers: Reversible solution viscosity change of poly(methyl methacrylate) having spirobenzopyran side groups." *Macromolecules* 12(6) (1979): 1176–1180.

114. J. Lv, Y. Liu, J. Wei, E. Chen, L. Qin, and Y. Yu. "Photocontrol of fluid slugs in liquid crystal polymer microactuators." *Nature* 537(7619) (2016): 179–184.

115. M. Wang, B.P. Lin, and H. Yang. "A plant tendril mimics soft actuator with phototunable bending and chiral twisting motion modes." *Nat. Commun.* 7 (2016): 13981.

116. A.H. Gelebart, D.J. Mulder, G. Vantomme, A.P.H.J. Schenning, and D.J. Broer. "A rewritable, reprogrammable, dual light-responsive polymer actuator." *Angew. Chem. Int. Ed.* 56(43) (2017): 13436–13439.

117. B. Tasic, W. Li, A. Sánchez-Ferrer, M. Copic, and I. Drevensek-Olenik. "Light-induced refractive index modulation in photoactive liquid-crystalline elastomers." *Macromol. Chem. Phys.* 214(23) (2013): 2744–2751.

118. E. Sungur, M.H. Li, G. Taupier, A. Boeglin, S. Mery, P. Keller, and K.D. Dorkenoo. "External stimulus-driven variable-step grating in a nematic elastomer." *Opt. Express* 15(11) (2007): 6784–6789.

119. Y. Liu, B. Xu, S. Sun, J. Wei, L. Wu, and Y. Yu. "Humidity- and photo-induced mechanical actuation of cross-linked liquid crystal polymers." *Adv. Mater.* 29(9) (2017): 1604792.

120. O.S. Bushuyev, M. Aizawa, A. Shishido, and C.J. Barrett. "Shape-shifting azo dye polymers: Towards sunlight-driven molecular devices." *Macromol. Rapid Commun.* 39(1) (2018): 1700253.

121. M. Yamada, M. Kondo, J.I. Mamiya, Y. Yu, M. Kinoshita, C.J. Barrett, and T. Ikeda. "Photomobile polymer materials: Towards light-driven plastic motors." *Angew. Chem. Int. Ed.* 47(27) (2008): 4986–4988.

122. R.Y. Yin, W.X. Xu, M. Kondo, C.C. Yen, J. Mamiya, T. Ikeda, and Y. Yu. "Can sunlight drive the photoinduced bending of polymer films?" *J. Mater. Chem.* 19(20) (2009): 3141–3143.

123. F. Cheng, R Yin, Y. Zhang, C.C. Yen, and Y. Yu. "Fully plastic microrobots which manipulate objects using only visible light." *Soft Matter* 6(15) (2010): 3447–3449.

124. F. Cheng, Y. Zhang, R. Yin, and Y. Yu. "Visible light induced bending and unbending behavior of crosslinked liquid-crystalline polymer films containing azotolane moieties." *J. Mater. Chem.* 20(11) (2010): 4888–4896.

125. A.H. Gelebart, D.J. Mulder, M. Varga, A. Konya, G. Vantomme, E.W. Meijer, R.L.B. Selinger, and D.J. Broer. "Making waves in a photoactive polymer film." *Nature* 546(7660) (2017): 632–636.

126. W.R. Brode, J.H. Gould, and G.M. Wyman. "The relation between the absorption spectra and the chemical constitution of dyes. XXV: Phototropism and cis-trans isomerism in aromatic azo compounds." *J. Am. Chem. Soc.* 74(18) (1952): 4641–4646.

127. M. Irie and R. Iga. "Photoresponsive polymers. 9: Photostimulated reversible sol-gel transition of polystyrene with pendant azobenzene groups in carbon disulfide." *Macromolecules* 19(10) (1987): 2480–2484.

128. R. Lovrien. "The photoviscosity effect." *Proc. Natl. Acad. Sci. U.S.A.* 57(2) (1967): 236–242.

129. X. Guo, D. Zhang, G. Yu, M. Wan, J. Li, Y. Liu, and D. Zhu. "Reversible photoregulation of the electrical conductivity of spiropyran-doped polyaniline for information recording and nondestructive processing." *Adv. Mater.* 16(7) (2004): 636–640.

130. Q. Shen, L. Wang, S. Liu, Y. Cao, L. Gan, X. Guo, M.L. Steigerwald, Z. Shuai, Z. Liu, and C. Nuckolls. "Photoactive gate dielectrics." *Adv. Mater.* 22(30) (2010): 3282–3287.

131. T. Wang, X. Li, Z. Dong, S. Huang, and H. Yu. "Vertical orientation of nanocylinders in liquid-crystalline block copolymers directed by light." *ACS Appl. Mater. Interfaces* 9(29) (2017): 24864–24872.

132. J. Cognard. "Alignment of nematic liquid crystals and their mixtures." *Mol. Cryst. Liq. Cryst. (Suppl. Ser.)* 1(1982): 1–78.

133. C. Jérôme. "Photoreversibility and biocompatibility of polydimethylsiloxane-coumarin as adjustable intraocular lens material." *Macromolecular Bioscience*, 17(7) (2017): 1600495.

134. H. Yoshida. "Vertically aligned nematic (VAN) LCD technology." In J. Chen, W. Cranton, and M. Fihn (Eds.), *Handbook of Visual Display Technology* Berlin: Springer, 2014, pp. 1485–1505.

135. J. Yao, J. Brenizer, R. Hui, and S. Yin. "Photonic fiber and crystal devices: Advances in materials and innovations in device applications VI." In *Proceedings of the* SPIE, San Diego, CA, 12–13 August 2012.

136. M. Liu, X.G. Zheng, S.M. Gong, L.L. Liu, Z. Sun, L.S. Shao and Y.H. Wang. "Effect of the functional diamine structure on the properties of a polyimide liquid crystal alignment film." *RSC Adv.* 5(32) (2015): 25348–25356.

137. C.J. Hsu, B.L. Chen, and C.Y. Huang. "Controlling liquid crystal pretilt angle with photocurable prepolymer and vertically aligned substrate." *Opt. Express* 24(2) (2016): 1463–1471.

138. C.M. Xue, J. Xiang, H. Nemati, H.K. Bisoyi, K. Gutierrez-Cuevas, L, Wang, M. Gao, S. Zhou, D.K. Yang, … and Q. Li. "Light-driven reversible alignment switching of liquid crystals enabled by azo thiol grafted gold nanoparticles." *ChemPhysChem* 16(9) (2015): 1852–1856.

139. F. Wang, L. Shao, Q. Bai, X. Che, B. Liu, and Y. Wang. "Photo-induced vertical alignment of liquid crystals via in situ polymerization initiated by polyimide containing benzophenone." *Polymers* 9(6) (2017): 233.

140. O. Yaroshchuk and Y. Reznikov. "Photoalignment of liquid crystals: Basics and current trends." *J. Mater. Chem.* 22(2) (2012): 286–300.

141. K. Miyachi, K. Kobayashi, Y. Yamada, and S. Mizushima. "The world's first photoalignment LCD technology applied to generation ten factory." *SID Symp. Digest Tech. Papers* 41(1) (2010): 579–582.

142. H. Akiyama, M. Momose, K. Ichimura, and S. Yamamura. "Surface-selective modification of poly(vinyl alcohol) films with azobenzenes for in-plane alignment photocontrol of nematic liquid crystals." *Macromolecules* 28(1) (1995): 288–293.

143. K. Sakamoto, K. Miki, M. Misaki, K. Sakaguchi, Y. Hijikata, M. Chikamatsu, and R. Azumi. "Highly polarized polymer-based light-emitting diodes fabricated by using very thin photoaligned polyimide layers." *J. Appl. Phys.* 107(11) (2010): 113108.

144. M. Luo and T.H. Epps. "Directed block copolymer thin film self-assembly: emerging trends in nanopattern fabrication." *Macromolecules* 46(19) (2013): 7567–7579.

145. M. Yamada, A. Hirao, S. Nakahama, T. Iguchi, and J. Watanabe. "Synthesis of side chain liquid crystalline homopolymers and block copolymers with well-defined structures by living anionic polymerization and their thermotropic phase behavior." *Macromolecules* 28(1) (1995): 50–58.

146. G. Mao and C.K. Ober. "Block copolymers containing liquid crystalline segments." *Acta Polymerica* 48(10) (1997): 405–422.

147. T. Seki. "New strategies and implications for the photoalignment of liquid crystalline polymers." *Polymer J.* 46(11) (2014): 751–768.

148. K. Fukuhara, S. Nagano, M. Hara, and T. Seki. "Free-surface molecular command systems for photoalignment of liquid crystalline materials." *Nat. Commun.* 5 (2014): 3321–3328.

149. K. Xiao, X. Kong, Z. Zhang, G. Xie, L. Wen, and L. Jiang. "Construction and application of photoresponsive smart nanochannels." *J. Photochem. Photobiol., C* 26 (2016): 31–47.

150. N. Shah, J. Hong, M. Hyder, and P. Hammond. "Osteophilic multilayer coatings for accelerated bone tissue growth." *Adv. Mater.* 24 (2012): 1445–1450.

151. M.V. Kiryukhin, S.R. Gorelik, S.M. Man, G.S. Subramanian, M.N. Antipina, H.Y. Low, and G.B. Sukhorukov. "Individually addressable patterned multilayer microchambers for site-specific release-on-demand." *Macromol. Rapid Commun* 34(1) (2013): 87–93.

152. M.N. Antipina, M.V. Kiryukhin, A.G. Skirtach, and G.B. Sukhorukov. "Micropackaging via layer-by-layer assembly: microcapsules and microchamber arrays." *Int. Mater. Rev.* 59(4) (2014): 224–244.

153. N.S. Sariciftci. "Plastic photovoltaic devices." *Materials Today* 7(9) (2004): 36–40.

154. Q. Ye and C. Chi. "Conjugated polymers for organic solar cells." In L.A. Kosyachenko (Ed.), *Solar Cells: New Aspects and Solutions*. Croatia: Intech Open, 2011, pp. 453–474.

155. J.Y. Kim, K. Lee, N.E. Coates, D. Moses, T-Q. Nguyen, M. Dante, and A.J. Heeger. "Efficient tandem polymer solar cells fabricated by all-solution processing." *Science* 317(5835) (2007): 222–225.

156. A. Ravve. "Photocrosslinkable polymers." In *Light-Associated Reactions of Synthetic Polymers*. New York: Springer, 2006, pp. 200–243.

157. S.E. Webber. "Photon-harvesting polymers." *Chem. Rev.* 90(8) (1990): 1469–1482.

158. A. Ikeda, A. Kameyama, T. Nishikubo, and T. Nagai. "Synthesis of new photoresponsive polyesters containing donor–acceptor norbornadiene (D–A NBD) residues by the polyaddition of D–A NBD dicarboxylic acids with bis(epoxide)s and their photochemical properties." *Macromolecules* 34(8) (2001): 2728–2734.

159. T. Nagai, M. Shimada, Y. Ono, and T. Nishikubo. "Synthesis of new photoresponsive polymers containing trifluoromethyl-substituted norbornadiene moieties." *Macromolecules*, 36(6) (2003): 1786–1792.

160. W. Kühlbrandt. "Many wheels make light work." *Nature* 374 (1995): 497–498.

161. G.R. Fleming and G.D. Scholes. "Physical chemistry: Quantum mechanics for plants." *Nature* 431(2004): 256–257.

162. M. Yamada, M. Kondo, R. Miyasato, J. Mamiya, M. Kinoshita, Y. Yu, C.J. Barrett, and T. Ikeda. "Photoresponsive behavior of laminated films composed of a flexible plastic sheet and a crosslinked azobenzene liquid-crystalline polymer layer with different initial alignment of mesogens." *Mol. Cryst. Liq. Crys.* 498(1) (2009): 65–73.

163. O.M. Wani, H. Zeng, and A. Priimagi. "A light-driven artificial flytrap." *Nat. Commun.* 8 (2017): 15546.

164. J. Küpfer, E. Nishikawa, and H. Finkelmann. "Densely crosslinked liquid single-crystal elastomers.". *Polym. Adv. Tech.* 5(2) (1994): 110–115.

165. Z. Jiang, M. Xu, F. Li, and Y. Yu. "Red light-controllable liquid-crystal soft actuators via low-power excited upconversion based on triplet-triplet annihilation." *J. Am. Chem. Soc.* 135(44) (2013): 16446–16453.

166. C. Huang, J. Lv, X. Tian, Y. Wang, Y. Yu, and J. Liu. "Miniaturized swimming soft robot with complex movement actuated and controlled by remote light signals." *Sci. Rep.* 5 (2015): 17414.

167. H. Zeng, P. Wasylczyk, C. Parmeggiani, D. Martella, M. Burresi, and D.S. Wiersma. "Light-fueled microscopic walkers." *Adv. Mater.* 27(26) (2015): 3883–3887.

168. D. Ryu and K.J. Loh. "Strain sensing using photocurrent generated by photoactive P3HT-based nanocomposites." *Smart Mater. Struct.* 21(6) (2012): 065016.

169. H.S. Kang, J.Y. Lee, S. Choi, H. Kim, H., Park, J.Y. Son, B.H. Kim, and S.D. Noh. "Smart manufacturing: past research, present findings, and future directions." *Int. J. Precis. Eng. Manuf.-Green Tech.* 3(1) (2016): 111–128.

170. S. Tibbits. "4D printing: multi-material shape change." *Architectural Design* 84(1) (2014): 116–121.

171. D-G. Shin, T-H. Kim, and D-E. Kim. "Review of 4D printing materials and their properties." *Int. J. Pr. Eng. Mangt.* 4(3) (2017): 349–357.

172. F. Cock, A. Cuadri, M. García-Morales, and P. Partal. "Thermal, rheological and microstructural characterisation of commercial biodegradable polyesters." *Polym. Test.* 32(4) (2013): 716–723.

173. O. Ivanova. A. Elliott, T. Campbell, and C. Williams. "Unclonable security features for additive manufacturing." *Addit. Manuf.* 1 (2014): 24–31.

174. X. Liu, Y. Zheng, S.R. Peurifoy, E.A. Kothari, and A.B. Braunschweig. "Optimization of 4D polymer printing within a massively parallel flow-through photochemical microreactor." *Polym. Chem.* 7(19) (2016): 3229–3235.

175. D. Rus and M.T. Tolley. "Design, fabrication, and control of soft robots." *Nature* 521 (2015): 467–475.

176. B. Jin, H. Song, R. Jiang, J. Song, Q. Zhao, and T. Xie. "Programming a crystalline shape memory polymer network with thermo- and photo-reversible bonds toward a single-component soft robot." *Sci. Adv.* 4(1) (2018): 3865.

177. O. Zhao, H.J. Qi, and T. Xie, T. "Recent progress in shape memory polymer: New behavior, enabling materials, and mechanistic understanding." *Prog. Polym. Sci.* 49–50 (2015): 79–120.

3

Smart Paints

Asit Baran Samui
Institute of Chemical Technology, Mumbai, India

Sushil S. Pawar
Naval Materials Research Laboratory, Ambernath, Maharashtra, India

CONTENTS

DOI: 10.1201/9781003037880-3

Abbreviations

TBTM	tributyltinmethacrylate
TBTA	tributyltinacrylate
PDMS	poly(dimethylsiloxane)
PEG	polyethylene glycol
AF	antifouling
NPs	nanoparticles
FR	foul release
MWCNT	multiwalled carbon nanotube
CA	contact angle
UV	ultraviolet
PEO	polyethylene oxide
PU	polyurethane
PSPM	poly (3-sulfopropyl methacrylate)
PCBMA	poly(carboxybetaine methacrylate)
AMPs	antimicrobial peptides
QAC	quaternary ammonium compounds
PVP	poly (4-vinyl pyridine)
MOFs	metal organic frameworks
PANI	polyaniline
ICP	intrinsically conducting polymer
MS (panels)	mild steel
DOPH	dioctyl phosphate
CSA	camphor sulfonic acid
TEOS	tetraethyl orthosilicate
APTS	3-aminopropyltriethoxysilane
GO	graphene oxide
RfA1	histidine-tagged reflectin A1
IR	infra-red
CIE	Commission on Illumination
TIR	thermal infrared radiation
NIR	near-infrared
Radar	radio detection and ranging
RAM	radar absorbing material
PPy	polypyrrole
LBL	layer-by-layer
CF	carbon fiber
CFRACs	carbon fiber radar absorbing coatings
CWA	chemical warfare agent
ROMP	living ring-opening metathesis polymerization
DCPD	dicyclopentadiene
HOPDMS	hydroxyl terminated polydimethylsiloxane
PDES	polydiethoxysiloxane
PSS	poly(styrene sulfonate)
PEI	poly(ethyleneimine)
BTE	benzotriazole

MSNs	mesoporous silica nanoparticles
PMAF17	poly(2-perfluorooctylethyl methacrylate)
MS	microgel sphere
CCF	coconut fiber
WW	wood waste
PHST	peach stone
SSA	self-cross-linked silicone acrylate
PVA	polyvinyl alcohol
SBR	styrene butadiene rubber
PVC	pigment volume concentration
F-POSS	fluorodecyl polyhedral oligomeric silsesquioxane

Symbols with Units

Symbol	Description	Unit
dB	Used to express one value of power to another	Dimensionless
GHz	Frequency of electromagnetic radiation	Cycles/sec

3.1 Introduction

Various types of paints have been used to decorate and protect materials and surfaces over the centuries. Modern paints and coatings are well suited for thousands of diverse applications, for example, coatings required for the marine environment, deserts, automobiles, underwater applications, industrial installations, etc. vary with the specific requirement dictated by the prevailing environment. Although the varieties of present-day paints are near-limitless, these are basically a "passive" type in nature. Smart paint is a kind of paint which will add something extra to the conventional function of the paint and will behave exactly similar in all target conditions. For example, a conducting polymer-based coating functions by first forming an impervious oxide layer, which is followed by isolation of the anodic and cathodic reaction, if there is any, by using the conducting nature of the paint.[1] Painting schemes comprise multiple coats with different roles to play. A smart action is to paint a surface once and have two layers with different functions. A self-stratified coating depends on the immiscibility of two binders that separate after applying the coating.[2] For antifouling coatings, toxins can be designed which are leached out to the surrounding water medium to maintain an appropriate dose that prevents the settlement of sea organisms while keeping the paint surface smooth.[3]

During the service life of a coating, a change in its mechanical properties occurs that leads to the formation of micro-cracks. These micro-cracks propagate slowly, resulting in atmospheric moisture and oxygen accessing the substrate that leads to accelerated disbanding of the coating. For micro-crack and mechanical damage, an autonomic repair function can be incorporated by using polymers with a self-healing ability.[4] The polymer used for self-healing can repair minor damage, which does not require any detection system with additional cost or any type of human participation.[5] Smart paint can be designed in such a way that a military vehicle changes color on the battlefield in response to stimuli that will create instant camouflage to make it indistinguishable from the background. A fire-retardant intumescent paint is designed by incorporating ingredients which react in the presence of fire to make carbon foam of about 100 times more height compared to coatings. The thick carbon foam, being a bad conductor of heat, protects the surface from further damage.[6] Thus, the addition of various attributes to conventional paint makes it exhibit smart behavior and protects the surface or performs better than conventional paints.

3.2 Antifouling Paint

Antifouling paint is used to protect underwater surfaces from fouling (hooking of marine organisms onto a ship's hull and other underwater surfaces to produce a hard and thick mass) so that fuel consumption and operational availability are not adversely affected (Figure 3.1). The smart toxic paint to prevent fouling is based on organotin polymer (tributyltinmethacrylate (TBTM)) resin as a binder.

The chemically linked organotin leaches in water in a controlled manner. After most of the organotin moiety is hydrolyzed in sea water and leached out to the surrounding water medium, the top microlayer of paint is left with a polymer containing carboxyl groups. This layer is highly hydrophilic and the moving sea water erodes the layer so that the next smooth layer comes into contact with the water, which starts leaching in the same way. Thus controlled leaching coupled with ablative action makes the system self-polishing in nature[3] (Figure 3.2).

Various copolymers have been designed to suit the requirements by having a specific mole percentage of TBTA/TBTM.[7, 8] The leaching rate above the optimum value of 0.4 μg.cm^{-2}day^{-1} is maintained by designing copolymers with TBTM content of about 30 mol% in the copolymer composition.

In 1999, the International Maritime Organization (IMO) adopted the resolution: "After 1 January 2003, the application of organotin-based paint on ships and after 1 January 2008, the presence of organotin in antifouling paint on ships are completely banned." Tributyltin-based resin was replaced with surrogate organometal and organic biocides with less toxicity on oceanic flora and fauna.

There are two effective strategies to prevent fouling using nontoxic coatings:

1. Easy removal of biofouling organisms mostly from hydrophobic surfaces (foul release).
2. Not allowing biofouling organisms to attach themselves by maintaining a hydrophilic coating.

The first category of coatings maintains low surface energy, which does not allow the marine organisms to build a strong interfacial bonding with the substrate. The loosely settled organisms get dislodged after the vessel picks up speed higher than the value established as a critical velocity,[9] which lies in the range of 10–20 knots.[10] To produce a surface with low adhesion for the adhesives produced by marine organisms, the following features need to be minimized:[11]

1. Mechanical intertwining, by maintaining a non-porous and smooth surface.
2. Wetting of the surface, by incorporating the suitable functional groups available on the surface.
3. Chemical bonding, by keeping the surface free from reactive functional groups.
4. Electrostatic interactions, by making a low polarity surface.
5. Diffusion of molecular chains from the marine adhesive onto the surface, by having closely packed functional groups.

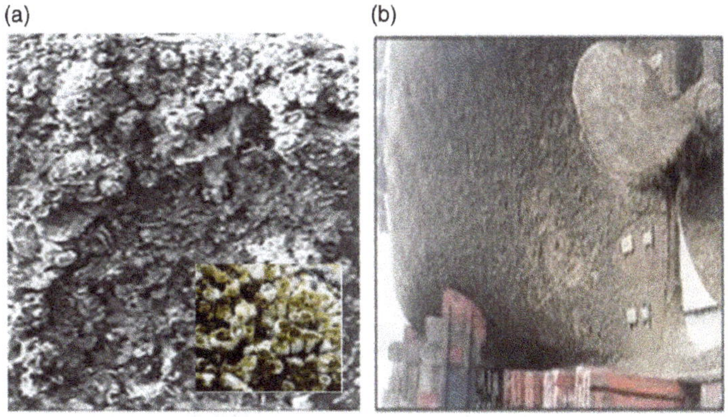

FIGURE 3.1 Fouling on (a) sea water exposed panel (inset: barnacle shell); (b) underwater surface of the ship.

Source: [31].

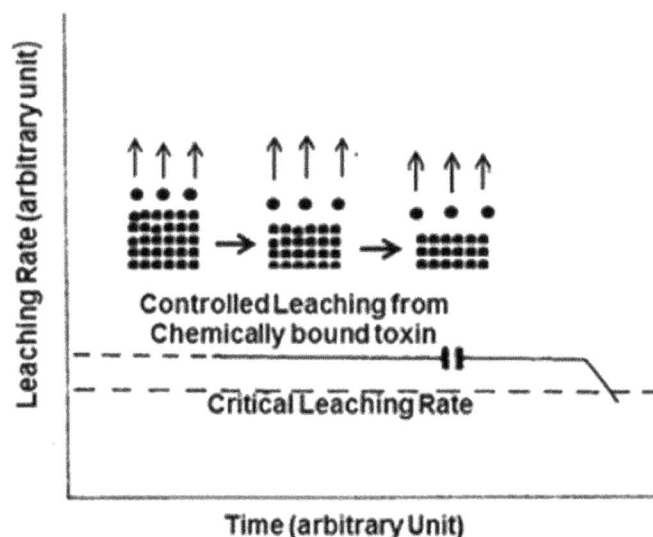

FIGURE 3.2   Representative plot for controlled leaching from chemically bound toxin vs. time.

Baier proposed that bio-adhesion should remain at a minimum by maintaining the surface tension in the range of 20–30 mN m[-1].[12] It was later discovered that the antifouling properties are not dependent on one parameter. Rather, they depend on the combined effect of surface energy and elastic modulus in the form of the square root of the product of surface energy and elastic modulus.[13, 14]

Poly(dimethylsiloxane) (PDMS), having an Si–O–Si backbone and a bond length and bond angle value higher than C–C–C bonds, has low surface energy due to –CH$_3$ side groups.[15] Because of the conformational mobility in PDMS, the pendant-CH$_3$ groups remain in a closed pack arrangement. The adhesion of marine organisms to this surface is low due to its mobile nature. These characteristics make the polymer maintain low surface energy and elastic modulus. Furthermore, it maintains a low glass transition temperature (T$_g$) and micro-roughness is practically absent.

3.2.1 Self-Replenishing Coating

It is a general observation that the leaching of the silicon-based component from the coating during service results in the variation of the surface chemical composition and topography, leading to inferior foul release (FR) performance. Three-dimensional vascular networks can be used to make self-replenishing surfaces by implanting them in polydimethylsiloxane (PDMS) that is filled with silicone oil so that an oil-infused matrix is generated.[16] The ability to be replenished with silicone oil from an outside source qualifies these materials to produce self-lubrication with the continuous renewal of the fouling-release layer.

3.2.2 Combined Fluorine–Silicone-Based Materials

Siloxane polymers, with a low critical surface tension and low elastic modulus, and fluoropolymers maintaining low surface tension, coupled with chain rigidity and a dense-packing ability, can be combined to get the best of both and a superior FR performance. A fluorinated-siloxane acrylic copolymer can be blended

with a PDMS matrix to make a coating so that the fluorinated copolymers segregate toward the air-interface, resulting in lower surface energy, while maintaining the usual mechanical properties.[17] Even with very low loading, the blend surface comprises – CH$_3$ terminals of siloxane chains and fluorinated side chains. This blended film exhibits extremely good FR properties with efficient removal of most barnacles that is superior to pure PDMS.[18]

3.2.3 Self-Stratified Coatings

The modified siloxane polymers attract the most attention as they offer possibilities of self-stratification. In self-stratifying coatings, the component-producing low surface energy migrates to the air interface during the formation of coating, while remaining an integral part of the polymer matrix[19] (Figure 3.3),

3.2.4 Amphiphilic Copolymers

The amphiphilic polymers are those types in which both hydrophilic and hydrophobic functionalities are combined, and the technique is considered a promising strategy to implement special interface functions in a variety of industries and technologies.[20] The presence of different functionalities at the top surface of the coatings offers molecular heterogeneity. The coexistence of two different functionalities on the top surface of a coating is expected to be advantageous by combining two different functions such as detachment of the biofouling organisms and inhibition of attachment, leading to high-performing FR surfaces.[21] The amphiphilic polymers, by maintaining a heterogeneous nanoscale mosaic chemical surface, with the coexistence of hydrophobic and hydrophilic domains, offer a deterrent to organisms during settlement and adhesion.[22]

3.2.5 PEG-Based Materials

Protein adsorption on the underwater surface creates a conditioning layer for microbial settlement and growth, leading to

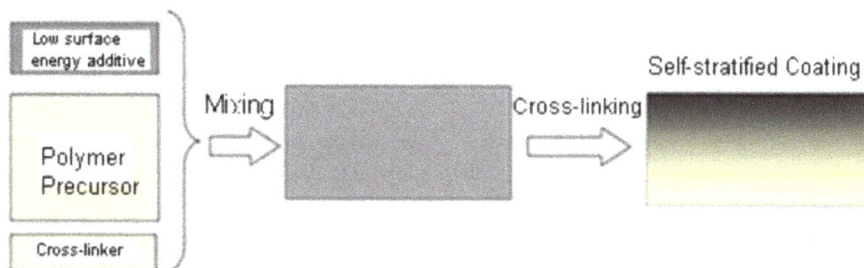

FIGURE 3.3 Schematics of the formation of self-stratifying coatings.

Source: [19]. Published with permission from The Royal Society of Chemistry.

Control panel   As removed panels   After gentle wash of as removed panels

FIGURE 3.4 Bimodal cross-linked PDMS nanocomposite coatings. Base Resin: hydroxy-terminated PDMS (M_n ~ 27,000); Cross-linker combination: *Tetra*ethoxysilane and aminopropyl triethoxy silane; Organoclay: Cloisite 10A (1% by weight).

Source: [31].

the formation of robust, surface-associated communities called biofilms. Although PEG film has high surface energy, PEG-coated surfaces resist protein adsorption and cell adhesion and also resist colonization of biofoulants.[23] The two main reasons for this are: (1) the elastic force with a repelling nature originates from the PEG chains undergoing compression during the movement of protein toward the surface; and (2) thermodynamically developed osmotic stress does not support the removal of the water molecules from the strongly hydrated PEG chains. Containing such a large amount of water, the extent of formation of the PEG–water complex is high, which cannot be interrupted because it is energetically and kinetically unfavorable.

3.2.6 Foul-Release Polymer Nanocomposites

It is known that PEG undergoes auto-oxidation in the presence of oxygen with transition metal ions as a catalyst that leads to degradation of the coatings. However, preventive action against decomposition is recorded after making PEG nanocomposites, and PEG-ZnO nanocomposites do not allow protein adsorption to a considerable extent and the bacteria hooked onto the PEG-ZnO nano-surfaces are easily removed after 4 hours of incubation.[24]

Hydrogel network structures, because they are porous and hold a large amount of water, are generally nontoxic, highly elastic, and do not allow bio-macromolecule adhesion as they prevent the coupling formation required for protein fouling. However, in spite of having excellent properties, such as allowing an efficient

mass transfer, maintaining hydrophilicity, and lack of response to stimulus and cell, the use of hydrogels is not widespread because of their inferior mechanical strength and because they become brittle upon dehydration.[25] High antifouling (AF) and mechanical properties of a hydrogel are achieved by making a nanocomposite, comprising a methacrylamide derivative, a hydroxyl derivative of methacrylate, and clay nanoparticles (NPs).[26]

The inherently superior FR properties of PDMS coatings can be improved by combining them with inorganic nano-additives, as this will enhance the mechanical strength of the coating, together with the enhancement of the hydrophobic character of the air interface. The NPs can be incorporated into a long-chain polymer matrix to form nanostructures with trapped NPs in polymer coils that alter the rheological behavior.[27] Multi-walled carbon nanotubes (MWCNTs), natural sepiolite, and organo-modified montmorillonite, etc. with hydrophobic nature, can be added to the PDMS-based coating to increase modulus, decrease roughness, increase the contact angle (CA), and thus the self-cleaning characteristics are improved.[28] Even a small amount of MWCNT incorporated into silicone-based coatings can augment the release characteristics of fouling organisms without much change in the mechanical properties.[29] Due to the strong CH–π interactions between the silicone methyl side chains and the π-electron-rich surface of MWCNTs, the extent of the residual surface energy is very low that contributes to the reduction of the adhesion of barnacles. Introduction of fluorine atoms into MWCNT also reduces the barnacle adhesion strength with the coating. There is a drastic reduction in pseudobarnacle attachments' strength on the nanocomposite coating just by replacing the MWCNT with fluorinated MWCNT.[30] Figure 3.4 presents the nanocomposite painted panels after they have been exposed in sea water.[31]

Nanoclay, being polar in nature and having high density, pulls the polar functionalities toward the substrate, while the surface contact angle increases to perform the enhanced foul release action. Super-hydrophobic coatings, with a CA more than 150°, are also suitable for use as an FR coating. In fact, the surface hydrophobicity is not the only criterion used to decide the FR performance, other contributory factors may be mentioned, such as the topology as well as the roughness length scale of the surface, and the extent of air entrapment in the coatings.[32]

Solar light-influenced FR paints for vessel bottoms are designed by combining photo-induced silicone/spherical single crystal TiO_2 to make nanocomposites. The incorporation of the latter in a silicone matrix modifies the roughness, wettability, and FR performance of the coating when it is exposed to UV-vis irradiation. A self-cleaning action is supported by the increased super-hydrophilicity of the FR coating.[33] The UV irradiation causes a decrease in the CA, which results in increased super-hydrophilicity. The UV-induced super-hydrophilicity will be difficult to perform in sunlight as the percentage of UV in sunlight reaching the Earth's surface is very low. The photocatalytic activity of TiO_2 NPs can be enhanced for visible light irradiation if noble metals such as Au or Ag are used as a dopant.[34] Core/shell nanofibers of TiO_2@carbon embedded in Ag nanoparticles (TiO_2@C/Ag NFs) have the ability to absorb light significantly and separate photo-generated electron-hole pairs excellently. This is considered the photosynergistic effect of the three components of TiO_2, carbon, and Ag. Many graphene-based nanocomposites with TiO_2, ZrO_2, ZnS, and CdS/TiO_2 exhibit higher photocatalytic performance compared to that using only metal oxide. Improved super-hydrophilic characteristics together with visible light-induced self-cleaning properties are observed with ZnO-rGO nanocomposite films.[35]

3.2.7 Engineered Micro-Topographies

3.2.7.1 Bio-Mimetic Antifouling Strategies

The antifouling strategies based on bio-mimetics fall into two main categories: microtopography and lubricant retention.[25] Microtopography, designed by mimicking nature without using any toxic component, has been used as an antifouling technique to counter biofouling.[36] Wettability, the fluid dynamics of a surface, and reduced attachment points are the factors responsible for improving antifouling performance.[37]

Larger fouling organisms find fewer attachment points due to the specific surface texture, and a topography that reduces the adhesion strength. The other antifouling strategy is the secretion and subsequent retention of a lubricant, which includes low-surface tension oils or hydrophilic proteoglycan aggregates that repel ions of the fouling agents and provide self-cleaning properties as well. Mollusc shells, fish skin, etc. are equipped with this type of antifouling strategy. The lubricants are held via soaking in topographical surface textures or by chemically

grafting to substrates or electromagnetic interactions. The microtopographic textures in a Nepenthes pitcher plant can hold an aqueous lubricant, which makes it easy for insects to slip inside it.[38] No sign of biofouling is observed on the surfaces of macroalgae, and this bio-inspired defense mechanism has been adopted to design an AF material.[39] The bio-inspired defense can be built up by using soft lithography using a PDMS negative mold, by pouring a PDMS solution over the macroalga sample and curing and removing the biomass. The lowest extent of biofouling is observed.

3.2.7.2 Bacteria-Resistant Polymer Brush Coatings

Polymer brushes are assemblies of macromolecular chains with a high density, linked at one end to a surface or an interface. Biopassive and anti-adhesion polymer brush coatings are designed to make the surface bacteria-resistant.[40] Along with high mechanical stability, the protection against the settlement of biomolecules and cells is ensured because of the adjustable thickness of the grafted polymer layer, the possibility of introducing selective functional groups, and because it has molecular exclusion properties.

3.2.7.3 Anti-Adhesion Coatings

Covalently bound polyethylene oxide (PEO) brushes as neutral hydrophilic coatings on glass and silica surfaces are able to nearly exclude the adhesion of *Staphylococci* and *Escherichia coli*, compared to their abundant settlement on an uncoated surface. The PEO brushes are thought to suppress the microbial adhesion by decreasing the Lifshitz-van der Waals attraction.[41] The zwitterionic polymer brush-derived coatings, based on ammonium betaine, on polyurethane (PU) vascular catheters, neatly prevent the adhesion of a broad range of microorganisms, as against bare catheters. Also, biofilm formation is prevented.[42, 43] Ammonium betaine-based polymer-coated catheters implanted in rabbits show much less inflammation and almost negligible adherence of bacteria, along with the reduction of platelet activation and accumulation of the thrombotic material on the catheter exterior.

3.2.7.4 Bioactive Coatings

Antimicrobial agents are either released from the surfaces or immobilized on the surfaces depending on their ability to kill the bacteria or inhibit their growth:

- *Bactericide-releasing coatings*: Because of the ability of polymer brushes to incorporate only a small amount of bactericidal agents, they have not received sufficient attention in the formulation of bactericide-releasing coatings. However, by growing poly (3-sulfopropyl methacrylate) (PSPM) brushes from gold and silicon surfaces and loading with silver ions, the growth of *S. aureus* and *P. aeruginosa* and biofilm formation can be inhibited.[44] The poly(carboxybetaine methacrylate) (PCBMA) brushes with an Ag-loaded coating kills the *E. coli* almost completely within 1 h.[45]

- *Contact killing coating*: Generally, contact killing depends on either the disruption of the bacterial membrane or by the specific interaction of the immobilized antibacterial agent with the biomolecule on the surface of the bacteria.

3.2.7.5 Polymer Brush-Based Contact Killing Coatings

The optimal interaction of the immobilized antimicrobial agents with the bacterial membrane depends on the spacer length, along with the accessibility of the antimicrobial agent. Coatings based on polymer brushes offer various advantages, such as: (1) sufficient conformational freedom for the immobilized biocidal agent due to the flexibility of the grafted chains to interact with the bacterial membrane; (2) a non-fouling polymer brush coating together with a fouling polymer brush reduces the bacterial adhesion and also prevents dead bacteria from being deposited on the surface; and (3) the hydrophilic nature of the polymer brush coating is responsible for the biocompatibility of the surfaces.[40]

- *Polymer brushes incorporating antimicrobial peptides*: Antimicrobial peptides (AMPs) belong to either natural or synthetic resources. By grafting a polymer brush system onto silica/paramagnetic silica microparticles and then conjugating it with cysteine-functionalized magainin, antibacterial activity is realized with 75% efficiency and more.[46] Magnetic field-controlled, localized, antibacterial action is possible by using the same system grafted onto magnetic particles.

- *Polymer brushes incorporating lysozyme*: Immobilization of lysozyme, a natural defense protein, on different substrates, such as polyvinyl alcohol and nylon 6/6, can lead to the development of antimicrobial surfaces.[47]

- *Synthetic bactericidal polyelectrolyte brushes*: The quaternary ammonium compound (QAC) polymeric system

is one of the few widely studied synthetic polymers. By securing poly(4-vinyl pyridine) (PVP) to glass slides via a covalent link and treating it with an alkyl bromide solution, positively charged poly(4-vinyl-N-alkyl pyridinium bromide) can be synthesized.[48] A hexyl-PVP-covered glass substrate is able to kill most of the sprayed bacteria, such as *S. aureus*, *S. epidermis*, *E. coli*, and *P. aeruginosa* as against less than 10% by covering with a long alkyl chain, and no activity of non-alkylated PVP chain-covered surfaces. This is due to the fact that the long alkyl chains stick to one another through hydrophobic interaction to provide compact coating (Figure 3.5). Thus, there exists a critical alkyl chain length that is optimized to formulate with superior antimicrobial activity.

3.2.8 Metal-Organic Frameworks (MOFs)-Based Antifouling Coatings

Metal-organic frameworks (MOFs), derived by combining metal or metal-oxo connectors and organic linkers, have important applications in molecular-level storage and as a carrier of drug in drug delivery.[49, 50] For example, surface-anchored metal-organic frameworks based on cupric ion show automatic release at a very controlled rate and the phenomenon has the potential to be exploited as a stimulus-responsive antifouling material.[51]

3.3 Corrosion-Resistant Coatings

Degradation of metal components, in the presence of moisture, salt spray, oxidation, or exposure to a variety of environmental or industrial chemicals, can be prevented by coatings designated as corrosion-resistant coatings. The primary role of the coatings is to maintain a barrier between the chemical compounds or corrosive

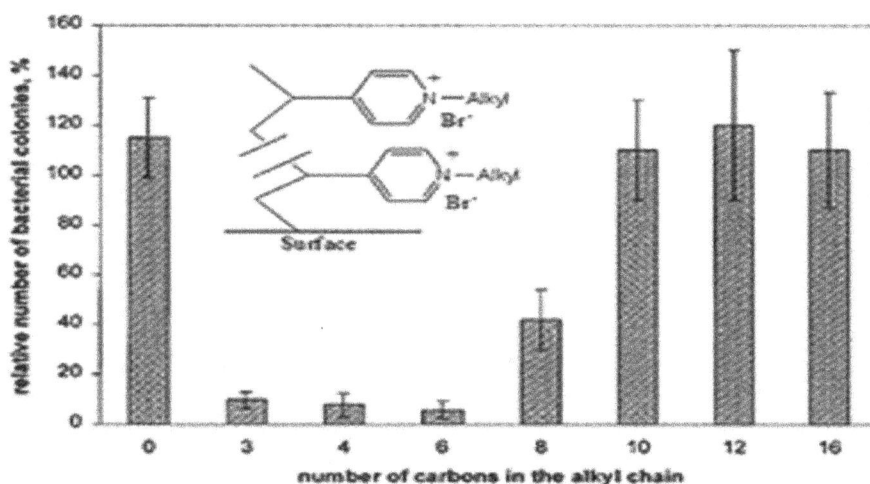

FIGURE 3.5 The percentage of S. aureus colonies grown on the infected surfaces of glass slides modified with PVP that was N-alkylated with different linear alkyl bromides relative to the number of colonies grown on a commercial NH_2-glass slide (used as a standard). The bacterial cells were sprayed from an aqueous suspension (106 cells per ml) onto the surfaces. All experiments were performed at least in quadruplicate, and the error bars indicate the standard deviations from the mean values obtained.

materials so that none are allowed to reach the substrate. In some cases, zinc or aluminum-based primers are used to allow these metals to protect the metal galvanically. There are large numbers of protective techniques developed in the smart paint category.

3.3.1 Conducting Polymer-Based Corrosion-Resistant Coating

The use of conducting polymer in a coating enables it to interact with the substrate metal during application and also with the surroundings during exposure, which makes it one of the smart anticorrosive coatings. The emergence of polyaniline (PANI) as an anticorrosive pigment was initiated by Wesseling, as he demonstrated that passivation occurs on the metal surface due to the redox reaction of PANI, resulting in the oxide layer formation.[1] Also, by dispersing PANI in the paint, the same oxide film appeared on the metal surface below the paint layer, and the coating qualified as an industrially viable corrosion prevention primer.[52] The properties of conducting polymer in coatings and its effect on the corrosion rate of ferrous alloys (iron, steel, and stainless steel) were reviewed for a further understanding of the corrosion prevention behavior.[53] The proposed mechanism of corrosion protection comprises a barrier formation to prevent ingress of corroding agents, anodic protection, and the mediation of oxygen reduction, respectively. An alternative mechanism is also proposed, that the conducting polymer becomes polarized via the formation of galvanic coupling with the base metal substrate at defects sites in the coating, which induces the intrinsically conducting polymer (ICP) to release an inhibiting anion.[54] Thus, the results of cathodic reduction of the conducting polymer or ion exchange with cathodically generated OH⁻, or both, release an anion dopant. The corrosion-inhibiting nature of the anionic dopant provides corrosion protection. The phosphonate dopant[55] and camphor sulfonate anion also show inhibiting action when used as a dopant with an ICP-based coating.[56] A styrene-butyl acrylate copolymer binder with PANI-HCl as a filler exhibits very good corrosion resistance properties even at a relatively low thickness.[57] Paints containing lower PANI-HCl (0.1–0.5 parts PANI-HCl) protect mild steel better than that of higher PANI-HCl loading, as confirmed by an accelerated weathering test, and other underwater exposure studies of painted mild steel panels

(Table 3.1). Low water vapor permeability confirms the very good barrier properties. Another dopant such as dioctyl phosphate incorporated in the epoxy coating (5 phr PANI–DOPH) exhibits protection for a longer exposure time than the other resins.[58] The corrosion protection study of epoxy-coated mild steel panels with DOPH doped PANI (5 phr), exposed in an extremely corrosive surf zone, exhibits the start of corrosion only after 18 months, whereas all the panels coated with other resins undergo heavy corrosion (Table 3.2).

3.3.2 Nanocomposite Coatings

Polyaniline camphor sulfonic acid (CSA)/zinc oxide nanoparticles (PANI-CSA/ZnO), organic-inorganic hybrid nanocomposites can be dispersed in the epoxy coating to act as a corrosion-inhibiting pigment and the resulting coating, when applied to carbon steel, has superior corrosion resistance than the coating without nanoparticles.[59] The coating exhibits excellent resistance to corrosion in synthetic sea water, confirming its higher barrier properties. Also, the open circuit potential of the conducting paint shifts to the noble region. The barrier and corrosion resistance properties are improved further in the presence of ZnO nanorods.

Layered structured nanomaterials, such as graphene oxide, when incorporated in a polymer to make a nanocomposite, provide very high barrier properties against gases and moisture.[60,61] SiO_2-GO nanohybrids can be prepared by reacting tetraethyl orthosilicate (TEOS) and 3-aminopropyltriethoxysilane (APTS) in a water-alcohol solution via a sol-gel method, which is used for making epoxy composites.[62] Even with 0.2 wt% SiO_2-GO nanohybrids, the epoxy coatings maintain a satisfactory barrier performance. The SiO_2-GO hybrids can also be prepared by heating a GO dispersion in water with a TEOS solution in toluene at 70–80°C for 24 h.[63] When the SiO_2-GO nanohybrid is incorporated into an epoxy coating at a very low concentration, the water contact angle of the nanocomposite increases along with its superior corrosion protection. The PANI–GO (1% GO) composite pigment, dispersed in epoxy resin, prevents corrosion for long time, with a corrosion rate of 6.5×10^{-5} mm year⁻¹, and maintains high impedance modulus and real impedance, high coating resistance and minimum coating capacitance after 4 days of immersion in artificial sea water.[64]

TABLE 3.1

Results of Salt Spray Exposure of PANI-HCl Containing Paint

| | Corrosion Performance (% Area Corroded) | | | | | | | |
| | Parts PANI-HCl in the Paint | | | | | | | |
Exposure Time (h)	0.0	0.1	0.5	1.0	3.0	5.0	10.0	20.0
210	Blister	–	–	–	Spot	20	5	5
310	Blister	–	–	–	5	30	7	35
410	Blister	–	–	–	15	50	12	55
510	5	Spot	Spot	12	35	75	25	75
610	25	10	10	25	50	85	85	95
710	80	75	70	65	85	95	100	100

Source: Data from [57].

TABLE 3.2

Results of Field Exposure Study of PANI–DOPH Containing a Coating on MS Panels

Coating	Thickness of coating (µm)	Exposure Time for Onset of Corrosion (months), Concentration of PANI–DOPH in the Coating (phr)				
		0.0	0.1	0.5	1.0	5.0
Silicone alkyd	100± 5	3	6	6	6	6
Soya alkyd	--do--	2	5	5	5	5
AC 80	--do--	4	8	8	7	6
Epoxy	--do--	4	15	15	16	No corrosion
PU	--do--	4	10	10	12	12
Epoxy red oxide	--do--	3	–	–	–	–
Amerlock 400	--do--	2	–	–	–	–

Source: Data from ref. [58].

3.3.3 Stimulated Protective Biofilms

Protective biofilms formation is triggered by some coatings. Biogenetically engineered bacteria release some species like polypeptides and polyphosphates that act as a corrosion inhibitor.[65] However, the possibility of controlled release has not yet been realized. Corrosion protection of Al 2024 in 3.5% NaCl solution is effected by the formation of a biofilm.[66] Since the bacteria have the ability to coat metals with a regenerative biofilm, this feature is becoming important in preventing corrosion.[67] Although homopolysaccharides behave interestingly for the protection of steel, they are rather slow in building up biopolymer layers on metal to make it completely protected. The L. fermentum Ts produces exopolysaccharides, functioning as a corrosion inhibitor for mild steel.[68]

3.4 Paints for Camouflage Purposes

3.4.1 Visible Camouflage

Visible camouflage, better known as active or adaptive camouflage, is the ability of an object to merge into the surroundings so that it cannot easily be identified. In theory, active camouflage enables an object to hide perfectly from visual detection. As an example, military vehicles need to be concealed from enemy detection by using visible camouflage. Similarly, animals use visible camouflage to avoid detection and being killed by other animals. The color-changing ability permits camouflage against different backgrounds. Octopuses, cuttlefish, and some terrestrial amphibians and reptiles, including chameleons, are very active in changing their color rapidly and also their pattern, mainly for signaling and also for camouflaging.[69]

In the animal kingdom, the color variety of creatures is widespread and that has an important biological function in camouflage, communication, and temperature control.[70, 71] The natural color mechanism inspired the scientific community to develop materials with a color-based functionality.[72, 73] These advances have benefited the development of both commercial and military technologies, including reconfigurable camouflage, advanced adaptable optics, and so on.[74-76] The two common aspects in adaptive color generation in nature and as well as artificially are: (1) the adaptive color can be varied by compacting or spreading pigment; and (2) efficient reflection can be maintained by using optical interference/diffraction.

The hydrogel-like behavior of reflectin thin film allows it to swell in the presence of water vapor. This, in turn, shifts the peak reflectance in the visible wavelength region.[77] Histidine-tagged reflectin A1 (RfA1)-covered silica substrates appear orange under white light illumination, whereas, under IR illumination (λ = 940 nm), they appear black.[78] The RfA1-coated substrate turns red in the presence of acetic acid, while reaching the maximum relative brightness. Under low or zero illumination conditions, the targets in military or surveillance actions are detected by using an optical detection system.[79]

Three types of night vision detection systems are used: (1) amplification of visible light from an object in a low lighting environment; (2) a reflection of IR emitted from the device is detected; and (3) the thermal IR (heat) emitted from the object is detected by the device. To suppress these signatures to avoid detection, several strategies have been adopted. Detection in the visible range, the near IR to low IR range, can be avoided by designing materials with high absorption and low reflection. Mid-IR camouflage is achieved by way of maximum absorption, minimum mid-IR reflections, and matching the optical emission of the surroundings in the mid-IR range. Visual camouflage plays an important role in military targeting. Normally, geometric-shaped camouflage patterns are used for tactical military equipment and clothing, which are able to hide their identification features, such as shape, shadow, highlights, and various types of lines that distinguish artificial objects from the natural environment. As this is a combination of art and science, the functioning of camouflage paint over a range of backgrounds needs help from appropriate computer techniques. The concealment of military objects, such as a building, is done by using camouflage painting, which blends the visual features of a building with the surroundings, and the camouflage can be done so effectively that it is hard to detect the building from the air.

3.4.2 Aircraft Camouflage

Aircraft camouflage is used to make it difficult for aircraft to be seen. One possible approach is the use of multiple coatings to tackle the background and lighting conditions. A common method is the application of aircraft coating using a disruptive pattern, comprising green and brown colors above and sky colors below. The color scheme is important. For example, when the aircraft is flying by night, it is coated black, but an observer sees it darker than the night sky, which needs to be replaced by color from the camouflage schemes. However, the schemes are used if definite requirements exist. In most camouflage approaches, various shades of green and blue colors are employed.

By combining a spectroradiometric model (for 3D signature prediction) with statistics of varying background fields and an International Commission on Illumination (CIE) color difference metric, camouflage color for helicopters is determined.[80] Using the color inventory, each paint is modeled with bidirectional reflectance distribution function scattering properties so that it matches the existing army paint and spectral reflectance. It is commonly observed that performance varies more in the desert terrain than in vegetative terrain. A marginal monocoat solution, required for joint vegetated/desert terrains, can also be formulated. CNT coating is applied to fabric to modify the optical properties for camouflaging purposes.[79] To match the optical reflection with the environment, the top CNT layer density is varied.

3.4.3 Camouflage Painting of Buildings

The main camouflage principles, such as color resemblance, obliterative shading, disruptive coloration, shadow elimination, are required to be satisfied during the camouflage painting of a building.[81] The applied camouflage pattern must be correct to create an effective blending of the building with the surrounding terrain. The building camouflage is different from personal camouflage in that the primary threat in the former is from aerial reconnaissance. The characteristic shape of the building is disrupted, shine is reduced to make the building difficult to identify even if it is spotted. The paint application is one of the main approaches, while patterns are designed to make it more difficult to visualize, and matte color is applied to reduce shine.[82] Digital camouflage is designed by using a number of small rectangular

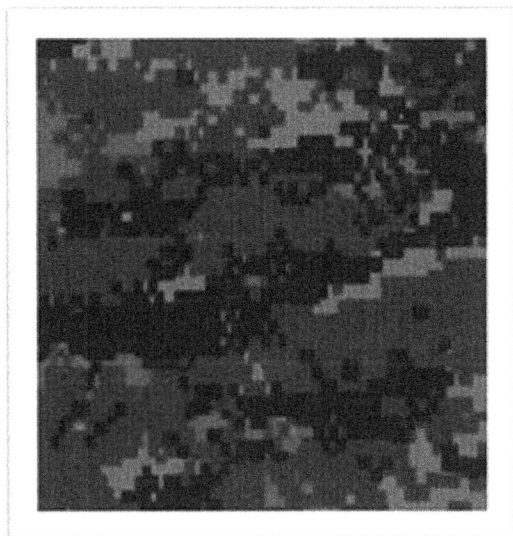

FIGURE 3.6 Digital CADPAT camouflage pattern.

pixels of color (Figure 3.6). Theoretically, this is a highly effective camouflage pattern, more effective than a standard uniform one, as it mimics the dappled textures and rough boundaries found in natural surroundings.

The typical value of a complex refractive index of a CNT forest is, $n_{eff} = 1.04+0.01i$ at visible wavelengths. Thus, if a thick layer of the CNT forest is used to design a conformal coating on a substrate, the impedance matching with air is achieved along with near perfect absorption.[83] The result is that practically no light reflection and scattering are emitted from the object.

3.4.4 Thermal Camouflage

Once IR detection systems are introduced in missile seekers, surveillance systems, search and track systems, the signature of military targets is needed to be minimized to avoid detection. Various techniques used to avoid detection are minimizing the heat radiation of hot spots in the military object and also applying low emissive paint to the body of the vehicle. The basic functions of the paints are to reduce the emission of thermal infrared radiation (TIR) and the wavelength shift for infrared emissions. Further, the reduction of emissions in the thermal region can be combined with reduced reflection in the near-infrared (NIR) bands, used for laser detection coatings. However, TIR signatures increase by using low NIR reflecting paints as they strongly absorb the NIR component of solar radiation, which increases the possibility of detection in the TIR band.

When infrared radiation falls on a coating, there is refraction at the interface before it reaches the pigment and filler particles. At the surface of these particles there is partial reflection and partial refraction, and multiple reflections and refractions in coatings with normal pigments diffuse the beam. The radiation partially propagates back to the coating surface and leaves in all directions, a process known as "scattering." While designing the coating on a military vehicle, evading detection has to be coupled with the requirement of keeping the temperature under control.

The selection of pigments is very important in providing thermal control by the coatings. The specific thermal requirements may be given as: (1) high solar absorption to minimize the NIR laser detectability; (2) low solar absorption so that illuminated equipment remains cool in hot environments; or, (3) low emission to minimize thermal detectability.[84] Low emission characteristics with solar absorption of any intensity are combined in a coating.

3.4.4.1 The Role of Pigments

Regarding the use of pigments, carbon black can be used to make dark colored paints, e.g. navy grays, which are strongly absorbent in a solar wavelength and can reduce their detectability in both photographic and NIR (by laser) mode. However, carbon black-based coatings suffer from the drawback of emitting strong thermal radiation. The metal substrates can be coated with metal oxide, which will exhibit high solar absorption and low emissivity in the TIR region. This is possible due to the absorption of short wavelength radiation by thin oxide layers, and no absorption for the infrared region and the result is the reduction of possibility of detection in both TIR and NIR regions. Heat build-up remains a problem due to the absorption of solar radiation. Organic perylene blacks exhibiting weak absorption in the NIR can be used as an alternative to carbon black and black metal oxide.

On the basis of reflection/absorption requirements, different size particles can be selected that have optimum reflectance in various wavelengths. Titanium dioxide of an average particle size of 10 μm is able to reflect in the wavelength region of 800–2300 nm, and is free from absorption in the 400–800 nm wavelength range. Chromium trioxide and iron oxides, the valuable pigments exhibiting strong absorption in the NIR wavelength, can only be used at an optimum level to achieve the desired effect of hiding. Conventional extender pigments, offering transparency and non-reflectance in the entire visible and NIR regions, are barytes, diatomaceous silica, and talc.

3.4.4.2 Spectral Properties of Resins

Resins are selected on the basis of their covering of pigments and non-interference with their inherent infrared properties and also because they remain transparent to the spectral region of interest. Mostly organic resins do not absorb in the NIR wavelengths. Conversely, the resins of the coatings have a strong role to play in their absorption in the TIR region. Thus, both pigment and resin decide the performance of the coating. Generally, the organic resins show strong absorption in the wavelength range of 3.3–8.0 μm, characteristic of their functional groups. One of the resins transparent in TIR is poly(vinylidene fluoride) that weakly absorbs in the TIR region.[85] Having excellent weather stability, the resin is used to formulate coatings for use in the IR region. Siloxane resins with methyl substitution have mostly low emissivity, which qualifies them to be used for low emittance coatings.[86] Based on the refractive indices and particle sizes, pigments can be chosen which scatter light in the wavelength band in which the resin absorbs. The strategy is very useful in the TIR region. One interesting strategy, to reduce absorption by the resin, is the use of leafy metallic pigments. The dispersion of leafy pigments practically makes a continuous film that minimizes the penetration of incident radiation into the underlying resin and its absorption. Low emissive coatings are designed to reflect in the TIR region and emit to a very low extent. Flaky pigments, such as aluminum flakes, are incorporated in such paints.

3.4.4.3 Camouflage Coatings

Camouflage coatings are made in such a way that their normal reflectance resembles that of the surroundings, which enables them to be indistinguishable from the surroundings under both visible light and infrared radiation.

In forests, the leaves have a green color due to chlorophyll, which exhibits low reflectance in the range of 400–700 nm and a very high value beyond 700 nm.[87] Most of the green pigments having a chlorophyll color absorb strongly in the wavelength region of NIR. Because of the green color of chromium trioxide, its reflectance characteristics are similar to that of chlorophyll, which made it the main pigment during the early development of camouflage coatings suitable for the NIR region. Various lead salts were also used. However, the toxicity of the salts prevented the continuation of their use.

Selection is made from a wide range of visual colors for modern camouflage coatings. For desert regions, the camouflage can be done by reproducing natural sand and earth colors in a range of gray to yellow and earth colors. Iron oxides, having color ranges from red to black, are successfully used. The use of carbon black in coating is restricted to trace level because of its absorption nature in the infrared region.

In the ocean, the hulls and superstructures of marine vessels are painted gray to ensure visual camouflage. This, in turn, exhibits high absorption of solar radiation. High solar absorption increases the surface temperature, which affects the comfort of the crew and increases the emission of thermal radiation. Due to the strong absorption of infrared by blue and black pigments, antimony sulfide can be used instead of carbon black. However, during long exposure, it degrades to antimony oxide, thus exhibiting poor weatherability and color retention. Also, perylene black, which absorbs in the visible region and is transparent throughout the infrared, can be used to replace carbon black.

3.4.5 Radar Camouflage

Radio detection and ranging (Radar) works by sending out electromagnetic waves and analyzing the reflected signals it receives back. In the military sector, particularly in marine and air defense strategies and also in the air traffic sector, the microwave radiations are mostly used to detect and image the target of interest.[88] Radar functions on the principle that when there is change in the properties of the medium, the electromagnetic waves are reflected. The parameters involved in the process are a combination of conductivity, permittivity, and permeability. In warfare strategies, the system is also required to be camouflaged from enemy radar detection systems so that the damage can be done from close range without being detected. A solid object, such as an incoming aircraft, having a metal surface, reflects the radar wave that is picked up by the radar receiver. To avoid radar detection, the design of the aircraft can be made such that the radar waves are deflected instead of reflected. A second avoidance method is the treatment of the surface to minimize the reflection of the radar wave or, in simple terms, the radar signature. Successful low detection technology uses aerodynamic geometry together with specific functions of the material used on the surface of the aircraft fuselage and particularly the leading edges, which make detection difficult. With further improvement in the material sciences, the material application have become a great contributing factor. Radar absorbing materials (RAMs) are normally magnetic, dielectric, or hybrid in nature. Such a classification comes from the wave-material interaction mechanism. The absorber materials successfully developed belong to both organic materials, such as carbon particles, carbon nanotubes, or electrically conducting polymers, and inorganic materials, such as ferrites and others. Conducting polymers, having the ability to modulate conductivity through the variation of doping, are expected to act as a microwave-absorbing center.[89] RAMs are also useful for application in aircraft, ships, and automobiles to protect the communication and safety systems. Polyaniline (PANI) can be used as a dielectric radiation absorber by incorporating it in a polyurethane (PU) matrix (15 wt%) followed by impregnation

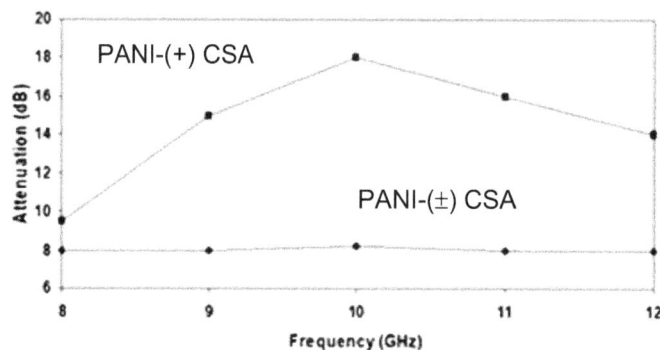

FIGURE 3.7   Microwave attenuation of PANI/CSA (65 parts loading) containing polyurethane paint film (thickness: 1.2 ± 0.2 mm).
Source: [90].

in nonwoven polymeric supports. PANI/PU resin can attenuate up to about 99% of the incident beam. The polyaniline powders form conduction paths that allow the loss of incident radiation via the mobility of the electrons. Figure 3.7 shows the microwave attenuation of PANI/camphor sulfonic acid-loaded PU film in an X-band.[90]

By manipulating the conductivity of RAM, the dielectric loss component can be controlled. The values of the real part, ε', and the imaginary part, ε'', of the complex permittivity, required to maintain minimum reflectivity at a particular film thickness, can be optimized by theoretical calculation. To match the optimum ε', ε'' for lowest reflectivity, the conductivity of the active component, e.g., PPy powder, is varied. For a thickness of 2.5 mm, about 2 wt% of PPy powder is required to maintain a reflectivity less than 10 dB in the Ku band (frequency range: 12–18 GHz).[91] PPy can be used to make a lightweight broadband RAM by coating it on rigid or flexible open cell foams.

Heterostructure composites such as carbon fiber (CF)/Co$_{0.2}$ Fe$_{2.8}$O$_4$/PANI is designed by using a layer-by-layer (LBL) structure so that the microwave attenuation of carbon fiber is enhanced.[92] The minimum reflection loss values in an absorber of thickness around 3.5 mm are observed below -20 dB in the Ku band. Thus, by introducing more phases into the CF, the loss values are enhanced due to the combined effect of magnetic loss and dielectric loss components. Ferrites, having microwave-absorbing (MW) properties, can be incorporated into ferrite-polymer composites so that the absorber can be made lightweight, and of low cost along with good design flexibility. The Ni-Zn ferrites have the desired properties for microwave absorption. The film formed by combining nano-sized ferrite and nano-sized polyaniline (PANI) with 2K bisphenol-A epoxy resin can absorb up to a maximum of 25 dB in the X-band (8–12 GHz).[93] The role of nano-sized PANI is more prominent as the increase in its proportion in the composite enhances the absorption considerably.

Carbon fiber alone can be used to design carbon fiber radar-absorbing coatings (CFRACs), comprising varying amounts of CF in polyurethane resin.[94] By varying the CF content and the thickness of the coating, the peak reflection of the CFRAC can be changed, so that it moves toward a low-frequency direction by increasing the parameters. At 0.8 wt% of CF content and the

coating thickness of 1.2 mm, the reflection loss of the coatings reaches a maximum of around 11 dB over X and Ku band.

Nanotubes perfectly absorb over a broad spectrum of radiation, including radio waves and UV-vis light. By making some space between nanotubes, their index of refraction can be made similar to that of the surrounding air. Thus, the scattering by the nanotubes-based substrate is minimized and absorption is maximized. The energy is converted to heat. Very low amounts of reflected waves are not enough to give any reliable radar signature.

3.4.6 Coating-Based Degradation of Chemical Warfare Agent (CWA)

Lethal chemicals like CWAs need to be captured and degraded for the safety of soldiers and the general public. Among many materials, metal-organic frameworks (MOFs) are able to adsorb efficiently and also act as catalysts in removing CWAs.[95, 96] Because they have a very high surface area and porosity, MOFs are found to be a very effective material for the absorption of CWAs.[97] In addition, the secondary building units containing metal in the structure act as catalytic sites of a Lewis-acidic type to destroy CWAs.[98] An atomic layer deposited TiO$_2$ coating on polyamide-6 nanofibers forms conformal Zr-based MOF thin films, including UiO66-NH$_2$, and others.[99] The MOF-functionalized nanofibers can also be designed for CWA simulant compound and nerve agents like soman, respectively.

3.5 Self-Healing Coatings

Organic coatings are vulnerable to mechanical damage during their service life. The damage allows corrosion to reach the metal surface via cracks and the result is coating failure. The failure of a coating makes the protection process costly. Although various remedial measures are available, the methods are costly and depend on precise detection. The development of smart coatings, with the ability to self-heal, suits applications requiring a long service life. Two main approaches are followed when designing self-healing coatings.[100] Polymer capsules are designed as the first self-healing generation. The composite nature of paints is well suited to incorporate self-healing agents. The polymer capsules containing reactive agents are incorporated into the coatings. Various healing agents, such as linseed oil, isocyanate, epoxy, dicyclopentadiene, etc., are encapsulated in the microcapsule. The role of the microcapsule is to protect the healing agent from degradation during the service life of the paint and also the damaging interaction of paint components with the healing agent is prevented. Apart from isolating the healing material from the paint matrix, the microcapsule wall thickness, brittleness, the interaction of the shell wall with the paint matrix and the healing chemistry are also considered when designing the encapsulation of specific healing agents. In the event of a fracture, the microcapsule breaks and liquid comes out, fills the crack and solidifies by curing of the liquid with the help of a catalyst dispersed in the medium. This prevents corrosion penetration. As an example, epoxy resin microencapsulated in the urea-formaldehyde shell is dispersed in the epoxy resin and catalyst mixture.[101] The coating self-heals during damage and the healed film exhibits a good corrosion prevention performance without any deterioration.

The living ring-opening metathesis polymerization (ROMP) used to meet the requirements for various polymerizations, along with the characteristics of low shrinkage upon polymerization, requires a transition metal catalyst (Grubbs' catalyst) for self-healing. Further, the catalyst is tolerant to the series of functional groups as well as the presence of oxygen and water, and is well suited for self-healing requirements. Microcapsules containing dicyclopentadiene (DCPD) and Grubbs' catalyst, incorporated in the epoxy matrix, break during fracture and heal the cracks under ambient conditions while demonstrating a recovery of 75% fracture load.[102]

Organic silane molecules react with moisture to form hydrolyzed intermediates that undergo cross-linking to form a solid film. This kind of behavior promotes this material to function in the self-healing arena. Moreover, the catalyst-free self-healing does not interfere with the corrosion performance and designing the microcapsule does not require any difficult formulations. The self-healing chemistry of siloxane is based on the reaction of hydroxyl terminated polydimethylsiloxane (HOPDMS) and polydiethoxysiloxane (PDES) catalyzed by di-n-butyl tin dilaurate to form a cross-linked product.[103] Thus, the material exhibits excellent healing chemistry while remaining stable in air and water and the activity is retained even after exposure to higher temperatures. This is useful for parent matrix curing at elevated temperatures without affecting the siloxane compounds. Generally, two approaches are followed. In the first approach, the catalyst is microencapsulated and the siloxanes are mixed with the polymer matrix as phase-separated droplets, while in the second approach, both the components are separated by encapsulating them separately.[104] In the first type, a matrix-initiated reaction with the catalyst is possible, while in the second method, both being microencapsulated, the healing ingredients remain stable during service. The self-healing performances of both methods are satisfactory.

Fluorinated silane compounds containing triethoxysilane, with the ability to hydrolyze in a wet environment and polymerize to form a hydrophobic film, is qualified for use in self-healing by encapsulating it to make an integral part of a coating[105] (Figure 3.8). The self-healed crack has similar corrosion-resistant properties. Of course, the siloxane, being hydrophobic in nature, prevents the contact of the substrate with the surrounding corrosion medium.

3.5.1 Multilayer Anticorrosion Systems

After the protective barrier breaks, the corrosion process starts that results in the formation of oxides, the movement of metal cations into the coating, local variation of pH and electrochemical potential and others. One of the self-healing strategies is designed by using a multilayer strategy. The polyelectrolyte multilayers have the ability to interfere with the corrosion process in the damaged area to slow down the process and also provide local corrosion protection. A combination of strong and weak polymeric acids and bases, such as poly(styrene sulfonate) (PSS) and poly(ethyleneimine) (PEI) polyelectrolyte, makes an ideal multilayer system.[106] The buffering capacities of PEI are used to neutralize the alkali generated during the onset of corrosion in the damaged area, and thus the self-healing action takes place. The positively charged imine groups capture the free hydroxide ions produced during the onset of the corrosion process that maintain a constant pH. Because of the neutralization, swelling of the polyelectrolyte multilayers occurs that enhances their mobility so that the polyelectrolyte chains diffuse faster toward the corrosive crack. The corrosion process in the affected

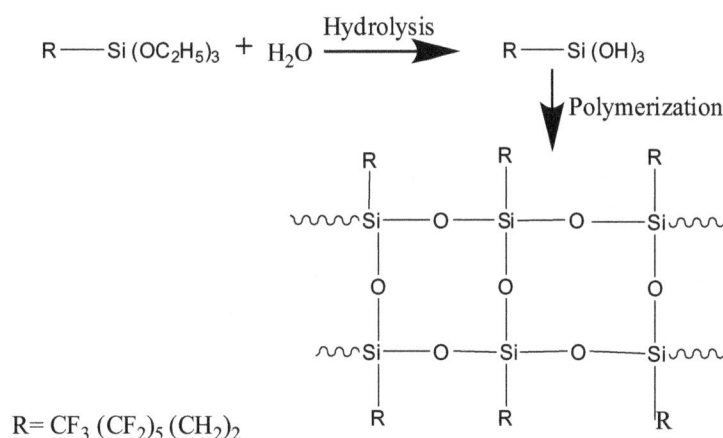

$$R\text{---}Si(OC_2H_5)_3 + H_2O \xrightarrow{\text{Hydrolysis}} R\text{---}Si(OH)_3$$

$$\downarrow \text{Polymerization}$$

$$R = CF_3(CF_2)_5(CH_2)_2$$

FIGURE 3.8 Schematics of self-healing behavior of fluorinated silanes.

area is thus minimized. However, the anticorrosion ability of polyelectrolyte multilayers drops to a very low value after it remains immersed for a long time. This is due to the absorption of many corrosive electrolytes to cause their swelling. The swelling causes its separation from the substrate. In spite of the diminishing ability with time, the multilayer structured coating, in which the mutually reactive components are incorporated, makes a smart anticorrosion system, comprising polyelectrolyte and inhibitor layers.[107] The general self-healing properties of the multilayer systems are dependent on three mechanisms: (1) the polyelectrolytes are required to have a pH-buffering activity so that the pH values at the metal surface in corrosive environment are restricted to between 5 and 7.5; (2) the inhibitors are required to be released from the multilayers strictly after the onset of the corrosion process; and (3) the polyelectrolytes forming the coating need to be mobile and should have the required ability to fill and seal the mechanical cracks that have developed in the coating. The multilayer coating can be applied by adopting a layerwise deposition technique using PEI and PSS as alternate layers with an inhibitor coating sandwiched between two consecutive multi-layer sequences. The strategy enables the coating to self-heal and prolongs the anticorrosive performance for a long time.

The improved corrosion resistance and the rapid self-healing ability can be combined by incorporating a cerium-based conversion layer in the multilayer structure, containing graphene oxide and branched PEI/PAA as polymeric electrolytes.[108] The cerium-based conversion layer is generally incorporated for application on metallic substrates like Mg alloy. Having a relatively higher aspect ratio, graphene oxide, which exhibits a good barrier property against the corrosion medium, is deposited above the cerium conversion layer. The Ce(IV), PEI/GO, and PEI/PAA combination can heal the damaged area within 10 minutes, which is possible by decreasing the cross-link density to allow higher swelling ability of the multilayer.

3.5.2 Antifouling Coatings

A self-healing coating with low toxicity is designed by combining the anticorrosion effect via the use of microcapsules filled with film formers (e.g., vegetable oil) and corrosion inhibitors together with an antifouling additive (silver compounds) filled in another microcapsule for slow release.[109] In the case of mechanical damage, the release of core materials takes place when the microcapsules break and fill the crack for the self-healing to occur, leading to the formation of the cured film. Slow release of the antifouling agent protects the paint surface from fouling. Antifouling coatings undergo degradation during their service life. For example, in polymer brushes, the degradation is followed by the detachment of the grafted polymeric chains, resulting in the loss of the antifouling effect. A coating can be designed with 3D grafting, i.e., grafting inside as well as on the surface of the host matrix, which ensures self-healing in case of detachment of a portion of the grafted polymers. The mechanism works as a few of the chains stored inside relocate from the film's interior to the outside under the chemical potential gradient. In one strategy, pH-responsive polymer films of 2-vinyl pyridine are cross-linked by the quaternization reaction using diiodomethane,

which is followed by grafting PEO onto the surface and inside the polymer films.[110] A long-lasting (fourfold) antifouling effect can be realized with 3D grafting, as compared to a limited period antifouling effect with surface grafting. The insoluble Cu-1 H-benzotriazole (BTA) complex, a corrosion inhibitor, can function as a pH and sulfide ion-sensitive nano valve to control the pore opening of BTA-loaded mesoporous silica nanoparticles (MSNs). Thus, a combination of BTA and benzalkonium chloride (biocide) can be incorporated into nanopores of MSNs to function as a pH/sulfide ion-responsive release system.[111] The spontaneous and premature release of active species can be prevented by this strategy. The anticorrosion self-healing property is exhibited by the hybrid coating in response to a decrease in pH and also the sulfide ions present maintain a high barrier level. The presence of biocide maintains the antifouling effect.

Under ambient conditions, the thiol radicals, produced due to the mechanical breakage of disulfide bonds, are able to quickly exchange with other disulfide bonds to prevent the catastrophic failure of the materials. A disulfide containing methacrylate cross-linker can be introduced to impart self-healing properties. It is possible to design self-healing coatings based on various functionalities, especially for antifouling and antibacterial coatings. As an example, PEG-based methacrylate, hydroxyl containing acrylamide and trimethylammonium chloride-based methacrylate, are used to prepare antifouling and antibacterial hydrogel coatings.[112] The healing is spontaneous under ambient conditions without any external agent. Very thick hydrogels induce faster healing.

In addition to higher stability, excellent antifouling and wetting properties, the fluoropolymer brush comprising poly(2-perfluorooctylethyl methacrylate) (PMAF17) maintains an extended service life as it can be repaired many times by simple heat treatment. This fluoropolymer brush can be covalently linked to a nanostructured silicon substrate to impart antifouling and self-healing properties.[113] Short duration heat treatment of the damaged surface restores the contact angle from a lower value to the original higher value of 152°. Above the T_g of the brush, the polymer chains display sufficient mobility to reorient and repair the damage. The process is facilitated by the low surface energy of the fluorinated materials, which permits the undamaged fluorinated tails to come to the air interface (top of the surface) during heating and the hydrophobicity and antifouling character of the surface are regained. Various hydrophilic polymers, developed for formulating self-healing coatings, suffer from lower mechanical strength, lower corrosion prevention ability, deposition of oily marine pollutants, etc. Also, a self-healing underwater oil-repellent and biofouling-resistant self-healing coating can be realized by adopting a three-dimensionally (3D) ordered structure via the self-assembly of ordered hybrid microgel spheres and microgel spheres with grafted hydrophilic copolymer chains. A typical design strategy can be given as:

1. Microgel spheres (MS) are synthesized via the polymerization of amide and acid containing monomers and a diacrylate cross-linker.

2. SiO$_2$ nanoparticles are deposited on MS via chemical linkage to obtain ordered microgel spheres.

3. A hydrophilic block copolymer, based on a silane-containing monomer, methyl acrylate and a phosphorylcholine-containing monomer, is synthesized.

4. The block copolymer is grafted onto the ordered MS to obtain modified ordered microgel spheres.

5. A solution of reactive acrylic resin and a diisocyanate trimer cross-linking agent is soaked into the voids of modified ordered MS and cross-linked at an elevated temperature to obtain a self-repairing underwater coating, having a super-oleophobic and anti-biofouling character.[114]

The coating has the unique properties of being hydrophilic and oleophilic in nature, while the oil contact angle is about 160.8° after the MS is immersed in water. After the coating is crushed, the original topographic features return after immersing in water and, as earlier, is able to exhibit oil-repellent property. The self-healing ability of the modified ordered MS-based coating is susceptible to variation in temperature as it is observed that faster self-repairing occurs at lower temperatures. The oil-repellent and biofouling-resistant properties do not show much variation in any type of environment of highly acidic, alkaline, and salty natures.

3.6 Intumescent Fire-Retardant Coatings

The danger of fire exists wherever we work or dwell. Fire is one of the most difficult things to control, in spite of the progress man has made through centuries. Every year, fires destroy assets worth billions and cause innumerable fatalities. Protection against fire is important for saving lives, as well for minimizing financial loss. In ships, fires of any intensity may break out due to an accident. Fire is also possible in war. Use of fire-retardant coatings in defense services arose largely from the destruction of ships during World War II.[115, 116] To save lives and assets from being destroyed by fire, the provision of a fire-retardant system is mandatory. The structures can be protected from fire by active and passive mechanisms. The active mechanism includes sprinklers, alarms, or gas release systems which are activated immediately after sensing the fire. The passive mechanism includes traditional fire retardants which can be divided into two types.[117] The first is the additive-type fire retardant and the second is the reactive type. The additive-type fire retardant has the advantage of low price, high availability, and wider use, and dominates the market.[117, 118] In reactive-type fire retardants, the fire-retardant segment is attached chemically to the polymer backbone to convert the polymer into an intrinsic fire-retardant material. This method is more promising and is becoming important in the preparation of fire-retardant coatings.[6, 119] The intumescence concept was developed by Gay-Lussac in 1821 by designing flame-retardant textiles and woven fabrics by applying a coating of a mixture of borate and ammonium phosphate to the cloth.[115] After a long gap, in 1934, a German patent proposed the development of flame-retardant wood by using a mixture of diammonium phosphate and formaldehyde. In 1938, after the first literature on flame-retardant intumescent coatings appeared, a discussion on the composition of intumescent systems vs. their function was held.[6]

3.6.1 Intumescent Fire-Retardant Paints

In case of fire, the coating swells up many times the original thickness and forms a stable, inert carbonaceous char (Figure 3.9).[120] This forms a physical barrier to isolate the flame and the flammable substrate. This char thermally insulates the surface and reduces the spread of the flames.

Because of the release of large amounts of smoke and toxic gases like halogen hydrides from the non-intumescent halogen-containing fire-retarded systems during burning, some of them have been banned by the European Commission.[6] As a halogen-free product, the intumescent formulations form a charred foam layer on the surface during heating to provide good heat insulation, which prevents both heat and mass transfer. For paint to function as an intumescent coating, four basic ingredients are required in the composition (Table 3.3).

The carbon source is an essential part of the fire-retardant additive to form a carbonaceous char which gives the coating better performance to protect against fire.[121] Natural products like coconut fiber (CCF), wood waste (WW), and peach stone (PHST) are also used as carbon sources in intumescent coatings formulations.[122] The optimum percentage of CCF, WW, and PHST varies in the range of 6–9% for the required performance in intumescent coatings. Similarly, the binder selection is also crucial in varying the properties of the intumescent coating. The resistance to stability of these coatings in the environment is also dependent on the nature of the binder systems. In earlier times, various types of thermoplastics, such as emulsion and latex, thermosets like epoxy, polyurethane, phenolic/amino resins were used as binders. The use of halogenated binders has been restricted in many countries due to stringent regulations on the release of large amounts of smoke and toxic gases. The binder has an important role in the foaming process of an intumescent composition. Inhibition of foaming can be performed by some polymer binders. Therefore, their contents should be minimized.[123] Usually, polymers and copolymers derived from vinyl acetate, vinyl ether, dibutyl maleic acid, styrene, acrylic acid ester, or vinyl toluene are used in the intumescent composition.

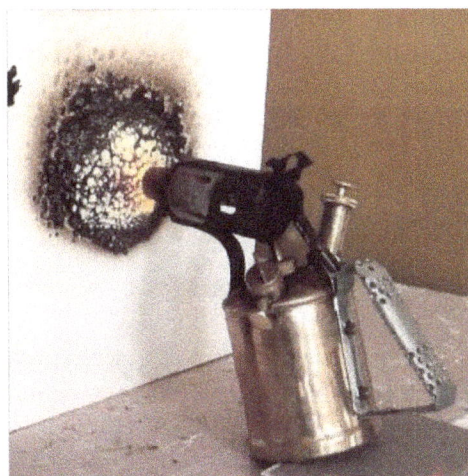

FIGURE 3.9 Photograph of panel exposed to fire at 1100°C.

TABLE 3.3

Ingredients of Intumescent Fire-Retardant Paint

Binder	Carbonific	Catalyst	Blowing Agents
Thermoplastic Resin (latex polymers)	Source of Carbon	Phosphorus	Releases Non-flammable Gases
1. PVA Emulsion	1. Pentaerythritol	1. Ammonium Polyphosphate	1. Urea
2. Vinyl Acetate Acrylic Emulsion	2. Carbohydrates (starch)	2. Melamine Phosphate	2. Melamine
3. SBR Emulsion	3. Sugars (Glucose, Maltose)	3. Urea phosphate	3. Chlorinated Paraffin
4. Acrylic Emulsion Solvent-based Vinyl Toluene Acrylate Polymer			4. Dicyandiamide
5. Chlorinated Rubber			
6. Epoxy			

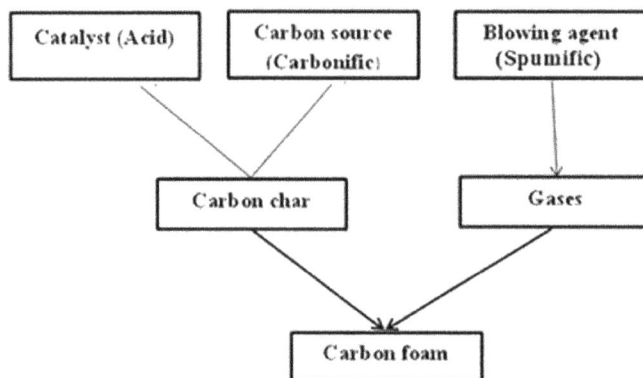

FIGURE 3.10 Schematics of intumescent action during a fire of intumescent fire-retardant paint.

Among different types of water-borne resins used in an intumescent paint formulation, the vinyl binders exhibit the best foam expansion and heat protection because of their substantial hardening after thermal degradation.[124]

In a typical formulation, an epoxy emulsion and self-cross-linked silicone acrylate (SSA) emulsion are used as a mixed binder in preparing a water-borne intumescent fire-retardant coating with improved fire protection and a foam structure of the coating.[125] The coatings with improved anti-oxidation properties can be realized by adding SSA to the mixed binders. The presence of epoxy and silicone acrylate binders increases the extent of cross-linking of the blend to improve the compactness of the coating, which minimizes the migration of fire-retardant additives and enhances the corrosion resistance by slowing down the water permeation.

3.6.2 The Mechanism of Intumescence

Several distinct reactions must take place simultaneously and in the proper sequence for intumescence to occur. This mechanism can be explained in five steps:

1. Phosphoric acid (H_3PO_4) is produced by decomposition of the catalyst.

2. The H_3PO_4 produced undergoes reaction with the carbon source.

3. The phosphate salt of the carbon source decomposes to form foamable carbon together with the release of acid.

4. Melting of the resinous material results in the formation of a film on top of the carbonaceous material.

5. Gases are released from the incorporated blowing agents that force carbon into foam, resulting in a thick foamed insulation layer.

For the reactions to take place in the proper order, the materials have to be carefully selected. The required temperature for the decomposition has to be determined. If the blowing agent decomposition temperature is too low, the gases will escape before the char formation. If it is too high, the char will be lifted off or blown apart by the released gases. To make a proper balance, the thermal behavior of the intumescent materials with that of the binder needs to be evaluated. The critical region of softening and decomposition of the binder should match that of the intumescent material. The char must be tough and adherent to resist the violent draughts arising in fires. For this to happen, certain other ingredients can be added to improve the char toughness. The addition of urea formaldehyde or melamine formaldehyde will contribute to increasing the toughness of the skin of the char. Smoke suppression can be achieved by the addition of zinc borate. Most of the ingredients fulfill more than one function during the formation of char due to the presence of carbon. The steps in the chemical reaction occurring during the char formation are shown in Figure 3.10.

3.6.3 Evaluation of the Compositions in the Fire Test

3.6.3.1 Two-Feet Tunnel Test

The standard fire-test cabinet and Monsanto's modification have shown the relationship of PVC to flame spread rate as determined in a 2-feet tunnel.[6] This apparatus is used effectively in the preliminary screening of experimental coatings to measure their intumescing fire-retardant performance. The paint is applied to the wooden panels (Himalayan fir) of size 76 cm x 10 cm x 1 cm to build up a film thickness of 500 ± 25 μm. After the paint is dry, the panel is positioned in the inclined holder through a slot on top and is marked in such a way that the "0" mark is positioned exactly above the tip of the burner and further markings are made at 2.5 cm succession up to the end of the panel. The panel is exposed to a flame for 10 minutes with a pressure of 10 Torr and the following observations are normally made:

1. *Char height*: Maximum height of char to a millimeter at a point directly above the burner.
2. *Flame spread length*: Only flame extending along the surface of the test panel should be recorded as the flame spread length.
3. *Afterglow*: The time in seconds for which the panel continues glowing after the source of fire is removed.
4. *After flaming*: The time in seconds for which the flame continues after the source of fire is removed.

In a typical fire-retardant composition, with additives like melamine polyphosphate/dipentaerythritol, nonflammable gases are produced after decomposition and, at a sufficiently high temperature, phosphorous oxide containing an intumescent char layer is formed.[126] The temperature range prevalent during char formation influences the average cell size and their distribution. For example, by increasing the temperature from 500°C to 800°, there is an increase of cell size and their distribution is wider. If the char layer is not uniform in nature, the fire protection of the coating is adversely affected. Several tests conducted with a fire-retardant coating under field conditions can be found in the literature.[120]

3.6.4 Hut Miniature Test

Miniature huts are constructed by using plywood sheets coated with fire-retardant intumescent paint and conventional fire-retardant paint respectively. Fire is simulated inside the huts by burning 300 ml alcohol inside the huts. The paint system and the observations are given in Table 3.4 and the condition of the hut exposed to fire is shown in Figure 3.11.

3.6.5 Fire Test for Electric Cables

Electric cables, coated with intumescent fire-retardant paint (500 μm), are exposed to fire at 750°C using a Bunsen burner for 7 minutes. The extent of damage is presented in Figure 3.12.

3.7 Self-Cleaning Coatings

Due to the awareness regarding the high potential in commercial products and their ability to reduce regular labor-intensive cleaning costs, self-cleaning coating technologies are attracting immense attention. These coatings are suitable for application on a large number of substrates, such as glasses, construction walls, textiles, and paints' top coat. Self-cleaning coatings can be designed with both a hydrophobic and a hydrophilic nature. The water contact angles for hydrophobic coatings are higher than 90°. When designing a self-cleaning coating, super-hydrophobicity is desired, which is defined as having a water contact angle above

TABLE 3.4

Hut Miniature Testing of Intumescent Paint System for Its Fire Performance

System	The Quantity of Fuel, ml	Condition after Alcohol Is Burnt off	Afterglow	The Event of Damage to the Hut
Hut coated with intumescent fire-retardant paint	300	The structure caught fire and burnt for 11.5 min	Nil	Provides good protection of the structured substrate
Hut coated with conventional fire-retardant paint	300	Reduced to ash	Yes	Reduced to ash

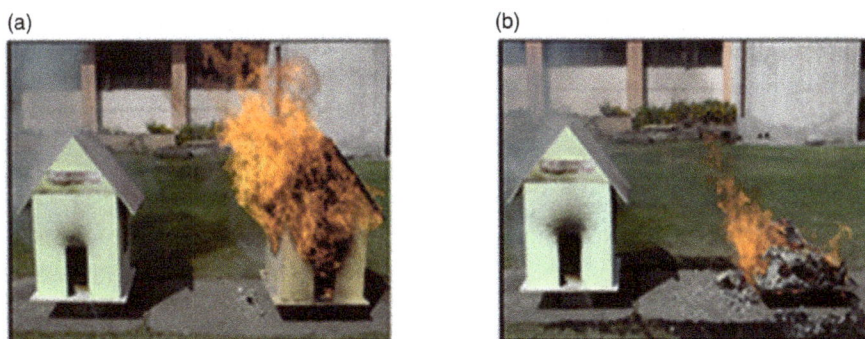

FIGURE 3.11 Performance of the fire-retardant paint after 7 minutes of the test; (a) Emulsion intumescent fire-retardant paint; (b) Conventional fire-retardant paint.

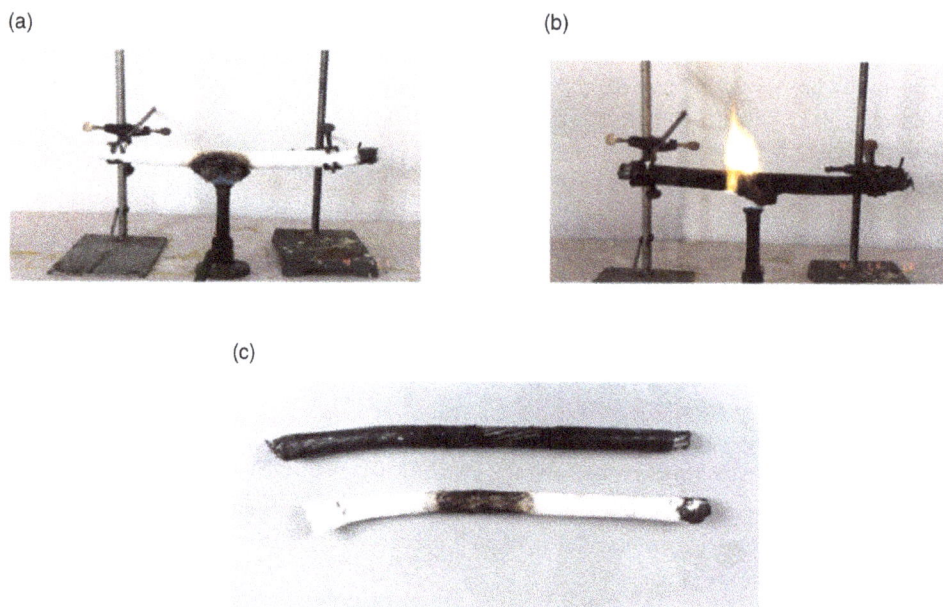

FIGURE 3.12 Photograph of intumescent paint coated cables before and after exposure to fire for 7 minutes. (a): Application of flame to cable coated with intumescent paint; (b) application of flame to uncoated cable; (c) condition of cables after flame exposure for 7 min. The protection against flame is established by applying intumescent paint.

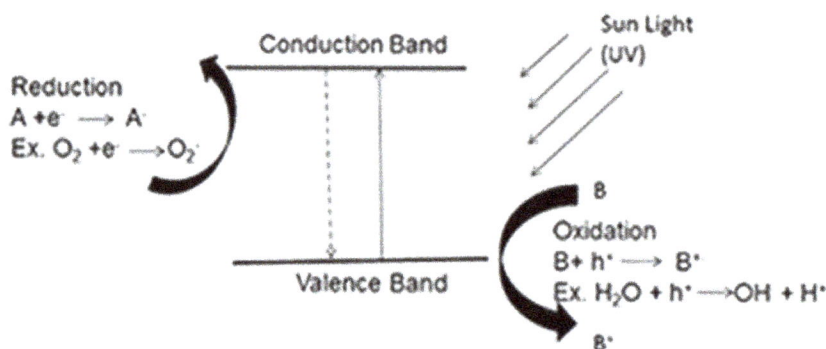

FIGURE 3.13 Photo-catalytic action of nano-TiO_2 during degradation of organic dirt.

150°. An example can be seen in nature, where water droplets are found to slide down the surface of a lotus leaf because of its super-hydrophobicity, which is called "the lotus effect." In addition, the spherical water droplets, as they roll down, carry dirt particles away. On the hydrophilic surface, the water forms a thin film, which prevents direct contact of dirt particles with the surface, making it easy to wash off the dirt.

3.7.1 Super-Hydrophobic Self-Cleaning Surfaces

Mainly there is the interplay of two factors involved in the self-cleaning action of lotus leaves:

1. the hydrophobic nature of the surface;
2. micro bumps present on the surface with the nanoscale protrusions spread on the surface.

By copying this lotus effect, a proper surface morphology and roughness can be developed to design various artificial super-hydrophobic self-cleaning surfaces.[127]

3.7.2 Self-Cleaning Surfaces

A self-cleaning surface has the ability to keep itself clean when exposed to sunlight and no human interference is required. This is possible by incorporating specific photoresponsive materials in the coating. When exposed to light, the photocatalysts can destroy organic pollutants and also the xenobiotic species present on the hydrophilic surface[128] and nano-TiO_2 are well suited for this purpose. For practical purposes, a thin layer of nano-TiO_2 is applied on the surface or incorporated into the coating. When exposed to radiation which has energy higher than its band gap, the electrons in TiO_2 are excited from the valance band to the conduction band, which results in the formation of an electron-hole pair on the photocatalyst surface (Figure 3.13).

These electron-hole pairs undergo reaction with the oxygen present in the air and water molecules present on the surface and form highly oxidative species as shown in Figure 3.14.

These highly oxidative species attack the organic pollutants, leading to their decomposition into CO_2 and H_2O. Together with organic dirt they destroy the micro-organisms. Thus, TiO_2 acting as a catalyst for photocatalytic degradation is not consumed at all.

Solar panels are adversely affected due to contamination by dust particles from the surroundings. Such dust particles reduce the amount of light reaching the working part of the solar cell and the output electrical power is much lower. This effect can be eliminated with nano TiO_2. If the TiO_2 content is maintained below 45% in a top coat for outdoor exposure, the self-cleaning coatings perform the dual role of photocatalytic degradation of the dirt as well as protecting the architectural coating underneath from UV-induced damage.[129] Because of the nano size of TiO_2, its incorporation into the coatings does not hamper the passage of light and the coating remains transparent to visible light. This clear coating can be used to make architectural silicone latex coating that will protect the inherent nature of the structure and can also be viewed by visitors without any change in its original appearance. The organic film produced has enhanced stability because of reduced transmission of the UV component in sunlight through the TiO_2-loaded clear coat, together with greatly reduced water permeability resulting from the siloxane binder.

The efficiency of nano TiO_2-based self-cleaning coating can be improved by using a thicker coating, which will help in producing an increased number of excited charge carriers. In a typical coating of 3 μm thick, there is strong absorption of near-UV light (quantum yields: approx. 0.15%), while for a shallow thickness (25 nm) coating, it is much less (quantum yields: approx. 0.04%).[130] Thus, the charge carriers have a sufficiently long half-life so that they can reach the surface easily and participate in the reaction with water and oxygen to generate charged species.

$$TiO_2 + h\Box \xrightarrow{\text{UV light}} h^+ + e^-$$

$$H_2O + h^+ \longrightarrow OH^{\cdot} + H^+$$

$$O_2 + e^- \longrightarrow \dot{O_2}^-$$

FIGURE 3.14 Schematics of generation of reactive species due to photocatalysis (h_0 to be replaced by h (nue)).

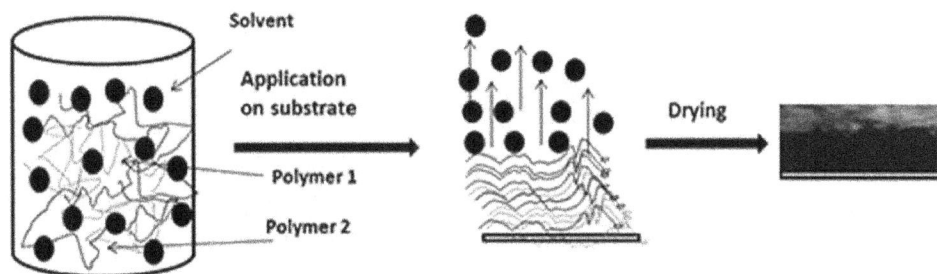

However, the thickness cannot be increased beyond its limit as other properties will be adversely affected.

3.7.3 Self-Cleaning Surfaces in Oil and Water

The super-hydrophobic self-cleaning surfaces, designed by creating micro/nano morphologies, have their deficiencies in terms of mechanical strength and non-functioning in the presence of oil. By using spraying, dipping, or extrusion methods a paint can be designed from a suspension of perfluorosilane-coated titanium dioxide nanoparticles. The paint can be applied to both hard and soft surfaces to create a self-cleaning surface that functions even under oil.[131] Clothes, paper, glass, and steel, etc. can be coated with the formulation for self-cleaning applications. The highly fluorinated small molecule, fluorodecyl polyhedral oligomeric silsesquioxane (F-POSS) blended with a fluorinated solvent to minimize crystallization, can be deposited as a smooth coating, leading to surfaces characterized by very low contact angle hysteresis in a broad range of liquids.[132] Mechanically durable omniphobic coatings for adhesion to a variety of substrates can be made by blending F-POSS with fluorinated polyurethane. Dip coating is another method which can be used with this formulation for non-planar substrates. The smooth, all-solid nature of the coating is inherently stable under pressure and more abrasion-resistant due to its smooth and all-solid nature.

3.8 Self-Stratified Coatings

Some incompatible polymer blends have found successful applications as plastics, rubber, films, etc. The heterophase structures in polymer/polymer are widely exploited in plastics, films, and membranes to improve their mechanical properties or control their permeability. However, with few exceptions, the incompatibility in coatings technology is considered unacceptable for multicomponent binders. A typical example can be given: rubber can be used to modify epoxy resins to make polymer/polymer composites that have good fracture toughness.[133] Thus, the self-stratification of polymer blends can be gainfully employed to make multilayer coating structures from a single coat application of a paint formulation.[2] The self-stratification is presented in Figure 3.15.

Heterophase composite-type polymer structures, derived from solvent-based self-stratified paints, impart high protective

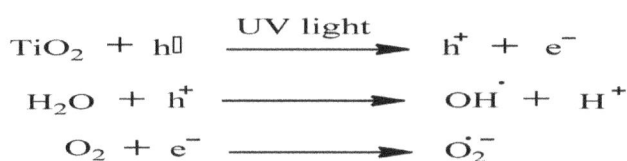

FIGURE 3.15 Schematics of self-stratification of a two-component coating.

properties and mechanical durability to the coatings.[134] With defined polymer compositions, the coating properties and other associated properties of a coating can be optimized by using self-stratification. Further, a double or multi-component self-stratifying coating structure solves the problem of uncertain adhesion of overlaying coats and the requirement of tie coats. The adhesion of the coating to a substrate has to be very good so that protection against corrosion of metallic substrates, and improvement of the chemical durability, weather and ultraviolet (UV) resistance are ensured. In addition, the application of single phase paint reduces emissions and energy consumption, thus making it sustainable in nature.[135]

3.8.1 Factors Affecting Self-Stratification

The factors affecting self-stratification can be listed as:[136]

- *Kinetic and thermodynamic conditions for stratification*: The polymer blend is required to undergo phase separation and the curing reaction (e.g., cross-linking) must occur only after the stratification process. Alternatively, the two processes must only partially overlap for the systems involving a curing agent. Thermodynamically, a polymer/polymer incompatibility has to exist, the base layer/substrate interfacial tension has to be very low and the total surface and interfacial surface energy must be very low.

- *Mechanisms of self-stratification*: The orientation in an assigned direction, necessary for self-stratification, cannot be done only by intermolecular forces. A separate driving force, which can be a surface tension gradient, a selective substrate wetting, contraction forces, etc., is needed by the binder. Different stratification mechanisms can be identified as: mechanisms driven by gravitation, selective wetting, pigment wetting, surface tension gradients, and phase contraction.

If driven by phase contraction, the several thermodynamic factors involved are:[137] ratio of polymer/polymer, thermodynamic affinity of the solvents toward the polymer partners, the presence of a compatibilizer, decompatibilizer, or surface tension controller, phase viscosities, pigment volume concentration (PVC) level, or film thickness.

3.8.2 Formulation of Coatings

- *Resins*: The choice of resin, particularly its type, molecular weight, glass transition temperature, cross-linking rate, and curing agent if used, is vital for the stratification to occur.

- *The molecular weight of polymers*: It has been established that the incompatibility increases with an increase in molecular weight that results in better phase separation.[137]

- *Glass transition temperature*: The Flory Huggins equation and the specific interactions between polymers suggest that a large difference in their T_g will support phase separation because of differences in the polymer chain flexibility.

- *Cross-linking rate and curing agent*: For proper stratification during the film-forming process, the layer has to develop before curing or partially overlap with the curing reaction. Curing before the gel point plays an important role as it increases the molecular weight of the chains. The cross-linking agent has to be chosen in such a way that a too reactive catalyst can hinder the self-stratification because of the gelation of the system. Because of a too reactive nature, the viscosity increases rapidly, which slows down the coalescence of the spherical particles of the separated phase into larger domains that obstruct the self-stratification.[138] The hardener can adversely affect the stratification if there is incompatibility between the hardener and the resin.[139] However, the selection and the effect of the curing agent depend on the resin used. For example, XE460 polyamine hardener helps in attaining stratification.

- Solvent and rate of evaporation: Self-stratification can be initiated by the evaporation of the solvent.[140] The stratification process of high surface energy resin can be optimized by adding a solvent with high volatility and polarity. On the other hand, the low surface energy component needs low volatility and polarity solvent.[138]

- *Pigments*: In the initial phases of development, the pigments of the right kind were used for stratification detection without bothering about the quality of the coating. The important properties of the pigment are its adsorption properties, particle size, and their affinity to a medium. If properly chosen, the pigments do not adversely affect the stratification process and even in a few cases this fares better than the pigment-free systems.

- *Other additives*: Additives are usually kept to a minimum in self-stratifying coatings as mostly they act in the direction of compatibilizing the phases rather than doing the reverse required for self-stratification. Sometimes decompatibilizers are added to make the phase separation process faster with moderate or negligible influence on the composition of equilibrium solutions. Silicon or fluorinated oligomers or polymers are commonly used for this purpose due to their complete incompatibility with conventional resins.[141] Wetting and dispersing agents also play important roles. Leveling agents do not have any role in modifying the film, but adversely affect the level of phase separation. Similar damage is done to the resulting coating by the flow control and surface active agents.

- *Solubility parameters and surface tensions*: These are very important parameters to be considered while dealing with stratification. Therefore, they are required to be ascertained for all the materials used in the paint formulation. The solubility characteristics of chemicals are usually determined experimentally by Hansen's solubility parameters.

- *The thickness of the coating*: Better stratification is possible with thick coatings (200 μm). However, if the coating is too thick, a non-uniform stratification occurs and less glossy coating results.[142] Approximately 30–40 μm is the minimum thickness for satisfactory stratification.

REFERENCES

1. B. Wesseling. "Passivation of metals by coating with polyaniline: Corrosion potential shift and morphological changes." *Adv. Mater.* 6(3) (1994): 226–228.
2. V.V. Verkholantsev. "Nonhomogeneous-in-layer coatings." *Prog. Org. Coat.* 13(2) (1985): 71–96.
3. R.V. Subramanian and K.N. Somasekharan. "Polymers for controlled release of organotin toxin." *J. Macromol. Sci. Chem.* 16(1) (1981): 73–93.
4. S.K. Ghosh. *Self-Healing Materials.* Weinheim: Wiley-VCH, 2009.
5. J.W.C. Pang and I.P. Bond. "A hollow fiber reinforced polymer composite encompassing self-healing and enhanced damage visibility." *Compos. Sci. Technol.* 65(11–12) (2005): 1791–1799.
6. G. Impallomeni, G. Montaudo, C. Puglisi, E. Scamporrino, and D. Vitalini. "The role of intumescence on the flammability of vinyl and vinylidene polymers." *Appl. Polym. Sci.* 31(5) (1986): 1269–1274.
7. N.A. Ghanem, N.N. Messiha, N.E. Ikladious, and A.F. Shaaban. "Organotin polymers. IV. Binary and ternary copolymerizations of tributyltin acrylate and methacrylate with styrene, allyl methacrylate, butyl methacrylate, butyl acrylate, and acrylonitrile." *J. Appl. Polym. Sci.* 26(1) (1981): 97–106.
8. P.C. Deb and A.B. Samui. "Studies on tributyl tin methacrylate iii; copolymerisation." *Die Angew. Makromol. Chem.* 112 (1983): 15–22.
9. R.F. Brady Jr. and I.L. Singer. "Mechanical factors favoring release from fouling release coatings." *Biofouling* 15(1–3) (2000): 73–81.
10. D.M. Yebra, S. Kiil, and K. Dam-Johansen. "Antifouling technology–past, present and future steps towards efficient and environmentally friendly antifouling coatings." *Prog. Org. Coat.* 50 (2004): 75–104.
11. A.A. Finnie and D.N. Williams. "Paint and coatings technology for the control of marine fouling". In S. Dürr, and J.C. Thomason (Eds.), *Biofouling.* Oxford: Blackwell, 2010, pp. 185–206.
12. R.E. Baier. "Influence of the initial surface condition of materials on bioadhesion." In R.F. Acker, B.F. Brown, J.R. DePalma, and W.P. Iverson (Eds.), *Proceedings of the 3rd International Congress on Marine Corrosion and Fouling, 1973.* Evanston, IL: Northwestern University Press, 1973, pp. 633–639.
13. J.G. Kohl and I.L. Singer. "Pull-off behavior of epoxy bonded to silicone duplex coatings." *Prog. Org. Coat.* 36 (1999): 15–20.
14. R.F. Brady. "Clean hulls without poisons: Devising and testing nontoxic marine coatings." *J. Coat. Technol.* 72(900) (2000): 45–56.
15. R.F. Brady. "Properties which influence marine fouling resistance in polymers containing silicon and fluorine." *Prog. Org. Coat.* 35(1–4) (1999): 31–35.
16. C. Howell, T.L. Vu, J.J. Lin, S. Kolle, N. Juthani, E. Watson, J.C. Weaver, J. Alvarenga, and J. Aizenberg. "Self-replenishing vascularized fouling-release surfaces." *ACS Appl. Mater. Interfaces* 6(15) (2014): 13299–13307.
17. J.A. Mielczarski, E. Mielczarski, G. Galli, A. Morelli, E. Martinelli, and E. Chiellini. "The surface-segregated nanostructure of fluorinated copolymer-poly(dimethylsiloxane) blend films." *Langmuir* 26(4) (2010): 2871–2876.
18. I. Marabotti, A. Morelli, L.M. Orsini, E. Martinelli, G. Galli, E. Chiellini, E.M. Lien, M.E. Pettitt, M. E. Callow, … and M. Jenko. "Fluorinated/siloxane copolymer blends for fouling release: Chemical characterization and biological evaluation with algae and barnacles." *Biofouling* 25(6) (2009): 481–493.
19. A.G. Nurioglu, A.C.C. Esteves, and G.J. deWith. "Nontoxic, non-biocide-release antifouling coatings based on molecular structure design for marine applications." *Mater. Chem. B* 3(32) (2015): 6547–6570.
20. P. Alexandridis and B. Lindman. "Amphiphilic molecules: Small and large." In P. Alexandridis and B. Lindman (Eds.), *Amphiphilic Block Copolymers: Self-Assembly and Applications*, 1st edn. Amsterdam: Elsevier Science, 2000, pp. 1–12.
21. S.H. Baxamusa and K.K. Gleason. "Random copolymer films with molecular-scale compositional heterogeneities that interfere with protein adsorption." *Adv. Funct. Mater.* 19(21) (2009): 3489–3496.
22. G. Galli and E. Martinelli. "Amphiphilic polymer platforms: Surface engineering of films for marine antibiofouling." *Macromol. Rapid Commun.* 38(8) (2017): 1600704.
23. R.A. Pullin, T.G. Nevell, and J. Tsibouklis. "Surface energy characteristics and marine antifouling performance of poly(1H, 1H, 2H, 2H-perfluorodecanoyl diitaconate) film structures." *Mater. Lett.* 39(3) (1999): 142–148.
24. N. Misdan, A.F. Ismail, and N. Hilal. "Recent advances in the development of (bio)fouling resistant thin film composite membranes for desalination." *Desalination* 380 (2016): 105–111.
25. Y. Higaki, M. Kobayashi, D. Murakami, and A. Takahara. "Anti-fouling behavior of polymer brush immobilized surfaces." *Polym.* 48 (2016): 325–331.
26. N.Y. Kostina, S. Sharifi, A. de Los, S. Pereira, J. Michálek, D.W. Grijpma, and C. Rodriguez-Emmenegger. "Novel antifouling self-healing poly(carboxybetaine methacrylamide-co-HEMA) nanocomposite hydrogels with superior mechanical properties." *J. Mater. Chem. B* 1(41) (2013): 5644–3650.
27. A. Derbalah, S.A. El-Safty, M.A. Shenashen, and N.A. Abdel Ghany. "Mesocage collector cavities as nanopockets for remediation and real assessment of carbamate pesticides in aquatic water." *Nano-Struct. & Nano-Objec.* 3 (2015): 17–27.
28. M.S. Selim, M.A. Shenashen, S.A. El-Safty, S.A. Higazy, M.M. Selim, H. Isago, and A. Elmarakbi. "Recent progress in marine foul-release polymeric nanocomposite coatings." *Prog. Mater. Sci.* 87 (2017): 1–32.
29. N. Roy and A. K. Bhowmick. "Novel in situ polydimethylsiloxane-sepiolite nanocomposites: Structure-property relationship." *Polymer* 51(22) (2010): 5172–5185.
30. F. Irani, A. Jannesari, and S. Bastani. "Effect of fluorination of multiwalled carbon nanotubes (MWCNTs) on the surface properties of fouling-release silicone/MWCNTs coatings." *Prog. Org. Coat.* 76 (2013): 375–383.
31. A.B. Samui, J.G. Chavan, and S.K. Rath. "Development of nanocomposites PDMS based foul release coating." Unpublished results, 2011.
32. A.J. Scardino, H. Zhang, D.J. Cookson, R.N. Lamb, and R. de Nys. "The role of nano-roughness in antifouling," *Biofouling* 25(8) (2009): 757–767.

33. M.S. Selim, M.A. Shenashen, S.A. El-Safty, S.A. Higazy, M.M. Selim, H. Isago, and A. Elmarakbi. "Recent progress in marine foul-release polymeric nanocomposite coatings." *Prog. Mater. Sci.* 87 (2017): 1–32.

34. P. Zhang, C. Shao, Z. Zhang, M. Zhang, J. Mu, Z. Guo, Y. Sun, and Y. Liu. "Core/shell nanofibers of TiO_2@carbon embedded by Ag nanoparticles with enhanced visible photocatalytic activity." *J. Mater. Chem.* 21(44) (2011): 17746–17753.

35. H-Y. He. "Photoinduced superhydrophilicity and high photocatalytic activity of ZnO–reduced graphene oxide nanocomposite films for self-cleaning applications." *Mater. Sci. Semicon. Proc.* 31 (2015): 200–208.

36. B. Bhushan. "Bioinspired structured surfaces." *Langmuir* 28(3) (2012): 1698–1714.

37. M. Lejars, A. Margaillan, and C. Bressy. "Fouling release coatings: A nontoxic alternative to biocidal antifouling coatings." *Chem. Rev.* 112(8) (2012): 4347–4390.

38. H.F. Bohn and W. Federle. "Insect aquaplaning: Nepenthes pitcher plants capture prey with the peristome, a fully wettable water-lubricated anisotropic surface." *Proc. Natl Acad. Sci., U.S.A.* 101(39) (2004): 14138–14143.

39. J. Chapman, C. Hellio, T. Sullivan, R. Brown, S. Russell, E. Kiterringham, L. Le Nor, and F. Regan. "Bioinspired synthetic macroalgae: Examples from nature for antifouling applications." *Int. Biodeter. Biodegr.* 86 (2014): Part A, 6–13.

40. N. Hadjesfandiari, K. Yu, Y. Mei, and J.N. Kizhakkedathu. "Polymer brush-based approaches for the development of infection-resistant surfaces." *Mater. Chem. B* 2(31) (2014): 4968–4978.

41. A. Roosjen, H.J. Busscher, W. Norde, and H.C. Van der Mei. "Bacterial factors influencing adhesion of Pseudomonas aeruginosa strains to a poly(ethylene oxide) brush." *Microbiology* 152(9) (2006): 2673–2682.

42. G. Cheng, Z. Zhang, S. Chen, J.D. Bryers, and S. Jiang. "Inhibition of bacterial adhesion and biofilm formation on zwitterionic surfaces." *Biomaterials* 28(29) (2007): 4192–4199.

43. R.S. Smith, Z. Zhang, M. Bouchard, J. Li, H.S. Lapp, G.R. Brotske, D.L. Lucchino, L. Weaver, L.A. Roth, … and C. Loose. "Vascular catheters with a nonleaching polysulfobetaine surface modification reduce thrombus formation and microbial attachment." *Sci. Transl. Med.* 4(153) (2012): 153RA132.

44. M. Ramstedt, N. Cheng, O. Azzaroni, D. Mossialos, H.J. Mathieu, and W.T.S. Huck. "Synthesis and characterization of poly(3-sulfopropylmethacrylate) brushes for potential antibacterial applications." *Langmuir* 23(6) (2007): 3314–3321.

45. R. Hu, G. Li, Y. Jiang, Y. Zhang, J.J. Zou, L. Wang, and X. Zhang. "Silver–zwitterion organic-inorganic nanocomposite with antimicrobial and antiadhesive capabilities." *Langmuir* 29(11) (2013): 3773–3779.

46. T. Blin, V. Purohit, J. Leprince, T. Jouenne, and K. Glinel. "Bactericidal microparticles decorated by an antimicrobial peptide for the easy disinfection of sensitive aqueous solutions." *Biomacromolecules* 12(4) (2011): 1259–1264.

47. P. Appendini and J.H. Hotchkiss. "Immobilization of lysozyme on food contact polymers as potential antimicrobial films." *Packag. Technol. Sci.* 10(5) (1997): 271–279.

48. J.C. Tiller, C.J. Liao, K. Lewis, and A.M. Klibanov. "Designing surfaces that kill bacteria on contact." *Proc. Natl. Acad. Sci. U.S.A.* 98(11) (2001): 5981–5985.

49. J. An, S.J. Geib, and N.L. Rosi. "Cation-triggered drug release from a porous zinc-adeninate metal-organic framework." *J. Am. Chem. Soc.* 131(24) (2009): 8376–8377.

50. P. Horcajada, C. Serre, G. Maurin, N.A. Ramsahye, F. Balas, M. Vallet-Regi, M. Sebban, F. Taulelle, and G. Ferey. "Flexible porous metal-organic frameworks for controlled drug delivery." *J. Am. Chem. Soc.* 130(21) (2008): 6774–6780.

51. M.P.A. Sancet, M. Hanke, Z. Wang, S. Bauer, C. Azucena, H.K. Arslan, M. Heinle, H. Gliemann, C. Wöll, and A. Rosenhahn. "Surface anchored metal-organic frameworks as stimulus-responsive antifouling coatings." *Biointerphases* 8(1) (2013): 29.

52. B. Wesseling. "Scientific and commercial breakthrough for organic metals." *Synth. Met.* 85(1–3) (1997): 1313–1318.

53. G.M. Spinks, A.J. Dominis, G.G. Wallace, and D.E. Tallman. "Electroactive conducting polymers for corrosion control." *J. Solid State Electrochem.* 6(2) (2002): 85–100.

54. M. Kendig, M. Hon, and L. Warren. "Smart corrosion inhibiting coatings." *Prog. Org. Coat.* 47(3–4) (2003): 183–189.

55. P.J. Kinlen, V. Menon, and Y.W. Ding. "A mechanistic investigation of polyaniline corrosion protection using the scanning reference electrode technique." *J. Electrochem. Soc.* 146(10) (1999): 3690–3695.

56. S. de Souza, J. Pereira, S.C. de Torres, M. Temperini, and R. Torresi. "Polyaniline based acrylic blends for iron corrosion protection." *Electrochem. Solid-State Lett.* 4(8) (2001): B27–B30.

57. A.B. Samui, A.S. Patankar, J. Rangarajan, and P.C. Deb. "Study of polyaniline containing paint for corrosion prevention." *Prog. Org. Coat.* 47(1) (2003): 1–7.

58. A.B. Samui and S.M. Phadnis. "Polyaniline-dioctyl phosphate salt for corrosion protection of iron." *Prog. Org. Coat.* 54(3) (2005): 263–267.

59. A. Mostafaei and F. Nasirpouri. "Epoxy/polyaniline–ZnO nanorods hybrid nanocomposite coatings: Synthesis, characterization, and corrosion protection performance of conducting paints." *Prog. Org. Coat.* 77(1) (2014): 146–159.

60. H. Di, Z. Yu, Y. Ma, F. Li, L. Lv, Y. Pan, Y. Lin, Y. Liu, and Y. He. "Graphene oxide decorated with Fe3O4 nanoparticles with advanced anticorrosive properties of epoxy coatings." *J. Taiwan Inst. Chem. Eng.* 64 (2016): 244–251.

61. X. Wang, W. Xing, L. Song, H. Yang, Y. Hu, and G.H. Yeoh. "Fabrication and characterization of graphene-reinforced waterborne polyurethane nanocomposite coatings by the sol-gel method." *Surf. Coat. Tech.* 206 (2012): 4778–4784.

62. B. Ramezanzadeh, H. Haeri, and M. Ramezanzadeh. "A facile route of making silica nanoparticles-covered graphene oxide nanohybrids (SiO_2-GO);fabrication of SiO_2-GO/epoxy composite coating with superior barrier and corrosion protection performance." *Chem. Eng. J.* 303 (2016): 511–528.

63. S. Pourhashem, M.R. Vaezi, and A. Rashidi. "Investigating the effect of SiO_2-graphene oxide hybrid as inorganic nanofiller on corrosion protection properties of epoxy coatings." *Surf. Coat. Technol.* 311 (2017): 282–294.

64. V.A. Mooss, A.A. Bhopale, P.P. Deshpande, and A.A. Athawale. "Graphene oxide-modified polyaniline pigment for epoxy based anti-corrosion coatings." *Chem. Pap.* 71(8) (2017): 1515–1528.

65. F. Mansfeld, C. Hsu, Z. Sun, D. Ornek, and T. Wood. "Technical note: Ennoblement—a common phenomenon." *Corrosion* 58(3) (2002): 187–191.

66. R. Zuo, E. Kus, F. Mansfeld, and T.K. Wood. "The importance of live biofilms in corrosion protection." *Corr. Sci.* 47(2) (2005): 279–287.

67. T. Ignatova-Ivanova, R. Ivanov, I. Iliev, and I. Ivanova. "Study of anticorrosion effect of EPS from now strains lactobacillus delbruecii." *Biotechnol & Biotechnol EQ.* 23(Suppl. 1) (2009): 705–708.

68. T. Ignatova-Ivanova and R. Ivanov. "Exopolysaccharides from lactic acid bacteria as corrosion inhibitors." *J. Life Sci.* 8 (2014): 940–945.

69. P. Forbes. *Dazzled and Deceived*. London: Yale University Press, 2009.

70. S. Kinoshita and S. Yoshioka. *Structural Colors in Biological Systems: Principles and Applications*. Osaka, Japan: Osaka University Press, 2005.

71. P. Ball. "Nature's color tricks." *Sci. Am.* 306(5) (2012): 74–79.

72. A. Saito. "Material design and structural color inspired by biomimetic approach." *Sci. Technol. Adv. Mater.* 12(6) (2011): 064709/1–064709/13.

73. Y. Zhao, Z. Xie, H. Gu, C. Zhu, and Z. Gu. "Bio-inspired variable structural color materials." *Chem. Soc. Rev.* 41(8) (2012): 3297–3317.

74. S.P. Mahulikar, H.R. Sonawane, and G.A. Rao. "Infrared signature studies of aerospace vehicles. " *Prog. Aerosp. Sci.* 43(7–8) (2007): 218–245.

75. S. Daehne, U. Resch-Genger, and O.S. Wolfbeis. *Near-Infrared Dyes for High Technology Applications*, vol. 52. Dordrecht: Kluwer Academic Publishers, 1998.

76. E. Kreit, I.M. Mäthger, R.T. Hanlon, P.B. Dennis, R.R. Naik, E. Forsythe, and J. Heikenfeld. "Biological versus electronic adaptive coloration: How can one inform the other?" *J. R. Soc. Interface* 10(78) (2012): 20120601.

77. G. Qin, P.B. Dennis, Y. Zhang, X. Hu, J.E. Bressner, Z. Sun, W.J. Crookes-Goodson, R.R. Naik, F.G. Omenetto, and D.L. Kaplan. "Recombinant reflection-based optical materials." *J. Polym. Sci. Polym. Phys.* 51(4) (2013): 254–264.

78. L. Phan, W.G. Walkup 4th, D.D. Ordinario, E. Karshalev, J.M. Jocson, A.M. Burke, and A.A. Gorodetsky. "Reconfigurable infrared camouflage coatings from a cephalopod protein." *Adv. Mater.* 25(39) (2013): 5621–5625.

79. X.S. Tang. "Carbon nanotube coatings for visible and IR camouflage." US Patent No. US 2013/0137324A1 (2013).

80. F.W. Bacon, F.J. Iannarilli Jr., J.A. Conant, T. Deas, and M. Dinning. "Quantitative camouflage paint selection for the CH-47F helicopter." *Col Res Appl* 34(6) (2009): 406–416.

81. R. Baušys and K.S. Danaitis. "Camouflage painting of buildings." Paper presented at Modern Building Materials, Structures, and Techniques: The 10th International Conference, Vilnius, Lithuania, 19–20 May, 2010.

82. "Camouflage for Military Vehicles," pp. 853–859. Available at: www.olivedrab.com/od_mvg_camo.php3 (accessed September 22, 2020).

83. H. Shi, J.G. Ok, H.W. Baac, and L.J. Guo. "Low-density carbon nanotube forest as an index-matched and near perfect absorption coating." *Appl. Phys. Lett.* 99(21) (2011): 211103.

84. L.V. Wake and R.F. Brady. "Formulating infrared coatings for defence applications." MRL Research Report: MRL-RR-1–93; March 1993.

85. P. Nallasamy and S. Mohan. "Vibrational spectroscopic characterization of form II poly (vinylidene fluoride)." *Ind. J. Pure Appl. Phys.* 43 (2005): 821–827.

86. H.Y.B. Mar and P.B. Zimmer. "Low infrared emissivity paints comprising an oxime cured silicone rubber." US Patent No. US 4,131,593–1978 (1978).

87. C. Buschmann, S. Lenk, and H.K. Lichtenthaler. "Reflectance spectra and images of green leaves with different tissue structure and chlorophyll content." *Israel J. Plant Sci.* 60 (2012): 49–64.

88. L. de Castro Folgueras and M. Cerqueira Rezende. "Multilayer radar absorbing material processing by using polymeric nonwoven and conducting polymer." *Materials Research* 11(4) (2008): 245–249.

89. S.M. Lee. *International Encyclopedia of Composites*. New York: Wiley-VCH Publishers, 1991.

90. A.B. Samui. "Development of polyaniline-based radar absorbing coatings." Unpublished results, 2009.

91. V-V. Truong, B.D. Turner, R.F. Muscat, and M.S. Russo. "Conducting-polymer-based radar-absorbing materials." Paper presented at The Far East and Pacific Rim Symposium on Smart Materials, Structures, and MEMS, , Adelaide, Australia, 1997.

92. J. Fang, Z. Chen, W. Wei, Y. Li, T. Liu, Z. Liu, X. Yue, and Z. Jiang. "A carbon fiber based three-phase heterostructure composite $CF/Co_{0.2} Fe_{2.8} O_4/PANI$ as an efficient electromagnetic wave absorber in the Ku band." *RSC Adv.* 5(62) (2015): 50024–50032.

93. E. Acikalin, K. Coban, and A. Sayinti. "Nanosized hybrid electromagnetic wave absorbing coatings." *Prog. Org. Coat.* 98 (2016): 2–5.

94. G. Ban, Z. Liu, S. Ye, H. Yang, R. Tao, and P. Luo. "Microwave absorption properties of carbon fiber radar absorbing coatings prepared by water-based technologies." *RSC Adv.* 7(43) (2017): 26658–26664.

95. Y. Liu, S-Y. Moon, J.T. Hupp, and O.K. Farha. "Dual-function metal-organic framework as a versatile catalyst for detoxifying chemical warfare agent simulants." *ACS Nano* 9(12) (2015): 12358–12364.

96. G.W. Peterson, S-Y. Moon, G.W. Wagner, M. G. Hall, J.B. DeCoste, J.T. Hupp, and O.K. Farha. "Tailoring the pore size and functionality of UiO-type metal-organic frameworks for optimal nerve agent destruction." *Inorg. Chem.* 54(20) (2015): 9684–9686.

97. V. Padial, E.Q. Procopio, C. Montoro, E. Lopez, J.E. Oltra, V. Colombo, A. Maspero, N. Masciocchi, S. Galli, … and J.A.R. Navarro. "Highly hydrophobic isoreticular porous metal-organic frameworks for the capture of harmful volatile organic compounds." *Angew. Chem. Int. Ed.* 52 (32) (2013): 8290–8294.

98. J.E. Mondloch, M.J. Katz, W.C.I. Iii, P. Ghosh, P. Liao, W. Bury, G.W. Wagner, M.G. Hall, J.B. DeCoste, … and O.K. Farha. "Destruction of chemical warfare agents using metal-organic frameworks." *Nat. Mater.* 14(5) (2015): 512–514.

99. J. Zhao, D.T. Lee, R.W. Yaga, M.G. Hall, H.F. Barton, I.R. Woodward, C.J. Oldham, H.J. Walls, G.W. Peterson, and G.N. Parsons. "Ultra-fast degradation of chemical warfare agents using MOF-nanofiber kebabs." *Nat. Mater.* 55(42) (2016): 13224–13228.

100. A.E. Hughes, I.S. Cole, T.H. Muster, and R.J. Varley. "Designing green, self-healing coatings for metal protection." *NPG Asia Mater.* 2(4) (2010): 143–151.

101. Z. Yang, Z. Wei, L. Le-ping, W. Hong-mei, and L. Wu-jun. "The self-healing composite anticorrosive coating." *Phys. Procedia.* 18 (2011): 216–221.

102. S.R. White, N.R. Sottos, P.H. Geubelle, J.S. Moore, M.R. Kessler, S.R. Sriram, E.N. Brown, and S. Viswanathan. "Autonomic healing of polymer composites." *Nature* 409 (2001): 794–797.

103. S.H. Cho, H.M. Andersson, S.R. White, N.R. Sottos, and P.V. Braun. "Polydimethylsiloxane-based self-healing materials." *Adv. Mater.* 18(8) (2006): 997–1000.

104. S.H. Cho, S.R. White, and P.V. Braun. "Self-healing polymer coatings." *Adv. Mater.* 21(6) (2009): 645–649.

105. M. Huang, H. Zhang, and J. Yang. "Synthesis of organic silane microcapsules for self-healing corrosion resistant polymer coatings." *Corr. Sci.* 65 (2012): 561–566.

106. D.V. Andreeva, D. Fix, H. Möhwald, and D.G. Shchukin. "Buffering polyelectrolyte multilayers for active corrosion protection." *J. Mater. Chem.* 18 (2008): 1738–1740.

107. D.V. Andreeva, D. Fix, H. Möhwald, and D.G. Shchukin. "Self-healing anticorrosion coatings based on pH-sensitive polyelectrolyte/inhibitor sandwichlike nanostructures." *Adv. Mater.* 20(14) (2008): 2789–2794.

108. F. Fan, C. Zhou, X. Wang, and J. Szpunar. "Layer-by-layer assembly of a self-healing anticorrosion coating on magnesium alloys." *ACS Appl. Mater. Interfaces* 7(49) (2015): 27271–27278.

109. T. Szabó, L. Molnár-Nagy, J. Bognár, L. Nyikos, and J. Telegdi. "Self-healing microcapsules and slow release microspheres in paints." *Prog. Org. Coat.* 72(1–2) (2011): 52–57.

110. H. Kuroki, I. Tokarev, D. Nykypanchuk, F. Zhulina, and S. Minko. "Stimuli-responsive materials with self-healing antifouling surface via 3D polymer grafting." *Adv. Funct. Mater.* 23(36) (2013): 4593–4600.

111. Z. Zheng, X. Huang, M. Schenderlein, D. Borisova, R. Cao, H. Möhwald, and D. Shchukin. "Self-healing and antifouling multifunctional coatings based on pH and sulfide ion sensitive nanocontainers." *Adv. Funct. Mater.* 23(26) (2013): 3307–3314.

112. W.J. Yang, X. Tao, T. Zhao, L. Weng, E-T. Kang, and L. Wang. "Antifouling and antibacterial hydrogel coatings with self-healing properties based on dynamic disulfide exchange reaction." *Polym. Chem.* 6(39) (2015): 7027–7035.

113. Z. Wang and H. Zuilhof. "Self-healing superhydrophobic fluoropolymer brushes as highly protein-repellent coatings." *Langmuir* 32(25) (2016): 6310–6318.

114. K. Chen, S. Zhou, and L. Wu. "Self-healing underwater superoleophobic and anti-biofouling coatings based on the assembly of hierarchical microgel spheres." *ACS Nano* 10(1) (2016): 1386–1394.

115. R. Talbert. *Paint Technology Handbook"*. 1st edn. Boca Raton, FL: CRC Press, 2007.

116. L.V. Wake. "Fire retardant performance of some interior shipboard paint schemes." MRL technical report; MRL-TR-91-31. Maribyrnong, Vic: DSTO Materials Research Laboratory, 1991, pp. 22–23.

117. S. Duquesne, S. Magnet, C. Jama, and R. Delobel. "Intumescent paints: Fire protective coatings for metallic substrates." *Surf. Coat. Technol.* 180–181 (2004): 302–307.

118. M. Hirose, J.H. Zhou, and K. Nagai. "The structure and properties of acrylic-polyurethane hybrid emulsions." *Prog. Org. Coat.* 38(1) (2000): 27–34.

119. M. Hirose, F. Kadowaki, and J.H. Zhou. "The structure and properties of core-shell type acrylic-polyurethane hybrid aqueous emulsions." *Prog. Org. Coat.* 31(1–2) (1997): 157–169.

120. S.S. Pawar, T.K. Mahato, and D. Kumar. "Development of intumescent paint for naval application." Unpublished results, 2007.

121. L.M.R. Mesquita, P.A.G. Piloto, M.A.P. Vaz, and T.M.G. Pinto. "Decomposition of intumescent coatings: Comparison between a numerical method and experimental results." *Acta Polytechnica* 49(1) (2009): 60–65.

122. M.M. de Souza, S.C. de Sá, A.V. Zmozinski, R.S. Peres, and C.A. Ferreira. "Biomass as the carbon source in intumescent coatings for steel protection against fire." *Ind. Eng. Chem. Res.* 55(46) (2016): 11961–11969.

123. R.N. McNair and J. Stepler. "Intumescent fabric coatings, investigation of, for protection against thermal radiation and flame." *American Dyestuff Rep.* 58 (1970): 27–36.

124. J.T. Pimenta, C. Gonçalves, L. Hiliou, J.F.J. Coelho, and F.D. Magalhães. "Effect of binder on the performance of intumescent coatings." *J. Coat. Technol. Res.* 13(2) (2016): 227–238.

125. G. Wang and J. Yang. "Influences of binder on fire protection and anticorrosion properties of intumescent fire resistive coating for steel structure." *Surf. Coat. Technol.* 204(8) (2010): 1186–1192.

126. G.J. Wang and J.X. Yang. "Thermal degradation study of fire resistive coating containing melamine polyphosphate and dipentaerythritol." *Prog. Org. Coat.* 72(4) (2011): 605–611.

127. P. Roach, N.J. Shirtcliffe, and M.I. Newton. "Progress in superhydrophobic surface 7evelopment." *Soft Matter.* 4 (2008): 224–240.

128. A. Fujishima, K. Hashimoto, and T. Watanabe. *TiO₂ Photocatalysis: Fundamentals and Applications.* Tokyo: BKC Inc., 1999.

129. F. Xu, T. Wang, H.Y. Chen, J. Bohling, A.M. Maurice, L. Wu, and S. Zhou. "Preparation of photocatalytic TiO_2-based self-cleaning coatings for painted surface without interlayer." *Prog. Org. Coat.* 113 (2017): 15–24.

130. I.P. Parkin and R.G. Palgravej. "Self-cleaning coatings." *Mater. Chem.* 15(17) (2005): 1689–1695.

131. Y. Lu, S. Sathasivam, J. Song, C.R. Crick, C.J. Carmalt, and I.P. Parkin. "Robust self-cleaning surfaces that function when exposed to either air or oil." *Science* 347(6226) (2015): 1132–1135.

132. M. Boban, K. Golovin, B. Tobelmann, O. Gupte, J.M. Mabry, and A. Tuteja. "Smooth, all-solid, low-hysteresis, omniphobic surfaces with enhanced mechanical durability." *ACS Appl. Mater. Interfaces* 10(14) (2018): 11406–11413.

133. R.A. Paerson and A.F. Yee. "Toughening mechanisms in thermoplastic-modified epoxies: Modification using poly (phenylene oxide)." *Polymer* 34(17) (1993): 3658–3670.

134. V.V. Verkholantsev. "Coatings based on polymer-polymer composites." *Prog. Org. Coat.* 18(1) (1990): 43–77.

135. J. Baghdachi, H. Perez, P. Talapatcharoenkit, and B. Wang. "Design and development of self-stratifying systems as sustainable coatings." *Prog. Org. Coat.* 78 (2015): 464–473.

136. A. Beaugendre, S. Degoutin, S. Bellayer, C. Pierlot, S. Duquesne, M. Casetta, and M. Jimenez. "Self-stratifying coatings: A review." *Prog. Org. Coat.* 110 (2017): 210–241.

137. V.V. Verkholantsev and M. Flavian. "Epoxy thermoplastic heterophase and self-stratifying coatings." *Mod. Paint Coat.* 85(1995): 100–106.

138. A. Toussaint. "Self-stratifying coatings for plastic substrates" (BRITE EURAM project RI 1B 0246 C(H)), *Prog. Org. Coat.* 28(3) (1996): 183–195.

139. P. Vink and T.L. Bots. "Formulation parameters influencing self-stratification ofcoatings." *Prog. Org. Coat.* 28(1996): 173–181.

140. P. Mokarian-Tabari, M. Geoghegan, J.R. Howse, S.Y. Heriot, R.L. Thompson, and R.A.L. Jones. "Quantitative evaluation of evaporation rate during spin-coating of polymer blend films: Control of film structure through defined-atmosphere solvent-casting." *Eur. Phys. J. E.* 33 (2010): 283–289.

141. V.V. Verkholantsev and M. Flavian. "Polymer structure and properties of heterophase and self-stratifying coatings." *Prog. Org. Coat.* 29(1–4) (1996): 239–246.

142. S. Liang, N.M. Neisius, and S. Gaan. "Recent developments in flame retardant polymeric coatings." *Prog. Org. Coat.* 76(11) (2013): 1642–1665.

4

Organic Phase Change Materials: Synthesis, Processing, and Applications

Swati Sundararajan
Zuckerberg Institute of Water Research, Jacob Blaustein Institutes for Desert Research, Ben-Gurion University of the Negev, Israel

Asit Baran Samui
Institute of Chemical Technology, Mumbai, India

CONTENTS

DOI: 10.1201/9781003037880-4

Abbreviations

TES	thermal energy storage
PCM	phase change material
PEG	poly(ethylene glycol)
SSPCM	solid-solid phase change material
PW	paraffin waxes
microPCM	phase change microcapsule
Poly(HEMA)	poly(2-hydroxyethyl methacrylate)
CNT	carbon nanotube
MWCNT	multi-walled carbon nanotube
EPDM	ethylene/propylene diene monomers
CNTS	carbon nanotube sponge
3D	three-dimensional
GF	graphene foam
HGF	hierarchical graphene foam
OMMT	organic montmorillonite
CDA	cellulose diacetate
PEO	Poly(ethylene oxide)
GNP	graphene nanoplatelet
PMMA	poly(methyl methacrylate)
PAA	poly(acrylic acid)
Eco-A	poly(ethylene-co-acrylic acid)
PU	polyurethanes
MDI	methylene diphenyl diisocyanate
AC	activated carbon
CMK-5	ordered mesoporous carbon
Brij58	polyethylene glycol hexadecyl ether
EG	expanded graphite
SG	sulphonated graphene
GO	graphene oxide
OA	oleylamine
HGA	hybrid graphene aerogel
PANI	polyaniline
RMS	radial mesoporous silica
TEOS	tetraethylorthosilicate
MCF	mesocellular foams
MCS	mesoporous calcium silicate
SWCNT	single-walled carbon nanotube
CNIC	carbon nitride intercalation compounds
BDO	1,4-butanediol
IPDI	isophorone diisocyanate
HMDI	hexamethylene diisocyanate
PHBV	poly(3-hydroxybutyrate-co-3-hydroxy valerate)

PVDF	polyvinylidene difluoride
NIR	near-infrared
rGO	reduced graphene oxide
BPEI	branched polyethyleneimine
DOX	doxorubicin
TPU	thermoplastic polyurethane

Symbols with Units

Symbol	Description	Unit
Q	Heat storage capacity	J/kg.K
η	Encapsulation efficiency	Dimensionless
T_g	Glass transition temperature	°C, K
T_m	Melting/fusion temperature	°C, K
m	Total mass	kg
ΔH	Enthalpy	J
V	Voltage	V
I	Current	A
t	Time	H

4.1 Introduction

The global energy crisis has been aggravated due to the depletion of fossil fuels, rapidly expanding economies, and growing population. The research on renewable energy sources has increased tremendously and one of the ways to meet these demands is by using efficient and economical thermal energy storage (TES) systems. TES can be realized by sensible, latent, and thermochemical energy storage. Among these methods, latent energy storage using phase change materials (PCM) is a highly efficient storage system due to high TES density under nearly isothermal conditions.[1,2] PCMs absorb heat from the surroundings during the melting process and heat is released back to the surroundings during the crystallization process (Figure 4.1).

As the material is heated, the PCM absorbs heat, causing the collapse of the solid structure and subsequently undergoes solid-to-liquid phase transformation. While cooling, this energy is released to the surroundings during the crystallization process, making the system reversible and maintenance-free. The crystallinity of PCMs is favored by a regular and symmetrical structure with fewer short branches and a high degree of stereoregularity.[3]

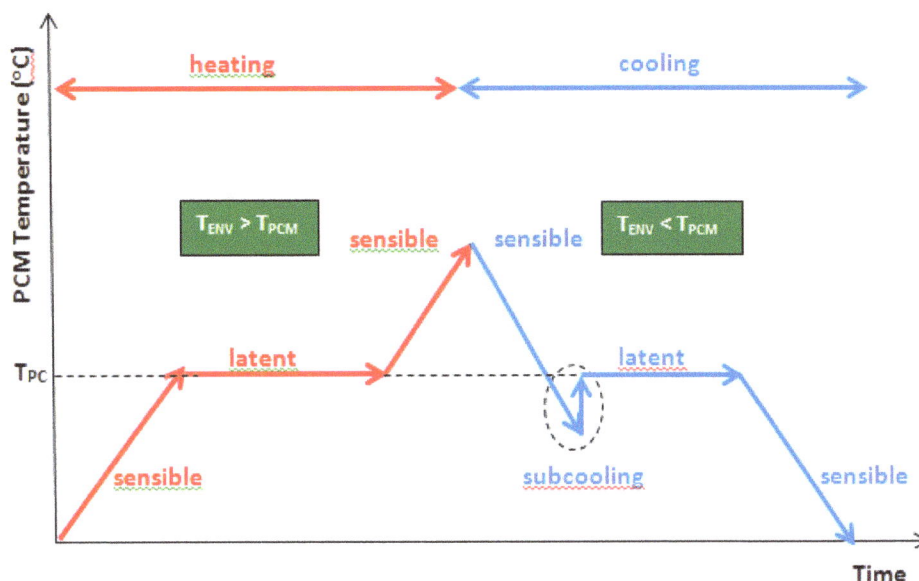

FIGURE 4.1 Schematic representation of the working of PCMs.

An ideal PCM should have the characteristics of high transition enthalpy, a suitable transition temperature, cycling stability, good thermal conductivity, low cost, non-toxic nature, and chemical and thermal stability. The heat storage capacity (Q) of a PCM is determined by using Equation 4.1:[4]

$$Q = \int_{T_i}^{T_m} mC_p dT + ma_m \Delta H_m + \int_{T_m}^{T_f} mC_p dT \qquad (4.1)$$

where m is the mass, C_p is the specific heat, a_m is the melt fraction of PCM, ΔH_m is the enthalpy of fusion and dT is the temperature difference. PCMs can be of three types, depending on the nature of the material: organic, inorganic, and eutectic PCMs, as presented in Figure 4.2.

Inorganic PCMs offer high storage capacity along with good thermal conductivity but can undergo phase segregation with repeated cycling, subcooling, and can corrode metallic containers. Organic PCMs have lower transition enthalpy and poor thermal conductivity. The organic PCMs offer advantages such as little or no subcooling, thermal stability, a non-corrosive nature, and chemical stability. Inorganic PCMs include mainly the hydrated salts, whereas the organic PCMs include paraffin, fatty acids, polyhydric alcohols, and sugar alcohols. Eutectics can be of the inorganic or organic type, using mixtures of two or more PCMs. Although these eutectics are stable from phase separation and offer good phase change enthalpy, extensive research is required to find eutectics with phase stability and suitable transition temperatures.[5,6] Some of the most commonly used PCMs are listed in Table 4.1.

The applications of PCMs can be extended to thermoregulating textiles, solar energy storage, thermal comfort in energy-efficient buildings, solar cookers, waste heat recovery, temperature-controlled greenhouses, temperature-sensitive transport of biopharmaceutical products and therapeutic packs. More recently, many novel applications have emerged in the field of nanomedicine, anti-icing coatings, barcoding, and biometric identification.[1,7–10]

Phase transition can be of solid-gas, liquid-gas, solid-liquid, and solid-solid categories. Phase transitions involving the gas phase are not preferred due to the large volume changes involved during the transition to the gas phase and the difficulty in its containment. Solid-liquid PCMs undergo limited volume change and have good thermal energy storage properties. Solid-liquid phase transitions are effective TES systems but pose the disadvantages of liquid phase containment and poor thermal behavior with cycling. The solid-solid PCMs (SSPCMs) have low thermal energy storage capacity compared to solid-liquid PCMs but can be used directly without any need for containment, so that no liquid leakage occurs.[11] The SSPCMs can be prepared by using physical and chemical methods. The physical methods include blending and impregnation, adsorbing and soaking, whereas the chemical methods include grafting, copolymerization, and cross-linking reactions with the solid-liquid PCM. The hard polymer segment acts as a supporting matrix and prevents leakage of the liquid phase. This chapter consists of three sections: Section 4.2 covers the use of paraffin as PCMs; Section 4.3 comprises PEG, fatty acids, and esters as PCMs, Section 4.4 elaborates on hybrid form-stable PCMs, Section 4.5 includes the chemical methods used for preparing solid-solid PCMs, and Section 4.6 focuses on the application of PCMs.

4.2 Paraffin as a Phase Change Material

Paraffin waxes (PW) are widely applied as solid-liquid PCMs due to their interesting properties, such as high phase transition energy, slight supercooling, non-toxicity, non-corrosiveness, and good chemical stability.[12] However, the practical applications are not so encouraging due to low thermal conductivity and seepage during the fusion process. PW can be used for potential applications by improvement in their heat transfer ability and

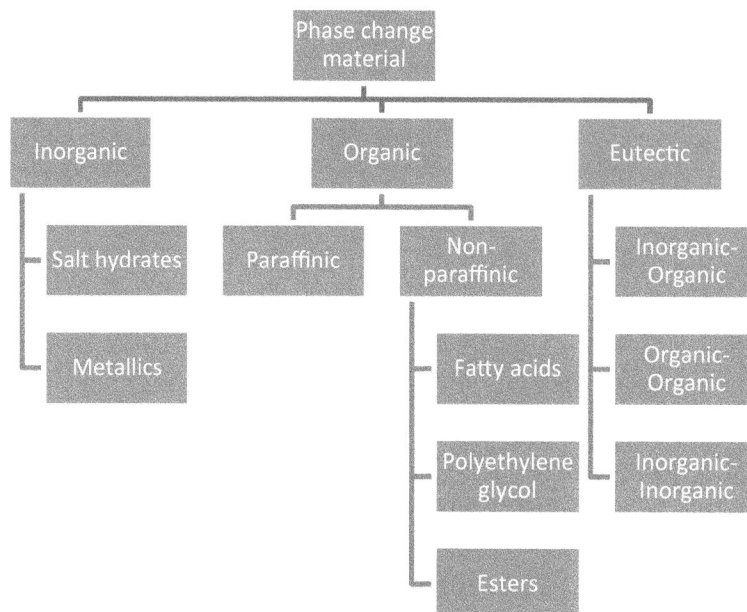

FIGURE 4.2 Classification of PCMs.

TABLE 4.1

Commonly Used PCMs

Inorganic PCMs	Organic PCMs	Eutectics
$CaCl_2.6H_2O$	Paraffin C_{14}	Capric + lauric acid (65 + 35 mol%)
$Na_2CO_3.10H_2O$	Paraffin C_{16}-C_{18}	Mystiric + capric acid (34 + 66 mol%)
$Mn(NO_3)_2.6H_2O$	Capric acid	Capric + lauric acid (45 + 55 mol%)
$LiClO_3.3H_2O$	Stearic acid	Urea (37.5%) + acetamide (63.5%)
$Na_2HPO_4.12H_2O$	Palmitic acid	$MgNO_3.6H_2O$ (58.7%) + $MgCl_2.6H_2O$ (41.3%)
$KF.4H_2O$	Oleic acid	$CaCl_2.6H_2O$ (66.6%) + $MgCl_2.6H_2O$ (33.3%)
H_2O	Polyethylene glycol 200–10000	$CaCl_2$ (48%) + NaCl (4.3%) + KCl (0.4%) + H_2O (47.3%)
$CaBr_2.6H_2O$	Erythritol	$Ca(NO_3)_2.4H_2O$ (47%) + $Mg(NO_3)_2.6H_2O$ (33%)
NaCl	1-tetradecanol	
KNO_3	Naphthalene	

the prevention of leakage with phase change. Various methods include the impregnation of PCMs into a polymer matrix,[13–15] use of a porous or layered material,[16,17] and microencapsulation. Phase-change microcapsules (microPCMs) have a typical core-shell structure; the PCM core is surrounded by a shell of organic or inorganic polymer. MicroPCMs have been applied in a variety of applications, such as solar energy storage,[18,19] air conditioning,[20] building energy conservation,[21] and thermal-regulating fibers and textiles.[22]

4.2.1 Microencapsulation of Paraffin Wax

Microencapsulation is an effective way to use PCM as the core, which is covered with a shell made from polymer materials. The resulting material, called a PCM microcapsule, can prevent the leakage of the interior PCMs and increase the heat transfer area.[23] Various strategies are followed to strengthen the shell, to increase the thermal conductivity, and add some other attributes to the microcapsule.

The microcapsules with silver nanoparticles can be fabricated by using in-situ polymerization, with aminoplast as the wall and bromo-hexadecane as the PCM core.[24] The distribution of silver nanoparticles on the surface increases the shell toughness, thermal stability, and mechanical strength of the microcapsules.

Microencapsulated PCMs using melamine-formaldehyde resin/SiO_2 shell are made by using 1-hexadecanol as the PCM core:[25] SiO_2 particles, organically modified by using dimethyldichlorosilane, stabilize the Pickering emulsion, and further, a hybrid shell is prepared by in-situ polymerization of melamine and formaldehyde. The employment of SiO_2 adds to the higher utility of the shell material and its impregnation in the shell wall strengthens the shell, improves the penetration resistance, and enhances the thermal reliability and conductivity.

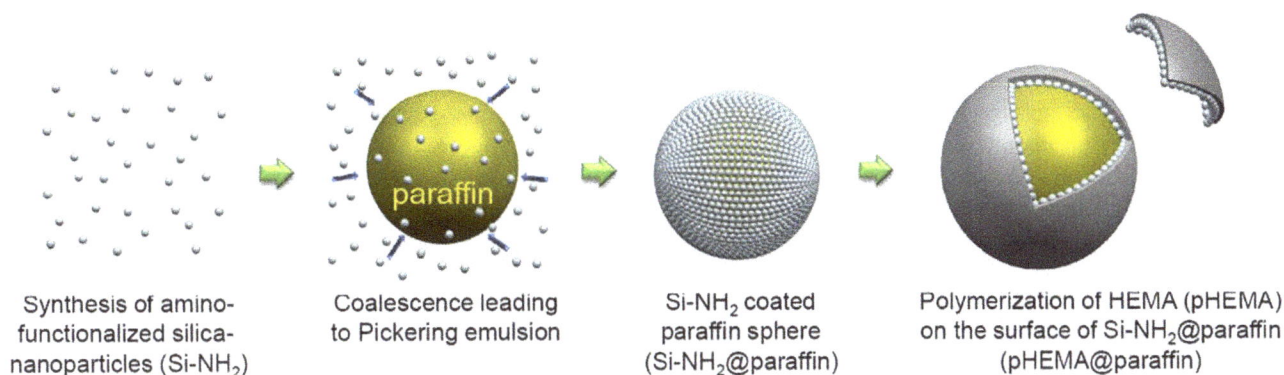

FIGURE 4.3 Schematic representation of the preparation of paraffin encapsulated poly(HEMA).

Source: [27]. Reprinted (adapted) with permission from American Chemical Society. Copyright (2015) American Chemical Society.

MicroPCMs possessing an n-octadecane (C_{18}) core with a melamine-urea-formaldehyde (MUF)/diatomite shell can be synthesized by using in-situ polymerization.[26] An improvement in mechanical properties can be achieved by the addition of moderate quantities of diatomite in these microPCMs. However, the enthalpy and encapsulation efficiency (η) decrease slightly. A water-absorbing polymer shell of poly(2-hydroxyethyl methacrylate) (poly(HEMA)) is used to coat the capsule of amino-functionalized silica particles containing paraffin (schematically presented in Figure 4.3) to endure the volume change with repeated cycling, ensure high sealing tightness and good flexibility.[27] As compared to dry poly(HEMA), a microPCM with 455 μm diameter has enhanced thermal conductivity of 0.47 W m^{-1} K^{-1} and latent heat of 167 J g^{-1}. Further, the PCM has high encapsulation ratio and efficiency of 97.7% and 97.3%, respectively.

The use of inorganic shells in microPCMs provides excellent mechanical properties. However, excess rigidity results in poor endurance when used in practical applications. The problem can be addressed by using an organic-inorganic hybrid shell which can be designed to have a synergetic combination of the mechanical properties of inorganic shells and the flexibility of organic shells. The mechanical strength of microPCMs can be improved by the addition of a moderate amount of modified carbon nanotubes (CNTs) in the polymer system. MicroPCMs based on an n-octadecane (C18) core and a MUF shell, along with O_2-plasma-modified multi-walled carbon nanotubes (MWCNTs), exhibit remarkable improvement in thermal and mechanical properties.[28] There are, however, slight decreases in enthalpy and encapsulated efficiency.

4.2.2 Paraffin Wax Composites and Nanocomposites

4.2.2.1 Paraffin Wax/Polymer Nanocomposites

Lightweight polymer composite materials with TES and high mechanical properties are useful in applications such as the automotive industry, where the diffusion of lightweight structures can disturb the thermal management of the cockpit environment, or in portable electronics field, where the space constraint does not permit the setting-up of a cooling system. Structural laminates can be made by combining an epoxy resin, with paraffin as PCM, which is stabilized with CNTs, and reinforcing carbon fibers.[17] The stabilized paraffin maintains an inherent ability to undergo fusion and crystallization in the laminates even after repeated thermal cycling. The transition enthalpy of the composite is proportional to the weight fraction of paraffin reaching a high value of 47.4 J cm^{-3}. Moreover, the thermal conductivity of the laminates' through-thickness direction increases with the content of CNT-stabilized PCM. Dynamic mechanical analysis exhibits a sharp drop of storage modulus at T_g as well as at T_m of paraffin. Similarly, there are two tan δ peaks corresponding to similar transitions as in the case of storage modulus. Paraffin-containing samples show a progressive failure during the application of load in the tensile test. A sequence of drops and plateaus in the load-displacement curves indicates the mechanical energy absorption capability of the laminates.

TES composites can be designed with a high-performance epoxy matrix together with a paraffinic PCM and the CNTs responsible for confining paraffin.[15] To confine paraffin in CNTs, the latter is added to molten paraffin under mechanical stirring at an rpm of 500 for 30 min and cooled to an ambient temperature by pouring into silicon molds. For shape stabilization, a minimum 10 wt% CNT is required to contain the paraffin. The CNT-confined paraffin, thus prepared is added to the epoxy resin and the mix is cured at an ambient temperature followed by post-curing at a high temperature. Blending does not affect the inherent melting and freezing peak temperatures. However, without CNT, the paraffin exudes out as the confinement action is missing in the blend. In accordance with the mixture rule, paraffin loading in epoxy reduces the elastic modulus and flexural strength of epoxy matrix. The electrical resistivity decreases with the loading of CNT.

4.2.2.2 Form-Stable Composite of Paraffin Wax

Form-stable composites of PW in an ethylene/propylene diene monomers (EPDM) rubber matrix can be prepared by melt mixing with varying degrees of PW concentration.[13] The compounds are then vulcanized to become shape-stabilized rubber materials for TES. There is a reduced rate of vulcanization of the EPDM matrix due to the presence of rubber. Homogeneous distribution of wax particles within the rubber allows the melting enthalpy to remain almost unchanged and remain so even at higher concentrations and after repeated thermal cycles. The mechanical properties of the EPDM matrix were positively affected by the incorporation of polyethylene wax (increasing proportionally to its content), as identified by dynamic and quasi-static tensile tests.

PCM blends based on soft PW and polyethylene are made by mixing with copper powder by first blending and then melt pressing techniques.[29] The melting peaks at lower wax loading are clearly defined as lower temperature shoulder and higher temperature peak. However, at higher wax loading, the peaks are exactly the opposite. Shorter low-density polyethylene chains and/or branches on co-crystallizing with wax make this kind of melting. This effect becomes less pronounced with an increase in the wax content. The transition temperature of the blend decreases with the wax loading as a result of the molten wax acting as a plasticizer in the linear low-density polyethylene matrix. The crystalline nature of the wax enhances the modulus of the PE. The presence of Cu micro-particles did not significantly change the thermal behavior of the blends. By increasing the copper content, the thermal stability of the composite is increased.

4.2.3 Wax in Porous Carbon Materials

4.2.3.1 CNT Sponges

PW can also be encapsulated in the porous scaffolds of carbon nanotube sponges (CNTS) to make electrically conductive composites with improved transition enthalpy and high thermal conductivity.[30]

Efficient TES can be realized in this composite via electrothermal conversion conversion or light absorption. The sponges can recover their original shape without any visible plastic deformation even after a large compressive strain or deformation strain for any arbitrary shape. Instead of molten wax, the wax solution is dripped on the sponge to infiltrate into the pores due to the high viscosity of the former as illustrated by Figure 4.4 (a). The porous CNTS has pores of the order of a few hundreds of nm to 1 μm (Figure 4.4 (b)) and this inter-nanotube spacing is filled by PW (Figure 4.4 (c) and Figure 4.4 (d)). The composite is extremely stable that by keeping it above melting temperature for two weeks, no liquid separates. Also, the enthalpy values remain slightly lower at PW loading below 87%.

However, with increasing PW content, the value becomes higher than pure PW. The six-fold increment of thermal conductivity of PW occurs after making a composite. There are many C-H bonds in PW and delocalized π-electrons on the surface of the nanotubes, which lead to extensive C-H… π-interactions at the PW-nanotube interface. This interaction affects the phase transition enthalpy of the paraffin. Further, the compressive stress applied by nanotubes on the melting PW during its volume expansion also inhibits the solid-to-liquid transition, resulting in high system enthalpy.

Vertically aligned CNT arrays are highly porous, and because of high absorption of sunlight, appear black in their original state. Further, the presence of CNT leads to enhanced photon absorption and, therefore, the PW in the composite can be heated. When current passes through the composite, the resistance of individual

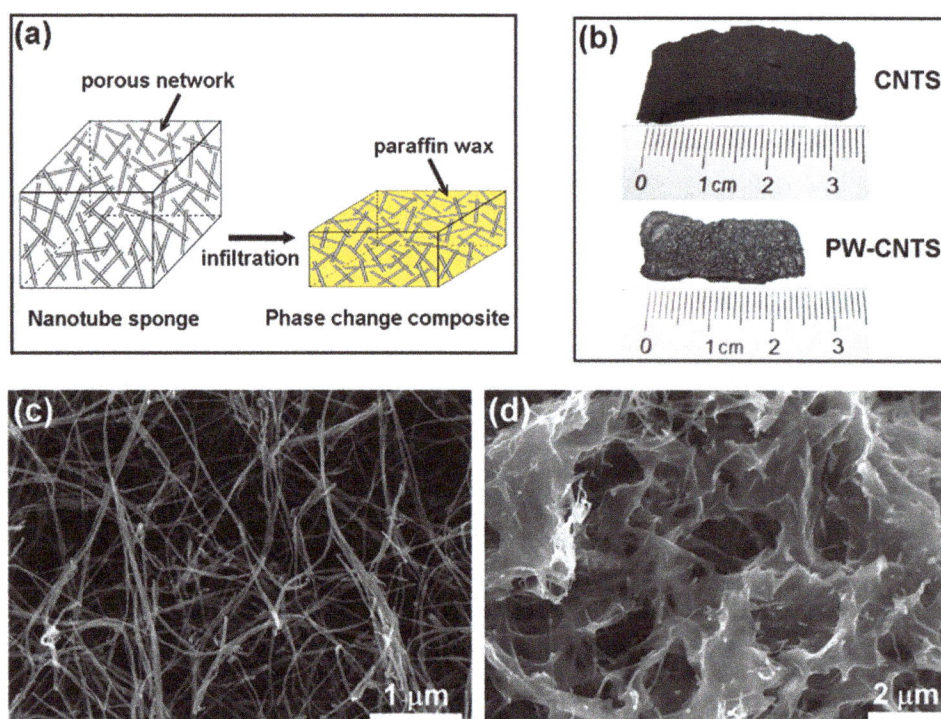

FIGURE 4.4   (a) Schematic for preparation of CNTS-PW composite, (b) Photograph of original CNTS and CNTS after PW infiltration, (c) SEM of sponge showing a porous structure; (d) SEM image of CNTS-PW composite.

nanotubes along with contact resistance in the network generates Joule heat. The main heat transfer mechanism comprises solid-to-solid heat conduction at the PW nanotube interface.[31]

The electro-heat storage efficiency (η) can be calculated by taking the ratio of stored heat in the PW [total mass (m) of PW enclosed in composite multiplied by its enthalpy (ΔH)] and the received electrical energy during the phase change [product of voltage (V), current (I) and time (t)]. Therefore, the energy storage efficiency in the system is defined as $\eta = (m\Delta H)/VIt$ on the basis of the assumption that all the encapsulated PW material participates in the phase change. Sunlight-to-heat storage efficiency ($\eta > 90\%$) is possible by simulating the irradiation of light at a wavelength (half-wave width of 80 nm) on dye-grafted PCM.[32]

4.2.3.2 Graphene Foam

A three-dimensional (3D) interconnected network of graphene foam (GF) exhibits enhanced thermal resistance for heat transfer from PCM to GF walls. A 3D hierarchical GF (HGF) can be obtained by filling the pores of GF with hollow graphene networks, which is used to prepare a PW-based composite PCM.[33] The thermal conductivity of the PW/HGF composite PCM is 87% and 744% higher than the PW/GF composite PCM and pure PW, respectively. Shape stability, negligible change in transition temperature, and high TES density (95% of pure PW), coupled with good chemical stability and thermal reliability characterize the composite. In addition, this PCM offers high light-to-thermal energy conversion and storage efficiency.

4.2.3.3 Intercalated Wax

Composite PCM comprising of organic montmorillonite (OMMT)/paraffin/grafted MWCNT, can be synthesized by using ultrasonic dispersion and liquid intercalation.[34] Paraffin is intercalated into the OMMT interlayer, and grafted MWCNT is well dispersed in the interlayers of OMMT. The phase change enthalpy of OMMT/paraffin/grafted MWCNT composite is around 47.7 J g^{-1}, which reduces slightly after extended thermal cycles. The thermal conductivity of OMMT/paraffin increases by 34% after adding grafted MWCNT.

4.2.4 Structural Laminates with Paraffin Wax

Structural laminates can be designed by combining PW, stabilized with CNTs, with an epoxy resin and reinforcing carbon fibers.[17] The stabilized paraffin can melt and crystallize in the laminates, and the melting enthalpy of the composite is proportional to the weight fraction of the paraffin (maximum value: 47.4 J cm^{-3}), which is stable in repeated thermal cycling. The presence of CNTs increases the thermal conductivity of the laminates' through-thickness direction. The flexural modulus is slightly lower due to the presence of PCM.

4.3 Polyethylene Glycol, Fatty Acid, and Esters as Phase Change Materials

Fatty acids possess higher enthalpy of fusion compared to that of PW. Fatty acids exhibit reproducible melting and crystallization behavior with minimum supercooling. The general formula of all fatty acids is given by $CH_3(CH_2)_{2n}$-COOH. However, their major drawback is their excessive cost which is about 2–2.5 times higher than PW. Also, they are mildly corrosive. PEG, having a general formula H-(O-CH$_2$-CH$_2$)$_n$-OH, has been exhaustively studied as a phase change material due to high phase change enthalpy, variable transition temperatures, ease of chemical modification, congruent transition behavior, chemical stability, thermal stability, non-corrosiveness, and biocompatibility.[35] The enthalpy and melting temperature of PEG vary with molecular weight. The PCM can be designed by physical blending or chemically attaching it to various organic and inorganic components.

4.3.1 Form-Stable PCMs

Form-stable composites can be prepared by physically blending a soft segment with a higher melting polymeric substrate so that the soft segment is fixed in the polymer matrix preventing its leakage during the solid-to-liquid phase transition while maintaining the temperature below the melting point of the polymer matrix.

4.3.1.1 Organic Form-Stable PCMs

4.3.1.1.1 Composites with Cellulose and Cellulose Derivatives

Constant efforts are made to use natural polymers for commercial applications so that non-biodegradable polymer pollution is minimized. Cellulose-based polymers are abundant in nature. Further, these are polar in nature, thermally and chemically stable, which are required for making form-stable PCM blends.

The earliest blends of PEG were prepared with cellulose. Cellulose extracted from wood pulp is blended with PEG in various proportions.[36] Intermolecular hydrogen bonding between PEG and cellulose prevents the flow of molten PEG that results in a solid-solid transition of PEG from crystalline to an amorphous state at temperatures above the melting point of PEG. The crystal size of PEG is usually smaller in the blend. This has a direct effect on the enthalpy of the transition and the transition temperature, which are lower at 120 J g^{-1} and 52°C respectively. Another renewable polymer cellulose diacetate (CDA) can be modified with PEG by both physical and chemical bonding methods.[37] By the chemical bonding method, the PEG molecules are constrained, which has a direct effect on their crystallization ability. The solid-solid PCM of CDA and PEG exhibits lower enthalpy of 74 J g^{-1}, which is due to its lower crystallinity. On the other hand, the physical blending method gives rise to solid-liquid PCM with much lower constraints on the crystallization of PEG. The blend has a higher enthalpy of 105 J g^{-1} and a phase transition temperature of 40°C. It is also the PEG content which decides the phase change pattern. By keeping the PEG content below 85%, it is possible to retain a solid-solid phase change during melting (155 J g^{-1} at 52°C).[38] PEG is blended with cellulose acetate in the presence of microwaves (Figure 4.5 (a)), reducing energy and solvent consumption and enhancing the encapsulation efficiency with 96.5 wt% PEG.[39] The latent enthalpy varies from 23.5–155.4 J g^{-1}. The network structure of the blend is visible at lower concentrations and globular morphology is prominent at higher concentrations of PEG (Figure 4.5 (b)).

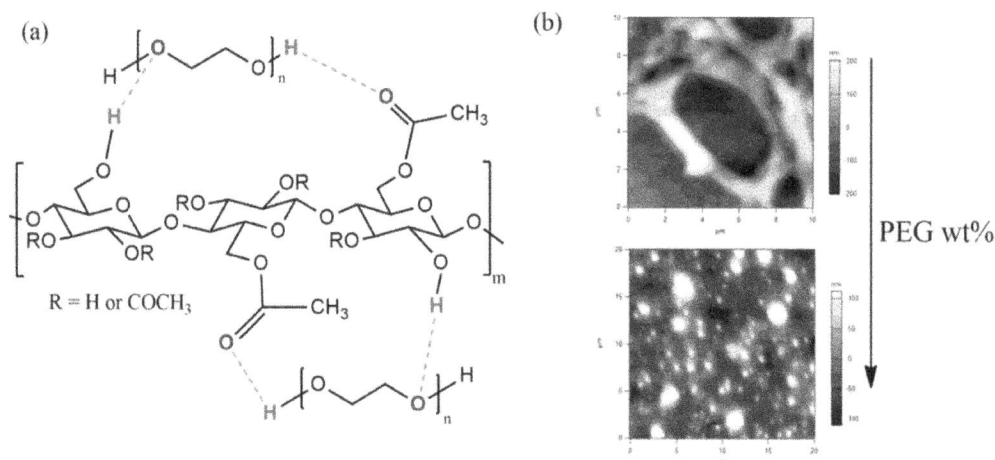

FIGURE 4.5   (a) Interaction of PEG and cellulose acetate in composite; (b) AFM images of form-stable PEG-CA blend with increasing concentration of PEG. Source: [39]. Reprinted with permission from Elsevier; Swati et al. (2016).

Also, the presence of cellulose acetate reduces the hydrophilic nature of PEG. The type of phase transition of polyethylene oxide (PEO) and cellulose polymers blends depends on the type of the latter.[40] Thus, the PEO-cellulose acetate blends show solid-liquid phase transition for all concentrations, whereas solid-solid phase transitions are possible for 25 and 50 wt% PEO in PEO-carboxymethyl cellulose blends, 25 wt% PEO in PEO-cellulose ether blends, and for the entire range of concentration of PEO (25–75 wt%) in PEO-cellulose blends. Strong hydrogen bonding interactions between PEO/cellulose and PEO/cellulose derivatives blends make the solid-solid phase transition possible. At 1:1 (w/w) compositions of PEO/cellulose acetate and PEO/carboxymethyl cellulose blends, synergistic transition enthalpy is possible. Blends of PEG with biodegradable natural polymers, such as cellulose, agarose, and chitosan, having strong hydrogen bonding interactions, form stable PCMs.[41] 3D porous cellulose/graphene nanoplatelets (GNPs) aerogels are fabricated in which PEG is vacuum impregnated to prepare phase change composites.[42] The GNPs, having a very high thermal conductivity, increase the heat-carrying capacity to a large extent (at 5.3 wt%, the thermal conductivity of 1.35 W m^{-1} K^{-1}). High fusion enthalpy of 156 J g^{-1} and the corresponding crystallization enthalpy of 149 J g^{-1} can be realized by loading 89.2 wt% PEG in the composite. With large surface to volume ratio and small diameter, the insoluble and fibrous carboxymethyl cellulose-1 fibers can be treated with a eutectic mixture of lauric and stearic acid.[43] The lightweight composite exhibits a latent heat of 115 J g^{-1} with very good thermal reliability.

4.3.1.1.2 Acrylate-Based Composites

PEG and acrylates, being moderately polar in nature, the blends/composites are compatible. The acrylates are thermally stable and do not impart any color or negative effect. Blends of PEG with acrylic polymers, such as polymethyl methacrylate (PMMA), Eudragit S and Eudragit E, can be prepared with enthalpy values ranging from 141–149 J g^{-1}.[44,45] These blends exhibit solid-solid phase transition till 80 wt% encapsulation of PEG. The blends

have good thermal and chemical reliability, as the enthalpy of fusion reduces by only 10% with no chemical degradation even after 3000 thermal cycles. The interactions of acid groups of acrylic acid polymers, such as poly(acrylic acid) (PAA) and poly(ethylene-co-acrylic acid) (Eco-A), with PEG, lead to the formation of interpolymer complexing blends, stabilized through hydrogen bonds.[46] For the PEG-PAA blend, latent enthalpy can be observed only at high concentrations of PEG (≥75 wt%), whereas in PEG-EcoA blends, the latent enthalpy values are observed at all compositions. With low molecular weight PEG (≤ 1000), the incompatibility in PEG-EcoA blends leads to seepage. A series of binary fatty acid eutectics of capric, lauric, myristic, and stearic acid can be developed to prepare form-stable PCM with a PMMA supporting matrix.[47] The combination of lauric and myristic acid with 70 wt% composition within the PMMA matrix is most suitable for use as a form-stable PCM with storage enthalpy of 113 J g^{-1}.

Although the versatility of the PEG-PMMA composite is established, its commercial feasibility depends on the thermal conductivity of the composite. To overcome the low thermal conductivity of the organic composite PCMs, thermal fillers can be used to improve the thermal conductivity. A metal-based thermally conductive filler, viz. aluminum nitride, is added during the in-situ polymerization of methyl methacrylate to prepare a composite withstanding maximum encapsulation of 70 wt% PEG.[48] By varying the mass fractions of aluminum nitride in the composite in the range of 5–30%, the thermal conductivity is increased from 0.253 to 0.389 W m^{-1} K^{-1}. However, increasing the mass fractions of aluminum nitride results in reduced latent enthalpy from 116 to 79 J g^{-1}. On the other hand, the volume resistivity increases from 0.26×10^{10} to 5.92×10^{10} Ω cm for the PEG/PMMA composite by adding 30 wt% aluminum nitride, which promotes the electric insulation capabilities of the PCM. Therefore, this composite PCM can be used for thermal management of electronic devices that have acceptable electrical insulation properties. Carbon-based filters are gaining in importance due to their excellent thermal and electrical properties, large surface area, low density, and significant chemical stability. Thus, GNPs are added as conductive fillers via ultrasonic treatment to form PEG-PMMA form-stable PCM networks that have improved thermal and electrical conductivity.[49] For PEG-PMMA composites, a maximum of 70 wt% PEG can be incorporated. Due to the high aspect ratio of GNP, a thermally conductive pathway is formed in the composite. As a result, the thermal conductivity of PEG-PMMA/GNP composite increases by about 20 times to 2.339 W m^{-1} K^{-1} from that of pure PEG-PMMA composite with the incorporation of 8 wt% GNP. The electrical conductivity of the composite is also improved due to the electrically conducting nature of GNP. The percolation threshold is reached with only 2 wt% GNP and the corresponding electrical conductivity is 10^{-4} S cm^{-1}. The addition of GNPs aids the crystallization of PEG and therefore, the supercooling extent reduces with an increase in thermal stability. This type of composite is considered potential material for the applications of EMI shielding, anti-electrostatic material and bipolar plates in a proton exchange membrane fuel cell.

4.3.1.1.3 Polyurethane Blends

Cross-linked polyurethanes (PU) are considered a suitable container for the PCMs with the aim of obstructing the migration of the PCM with repeated use and giving way to a synergistic phase change effect, allowing a wider transition range. The solid-solid PCM with the urethane linkage is prepared by synthesizing the prepolymer from PEG and methylene diphenyl diisocyanate (MDI), which is further reacted with glucose. The reaction mixture is then blended with extra PEG and cured with heat over an extended period to obtain form-stable PCM.[50] The synthesized polymeric PCM exhibits a solid-solid phase change and a synergistic effect is observed due to the presence of PEG in the cross-linked polymeric chain as well as in the blend. An improvement in enthalpy of 21.3% over the polymeric PCM can be achieved. An environmentally friendly, cost-effective method is adopted to prepare polyurethane/paraffin composites by avoiding the use of toxic organic solvents.[51] This method allows the maximum encapsulation of a series of paraffin waxes in the hard PU segment made with PEG 10,000. Fusion enthalpy of 141 J g^{-1} and transition temperatures ranging from 20 to 65°C, exhibited by this combination, are suitable for multiple energy storage applications. In these polyurethane composites, the chain extender is a vital component that is responsible for increasing the phase separation that enhances the crystallization of soft segments.[52] However, with PU-based graphene composites of SSPCMs without a chain extender, higher heat of fusions, higher thermal stability, and better thermal reliability have been observed.[53]

In cross-linked polyurethane prepared by using PEG and a branching unit, xylitol can be used as a supporting matrix to prepare a lauric acid/PU composite having a high melting enthalpy of 125 J g^{-1} and reduced supercooling as a result of the overlap of phase transitions of PU and lauric acid.[54] Using the same cross-linked polyurethane system with PEG as supporting material, paraffin can be used as the working material.[55] The system exhibits synergistic phase change with enthalpy reaching 210.6 J g^{-1} at an encapsulation percentage of 74 wt%. Paraffin promotes microphase separation by reducing the hard-soft segment compatibility. Also, by integrating dye molecules with the polyurethane matrix, direct conversion of visible light to thermal energy can be achieved and this energy is stored by the PCM.[56] By using hexadecanol as a PCM encapsulated by the PU matrix, very high enthalpy of the order of 229.5 J g^{-1} is possible.

4.3.1.1.4 Carbon-Based Composites

Carbon-based materials, such as activated carbon (AC), carbon nanotubes, or ordered mesoporous carbon (CMK-5), have gained significance, not only as thermally conductive fillers but also as containment matrices. Along with improved thermal conductivity of the composite, the porous nature of these materials allows for maximum accommodation of the PCM material and offers structural stability without compromising the transition enthalpy significantly. Physical blending and impregnation method has been most commonly applied for the preparation of these composites. Due to the porous nature of these materials, there is a minimum threshold for the accommodation of working material below which the soft segment is well absorbed into the pores, thereby hindering the crystallization and thermal behavior. For the PEG-mesoporous active carbon form-stable PCM, this threshold limit is 30 wt% PEG.[57] The phase change properties of these PEG/AC PCMs are affected by the adsorption confinement of PEG segments in the porous structure of the AC as well as the extent of the interference by the AC as an impurity with the normal crystallization of the PEG. The phase change activation energy increases with the increasing AC content and the decreasing PEG content. For larger chain polyethylene glycol hexadecyl ether (Brij58) working material, this threshold limit is 25 wt% PEG with a porous activated carbon supporting framework.[58] Porous AC positively influences the crystallization and nucleation process of Brij58. Confined non-isothermal crystallization of Brij58 as a result of nanoporous activated carbon matrix is associated with an increase in half-time and the relative degree of crystallinity. The pore structure of these porous materials is also important as it affects the phase change behavior of composites.[59] The irregular pores of the AC have a pore diameter of 4 nm whereas the hexagonally ordered pores of CMK-5 have a pore diameter of 3.5 nm. The mesoporous pores of expanded graphite (EG) have pore diameters around 13 μm. The phase change enthalpies of porous carbon composites and the crystallinity of PEG increase in the order of PEG-AC < PEG-CMK-5 < PEG-EG. Thus, higher pore volumes and ordered pore geometries are essential for shape stabilization. The large pore volume of the EG supports crystallization, whereas the small pore diameter of the AC and CMK-5 hinders the crystallization of the PEG.

MWCNTs with (-COOH, -NH$_2$, -OH functional groups) and without functionalization, show different phase change behavior of the PEG in the PEG/MWCNT composite.[60] PEG/functionalized MWCNTs exhibit lower phase transition enthalpy and transition temperatures, compared to pure PEG/MWCNT PCM. The functional groups affect the enthalpy values of the composite PCM in the order of MWCNT-COOH > MWCNT-NH$_2$ > MWCNT-OH > MWCNT. This order is a result of the influence of hydrogen bonding, capillary forces, and surface adsorption.

Along with direct use of commercially available carbon materials, porous carbon is also synthesized through various routes, using different raw materials. By carrying out controlled carbonization of metal-organic frameworks at different temperatures, highly porous carbons, with an exceptionally high surface area and a large pore volume reaching 2551 m^2 g^{-1} and 3.1 cm^3 g^{-1} respectively, have been used as a supporting matrix in PEG-based shape-stabilized PCMs.[61] Carbonization of metal-organic frameworks at high temperatures, leading to the migration and evaporation of zinc oxide particles, forms a highly porous matrix with an adsorption capacity of 92.5 wt% PEG which can afford high latent heat of 162 J g^{-1}. Additionally, the thermal conductivity is 50% higher, compared to pure PEG. Potato is also used as raw material to prepare porous carbon by using consecutive freeze drying and heat treatment, thereby utilizing the renewable plant resources/agricultural feedstock.[62] The prepared porous carbons exhibit an average pore diameter of 204.7 nm with 73.4% porosity while providing reasonable mechanical strength to retain their shape stability and prevent leakage of the melted PEG. A maximum loading of 50 wt% PEG can be attained. Melted PEG with good wettability to carbon confirms its exudation stability above melting temperatures.

4.3.1.1.5 Graphite and Graphite Derivatives-Based Composites

The importance of graphite and its derivatives lies in its two-dimensional nature, high thermal conductivity, and superior electronic and mechanical properties, along with the added advantage of low density. Similar to carbon composites, these composites of graphite are also commonly prepared by using the blending and impregnation process. Blends of PEG and EG contain 90 wt% PEG with a high enthalpy of 161 J g^{-1} at 61°C.[63] Further, the thermal conductivity improves with an increase in percentage incorporation of EG with maximum values reaching 1.324 W m^{-1} K^{-1} at 10 wt% of EG. The 3D netlike architecture of graphene/PEG composites, containing 93 wt% PEG, has fusion enthalpies reaching 166 J g^{-1}.[64] This interconnected architecture of composite is used as a source of heat for thermoelectric device fabrication, allowing a long steady-state output time with increasing weight percentage of PEG. Another environmentally-friendly derivative of graphene, sulphonated graphene (SG), is used as a nanocomposite matrix to incorporate 96 wt% PEG by using solution processing in an aqueous medium.[65] The thermal conductivity increases fourfold from 0.263 W m^{-1} K^{-1} for pristine PEG to 1.042 W m^{-1} K^{-1} for the composite. However, the latent heat value is 12.9% lower than pure PEG.

The oxygen functional groups of graphene oxide (GO) undergo strong hydrogen bonding interactions with PEG up to 90 wt% incorporation.[66] GO is effective in lowering the transition temperature without much impact on the phase change enthalpy in comparison to other carbon-based porous materials, including AC and CMK-5. In a typical procedure, 96 wt% PEG/GO nanocomposites are prepared by using both blending and impregnation and the microwave method.[67,68] The interlayer spacing between the GO sheets is enhanced by using the microwave method, which helps to efficiently disperse the PEG molecular chains. Thus, the TES of the composite prepared by the microwave method is 175 J g^{-1}, while the composite prepared by the blending and impregnation method has a latent heat of 143 J g^{-1}.

With an increase in the concentration of GO, the photo-thermal energy conversion efficiency is improved. The surface of the GO can be modified by carboxylation and reduction to prepare GO-COOH and reduced GO respectively; which is used for the preparation of PEG-based nanocomposites.[69] The carboxylic functionalization leads to a random distribution of carboxyl groups which can interact strongly with PEG chains whereas the reduction with NaBH$_4$ removes surface oxygen functional groups, responsible for weak interactions with PEG chains. Therefore, melting enthalpy and phase change temperature reduction follow the trend of PEG/GO-COOH < PEG/GO < PEG/reduced GO. Apart from surface functionalization, the large interface area is also a contributory factor in the phase change behavior of PEG. The reduction and functionalization of GO with oleylamine (OA) (Figure 4.6(a)), followed by adsorption of palmitic acid and simultaneous self-assembly into the 3D structure (Figure 4.6(b)), result in a phase change composite.[70]

With very low loading of 0.6 wt% of GO, the composite retains 99.6% of the latent enthalpy of palmitic acid with a value of 196.6 J.g^{-1}. The enthalpy improves with thermal cycling after 500 thermal cycles to 202 J.g^{-1}. The composite exhibits high photo-to-thermal conversion efficiency.

3D hybrid graphene aerogels (HGA) act as both nucleating agent and supporting material. By allowing the incorporation of 2 wt% GO and 4 wt% GNP in hybrid graphene aerogels and introducing them in PEG by using vacuum impregnation and physical blending methods, composite PCMs can be produced.[19,71] These composites exhibit high latent heat values in the range of 170–186 J g^{-1} and also the 3D supporting framework leads to a significant increase in thermal conductivity. The vacuum impregnated composite exhibits an increased thermal conductivity from 0.3 W m^{-1} K^{-1} of pure PEG to a value of 1.43 W m^{-1} K^{-1} by incorporating aerogel, having a composition of 0.45 wt% GO and 1.8 wt% GNP. Additionally, this composite PCM shows light-to-thermal energy conversion with 91.9% efficiency. For the composite prepared by using physical blending method, the thermal conductivity reaches a maximum value of 1.7 W m^{-1} K^{-1} and electrical conductivity reaches a value of 2.5 Sm^{-1}. With ultralow filler concentrations, this composite exhibits a high latent heat value of 178 J g^{-1}, which is 98.2% of that of pure PEG.

4.3.2 Phase Change Polymer Blends with Engineering Polymers

Engineering polymers, such as epoxy and acrylonitrile, offer good chemical resistance, toughness, and mechanical strength and, therefore, have been used as supporting matrices to prepare form-stable PCM. A composite of PEG and epoxy resin of diglycidyl ether of bisphenol-A, prepared by casting/molding, contains 75% of PEG-4000.[72] The epoxy resin with a 3D network acts as the supporting material and imparts mechanical and thermal stability to the PCM. In a similar way, a cross-linked polymer matrix of poly(acrylonitrile-co-itaconate) can be blended with PEG up to maximum incorporation of 73 wt% PEG, which has good fusion enthalpy of 119 J g^{-1}.[73] Due to steric hindrance, the movement of PEG is limited, leading to a reduced size of the crystallites and therefore, lower phase transition enthalpies. Due to concern over a fire in an

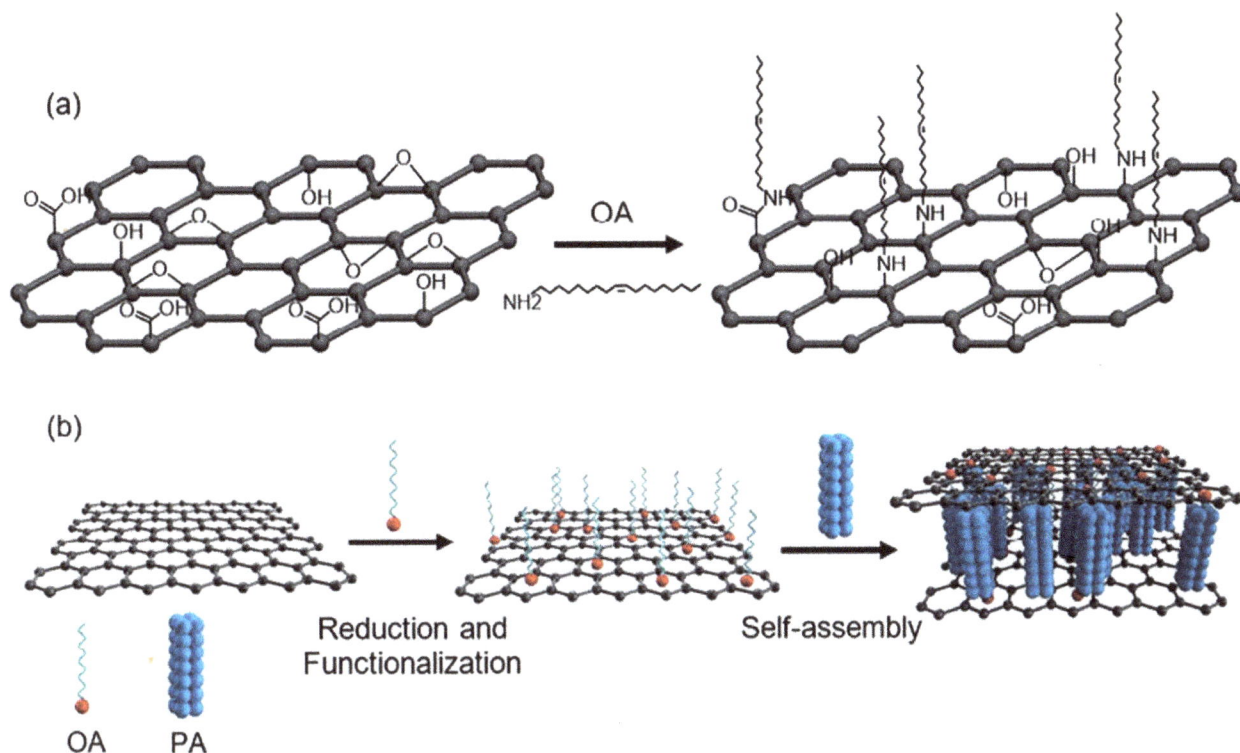

FIGURE 4.6 (a) Functionalization of GO; (b) Schematic illustration of the assembly of functionalized GO and palmitic acid followed by self-assembly into a 3-D composite.

Source: [70]. Reprinted (adapted) with permission from *Journal of Physical Chemistry C* 119, 40, 22787–22796. Copyright (2015) American Chemical Society.

enclosed area, fire-retardant PCM has been considered. The preparation of halogen-free, flame-retardant, form-stable, organic PCM can be done by using the twin screw extruder technique with paraffin as the working material and high-density polyethylene as the supporting material.[74] With latent heat energy in the range of 79–94 J g^{-1}, these PCMs have improved thermal stability and produce a large amount of charred residue during burning. Highly porous, lightweight, superhydrophobic polypropylene aerogel prepared from industrial wastes is considered a supporting material to store paraffin.[75] Extremely high PCM loading up to 1060 wt% can be realized with a heat storage capacity in the range of 141–160 J g^{-1}.

4.3.3 Phase Change Polymer Blends with Conducting Polymers

A conducting polymer, polyaniline (PANI) composite with tetradecanol is prepared using in-situ polymerization.[76] This composite has a melting enthalpy of 163 J g^{-1} and possesses the ability to endure heat shock. However, due to processing difficulty, more suitable fatty acids have been preferred when preparing PCM composites with PANI. PANI has been used as supporting material to encapsulate 82 wt% of myristic acid by using a surface polymerization method.[77] Keeping in mind the importance of thermal conductivity, exfoliated GNP has been used as a filler for the PANI/palmitic acid composite by using ultrasonication.[78] With a low loading of 7.8 wt% of GNP, high enthalpy of 158 J g^{-1} and thermal conductivity, an improvement of 237.5% can be achieved.

4.3.4 Miscellaneous Organic Composites

Many studies have dealt with unconventional hard segments in an attempt to understand the feasibility of various materials as a containment matrix. The blends of PEO and potato starch with weight concentrations of 3:1, 1:1, and 1:3 w/w are held together by the strong intermolecular hydrogen bonding interactions.[79] The solid-solid phase transition is observed for 1:3 and 1:1 w/w PEO/potato starch blend with melting enthalpies 43 J g^{-1} and 47 J g^{-1}, respectively, whereas the 3:1 w/w PEO/potato starch blend exhibits solid-solid phase change with partial melting and melting enthalpy of 97 J g^{-1}. For this biodegradable PCM, transition enthalpy is dependent on the strength of the hydrogen bonds between the PEO and the starch. Shape-stabilized PCMs of PEG with sugars such as glucose, lactose, and fructose allow maximum incorporation of 90 wt% PEG.[80] The blends are held together by hydrogen bonds between ether oxygen and hydroxyl end groups as a result of which the latent heat reduces by only 10–20%, compared to pristine PEG. Both these blends are prepared using the cost-effective method and the product is biodegradable in nature.

4.4 Hybrid Form-Stable PCMs

Various inorganic hard segments, such as silica, montmorillonite, diatomite, calcium silicate, clay, gypsum, vermiculite, and bentonite, etc. perform well as supporting materials along with organic soft segments. The incorporation of an inorganic segment adds to the chemical stability, thermal conductivity, fire resistance, and mechanical strength of the composites.

4.4.1 Silica Composites

Silicon dioxide is a highly promising carrier material due to its high surface area, porosity, excellent thermal stability, and non-toxic nature. The blend preparation method has been extensively studied using an array of techniques, starting from simple solution blending,[81] the conventional sol-gel method,[82] to the ultrasound-assisted sol-gel method,[83] the temperature-assisted sol-gel method,[84,85] and the coagulant-assisted sol-gel method.[85] All the methods can accommodate around 80 wt% PEG as a result of hydrogen bonding interactions, capillary forces, and the surface tension forces between the PEG and SiO_2.

A high melting enthalpy of 138 J g^{-1} is possible with the solution-blended PCM, whereas a much lower value of 75 J g^{-1} is exhibited by the PCM prepared by the conventional sol-gel method. Further, the type of sol-gel method affects the enthalpy of the composite. As an example, a PEG/SiO_2 composite prepared by the temperature-assisted sol-gel method has a higher enthalpy of 122 J g^{-1}. compared to that for calcium chloride-assisted composite (91 J g^{-1}).[85] The complex formation between PEG and calcium ion reduces its crystallization ability and leads to lower enthalpy. For the ultrasonic-assisted sol-gel method, using ultrasonic power of 300 W and a temperature below 60°C, a good phase change enthalpy of 103 J g^{-1} can be realized.[83] A low molecular weight PEG/SiO_2 composite exhibits poor thermal behavior due to the confinement of the PEG chains by the silica framework.[82] The enthalpy and crystallization behavior of the composite improve with an increase in the molecular weight of PEG used in composite preparation.

Waste raw materials are also used for making form-stable PEG composites. Oil-shale ash is used to extract sodium silicate by using the calcination-alkali leaching method under optimized conditions.[86] The PEG-SiO_2 composite, prepared by using the temperature-assisted sol-gel method in the absence of a surfactant or co-solvent, exhibits maximum latent enthalpy of 152 J g^{-1} at 58°C. The supercooling extent, melting, and solidifying time are lower by 22%, 27%, and 23% respectively than that of pure PEG.

The enhancement of thermal conductivity is a prerequisite for the practical application of PCMs. The thermal conductivity of the PEG/SiO_2 composites can be enhanced by certain metal and carbon-based fillers. The addition of β-aluminum nitride enhances the thermal conductivity from 0.3847 to 0.7661 W m^{-1} K^{-1}, while the loading is varied from 5% to 30 wt%.[87] Together with the increase in thermal conductivity, the enthalpy values decrease from 161 to 130 J g^{-1}. Copper can also be introduced in situ via a chemical reduction of copper sulfate by applying the ultrasound-assisted sol-gel method.[88] With 2.1 wt% copper, the thermal conductivity of the composite reaches 0.414 W m^{-1} K^{-1} with a corresponding melting enthalpy of 110 J g^{-1}. The introduction of graphite to the PEG-SiO_2 composite enhanced the thermal conductivity substantially (0.558 W m^{-1} K^{-1} with 2.7 wt% graphite).[84] By dispersing MWCNT in a PEG/SiO_2 composite,

thermal conductivity can be improved (0.463 W m^{-1} K^{-1} with 3 wt% MWCNT).[89] The presence of MWCNT qualifies the PCM to exhibit high photo-to-thermal conversion.

4.4.1.1 Mesoporous Matrices

Mesoporous matrices are attractive supporting materials due to their unique pore structures, high adsorption capacities, surface permeability, and fire-retardant properties. The phase transition properties and shape stabilization are directly related to the average pore size of the material.[90] If the pore size is too small, then PCM crystallization will be impeded, and if the pores are large, then the capillary force might not be sufficient to hold the liquid PCM.

A simple physical blending and impregnation process has been used in the preparation of composites of both PEG and its derivatives (polyethylene glycol octadecyl ether and Brij58) with mesoporous silica.[91,92] PEG/silica PCMs are bound by a combined effect of hydrogen bonding, capillary forces, and surface adsorption phenomena to exhibit higher enthalpy with regard to mesoporous silica molecular sieves (MCM-41 and SBA-15).[91]

The preparation of radial mesoporous silica (RMS) was carried out using tetraethylorthosilicate (TEOS) as SiO_2 precursor along with a cetyltrimethylammonium bromide template and this RMS can be vacuum impregnated with PEG.[93] The immersion time of 50 minutes with an immersion temperature of 70°C is optimum for the preparation of this PCM. Maximum enthalpy of 130 J g^{-1} can be realized by using 80 wt% PEG/RMS composite. The extent of supercooling reduces significantly by 4.8–19.7%.

Mesoporous silica nanoparticles including SBA-15, MCM-41, and mesocellular foams (MCF) composites with lauric acid are prepared using an evaporative solution impregnation to achieve dual thermal responsive PCMs.[94] MCF, with high pore volume, can accommodate 83 wt% lauric acid and had enthalpies reaching up to 124 J g^{-1}. The dual temperature response of MCF is possible due to the nanoconfined phase of the organic material inside the mesoporous material and the interparticle voids filling on the external silica surface. Hexagonal ordered silica, viz. MCM-41 and SBA-15, have a high fraction of mesopore volume which is occupied by the non-melting layer and, therefore, are less likely to have a dual temperature response (Figure 4.7). This dual temperature range is helpful in achieving a large operating window and to improve the heat transfer rate.

The modification of surface functional groups of mesoporous silica supporting material, namely SBA-15, can be engineered to control the thermal properties of the composite and crystallization behavior of the PCM.[69] The modification of terminal end-groups of PEG/HO-SBA-15-OH to amino and methyl groups yield PEG H_2N-SBA15-NH_2 and H_2N-SBA15-CH_3, respectively. The amino groups are fabricated on the internal surface of the SBA-15 channels whereas the methyl groups are placed on the external surface, thereby inducing opposing polarities, which prevent the spilling of PEG from the SBA-15 channels and promotes the crystallization of PEG. The grafting of –NH_2 groups in channels of silica reduces the adsorption sites of the PEG chain and changes the adsorption confirmation from the train to a loop structure, effectively enhancing the crystallization of PEG. The melting and

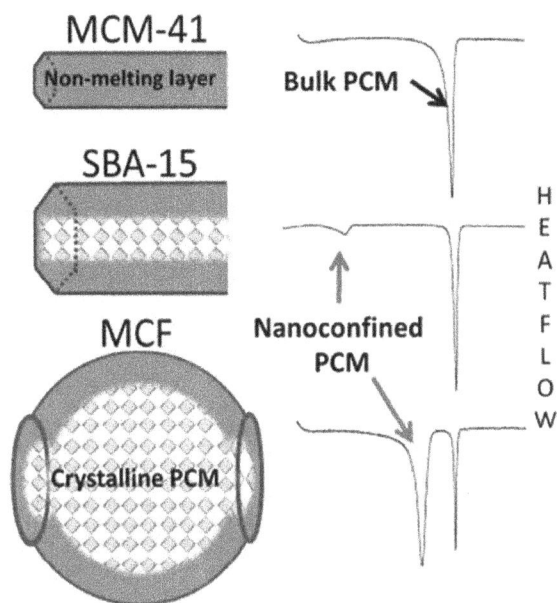

FIGURE 4.7 Dual temperature response by mesocellular foams as compared to hexagonally ordered mesoporous silica.

Source: [94]. Reprinted (adapted) with permission from *Journal of Physical Chemistry C* 119, 27, 15177–15184. Copyright (2015) American Chemical Society.

crystallization enthalpies of PEG/H$_2$N-SBA15-CH$_3$ reach 88 J g^{-1} and 82 J g^{-1}, respectively with 70 wt% PEG. The PEG/HO-SBA-15-OH composite, having only –OH grafting, cannot crystallize effectively and therefore, has zero enthalpies. Similarly, in the modification of surface functional groups of the MCM-41 silica support, good enthalpy of 80 J g^{-1} is possible with PEG. Thus, the surface functionalization behavior of various silica supports can be modified effectively to enhance the crystallization of PEG.

4.4.2 Building Material Composites

The fabrication of composite PCMs by using porous building materials as a supporting matrix for the incorporation of liquid PCM offers the advantages of thermal comfort and energy-efficient maintenance of indoor temperature. Simple blending impregnation and vacuum impregnation processes have been used for the preparation of building composite PCMs with maximum incorporation of the soft segment.

Both ordinary Portland cement[95] and mesoporous calcium silicate (MCS)[96] can be impregnated with PEG, using a simple physical blending process, allowing incorporation up to 70 wt% respectively. The PEG molecules are confined in the pores by the combined effect of the surface tension and capillary forces. Vacuum impregnation of gypsum and natural clay can accommodate a maximum of 18 and 22 wt% PEG, respectively and therefore, yield low enthalpies of 24 J g^{-1} for PEG/gypsum and 29 J g^{-1} for PEG/clay composite.[97] While studying the thermal performance of these building PCMs, by fabricating wallboards with both the composites using cubicle systems, the inside center temperatures of cubicles made of PEG/natural clay and PEG/gypsum are found to be lower by 2.08°C for 60 minutes and

1.47°C for 120 minutes, respectively as compared to the outside temperature. Low-cost clay mineral, bentonite exhibits a melting temperature and melting enthalpy varied in the range of 4–30°C and 38–74 J g^{-1} respectively when its blend with capric acid, dodecanol, PEG 600, and heptadecane is prepared by vacuum impregnation.[98] The transition temperature range is suitable for passive solar heating and cooling in building envelopes. By adding 5 wt% EG, the thermal conductivity of bentonite-capric acid, bentonite-PEG 600, bentonite-heptadecane, and bentonite-dodecanol composites can be increased by 65%, 63%, 39%, and 47% respectively. Also, the heating times of the bentonite-PCM composites with EG can be reduced by 12–22 seconds.

Natural Na-montmorillonite clay, having an average size of 936–1190 nm, is used to produce PEG-montmorillonite composites using ultrasonic impregnation, incorporating up to 44 wt% of PEG.[99] With the original heat storage capacity of 104.8 J g^{-1} remaining almost constant till 100 thermal cycles, these composites are considered promising candidates for the design of thermal insulation systems. The montmorillonite nanoclay can be used only for the improvement of thermal properties of the nanocomposite over a neat PEG/epoxy blend.[100] The presence of nanoclay maintains the heat transfer barrier properties, which is manifested by a 31% reduction in the top surface temperature.

4.4.3 Composites with Diatomite

Another economical, porous building material, diatomite has gained significant attention as a hard segment in the preparation of solid-solid PCM due to its chemical inertness, high surface area, high adsorption capability, and low density. It can be directly incorporated[101] or vacuum impregnated[101–103] to prepare composite PCM. For the preparation of composite by the direct impregnation method, an immersion time of 90 minutes at an immersion temperature of 75°C is ideal.[101] A loading of 55 wt% PEG can be realized with transition enthalpies ranging from 78–107 J g^{-1}. By using the vacuum impregnation method, the maximum incorporation of 58 wt% PEG can be achieved.[103] As an example, the enthalpy of PEG 4000/diatomite PCM is 106 J g^{-1}, with a transition temperature of 58°C, and it is highly stable after repeated thermal cycles. PEG/diatomite/single-walled CNTs (SWCNTs) form-stable composite PCMs, containing PEG of 60 wt%, and exhibit a shift of the melting point to a lower temperature whereas the solidification peak shifts to a higher temperature in the presence of SWCNTs.[104] The melting and solidification time of a SWCNTs-loaded matrix decreases by 54.7% and 51.1%, respectively. The thermal conductivity increases 2.8 times compared to composite PCM without SWCNTs. No leakage occurs in the extensive thermal cycling study.

Raw diatomite is normally leached with alkali to dredge diatomite pores and, thereby, improve the surface area and pore size.[105] Spherical, crystalline silver nanoparticles (3–10 nm) can be uniformly deposited on the surface of diatomite, which is blended with PEG to enhance the heat transfer properties of PCM. About 63 wt% PEG can be loaded to exhibit a melting enthalpy of 111 J g^{-1} at 60°C. With 7.2 wt% of Ag, an improvement in thermal conductivity of PEG/diatomite composite by 127% can be achieved, reaching a value of 0.82 W m^{-1} K^{-1}.

4.4.4 Miscellaneous Hybrid Composites

Among the hybrid composites, a number of unconventional inorganic hard segments have been employed to prepare SSPCMs with a wide array of properties.

A phosphamide-containing silsesquioxane matrix, with an average pore diameter of 4.1 nm, is used as a supporting matrix for the preparation of fire-retardant PCM. With 82 wt% PEG, the melting enthalpy can be as high as 125 J g^{-1} at a melting temperature of 56°C.[106] Bischofite, a by-product of the non-metallic mining industry, is obtained during potassium chloride production with a major component of $MgCl_2.6H_2O$.[107] $MgCl_2.6H_2O$ itself is a well-recognized inorganic PCM. Its cycling performance can be most satisfactorily improved by adding 5% PEG. However, high supercooling of 37°C remains a disadvantage.

Bulk carbon nitride (bulk-C_3N_4) and carbon nitride intercalation compounds (CNIC) can be used as shape stabilizers for PEG by using the blending and impregnation method.[108] The higher specific area of CNIC in PEG/CNIC blends than that of bulk C_3N_4 aids the crystallization of PEG. The PEG/CNIC blend possesses a fusion enthalpy of 46 J g^{-1} for 60 wt% PEG with a reduction of the melting temperature by 24°C from that of pure PEG. Thus, the use of graphitic carbon nitride as a shape stabilizer helps in lowering the phase change enthalpy and extent of supercooling.

To a PEG/expanded vermiculite composite, silver nanowire is added as a thermal conductivity enhancer.[109] The silver nanowire, of length 5–20 μm and diameter 50–100 nm, is wrapped by PEG and is well infused into the pores of expanded vermiculite. A maximum of 66 wt% PEG can be used for shape stabilization. On the other hand, increasing the concentration of silver nanowire decreases the latent enthalpy and increases the thermal conductivity (enthalpy of 99 J g^{-1} and thermal conductivity at 0.68 W m^{-1} K^{-1} at 19.3 wt% Ag nanowire). Expanded vermiculite acts as a nucleating agent in promoting the crystallization of the PEG chains and causes a reduction of the supercooling extent by 7°C. By adding silicon carbide nanowires as conductive fillers to vermiculite composites, the conductivity can be enhanced to 0.53 W m^{-1} K^{-1} with a melting enthalpy of 65 J g^{-1} at 3.28 wt% of the former.[110] Hybrid composite PCMs can be prepared by using inherently thermally conductive inorganic supporting materials, such as flower-like TiO_2 nanostructures with a unique pore structure and a high surface area that is obtained by the hydrothermal method, which contain 50 wt% PEG with a melting enthalpy of 86 J g^{-1}.[111] With 4 wt% GO and 30 wt% boron nitride serving as hard segments and thermally conductive fillers, respectively, PEG can be constricted to undergo crystalline to amorphous phase transition without any liquid leakage.[112] The melting enthalpy and thermal conductivity of the composite are 107 J g^{-1} and 3 W m^{-1} K^{-1} respectively and have the capability to exhibit light to electric energy conversion.

4.5 Chemical Methods for Preparing Solid-Solid PCMs

Various chemical methods, including grafting, cross-linking, and copolymerization have been used in the preparation of SSPCMs,

wherein the working material is linked chemically to the supporting material through different linkages, such as urethane, epoxy, ester, and ether modification. The chemical methods of modification yield more chemically and thermally stable PCMs but compromise the TES capacity.

4.5.1 Polyurethane Modification

Among the chemical methods, urethane modification is most commonly applied for the preparation of chemical SSPCMs. A wide array of multifunctional chain extenders with MDI have been used in the preparation of polyurethane PCM, including 1,4-butanediol (BDO),[113] pentaerythritol,[114] polyvinyl alcohol,[115] β-cyclodextrin,[116] trihydroxy surfactants,[117] and hexahydroxy polyols.[118] Undergoing a crystalline to amorphous phase transition with a reduction of crystal size of PU PCM compared to that of pure PEG, there is also an increase in thermal resistance due to cross-linking and incorporation of phenyl groups. In SSPCMs with low cross-link density, the PEG soft segments are less constrained from moving freely and maintain a higher crystallization ability.[116]

Thermoplastic polyurethane exhibiting solid-solid phase transition can be prepared by using PEG as a soft segment and MDI, along with a novel tetrahydroxy compound as a hard segment.[119] The lower phase change enthalpy of 137 J g^{-1} at 56°C is due to the lower crystallization ability of PUPCM than that of pure PEG. However, these materials are thermally stable with good mechanical properties and processing abilities. PU SSPCM is synthesized by a two-step condensation process by reaction of PEG and MDI with a novel tetrahydroxy compound, namely terephthalic acid bis-(2-hydroxy-1-hydroxymethyl-ethyl) ester, to get PUPCM with a phase change temperature of 48°C and latent heat of 154 J g^{-1}.[120]

By employing PEG as a soft segment with isophorone diisocyanate (IPDI) and BDO as hard segments, PEG-PU shape memory polymer can be prepared, which has an enthalpy value of 100 J g^{-1}.[11] The hydrogen bonded hard segments restrict the movement of the soft segments when heated above the phase transition temperature of 47°C. The PU prepared with PEG 1000 and coupling reagents, IPDI and toluene diisocyanate, exhibit solid-liquid PCM behavior and all other polyurethane PCMs (PEG mol. wt. 6000, 10000) possess solid-solid transition behavior.[121] Due to the linear and symmetric backbone, PEG-hexamethylene diisocyanate (HMDI) PCMs have higher phase change enthalpies than PEG-IPDI and PEG-toluene diisocyanate PCMs. With higher molecular weight of PEG (PEG 6000 and 10000), the transition temperatures remain unaffected with a variation in the type of diisocyanate. The enthalpy values are significantly lower and overcooling more pronounced for low molecular weight PEG (PEG 1000). The melting enthalpy and thermal efficiency of the biodegradable polymeric PCM from an HMDI/PEG/castor oil combination are better than the PCM with MDI, due to the steric hindrance caused by the rigid benzene rings of the MDI.[122] Novel comb-polyurethane can be prepared by reaction of the soft segment, monomethoxy PEG modified by diethanolamine, with the hard segment consisting of IPDI and BDO.[123] As the soft segment is linked to the main PU chain only by one end, compared to the restriction of the chain from both ends in PEG-PU, the comb-PU shows higher phase transition enthalpy of 122 J g^{-1} and better crystallization properties than PEG-PU.

Two different PEG-based components, viz. isocyanate-terminated prepolymer and tetrahydroxy-terminated prepolymer can be cross-linked to ease the processing of thermosetting SSPCMs in a solvent-free medium, resulting in the formation of flexible PCMs.[124] Phase change enthalpy of 98 J g^{-1} is exhibited by the flexible PCM having a thermal stability higher than 300°C. Ionic polyurethanes are synthesized by using PEG as a soft segment with MDI, N-methyl diethanolamine and 1,3-propane sulfonate as the hard segments.[125] This PCM exhibits high phase change enthalpy of 152 J g^{-1} when PEG 10,000 (ΔH_m = 187 J g^{-1}) is used to form the ionomer. The introduction of ionic groups results in better phase separation, improved thermal stability, and reduced interference of hard segments during the crystallization of the soft PEG segments. Using a grafting-to method, PVA-grafted-octadecanol copolymers are prepared using urethane bonds. Compared to pure octadecanol, the melting enthalpy is much reduced for the copolymer but can be improved with an increasing grafting ratio.[126]

HMDI biuret, a cross-linking agent and a hard segment are reacted with PEG along with the addition of halloysite nanotubes.[127] These nanotubes act as cross-linking and nucleating agents, increasing the rate of crystallization and, hence, the TES density of the system. The excessive cross-linking by HMDI biuret leads to a reduction in the phase change enthalpy, which is improved by the presence of halloysite nanotubes due to the heterogeneous nucleation effect enhancing it to a value of 118.7 J g^{-1}.

4.5.2 Hyperbranched Polymers

Hyperbranched polyurethanes are being explored as solid-solid PCMs due to their many advantages, including easy synthesis, a large number of reactive end groups, a highly branched structure, minimal chain entanglement, and maximum chain rearrangement. Initial reports point toward the preparation of hyperbranched polyurethane as solid-solid PCM using PEG, diisocyanate and hyperbranched polyester as a chain extender.[128–131] Due to the presence of strong covalent bonds, the hard polyurethane segment prevents the PEG from melting to a liquid state during the crystalline to an amorphous phase transition.[129] The hyperbranched polyurethanes maintain a solid-solid phase transition when the molecular weight of PEG is maintained beyond 2000.[130] Also, the phase transition temperatures and enthalpy reduce with reducing the soft segment concentration and increase with increasing the soft segment molecular weight. The strategy of doping the hyperbranched polyurethane with pentaerythritol improves the phase change enthalpy and thermal resistance of the material.[131] PEG is also used in the preparation of hyperbranched polyol, which is used later as a chain extender.[132] For easy synthesis and processing on a large scale, hyperbranched polyurethane can be synthesized from PEG soft segment and trifunctional aromatic (phloroglucinol) (Figure 4.8) and aliphatic (trimethylolpropane) core moieties as branching units.[133,134]

4.5.3 Natural Polymer Grafted SSPCMs

Natural polymers impart properties such as biodegradability, renewability, biocompatibility, and non-toxicity, which make them attractive for use as hard segments in the preparation of

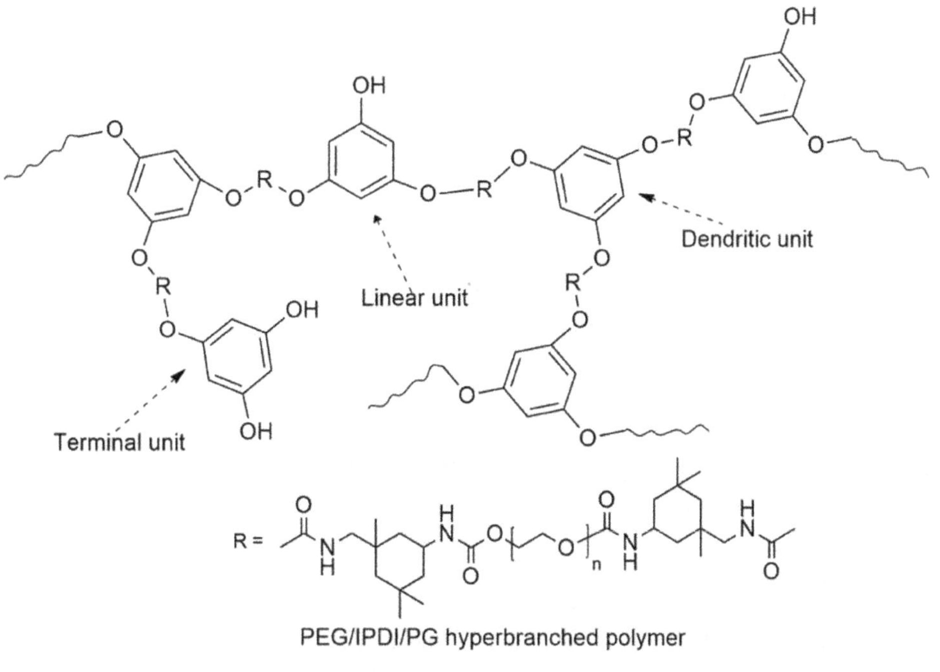

FIGURE 4.8 Structure of hyperbranched polyurethane with aromatic branching core unit, phloroglucinol.

Source: [134]. Reprinted with permission from *Industrial and Engineering Chemistry Research* 56, 49, 14401–14409. Copyright (2017) American Chemical Society.

solid-solid PCMs. Cellulose from MMTritylcellulose can be used to prepare cellulose-*g*-PEG copolymers using ether linkage with a thermal storage capacity greater than 120 J g^{-1} and a phase transition varying from room temperature to 50°C.[135]

The PCM undergoes solid-solid transition only when cellulose composition is more than 15%. Graft copolymers with a polyethylene glycol monomethyl ether soft segment and a cellulose hard segment are prepared via the formation of urethane bonds in the ionic liquid.[136] The thermal energy storage capacity can reach 154 J g^{-1}. Cotton with a major cellulose constituent is also bound to PEG with urethane linkage.[137] With a grafting percentage of nearly 30%, the material has the potential for the development of smart fabric, as the inherent properties of cotton cloth remain mostly unchanged after grafting and the resistance to degradation increases during the wash. A derivative of cellulose, CDA-based PCM with PEG is prepared by urethane bonding.[37,138] The PEG molecular weight of 10,000 exhibits maximum enthalpy.

4.5.4 Modified PEG Polymers

PEG derivatives are appealing as SSPCMs due to their similar crystal structure and crystallization behavior as PEG, along with the added advantage of designing polymerizable monomers. Derivatives of PEG have high phase change enthalpy with adjustable transition temperatures over a wide range of molecular weights. Copolymerization reactions have been used to prepare PEG-based copolymers with parallel thermal energy storage capacity. Styrene can be copolymerized with a vinylic end group of PEG monomethyl ether, yielding a TES material with a maximum storage capacity of 109 J g^{-1} at phase change temperature of 58°C.[139] The crystalline properties of PEG monomethyl ether remain unaffected during synthesis. Polydecaglycerol acrylate and PEG acrylate are copolymerized together to prepare SSPCM having a melting enthalpy of 163 J g^{-1}.[140] A high enthalpy value is facilitated with extensive hydrogen bonding from polydecaglycol chains. Poly(PEG methyl ether methacrylate) homopolymer can be prepared via free radical bulk polymerization of PEG methyl ether methacrylates, which exhibits good crystallinity.[141]

PEG-acrylate-based cross-linked copolymer can be synthesized to enhance thermal stability while retaining the high value of enthalpy. This is possible as the PEG is present as a side chain. The cross-linked copolymer PCMs possess thermal stability beyond 300°C with maximum fusion enthalpy of 145 J g^{-1}.[142] PEG acrylate can be copolymerized with methyl methacrylate to realize maximum enthalpy of 165 J g^{-1}. The monomer reactivity ratio calculation indicates a strong tendency to undergo cross-propagation.[143] PEG octadecyl ether monomer can be synthesized by varying the PEG chain length, which has transition enthalpy varying in the range of 143–183 J g^{-1}.[144] Polymeric PCM can be prepared by copolymerization of PEG octadecyl ether monomer with methyl methacrylate to prevent the leakage of the liquid phase.[145] Copolymer poly (PEG alkyl ether vinyl ether) is prepared with varying lengths of PEG spacers having the highest enthalpy of 103 J g^{-1}. The PCM is thermally stable up to 295°C and, thus, can be used for the preparation of melt-spinning thermoregulated fiber.

4.5.5 Miscellaneous Polymeric Supports

Using diverse polymeric supporting materials, SSPCMs can be synthesized chemically. Polyethylene terephthalate-PEG block copolymers[146] and a coupling blend of PEG with polyacrylamide[147] can be synthesized using chemical methods. PEG with carboxyl groups can be linked with Eu^{3+} to prepare a phase change luminescent material with phenanthroline as a secondary ligand.[148] The Eu-PEG polymeric SSPCM has phase change enthalpy of 97 J g^{-1}, possessing luminescent properties with UV absorption at 309 nm.

PEG modified with maleic anhydride can be grafted onto a polyacrylonitrile chain, yielding a comb graft architecture SSPCM exhibiting phase change enthalpy of 74 J g^{-1}.[149] PEG can be grafted onto a poly(styrene-co-acrylonitrile) matrix for the preparation of a solid-solid PCM.[150] The architecture has thermal stability up to 350°C with thermal reliability beyond 1000 cycles. Form-stable PCM of cross-linked poly(acrylonitrile-co-itaconate)/PEG blend exhibits thermal reliability and heat storage durability for 1000 thermal cycles.[73]. The phase change temperature, being in the range of 23–53°C with good phase change enthalpy of 119 J g^{-1}, is well suited for commercial applications.

The cross-linking reaction of PEG with melamine-formaldehyde[151] and carboxylated PEG with epoxy groups of poly(glycidyl methacrylate)[152] results in PCM, wherein the hard segment restricts the flow of working material, resulting in low enthalpy. Polystyrene-graft-PEG copolymers with a different mass percentage of PEG content exhibit latent heat values ranging between 116–174 J g^{-1} with a transition temperature varying from 55 to 58°C.[153] Palmitic acid is grafted onto a polystyrene backbone using palmitoyl chloride to prepare solid-solid PCMs with maximum incorporation of 75 wt% palmitic acid.[154] Although the graft polymer has a low storage enthalpy of 40 J g^{-1}, the thermal reliability for 5000 thermal cycles is quite high. Environmentally-friendly PCM containing poly(3-hydroxybutyrate-co-3-hydroxy valerate) (PHBV)/PEG copolymer has a latent heat of fusion in the range of 126–135 J g^{-1} above 20% PHBV fraction.[155] 3D semi-interpenetrating polymer networks can be prepared by using the simultaneous interpenetrating technique.[156] One network is formed by the condensation polymerization of TEOS with PEG and the second network is prepared by free-radical polymerization of HEMA to form poly(HEMA). Although interpenetrated, a high transition enthalpy of 145 J g^{-1} can be achieved. Further, the relative hydrophilic nature of PEG can be reduced with increasing HEMA concentration.

4.6 Applications of PCMs

Phase change polymers are being evaluated for a large number of applications and many have been inducted into commercial use as a maintenance-free heating/cooling system. The possible application areas are air conditioning or telecom shelters, as the interruption of the power supply can be taken care of by PCM. Transportation of perishable foods, sundry electronics, temperature-sensitive pharmaceuticals and chemicals (explosives) require cooling, which can be done by PCM at a much cheaper rate compared to conventional methods. The automobile sector offers a tremendous application of PCM for better utilization and comfort. In construction, PCMs are embedded to make a building more thermally comfortable than it would be without PCM. Heating or cooling houses in different climates can be done with PCM to a great extent. PCM can be used to keep caterers' food hot during transportation. The greenhouse is another area, where PCM can be comfortably used to maintain temperature variation at a minimum level.

4.6.1 The Textile Industry

The first application of PCMs was explored 30 years ago by National Aeronautics and Space Administration (NASA), in spacesuits to protect astronauts from temperature fluctuations in outer space. Thus, the concept of smart textiles for human comfort came into existence in the form of parkas, gloves, sports apparel, footwear, and thermal undergarments. These smart textiles adjust the body temperature by regulating the heat uptake or release, depending on the external environment. PCM can be incorporated into fabrics by coating, impregnation of hollow fibers with PCMs, microencapsulation, or electrospinning.

In the initial periods of product development, various fabrics, including polyester, nylon 66, wool, and cotton were treated with aqueous PEG solutions using a conventional pad-dry technique to enhance the thermal properties by a minimum of 2–2.5 times that of untreated fabric.[157] The interactions between PEG and the amide groups of nylon and wool are the reason for exhibiting the highest enthalpy among various fabrics. The thermal values are also quite stable even after 50 thermal cycles.

PEG nanocapsules with an urea-formaldehyde shell are fixed onto a cotton fabric using the pad-dry-cure method.[158] Increasing the binder agent improves the tensile strength and abrasion resistance of the treated fabric.

Solid-solid PCM can be produced by grafting PEG 1000 onto cotton waste fibers with glutaraldehyde as a cross-linker.[159] A difference of 1–1.5°C in temperature is observed for the box containing the PCM for 23–25 min. Poor adhesion properties of PEG to non-cellulosic fiber [poly(lactic acid)] is improved by using a highly cross-linked PEG system consisting of PEG-dimethyloldihydroxyethyleneurea.[160] This strategy helped in improving wash fastness while the enthalpy value is mostly unaltered. The PLA fabric shows a weight gain of 44% after cross-linking, allowing less air to permeate through the fabric, thus improving the thermal properties, allowing better static charge dissipation with lower surface tension compared to the polyethylene terephthalate fabric used. The improvement in adhesion of PEG to wool fibers is done by using fluorinated and chlorine-Hercosett pretreated wool fabrics followed by reaction with PEG-dimethyloldihydroxyethyleneurea.[161] The fluorinated fabric shows the highest thermal properties with a melting enthalpy value of 17 J g^{-1} for washed fabric, together with better dimensional stability, enhanced felting performance, reduced water vapor permeability, and increased air permeability.

Non-woven mats with a PEG core and polyvinylidene difluoride (PVDF) sheath have been prepared by coaxial electrospinning.[162] The use of non-woven mats adds to the mechanical strength and specific surface area with desirable dimensions over conventional electrospun fibers and eliminates the need for any further fiber processing. The optimized nanofibers, having an average diameter of 721 nm and latent fusion enthalpy of 34.8 J g^{-1}, have an encapsulation ratio of 20 wt% PEG. These fibers exhibit a reduced degree of elongation with the introduction of PEG into PVDF, but satisfactory mechanical strength is retained. To improve the mechanical strength of the PEG-PVDF nanofibers, fumed silica (2 wt%; PEG 1000) is added to the spinning solution.[163] The diameter of the nanofibers reduces with an increase in fusion enthalpy to 59 J g^{-1} and improvement of tensile strength.

Flexible, porous membranes can be prepared by diffusing two PEGs of varying molecular weights into a PU solution.[164] The membrane preserves its solid state on heating above the PEG transition temperature with enthalpies reaching up to 129 Jg^{-1}. The PU/PEG flexible membrane has a porous structure, a suitable phase transition temperature, and high transition enthalpy. The introduction of porous technology into functional textile formation contributes to the development of new ways to improve the transition enthalpy for adjustable textiles.

Direct grafting of PEG onto the cotton framework, as indicated by POM images in Figure 4.9, is done up to a grafting percentage

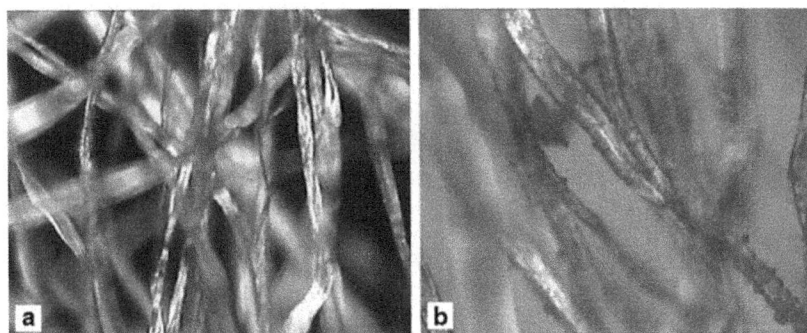

FIGURE 4.9 POM images of (a) pure cotton and (b) modified cotton. The roughness in morphology indicates the grafting of PEG and, therefore, the modified fiber can directly be applied for textile production.

of 30 wt%.[137] The grafted cotton has improved tensile strength along with an increase in gram per square meter value. A reduction in tear strength and an increase in abrasion resistance from 33,000 rubs to 42,000 rubs for thread breaking is directly related to the strong chemical bonds of PEG on the cotton. The water, perspiration, and rubbing fastness of the fabric remains almost the same.

4.6.2 Building Applications

As the demands for comfortable living environments are increasing, buildings are found to have high energy consumption and contribute to 40% of total world energy consumption. PCM wallboards, underfloor heating systems, shutters, ceiling boards, and heat exchangers find applications in the heating or cooling of buildings.[44] PCMs can be applied by adding PCM pellets to wallboards during manufacturing or impregnating wallboards with PCM, using PCM-enhanced concrete or PCM-incorporated PU foam.[165]

Diatomite or diatomaceous earth is a natural amorphous silicate possessing high porosity, excellent absorption capacity, chemical stability, and is of low cost.[102] This lightweight building material can be integrated with a maximum of 50 wt% PEG to prepare shape-stabilized PCMs. The addition of EG improves the heat transfer rate of the PEG/diatomite composites. The wallboards fabricated with composite PCM, exhibit a temperature difference of 2.3°C on the inner surface.

PEG-cement composites are used directly for building applications.[95] When the mass ratio of PEG and cement is raised to 1:3, the corresponding fusion temperature and enthalpy of the PCM are 24°C and 24 J g^{-1}. PEG is easily impregnated into porous cement showing no leakage during the phase transformation. In the PEG-MCS shape-stabilized composite, 70 wt% of PEG can be accommodated, resulting in melting and solidification enthalpies of 122 and 107 J g^{-1} and the corresponding heat storage and heat release are improved by 28.2% and 27.3% respectively, compared to pure PEG.[96] The direct impregnation of PEG into a building material, MCS, with a transition temperature in the range of 50–70°C, makes it suitable for building envelopes during hot summers.

Building materials, gypsum, and natural clay have been used as porous supporting materials for the preparation of PEG-containing shape-stabilized composites using the vacuum impregnation method.[166] A maximum loading of 18 and 22 wt% PEG in PEG/gypsum and PEG/natural clay composite can be realized. The thermal performance wallboards, fabricated from PEG/natural clay and PEG/gypsum PCMs, are able to maintain a temperature difference of 2.08°C for 60 min and 1.47°C for 120 min, respectively. Low-cost clay mineral, bentonite, impregnated with four different kinds of PCM, including capric acid, PEG 600, dodecanol, and heptadecane, exhibits a wide range of melting temperature and melting enthalpy (4–30°C and 38–74 J g^{-1} respectively).[98] The transition temperature chosen makes it suitable for passive solar heating and cooling in building envelopes. Thermal conductivity is also increased by adding EG (5 wt%), so that response times are lowered.

4.6.3 Solar Energy Storage

Solar energy is the most abundantly available form of renewable energy. Several techniques have been explored and used to effectively capture solar radiation. Solar energy conversion to thermal energy is an efficient method to retain a large quantity of thermal energy in a small volume. The inherent low thermal conductivity of organic PCMs acts as a major drawback in its application, but considerable research is now focused upon the development of effective TES materials with high thermal conductivity and high solar energy absorption efficiency.

SSPCM with a PEG soft segment and a supporting framework composed of an organic dye, 1,4-bis((2-hydroxyethyl)amino) anthracene-9,10-dione was prepared for functioning as a solar energy harvester.[32] This PEG 20000-dye PCM exhibits melting enthalpy of 143 J g^{-1}, shows absorption in the visible range at 555 nm and has high light-to-thermal energy conversion with an efficiency of 0.937. The temperature of this PCM composite increases to 70°C in 34 min under sunlight irradiation, thereby, effectively absorbing visible sunlight through a non-radiative process (Figure 4.10).

The heat supply for a thermoelectric device is generated by using graphene/PEG composites with a 3D netlike architecture.[64] This interconnected architecture enables efficient heat transfer through thermally conductive pathways. Graphene/PEG composites have superior thermal and electrical conductivity together with high structural stability. A high PEG loading of 93 wt% and a corresponding high melting enthalpy of 166 J g^{-1} with 93.3% efficiency ensure longer steady-state output time and therefore, higher heat output is generated by the thermoelectric device.

For effective light harvesting, MWCNTs can be blended with PEG-SiO$_2$ to make a composite having improved thermal conductivity, absorption capacity over a broad range of sunlight with high light to heat energy conversion and storage efficiency.[89] The maximum latent heat with 0.5 wt% of MWCNT remains 139 J g^{-1} at 54°C and the thermal conductivity at 0.389 W m^{-1} K^{-1}. This PCM shows a high efficiency of 0.918 for light-to-heat conversion and thermal storage.

SWCNT (diameter: < 2 nm and length of 5–20 μm), modified by introducing nitrophenyl groups and stabilized in toluene, can be added to form-stable composite PCM derived from PEG, diisocyanate, and N, N'-dihydroxyethyl aniline.[167] SWCNT functions as an effective photon capturer and shows light to thermal conversion with energy storage efficiency in the range of 0.845–0.913 upon solar irradiation. The enthalpy is higher than 100 J g^{-1} and thermal conductivity is higher by 25.1%, which results in a low heating time of 870 sec. Due to strong absorbance in the near-infrared zone, these composite PCMs can be used in military stealth and smart textile applications. Smart TES composites have efficient electro/photo-to-heat conversion and storage with high thermal conductivity can be developed by carrying out cross-linking copolymerization of PEG and IPDI which has been injected into graphite foam.[168] This material is found to have thermal conductivity of 3.5 W m^{-1} K^{-1}. By incorporation of more than 80 wt% PEG 8000 in the graphite foam, the electro-to-heat storage efficiency can reach a maximum of 80% at low voltages with photo-to-heat conversion efficiency of 67%.

The 3D heat conductive pathway formed in PEG10000/HGA composite (GO and GNP amount varied) results in

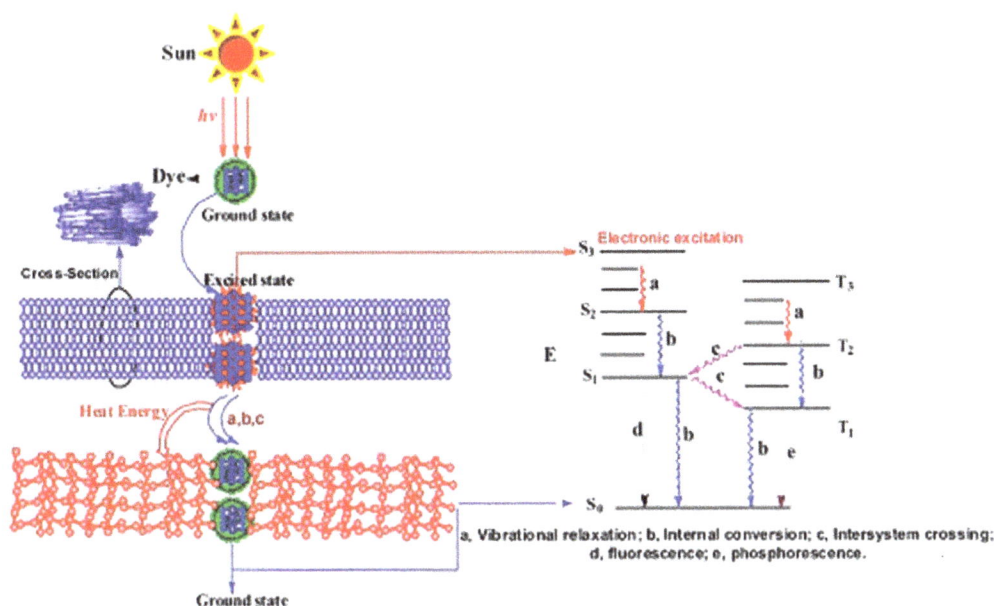

FIGURE 4.10 Proposed mechanism for solar-thermal conversion by PEG incorporated organic dye through nonradiative decay and generated latent heat is stored by PCM.

Source: [32]. Reproduced with permission from the Royal Society of Chemistry.

high thermal conductivity of 1.43 W m^{-1} K^{-1}, a high transition enthalpy ranging from 178–182 J g^{-1}, and crystallization enthalpy from 170–174 J g^{-1}.[19] The energy storage efficiency from solar simulator (a xenon lamp) reaches 80–92% using the tangential method and effective light-to-heat conversion is achieved within 33 min.

4.6.4 Innovative Applications of PCMs

The conventional applications of PCM for TES have been recognized widely and commercially applied by a number of companies, such as Outlast technologies, DuPont, Rubitherm, Microtek, Teappcm, and Sunamp, PCM Thermal Solutions, and counting. But very recently, the potential of PCMs is being explored for high-end applications, including drug delivery, thermal barcoding, solar-powered refrigerators, etc.

4.6.4.1 Reduction of Cooling Costs in Warehouse-Scale Computers

With the increasing computing density of the computing infrastructure, a significant portion of initial capital expenditures and recurring operating expenditures are dedicated to cooling.[169] To prevent high server failure, the cooling system should have the ability to handle the peak demand of the data center. Further, the cooling system also may become insufficient as servers undergo upgrading or replacement. To mitigate these challenges, the PCM can be used to temporarily store the heat generated during periods of peak load and release the absorbed heat when there is an excess of cooling capacity. The advantages of this tactic may not be immediately realized as heat is stored on a temporary basis and then released at a later time. However, the understanding gained from this work is that the thermal behavior of the data

center can be designed so that the heat is released only when it economically feasible.

4.6.4.2 Solar Power Refrigerators

Dulas, a Welsh company renowned in renewable energy technology, is replacing batteries by using PCM as an essential component of solar-powered direct-drive refrigerators for off-grid vaccine storage in developing countries.[170] As battery-based fridges count on traditional evaporators to remain cool, there can be parts of refrigerator where the temperature is below 0°C, so there exists a risk of freezing the stored vaccines, resulting in their degradation. In order to improve vaccination rates, the whole fridge must be maintained at a constant temperature ranging between 2 and 7°C. The problem is solved by having a lining of PW-based PCM that freezes at 5°C. When there is no solar power (at night), the PCM melts and absorbs heat without any temperature variations, until all the PCM is turned into liquid. This process ensures that the vaccines stay within their optimum temperature range. Another advantage of PCM design is that the thermal storage medium is capable of undergoing over 10,000 cycles with minimum degradation.

4.6.4.3 Thermal Packaging

PCMs can be used in the thermal packaging industry to maintain temperature-sensitive products within the prescribed temperature range during transport.[171] As the PCM has a phase transition temperature within the required temperature range, it undergoes a change in its phase (for example, solid-liquid), so it results in effectively extending the duration of temperature control via latent heat, by cooling the product. The use of electricity-free, PCM packets, as an incubator has been approached to isolate bacteria for the detection of typhoidal fever.[172] Such a low-cost system can be applied effectively in remote areas where sophisticated lab equipment and trained personnel are lacking.

4.6.4.4 Temperature-Sensitive Drug Delivery

The field of nano-drug delivery systems using external stimuli such as light, temperature, magnetic field, and ultrasound is constantly being enriched with much pioneering research. As shown in Figure 4.11, near-infrared (NIR) irradiation can effectively deliver the drug by encapsulation of photothermal agents such as gold or carbon nanomaterials in the thermoresponsive polymer system.[173] Biocompatible and biodegradable organic PCMs can be used in temperature-sensitive drug delivery, which can be particularly useful for chemotherapeutics, providing a pathway for localized delivery and posing minimum side-effects.[77] Ongoing

research data suggest the use of PCM microcapsules as a drug carrier and allowing its release above the melting point of the PCM. Even multiple components can be released as a function of temperature.

4.6.4.5 Thermal Barcoding

Due to the problem of counterfeiting associated with conventional barcodes, covert barcodes need to be developed for tracking objects, tracing documents, and even identifying criminals or terrorists.[174] Phase change nanoparticles and their eutectic mixtures offer high loading capacity with sharp melting peaks, providing a robust system for forensic investigations. The detection of the solid-liquid phase change can be identified using infrared thermal imaging (Figure 4.12) instead of DSC to provide

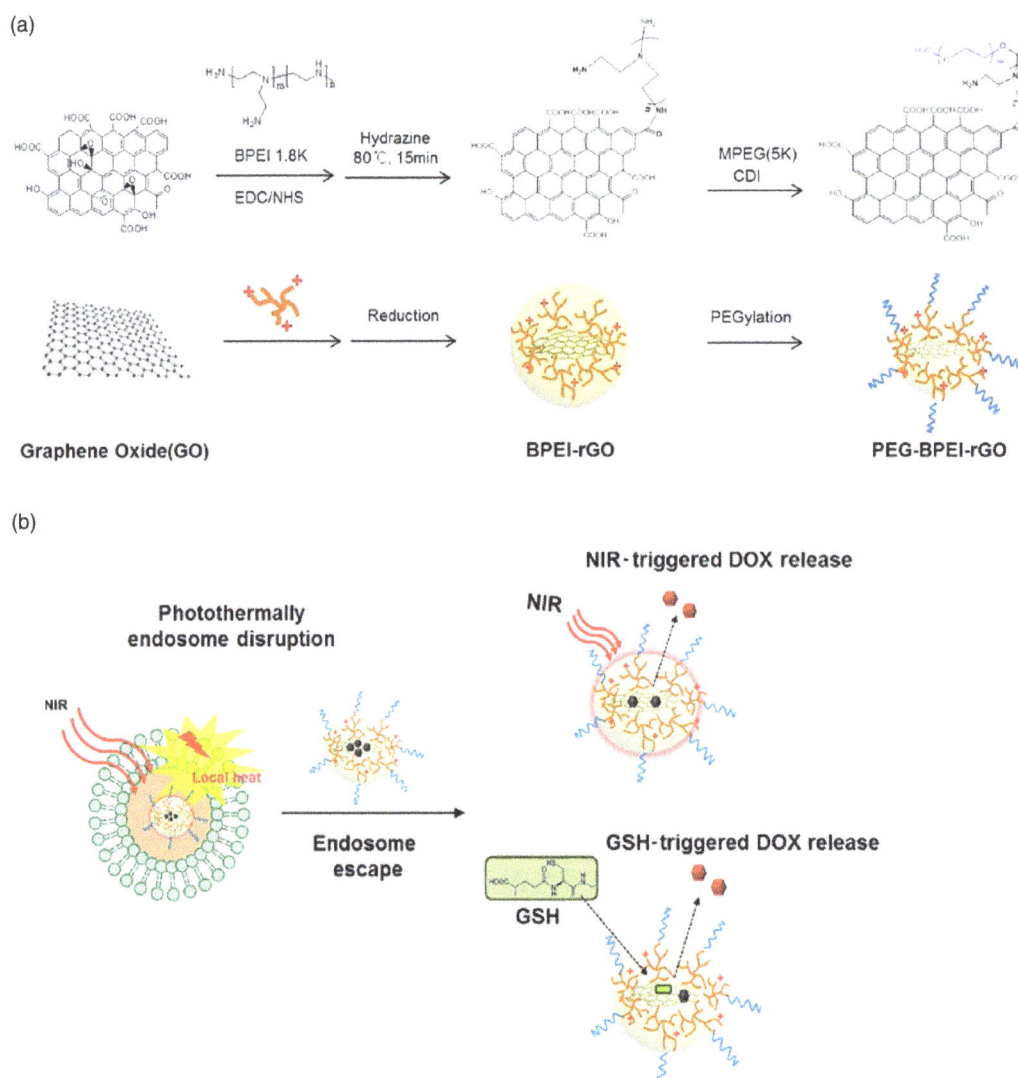

FIGURE 4.11   (a) Functionalization of reduced graphene oxide (rGO) with PEG and branched polyethyleneimine (BPEI); (b) Schematic for PEG-BPEI-rGO composite loaded with an anti-cancer drug, doxorubicin (DOX) and photothermally triggered drug delivery by endosomal disruption using NIR.

Source: [173]. Reprinted (adapted) with permission from *ACS Nano* 7, 6735–6746. Copyright (2013) American Chemical Society.

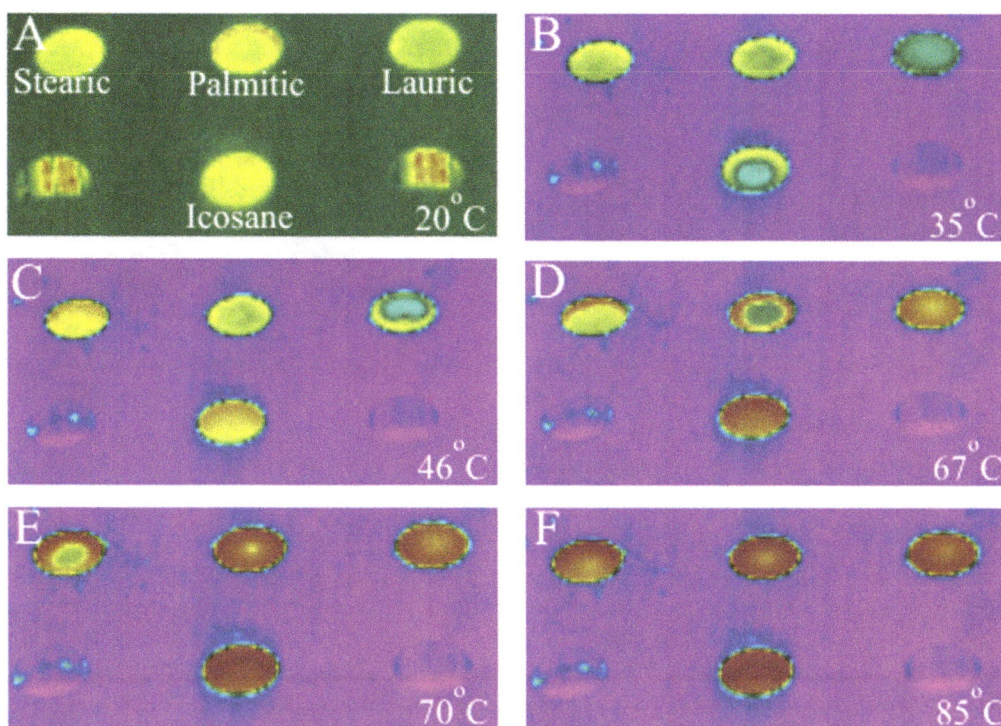

FIGURE 4.12 Infrared imaging of four different PCMs acquired over time indicating different emissivity of each material forming a code depending upon the difference in melting temperature. Infrared imaging of four different PCMs acquired over time indicating different emissivity of each material forming a code depending upon the difference in melting temperature. (A) Image of all four PCMs at the start of test (20 °C); (B-E) melting of samples with increasing temperature wherein purple color indicates relatively low temperatures and red color indicates higher temperatures; and (F) complete melting of samples observed.

Source: [8]. Reprinted (adapted) with permission from *Journal of Physical Chemistry C* 120, 38, 22110–22114. Copyright (2016) American Chemical Society.

a novel non-contact, sensitive technique for decrypting thermal barcodes.[88]

4.6.4.6 Biometric Identification

Fluorescent, thermoresponsive sensor systems were developed by impregnating common filter paper with molecular rotor and PCM by absorption in the molten state.[99] This hybrid material has the ability to respond to fluorescence change reversibly (Figure 4.13). A critical change in the fluorescence intensity is observed at the melting and crystallization temperatures of PCM and can be effectively used for biometric identification.

4.6.4.7 Anti-Icing Coating

Ice accumulation on aircraft or runways can lead to fatal accidents and the currently available mechanical and chemical methods are both labor- and time-intensive. An ingenious method was developed by Bhamidipati of E-Paint Company to prevent the accumulation of ice using bio-inspired anti-icing coatings comprising silicone-based PCMs, preferably a hydrophobic resin.[175] When the ice forms, the phase change occurs with the expansion of the PCM along with substrate contraction. The latent heat released from the freezing of water is passed on to the PCM near the surface, thereby experiencing local shear stress due to volume change and ultimately resulting in the disruption of the ice-surface bond.

4.6.4.8 Thermal Peak Management Using Organic Phase Change Materials

A composite coating on printed circuit boards or electronic assemblies for electronic components is expected to buffer a certain amount of thermal energy, dissipated from a device.[176] To prevent damage to the instrument during temperature peaks in electronic components, a phase change material can be used. PCMs can be coated directly onto the chip package or printed circuit board using different mechanical retaining jigs. Sugar alcohols as PCM and epoxy or acrylate as a resin matrix, together with metal nanoparticles or CNT as thermal conductivity enhancer are used to make the coating.

4.6.4.9 Paraffin Wax in 3D Printing

For potential applications in winter sports equipment, 3D printable blends with TES capabilities can be developed by incorporating paraffin in thermoplastic polyurethane (TPU).[177] Thoroughly mixed encapsulated paraffin and TPU granules are first extruded and then the filaments are used for 3D printing using the set-up: Layer height 0.20 mm, bed temperature 40°C, nozzle temperature 240°C, deposition rate 40 mm s^{-1}. Homogeneous distribution of PCM capsules in the polymer matrix and good adhesion between the layers in the 3D printed parts can be achieved by this technique. Reasonable energy storage/release capability is obtained in

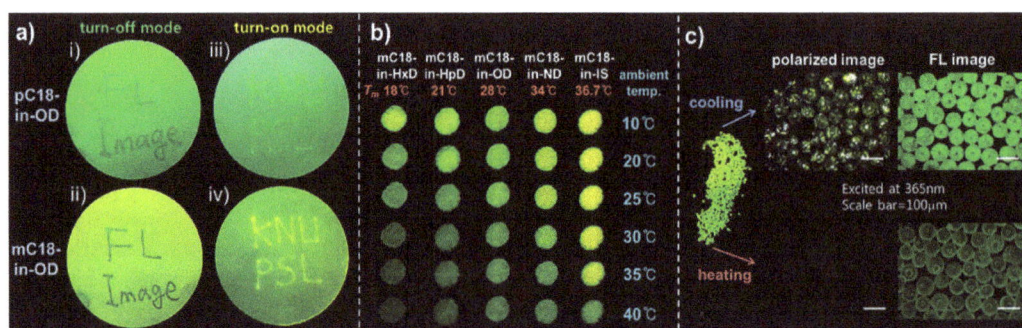

FIGURE 4.13   (a) Fluorescence imaging of characters on PCM incorporated filter paper on cooling ((i) and (ii)) and heating ((iii) and (iv)); (b) thermoresponsive sensor deposited on PCM impregnated filter paper with different melting temperatures, used as an array system; (c) polarized and corresponding fluorescent microscopy images of the PCM microcapsules.

Source: [9]. Reprinted (adapted) with permission from *ACS Applied Materials & Interfaces* 7, 26, 14485–14492. Copyright (2015) American Chemical Society.

the 3D printed parts, with fusion enthalpy values reaching up to 70 J g⁻¹. The elastic modulus increases with the microcapsule content and the higher elastic modulus of melamine formaldehyde shell of the capsules, compared to the TPU matrix, generate the stiffening of the matrix. The hard microcapsules also increase creep resistance and Shore A hardness of the material.

4.7 Environmental Impact of PCMs

In the literature, numerous PEG-based SSPCMs have been explored with respect to their chemical, thermal, and morphological characteristics. One important aspect to be taken into consideration during PCM preparation is the need to develop environmentally and user-friendly materials of a biodegradable and non-toxic nature.

PEG, a synthetic biomaterial, is innocuous and widely used in the production of cosmetics and pharmaceuticals. Even high molecular weight PEGs are susceptible to biodegradation.[178] Due to the abundant availability, renewability, low cost and biodegradability of the natural polymers, many researchers have integrated soft segment PEG with natural polymers as hard segments and achieved solid-solid phase transformation with satisfactory thermal properties. Composites of PEG with natural polymers, including cellulose,[136] cellulose acetate,[179] carboxymethyl cellulose,[40] cellulose ether,[40] agarose,[41] chitosan,[41] starch,[79] sugars (glucose, fructose, and lactose),[80] have been prepared by using both physical blending and chemical grafting methods. β-cyclodextrin with α-glucopyranose units, widely used for biomedical applications, is cross-linked with PEG in the presence of MDI to prepare SSPCM.[116] Similarly, trihydroxy surfactants based on natural fatty acids, Span 80 and tween 80 are used as cross-linking agents in the solventless preparation of PEG-based polyurethanes.[117] The thermoplastic polyester, PHBV, produced naturally by bacteria, can be copolymerized with PEG diacrylate to form environmentally-friendly copolymer PCM.[155] Using cotton waste in the development of thermoregulating fabric by grafting of PEG will aid in industrial waste management.[159]

Solvent-free synthesis of PEG-based SSPCM, with vegetable oil as a hard segment linked by urethane bonds, can be carried out, leading to advanced energy storage material.[122] It is believed that the use of biodegradable starting materials in the preparation of PCM can benefit the biodegradability of the prepared SSPCM. The microwave route can be adopted for green synthesis of PEG-CDA composites with minimum use of a solvent.[39] The energy- and time-efficient techniques follow the principles of green chemistry. All these TES materials are biocompatible, biodegradable, and environmentally safe macromolecules which should be used in the development of advanced SSPCMs.

4.8 Conclusion

A wide array of polymers has been discussed which can be prepared by using both physical and chemical methods, capable of undergoing thermal energy storage. Among the physical methods, blending has been most commonly used to prepare solid-solid PCMs whereas urethane modification has been most commonly applied to prepare chemically bonded PCMs. Several methods of modification to prepare polymers for thermal energy storage have been discussed, including grafting, cross-linking, copolymerization, and hyperbranching. The use of several modification techniques listed offer a wide array of transition temperatures and enthalpies which can accommodate the user's requirements.

REFERENCES

1.   M. M. Farid, A. M. Khudhair, S. A. K. Razack and S. Al-Hallaj, "A review on phase change energy storage: Materials and applications." *Energy Convers. Manag.* 45 (2004): 1597–1615.

2.   K. Pielichowska and K. Pielichowski, "Phase change materials for thermal energy storage." *Prog. Mater. Sci.* 65 (2014): 67–123.

3.   G. Odian, *Principles of Polymerization*, 4th edn. New York: Wiley-Interscience, 2004.

4.   A. Sharma, V. V. Tyagi, C. R. Chen and D. Buddhi, "Review on thermal energy storage with phase change materials and applications." *Renew. Sustain. Energy Rev.* 13 (2009): 318–345.

5. H. Mehling and L. F. Cabeza, "Solid-liquid phase change materials," in H. Mehling and L. F. Cabeza, *Heat and Cold Storage with PCM: An Up to Date Introduction into Basics and Applications.* Berlin: Springer, 2008, pp. 11–55.

6. B. Zalba, J. M. Marín, L. F. Cabeza and H. Mehling, "Review on thermal energy storage with phase change: materials, heat transfer analysis and applications." *Appl. Therm. Eng.* 23 (2003): 251–283.

7. D. C. Hyun, N. S. Levinson, U. Jeong and Y. Xia, "Emerging applications of phase-change materials (PCMs): Teaching an old dog new tricks." *Angew. Chemie Int. Ed.* 53 (2014): 3780–3795.

8. S. Hou, W. Zheng, B. Duong and M. Su, "All-optical decoder for rapid and noncontact readout of thermal barcodes." *J. Phys. Chem. C* 120 (2016): 22110–22114.

9. Y.-J. Jin, R. Dogra, I. W. Cheong and G. Kwak, "Fluorescent molecular rotor-in-paraffin waxes for thermometry and biometric identification." *ACS Appl. Mater. Interfaces* 7 (2015): 14485–14492.

10. M. M. Kenisarin and K. M. Kenisarina, "Form-stable phase change materials for thermal energy storage." *Renew. Sustain. Energy Rev.* 16 (2012): 1999–2040.

11. Q. Meng and J. Hu, "A poly(ethylene glycol)-based smart phase change material." *Sol. Energy Mater. Sol. Cells* 92 (2008): 1260–1268.

12. H. A. Hussein, A. H. Abed and A. R. Abdulmunem, "An experimental investigation of using aluminum foam matrix integrated with paraffin wax as a thermal storage material in a solar heater," in *Proceeding 2nd Sustainable and Renewable Energy Conference*, Baghdad, Iraq, 2016.

13. A. Dorigato, M. V. Ciampolillo, A. Cataldi, M. Bersani and A. Pegoretti, "Polyethylene wax/EPDM blends as shape-stabilized phase change materials for thermal energy storage." *Rubber Chem. Technol.* 90 (2017): 575–584.

14. G. Fredi, A. Dorigato and A. Pegoretti, "Multifunctional glass fiber/polyamide composites with thermal energy storage/release capability." *Express Polym. Lett.* 12 (2018): 349–364.

15. G. Fredi, A. Dorigato, L. Fambri and A. Pegoretti, "Wax confinement with carbon nanotubes for phase changing epoxy blends." *Polymers (Basel).* 9 (2017): 405.

16. Y. Yang, J. Kuang, H. Wang, G. Song, Y. Liu and G. Tang, "Enhancement in thermal property of phase change microcapsules with modified silicon nitride for solar energy." *Sol. Energy Mater. Sol. Cells* 151 (2016): 89–95.

17. G. Fredi, A. Dorigato, L. Fambri and A. Pegoretti, "Multifunctional epoxy/carbon fiber laminates for thermal energy storage and release." *Compos. Sci. Technol.* 158 (2018): 101–111.

18. Z. Liu, Z. Chen and F. Yu, "Microencapsulated phase change material modified by graphene oxide with different degrees of oxidation for solar energy storage." *Sol. Energy Mater. Sol. Cells* 174 (2018): 453–459.

19. J. Yang, G. Qi, Y. Liu, R. Bao, Z. Liu, W. Yang, B. Xie and M. Yang, "Hybrid graphene aerogels/phase change material composites: Thermal conductivity, shape-stabilization and light-to-thermal energy storage." *Carbon N. Y.* 100 (2016): 693–702.

20. L. Zheng, W. Zhang and F. Liang, "A review about phase change material cold storage system applied to solar-powered air-conditioning system." *Adv. Mech. Eng.* 9 (2017): 168781401770584.

21. J. Lei, J. Yang and E.-H. Yang, "Energy performance of building envelopes integrated with phase change materials for cooling load reduction in tropical Singapore." *Appl. Energy* 162 (2016): 207–217.

22. Y. Shin, D.-I. Yoo and K. Son, "Development of thermo-regulating textile materials with microencapsulated phase change materials (PCM). IV. Performance properties and hand of fabrics treated with PCM microcapsules." *J. Appl. Polym. Sci.* 97 (2005): 910–915.

23. Z. Zhang, G. Shi, S. Wang, X. Fang and X. Liu, "Thermal energy storage cement mortar containing n-octadecane/expanded graphite composite phase change material." *Renew. Energy* 50 (2013): 670–675.

24. Q. Song, Y. Li, J. Xing, J. Y. Hu and Y. Marcus, "Thermal stability of composite phase change material microcapsules incorporated with silver nano-particles." *Polymer (Guildf).* 48 (2007): 3317–3323.

25. D. Yin, H. Liu, L. Ma and Q. Zhang, "Fabrication and performance of microencapsulated phase change materials with hybrid shell by in situ polymerization in Pickering emulsion." *Polym. Adv. Technol.* 26 (2015): 613–619.

26. Y. Sun, R. Wang, X. Liu, E. Dai, B. Li, S. Fang and D. Li, "Synthesis and performances of phase change microcapsules with a polymer/diatomite hybrid shell for thermal energy storage." *Polymers (Basel).* 10 (2018).

27. T. Do, Y. G. Ko, Y. Chun and U. S. Choi, "Encapsulation of phase change material with water-absorbable shell for thermal energy storage." *ACS Sustain. Chem. Eng.* 3 (2015): 2874–2881.

28. Y. Sun, R. Wang, X. Liu, M. Li, H. Yang and B. Li, "Improvements in the thermal conductivity and mechanical properties of phase-change microcapsules with oxygen-plasma-modified multiwalled carbon nanotubes." *J. Appl. Polym. Sci.* 134 (2017): 45269.

29. J. A. Molefi, "Investigation of phase change conducting materials prepared from polyethylenes, paraffin waxes and copper." PhD thesis, University of the Free State (2008).

30. L. Chen, R. Zou, W. Xia, Z. Liu, Y. Shang, J. Zhu, Y. Wang, J. Lin, D. Xia and A. Cao, "Electro- and photodriven phase change composites based on wax-infiltrated carbon nanotube sponges." *ACS Nano* 6 (2012): 10884–10892.

31. Z. Liu, R. Zou, Z. Lin, X. Gui, R. Chen, J. Lin, Y. Shang and A. Cao, "Tailoring carbon nanotube density for modulating electro-to-heat conversion in phase change composites." *Nano Lett.* 13 (2013): 4028–4035.

32. Y. Wang, B. Tang and S. Zhang, "Visible light-driven organic form-stable phase change materials for solar energy storage." *RSC Adv.* 2 (2012): 5964.

33. G. Qi, J. Yang, R. Bao, D. Xia, M. Cao, W. Yang, M. Yang and D. Wei, "Hierarchical graphene foam-based phase change materials with enhanced thermal conductivity and shape stability for efficient solar-to-thermal energy conversion and storage." *Nano Res.* 10 (2017): 802–813.

34. M. Li, Q. Guo and S. Nutt, "Carbon nanotube/paraffin/montmorillonite composite phase change material for thermal energy storage." *Sol. Energy* 146 (2017): 1–7.

35. S. Sundararajan, A. B. Samui and P. S. Kulkarni, "Versatility of polyethylene glycol (PEG) in designing solid–solid phase change materials (PCMs) for thermal management and their application to innovative technologies." *J. Mater. Chem. A* 5 (2017): 18379–18396.

36. X.-H. Liang, Y.-Q. Guo, L.-Z. Gu and E.-Y. Ding, "Crystalline-amorphous phase transition of poly(ethylene glycol)/cellulose blend." *Macromolecules* 28 (1995): 6551–6555.

37. E.-Y. Ding, Y. Jiang and G.-K. Li, "Comparative studies of the structures and transition characteristics of cellulose diacetate modified with polyethylene glycol prepared by chemical bonding and physical blending methods." *J. Macromol. Sci. Part B* 40 (2001): 1053–1068.

38. Y. Guo, Z. Tong, M. Chen and X. Liang, "Solution miscibility and phase-change behavior of a polyethylene glycol-diacetate cellulose composite." *J. Appl. Polym. Sci.* 88 (2003): 652–658.

39. S. Sundararajan, A. B. Samui and P. S. Kulkarni, "Shape-stabilized poly(ethylene glycol) (PEG)-cellulose acetate blend preparation with superior PEG loading via microwave-assisted blending." *Sol. Energy* 144 (2017): 32–39.

40. K. Pielichowska and K. Pielichowski, "Biodegradable PEO/cellulose-based solid-solid phase change materials." *Polym. Adv. Technol.* 22 (2011): 1633–1641.

41. S. B. Şentürk, D. Kahraman, C. Alkan and İ. Gökçe, "Biodegradable PEG/cellulose, PEG/agarose and PEG/chitosan blends as shape stabilized phase change materials for latent heat energy storage." *Carbohydr. Polym.* 84 (2011): 141–144.

42. J. Yang, E. Zhang, X. Li, Y. Zhang, J. Qu and Z. Yu, "Cellulose/graphene aerogel supported phase change composites with high thermal conductivity and good shape stability for thermal energy storage." *Carbon N. Y.* 98 (2016): 50–57.

43. L. Cao, Y. Tang and G. Fang, "Preparation and properties of shape-stabilized phase change materials based on fatty acid eutectics and cellulose composites for thermal energy storage." *Energy* 80 (2015): 98–103.

44. C. Alkan, A. Sari and O. Uzun, "Poly(ethylene glycol)/acrylic polymer blends for latent heat thermal energy storage." *AIChE J.* 52 (2006): 3310–3314.

45. A. Sari, C. Alkan, A. Karaipekli and O. Uzun, "Poly(ethylene glycol)/poly(methyl methacrylate) blends as novel form-stable phase-change materials for thermal energy storage." *J. Appl. Polymer Sci.* 116 (2010): 929–933.

46. C. Alkan, E. Günther, S. Hiebler and M. Himpel, "Complexing blends of polyacrylic acid-polyethylene glycol and poly(ethylene-co-acrylic acid)-polyethylene glycol as shape stabilized phase change materials." *Energy Convers. Manag.* 64 (2012): 364–370.

47. L. Wang and D. Meng, "Fatty acid eutectic/polymethyl methacrylate composite as form-stable phase change material for thermal energy storage." *Appl. Energy* 87 (2010): 2660–2665.

48. L. Zhang, J. Zhu, W. Zhou, J. Wang and Y. Wang, "Characterization of polymethyl methacrylate/polyethylene glycol/aluminum nitride composite as form-stable phase change material prepared by in situ polymerization method." *Thermochim. Acta* 524 (2011): 128–134.

49. L. Zhang, J. Zhu, W. Zhou, J. Wang and Y. Wang, "Thermal and electrical conductivity enhancement of graphite nanoplatelets on form-stable polyethylene glycol/polymethyl methacrylate composite phase change materials." *Energy* 39 (2012): 294–302.

50. C. Chen, W. Liu, Z. Wang, K. Peng, W. Pan and Q. Xie, "Novel form stable phase change materials based on the composites of polyethylene glycol/polymeric solid-solid phase change material." *Sol. Energy Mater. Sol. Cells* 134 (2015): 80–88.

51. K. Chen, X. Yu, C. Tian and J. Wang, "Preparation and characterization of form-stable paraffin/polyurethane composites as phase change materials for thermal energy storage." *Energy Convers. Manag.* 77 (2014): 13–21.

52. I. Yilgor and E. Yilgor, "Structure-morphology-property behavior of segmented thermoplastic polyurethanes and polyureas prepared without chain extenders." *Polym. Rev.* 47 (2007): 487–510.

53. K. Pielichowska, M. Nowak, P. Szatkowski and B. Macherzyńska, "The influence of chain extender on properties of polyurethane-based phase change materials modified with graphene." *Appl. Energy* 162 (2016): 1024–1033.

54. W. Kong, X. Fu, Y. Yuan, Z. Liu and J. Lei, "Preparation and thermal properties of crosslinked polyurethane/lauric acid composites as novel form stable phase change materials with a low degree of supercooling." *RSC Adv.* 7 (2017): 29554–29562.

55. Y. Zhang, L. Wang, B. Tang, R. Lu and S. Zhang, "Form-stable phase change materials with high phase change enthalpy from the composite of paraffin and cross-linking phase change structure." *Appl. Energy* 184 (2016): 241–246.

56. B. Tang, L. Wang, Y. Xu, J. Xiu and S. Zhang, "Hexadecanol/phase change polyurethane composite as form-stable phase change material for thermal energy storage." *Sol. Energy Mater. Sol. Cells* 144 (2016): 1–6.

57. L. Feng, J. Zheng, H. Yang, Y. Guo, W. Li and X. Li, "Preparation and characterization of polyethylene glycol/active carbon composites as shape-stabilized phase change materials." *Sol. Energy Mater. Sol. Cells* 95 (2011): 644–650.

58. L. Zhang, H. Shi, W. Li, X. Han and X. Zhang, "Thermal performance and crystallization behavior of poly(ethylene glycol) hexadecyl ether in confined environment." *Polym. Int.* 63 (2014): 982–988.

59. C. Wang, L. Feng, W. Li, J. Zheng, W. Tian and X. Li, "Shape-stabilized phase change materials based on polyethylene glycol/porous carbon composite: The influence of the pore structure of the carbon materials." *Sol. Energy Mater. Sol. Cells* 105 (2012): 21–26.

60. L. Feng, C. Wang, P. Song, H. Wang and X. Zhang, "The form-stable phase change materials based on polyethylene glycol and functionalized carbon nanotubes for heat storage." *Appl. Therm. Eng.* 90 (2015): 952–956.

61. J. Tang, M. Yang, W. Dong, M. Yang, H. Zhang, S. Fan, J. Wang, L. Tan and G. Wang, "Highly porous carbons derived from MOFs for shape-stabilized phase change materials with high storage capacity and thermal conductivity." *RSC Adv.* 6 (2016): 40106–40114.

62. B. Tan, Z. Huang, Z. Yin, X. Min, Y. Liu, X. Wu and M. Fang, "Preparation and thermal properties of shape-stabilized composite phase change materials based on polyethylene glycol and porous carbon prepared from potato." *RSC Adv.* 6 (2016): 15821–15830.

63. W. Wang, X. Yang, Y. Fang, J. Ding and J. Yan, "Preparation and thermal properties of polyethylene glycol/expanded graphite blends for energy storage." *Appl. Energy* 86 (2009): 1479–1483.

64. Y. Jiang, Z. Wang, M. Shang, Z. Zhang and S. Zhang, "Heat collection and supply of interconnected netlike graphene/polyethyleneglycol composites for thermoelectric devices." *Nanoscale* 7 (2015): 10950–10953.

65. H. Li, M. Jiang, Q. Li, D. Li, Z. Chen, W. Hu, J. Huang, X. Xu, L. Dong, H. Xie and C. Xiong, "Aqueous preparation of polyethylene glycol/sulfonated graphene phase change composite with enhanced thermal performance." *Energy Convers. Manag.* 75 (2013): 482–487.

66. C. Wang, L. Feng, H. Yang, G. Xin, W. Li, J. Zheng, W. Tian and X. Li, "Graphene oxide stabilized polyethylene glycol for heat storage." *Phys. Chem. Chem. Phys.* 14 (2012): 13233–13238.

67. G.-Q. Qi, C.-L. Liang, R.-Y. Bao, Z.-Y. Liu, W. Yang, B.-H. Xie and M.-B. Yang, "Polyethylene glycol based shape-stabilized phase change material for thermal energy storage with ultra-low content of graphene oxide." *Sol. Energy Mater. Sol. Cells* 123 (2014): 171–177.

68. W. Xiong, Y. Chen, M. Hao, L. Zhang, T. Mei, J. Wang, J. Li and X. Wang, "Facile synthesis of PEG based shape-stabilized phase change materials and their photo-thermal energy conversion." *Appl. Therm. Eng.* 91 (2015): 630–637.

69. C. Wang, W. Wang, G. Xin, G. Li, J. Zheng, W. Tian and X. Li, "Phase change behaviors of PEG on modified graphene oxide mediated by surface functional groups." *Eur. Polym. J.* 74 (2016): 43–50.

70. A. R. Akhiani, M. Mehrali, S. Tahan Latibari, M. Mehrali, T.M.I. Mahlia, E. Sadeghinezhad and H.S.C. Metselaar, "One-step preparation of form-stable phase change material through self-assembly of fatty acid and graphene." *J. Phys. Chem. C* 119 (2015): 22787–22796.

71. G.-Q. Qi, J. Yang, R.-Y. Bao, Z.-Y. Liu, W. Yang, B.-H. Xie and M.-B. Yang, "Enhanced comprehensive performance of polyethylene glycol based phase change material with hybrid graphene nanomaterials for thermal energy storage." *Carbon N. Y.* 88 (2015): 196–205.

72. Y. Fang, H. Kang, W. Wang, H. Liu and X. Gao, "Study on polyethylene glycol/epoxy resin composite as a form-stable phase change material." *Energy Convers. Manag.* 51 (2010): 2757–2761.

73. S. Mu, J. Guo, Y. Yu, Q. An, S. Zhang, D. Wang, S. Chen, X. Huang and S. Li, "Synthesis and thermal properties of cross-linked poly(acrylonitrile-co-itaconate)/polyethylene glycol as novel form-stable change material." *Energy Convers. Manag.* 110 (2016): 176–183.

74. Y. Cai, Q. Wei, F. Huang and W. Gao, "Preparation and properties studies of halogen-free flame retardant form-stable phase change materials based on paraffin/high density polyethylene composites." *Appl. Energy* 85 (2008): 765–775.

75. H. Hong, Y. Pan, H. Sun, Z. Zhu, C. Ma, B. Wang, W. Liang, B. Yang and A. Li, "Superwetting polypropylene aerogel supported form-stable phase change materials with extremely high organics loading and enhanced thermal conductivity." *Sol. Energy Mater. Sol. Cells* 174 (2018): 307–313.

76. J. L. Zeng, J. Zhang, Y. Y. Liu, Z. X. Cao, Z. H. Zhang, F. Xu and L. X. Sun, "Polyaniline/1-tetradecanol composites." *J. Therm. Anal. Calorim.* 91 (2008): 455–461.

77. J. L. Zeng, F. R. Zhu, S. B. Yu, Z. L. Xiao, W. P. Yan, S. H. Zheng, L. Zhang, L. X. Sun and Z. Cao, "Myristic acid/polyaniline composites as form stable phase change materials for thermal energy storage." *Sol. Energy Mater. Sol. Cells* 114 (2013): 136–140.

78. J. L. Zeng, S. H. Zheng, S. B. Yu, F. R. Zhu, J. Gan, L. Zhu, Z. L. Xiao, X. Y. Zhu, … and Z. Cao, "Preparation and thermal properties of palmitic acid/polyaniline/exfoliated graphite nanoplatelets form-stable phase change materials." *Appl. Energy* 115 (2014): 603–609.

79. K. Pielichowska and K. Pielichowski, "Novel biodegradable form stable phase change materials: Blends of poly(ethylene oxide) and gelatinized potato starch." *J. Appl. Polym. Sci.* 21 (2010): 1725–1731.

80. C. Alkan, E. Günther, S. Hiebler, Ö. F. Ensari and D. Kahraman, "Polyethylene glycol-sugar composites as shape stabilized phase change materials for thermal energy storage." *Polym. Compos.* 33 (2012): 1728–1736.

81. W. Wang, X. Yang, Y. Fang and J. Ding, "Preparation and performance of form-stable polyethylene glycol/silicon dioxide composites as solid-liquid phase change materials." *Appl. Energy* 86 (2009): 170–174.

82. H. Yang, L. Feng, C. Wang, W. Zhao and X. Li, "Confinement effect of SiO_2 framework on phase change of PEG in shape-stabilized PEG/SiO_2 composites." *Eur. Polym. J.* 48 (2012): 803–810.

83. B. Tang, J. Cui, Y. Wang, C. Jia and S. Zhang, "Facile synthesis and performances of PEG/SiO_2 composite form-stable phase change materials." *Sol. Energy* 97 (2013): 484–492.

84. J. Li, L. He, T. Liu, X. Cao and H. Zhu, "Preparation and characterization of PEG/SiO_2 composites as shape-stabilized phase change materials for thermal energy storage." *Sol. Energy Mater. Sol. Cells* 118 (2013): 48–53.

85. L. He, J. Li, C. Zhou, H. Zhu, X. Cao and B. Tang, "Phase change characteristics of shape-stabilized PEG/SiO_2 composites using calcium chloride-assisted and temperature-assisted sol gel methods." *Sol. Energy* 103 (2014): 448–455.

86. T. Qian, J. Li, H. Ma and J. Yang, "The preparation of a green shape-stabilized composite phase change material of polyethylene glycol/SiO_2 with enhanced thermal performance based on oil shale ash via temperature-assisted sol–gel method." *Sol. Energy Mater. Sol. Cells* 132 (2015): 29–39.

87. W. Wang, X. Yang, Y. Fang, J. Ding and J. Yan, "Enhanced thermal conductivity and thermal performance of form-stable composite phase change materials by using β-Aluminum nitride." *Appl. Energy* 86 (2009): 1196–1200.

88. B. Tang, M. Qiu and S. Zhang, "Thermal conductivity enhancement of PEG/SiO_2 composite PCM by in situ Cu doping." *Sol. Energy Mater. Sol. Cells* 105 (2012): 242–248.

89. B. Tang, Y. Wang, M. Qiu and S. Zhang, "A full-band sunlight-driven carbon nanotube/PEG/SiO_2 composites for solar energy storage." *Sol. Energy Mater. Sol. Cells* 123 (2014): 7–12.

90. A. Elgafy and K. Lafdi, "Effect of carbon nanofiber additives on thermal behavior of phase change materials." *Carbon N. Y.* 43 (2005): 3067–3074.

91. L. Feng, W. Zhao, J. Zheng, S. Frisco, P. Song and X. Li, "The shape-stabilized phase change materials composed of polyethylene glycol and various mesoporous matrices (AC, SBA-15 and MCM-41)." *Sol. Energy Mater. Sol. Cells* 95 (2011): 3550–3556.

92. L. Zhang, H. Shi, W. Li, X. Han and X. Zhang, "Structure and thermal performance of poly(ethylene glycol) alkylether (Brij)/porous silica (MCM-41) composites as shape-stabilized phase change materials." *Thermochim. Acta* 570 (2013): 1–7.

93. X. Min, M. Fang, Z. Huang, Y. Liu, Y. Huang, R. Wen, T. Qian and X. Wu, "Enhanced thermal properties of novel shape-stabilized PEG composite phase change materials with radial mesoporous silica sphere for thermal energy storage." *Sci. Rep.* 5 (2015): 12964.

94. R. Mitran, D. Berger, C. Munteanu and C. Matei, "Evaluation of different mesoporous silica supports for energy storage in shape-stabilized phase change materials with dual thermal responses." *J. Phys. Chem. C* 119 (2015): 15177–15184.

95. H. Li and G.-Y. Fang, "Experimental investigation on the characteristics of polyethylene glycol/cement composites as thermal energy storage materials." *Chem. Eng. Technol.* 33 (2010): 1650–1654.

96. T. Qian, J. Li, X. Min, Y. Deng, W. Guan and H. Ma, "Polyethylene glycol/mesoporous calcium silicate shape-stabilized composite phase change material: Preparation, characterization, and adjustable thermal property." *Energy* 82 (2015): 333–340.

97. A. Sarı, "Composites of polyethylene glycol (PEG600) with gypsum and natural clay as new kinds of building PCMs for low temperature-thermal energy storage." *Energy Build.* 69 (2014): 184–192.

98. A. Sarı, "Thermal energy storage characteristics of bentonite-based composite PCMs with enhanced thermal conductivity as novel thermal storage building materials." *Energy Convers. Manag.* 117 (2016): 132–141.

99. E. Onder, N. Sarier, G. Ukuser, M. Ozturk and R. Arat, "Ultrasound assisted solvent free intercalation of montmorillonite with PEG1000: A new type of organoclay with improved thermal properties." *Thermochim. Acta* 566 (2013): 24–35.

100. A. R. Bahramian, L. S. Ahmadi and M. Kokabi, "Performance evaluation of polymer/clay nanocomposite thermal protection systems based on polyethylene glycol phase change material." *Iran. Polym. J.* 23 (2014): 163–169.

101. T. Qian, J. Li, H. Ma and J. Yang, "Adjustable thermal property of polyethylene glycol/diatomite shape-stabilized composite phase change material." *Polym. Compos.* 37 (2016): 854–860.

102. S. Karaman, A. Karaipekli, A. Sarı and A. Biçer, "Polyethylene glycol (PEG)/diatomite composite as a novel form-stable phase change material for thermal energy storage." *Sol. Energy Mater. Sol. Cells* 95 (2011): 1647–1653.

103. T. Qian, J. Li, X. Min, Y. Deng, W. Guan and L. Ning, "Diatomite: A promising natural candidate as carrier material for low, middle and high temperature phase change material." *Energy Convers. Manag.* 98 (2015): 34–45.

104. T. Qian, J. Li, W. Feng and H. Nian, "Enhanced thermal conductivity of form-stable phase change composite with single-walled carbon nanotubes for thermal energy storage." *Sci. Rep.* 7 (2017): 44710.

105. T. Qian, J. Li, X. Min, W. Guan, Y. Deng and L. Ning, "Enhanced thermal conductivity of PEG/diatomite shape-stabilized phase change materials with Ag nanoparticles for thermal energy storage." *J. Mater. Chem. A* 3 (2015): 8526–8536.

106. Y. Qian, P. Wei, P. Jiang, Z. Li, Y. Yan and J. Liu, "Preparation of a novel PEG composite with halogen-free flame retardant supporting matrix for thermal energy storage application." *Appl. Energy* 106 (2013): 321–327.

107. A. Gutierrez, S. Ushak, H. Galleguillos, A. Fernandez, L.F. Cabeza and M. Grágeda, "Use of polyethylene glycol for the improvement of the cycling stability of bischofite as thermal energy storage material." *Appl. Energy* 154 (2015): 616–621.

108. L. Feng, P. Song, S. Yan, H. Wang and J. Wang, "The shape-stabilized phase change materials composed of polyethylene glycol and graphitic carbon nitride matrices." *Thermochim. Acta* 612 (2015): 19–24.

109. Y. Deng, J. Li, T. Qian, W. Guan, Y. Li and X. Yin, "Thermal conductivity enhancement of polyethylene glycol/expanded vermiculite shape-stabilized composite phase change materials with silver nanowire for thermal energy storage." *Chem. Eng. J.* 295 (2016): 427–435.

110. Y. Deng, J. Li and H. Nian, "Polyethylene glycol-enwrapped silicon carbide nanowires network/expanded vermiculite composite phase change materials: Form-stabilization, thermal energy storage behavior and thermal conductivity enhancement." *Sol. Energy Mater. Sol. Cells* 174 (2018): 283–291.

111. Y. Deng, J. Li, H. Nian, Y. Li and X. Yin, "Design and preparation of shape-stabilized composite phase change material with high thermal reliability via encapsulating polyethylene glycol into flower-like TiO_2 nanostructure for thermal energy storage." *Appl. Therm. Eng.* 114 (2017): 328–336.

112. J. Yang, L. S. Tang, R. Y. Bao, L. Bai, Z. Y. Liu, B. H. Xie, M. B. Yang and W. Yang, "Hybrid network structure of boron nitride and graphene oxide in shape-stabilized composite phase change materials with enhanced thermal conductivity and light-to-electric energy conversion capability." *Sol. Energy Mater. Sol. Cells* 174 (2018): 56–64.

113. J.-C. Su and P.-S. Liu, "A novel solid–solid phase change heat storage material with polyurethane block copolymer structure." *Energy Convers. Manag.* 47 (2006): 3185–3191.

114. W.-D. Li and E.-Y. Ding, "Preparation and characterization of cross-linking PEG/MDI/PE copolymer as solid–solid phase change heat storage material." *Sol. Energy Mater. Sol. Cells* 91 (2007): 764–768.

115. X.-M. Zhou, "Preparation and characterization of PEG/MDI/PVA copolymer as solid-solid phase change heat storage material." *J. Appl. Polym. Sci.* 113 (2009): 2041–2045.

116. K. Peng, C. Chen, W. Pan, W. Liu, Z. Wang and L. Zhu, "Preparation and properties of β-cyclodextrin/4,4′-diphenylmethane diisocyanate/polyethylene glycol (β-CD/MDI/PEG) crosslinking copolymers as polymeric solid–solid phase change materials." *Sol. Energy Mater. Sol. Cells* 145 (2016): 238–247.

117. X. Fu, W. Kong, Y. Zhang, L. Jiang, J. Wang and J. Lei, "Novel solid-solid phase change materials with biodegradable trihydroxy surfactant for thermal energy storage." *RSC Adv.* 5 (2015): 68881–68889.

118. C. Chen, W. Liu, H. Wang and K. Peng, "Synthesis and performances of novel solid–solid phase change materials with hexahydroxy compounds for thermal energy storage." *Appl. Energy* 152 (2015): 198–206.

119. P. Xi, Y. Duan, P. Fei, L. Xia, R. Liu and B. Cheng, "Synthesis and thermal energy storage properties of the polyurethane solid–solid phase change materials with a novel tetrahydroxy compound." *Eur. Polym. J.* 48 (2012): 1295–1303.

120. P. Xi, F. Zhao, P. Fu, X. Wang and B. Cheng, "Synthesis, characterization, and thermal energy storage properties of a novel thermoplastic polyurethane phase change material." *Mater. Lett.* 121 (2014): 15–18.

121. C. Alkan, E. Günther, S. Hiebler, Ö. F. Ensari and D. Kahraman, "Polyurethanes as solid–solid phase change materials for thermal energy storage." *Sol. Energy* 86 (2012): 1761–1769.

122. Z. Liu, X. Fu, L. Jiang, B. Wu, J. Wang and J. Lei, "Solvent-free synthesis and properties of novel solid–solid phase change materials with biodegradable castor oil for thermal energy storage." *Sol. Energy Mater. Sol. Cells* 147 (2016): 177–184.

123. X. Du, H. Wang, X. Cheng and Z. Du, "Synthesis and thermal energy storage properties of a solid–solid phase change material with a novel comb-polyurethane block copolymer structure." *RSC Adv.* 6 (2016): 42643–42648.

124. X. Fu, Y. Xiao, K. Hu, J. Wang, J. Lei and C. Zhou, "Thermosetting solid–solid phase change materials composed of poly(ethylene glycol)-based two components: Flexible application for thermal energy storage." *Chem. Eng. J.* 291 (2016): 138–148.

125. K. Chen, R. Liu, C. Zou, Q. Shao, Y. Lan, X. Cai and L. Zhai, "Linear polyurethane ionomers as solid–solid phase change materials for thermal energy storage." *Sol. Energy Mater. Sol. Cells* 130 (2014): 466–473.

126. H. Shi, J. Li, Y. Jin, Y. Yin and X. Zhang, "Preparation and properties of poly(vinyl alcohol)-g-octadecanol copolymers based solid-solid phase change materials." *Mater. Chem. Phys.* 131 (2011): 108–112.

127. Y. Zhou, D. Sheng, X. Liu, C. Lin, F. Ji, L. Dong, S. Xu and Y. Yang, "Synthesis and properties of crosslinking halloysite nanotubes/polyurethane-based solid-solid phase change materials." *Sol. Energy Mater. Sol. Cells* 174 (2018): 84–93.

128. Q. Cao and P. Liu, "Hyperbranched polyurethane as novel solid–solid phase change material for thermal energy storage." *Eur. Polym. J.* 42 (2006): 2931–2939.

129. Q. Cao and P. Liu, "Crystalline-amorphous phase transition of hyperbranched polyurethane phase change materials for energy storage." *J. Mater. Sci.* 42 (2007): 5661–5665.

130. Q. Cao, L. Liao and H. Xu, "Study on the influence of thermal characteristics of hyperbranched polyurethane phase change materials for energy storage." *J. Appl. Polym. Sci.* 115 (2010): 2228–2235.

131. L. Liao, Q. Cao and H. Liao, "Investigation of a hyperbranched polyurethane as a solid-state phase change material." *J. Mater. Sci.* 45 (2010): 2436–2441.

132. X. Du, H. Wang, Y. Wu, Z. Du and X. Cheng, "Solid-solid phase-change materials based on hyperbranched polyurethane for thermal energy storage." *J. Appl. Polym. Sci.* 134 (2017): 1–8.

133. S. Sundararajan, A. B. Samui and P. S. Kulkarni, "Synthesis and characterization of poly(ethylene glycol) (PEG) based hyperbranched polyurethanes as thermal energy storage materials." *Thermochim. Acta* 650 (2017): 114–122.

134. S. Sundararajan, A. B. Samui and P. S. Kulkarni, "thermal energy storage using poly(ethylene glycol) incorporated hyperbranched polyurethane as solid–solid phase change material." *Ind. Eng. Chem. Res.* 56 (2017): 14401–14409.

135. Y. Li, R. Liu and Y. Huang, "Synthesis and phase transition of cellulose-graft -poly(ethylene glycol) copolymers." *J. Appl. Polym. Sci.* 110 (2008): 1797–1803.

136. Y. Li, M. Wu, R. Liu and Y. Huang, "Cellulose-based solid–solid phase change materials synthesized in ionic liquid." *Sol. Energy Mater. Sol. Cells* 93 (2009): 1321–1328.

137. A. Kumar, P. S. Kulkarni and A. B. Samui, "Polyethylene glycol grafted cotton as phase change polymer." *Cellulose* 21 (2014): 685–696.

138. Y. Jiang, E. Ding and G. Li, "Study on transition characteristics of PEG/CDA solid–solid phase change materials." *Polymer (Guildf).* 43 (2002): 117–122.

139. P. Xi, X. Gu, B. Cheng and Y. Wang, "Preparation and characterization of a novel polymeric based solid–solid phase change heat storage material." *Energy Convers. Manag.* 50 (2009): 1522–1528.

140. J. Guo, H. Xiang, Q. Wang, C. Hu, M. Zhu and L. Li, "Preparation of poly(decaglycerol-co-ethylene glycol) copolymer as phase change material." *Energy Build.* 48 (2012): 206–210.

141. B. Tang, Z. Yang and S. Zhang, "Poly(polyethylene glycol methyl ether methacrylate) as novel solid-solid phase change material for thermal energy storage." *J. Appl. Polym. Sci.* 125 (2012): 1377–1381.

142. H. Zhang, D. Sun, Q. Wang, J. Guo and Y. Gong, "Synthesis and characterization of polyethylene glycol acrylate crosslinking copolymer as solid-solid phase change materials." *J. Appl. Polym. Sci.* 131 (2014): 39755.

143. S. Sundararajan, A. B. Samui and P. S. Kulkarni, "Synthesis and characterization of poly(ethylene glycol) acrylate (PEGA) copolymers for application as polymeric phase change materials (PCMs)." *React. Funct. Polym.* 130 (2018): 43–50.

144. J. Y. Meng, X. F. Tang, W. Li, H. F. Shi and X. X. Zhang, "Crystal structure and thermal property of poly-ethylene glycol octadecyl ether." *Thermochim. Acta* 558 (2013): 83–86.

145. J. Meng, X. Tang, Z. Zhang, X. Zhang and H. Shi, "Fabrication and properties of poly(polyethylene glycol octadecyl ether methacrylate)." *Thermochim. Acta* 574 (2013): 116–120.

146. J. Hu, H. Yu, Y. Chen and M. Zhu, "Study on phase-change characteristics of PET-PEG copolymers." *J. Macromol. Sci. Part B* 45 (2006): 615–621.

147. X.-M. Zhou, "Study on phase change characteristics of PEG/PAM coupling blend." *J. Appl. Polym. Sci.* 21 (2010): 1591–1595.

148. X. Gu, P. Xi, B. Cheng and S. Niu, "Synthesis and characterization of a novel solid-solid phase change luminescence material." *Polym. In.* 59 (2010): 772–777.

149. J. Guo, P. Xie, X. Zhang, C. Yu, F. Guan and Y. Liu, "Synthesis and characterization of graft copolymer of polyacrylonitrile- g -polyethylene glycol-maleic acid monoester macromonomer." *J. Appl. Polym. Sci.* 131 (2014): 40152.

150. S.-Y. Mu, J. Guo, Y.-M. Gong, S. Zhang and Y. Yu, "Synthesis and thermal properties of poly(styrene-co-acrylonitrile)-graft-polyethylene glycol copolymers as novel solid–solid phase change materials for thermal energy storage." *Chinese Chem. Lett.* 26 (2015): 1364–1366.

151. L. Yanshan, W. Shujun, L. Hongyan, M. Fanbin, M. Huanqing and Z. Wangang, "Preparation and characterization of melamine/formaldehyde/polyethylene glycol crosslinking copolymers as solid–solid phase change materials." *Sol. Energy Mater. Sol. Cells* 127 (2014): 92–97.

152. C. Chen, W. Liu, H. Yang, Y. Zhao and S. Liu, "Synthesis of solid–solid phase change material for thermal energy storage by crosslinking of polyethylene glycol with poly (glycidyl methacrylate)." *Sol. Energy* 85 (2011): 2679–2685.

153. A. Sarı, C. Alkan and A. Biçer, "Synthesis and thermal properties of polystyrene-graft-PEG copolymers as new kinds of solid–solid phase change materials for thermal energy storage." *Mater. Chem. Phys.* 133 (2012): 87–94.

154. A. Sarı, C. Alkan, A. Biçer and A. Karaipekli, "Synthesis and thermal energy storage characteristics of polystyrene-graft-palmitic acid copolymers as solid–solid phase change materials." *Sol. Energy Mater. Sol. Cells* 95 (2011): 3195–3201.

155. H. Xiang, S. Wang, R. Wang, Z. Zhou, C. Peng and M. Zhu, "Synthesis and characterization of an environmentally friendly PHBV/PEG copolymer network as a phase change material." *Sci. China Chem.* 56 (2013): 716–723.

156. S. Sundararajan, A. B. Samui and P. S. Kulkarni, "Interpenetrating phase change polymer networks based on crosslinked polyethylene glycol and poly(hydroxyethyl methacrylate)." *Sol. Energy Mater. Sol. Cells* 149 (2016): 266–274.

157. T. L. Vigo and C. M. Frost, "Temperature-adaptable fabrics." *Text. Res. J.* 55 (1985): 737–743.

158. M. Karthikeyan, T. Ramachandran and O. L. S. Sundaram, "Nanoencapsulated phase change materials based on polyethylene glycol for creating thermoregulating cotton." *J. Ind. Text.* 44 (2013): 130–146.

159. A. Kuru and S. A. Aksoy, "Cellulose-PEG grafts from cotton waste in thermo-regulating textiles." *Text. Res. J.* 84 (2014): 337–346.

160. A. Khoddami, O. Avinc and F. Ghahremanzadeh, "Improvement in poly(lactic acid) fabric performance via hydrophilic coating." *Prog. Org. Coatings* 72 (2011): 299–304.

161. F. Ghahremanzadeh, A. Khoddami and C. M. Carr, "Improvement in fastness properties of phase-change material applied on surface modified wool fabrics." *Fibers Polym.* 11 (2010): 1170–1180.

162. T. T. T. Nguyen, J. G. Lee and J. S. Park, "Fabrication and characterization of coaxial electrospun polyethylene glycol/polyvinylidene fluoride (core/sheath) composite non-woven mats." *Macromol. Res.* 19 (2011): 370–378.

163. T. T. T. Nguyen and J. S. Park, "Fabrication of electrospun nonwoven mats of polyvinylidene fluoride/polyethylene glycol/fumed silica for use as energy storage materials." *J. Appl. Polym. Sci.* 121 (2011): 3596–3603.

164. G. Z. Ke, H. F. Xie, R. P. Ruan and W. D. Yu, "Preparation and performance of porous phase change polyethylene glycol/polyurethane membrane." *Energy Convers. Manag.* 51 (2010): 2294–2298.

165. R. Baetens, B. P. Jelle and A. Gustavsen, "Phase change materials for building applications: A state-of-the-art review." *Energy Build.* 42 (2010): 1361–1368.

166. A. Sari, "Composites of polyethylene glycol (PEG600) with gypsum and natural clay as new kinds of building PCMs for low temperature-thermal energy storage." *Energy Build.* 69 (2014): 184–192.

167. Y. Wang, B. Tang and S. Zhang, "Single-walled carbon nanotube/phase change material composites: Sunlight-driven, reversible, form-stable phase transitions for solar thermal energy storage." *Adv. Funct. Mater.* 23 (2013): 4354–4360.

168. R. Chen, R. Yao, W. Xia and R. Zou, "Electro/photo to heat conversion system based on polyurethane embedded graphite foam." *Appl. Energy* 152 (2015): 183–188.

169. M. Skach, M. Arora, C. Hsu, Q. Li, D. Tullsen, L. Tang and J. Mars, "Thermal time shifting," in *Proceedings of the 42nd Annual International Symposium on Computing Architecture – ISCA '15*. New York: ACM Press, 2015, pp. 439–449.

170. J. Deign, "How phase-change materials are changing lives" (2016). Available at: http://energystoragereport.info/phase-change-materials-saving-lives/

171. R. M. Formato, "The advantages & challenges of phase change materials (PCMs) in thermal packaging" (2013). Available at: www.coldchaintech.com/assets/Cold-Chain-Technologies-PCM-White-Paper.pdf

172. J. R. Andrews, K. G. Prajapati, E. Eypper, P. Shrestha, M. Shakya, K. R. Pathak, N. Joshi, P. Tiwari, M. Risal, … and A. Arjyal, "Evaluation of an electricity-free, culture-based approach for detecting typhoidal salmonella bacteremia during enteric fever in a high burden, resource-limited setting." *PLoS Negl. Trop. Dis.* 7 (2013): e2292.

173. H. Kim, D. Lee, J. Kim, T. Kim and W. J. Kim, "Photothermally triggered cytosolic drug delivery via endosome disruption using a functionalized reduced graphene oxide." *ACS Nano* 7 (2013): 6735–6746.

174. B. Duong, H. Liu, L. Ma and M. Su, "Covert thermal barcodes based on phase change nanoparticles." *Sci. Rep.* 4 (2014): 5170.

175. M. V. Bhamidipati, "Methods and compositions for inhibiting surface icing." US Patent No. US 20100322867A1-2009 (2009).

176. J. Maxa, A. Novikov and M. Nowottnick, "Thermal peak management using organic phase change materials for latent heat storage in electronic applications." *Materials (Basel)* 11 (2018): 31.

177. D. Rigotti, A. Dorigato and A. Pegoretti, "3D printable thermoplastic polyurethane blends with thermal energy storage/release capabilities." *Mater. Today Commun.* 15 (2018): 228–235.

178. Y.-L. Huang, Q.-B. Li, X. Deng, Y.-H. Lu, X.-K. Liao, M.-Y. Hong and Y. Wang, "Aerobic and anaerobic biodegradation of polyethylene glycols using sludge microbes." *Process Biochem.* 40 (2005): 207–211.

179. C. Chen, L. Wang and Y. Huang, "Electrospinning of thermo-regulating ultrafine fibers based on polyethylene glycol/cellulose acetate composite." *Polymer (Guildf).* 48 (2007): 5202–5207.

5

Smart Hydrogel Materials

Ramavatar Meena
CSIR-Central Salt & Marine Chemicals Research Institute, Bhavnagar, Gujarat, India
Academy of Scientific and Innovative Research (AcSIR), Ghaziabad, Uttar Pradesh, India

Faisal Kholiya
CSIR-Central Salt & Marine Chemicals Research Institute, Bhavnagar, Gujarat, India

CONTENTS

Abbreviations

PVP	polyvinyl pyrrolidone
SA	sodium alginate
PVA	polyvinyl alcohol
CMA	carboxymethyl agarose
BA	boric acid
P(MAA-g-EG)	poly(methacrylic acid)-grafted-poly(ethylene glycol)
Poly(NIPAAm-co-MA)	poly(n-isopropylacrylamideco-methacrylic acid)
MMA	methyl methacrylate
AA	acrylic acid
CMCT	carboxymethyl chitosan
PEO	poly (ethylene oxide)
semi-IPN	semi-interpenetrating polymer network
PAsp	polyaspartic acid
HEAA	N-(2-hydroxyethyl) acrylamide
METAC	[2-(methacryloyloxy)ethyl]trimethylammonium chloride
MBAA	N,N'-methylene bisacrylamide

VBIPS	3-(1-(4- vinyl benzyl)-1H-imidazole-3-ium-3-yl)propane-1-sulfonate
NaAA	sodium acrylic acid
PVA	poly (vinyl alcohol)
LCST	lower critical solution temperature
UCST	upper critical solution temperature
AQT	*N*-acryl-*N'*-(quinolin-8-yl)thiourea
NIPAAm	*N*-isopropylacrylamide
PPS-*b*-PDMA-*b*- PNIPAAM	poly[(propylene sulfide)-*block*-(*N*,*N*-dimethylacrylamide)-*block*-(*N*-isopropylacrylamide)]
RAFT	reversible addition-fragmentation chain transfer
ROS	reactive oxygen species
bFGF	basic fibroblast growth factor
GO–GMA	glycidyl methacrylate functionalized graphene oxide
GOD	glucose oxidase
PEGDA	poly (ethylene glycol) diacrylate
MMP	matrix metalloproteinase
CD	cyclodextrin
SEM	scanning electron microscopy
FTIR	Fourier transform infrared spectroscopy
STM	scanning tunneling microscopy
AFM	atomic force microscopy
PEG	polyethylene glycol

5.1 Introduction

In general, polymers which swell up in water instead of dissolving in it are called polymer-based hydrogels. The hydrogel can be defined in various ways, but the commonest definition of a hydrogel is a water-swollen and cross-linked polymeric material obtained by one or more monomers. This class of material shows swelling properties due to the thermodynamic attraction toward the solvent (Kopeček 2002). This swelling property makes hydrogels highly versatile and highly tunable in their properties which renders them an attractive research area (Figure 5.1).

There is a large amount of water in a hydrogel, and it also possesses a degree of flexibility which is similar to natural tissue and gives a promising wide range of applications in superabsorbent materials, matrix chemistry, and biology, as a medium for storage and delivery of substances in biomedicine, and as highly promising scaffolds to reconstitute artificial extracellular matrix environments (Kiyonaka et al. 2003; Lee and Mooney 2001; Lutolf 2009; Stuart et al. 2010). For the past three decades, research has increased on the uses of natural polymer-based hydrogels, as an alternative to synthetic polymers, due to their long service life, high capacity of water absorption, high gel strength, and non-toxic nature; consequently, they form a smart hydrogel (Ahmed 2015; Ebara et al. 2014).

Hydrogels are mainly classified into two major groups based on their source: (1) natural hydrogels; and (2) synthetic hydrogels (Zhao et al. 2013). Furthermore, hydrogels are also classified into two types by their polymeric composite: (1) homopolymer-based hydrogels; (2) copolymer-based hydrogels, and (3) multi-polymer interpenetrated hydrogels. In the homopolymer hydrogel, the formation of the network in the hydrogel is due to the single species of the monomer which is the basic unit of the correspondence polymer, while the copolymer hydrogel forms a network with two or more monomer spices with at least one hydrophilic monomer of correspondence polymer (Yang et al. 2002; Iizawa et al. 2007). Multi-polymer interpenetrated hydrogels are an important class of polymer hydrogel in which two or more independent cross-linked natural or synthetic polymers form a gel network (Maolin et al. 2000). As hydrogels have swelling and de-swelling properties, they have numerous fields of application, such as hygienic products (Singh et al. 2010), agriculture (Saxena 2010), drug delivery systems (Hamidi et al. 2008; Singh et al. 2010), sealing (Singh et al. 2010), coal dewatering (Park 2001), artificial snow (Singh et al. 2010), food additives (Chen et al. 1995), pharmaceuticals (Kashyap et al. 2005), biomedical applications (Kaihara et al. 2008; Stamatialis et al. 2008), tissue engineering and regenerative medicines (Saul and Williams 2011; Zhang et al. 2011), diagnostics (Plunkett and Moore 2004), wound dressings (Sikareepaisan et al. 2011), separation of biomolecules or cells (Wang et al. 2010), barrier materials to regulate biological adhesions (Roy et al. 2010), and biosensors (Krsko et al. 2009).

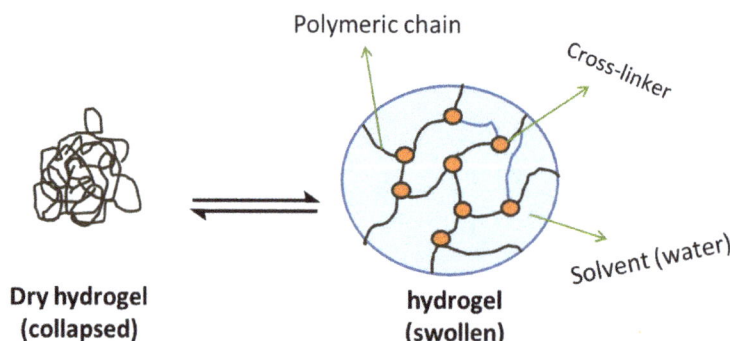

FIGURE 5.1 Hydrogel (collapsed and swollen).

5.2 Smart Hydrogels

Polymers that show dramatic changes in their properties responding to small changes in the environment, such as temperature, pH, solvents, etc., are called smart hydrogels or a stimuli-responsive polymeric system. Smart hydrogels are wide-ranging hydrogels undergoing switchable gel-to-solution or gel-to-solid transitions depending upon the application of external prompts. The external prompts include thermal, magnetic, ultrasonic, electrochemical, or light stimuli as physical prompts, and pH, redox reactions, supramolecular complexes, and bio-catalytically driven reactions as chemical prompts (Aguilar and San Román 2014). Numerous applications are suggested for stimuli-responsive hydrogels, such as functional matrices for sensing, actuators, biomedical applications, controlled drug release, tissue engineering, and imaging. Also, stimuli-triggered hydrogels are used to construct catalytic switches, logic-gate operations, surfaces for controlled growth of cells, and more. Methods to immobilize and pattern stimuli-responsive hydrogels on surfaces have been developed, and the fabrication of surfaces revealing signal-triggered stiffness and switchable interfacial electron transfer properties have been demonstrated (Willner 2017).

The stimuli-responsive or smart hydrogels can be divided into three types on the basis of their functions: (1) mechanical motion; (2) mass transport; and (3) conversion and transmission of information.

5.2.1 Function of Mechanical Motion

The first mechanical gels were prepared in 1980 and the 1990s. This hydrogel was used as an artificial hand to lift an object and also as an artificial fish that swam in a repeating flexing motion which was controlled by changing the temperature or an electric field.

5.2.2 Mass Transport

These kinds of hydrogels are widely used for biomedical applications as their function is to capture or release a chemical or biological substance and to separate or purify substances.

5.2.3 Conversion and Transmission of Information

Memorizing or converting information is the main function of stimuli devices. These types of hydrogels are prepared in such a way that the molecular designs memorize the structure at the molecular level in a polymer network.

5.3 Smart/Stimuli-Responsive Hydrogels and Their Applications

From the past five decades, formation of hydrogels has been focused on biopolymers instead of synthetic polymers, because they had a renewable source, and were eco-friendly, cost-effective, and non-toxic. Due to the non-toxicity of the biopolymers, they are widely used in wound dressings and the preparation of smart gels in biomedical applications (Gopi and Amalraj 2016; Van

Vlierberghe et al. 2011). There are several naturally occurring biopolymers, including polysaccharides, such as agarose, cellulose, alginate, chitosan, carrageen, etc., and proteins, such as gelatin, collagen, and DNA, which form a three-dimensional network that can hold an enormous amount of water which leads to a naturally occurring hydrogel (Singh et al. 2010).

Seaweed polysaccharides are well known for the ability to form gels in water and generally are called hydrogels. The gelling seaweed polysaccharides, namely agar, carrageenan, agarose, and alginate, are mainly obtainable only from renewable seaweed resources, except for alginate. They are hydrophilic in nature and used in wider applications, such as foods, textiles, pharmaceuticals, and biomedical applications. Due to their hydrophilic nature, aqueous applications are limited, but they are gifted with plentiful hydroxyl (-OH) groups in their structure, which are available to modify these polymers to obtain the desired functional materials for applications by physical and chemical interventions. Hence, this chapter considers some of our recent works (Chaudhary, Chejara et al. 2014; Chaudhary, Chejara et al. 2015; Chaudhary, Kondaveeti et al. 2014; Chaudhary, Kumar et al. 2015; Chaudhary, Nataraj et al. 2014; Chaudhary, Vadodariya et al. 2015; Mondal et al. 2013), related to the development of seaweed polysaccharides-based sustainable hydrogel materials for wider applications. Some of the seaweed polysaccharides-based functional hydrogel materials were successfully tested for emulsion (oil-water) separation (Figure 5.2 (a)) (Chaudhary, Chejara et al. 2015; Chaudhary, Kondaveeti et al. 2014) multifunctional material (Figure 5.2 (b)) (Chaudhary, Chejara et al. 2015),· mercury detection (ibid.), controlled release (Figure 5.2 (c)) (Chaudhary, Siddhanta, and Meena 2014), DNA resolution and for self-healing (Figure 5.2 (d)) applications (Chaudhary, Kondaveeti et al. 2014). Hence in this chapter, we briefly describe the seaweed polysaccharides-based functional hydrogel materials and their properties and stimuli-responsive/smart hydrogel.

5.4 Seaweed Polysaccharide-Based Hydrogels

Biopolymer-based hydrogels are of substantial interest in various research fields due to their high safety level and unique properties, e.g., they are biocompatible, hydrophilic, and biodegradable. Currently, various biopolymer-based hydrogels have been synthesized, such as seaweed polysaccharide-based hydrogels, protein-based hydrogels, and DNA-based hydrogels. In this section. brief information about seaweed polysaccharide hydrogels is given.

Hydrogels can be synthesized using mainly two methods: (1) chemically cross-linked hydrogels; and (2) physically cross-linked hydrogels (Figure 5.3).

5.5 Chemically Cross-Linked Hydrogels

Chemically cross-linked hydrogels can be synthesized by various methods, such as chain growth polymerization, addition, and condensation polymerization and gamma and electron beam

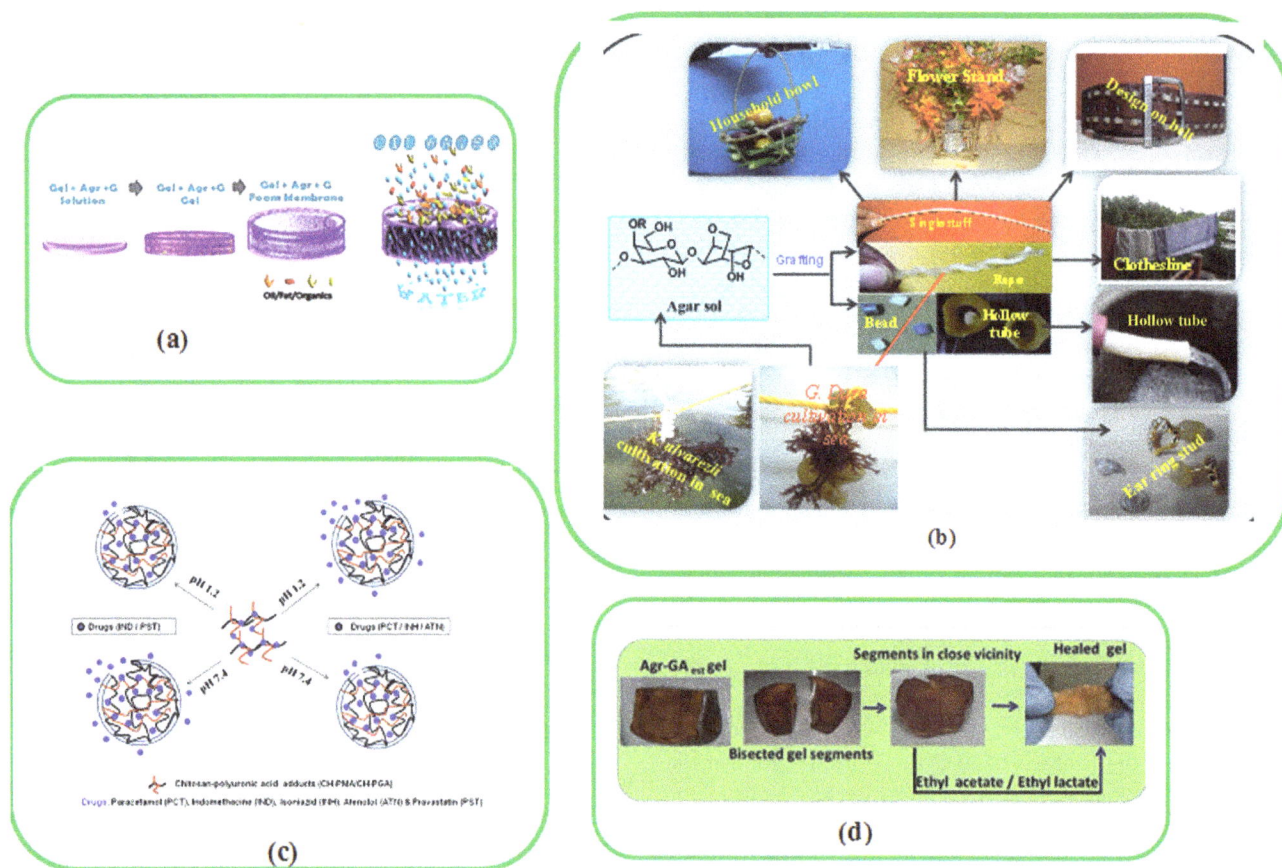

FIGURE 5.2 Seaweed polysaccharides-based functional hydrogel for (a) emulsion (oil-water) separation; (b) multifunctional material; (c) controlled release; and (d) self-healing.

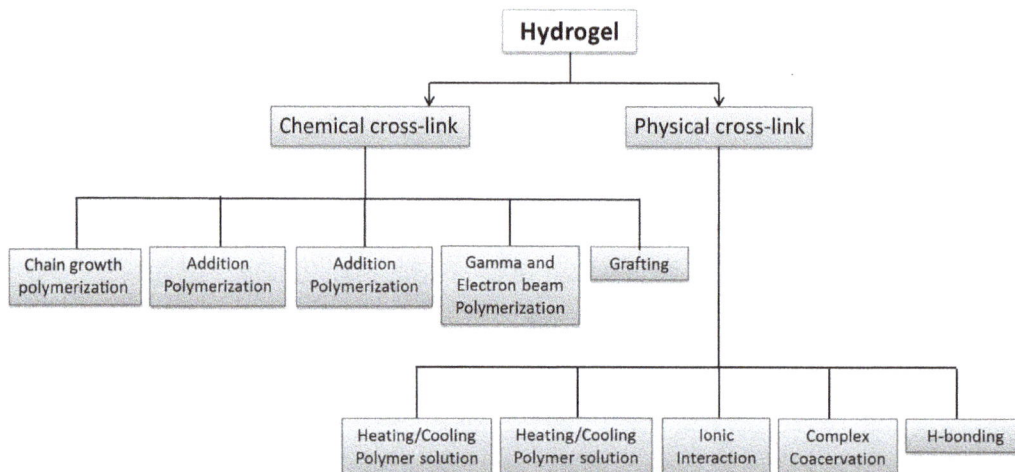

FIGURE 5.3 Types of hydrogels.

polymerization. In chain growth polymerization, the monomers react with the active side of the polymer to form the hydrogel. There are three steps in the chain growth polymerization process: initiation, propagation, and termination. After initiation, a free radical active site is generated which adds monomers in a chain link-like fashion. The addition and condensation cross-linking include the

stepwise addition of the poly-functional cross-linker to the functional monomer. By this addition, a cross-linked reaction, water-soluble polymer can be converted into the hydrogel. Furthermore, the gamma and electron beam polymerization implicates high energy electromagnetic irradiation as a cross-linker. The chemically cross-linked hydrogel also is prepared by adding a chemical cross-linker,

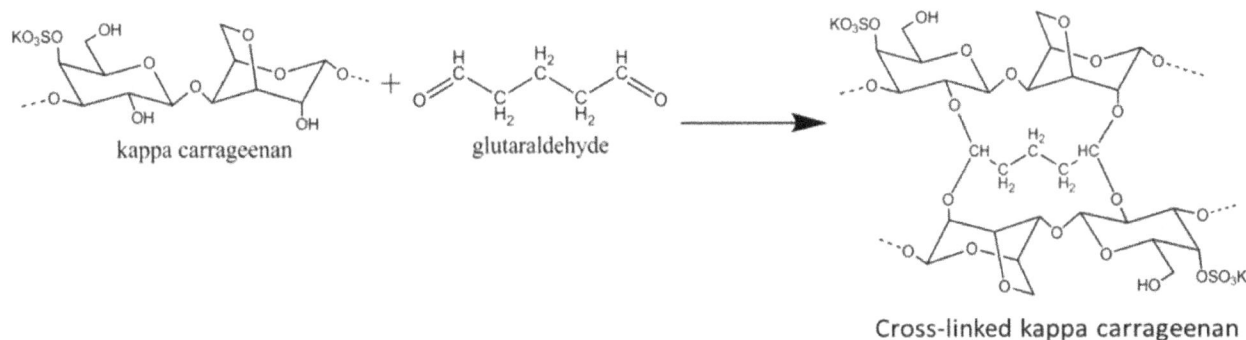

FIGURE 5.4 Chemically cross-linked kappa carrageenan and glutaraldehyde hydrogel.

such as glutaraldehyde, epichlorohydrin. In this technique, the new monomer or polymer is introduced into the native polymer to form a hydrogel (Chang et al. 2010; Distantina et al. 2013). Grafting is also one type of chemical cross-linker reaction for the formation of the hydrogel. In grafting, the monomer is introduced to the preformed polymer. The polymeric chain is activated by the chemical reagent or high energy radiation treatment (Gulrez et al. 2011). Here are some seaweed polysaccharides-based hydrogel syntheses using a chemical cross-link.

Carrageenans derived from red seaweed are high-molecular weight polysaccharides made up of repeating galactose units and 3, 6 anhydrogalactose (3, 6-AG), both sulfated and nonsulfated. The units are joined by alternating α-1, 3 and β-1, 4 glycosidic linkages. Carrageenans are large, highly flexible molecules that curl, forming helical structures. This gives them the ability to form a variety of different gels at room temperature. They are widely used in the food industries as thickening and stabilizing agents. The carrageenan is cross-linked with the glutaraldehyde and HCl is used as a catalyst. This as-prepared hydrogel was pH-responsive which has potential to be a drug delivery system (Figure 5.4) (Distantina et al. 2013).

In another report, Singh, Singh, and Singh (2015) prepared antimicrobial hydrogel for wound dressings, using carrageenan and polyvinyl pyrrolidone (PVP) with nanosilver. This hydrogel was prepared by gamma irradiation and shows potential microbicidal activity (≥3 log_{10} decreases in CFU/ml) against wound pathogens, *P. aeruginosa*, *S. aureus*, *E. coli*, and *C. albicans*. This PVP-carrageenan hydrogel could be used for wound dressings to control infection and offer a smooth healing process for burns and other skin injuries (ibid.).

Agarose/agar is a hydrophilic polymer widely used in bio-medicinal applications and bioengineering. Agarose has (1→ 3) linked β-D-galactose and (1→ 4) linked α-L-3, 6-anhydrogalactose disaccharide repeating units. Agarose is a low temperature, water-soluble, gelling polysaccharide, which is widely used in the food industries as a thickener as well as for stabilization. Agarose is also used in various industries, such as toothpaste, laxatives, diet pills, water-based paints, detergents, textile sizing, and various paper products (López-Simeon et al. 2012). Furthermore, Chaudhary, Chejara et al. (2015) used agarose for the preparation of an aerogel membrane through cross-linking with a natural cross-linker, namely genipin, with an amino-based polysaccharide chitosan. The aerogel was used to separate an oil-water emulsion. The separated water, through the

agarose-based aerogel, had ≥ 99% purity with 600 L.m^{-2}.h^{-1}.bar^{-1} (ibid.) (Figure 5.5).

Healing enhancement and pain control are critical issues in wound management. So far, different wound dressings have been developed. Among them, hydrogels are the most applied. Miguel et al. (2014) produced a thermoresponsive hydrogel using two biopolymers, agarose and chitosan. As prepared hydrogel shows improved healing properties on the wound, and the lack of a reactive or a granulomatous inflammatory reaction in skin lesions treated with hydrogel, this demonstrates its suitability for use in the near future as a wound dressing (ibid.). Also in one report, an agarose-based, chemically cross-linked, self-healing hydrogel (Agr-GAEst) was synthesized using gallic acid, which is organic acid found in many plants. The hydrogel was prepared in one pot and through microwave irradiation. The obtained gel showed a substantial degree of thixotropic (hysteresis loop area = 38.73%), quick self-healing ability (12 min) upon complete cleavage of the gel and excellent stretching ability (>20 times of its original length). These types of multifunctional gels can find applications in food and personal health care industries (Figure 5.6) (Chaudhary, Kondaveeti, et al. 2014).

Sodium alginate is a sodium salt of alginic acid which is derived from brown seaweed. Alginic acid is a linear copolymer with homopolymeric blocks of (1–4)-linked β-D-mannuronate (M) and its C-5 epimer α-L-guluronate (G) residues, respectively, covalently linked together in different sequences or blocks. It is reported that a bio-based hydrogel using sodium alginate (SA) and polyvinyl alcohol (PVA) has been developed. As prepared hydrogels offer good mechanical advantage and are highly conductive, they are of great importance for their excellent biocompatibility and biodegradability. PVA-SA-based hydrogels show significant high gel strength and improved conductivity when immersed in an NaCl solution (Figure 5.7) (Jiang et al. 2018). Also, other seaweed polysaccharides, such as fucoidan, laminarin, and ulvan-based hydrogel, were prepared, which are used widely for biomedical application due to their high biocompatibility and easy availability (Venkatesan et al. 2015).

5.6 Physically Cross-Linked Hydrogels

In the past, research was interested in physically cross-linked hydrogels, because they avoided the cross-linking agent. The

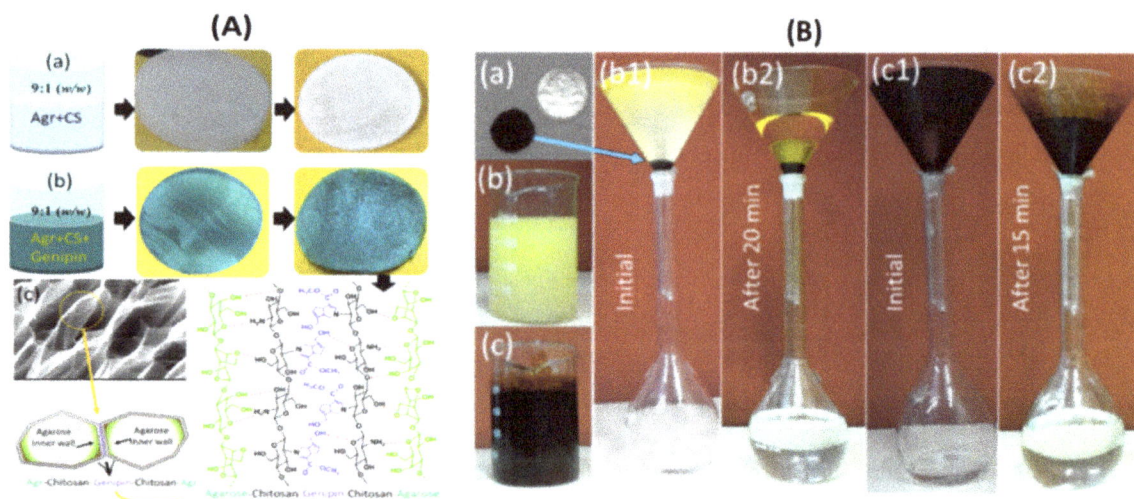

FIGURE 5.5 (A) Schematics of preparing chitosan-based aerogel membrane (a) control; (b) genipin cross-linked chitosan aerogel; and (c) genipin–chitosan cross-linked chemical structure with inner walls of CS linked with agarose. (B) (a) Photograph of coin-sized aerogel membrane used to separate (b) biodiesel/ water emulsion; (c) oil-spill wastewater emulsion collected from ship breaking yard; (b1 and b2) biodiesel/ water emulsion before and after separation and (c1 and c2) oil-spill wastewater emulsion before and after separation.

Source: Chaudhary, Vadodariya et al. (2015).

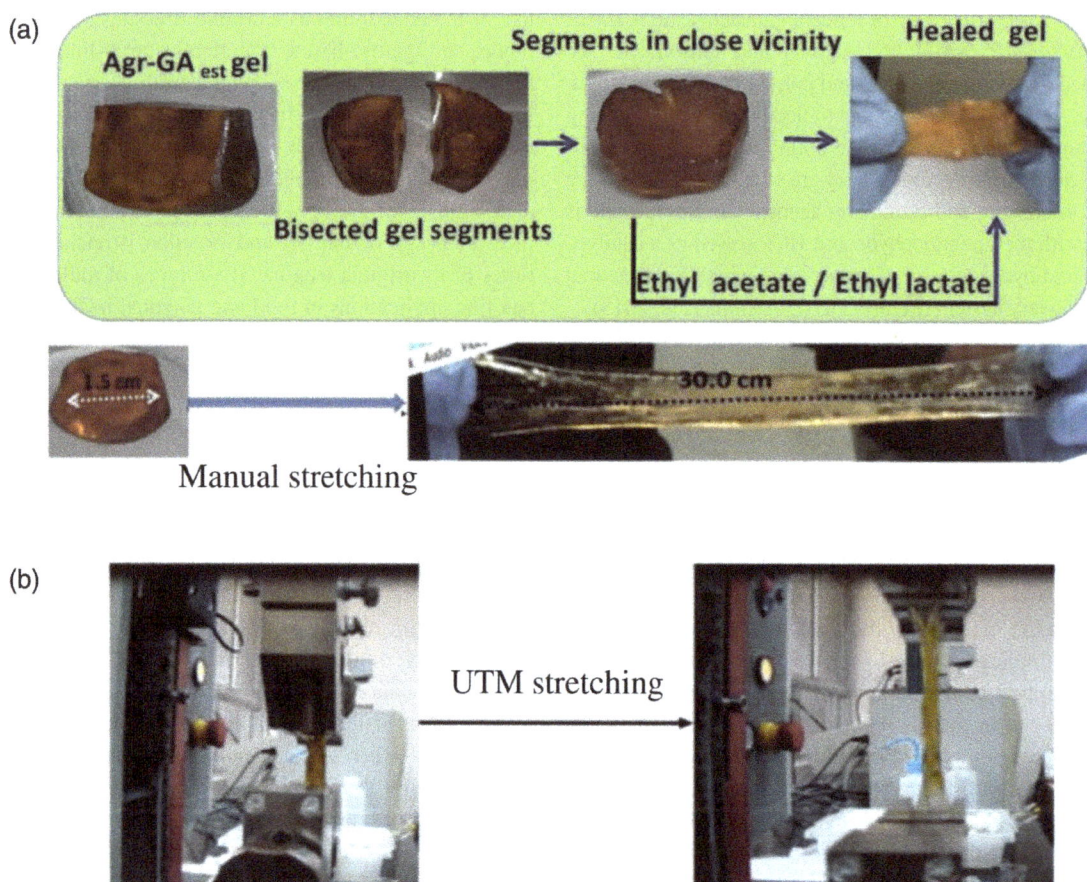

FIGURE 5.6 (a) Pictorial demonstration of self-healing and solvent responsive healing ability of Agr-GAEst (1:0.5) gel prepared in ethylene glycol; (b) pictorial demonstration of the physical state of agarose ester derivative gel prepared in ethylene glycol upon manual stretching and universal tensile testing machine (UTM).

Source: Chaudhary, Chejara et al. (2014).

physically cross-linked hydrogels are reversible and are easier to synthesize than chemical cross-linked hydrogels. Careful selection of hydrocolloid type, concentration, and pH can lead to the formation of a broad range of gel textures. Several methods are available to produce physically cross-linked hydrogels, such as heating/cooling a polymer solution, ionic interaction, complex coacervation, H-bonding, etc. In the heating/cooling polymer

solution, the gel is formed by cooling the solution, which is prepared by heating the polymeric solution. The gel formation is due to the helix formation, an association of the helices, and forming junction zones (Funami et al. 2007). For example, carrageenan forms hydrogels below the melting transition temperature and becomes a solution above it (Figure 5.8) (Meena et al. 2007).

In ionic interaction, ionic polymers are cross-linked by the addition of di- or tri-valent counterions. The principle of the gelling formation is based on the polyelectrolyte solution interacting with the multivalent ion of the opposite charge (Bajpai et al. 2008; Hennink and van Nostrum 2012) (Figure 5.9). The sodium alginate hydrogel shows significantly high permeability in the presence of the divalent ion like Ba^{2+}, and Sr^{2+} then Ca^{+2} (Mørch et al. 2006). In complex coacervates, gelling forms because of mixing polycation with polyanion and forms a soluble or insoluble complex, depending upon the pH and the temperature of the respective solution (Magnin et al. 2004) (Figure 5.9). Marsich et al. (2008) used the complex coacervate gelling technique and used alginate as a polyanion and lactose-modified chitosan as a polycation polysaccharide. This hydrogel shows better mechanical properties than the alginate. Also biochemical and biological studies showed that these 3D scaffolds are able to maintain chondrocyte phenotype and, in particular, to significantly stimulate and promote chondrocyte growth and proliferation.

The hydrogel is also prepared by H-bonding which is one type of physical cross-link. In such a method, the hydrogel is formed by lowering the pH of the aqueous of the carboxyl group containing the polymer. Chaudhary et al. (2018) used such a method for the synthesis of the multifunctional hydrogel (Figures 5.10, 5.11). For this, they used carboxymethyl agarose (CMA) and polyvinyl alcohol (PVA) cross-link by boric acid (BA) for the formation of the hydrogel through hydrogen bonding. This hydrogel has superior versatile properties, such as being stretchable, self-healing, and able to form films and fibers.

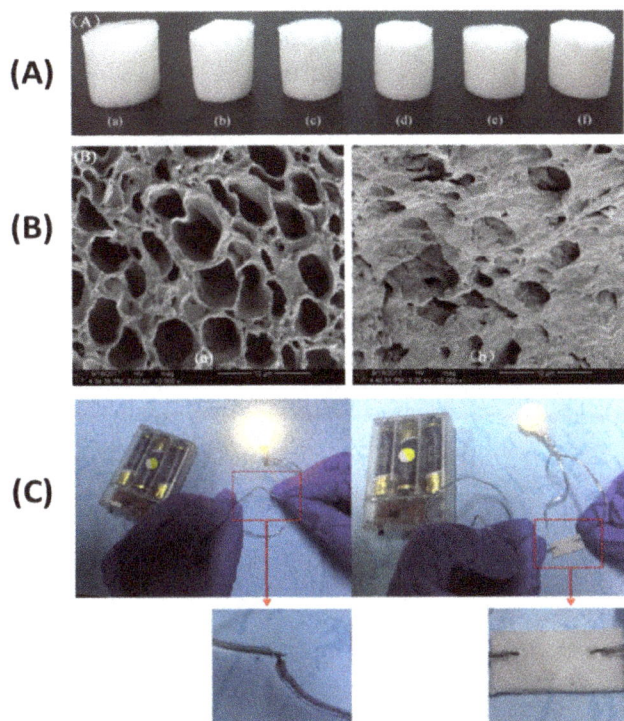

FIGURE 5.7 (A) The appearance change of PVA/SA hydrogel during immersing in saturated NaCl aqueous solution (a) PVA/SA; (b) PVA/SA-90; (c) PVA/SA-120; (d) PVA/SA-150; (e) PVA/SA-180; and (f) PVA/SA-210; (B) the SEM images of the cross-section of (a) PVA/SA and (b) PVA/SA-150; and (C) The conductivity photos of PVA/SA-150 hydrogel.

Source: Jiang et al. (2018).

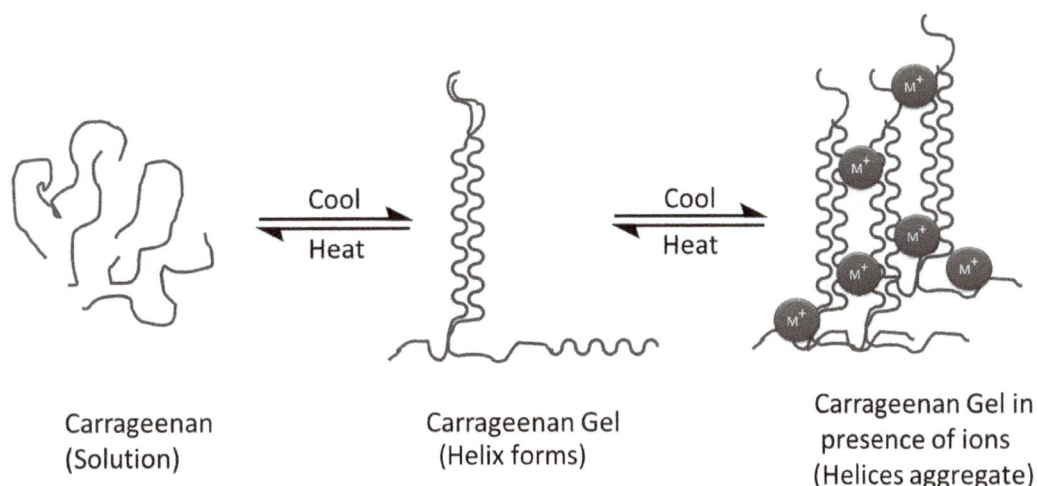

FIGURE 5.8 Carrageenan gel formation on cooling and aggregation of formed helices in present of ions.

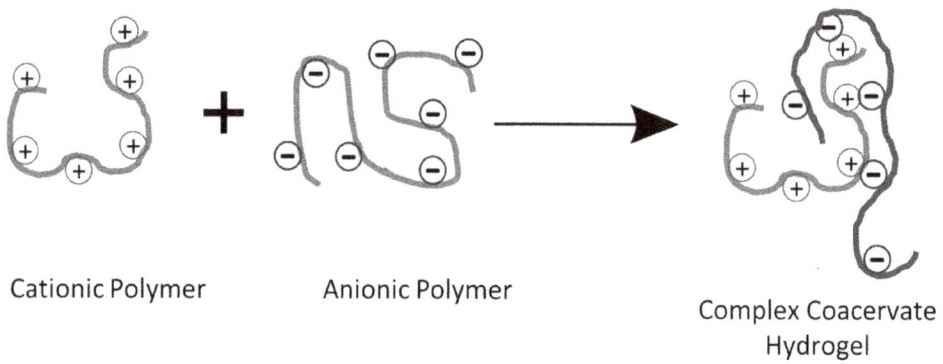

FIGURE 5.9 Complex coacervate hydrogel formation by cationic and anionic polymer.

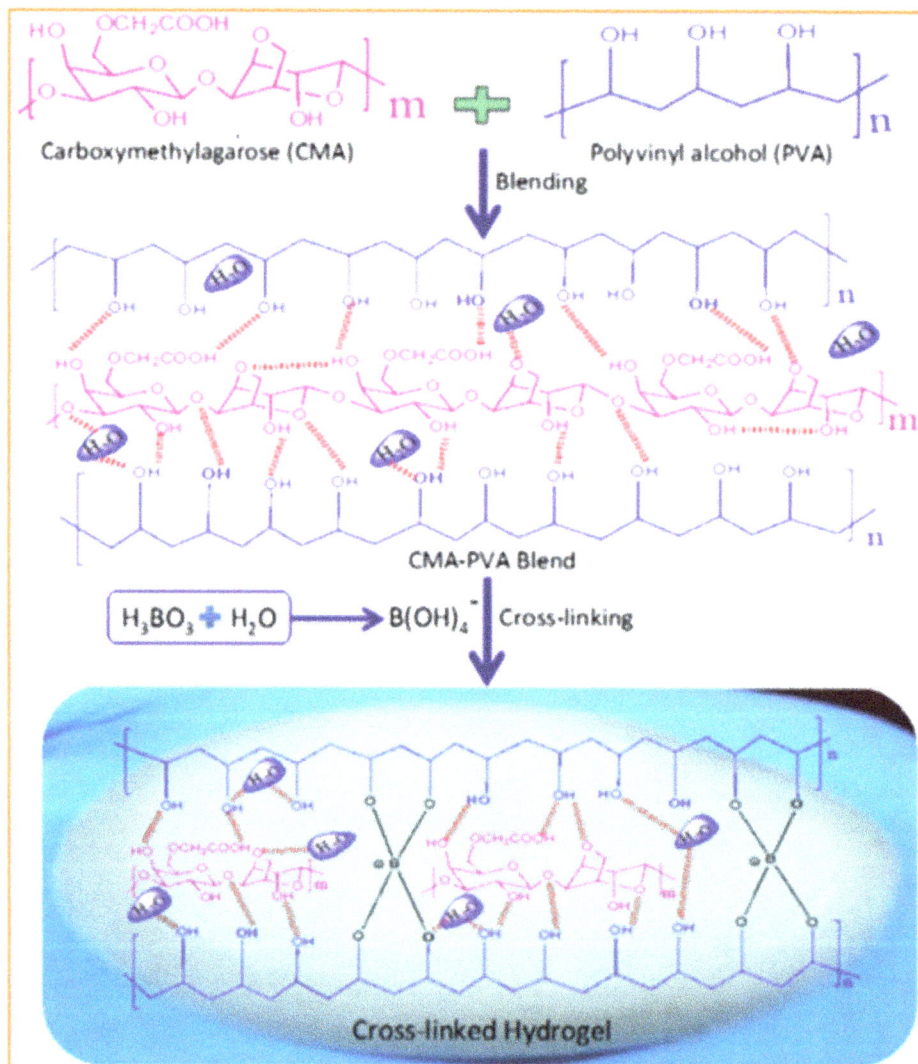

FIGURE 5.10 A plausible mechanism for the preparation of CMA-PVA cross-linked hydrogel and the interactions Inter/intramolecular hydrogen bonding taking place between CMA and PVA during the reaction.

Source: Chaudhary et al. (2018).

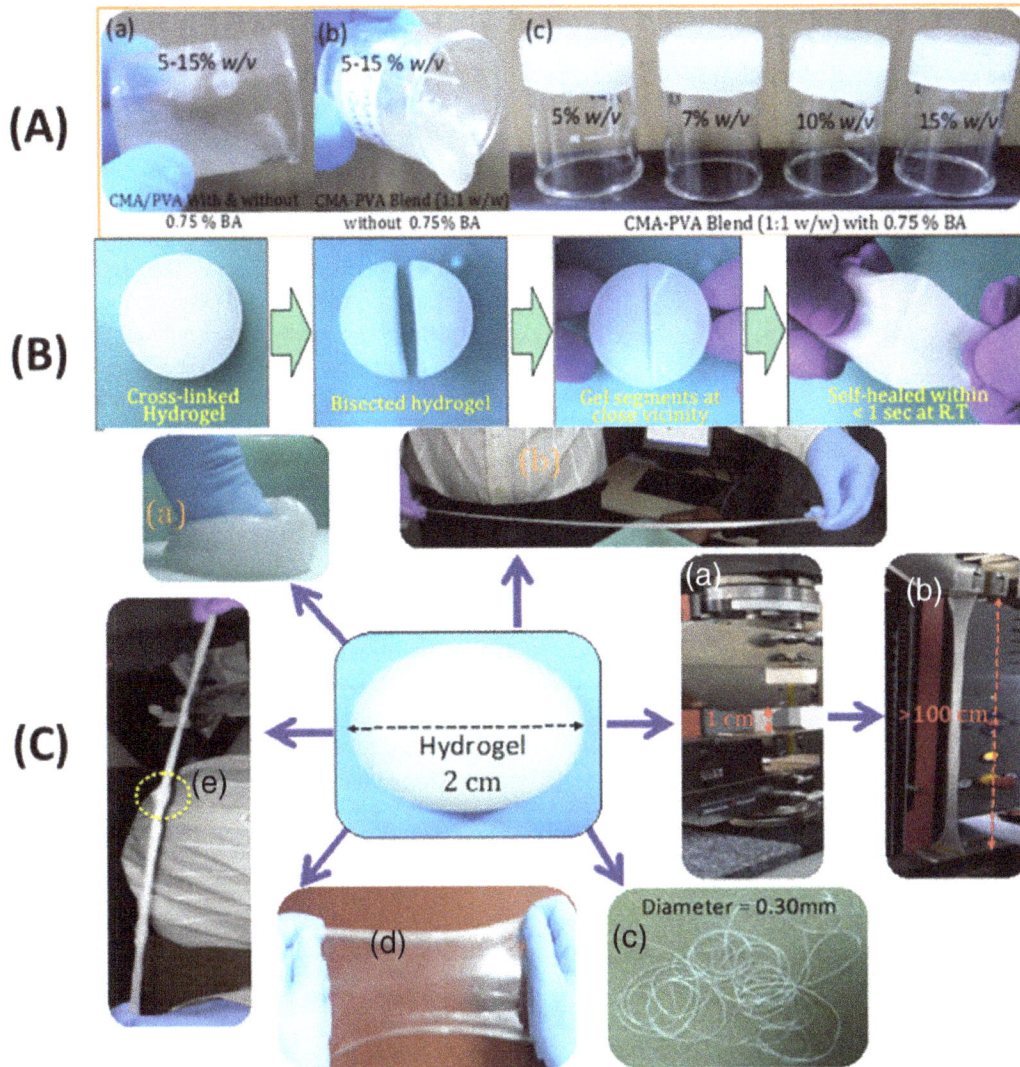

FIGURE 5.11 (A) (a) non-gelling nature of CMA/PVA solutions with or without boric acid; (b) non-gelling nature of CMA-PVA blends without boric acid; (c) gelling nature of CMA-PVA with boric acid cross-linker. (B) Self-healing behavior of hydrogel; and (C) (a) compressible nature (shown in two examples); (b) manual stretching >80 times of original length (c-c00) UTM stretching >100 times of original length (shown in two examples); (c) photographic image of prepared fiber; (d) notches insensitive stretching; (e) film forming nature of cross-linked hydrogel.

Source: Chaudhary et al. (2018).

5.7 Stimuli-Responsive/Smart Hydrogels

A stimuli-responsive polymeric hydrogel system can be made with a responsive polymer or the polymer combining with the responsive compound in which the polymer is used as a template for that compound. In this section, we will focus only on the stimuli-responsive polymer hydrogel system that is also implicated as a smart material. Stimuli are generally classified into three categories: chemical, physical, and biological (Figure 5.12) (Gil and Hudson 2004; Delcea et al. 2011). The chemical stimuli, i.e., electrochemical, pH, ionic strength, solvent, etc. mainly temper a molecular interaction between the polymer and the solvent and/or between polymer chains, while the physical stimuli, i.e., light, temperature, ultrasound, magnetic, mechanical, electrical, etc. modify the dynamics of the polymeric chain (Liechty et al.

2010). Biological stimuli, i.e., proteins, peptide, enzymes, sugar, etc. are used to introduce biodegradability, temperature-induced phase transformation, and the sensitivity of the presence of biologically active compounds (Kopeček 2007; Miyata et al. 1999; Ulbrich et al. 1982; Wang et al. 2001). Furthermore, there are also dual stimuli hydrogel systems, which are responsive to more than one stimulus.

5.7.1 Chemical Stimuli Hydrogels

5.7.1.1 pH-Responsive Hydrogels

pH is an important parameter for biomedical applications as pH changes occur in many specific or pathological compartments. For example, the human body exhibits different pH for different body part, i.e., the pH of blood shows neutral pH 7.3–7.5 while the stomach exhibits acidic pH 1.0–3.0 (Yu et al. 2015). Due to the different pH in tissues, this attracts biomedical and pharmaceutical applications by which pH-responsive hydrogels respond to a

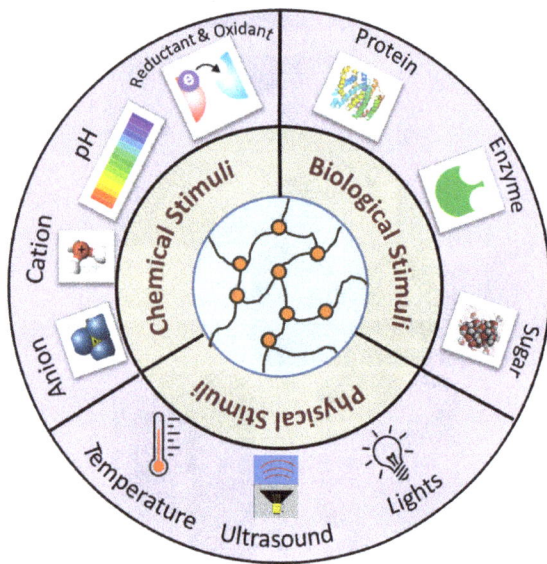

FIGURE 5.12   Classification of the smart/stimuli-responsive hydrogel.

dynamic environment. Furthermore, the pH of the tumor tissue is acidic, so an anti-cancer drug needs pH-responsive material to avoid waste and the side effect of the drug in the human body (Rofstad et al. 2006; Vaupel et al. 1989).

Researchers used synthetic polymer as well as a natural polymer for the preparation of pH-stimuli hydrogel. Synthetic polyacid monomers like acrylic acid, methacrylic acid, maleic anhydride, N,N-dimethylaminoethyl, methacrylate, and sulfonamide-containing polymers have also been used. For example, Horava and Peppas prepared pH-responsive hydrogels based on poly(methacrylic acid)-grafted-poly(ethylene glycol) (P(MAA-g-EG)). This prepared hydrogel is used in the delivery of vehicles of hematological factor IX (FIX) (Figure 5.13) (Horava and Peppas 2016). In other reports, poly(N-isopropylacrylamideco-methacrylic acid) (poly(NIPAAm-co-MA))-based pH-stimuli hydrogel synthesis was used for hydrophobic drugs, such as propranolol, lidocaine or a metoclopramide delivery system (Constantin et al. 2014). Also, methyl methacrylate (MMA) and acrylic acid (AA) monomers, used in the preparation of pH-dependent hydrogels, are also employed in the controlled delivery of small molecular-weight hydrophobic as well as hydrophilic agents.

There are also reports available for the pH-stimuli hydrogels with biopolymers. The main advantage is that the biopolymer is non-toxic and biodegradable (Guo and Kaletunç 2016). Alginate

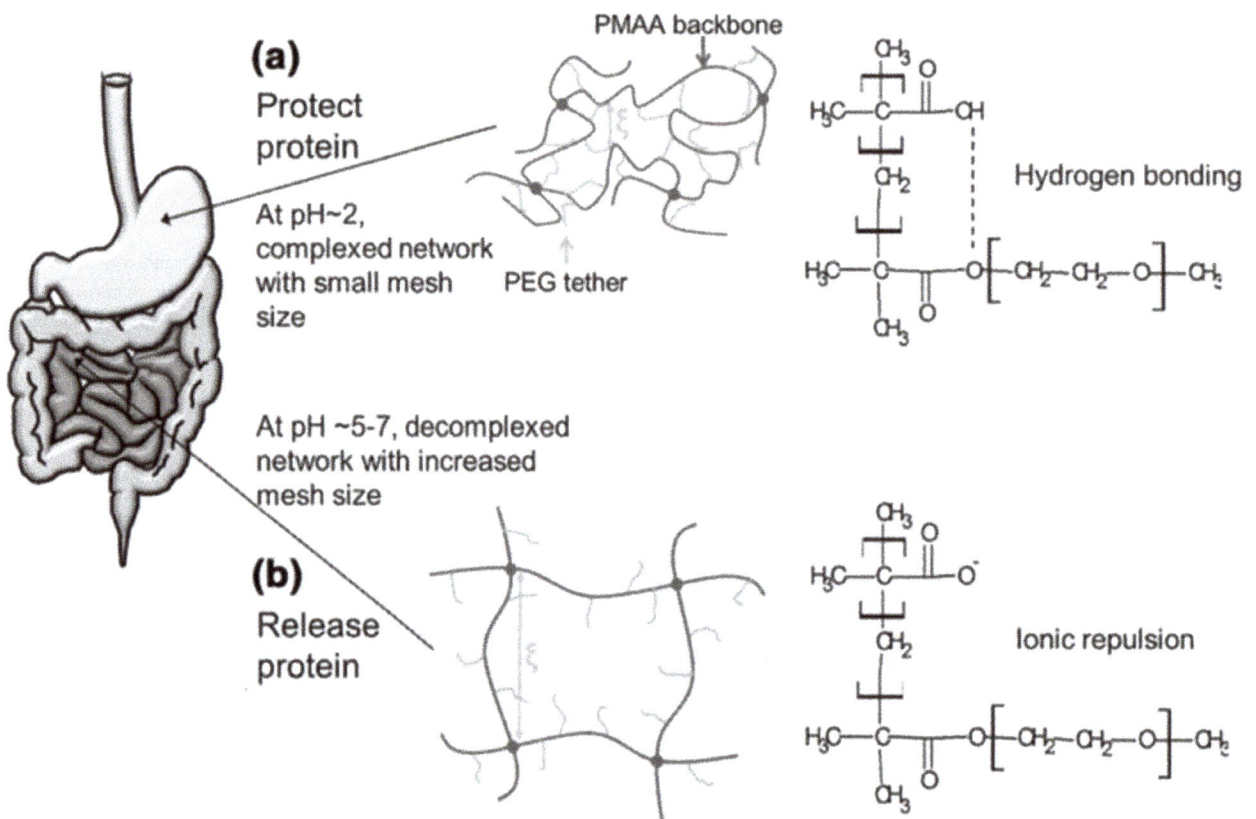

FIGURE 5.13   Complexation hydrogels are engineered to exploit the pH changes in the GI in order to (a) protect proteins from the harsh gastric conditions and (b) deliver them to the small intestine, where they can be absorbed into the bloodstream.

Source; Horava and Peppas (2016).

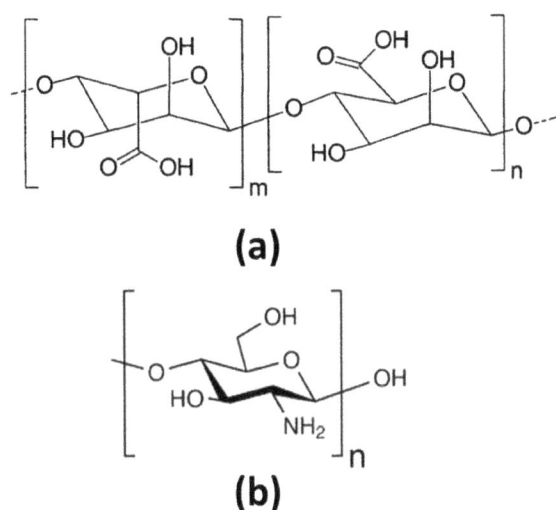

FIGURE 5.14 Structure of (a) alginate; (b) chitosan.

and chitosan-based pH-stimuli hydrogel prepared due to weak ionization shows a pH-responsive phase transition (Figure 5.14). For example, a biopolymer mixture was developed, composed of alginate and pectin that can form hydrogel (Al-P) when the pH is below 3.0, When the pH increases, the hydrogel undergoes solid-to-gel transition and the dissolution of the hydrogel dominates the bioactive compound release in the intestine (ibid.).

In another research study, chitosan was reacted with mono-chloroacetic acid under alkaline conditions to prepare carboxymethyl chitosan (CMCT). This prepared CMCT shows less swelling at pH 1.2 (for the first 2 hours) and quick swelling at pH 6.8 (for the next 3 hours) followed by linear swelling at pH 7.4 (for the next 7 hours) with a slight increase. The in vitro release profile depends on the swelling behavior, i.e., it is pH-dependent. These studies revealed that no chemical change was found in nateglinide during the preparation of hydrogel formulations (Vaghani et al. 2012). In another study, researchers developed novel drug delivery systems with pH-sensitive swelling and drug release properties for localized antibiotic delivery in the stomach by cross-linking chitosan and poly(ethylene oxide) (PEO) in a blend to form a semi-interpenetrating polymer network (semi-IPN). The results of this study suggest that freeze-dried chitosan-PEO semi-IPN could be useful for localized delivery of antibiotics in the acidic environment of the gastric fluid (Patel and Amiji 1996). As another example, researchers prepared an albumin (bio-polysaccharide-based pH-sensitive drug delivery system. Gel, which was made from the albumin solution whose pH was adjusted to a high value, was found to have high pH sensitivity in swelling (Park et al. 1998).

5.7.1.2 Salt- or Ionic Strength-Responsive Hydrogels

Salt or ionic concentration in hydrogel changes the structure of the polymer and makes the polymeric a hydrogel salt or ionic stimuli system (Buenger et al. 2012). Salt-stimuli hydrogel works similarly to pH stimuli. Varying the salt concentration by increasing or decreasing the hydrophobic interaction in the polymeric hydrogel system results in an electrostatic interaction

reduced between the copolymers and the enabling network precipitations (Gil and Hudson 2004; Park and Hoffman 1993). Also, some salts like a sodium phosphate or sodium sulfate buffer reduce the hydrogel swelling which is not due to phase transitions (Buenger et al. 2012). Salt-stimuli hydrogel is a very important term for the biological processes, such as cell locomotion, muscle contraction, and nerve excitation, which involve ionic strength modifications (Horkay et al. 2000). Salt-stimuli hydrogels have not been explored as smart drug delivery systems, compared to pH- and temperature-stimuli hydrogels as the salt concentration in the bulk medium affects the inter- and intramolecular hydrogen-bonding and polar interactions as well as hydrophobic interactions associated with water molecules. Thus, it can be anticipated they will not be suitable for drug release applications (Park and Hoffman 1993; Samchenko et al. 2011).

The literature shows some polyacids, such as poly (acrylic acid) and methacrylic acid, become viscous with the addition of lithium salt due to ionization (Horkay et al. 2000). Polyaspartic acid (PAsp) has a free carboxylic acid group or an amino group which shows high water absorber capacity. Zhao and coworkers studied the effect of salt on the water absorption capacity of polyaspartic acid. The results show the monovalent ion (Na^+) absorption capacity is more than the bivalent ion (Ca^{+2}). This is because the cross-linking density is enhanced in the presence of bivalent Ca^{2+}. The swelling capacity of PAsp reduces in the Ca^{+2} solution and shows reverse properties in the Na^+ solution (Zhao et al. 2005).

Xiao and coworkers prepared shape-transformable hydrogel, which is salt-responsive. They prepared hydrogel from the N-(2-hydroxyethyl) acrylamide (HEAA) cross-link and [2-(methacryloyloxy)ethyl]trimethylammonium chloride (METAC) cross-link with N,N′-methylene bisacrylamide (MBAA) and 3-(1-(4-vinyl benzyl)-1H-imidazole-3-ium-3-yl)propane-1-sulfonate (VBIPS). In this study, they designed poly METAC/HEAA–polyVBIPS bilayer hydrogels to exhibit bidirectional bending in response to salt solutions, salt concentrations, and counterion types. Such bidirectional bending of this bilayer hydrogel is fully reversible and triggered between the salt solution and pure water many times (Figure 5.15) (Xiao et al. 2017).

Lin et al. (2010) studied the salt effect on biopolymers, namely sodium alginate-based hydrogel. In their study, they prepared sodium alginate (ESA), sodium acrylic acid (NaAA), and poly (vinyl alcohol) (PVA) hydrogel by the polymerization method in an aqueous medium. The results show that the swelling rate of the hydrogel is higher in a multivalent anion salt solution than in a monovalent anion salt solution, i.e., $SR_{K2SO4} > SR_{KCl}$ and $SR_{Na2SO4} > SR_{NaCl}$ (ibid.).

5.7.2 Physical Stimuli Hydrogels

5.7.2.1 Temperature-/Thermo-Stimuli Hydrogels

Temperature-/thermo-stimuli hydrogels are explained through their ability to swell and shrink. Gel-to-solid shape changes when the temperature changes in the environment surrounding the hydrogel. The response of the hydrogel to temperature can

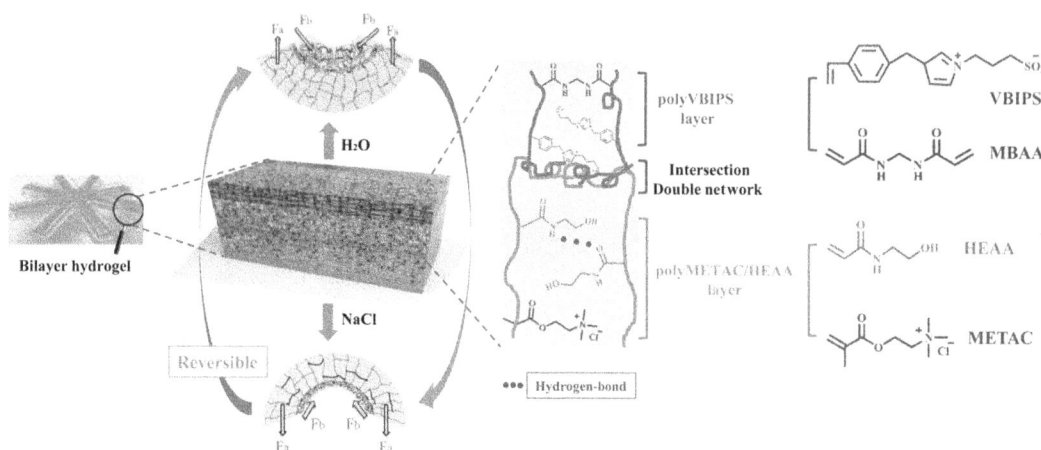

FIGURE 5.15 Schematic illustration of the microstructure and bending behavior of the poly METAC/HEAA–polyVBIPS bilayer hydrogel in water and salt solutions.

Source: Xiao et al. (2017).

be positive as well as negative (Laftah et al. 2011; Richter 2010). The temperature-stimuli hydrogel is the most studied class among all the stimuli class of hydrogels in tissue engineering and drug delivery systems because of the temperature swing in the physiological and pathological conditions (Klouda and Mikos 2008; Lin et al. 2014; Qiu and Park 2001). Thermo-stimuli hydrogels are divided into two classes: (1) lower critical solution temperature (LCST); and (2) upper critical solution temperature (UCST). In the LCST class, the hydrogels exhibit non-linear behavior in which the polymer is soluble upon a decrease in the temperature. On the other hand, the UCST class of hydrogels shows the opposite reaction in which polymer is soluble upon heating.

Liu and coworkers synthesized monomeric fluorophore using *N*-acryl-*N'*-(quinolin-8-yl)thiourea (AQT) and *N*-isopropylacrylamide (NIPAAm) by free radical copolymerization. This prepared polymeric system shows a lower critical solution temperature (LCST) ranging from 32.5–37.5°C. Further, it is used as a sensor for metal cations in aqueous solution (Liu et al. 2008). In another report, a triblock polymer poly [(propylene sulfide)-*block*-(*N,N*-dimethylacrylamide)-*block*-(*N*-isopropylacrylamide)] (PPS-*b*-PDMA-*b*-PNIPAAM) base hydrogel was synthesized by anionic and RAFT polymerization. This prepared hydrogel was in an aqueous solution at room temperature, the triblock polymer dissolved into a clear solution and assembled into stable micelles that, at relatively low concentrations, underwent sharp, reversible thermo-gelation when heated from room to physiologic temperature. The triblock copolymer hydrogels also demonstrated a controlled, sustained, and reactive oxygen species (ROS) concentration-dependent release of the model drug Nile red (Figure 5.16) (Gupta et al. 2014).

Some natural polymers, such as alginate, chitosan, cellulose derivatives, etc., are biodegradable and soluble on heating while making an opaque gel at low temperatures. The main reason for the formation of the gel is due to the polymeric chain being solvated at low temperatures while, on raising the temperature, the polymer molecules lose their hydrating water and have a

strong interaction between polymer chains to form a viscose solution. This transition temperature of polymers is not applicable to the physiological application but some approaches can apply to lower the LCST. For example, the transition temperature of methylcellulose is 40°C while adding NaCl decreases the transition temperature to 32–34°C (Ruel-Gariepy and Leroux 2004; Sarkar 1979; Silva et al. 2008). Wang, Li and coworkers prepared a temperature-responsive chitosan hydrogel and used it as an injectable scaffold to slowly release basic fibroblast growth factor (bFGF) within the ischemic myocardium, as the prepared hydrogel enhanced the positive effects of bFGF on angiogenesis, myocardial fibrosis, ventricular remodeling, and cardiac function in a rat infarction model (Wang, Li, et al. 2010). In another report, Yang and coworkers prepared hemicellulose obtained from acetic acid pulping of eucalyptus and N-isopropylacrylamide (NIPAAm) through UV photo-cross-linking-based temperature-responsive hydrogel. The LCST of the hydrogel increased with increasing the Hce-MA content, and a faster water uptake rate appeared. These types of hydrogel are favorable material for biomedical applications (Figure 5.17) (Yang et al. 2011).

5.7.2.2 Light-Stimuli Hydrogels

Light is the main external physical stimuli which can be used for the tuning of drug release kinetics. Here are some examples of some light-stimuli hydrogels prepared by researchers. Lo and coworkers prepared poly (*N*-isopropylacrylamide) (PNIPAAm) light-stimuli hydrogels integrating with glycidyl methacrylate functionalized graphene oxide (GO–GMA). The prepared hydrogel shows a volumetric change in response to infrared (IR) light illumination, due to the highly efficient photothermal conversion of GO–GMA (Lo et al. 2011). In another report, Shi and coworkers prepared a near-infrared light stimuli hydrogel using poly (N-isopropylacrylamide)/graphene oxide (PNIPAM-GO) by physical cross-linking and ultrahigh tensibility, achieved by adding a very low amount of chemical cross-linker N,N′-methylene bis(acrylamide). The prepared PNIPAM-GO nanocomposite

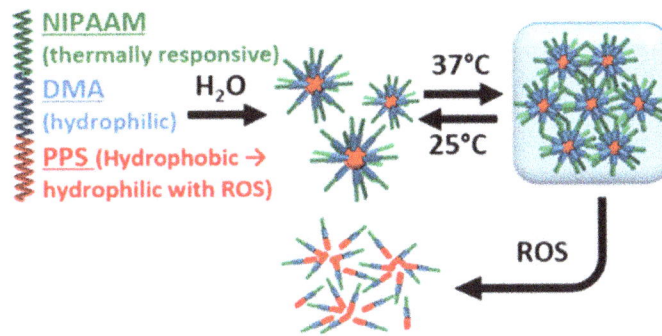

FIGURE 5.16 Temperature stimuli triblock polymer hydrogel.

Source: Gupta et al. (2014).

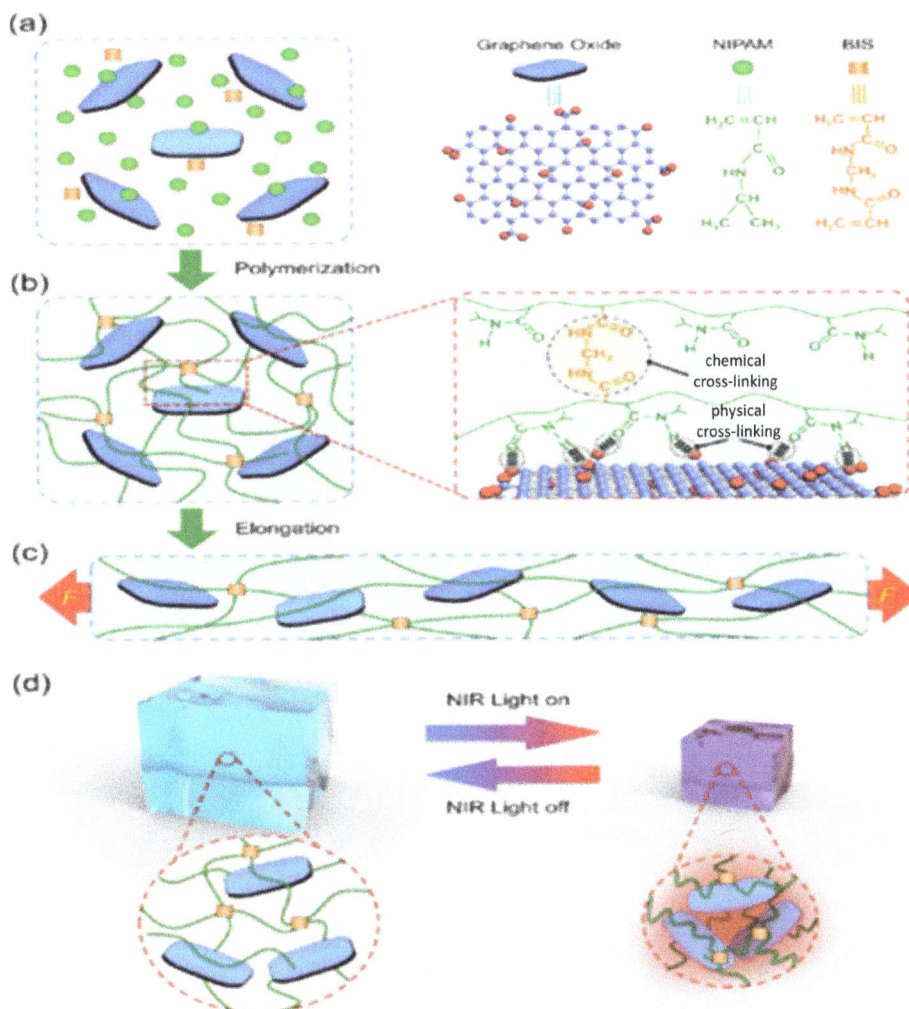

FIGURE 5.17 Schematic illustration of the fabrication process and performance mechanism of the proposed PNIPAM-GO nanocomposite hydrogels. (a) GO nanosheets are homogeneously dispersed in the monomer solution; (b) the PNIPAM-GO nanocomposite hydrogels are formed by both chemical and physical cross-linking, in which the PNIPAM chains are chemically cross-linked by BIS, and the hydrogen bond interactions between GO nanosheets and PNIPAM chains result in the physical cross-linking; (c, d) the PNIPAM-GO hydrogels exhibit ultrahigh tensibility (c) and reversible NIR light-responsive property (d).

Source: Shi et al. (2015).

hydrogels with ultrahigh tensibility exhibit rapid, reversible, and repeatable NIR light-responsive properties, which are highly promising for fabricating remote light-controlled devices, smart actuators, artificial muscles, and so on (Shi et al. 2015).

Furthermore, the photo-switchable cyclodextrin and azobenzene-based hydrogel was prepared after the use of curdlan (β-1,3 glucan, CUR) as a backbone. This hydrogel showed gel formation under vis-light (430 nm) and a solution form under UV-light (365 nm) (Tamesue et al. 2010). Also, plant-origin polyuronic acids, alginate and pectate-based visible-light responsive hydrogels were prepared which are coordinated to Fe (III) ions. The physical changes were observed in gel-to-solid form upon irradiation with visible (405 nm) LED light. This prepared hydrogel is a promising drug release vehicle at 37°C in the biological buffer and a neutral aqueous solution (Giammanco et al. 2015).

5.7.3 Biological Stimuli Hydrogels

For biological stimuli, the hydrogel is sensitive to biomolecules like glucose, enzyme, protein, receptors, and overproduced metabolites in inflammation. Glucose-stimuli hydrogels are responsive to sugars which are used to make a self-regulated system to control insulin in the blood. These types of hydrogels are achieved by incorporating glucose oxidase (GOD) (Albin et al. 1987; Brownlee and Cerami 1979; Ishihara and Matsui 1986). The first GOD immobilized hydrogel was prepared in 1987 by Albin and coworkers (Albin et al. 1987). In glucose-stimuli hydrogels, the GOD is aggregated by a pH-responsive hydrogel in which the blood sugar increase causes the glucose to be drawn out into a membrane where the glucose is converted to gluconic acid. Gluconic acid lowers the pH so that the hydrogel starts swelling and releases the insulin (Seminoff et al. 1989).

Bacteria located in the colon produce enzymes, in particular a reductive enzyme, i.e., azoreductase or hydrolytic, which is capable of degrading polysaccharides, such as pectin, chitosan, amylase/amylopectin, cyclodextrin, and dextrin (Chambin et al. 2006; Sinha and Kumria 2001; Vandamme et al. 2002), so that an enzyme-stimuli hydrogel is mostly used to destroy the polymer or its assembly. Secret and coworkers developed poly (ethylene glycol)-based hydrogel microparticles for pulmonary drug delivery. To make an enzyme-stimuli hydrogel, they used a poly (ethylene glycol) diacrylate (PEGDA) precursor to incorporate peptides in the polymer chain. The peptide is chosen so that it will be cleaved by a collagenase matrix metalloproteinase (MMP-1), as the model enzyme (Secret et al. 2014).

5.8 Properties of Smart Hydrogels

Smart hydrogels are characterized by various properties like swelling, mechanical properties, and toxicity.

5.8.1 Swelling

The hydrogel is formed due to the polymeric chain by cross-linking either physically or chemically and being considered as one molecule. These polymeric chains interact with the solvent and tend to expand to a full solvated state at the same time the cross-linked structure of the hydrogel applies a retractive force to pull the chain inside. Swelling ratio or water content in the hydrogel can be identified by Equations 5.1 or 5.2 (Ebara et al. 2014):

$$Water\ content = \frac{weight\ of\ water}{weight\ of\ water + weight\ of\ gel} \times 100 \ldots$$
$$(5.1)$$

$$Swelling\ ratio = \frac{weight\ of\ swollen\ gel}{weight\ of\ dry\ gel} \ldots \qquad (5.2)$$

The swelling properties of the hydrogel are important in pharmaceutical and biomedical fields since the equilibrium swelling ratio affects the solute diffusion coefficient, the surface wettability, and the mobility and optical and mechanical properties of the hydrogel. The swelling properties are determined by many factors, including the type and composition of the monomers, the cross-linking density, and other environmental factors, such as temperature, pH, and ionic strength (Chirani et al. 2015; Ebara et al. 2014).

5.8.2 Mechanical Properties

The mechanical properties of the hydrogel mainly depend on the structure and the composition of the hydrogel (Shibayama 2012). Mostly, shearing and the compression method are used to evaluate the mechanical properties of the hydrogel instead of stretching due to poor deformability (Vedadghavami et al. 2017). Biopolymers, such as gelatin gels and polysaccharides, have been extensively investigated because of their variety of applications in products such as cosmetics and foods. The mechanical performance of conventional hydrogels can be expressed as an elastic modulus. The elastic modulus can range from kPa to MPa, e.g., gelatin gel and agarose. The mechanical properties of the hydrogel are affected by the comonomer composition, cross-linking density, polymerization conditions, and degree of swelling. Okumura and Ito prepared a hydrogel with unique mechanical and swelling properties through cross-linking polyrotaxane, which consists of PEG threaded with a ring molecule of cyclodextrin (CD). The prepared hydrogel slide-ring gel is also called a topological gel (Okumura and Ito 2001). These unique properties of the hydrogel arise because the CDs are not covalently bonded to the axis polymer, so the cross-links can slide along the axial chain (Gong et al. 2003). The evaluation of the mechanical property is crucial in various biomedical applications, namely, ligament and tendon repair, wound dressing material, the matrix for drug delivery, tissue engineering, and as cartilage replacement material. The mechanical properties of hydrogels should be such that they can maintain their physical texture during the delivery of therapeutic moieties for a predetermined period of time. For the desired mechanical properties of the hydrogel, the degree of cross-linking should vary. By increasing the degree of cross-linking, the strength of the hydrogel increases. A higher degree of cross-linking makes the hydrogel more brittle by decreasing the percentage of elongation. Therefore, the degree of cross-linking for the hydrogel should be optimized to achieve a strong yet elastic hydrogel. Furthermore, the desired mechanical properties of the hydrogel were also accomplished by the researchers using copolymerization of the monomer with the native polymer. This comonomer may attach to the hydrogen bond in the hydrogel. Grassi et al. (2009) determined the mechanical properties of calcium alginate hydrogel. The mechanical characterization consisted of relaxation experiments (normal stress relaxation at constant deformation) to determine the hydrogel linear visco-elastic range and to define the relaxation spectra and the Young modulus by using the generalized Maxwell model. On the basis of the Young modulus and Flory's theory, it was possible to determine the hydrogel's cross-linking density. This value was then used to estimate the average polymeric mesh size according to the equivalent network theory (ibid.).

5.8.3 Rheology

The rheology of the hydrogel mainly depends on the hydrogel structure (e.g., association, entanglement, and cross-links) present in the system. Polymer solutions are essentially viscous at low frequencies and tend to fit the scaling laws: $G' \sim \omega^2$ and $G'' \sim \omega$. At high frequencies, elasticity dominates ($G' > G''$). This corresponds to Maxwell-type behavior with a single relaxation time, which may be determined from the cross-over point, and this relaxation time increases with concentration. Cross-linked microgel dispersions exhibit G' and G'' that are almost independent of oscillation frequency (Meena et al. 2013; Meena et al. 2014).

5.8.4 Surface Properties

The surface of the hydrogel could be rough, smooth, or stepped. Procedures to determine the surface properties of the hydrogel include electron spectroscopy, secondary ion mass spectrometry, scanning electron microscopy (SEM), Fourier transform infrared spectroscopy (FTIR), scanning tunneling microscopy (STM), and atomic force microscopy (AFM). FTIR is a useful technique for investigating the chemical structure of the hydrogel. Comparison of the FTIR spectra with the raw material gives information about the structural arrangement of the hydrogel. The SEM images indicate the surface changes of the hydrogel; they also provide information about the composition and other properties, such as conductivity of the hydrogel. To investigate the network structure of the hydrogel, SEM is a powerful technique and also widely used to visualize the structure.

5.8.5 Biocompatible Properties

Nowadays hydrogels are widely used in the biomedical field so it is important that hydrogel should be biocompatible and non-toxic. Biocompatibility consists basically of two elements: (1) bio-safety, i.e., the appropriate host response must be not only systemic but also local (the surrounding tissue), the absence of cytotoxicity, mutagenesis, and/or carcinogenesis; and (2) bio-functionality, i.e., the ability of a material to perform the specific task for which it is intended. The polymers which are used for the hydrogel must pass the cytotoxicity assay as well as the *in-vivo*

toxicity test. If toxic chemicals are used to prepare hydrogels, the conversion of the toxic chemical should be 100% to pass the *in-vivo* biocompatibility test which is challenging. To remove such toxic chemicals from the hydrogel, various purification methods should be followed, such as solvent washing and/or dialyzing the sample. To overcome that problem, biopolymers should be used instead of those synthetic polymers (Das 2013). So in this section, we focus only on applications of biopolymer-based hydrogels.

5.9 Applications

Hydrogels are able to hold a high water content and also are bio-degradable so that mostly they have been used in the biomedical field, such as wound dressings, drug delivery, agriculture, sanitary pads, as well as trans-dermal systems, dental materials, implants, injectable polymeric systems, ophthalmic applications, or hybrid-type organs (encapsulated living cells). Also, hydrogels have been used for water purification, in the food industries, etc. (see Figure 5.16) (Das and Pal 2015).

Agarose forms a gel at 1.5–2% concentration in an aqueous media which is directly used for electrophoresis, immunoelectro-phoresis, and immunodiffusion. The derivative of the agarose, namely, carboxymethylagarose, is used for the preparation of a self-healing and super-stretchable hydrogel (Chaudhary, Kondaveeti et al. 2014; Chaudhary et al. 2018). Agar/agarose is insoluble in water at room temperature but soluble in hot water and forms a gel on cooling which is thermally reversible (Norziah et al. 2006). This type of hydrogel is useful for quick wound healing (Zhao et al. 2017). Agar is an ionic polysac-charide so it is easily made into a complex with the protein so that it is useful to remove protein impurities from the vine, juice, and vinegar (Laurienzo 2010). Pharmaceutical graded agar is used in molecule microbiology to get DNA information (ibid.). Also, the agar/agarose beads are also investigated for a sustainable drug delivery system (Nakano et al. 1979). The agar and carrageenan are grafted with polyvinylpyrrolidone (PVP) which shows superior properties such as better spreadability and water-holding capability (Prasad et al. 2006). This prepared grafted hydrogel could be used as moisturizer formulations and as active carriers of drugs. Also such a gel is useful in dressings in bio-medical applications (Lugão et al. 1998). Agar/agarose gel is also used in the food industries for gel formation and food gums, as well as food additives, thanks to its properties as an emulsifying and gelling agent (Suleria et al. 2015).

An alginate structure can easily entrap a biological agent like a drug, protein, DNA, etc., so it is used for loading and release of such molecules without them losing their biological activity because of the relatively mild gelation process (D'Ayala et al. 2008). Alginate can produce a gel in the presence of the metal ion like Ca^{+2}, Zn^{+2}, and Ba^{+2}, which is widely used for the microencapsulation of drugs (Russo et al. 2007). Also, there is one report on the alginate in which alginate interacts with poly (ethylene glycol) (PEG). This formed hydrogel exhibited superior properties such as protein resistance, low toxicity, and immuno-genicity (Han et al. 1989). The chitosan and PEG alginate-based hydrogel encapsulates biomolecules such as albumin and hirudin, which is a good carrier for the oral delivery of such molecules (Chandy et al. 1998).

Acknowledgments

We thank the Science and Engineering Research Board (SERB), DST, the Government of India (EMR/2016/004944); CSIR (for awarding CSIR-SRF to FK), and CSIR-CSMCRI (OLP-088) for financial support. Thanks also to the Analytical Discipline and Centralized Instrumental Facilities for providing instrumentation facilities.

REFERENCES

Aguilar, M.R. and San Román, J. (2014). "Introduction to smart polymers and their applications." In *Smart Polymers and Their Applications*. Cambridge: Woodhead Publishing, pp. 1–11.

Ahmed, E.M. (2015). "Hydrogel: Preparation, haracterization, and applications: A review." *Journal of Advanced Research* 6(2): 105–121.

Albin, G.W., Horbett, T.A., Miller, S.R., and Ricker, N.L. (1987). "Theoretical and experimental studies of glucose sensitive membranes." *Journal of Controlled Release* 6(1): 267–291.

Bajpai, A.K., Shukla, S.K., Bhanu, S., and Kankane, S. (2008). "Responsive polymers in controlled drug delivery." *Progress in Polymer Science* 33(11): 1088–1118.

Brownlee, M. and Cerami, A. (1979). "A glucose-controlled insulin-delivery system: semisynthetic insulin bound to lectin." *Science* 206(4423): 1190–1191.

Buenger, D., Topuz, F., and Groll, J. (2012). "Hydrogels in sensing applications." *Progress in Polymer Science* 37(12): 1678–1719.

Chambin, O., Dupuis, G., Champion, D., Voilley, A., and Pourcelot, Y. (2006). "Colon-specific drug delivery: Influence of solution reticulation properties upon pectin beads performance." *International Journal of Pharmaceutics* 321(1–2): 86–93.

Chandy, T., Mooradian, D.L., and Rao, G.H.R. (1998). "Chitosan/polyethylene glycol–alginate microcapsules for oral delivery of hirudin." *Journal of Applied Polymer Science* 70(11): 2143–2153.

Chang, C., Zhang, L., Zhou, J., Zhang, L., and Kennedyv J.F. (2010). "Structure and properties of hydrogels prepared from cellulose in NaOH/urea aqueous solutions." *Carbohydrate Polymers* 82(1): 122–127.

Chaudhary, J.P., Chejara, D.R., Eswaran, K., Meena, R., and Ghosh, P.K. (2015). "Seaweed-derived polymeric materials for multi applications including marine algal cultivation." *RSC Advances* 5(25): 19426–19431.

Chaudhary, J.P., Chejara, D.R., Makwana, D., Prasad, K., and Meena, R. (2014). "Agarose-based multifunctional materials: Evaluation of thixotropy, self-healability, and stretchability." *Carbohydrate Polymers* 114: 306–311.

Chaudhary, J.P., Kholiya, F., Vadodariya, N., Budheliya, V.M., Gogda, A., and Meena, R. (2018). "Carboxymethylagarose-based multifunctional hydrogel with super stretchable, self-healable having film and fiber forming properties." *Arabian Journal of Chemistry*. doi: https://doi.org/10.1016/j.arabjc.2017.12.034.

Chaudhary, J.P., Kondaveeti, S., Gupta, V., Prasad, K., and Meena, R. (2014). "Preparation and functional evaluation of agarose derivatives." *Journal of Applied Polymer Science* 131(16): 40630.

Chaudhary, J.P., Kumar, A., Paul, P., and Meena, R. (2015). "Carboxymethylagarose-AuNPs generated through green route for selective detection of Hg$_2$+ in aqueous medium with a blue shift." *Carbohydrate Polymers* 117: 537–542.

Chaudhary, J.P., Nataraj, S.K., Gogda, A., and Meena, R. (2014). "Bio-based superhydrophilic foam membranes for sustainable oil-water separation." *Green Chemistry* 16(10): 4552–4558.

Chaudhary, J. P., Siddhanta, A. K., and Meena R. (2014). "Microwave assisted synthesis of chitosan-polyuronic acid adducts and their drug release performance." 20th ISCB International Conference (ISCBC-2014) on Chemistry and medicinal plants in translational medicine for healthcare, Delhi University.

Chaudhary, J.P., Vadodariya, N., Nataraj, S.K., and Meena, R. (2015). "Chitosan-based aerogel membrane for robust oil-in-water emulsion separation." *ACS Applied Materials & Interfaces* 7(44): 24957–24962.

Chen, X., Martin, B.D., Neubauer, T.K., Linhardt, R.J., Dordick, J.S., and Rethwisch, D.G. (1995). "Enzymatic and chemoenzymatic approaches to synthesis of sugar-based polymer and hydrogels." *Carbohydrate Polymers* 28(1): 15–21.

Chirani, N., Yahia, L., Gritsch, L., Motta, F.L., Chirani, S., and Faré, S. (2015). "History and applications of hydrogels." *Journal of Biomedical Sciences* 4(2:13): 1–23.

Constantin, M., Bucatariu, S., Harabagiu, V., Popescu, I., Ascenzi, P., and Fundueanu, G. (2014). "Poly(N-isopropylacrylamide-co-methacrylic acid) pH/thermo-responsive porous hydrogels as self-regulated drug delivery system." *European Journal of Pharmaceutical Sciences* 62: 86–95.

Das, D. and Pal, S. (2015). "Modified biopolymer-dextrin based crosslinked hydrogels: Application in controlled drug delivery." *RSC Advances* 5(32): 25014–25050.

Das, N. (2013). "Preparation methods and properties of hydrogel: A review." *International Journal of Pharmacy and Pharmaceutical Sciences* 5(3): 112–117.

D'Ayala, G.G., Malinconico, M., and Laurienzo, P. (2008). "Marine derived polysaccharides for biomedical applications: Chemical modification approaches." *Molecules* 13(9): 2069–2106.

Delcea, M., Möhwald, H., and Skirtach, A.G. (2011). "Stimuli-responsive LbL capsules and nanoshells for drug delivery." *Advanced Drug Delivery Reviews* 63(9): 730–747.

Distantina, S., Rochmadi, R., Fahrurrozi, M., and Wiratni, W. (2013). "Preparation and characterization of glutaraldehyde-crosslinked kappa carrageenan hydrogel." *Engineering Journal* 17(3): 57–66.

Ebara, M., Kotsuchibashi, Y., Uto, K., Aoyagi, T., Kim, Y-J., Narain, R., Idota, N., and Hoffman, J.M. (2014). "Smart hydrogels." In *Smart Biomaterials*. NIMS Monograph. Tokyo: Springer, pp. 9–65.

Funami, T., Hiroe, M., Noda, S., Asai, I., Ikeda, S., and Nishinari, K. (2007). "Influence of molecular structure imaged with atomic force microscopy on the rheological behavior of carrageenan aqueous systems in the presence or absence of cations." *Food Hydrocolloids* 21(4): 617–629.

Giammanco, G.E., Sosnofsky, C.T., and Ostrowski, A.D. (2015). "Light-responsive iron(III)–polysaccharide coordination hydrogels for controlled delivery." *ACS Applied Materials & Interfaces* 7(5): 3068–3076.

Gil, E.S. and Hudson, S.M. (2004). "Stimuli-responsive polymers and their bioconjugates." *Progress in Polymer Science* 29(12): 1173–1222.

Gong, J.P., Katsuyama, Y., Kurokawa, T., and Osada, Y. (2003). "Double-network hydrogels with extremely high mechanical strength." *Advanced Materials* 15(14): 1155–1158.

Gopi, S., Amalraj, A., and Thomas, S. (2016). "Effective drug delivery system of biopolymers based on nanomaterials and hydrogels – a review." *Drug Designing: Open Access* 5: 129.

Grassi, M., Chiara, S., Danilo, P., Tommasina, C., Romano, L., and Grassi, G. (2009). "Structural characterization of calcium alginate matrices by means of mechanical and release tests." *Molecules* 14(8): 3003–3017.

Gulrez, S.K.H, Al-Assaf, S., and Phillips, G.O. (2011). "Hydrogels: Methods of preparation, characterisation, and applications." In A. Carpi (Ed.), *Progress in Molecular and Environmental Bioengineering - From Analysis and Modeling to Technology Applications*. Rijeka: InTech.

Guo, J. and Kaletunç, G. (2016). "Dissolution kinetics of pH-responsive alginate-pectin hydrogel particles." *Food Research International* 88: 129–139.

Gupta, M.K., Martin, J.R., Werfel, T.A., Shen, T., Page, J.M., and Duvall, C.L. (2014). "Cell protective, ABC triblock polymer-based thermoresponsive hydrogels with ROS-triggered degradation and drug release." *Journal of the American Chemical Society* 136(42): 14896–14902.

Hamidi, M., Azadi, A., and Rafiei, P. (2008). "Hydrogel nanoparticles in drug delivery." *Advanced Drug Delivery Reviews* 60(15): 1638–1649.

Han, D.K., Park, K.D., Ahn, K-D., Jeong, S.Y., and Kim, Y.H. (1989). "Preparation and surface characterization of PEO-grafted and heparin-immobilized polyurethanes." *Journal of Biomedical Materials Research* 23(A1): 87–104.

Hennink, W.E. and van Nostrum, C.F. (2012). "Novel crosslinking methods to design hydrogels." *Advanced Drug Delivery Reviews* 64: 223–236.

Horava, S.D. and Peppas, N.A. (2016). "Design of pH-responsive biomaterials to enable the oral route of hematological factor IX." *Annals of Biomedical Engineering* 44(6): 1970–1982.

Horkay, F., Tasaki, I., and Basser, P.J. (2000). "Osmotic swelling of polyacrylate hydrogels in physiological salt solutions." *Biomacromolecules* 1(1): 84–90.

Iizawa, T., Taketa, H., Maruta, M., Ishido, T., Gotoh, T., and Sakohara, S. (2007), "Synthesis of porous poly(N-isopropylacrylamide) gel beads by sedimentation polymerization and their morphology." *Journal of Applied Polymer Science* 104: 842–850.

Ishihara, K. and Matsui, K. (1986). "Glucose-responsive insulin release from polymer capsule." *Journal of Polymer Science Part C: Polymer Letters* 24(8): 413–417.

Jiang, X., Xiang, N., Zhang, H., Sun, Y., Lin, Z., and Hou, L. (2018). "Preparation and characterization of poly(vinyl alcohol)/sodium alginate hydrogel with high toughness and electric conductivity." *Carbohydrate Polymers* 186: 377–383.

Kaihara, S., Matsumura, S., and Fisher, J.P. (2008). "Synthesis and characterization of cyclic acetal based degradable hydrogels." *European Journal of Pharmaceutics and Biopharmaceutics* 68(1): 67–73.

Kashyap, N., Kumar, N., and Ravi Kumar, M.N.V. (2005). "Hydrogels for pharmaceutical and biomedical applications." 22(2): 107–150.

Kiyonaka, S., Sada, K., Yoshimura, I., Shinkai, S., Kato, N., and Hamachi, I. (2003). "Semi-wet peptide/protein array using supramolecular hydrogel." *Nature Materials* 3: 58–64.

Klouda, L. and Mikos, A.G. (2008). "Thermoresponsive hydrogels in biomedical applications." *European Journal of Pharmaceutics and Biopharmaceutics* 68(1): 34–45.

Kopeček, J. (2002). "Swell gels." *Nature* 417: 388–391.

Kopeček, J. (2007). "Hydrogel biomaterials: A smart future?" *Biomaterials* 28(34): 5185–5192.

Krsko, P., McCann, T.E., Thach, T-T., Laabs, T.L., Geller, H.M., and Libera, M.R. (2009). "Length-scale mediated adhesion and directed growth of neural cells by surface-patterned poly(ethylene glycol) hydrogels." *Biomaterials* 30(5): 721–729.

Laftah, W.A., Hashim, S., and Ibrahim, A.N. (2011). "Polymer hydrogels: A review." *Polymer-Plastics Technology and Engineering* 50(14): 1475–1486.

Laurienzo, P. (2010). "Marine polysaccharides in pharmaceutical applications: An overview." *Marine Drugs* 8(9): 2435–2465.

Lee, K.Y. and Mooney, D.J. (2001). "Hydrogels for tissue engineering." *Chemical Reviews* 101(7): 1869–1880.

Liechty, W.B., Kryscio, D.R., Slaughter, B.V., and Peppas, N.A. (2010). "Polymers for drug delivery systems." *Annual Review of Chemical and Biomolecular Engineering* 1: 149–173.

Lin, H., Zhou, J., Yingde, C., and Gunasekaran, S. (2010). "Synthesis and characterization of pH- and salt-responsive hydrogels based on etherificated sodium alginate." *Journal of Applied Polymer Science* 115(6): 3161–3167.

Lin, Z., Gao W., Hu, H., Ma, K., He, B., Dai, W., Wang, X., Wang, J., Zhang, X., and Zhang, Q. (2014). "Novel thermo-sensitive hydrogel system with paclitaxel nanocrystals: High drug-loading, sustained drug release and extended local retention guaranteeing better efficacy and lower toxicity." *Journal of Controlled Release* 174: 161–170.

Liu, Y., Meng, L., Lu, X., Zhang, L., and He, Y. (2008). "Thermo and pH-sensitive fluorescent polymer sensor for metal cations in aqueous solution." *Polymers for Advanced Technologies* 19(2): 137–143.

Lo, C-W., Zhu, D., and Jiang, H. (2011). "An infrared-light responsive graphene-oxide incorporated poly(N-isopropylacrylamide) hydrogel nanocomposite." *Soft Matter* 7(12): 5604–5609.

López-Simeon, R., Campos-Terán, J., Beltrán, H.I., and Hernández-Guerrero, M. (2012). "Free-lignin cellulose obtained from agar industry residues using a continuous and minimal solvent reaction/extraction methodology." *RSC Advances* 2(32): 12286–12297.

Lugão, A.B., Machado, L.D.B., Miranda, L.F., Alvarez, M.R., and Rosiak, J.M. (1998). "Study of wound dressing structure and hydration/dehydration properties." *Radiation Physics and Chemistry* 52(1): 319–322.

Lutolf, M.P. (2009). "Spotlight on hydrogels." *Nature Materials* 8: 451.

Magnin, D., Lefebvre, J., Chornet, E., and Dumitriu, S. (2004). "Physicochemical and structural characterization of a polyionic matrix of interest in biotechnology, in the pharmaceutical and biomedical fields." *Carbohydrate Polymers* 55(4): 437–453.

Maolin, Z., Jun, L., Min Y., and Hongfei, H. (2000). "The swelling behavior of radiation prepared semi-interpenetrating polymer networks composed of polyNIPAAm and hydrophilic polymers." *Radiation Physics and Chemistry* 58(4): 397–400.

Marsich, E., Borgogna, M., Donati, I., Mozetic, P., Strand, B.L., Gomez Salvador, S., Vittur, F., and Paoletti, S. (2008). "Alginate/lactose-modified chitosan hydrogels: A bioactive biomaterial for chondrocyte encapsulation." *Journal of Biomedical Materials Research Part A* 84A(2): 364–376.

Meena, R., Lehnen, R., and Saake, B. (2014). "Microwave-assisted synthesis of kC/Xylan/PVP-based blend hydrogel materials: physicochemical and rheological studies." *Cellulose* 21(1): 553–568.

Meena, R., Lehnen, R., Schmitt, U., and Saake, B. (2013). "Physicochemical and rheological properties of agarose/xylans composite hydrogel materials." *Polymer Composites* 34(6): 978–988.

Meena, R., Prasad, K., and Siddhanta, A.K. (2007). "Effect of genipin, a naturally occurring crosslinker on the properties of kappa-carrageenan." *International Journal of Biological Macromolecules* 41(1): 94–101.

Miguel, S.P., Ribeiro, M.P., Brancal, H., Coutinho, P., and Correia, I.J. (2014). "Thermoresponsive chitosan–agarose hydrogel for skin regeneration." *Carbohydrate Polymers* 111: 366–373.

Miyata, T., Asami, N., and Uragami, T. (1999). "A reversibly antigen-responsive hydrogel." *Nature* 399(6738): 766.

Mondal, D., Sharma, M., Maiti, P., Prasad, K., Meena, R., Siddhanta, A.K., Bhatt, P., Ijardar, S., Mohandas, V.P., and Ghosh, A. (2013). "Fuel intermediates, agricultural nutrients and pure water from *Kappaphycus alvarezii* seaweed." *RSC Advances* 3(39): 17989–17997.

Mørch, Ý.A., Donati, I., and Strand, B.L. (2006). "Effect of Ca2+, Ba2+, and Sr2+ on alginate microbeads." *Biomacromolecules* 7(5): 1471–1480.

Nakano, M., Nakamura, Y., Takikawa, K., Kouketsu, M., and Arita, T. (1979). "Sustained release of sulphamethizole from agar beads." *Journal of Pharmacy and Pharmacology* 31(1): 869–872.

Norziah, M.H., Foo, S.L., and Abd Karim, A. (2006). "Rheological studies on mixtures of agar (*Gracilaria changii*) and κ-carrageenan." *Food Hydrocolloids* 20(2): 204–217.

Okumura, Y. and Ito, K. (2001). "The polyrotaxane gel: A topological gel by figure-of-eight cross-links." *Advanced Materials* 13: 485–487.

Park, H.-Y., Song, I-H., Kim, J-H., and Kim, W-S. (1998). "Preparation of thermally denatured albumin gel and its pH-sensitive swelling." *International Journal of Pharmaceutics* 175(2): 231–236.

Park, J.H. and Kim, D. (2001). "Preparation and characterization of water-swellable natural rubbers." *Journal of Applied Polymer Science* 80: 115–121.

Park, T.G. and Hoffman, A.S. (1993). "Sodium chloride-induced phase transition in nonionic poly(N-isopropylacrylamide) gel." *Macromolecules* 26(19): 5045–5048.

Patel, V.R. and Amiji, M.M. (1996). "Preparation and characterization of freeze-dried chitosan-poly(ethylene oxide) hydrogels for site-specific antibiotic delivery in the stomach." *Pharmaceutical Research* 13(4): 588–593.

Plunkett, K.N. and Moore, J.S. (2004). "Patterned dual pH-responsive core-shell hydrogels with controllable swelling kinetics and volumes." *Langmuir* 20(16): 6535–6537.

Prasad, K., Mehta, G., Meena, R., and Siddhanta, A.K. (2006). "Hydrogel-forming agar-graft-PVP and κ-carrageenan-graft-PVP blends: Rapid synthesis and characterization." *Journal of Applied Polymer Science* 102(4): 3654–3663.

Qiu, Y. and Park, K. (2001). "Environment-sensitive hydrogels for drug delivery." *Advanced Drug Delivery Reviews* 53(3): 321–339.

Richter, A. (2010). "Hydrogels for actuators." In G. Gerlach and K-F. Arndt (Eds.), *Hydrogel Sensors and Actuators: Engineering and Technology*. Berlin: Springer, pp. 221–248.

Rofstad, E.K., Mathiesen, B., Kindem, K., and Galappathi, K. (2006). "Acidic extracellular pH promotes experimental metastasis of human melanoma cells in athymic nude mice." *Cancer Research* 66(13): 6699–6707.

Roy, D., Cambre, J.N., and Sumerlin, B.S. (2010). "Future perspectives and recent advances in stimuli-responsive materials." *Progress in Polymer Science* 35(1): 278–301.

Ruel-Gariepy, E. and Leroux, J-C. (2004). "In situ-forming hydrogels—review of temperature-sensitive systems." *European Journal of Pharmaceutics and Biopharmaceutics* 58(2): 409–426.

Russo, R., Malinconico, M., and Santagata, G. (2007). "Effect of cross-linking with calcium ions on the physical properties of alginate films." *Biomacromolecules* 8 (10): 3193–3197.

Samchenko, Y., Ulberg, Z., and Korotych, O. (2011). "Multipurpose smart hydrogel systems." *Advances in Colloid and Interface Science* 168(1): 247–262.

Sarkar, N. (1979). "Thermal gelation properties of methyl and hydroxypropyl methylcellulose." *Journal of Applied Polymer Science* 24(4): 1073–1087.

Saul, J.M. and Williams, D.F. (2011). "Hydrogels in regenerative medicine." In R. Lanza, J.A. Thomson and R. Nerem (Eds.), *Principles of Regenerative Medicine*, 2nd edn. San Diego, CA: Academic Press, pp. 637–661.

Saxena, A.K. (2010). "Synthetic biodegradable hydrogel (PleuraSeal) sealant for sealing of lung tissue after thoracoscopic resection." *The Journal of Thoracic and Cardiovascular Surgery* 139(2): 496–497.

Secret, E., Kelly, S.J., Crannell, K.E., and Andrew, J.S. (2014). "Enzyme-responsive hydrogel microparticles for pulmonary drug delivery." *ACS Applied Materials & Interfaces* 6(13): 10313–10321.

Seminoff, L.A., Olsen, G.B., and Kim, S.W. (1989). "A self-regulating insulin delivery system. I. Characterization of a synthetic glycosylated insulin derivative." *International Journal of Pharmaceutics* 54(3): 241–249.

Shi, K., Liu, Z., Wei, Y-Y., Wang, W., Ju, X-J., Xie, R., and Chu, L-Y. (2015). "Near-infrared light-responsive poly (N-isopropylacrylamide)/graphene oxide nanocomposite hydrogels with ultrahigh tensibility." *ACS Applied Materials & Interfaces* 7(49): 27289–27298.

Shibayama, M. (2012). "Structure-mechanical property relationship of tough hydrogels." *Soft Matter* 8(31): 8030–8038.

Sikareepaisan, P., Ruktanonchai, U., and Supaphol, P. (2011). "Preparation and characterization of asiaticoside-loaded alginate films and their potential for use as effectual wound dressings." *Carbohydrate Polymers* 83(4): 1457–1469.

Silva, S.M.C., Pinto, F.V., Antunes, F.E., Miguel, M.G., Sousa, J.S., and Pais, A.A.C.C. (2008). "Aggregation and gelation in hydroxypropyl methyl cellulose aqueous solutions." *Journal of Colloid and Interface Science* 327(2): 333–340.

Singh, A., Sharma, P.K., Garg, V.K., and Garg, G. (2010). "Hydrogels: A review." *International Journal of Pharmaceutical Sciences Review and Research* 4(2): 97–105.

Singh, D., Singh, A., and Singh, R. (2015). "Polyvinyl pyrrolidone/carrageenan blend hydrogels with nanosilver prepared by gamma radiation for use as an antimicrobial wound dressing." *Journal of Biomaterials Science, Polymer Edition* 26(17): 1269–1285.

Sinha, V.R. and Kumria, R. (2001). "Polysaccharides in colon-specific drug delivery." *International Journal of Pharmaceutics* 224(1–2): 19–38.

Stamatialis, D.F., Bernke, J. Papenburg, M.G., Saiful, S., Srivatsa, N.M., Bettahalli, S.S., and Wessling, M. (2008). "Medical applications of membranes: Drug delivery, artificial organs, and tissue engineering." *Journal of Membrane Science* 308(1): 1–34.

Stuart, M.A.C., Huck, W.T.S., Genzer, J., Müller, M., Ober, C., Stamm, M., Sukhorukov ,GB, Szleifer, I., Tsukruk, V.V., ... and Minko, S. (2010). "Emerging applications of stimuli-responsive polymer materials." *Nature Materials* 9: 101.

Suleria, H.A.S., Osborne, S., Masci, P., and Gobe, G. (2015). "Marine-based nutraceuticals: An innovative trend in the food and supplement industries." *Marine Drugs* 13(10): 6336.

Tamesue, S., Takashima, Y., Yamaguchi, H., Shinkai, S., and Harada, A. (2010). "Photoswitchable supramolecular hydrogels formed by cyclodextrins and azobenzene polymers." *Angewandte Chemie* 122(41): 7623–7626.

Ulbrich, K., Strohalm, J., and Kopeček, J. (1982). "Polymers containing enzymatically degradable bonds. VI. Hydrophilic gels cleavable by chymotrypsin." *Biomaterials* 3(3):150–154.

Vaghani, S.S., Patel, M.M., Satish, C.S., Patel, K.M., and Jivani, N.P. (2012). "Synthesis and characterization of carboxymethyl chitosan hydrogel: Application as pH-sensitive delivery for nateglinide." *Current Drug Delivery* 9(6): 628–636.

Vandamme, T.F., Lenourry, A., Charrueau, C., and Chaumeil, J.C. (2002). "The use of polysaccharides to target drugs to the colon." *Carbohydrate Polymers* 48(3): 219–231.

Van Vlierberghe, S., Dubruel, P., and Schacht, E. (2011). "Biopolymer-based hydrogels as scaffolds for tissue engineering applications: A review." *Biomacromolecules* 12(5): 1387–1408.

Vaupel, P., Kallinowski, F., and Okunieff, P. (1989). "Blood flow, oxygen and nutrient supply, and metabolic microenvironment of human tumors: A review." *Cancer Research* 49(23): 6449–6465.

Vedadghavami, A., Minooei, F., Mohammadi, M.H., Khetani, S., Rezaei Kolahchi, A., Mashayekhan, S., and Sanati-Nezhad, A. (2017). "Manufacturing of hydrogel biomaterials with controlled mechanical properties for tissue engineering applications." *Acta Biomaterialia* 62: 42–63.

Venkatesan, J., Lowe, B., Anil, S., Manivasagan, P., Al Kheraif, A.A., Kang, K-H., and Kim, S-K. (2015). "Seaweed polysaccharides and their potential biomedical applications." *Starch - Stärke* 67(5–6): 381–390.

Wang, C., Kopeček, J., and Stewart, R.J. (2001). "Hybrid hydrogels cross-linked by genetically engineered coiled-coil block proteins." *Biomacromolecules* 2 (3): 912–920.

Wang, F., Li, Z., Khan, M., Tamama, K., Kuppusamy, P., Wagner, W.R., Sen, C.K., and Guan, J. (2010). "Injectable, rapid gelling and highly flexible hydrogel composites as growth factor and cell carriers." *Acta Biomaterialia* 6(6): 1978–1991.

Wang, H., Zhang, X., Li, Y., Ma, Y., Zhang, Y., Liu, Z., Zhou, J., Lin, Q., … and Wang, C. (2010). "Improved myocardial performance in infarcted rat heart by co-injection of basic fibroblast growth factor with temperature-responsive chitosan hydrogel." *The Journal of Heart and Lung Transplantation* 29(8): 881–887.

Willner, I. (2017). "Stimuli-controlled hydrogels and their applications." *Accounts of Chemical Research* 50(4): 657–658.

Xiao, S., Yang, Y., Zhong, M., Chen, H., Zhang, Y., Yang, J., and Zheng, J. (2017). "Salt-responsive bilayer hydrogels with pseudo-double-network structure actuated by polyelectrolyte and antipolyelectrolyte effects." *ACS Applied Materials & Interfaces* 9(24): 20843–20851.

Yang, J.Y., Zhou, X.S., and Fang, J. (2011). "Synthesis and characterization of temperature-sensitive hemicellulose-based hydrogels." *Carbohydrate Polymers* 86(3): 1113–1117.

Yang, L., Chu, J.S., and Fix, J.A. (2002). "Colon-specific drug delivery: New approaches and in vitro/in vivo evaluation." *International Journal of Pharmaceutics* 235(1): 1–15.

Yu, F., Cao, X., Du, J., Wang, G., and Chen, X. (2015). "Multifunctional hydrogel with good structure integrity, self-healing, and tissue-adhesive property formed by combining Diels–Alder click reaction and acyl hydrazone bond." *ACS Applied Materials & Interfaces* 7(43): 24023–24031.

Zhang, L., Li, K., Xiao, W., Zheng, L., Xiao, Y., Fan, H., and Zhang, X. (2011). "Preparation of collagen-chondroitin sulfate–hyaluronic acid hybrid hydrogel scaffolds and cell compatibility in vitro." *Carbohydrate Polymers* 84(1): 118–125.

Zhao, W., Jin, C.Y., Liu, Y., and Fu, J. (2013). "Degradable natural polymer hydrogels for articular cartilage tissue engineering." *Journal of Chemical Technology & Biotechnology* 88(3): 327–339.

Zhao, X., Wu, H., Guo, B., Dong, R., Qiu, Y., and Ma, P.X. (2017). "Antibacterial anti-oxidant electroactive injectable hydrogel as self-healing wound dressing with hemostasis and adhesiveness for cutaneous wound healing." *Biomaterials* 122: 34–47.

Zhao, Y., Su, H., Li, F., and Tan, T. (2005). "Superabsorbent hydrogels from poly(aspartic acid) with salt-, temperature- and pH-responsiveness properties." *Polymer* 46(14): 5368–5376.

6

Self-Healing Polymers

Asit Baran Samui
Institute of Chemical Technology, Mumbai, India

CONTENTS

DOI: 10.1201/9781003037880-6

Abbreviations

T_g	glass transition temperature
PMMA	poly (methyl methacrylate)
BA	butyl acrylate
AMPS	2-acrylamido-2-methyl-1-propane sulfonic acid
ATRP	atom transfer radical polymerization
DA	Diels–Alder
BM	bismaleimide
PU	polyurethane
PUA	polyurea
PEG	polyethylene glycol
HDI	hexamethylene diisocyanate
MGP	methyl-α-D-glucopyranoside
PEI	polyethyleneimine
MWCNT	multi-walled carbon nanotube
POSS	polyhedral oligomeric silsesquioxane
HEA	2-hydroxyethyl acrylate
PDMS	poly (dimethylsiloxane)
PUF	poly (urea-formaldehyde)
PVDF	polyvinylidene fluoride
PC	polycarbonate
DETA	diethylenetriamine
UDETA	2-aminoethylimidazolidone
TPEs	thermoplastic elastomers
T_m	melting temperature
HBP	hydrogen-bonding brush polymer
PA	polyacrylate
PAA	polyacrylic acid
PEO	polyethylene oxide
BIIR	bromobutyl rubber
OSA	oxidized sodium alginate
ADH	adipic acid dihydrazide
CEC	N-carboxyethyl chitosan
EDHs	enzyme-assisted dual-network self-healing hydrogels
EPL	ε-poly-L-lysine
PAO	plasma amine oxidase
PEGDA	dibenzaldehyde-terminated poly (ethylene glycol)
DF-PEG	difunctional (benzaldehyde) poly(ethylene glycol)
DN	double networks
GOX	glucose oxidase
CAT	catalase
BSA	bovine serum albumin

NVP	1-vinyl-2-pyrrolidinone
AM	acrylamide
NIPAM	N-isopropylacrylamide
O-CMC	O-carboxymethyl chitosan
NHDF	normal human dermal fibroblast
r-DA	reversible Diels-Alder
P(MVE-alt-MA)	poly[(methyl vinyl ether)-alt-(maleic acid)]
MAc	maleic acid
β-CD	β-cyclodextrin
Ad	adamantane
TIPS	thermally induced phase separation technique
PU	polyurethane
ODA	octadecylamine
HTPDMS	hydroxyl terminated polydimethoxy silane
PDES	polydiethoxy siloxane
LbL	layer-by layer
SRT	squid ring teeth
CP	conducting polymer
PANI	polyaniline
PA	phytic acid
FFF	fused filament fabrication
GO	graphene oxide
DAC	2-(dimethylamino) ethyl acrylate methochloride
FGNS	functionalized graphene nanosheets
PET	polyethylene terephthalate
SWCNT	single walled carbon nanotube
SHM	structural health monitoring

6.1 Introduction

Self-healing materials are substances that can heal any damage caused by any external factors and recover their original properties partially or completely. In nature, some biological materials have the ability to self-repair, such as a cut in the body, bone fracture, etc. However, man-made systems do not have such kind of self-repairing ability. A self-healing property gives materials longer service, particularly in places where accessibility is restricted or places which are prone to damage. The basic idea is to continue the performance of the product as it was before the damage. The possibility of self-healing has propelled the development of a large number of strategies for three categories of materials: metals, ceramics, and polymers. Although many strategies have been developed over the years, the basic requirement remains the same; the generation of a mobile phase in response to damage, which fills the damaged area.[1] The temperature of self-healing varies according to the matrix, which increases in the order of

concrete< polymers< metals< ceramics

Depending on the self-healing nature, two types of processes are observed: automatic and non-automatic. The former occurs by using the damage as the trigger. This is evident in adaptive structures. External energy sources, including heat and light, etc. are used for the non-automatic process. Prevailing environmental conditions or a laser source act as the external energy. A subdivision of this classification considers intrinsic and extrinsic self-healing processes. The intrinsic process does not depend on the presence of an external agent in the matrix. Rather, it depends on the physical interactions and the formation of primary or secondary chemical bonds between the interfaces of cracks. The extrinsic process depends on the presence of micro or nanocapsules, hollow fibers, etc., which release the healing agents during crack formation.

Self-healing polymers and composites have been studied mostly because of their ability to self-heal at a lower temperature compared to metals and ceramics. Conventionally, there are two types of design approaches: (1) confinement of the healing agents that are released when damaged; and (2) use of a material that is decorated with reversible bonds, which become activated when damaged. The microcapsule is filled with a monomer or reactive prepolymer, which is released due to mechanical impact-induced damage. The released monomer or reactive prepolymer undergoes polymerization in the crack plane triggered by the catalyst dispersed in the polymer. This results in the healing of the crack.[2] The confinement of the healing agent occurs also in hollow fibers, which facilitate its transport over long distances.[3] Significant restoration of the original flexural properties is possible using this approach. Following the idea of multiple functions being performed by vascular networks in biological systems, interpenetrating microvascular networks have been designed. One of these is filled with one component of the healing agent, e.g., one part of a two-component epoxy resin while the other part is filled in another network and both are embedded in an epoxy substrate, which enables the system to heal during multiple damage cycles in the coating.[4] The two-part epoxy components, loaded separately in microvascular networks, are embedded in a polymeric matrix that is released during crack formation and self-heals the damaged portion.[5]

A numerical study is also used to find out the exact nature of microcapsules on the healing properties.[6] The elongated capsules maintain a high aspect ratio and, if properly oriented, release more of the liquid healing agent per unit crack area compared to spherical microcapsules.

In the simplest case of self-healing, heating is applied to facilitate the polymer chains' diffusion into one another so that new entanglements are formed that lead to the closure of the crack.[7] The crack healing occurs by keeping the operating temperature higher than the effective glass transition temperature (T_g). As an example, methanol treatment is used to reduce the T_g of a poly

PMMA matrix, so that the healing can occur at a lower temperature than the original T_g. If the damage is associated with the cleavage of the covalent bond in these thermoplastic polymer chains, new chemical bonds are required to be formed for self-healing to occur. Polycarbonates, having reactive end groups, can be used for the formation of polymer chains after the damage.[8] A thermally reversible Diels-Alder cycloaddition reaction can be used to design thermally reversible cross-linking via a side group of polymers; e.g., a cyclopentadiene core can function as diene as well as dienophile.[9] Other mechanisms have also been developed. The drawback associated with self-healing via covalent bond formation is that it needs a higher temperature. If some of the cross-links in a matrix are replaced by hydrogen bond-forming moieties, it acquires a facile reversible cross-linking ability.[10] The S-S bond, having favorable dynamics and cleavage reversibility, helps in designing self-healing polymers, the aromatic disulfides being the most common.[11] An oligomeric, fatty acids-based thermoplastic elastomer, comprising diethylene triamine, functionalized with urea, exhibits the breaking of mainly hydrogen bonds during damage and their rebuilding after pressing together the separated portions.[12] Ionic interactions,[13] metal complexes interactions,[14] and $\pi - \pi$ interactions[15] are also studied as non-covalent interactions for developing materials with self-healing characteristics. Mechanical stress can activate many physical and chemical processes, and a catalyst accelerating a chemical reaction can be used.

Self-healing is a very common event for living things. Taking this cue from nature, various self-healing strategies in polymers have been developed. This chapter deals with the concept, to effect self-healing.[16]

6.2 Strategies for Self-Healing

Two main approaches are followed for self-healing, which are subdivided into intrinsic and extrinsic in nature. Further, there are subdivisions such as hydrogen bonding, polar interactions, microcapsule and vesicle-based delivery of chemicals or solvents, etc. Various types of polymeric architectures are discussed and different ways of their healing are explored and discussed with examples.

6.3 Self-Healing of Methacrylate Polymers

6.3.1 Methanol-Induced Healing of Polymethylmethacrylate (PMMA)

Crack healing in PMMA can be done with the help of methanol. Crack healing is observed only in a working environment with a temperature higher than the T_g.[7] However, the self-healing of the methanol diffused PMMA occurs at much lower temperature than the normal T_g. This is possible as the effective T_g in this condition is quite low, e.g., 39.2°C and 21.8°C at a surrounding temperature of 40°C and 60°C respectively. The healing occurs in two stages. Stage I is characterized by wetting corresponding to progressive healing, observed by constant crack closure. Methanol transport behavior in PMMA controls the process indirectly. After stage I, stage II starts with diffusion continuing that improves the healing characteristics. Since the crack healing occurs during methanol transport over a short duration, the molecular chains are able to continue self-diffusion around the crack surface. This zone is known as the solvent-affected zone. The effective glass transition temperature controls the diffusion and healing and its extent is directly proportional to the temperature of the methanol treatment. After stage I of crack healing, the fracture surface remains in the same plane as that of the original crack surface. In stage II, the original crack surface disappears to regain the virgin fracture surface.

6.3.2 PMMA-Based Self-Healing Shape Memory Polymers

The vitrification-induced PMMA domains with microphase separation function as the physical cross-link for the diblock polymer of PMMA and butyl acrylate acrylamide sulfonic acid derivative-based copolymer P(BA-co-AMPS). The AMPS segments contribute to the reinforcement of the matrix through their ionic interaction. The two networks produced by the two interactions provide a shape memory property to the matrix. The rubbery BA component in the copolymer, aided by superior ductility, makes these materials self-healable.[17] Quick self-healing via entangling across the broken interface is made possible by the high mobility of the segments of P(BA-co-AMPS) with low T_g.

6.3.3 Methacrylate-Based Copolymers Healing by Photochemical [2+2] Cycloaddition

Photo-healing has the added advantage of using light, which is fast, simple, and environmentally-friendly clean energy. The double bond-containing compounds such as photo-cross-linkable cinnamate monomer can be cross-linked with UV radiation with a wavelength above 280 nm[18] (Figure 6.1). The recovery of mechanical strength is almost near the initial value. The recovery is faster at a higher temperature.

Cinnamoyl substituted ethane (cinnamate) monomer, having a photo-cross-linking ability, can be photopolymerized along with methacrylate monomers via the cyclobutane ring formation. During crack generation the cyclobutane ring of the cross-linked structure breaks, which is self-healed by irradiating it again by UV radiation above 280 nm.

6.3.4 Polymethacrylate with Dynamic Urea-Based Cross-Linking

Sterically hindered ureas have the ability to reverse the urea bond formation. Reversible reactions take place by applying heat or exposure to water to reform the isocyanate and the corresponding co-reactant.[19] The cross-link density of the copolymer with the urea bonds, formed in situ during photopolymerization, determines the self-healing ability. Thus, excellent scratch-healing and acceptable bulk-healing properties can be achieved with only 5% cross-link density at temperatures above 70°C. These urea-based systems are hard under ambient conditions and even in the healing environment, and the remaining mechanical properties are superior compared to many self-healing systems.[20]

6.4 Self-Healing of Styrene-Based Polymers

6.4.1 Disulfide-Based Materials

The disulfide-based materials undergo self-healing via a [2+1] radical-mediated mechanism; the sulfenyl radicals are generated in the first step by the homolytic cleavage of the S-S bond, which attacks the nearby disulfide bonds to induce the interchange of sulfur atoms via a three-membered transition state.[21] The generation of sulfenyl radicals and the interchange of sulfur atoms depend on:

1. the sulfenyl radical concentration that is chemically regulated by adjusting the S-S bond strength;
2. the dynamics of the polymeric chains, which is associated with the flow and reorganization characteristics;
3. the energy barrier of the exchange process.

Using an analytical tool, the above parameters can be studied. The S-S bonds are weakened by the presence of electron-donating groups in the phenyl rings, particularly the amino group and its derivatives, that facilitate the formation of sulfenyl radicals. Between the two most important non-covalent interactions, the π–π stacking influences the polymer chain mobility to some extent, whereas the hydrogen bonding is important in the reaction process. For the reaction to occur at room temperature, the reaction barrier should be sufficiently low.

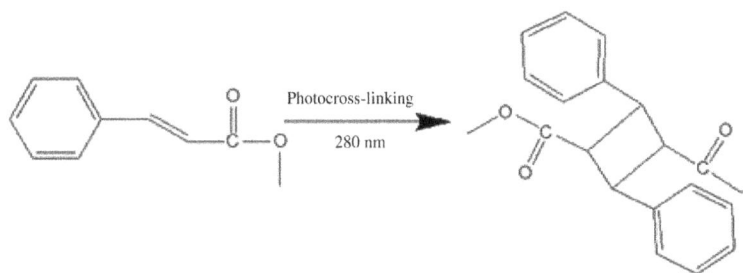

FIGURE 6.1 Photo-cross-linking of cinnamate esters via (2+2) cycloaddition.

6.4.2 Polystyrene with Disulfide Linkage

The introduction of thiol functionality into a polymer can be done by using an appropriate sulfur-based initiator. However, due to the large transfer constants of thiols, the method cannot be used. The use of protected thiols can solve the problem, as the presence of protective groups maintains a lower transfer constant and the groups can be removed after the polymerization. Also, a sulfur-containing nucleophilic precursor of the thiol (e.g., thiourea) can be reacted with the halogen end groups of the polymers to form thiol end-functionalized polymers.[22] The diester, derived from a hydroxyl group containing disulfide and bromo-derivative of carboxylic acid, is used as an initiator in the polymerization of styrene via ATRP, that results in polystyrene with an internal disulfide bond.[23] The disulfide bond may undergo reversible reductive cleavage in the presence of dithiothreitol to produce a thiol-ended polymer. This polymer reacts with $FeCl_3$, resulting in the oxidative coupling of the thiol groups so that by the regeneration of an internal disulfide bridge the original polymer is obtained. The self-healing concept is also extended to the epoxy network by incorporating disulfide links.

6.5 Self-Healing of the Terpolymer

6.5.1 Diels-Alder-Based, Self-Healable Triblock Copolymers

The chemical reaction between a conjugated diene and a substituted alkene forming a substituted cyclohexene derivative is called the Diels-Alder reaction (Figure 6.2). The Diels-Alder reaction can be reversible under certain conditions, which is called the retro-Diels-Alder reaction.

A polymeric network via a DA reaction can be made by mixing the triblock copolymers, formed by introducing a polydimethylsiloxane block between the furfuryl methacrylate blocks, with bismaleimide (BM) derivative (1:1 mole ratio) in chloroform solution at 50°C for more than 20 h in an N_2 environment.[24] When the reversible DA reaction in the scratched DA polymer film at 150°C is followed by heating at 50°C for 14 h, the scratch is completely healed, as observed in SEM micrographs. However, the healing does not occur at 100°C, as the temperature is above the melting point of the block copolymer. Thus, the flow behavior has no role in self-healing as this is a function of the DA reaction only.

6.5.2 Dynamic Covalent Bonds-Based Self-Healing

Self-healing materials can be prepared by the terpolymerization of a phenylboronic acid-based monomer and an acrylamide (AAm)

in the presence of a polyrotaxane derivative. The dynamic covalent bonds between the polyrotaxane derivative and the polymers are responsible for its flexibility and self-healing ability.[25] Reversible covalent bonds allow the material to have both a "physical" self-healing ability and a "chemical" self-healing ability, even when they are reattached after complete separation. Further, self-healing is possible in both a dry and a wet state. The terpolymer film self-heals efficiently as it reaches approximately at 100% recovery within 30 min at 60°C under humid conditions.

6.6 Self-Healing of Polyurethane

6.6.1 Carbon Dioxide-Induced Healing

Carbon dioxide has been used to produce a large number of small molecules as well as polymers. As an example, using a catalyst, urethanes and ureas can be formed by reacting CO_2 with primary or secondary amines.[26] During the synthesis of the polyurethanes (PU) networks from the reaction of a diol (polyethylene glycol (PEG)) with a diisocyanate (hexamethylene diisocyanate (HDI)) in the presence of H_2O vapors, gaseous CO_2, as well as polyurea (PUA), are generated. During mechanical damage of this PU network, there are cleavages of C-O, C-N, and other linkages. The agents such as CO_2 and H_2O do not help in the self-repair of the damaged PU film. However, in the presence of carbohydrates with multiple OH groups in PU, mechanically damaged network bonds are regenerated by CO_2 and H_2O.[27] As an example, PU networks formed from methyl-α-D-glucopyranoside (MGP), HDI, PEG, and H_2O regenerate in the air after being mechanically damaged. The repair is possible due to physical diffusion of cleaved network segments leading to the formation of carbonate and urethane linkages, respectively.

6.6.2 Disulfide-Based Polyurethane

The healing process via the reorganization of chemical bonds is always preferred to occur under ambient conditions, which make disulfide bonds appropriate candidates. Aromatic disulfides are potentially promising candidates as the exchange reactions can occur under ambient conditions in solution[28] and in the solid state.[29] Exchange of the disulfide compounds comprises [2+1] radical-mediated process; the first step involves the generation of sulfenyl radicals via homolytic cleavage of the S-S bond that attacks the disulfide bonds to produce an interchange of sulfur atoms.[30, 31] The calculated hydrogen bonding energies are almost similar to the S-S bond energies. The use of catalysts enhances the disulfide exchange reaction as it helps in the formation of S-based anions that add to the radical-mediated mechanism.

FIGURE 6.2   Schematics of Diels-Alder and retro-Diels-Alder reaction.

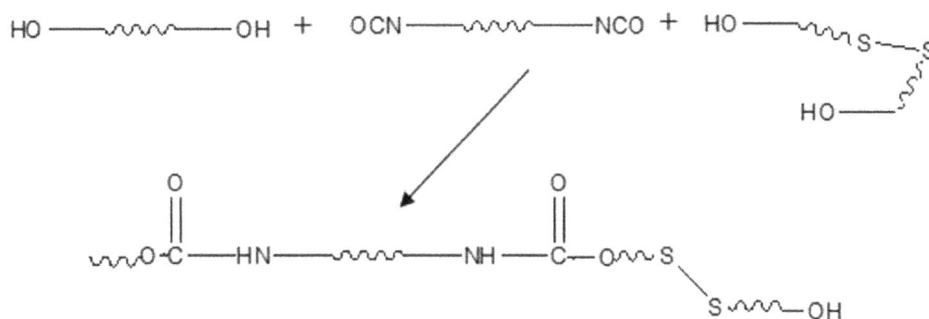

FIGURE 6.3 Representative synthesis of polyurethane with disulfide bonds.

A self-healing polyurethane system with a combination of S-S bond and hydrogen bonding can be realized by introducing a chain extender with disulfide bonds[32] (Figure 6.3).

Almost 50% of the initial mechanical strength is regained within 2 h of healing at 60°C and 100% is recovered within 6 h. At lower temperatures the healing also occurs, however, at a much slower rate. In the case of the PU matrix without an S-S bond, about 46% healing is possible, which indicates both the H-bond and S-S bonds are responsible for self-healing.

The UV-curing technology can be combined with a self-healing concept to design UV-curable self-healing oligomers that are formed by reacting hydrogen bonds forming compounds, such as polycarbonate diol and a multi-arm polyol-based photosensitive monomer.[33] The healing performance of a PU system is quite satisfactory. Using a hot air gun, the polymer can quickly be repaired with high repair efficiency. The self-healing can go as deep inside as 4 μm below the surface. The high healing temperature is permissible for such kind of architecture due to its good thermal stability.

6.7 Self-Healing Epoxy

6.7.1 Recyclable Epoxy Adhesive with Diels-Alder Reaction-Based Self-Healing

Epoxy adhesive with self-healing and a recycling ability can be made by incorporating MWCNTs modified with polyethyleneimine (PEI) into it with Diels-Alder (DA) bonds.[34] The Diels-Alder reaction between furan and maleimide in the epoxy monomer, being thermally reversible, imparts self-healing and recyclable ability to the adhesive. Self-healing is possible in this system when heated above the cleaving temperature for the unbroken bonds to separate into two individual molecules, which provides a greater chance for the DA reaction to occur to heal the fracture planes. The healing efficiency depends on the viscous flow and the DA healing effect. In fact, at temperatures higher than T_g, the furans with a very high mobility form a bond with the maleimides at the interface. Thus, the strategies of increasing the temperature and decreasing the T_g act to increase the bonding at the interface. However, due to the bonding of furans and maleimides through a thermo-reversible reaction, the equilibrium toward Diels-Alder adduct decreases with the increasing temperatures. This suggests that the strategy of lowering the T_g is highly suitable to increase healing efficiency.

6.7.2 Disulfide-Based Cross-Linked Epoxy

Epoxy resin cross-linked with thiols flows at a higher temperature, which is due to the breakage of the disulfide linkage.[11] Even in a solvent, the cross-linked networks break at the disulfide linkage and become soluble. The healing process is also efficient, as within 15 min sufficient healing occurs and is completed within 1 h at 60°C. Generally, longer healing times are required for better healing. The stress-strain curves for healed samples superimpose irrespective of healing times, and differ only in their elongations at the break.

Building blocks such as polyhedral oligomeric silsesquioxane (POSS) are very effective in the preparation of nanomaterials. However, they suffer from deficiencies, such as brittleness and cracking, which can be tackled by using self-healing attributes. Disulfide linkages can be used to prepare a self-healing composite, which undergoes a disulfide bond exchange reaction at a lower temperature. By reacting elemental sulfur (S_8) with methacrylated POSS compound (MMA-POSS) via "inverse vulcanization," a building block in the form of S-MMA-POSS can be produced to impart the self-healing ability. This can be incorporated into thermally cross-linked POSS-containing nanocomposites through a self-curing reaction and co-curing reaction with conventional thermosetting resins. Thermal treatment at 120°C breaks the S-S bond and the molecular system becomes mobile to heal. After cooling, the S-S bonds are re-established.

6.8 Polymers with Retro Thiol-Michael Reaction-Based Healing

6.8.1 Dual Stimuli-Responsive Self-Healing

Michael adduct formation can be made reversible at elevated temperatures as well as at higher pH values. A thiol-maleimide Michael reaction is known for its ability to maintain very high conversion under ambient conditions in the presence of air and water. Thiol-maleimide (TM) adducts can be used to cross-link mechanically stable polymeric materials to make them self-healable under elevated thermal and pH conditions.[35] The retro thiol-Michael reaction-based healing mechanism depends on the dissociation of the thiol-Michael adduct into the maleimide and thiol. Subsequent to this, new thiol-Michael adducts are formed via the reaction of the maleimide formed with another free thiol to effect a dynamic exchange. Lower temperatures and pH values assist only for a negligible dynamic exchange in the thiol-Michael adducts, as full creep recovery takes about 24 hours.

6.8.2 Cross-Linked Acrylate

A thiol-acrylate-based cross-linker can be used to cross-link a hydroxyl containing acrylate (HEA) polymer that displays self-healing and malleability properties.[36] Because of its stability under stress and strain at ambient conditions, the thiol-Michael linkage in the cross-linked polymers can only be activated by applying heat. The strategy is applicable for coatings, sealants, and high-performance elastomers, particularly those in which creep is not desired to occur at room temperature.

6.9 Polydimethylsiloxane Self-Healing

6.9.1 PDMS-Based Self-Healing Microcapsules

In automobiles, PU resin with a urethane bond (NHCOO) is mostly used as a skin layer because of its appearance and intrinsic elasticity. However, scratches during damage cannot be avoided. Normally, two approaches are feasible to avoid the scratches: the first is the enhancement of scratch resistance by using nanoceramics as a filler and the second is self-healing. Two different microcapsules need to be incorporated into the matrix to design a self-healing PU-based matrix. Poly (dimethylsiloxane) (PDMS) with a hydride group mixed with the platinum catalyst and the cross-linker are encapsulated in two different microcapsules with a poly (urea-formaldehyde) (PUF)-based shell.[37] The healing occurs via a Pt-catalyzed reaction of hydride with vinyl-terminated PDMS resin, which are provided by the broken microcapsules containing active core materials (PDMS, Pt, and a cross-linker).

6.9.2 Liquid Encapsulation for Stress Transfer-Based Self-Healing

The efficiency of the stress transfer from the matrix to the reinforcement is the key to the mechanical properties of a composite. This can be improved by enhancing the interfacial strength and load-bearing capability that are achieved by carrying out surface modification, morphological engineering of the reinforcement, the addition of surface binders, etc. Alternatively, a solid-liquid system can be designed for stress transfer. A unique load transfer mechanism, very different from conventional composites, can be designed with a liquid-filled solid shell in which the viscosity of the liquid is important as it is the determinant of the shape of the shell under load.[38] The mechanical properties depend on the interface between the liquid and the solid and the size and surface tension of the reinforcement. The liquid-solid interface can be designed by synthesizing liquid-filled microspheres, which comprise two polymer systems, such as a PDMS liquid filler in a polyvinylidene fluoride (PVDF) shell.[39] When load is applied, the spheres undergo temporary deformation so that there is even distribution of load. Interfacial failure is common in solid-solid composites due to modulus mismatch. On the other hand, the liquid in the solid-liquid composites flows according to the solid shape and a large amount of strain is accommodated without any damage at the interface. After removal of the load, the liquid flows back to attain the minimum energy shape.

6.10 Polycarbonate Self-Healing

By using sodium carbonate and an organic phosphate, the self-healing of polycarbonate (PC) can be done at 120°C, which is followed by treatment at 130°C with nitrogen flow.[8] The subsequent treatment increases the molecular weight of the specimen. The broken carbonate link is restored during self-healing.

6.11 Self-Healing Elastomers

Rubbers (elastomers) originating from natural or synthetic resources have been used for centuries in diverse applications because of their combined properties such as strength, toughness, and elasticity. There are constant efforts to develop better and better materials with enhanced mechanical strength or improved elasticity. Self-healing capabilities have been added to the required quality of rubbers with the aim of extending their lifetime and sustainability.

6.11.1 Hydrogen Bond-Mediated Healing

Covalent cross-linking, connected by physical associations of small glassy or crystalline domains, ionic aggregates or multiple hydrogen bonds in a macromolecular system exhibits rubber elasticity. Generally, flow and creep in a polymer system can be prevented by the presence of covalent cross-links or strong physical associations. The flexibility of the rubber chains, modified with self-healing function, is advantageous for rebuilding or redistributing the noncovalent interactions to self-heal any mechanical damage. The supramolecular rubbers with little sticky equilibrated surfaces exhibit a weak self-adhesion. The presence of a large density of free moieties on a surface makes the healing of damaged surfaces possible because of their ability to recombine across an interface. Maintaining a strong directional interaction together and avoiding crystallization during the preparation of supramolecular rubbers from small molecules form a challenging task. A variety of molecular architectures mixed together makes it difficult for crystallization to occur and the entropy of mixing and directional specific interactions are balanced so that the macroscopic phase separation of different species is avoided. As an example, the fatty dimer reacts with substituted imidazolidone, substituted ureas, in such a way as to form multiple bonds.[40] Hydrogen bonds play an active part in the formation of both chains and cross-links. Strong and long-lived supramolecular associations are desired while the fraction of non-associated groups in the network is kept low at equilibrium. It is also required that the strength of the associations is kept lower than that of covalent bonds so that many non-associated groups are available near the fracture surface. These "eager" to link groups present near the surface perform efficient self-healing and, therefore, repeated self-healing at room temperature is done just by bringing together the fractured surfaces. The composition can be varied slightly by using a combination of vegetable oil-derived fatty acid derivatives, diethylenetriamine, and urea so that the elastomer formed comprises oligomers.[41] The wide molecular weight distribution of the randomly branched oligomers with a broad molecular weight distribution and the presence of hydrogen bonding groups make the elastomer self-healing in nature.

Organized solid and viscoelastic supramolecular materials can be designed from the reaction of a mixture of fatty acid with diethylenetriamine (DETA) and 2-aminoethylimidazolidone (UDETA).[42] The UDETA acts as the hydrogen bonding moiety with a self-complementary nature. Additional complementary H-bonding units can be incorporated by grafting urea onto all the secondary amines of DETA. The self-healing behavior originates from the breaking and formation of the hydrogen bonds.

6.11.2 Thermoplastic Elastomers Healing by Hydrogen Bonding

Stiffness and spontaneous healing can be combined in a multiphase design such as a thermoplastic elastomeric system. The hydrogen bonds acting as the dynamic healing motifs are placed in the soft phase in a multiphase system. Thermoplastic elastomers (TPEs) undergo microphase separation into "hard" glassy or crystalline domains embedded in a "soft" rubbery matrix. A dual role is performed by the hard domains, such as the formation of physical cross-links to introduce rubber elasticity and maintain the high modulus and stiffness of TPEs. The soft segments connected via the covalent bonds with the hard domains are not able to undergo self-healing below the glass transition (T_g) or melting temperature (T_m) of the hard phase. For the self-healing character to be incorporated, the covalent connected soft segments need to be replaced with a noncovalent, supramolecular soft matrix. A perfect TPE is constituted by the hydrogen bonding brush polymer that self-assembles into a two-phase morphology with high mechanical strength as well as extensibility.[43] The HBP system self-heals as a single-component solid material without the need for any external stimuli, healing agents, or others.

6.11.3 Block and Graft Copolymers Healing by Hydrogen Bonding

Triblock copolymers comprising a polymethylmethacrylate (PMMA) central block and polyacrylate-amide (PA-amide) terminal blocks exhibit good self-healing capability while maintaining higher mechanical properties.[44] The polar groups in both the soft PA-amide and the hard PMMA microphases, having strong interactions, enhance the mechanical properties. The hard and soft phase components (PMMA and PA-amide) can also be arranged in a brush macromolecular architecture, with the PA-amide graft on a PMMA backbone, by using atom transfer radical polymerization (ATRP) with PMMA macroinitiators.[45] The brush copolymers have a spherical morphology and the highly mobile nature and extensive dynamic hydrogen bonding of PA-amide brushes are responsible for self-healing properties at room temperature without the need for any external stimulus.

6.11.4 Hydrogen-Bonded Elastomer Complexes

Healable elastomers with high mechanical strength, high toughness, and good elasticity can be generated by complexing poly (acrylic acid) (PAA) and poly (ethylene oxide) (PEO).[46] The hydrogen-bonded PAA-PEO complex-based elastomers exhibit reversible elongation more than 30 times their initial length. Good elasticity is manifested due to the cross-link formation by reversible hydrogen bonds. During stretching the dynamic hydrogen bonds break and reform after the stretching is removed, while their original mechanical properties are restored. Simple preparation methods include mixing of an aqueous solution of two polymers to obtain precipitate, which is separated by centrifuge and compression molded for two days and finally air-dried. A larger proportion of PAA enhances the extent of hydrogen bonding, leading to higher rigidity at room temperature. When the cut pieces are placed to touch each other, the self-healing occurs at room temperature over time during incubation in a humid environment (RH>90%). Complete healing takes about 24 h. The healing at lower humidity takes longer at room temperature.

6.12 Self-Healing Ionomers

6.12.1 Ethylene/Sodium Methacrylate Copolymer Healing

Ionomers, formed by incorporating up to 20 mol% ionic species to induce interactions or aggregations, belong to a class of polymers that exhibits self-healing behavior for a large number of cycles. Mechanical and physical properties are highly influenced by this behavior. The aggregates are formed by several ion pairs and are known as physical cross-linking. Thus, the ionic clusters can be dispersed in a continuous semi-crystalline polymer matrix. With increasing temperature, the polymer goes from an ordered to a disordered state and, as a result, loses its mechanical strength. Further increase in temperature causes the semi-crystalline polymer to melt. Of course, the disordered clusters contribute to increased melt strength. The ionic nature of the network is known to enable the segments to rearrange, which induces healing after mechanical damage. The impact energy is distributed over the ionomer components, which melt and penetrate into each other. This way, the intermolecular ionic attractions are maintained by the ionic regions that help in their elastic recovery and repair the damage. It is thus apparent that, for self-healing, the ionic character, along with its unique interaction, is most important. For example, simple ionomers can be synthesized by copolymerizing ethylene and sodium salt of methacrylic acid. The ionomers can heal rapidly at room temperature. However, by increasing the temperature of the film before puncture, the healing response is prevented by a more residual deformation.[47] In this condition, the dispersion of impact energy over a larger area does not support the elastic recovery of the hole. The stages of the healing process in the case of projectile penetration can be depicted as: in the first stage, the localized molten polymer exhibits an elastic response by using the rigid film perimeter as a framework, while reverting back to fill the hole. In the next stage, the interdiffusion occurs in the molten surfaces to fuse them together and the strength is regained as the polymer cools. In the third stage, the strength regeneration occurs through the processes of continued interdiffusion, crystallization, and long-term relaxation of the polymer chains.

The ballistic penetration behavior also confirms that throughout the impact regions both elastomeric and viscous behavior remain prominent, while the ductile/elastic behavior is characteristic of the outer impact regions, and elastomeric and viscous behavior are in play in the region near the impact cavity.[48] A smaller diameter, lower impact energy bullet, producing a smaller crater, induces enhanced elastomeric behavior at the impact site. The repeatability of the extent of ionomeric healing decreases with the number of repetitions. Room temperature crystallinity and storage modulus can be increased by introducing the zinc stearate. On the other hand, the elevated temperatures induce a decrease in viscosity and loss modulus. The above properties reduce the elastomeric properties, which impede the healing capability. The zinc cations from the zinc stearate solvate the ionic portions, while the aliphatic long chain molecules remain miscible in the continuous matrix that increases the cluster size, resulting in reduced attraction and their strength.

6.12.2 Ion-Mediated Self-Healing of Siloxane Elastomers

Quaternary ammonium silanolate-initiated copolymerization of octamethyl- cyclotetrasiloxane (D4) and substituted cyclotetrasiloxanyl-ethane (bis-D4) produces a cross-linked ethylene bridge containing polymer networks with active silanolate groups at the end, which performs self-healing (Figure 6.4).[49]

These "living" reactive anionic species maintain thermally activated equilibration among different network isomers and cyclic oligomers. Due to the volatility of the cyclic oligomers, the system undergoes self-healing in a sealed environment at a high temperature.

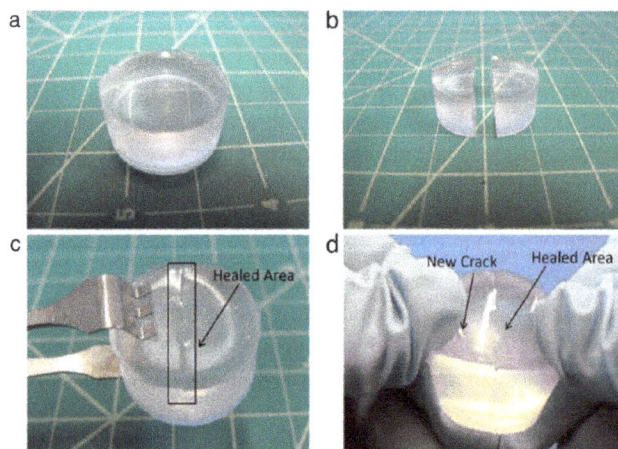

FIGURE 6.4 The original cylindrical sample (a) was cut in half using a razor blade (b). 24 hours after healing at 90°C (c), the healed sample was deformed by hand, and cracking occurred at a place other than the healed area (d).

6.12.3 Bromobutyl Rubber Healing

Because of the poor cross-linking efficiency of butyl rubber, bromobutyl rubber (BIIR) can be used because of its higher reactivity. By reacting allylic bromide groups with primary and secondary amines, various functionalities can be introduced into BIIR.[50] The imidazolium bromide groups, that form ionic associates, are generated via vulcanization of BIIR with butylimidazole at elevated temperatures in bulk.[51] The ionic groups exhibiting self-healing behavior demonstrate reversible interactions. The high flexibility of the elastomer facilitates the self-healing character.

6.12.4 Metal Methacrylate Cross-Linking in Non-Polar Rubbers

For non-polar rubbers, the strategies to introduce ionic groups play an important role in gaining the maximum benefit. Metal methacrylate monomers possess an ionic character. For example, zinc dimethacrylate (ZDMA) reacts with rubber molecules during its vulcanization. A large number of ion pairs produce reduced mobility. The grafting of ZDMA acts as a new type of cross-linking. Ionic cross-links can be introduced into the NR network by way of graft-polymerization of ZDMA, which maintains the mobility of NR chains while ensuring electrostatic interaction (Figure 6.5).[52] After the damaged parts are brought into contact, the transport of ionic associates occurs for reconstruction and redistribution of ionic cross-links for self-repairing.

6.13 Self-Healing Hydrogels

Self-healing hydrogels can be used over a long period with the desired functionality as they can maintain the integrity of network structures and mechanical properties. The approaches followed for self-healing are either dynamic covalent bonds or noncovalent bonds.

6.13.1 Schiff Base-Based Hydrogels
6.13.1.1 Polysaccharide-Based Hydrogels

The dynamic covalent bond approach is well established for self-healing. However, the non-autonomous self-healing nature of this approach cannot be applied in vivo. Further, the hydrogel synthesis cannot be interfered with by adding a

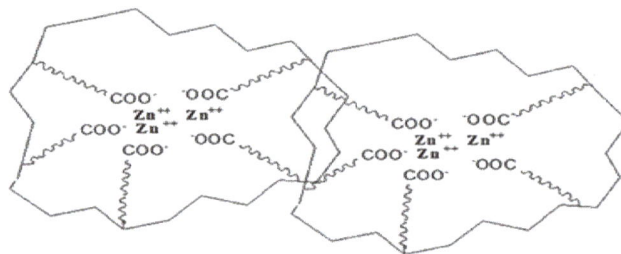

FIGURE 6.5 Schematics of physical cross-linking via ionic groups and metal ions.

dynamic covalent bond to the system and also the final system must be biocompatible. The biocompatible hydrogel containing a dynamic covalent bond, acylhydrazone, which is obtained by the condensation reaction between oxidized sodium alginate (OSA) and adipic acid dihydrazide (ADH), has the potential for application in vivo.[53] However, a slightly acidic environment (pH = 4.0–6.0) is required for the reversible reaction of the dynamic acylhydrazone bonds to occur.[54] A polysaccharide-based self-healing hydrogel can be formed by the reaction of the OSA solution with N-carboxyethyl chitosan (CEC) and ADH. The self-healing capability of the hydrogel is possible due to the coexistence of imine bonds derived from the reaction of the OSA with CEC and the acylhydrazone bonds derived from reaction between the OSA and ADH. Because of the more active nature of imine bonds compared to the acylhydrazone bonds, it can function as self-healing under neutral conditions.[55] Higher molecular weight CEC can add to the mechanical properties of the hydrogel. The self-healing by the hydrogel under physiological conditions works well without any external stimulus together with good cytocompatibility. This type of hydrogel is well suited for cell encapsulation and drug release

6.13.1.2 Enzyme-Induced Dual-Network Hydrogels

The pyrrolidinone-based monomer and substituted acrylamide react in the presence of ε-poly-L-lysine (EPL) to form enzyme-assisted, dual-network, self-healing hydrogels (EDHs) with autonomous self-healing and antimicrobial ability.[56] This is followed by the introduction of plasma amine oxidase (PAO) so that the oxidation of primary amines of EPL occurs in the air to form imine bonds (–CH = N–) and thus a secondary cross-linking network is formed. This enzyme-induced EDH system, having the dynamic covalent bonds of the Schiff base as well as the hydrogen-bonding interactions, exhibits autonomous healing without the need for any external stimuli. Broken gels combine together without any visible joint via self-healing over 6 h at 25°C and the mechanical properties at the weld line are the same as the original sample. The higher amount of EPL enhances the enzyme-induced Schiff base reaction rate, which, in turn, forms more dynamic imine bonds (–CH = N–). The imine bonds accelerate the self-healing rate. The presence of an ionic attraction between EPL and the target tissue allows it to exhibit adhesive properties. The materials have the property of fast wound dressing also.

6.13.1.3 Injectable Hydrogels

Hydrogel constitutes a drug carrier because of its exceptional pharmacokinetic property and its use as an injectable drug and cell delivery carrier is gaining popularity. To make it non-vulnerable to the stress-induced formation of cracks, the self-healing hydrogels concept is expected to prolong their lifetime. Self-healing hydrogels have the advantages of injecting the already formed hydrogels through a needle without clogging, due to their shear thinning behavior, which is followed by recovery of the original state. In situ formability and characteristics such as the ability to self-mend damage allow injectable hydrogels to encapsulate drugs effectively and are useful in invasive surgery. It is also known that acidosis is an important characteristic of the surrounding areas of tumor and adjacent tissues.[57] This requires hydrogels to be designed with a pH-responsive ability so that the drug is efficiently released in an acid environment, while negligible release occurs in a normal physiological environment. Typically, pH-responsive self-healing injectable hydrogels can be derived by reacting CEC with dibenzaldehyde-terminated PEG (PEGDA).[58] Hydrogels are synthesized by reacting solutions of CEC with PEGDA thoroughly and holding the homogeneous solution without shaking to form a gel at 37°C. The presence of the dynamic Schiff base allows rapid self-healing. Also, in vivo gel formation after injection is very rapid.

6.13.1.4 Chitosan-Based Fission Product Isolating Agent

A transparent 3D hydrogel framework based on chitosan with rapid self-healing activity can be employed for high purity separation of fission products 152Eu from others.[59] The chitosan to formaldehyde ratio can properly be selected to obtain a product with high separation activity and suitable visco-elastic properties. The Schiff base linkage and the aldehyde and amine functional groups maintain dynamic equilibrium so that the sol-gel transition of polymeric hydrogel can be repeated over several cycles.

6.13.2 Magnetic Hydrogels

By mixing carboxy modified Fe_3O_4 nanoparticles, difunctional (benzaldehyde) poly(ethylene glycol) (DF-PEG) and chitosan, a magnetic hydrogel can be prepared (Figure 6.6).[60]

Good dispersion of Fe_3O_4 nanoparticles in chitosan solution is possible due to the interaction between NH_2 on chitosan and carboxyl groups on the Fe_3O_4 surface that leads to the formation of a stable ferrofluid and finally a magnetic hydrogel by reaction with an aqueous solution of DF-PEG. Due to the magnetic feature of the hydrogel, it can be hung pendulously by a magnet, driven like a laboratory stirrer. The dynamic structure provides shape-changing ability and the hydrogel can be manipulated from a remote location by a magnet, forcing it to pass through a narrow channel with an obstacle in the middle. During the process, there is a change of shape of the hydrogel while maintaining the integral appearance. The dynamic Schiff-base linkage, considered a quasi-covalent linkage, allows the cleavage and regeneration

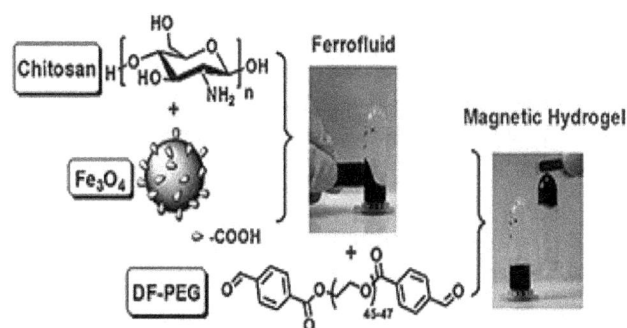

FIGURE 6.6 Synthetic scheme for the preparation of chitosan–Fe_3O_4 ferrofluid and magnetic dynamic hydrogel.

of the imine bond in the hydrogel network. Thus, the hydrogel is capable of self-healing automatically without the need for an external stimuli.[61]

6.13.3 Ion Interaction-Based Hydrogels
6.13.3.1 Acrylic Acid/Metal Ion-Based Hydrogels

Hydrogels, commonly used for soft tissues, can be designed with high mechanical properties, including high toughness, resilience, etc. A dynamic ionic interaction between the carboxyl group pertaining to polyacrylic-based hydrogel and Fe (III) and Ce (IV) can be the basis of self-healing.[62, 63] A unique cross-linking structure with tough properties and self-healing ability is achievable from the dual cross-linking of the triblock copolymer via modified micelles and ionic interactions.[64] The first cross-linking points are generated from the modification of the copolymer micelles modified with a double bond and the second originate from the ionic interaction between Fe(III) and carboxyl groups. Self-healing action occurs by the migration of Fe(III) ions and the dynamic interactions between the carboxyl groups and Fe(III) ions induce the self-healing action, which encourages the cross-linking of polymer chains around the damaged area. By bringing two fracture surfaces into contact, a distinct boundary between the two is diffused due to merging and almost disappears after 9 h. When stretched, the healed crack does not break.

6.13.3.2 Double Network Hydrogels

For hydrogels, the increases in strength and efficiency in self-healing go in a reverse direction. One of the strategies is the preparation of a double network (DN). The two interpenetrating networks improve the mechanical properties and allow the dissipation of energy efficiently, while self-healing is maintained by the construction of a highly reversible network facilitated by physical cross-linkers in one or both the networks. Tough and self-healing DN hydrogels, such as physically cross-linked agar/polyacrylic acid (PAA)-Fe^{3+}, comprise the first network as physically linked agar gel and the second network as physically linked PAA.[65] With optimum composition, the agar/PAA-Fe^{3+} gels exhibit high stiffness, high strength, good extensibility, and high toughness. The agar/PAA-Fe^{3+} DN gels, with physically linked networks via reversible hydrogen bonds and ionic bonds, exhibit 100% rapid recovery of stiffness and toughness in 15 min in a ferric ion solution and double the time under ambient conditions without the presence of any external stimuli. The easier movement of Fe^{3+} within the ionic solution promotes the bonding between Fe^{3+} and the PAA chains, and hence faster recovery. During cyclic loading-unloading, there is internal fracture of agar/PAA-Fe^{3+} DN gels at far below the yield point that spreads during the elongation process.

6.13.4 Hydrogen Bond-Based Hydrogels
6.13.4.1 Protein Hydrogels

By maintaining the synergy of two enzymes, viz., glucose oxidase (GOX) and catalase (CAT), protein hydrogels can be designed, which can undergo enzyme-mediated self-healing of

dynamic covalent bonds.[66] Glutaraldehyde undergoes a reversible covalent attachment to lysine residues of GOX, CAT, and bovine serum albumin (BSA) that is suitable for the formation and functionalization of the self-healing protein hydrogel system. Its role to support the hydrogel system is aided by a BSA scaffold whereas the pH of the system is adjusted by GOX through the addition of extra traces of glucose. Through the GOX catalysis, glucose is converted to gluconolactone, which is followed by its hydrolysis to gluconic acid so that the pH value of the hydrogel system is decreased. The oxidation of imine bonds by the H$_2$O$_2$ generated by the catalytic process is avoided by decomposing it to H$_2$O and O$_2$ by the enzyme CAT. Furthermore, GOX reuses the generated O$_2$ to accelerate the whole catalytic reaction. The imine bonds help in healing the protein hydrogel during the change of pH. As free H$^+$ facilitates this reaction, the pH range of 5–8 enhances the reaction rate, while at too high a H$^+$ concentration, the amino group is protonated. As the nucleophilic attack is too slow, the formation of the imine bond could not be kept at a minimum, resulting in reaction inefficiency.

6.13.4.2 Cell-Loaded Hydrogels

Self-healing and biocompatible dually cross-linked hydrogels are prepared by the introduction of a cell-loaded healing route. 1-vinyl-2-pyrrolidinone (NVP), acrylamide (AM), N-isopropylacrylamide (NIPAM), and O-carboxymethyl chitosan (O-CMC) are copolymerized via the radical polymerization route. This is followed by injecting normal human dermal fibroblast (NHDF) cells into the hydrogels and then incubating them in a 5% CO$_2$ environment for 48 h at 37°C.[67] The wound healing and tissue regeneration in vivo experiment is accelerated due to its enhanced hydrogen bonding ability. Dual cross-linking through hydrogen-bonding interactions between secretory proteins and polymer chains occurs due to the excretion of secretory proteins from the cells that have a large number of dangling carboxyl and amino groups. The nature of the matrix maintains excellent biocompatibility, antibacterial property, and nontoxicity.

6.13.5 Diels-Alder Reaction-Based Self-Healing
6.13.5.1 Furan Functionalized Polymer/Bis-Maleimide Gel Healing by the Diels-Alder Reaction

Self-healing has been found to be possible with a thermally reversible Diels-Alder (r-DA) cycloaddition reaction and a photoreversible [2+2] cycloaddition reaction. Maleimide (PA-MI) and furan (PA-F) pendent groups decorated polyamide undergo cross-linking via an r-DA cycloaddition reaction (Figure 6.7).[68] This strategy increases toughness, T$_g$, and the mechanical property together with self-healing behavior. Similarly, a gel is formed by cross-linking furan-functionalized polyketones (PK-furan) with bis-maleimide.[69] The polymer gel is completely turned back to clear and fluid solutions as before cross-linking. The thermally remendable nature of cross-linked PK-furan can be confirmed by first converting the cross-linked PK-furan into uniform bars by compression molding, at an elevated temperature, that behave like linear thermoplastics during heating. This behavior, such as remeltability, reprocessability, and recyclability, originates from

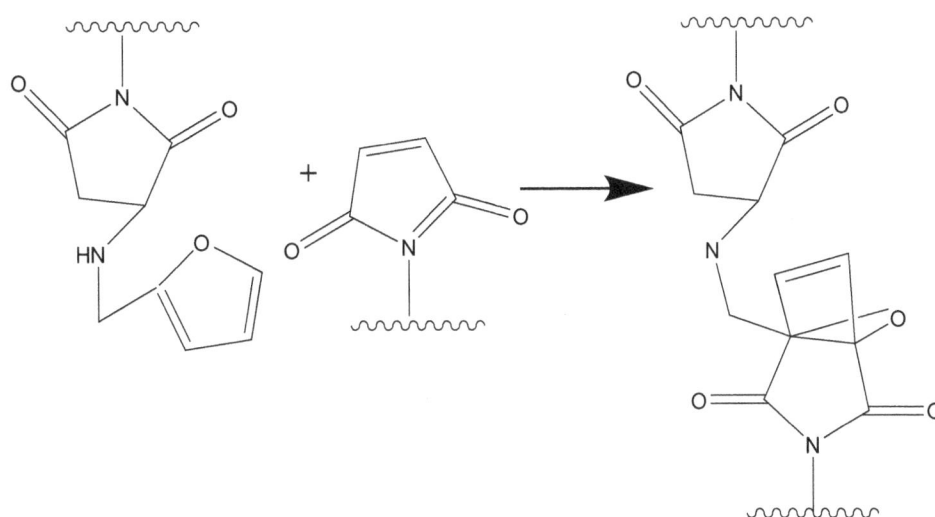

FIGURE 6.7 Schematic representation of Diels-Alder reaction of pendant furan and maleimide groups of polyamide chain.

the opening of the DA adduct. By slow cooling, a rigid polymer network can be obtained due to the regeneration of the DA adduct. The strategy, allowing the recyclability and reworkability of this system, can have a tremendous impact on waste plastic management.

6.13.6 Supramolecular Hydrogels

6.13.6.1 pH-Sensitive Supramolecular Hydrogels

The cell adhesive and biocompatible self-healing supramolecular hydrogels have several applications in biomedical fields, such as cell culture, transmission, and treatment, etc. Polycarboxylic acid containing polymer, such as poly[(methyl vinyl ether)-alt-(maleic acid)] (P(MVE-alt-MAc)) and its derivatives, with a hydrophilic, biocompatible, and bioadhesive nature is suitable for the purpose. In biotechnology, pharmacology and health care, such type of hydrogel act as a thickening and suspending agent, denture adhesive, mouthwash, etc.[70] P(MVE-alt-MAc) has the ability to exhibit polyanionic characteristics to the polymer backbone, and also can undergo chemical modification to form an amide linkage with amine and an ester linkage with hydroxyl groups. Supramolecular hydrogels can be obtained by mixing the aqueous solution of P(MVE-alt-MAc)-g-β-cyclodextrin (β-CD)/P(MVE-alt-MAc)-g-adamantane (Ad) that possesses pH-sensitivity and also is able to repair itself after being damaged.[71] The protons of Ad and those in the cavities of β-CD, being in immediate proximity, lead to the entrapment of the former in the cavities of the latter, forming inclusion complexes. The self-healing process takes place via the formation of supramolecular networks; the fastest occurs at pH2.

6.13.7 3D Printing of Hydrogels

3D printing has become a popular tool for the preparation of materials in the field of biotechnology, sensing, actuation, etc. Hydrogels with various functional properties for applications as an actuator, a scaffold for tissue engineering, should have a self-healing capability for wide area application. The 3D printing technology is associated with secondary processing to reinforce the soft 3D-printed object, which includes cross-linking by heat or UV radiation, physical cross-linking, thermally-induced gelation, etc. However, additional materials are added to the matrix in the process. Keeping the disadvantages in mind, macroporous gels that are doubly dynamic, self-healing, 3D-printable can be developed via a thermally-induced phase separation technique (TIPS).[72] After the 3D printing, the thermodynamically-driven cross-linking of the polymer-rich phase is accomplished by applying freeze-thaw cycles that result in the mechanical stabilization of macroporous soft gels. The formation of hydrogen bonds between various chains generates macroscopic interconnected pores. The hydrogen bond formation is facilitated by the TIPS treatment and there is improvement of the mechanical properties of the soft objects. Variation in the conditions of the freeze-thaw process can be used to manipulate the mechanical properties of the self-healing cryogels. As an example, the self-healing networks of copolymers of substituted acrylamide and methyl vinyl ketone form oxime bonds with a hydroxylamine cross-linker, which helps in the formation of stable dynamic covalent bonds, make the gel self-healing in nature.

6.14 Self-Healing Coatings

Functional coatings, such as anticorrosion and antimicrobial coating, superhydrophobic/superhydrophilic coating, UV-absorbing coating, etc. have become very important for the whole system to run smoothly. However, the coatings mostly suffer from reduced service life in the case of mechanical damage. As an example, damage of anticorrosive coatings allows the corrosion-causing agents to penetrate through cracks to cause rapid corrosion, resulting in reduced service life. By introducing a self-healing ability into the coating, the service life of the coating can be extended. Porous nano/microcarriers and core-shell capsules have been developed as paint additives for the prevention of corrosion, as corrosion inhibitors or monomers can be loaded into the pores. During mechanical damage, the released inhibitors/monomers form a coating on the bare substrate so that it is isolated from the corroding media and this prevents corrosion. Porous nanoparticles function when there is a change in pH due to the onset of corrosion. The nanoparticles do not weaken the coating and are effective as long as it has the ability to release active reagents. On the other hand, the special core materials of the core-shell capsules form a coating layer that does not allow the corrosive media to come into contact with the substrate. However, the mechanical properties of the microcapsules-containing coatings are mostly weaker.

6.14.1 Microcapsule-Based Self-Healing

6.14.1.1 Linseed Oil-Filled Microcapsules

Microencapsulated drying oil (e.g., linseed oil) and corrosion inhibitors with urea-formaldehyde as a shell are synthesized by in-situ polymerization and are added to both the primer and topcoat formulations.[73] Normally, the rough morphology of microcapsules ensures sufficient bonding with the matrix. Mechanical damage creates cracks, together with the breaking of the microcapsules. The released active agent, the drying oil, together with corrosion inhibitors, comes out and fills the crack within 90 s and cures slowly. For better curing behavior, the microcapsules are loaded with linseed oil together with cobalt octoate as the dryer material and/or octadecylamine (ODA) as a corrosion inhibitor.[74] Although the presence of cobalt octoate decreases the healing time and ODA increases it, the mixture works favorably, to cure in a short time. The presence of ODA improves the electrochemical resistance, indicating higher corrosion resistance.

6.14.1.2 Epoxy-Filled Microcapsules

Microencapsulated epoxy resin with a urea-formaldehyde shell can be dispersed in an epoxy resin matrix together with a catalyst to function as a self-healing coating.[75] Experimental damage can be done with a blade and self-healing can be observed to occur by maintaining the samples at 60°C for 72 h. During crack formation, the microcapsules break and the encapsulated epoxy resin flows to the crack plane. Healing takes place with the help of a catalyst dispersed in the matrix. In the corrosion test, the self-healing effect can be understood, as the current flow is too small compared to non-healed samples.

6.14.1.3 Polyurethane Paint in Microcapsules

Water-borne polyurethane (PU) paint-based microcapsules can be used for self-repairing protective coatings.[76] The microcapsules are produced in two steps: the first step comprises the synthesis of the prepolymer from butanediol and diisocyanate, which is used in the second stage to react with butanediol to encapsulate water-soluble PU via interfacial polymerization. The capsules possess the required long-term storage stability. During a study on self-healing, the self-curing for 3–4 h leads to significant self-repairing, which displays protection performance with

47–97% efficiency. Larger diameter capsules have higher healing efficiency.

6.14.1.4 Hydroxyl-Terminated Polydimethylsiloxane

Self-healing coatings can be made by incorporating microencapsulated catalysts (tin compound) and a phase-separated or encapsulated healing agent [hydroxyl terminated polydimethylsiloxane (HTPDMS) and polydiethoxy siloxane (PDES)] droplets in a matrix, on a metallic substrate.[77] Practically no reactions take place between the HTPDMS and PDES in the absence of a catalyst. When the self-healing coating layer is damaged by cracking or scratches, the contents of the microcapsules come out and undergo diffusive mixing in the damaged region. Finally, the healing of the damaged portion occurs by cross-linked HTPDMS to provide protection to the substrate from the environment. Enhancement of the adhesion of the silicone-based healing agent can be performed by using a methylacryloxy propyl triethoxy silane. A self-healing coating formulation of epoxy vinyl ester matrix is made by incorporating a phase-separated healing agent (96 vol% HOPDMS and 4 vol% PDES), PU-microencapsulated dimethyldineodecanoate tin catalyst solution and methylacryloxy propyl triethoxy silane adhesion promoter, respectively. The self-healing occurs at 50°C over 24 h.[78] A silane-based catalyst is used to cure PDMS efficiently at ambient temperature and does not require moisture for activation.[79] In fact, for aerospace or buried interface applications, the catalyst is very useful. For a coating containing dual microcapsules for both catalyst and curing agent, the self-healing is also quite satisfactory. In fact, for non-compatible components, the strategy works well. It cures at an ambient temperature over 24 h and provides an excellent self-healing performance.

6.14.1.5 Silyl Ester

Self-healing by using the double capsule method or the single capsule method containing a corrosion inhibitor suffers from several drawbacks: the first is the lack of the exact probability that both the components are available at the same time, while the second is the uncontrolled release of the corrosion inhibitor. To make a system without the above drawbacks, an encapsulated, water-reactive healing agent based on a silyl ester has been developed, which exhibits healing performance when incorporated in a coating.[80] The healing mechanism originates from the release of silyl ester at the damage site to form an adhered metal barrier hydrophobic system by reacting with moisture or water in the environment. Thus, it protects the metal from further corrosion attack as well as avoiding the unnecessary release of corrosion inhibitors from the crack sides when inhibitors are present. The octyldimethylsilyloleate (silyl ester) healing agent incorporated in one single capsule can theoretically give protection to a scratch of about 100 μm width and 0.7 mm length.

6.14.1.6 Solvents as Healing Agents

For self-healing of microcracks, a solvent can be used as a healing agent that reacts with the polymer to heal the microcracks. The halloysite nanotubes are used to encapsulate the solvents to impart self-healing ability to epoxy.[81] During crack propagation through the embedded nanotubes in the polymer, liquid solvent, such as dimethylsulfoxide and nitrobenzene, is released into the crack plane. Thus, by using this strategy, both strengthening the matrix and self-healing can be realized. The same behavior is exhibited by the CNT-based system.

6.14.2 Layer-by-Layer Coating

Fabrication via the layer-by-layer (LbL) technique is a unique method to design a nanocontainer for corrosion-inhibiting agents, which is released in a controlled fashion to function as a self-healing agent. This method is unique by design as the number of layers and ingredients can be regulated as desired. The electrostatic interaction is used to deposit the layers comprising charged species of inhibitors, nanoparticles, and polyelectrolytes.[82] In this type of coating, the release behavior and permeability for the inhibitor are controlled by manipulating the polyelectrolytes nature, surrounding stimuli, and multiple functionalities. The important advantage of the multilayer strategy of incorporation of the corrosion inhibitors in the polyelectrolyte for use as self-healing protective coatings is the isolation of the inhibitor, which does not interfere with the curing or the coating behavior and provides its controllable release according to changes in the local humidity and the pH.[83] Normally the trigger is the changing pH during corrosion. Capsules can be opened and closed by varying the pH of the surrounding medium. By increasing the ionic strength, the polyelectrolyte capsule can be opened. Also, a non-polar solvent damages the integrity of the polyelectrolyte and opens the capsule. By introducing photoresponsive material into the composition, the capsule can be opened by irradiation. Ag nanoparticles or IR dye can be incorporated into the polyelectrolyte shells of sulfonated polymer/amine hydrochloride polymer capsules to make it photoresponsive.[84] Under laser illumination, the capsules are deformed or cut, that triggers the release of encapsulated materials. A low power near-infrared laser diode with continuous-wave configuration can be used for the process.

Similarly, introducing magnetic component into the capsule allows it to be opened by magnetic treatments. Gold-coated cobalt (Co@Au) ferromagnetic nanoparticles are incorporated inside the capsule walls.[85] By applying external alternating magnetic fields, the Co@Au nanoparticles are rotated that creates disturbance and distortion in the capsule wall, leading to significant increase in its permeability.

Capsule opening is also possible by introducing an oxidation/reduction phenomenon of the capsule shell. Redox materials, such as a conducting polymer, can be incorporated into the polyelectrolyte microcapsules as micro-containers. Incorporating the capsules inside the conducting polymer (polypyrrole, PPy) film results in a composite electrode that can be designed to provide the combined properties of electrocatalytic and conducting PPy with the storage and release properties of the capsules.[86] Switching between the open and closed states of the capsule is now controlled electrochemically with the help of this electrode.

The release from microcapsules can also be operated by using an enzyme. Opening of capsules via enzymatic treatment depends on the biodegradable components in the shell. Ultrasonic treatment is another way to irreversibly open the capsules with a nanoparticles loaded shell.

6.14.3 Supramolecular Cross-Links

Some of the cross-links in a polymer can be replaced by reversible cross-links that are provided by dimers of hydrogen-bonding ureidopyrimidinone (UPy) moieties, resulting in superior relaxation of stresses even below T_g. The technique can be applied in coatings technology (Figure 6.8).

For a coating on a metal substrate, the stress is experienced due to the difference in thermal expansion/contraction between the coating and the substrate and also due to the increase in density upon cross-linking and shrinkage after evaporation of the solvent. UPy units, incorporated into a polyester-polyurethane networks-based coating on metals, can combine several properties of thermoset materials and thermoplastic materials.[87] Together with improved material strength and toughness, an additional relaxation mechanism is induced by UPy, leading to the softening of materials above T_g. By incorporating 20% supramolecular cross-links, there is a substantial increase in the fracture energy of the material. These materials are ideally suited for industrial application because of their higher creep compliances below T_g. Completely autonomous healing is possible as no external stimuli are required for materials to recover from stress and they also have the ability to be repeated indefinitely.

6.14.4 Shape Recovery Coating

Shape memory (SM) polymers can be used as self-healing protective coatings for metals and alloys and their ability depends on storing a new shape, due to deformation and then restoring the original shape with exposure to external stimulants, such as light or heat. Elastic polymeric networks, of a suitable stimuli-sensitive nature, can be used in the textile industries, in electronic devices, in packaging or medical devices. If the film is damaged, the film can be heated to a temperature higher than the lowest thermal transition (T_{trans}), acting as a switching segment.[88] The temperature above T_{trans} allows the relaxation of the switching segment for shape recovery.

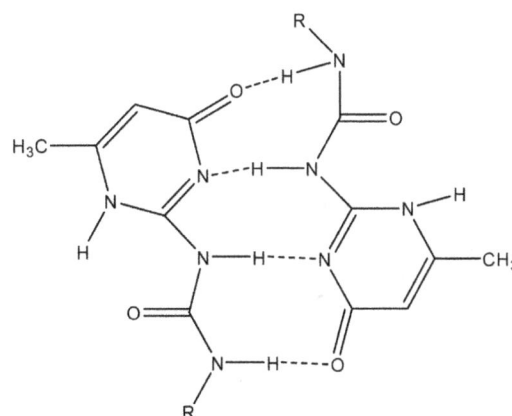

FIGURE 6.8 Hydrogen bonded dimer of ureidopyrimidinones.

6.14.4.1 Cross-Linked/Linear Polycaprolactone Self-Healing Coating

A cross-linked-polycaprolactone (PCL) network can be blended with linear PCL to exhibit shape recovery properties.[89] Self-healing, assisted by shape memory, occurs via the closure of the scratch by the cross-linked PCL network followed by self-healing assisted by the linear PCL chain.

6.14.4.2 Epoxy/PCL Fiber-Based Self-Healing Coating

A slightly modified epoxy-based shape memory coating is made by incorporating electrospun thermoplastic PCL fibers in a random distribution.[90] With the application of heat, the shape memory effect of the matrix acts to close the crack and simultaneously bond the crack surfaces through the melting and flow of the PCL fibers into the damaged region. However, the PCL fibers allow an application below 60°C due to their low melting point. The Diels-Alder adduct-based system can be used to enhance the operating temperature above 60°C.

6.14.5 Microvascular Networks

Similar to biological systems, the interpenetrating micro-vascular networks, carrying reactive components, can be embedded in synthetic materials for a self-healing performance. The circulation and delivery of reactive components are performed via microvascular networks that impart self-healing properties and others. The bio-inspired coating/substrate can be designed by incorporating a three-dimensional micro-vascular network to deliver a healing agent to the cracks.[91] By this strategy, a single network is sufficient to heal a single crack up to seven times.

6.14.5.1 Dicyclopentadiene/Grubbs' Catalyst-Based Microvascular Self-Healing

The dicyclopentadiene monomer is loaded into the embedded network, while the second component, a first-generation Grubbs' catalyst, belongs to the coating matrix. However, this strategy suffers from the drawback of depleted availability of the solid-phase catalyst, which reduces with repeated healing. This limitation can be tackled by using a multi-network architecture for delivery of both parts of the healing system. An isolated microvascular network can be produced that can be independently filled with resins and curing agents respectively. During crack formation, the active ingredients are released, which fill the crack plane and heal the crack around the ambient temperature.[5]

6.14.6 Silicone Coating for Marine Anti-Biofouling

Poly(dimethylsiloxane) (PDMS)-based coatings are known to provide a low surface energy surface, which allows only weak adhesion of marine organisms. However, the PDMS-based coatings are mechanically weak and adhesion to the substrates is poor and it is hard to repair the damaged area because of their cross-linked network structure. The mechanical properties of PDMS-based coatings are chemically modified with epoxy[92] and polyurethane.[93]

6.14.6.1 Polydimethyl Siloxane-Based Polyurea Self-Healing

By combining the urea groups and dimethyl siloxane segments in PDMS-based polyurea, a self-repairing action can be imparted to the matrix, along with good adhesion to the substrates.[94] The breaking and reconstruction of the hydrogen bonds are the main action during the self-repair of the PDMS-polyurea and by nature they are sensitive to temperature. PDMS-based polyurea is also used to hold the organic antifouling agent and release it in a controlled manner even when the surface remains stationary. When the two cut pieces are put together, they repair completely within 48 h under ambient conditions. In fact, the self-healing is also evident in artificial seawater for both PDMS-based polyurea and its formulation with antifoulant.

6.14.7 UV-Blocking Coatings

Self-healing UV-protective coatings are always a tricky issue as the compatibility of absorbers, self-healing of cracks, and maintenance of functions are to be optimized by the addition of UV absorbers. Generally, polymeric UV-protecting coatings contain inorganic particles or organic molecules as UV absorbers. However, a self-healing action can be added to the coating by connecting the UV absorbers and the matrix through host-guest interactions. A typical system comprises TiO_2 nanoparticles (stabilized by β-cyclodextrin) and acrylate copolymers as UV absorbers and as a polymeric matrix, respectively.[95] The resulting coating has both properties, e.g., blocking UV irradiation and healing the crack in the presence of water. Further, the UV-blocking property of the coating is completely recovered after healing. The coating shows repeatable self-healing behavior. Because of the dynamic interactions between the UV absorber and the polymer, there is no need to carry out any modification of the basic chemical structure of the polymer backbone. This adds to the versatility of the approach, as a variety of polymers can be used to formulate the coating. Self-healing behavior can be observed within 24 h for an abrasion-induced rough surface after putting a drop of water on it. The polymer chain entanglement and host-guest interactions play important roles in the self-healing process. When the coating is abraded, β-cyclodextrin (β-CD)–Adamantane (Ad) links break at the fracture surface. After adding water, the polymer chain movement increases, resulting in expansion of the coating volume. This facilitates both sides of the scratches to come into contact so that the exposed β-CD molecules and the Ad groups, "eager" to re-associate across the interface, do the self-healing.

6.14.8 Coaxial Electrospinning of Coatings

An electrospun coaxial healing agent incorporated into a self-healing polymer coating can be successfully used on metal substrates. Encapsulation of the healing agent or a catalyst-solvent mixture in a core-shell bead-on-string morphology can be performed via electrospinning onto a substrate.[96] The advantage of this system is that only physical forces are used to form the core/sheath structure, as against chemical methods where minimum emulsion stability is required for chemical reactions to occur and hold the capsule together.[97] Also, by manipulating the experimental design, the diameter of the microcapsules and connecting ligaments can be controlled. The advantages of electrospinning enable it to apply self-healing treatment to substrates across a large area under mild conditions. PVP can be used for encapsulation of polysiloxane-based healing agents in a sheath by taking the solution of the latter in a syringe with a metallic needle and the former in another syringe fitted with the inner polyethylene capillary. A polyethylene capillary tube is inserted into the stainless steel needle to make the spinneret nozzle.

6.15 Self-Healing Textiles

For long-term use of woven or non-woven textiles, incorporating self-healing materials is beneficial. Layer-by-layer films of polyelectrolytes are considered to be one of the foremost approaches for designing coatings with self-healing ability due to the mobility of the components. Films can be comfortably deposited on textiles by using environmentally safe water as a solvent. A large variety of materials can be incorporated to carry out the self-healing. In a typical process, polystyrene sulfonate (PSS) and squid ring teeth (SRT) are alternatively adsorbed to build a multilayer coating.[98] The soft amorphous regions of the protein, having the ability to form a supramolecular self-assembly, enable the matrix to exhibit self-healing because of its ability to form a strong adhesive bond in the presence of water and to induce a reversible glass-to-rubber transition. During self-healing, the SRT protein deforms and softens in water above its T_g, while the hydrogen bonds are reversibly maintained in the amorphous region. The polyelectrolyte chain diffusion rate increases in water or diluted saline water, which enhances the speed of the healing process. By applying pressure and warm water, SRT-coated textile samples can be repaired.

6.16 Self-Healing of Polymer Blends

6.16.1 Ionomer-Polycyclooctene Cross-Linked Polymer Blends

Cross-linked polycyclooctene is an outstanding shape memory polymer. However, it does not show self-healing behavior. The strategy of blending this polymer with others which have self-healing behavior is considered useful. Ethylene-methacrylic acid (EMAA) ionomers are known to form ionic clusters after acid is neutralized with metal salts, including that of sodium, zinc, and others. The resulting supramolecular interactions lead to unique mechanical properties, together with high healing degrees, while maintaining excellent mechanical integrity. Thus, the self-healing and multiple shape memory events are exhibited by a blend of a self-healing ionomer and a cross-linked polymer, such as polycyclooctene with a composition of (70/30 by wt).[99] The stiffness recovery after healing in standard ionomer healing conditions is almost 100%.

6.17 Conductive Polymer Composites

Development of stretchable electronic materials with a self-healing capability has potential for application in skin-like electronics and soft robotics. Loading conductive fillers into a self-healing network is a common approach, which is not foolproof as it interferes with the stretching characteristics. A better approach is the development of self-healable electronic materials with reasonable stretching characteristics. For applications in skin-like electronics, commonly used conducting polymers (CP) such as polyaniline (PANI), polypyrrole and derivatized polythiophene can be employed via structural design and molecular engineering.[100] As an example, aniline can be polymerized to PANI in the presence of a soft polymer with the carboxylic acid group, e.g., polyacrylic acid (PAA) and phytic acid (PA).[101] The PAA chains provide hydrogen bonding, electrostatic interaction, doping, and flexibility, while the PA provides doping and electrostatic interaction, and hydrogen bonding. It is apparent that for self-healing, via the restoration of electrical and mechanical properties in ambient conditions, hydrogen bonding and electrostatic interactions are indispensable. Additionally, they participate in the dissipation of mechanical energy under stretching. The film formation of polymer results in a porous film, which helps in increasing the sensitivity and deformability in mechanical sensors. Even after stretching beyond 200%, the film remains intrinsically conducting. During self-healing, the electrical properties recovered faster than the mechanical properties.

When designing self-healing conducting polymers, recovery of mechanical damage and electrical conductivity are given attention. After damage and self-healing, an additional strategy is needed to restore the optical properties. Irradiation-induced bond breaking, mostly photooxidation, is healed in a system designed by imine-based donor-acceptor CP.[102] Restoration of their absorption properties is done via imine metathesis under heating, which is possible without the need of an acid catalyst. In a solid state, the chromophores cannot move freely due to steric hindrance. However, the exchange takes place between imine bonds in close proximity. At a higher temperature, the mobility of chromophores increases, which, in turn, increases the extent of the reactions.

6.18 3D-Printed Polylactic Acid Blends

One of the most versatile and cost-effective methods for 3D printing employs fused filament fabrication (FFF). However, lack of adhesion at the interfilament junctions is likely to result in non-uniform mechanical strength. Even with various modifications, the problems remain partially unsolved. Among various design approaches, the dynamic covalent chemistry of the Diels-Alder reaction is known to enhance the strength as well as the toughness of FFF 3D printable materials.[103] During the FFF 3D printing process, the partially cross-linked terpolymers with furan/maleimide Diels-Alder (fmDA) functionality undergo depolymerization at elevated temperatures, due to the retro-DA reaction.[104] Upon cooling at low temperatures, the repolymerization occurs via fmDA adduct formation. Thus, reversible Diels-Alder chemistry is used to improve the mechanical properties to a considerable extent. A conventional FFF 3D printer can be used to process a

3D printable polymer ink, comprising fmDA and polylactic acid, without much decrease in its mechanical strength after printing.[105] The polymers, thus obtained, retain their remending nature after printing along with improved strength. When two broken ends are connected and heated at 120°C for 5 s and brought into contact and cooled and maintained at 65°C for 15 min and cooled, the blend (100% cross-link density in fmDA polymer) recovers 77% of its ultimate strength. The fmDA linkages remain active even after many thermal cycles.

6.19 Self-Healing Polymer Composites

Unlike traditional tough and static composites, the self-healing composites are able of adapt to environmental changes as well as undergo automatic recovery. The safety and reliability will improve through self-healing, as no catastrophic failure will occur. Also, the artificial composites will be maintainable at lower cost and the material life will be longer. Of the current self-healing composites, intrinsic self-healing materials are considered the most important. For capsule-based/vascular designs, various aspects such as healing agents, or considerations of various processes, such as rupture, mixing and microstructure fabrication techniques are required. For intrinsic self-healing composites, developing new healing mechanisms to achieve higher healing strength are the only consideration.[106]

Because of the stability of the components of epoxy resin, its easy encapsulation and straightforward reaction during damage to heal cracks, the epoxy-based healing reaction has gained in popularity. In areas such as construction, the capsule-based structures have become popular.[107] To heal the cracks in concrete, cyanoacrylate-filled cylindrical glass capsules are used.[108] For cement-based composite materials, a two-part epoxy system is also employed for self-healing.[109]

6.19.1 Vascular Self-Healing Composites

Healing of the vascular and circulation systems in animals is mimicked in the vascular self-healing composite which can be refilled to provide continuous delivery of a healing agent and flow control, so that the damage is immediately taken care of. In this system, the main effort is the fabrication of the vascular system. As a general strategy, the hollow fibers are aligned similarly to normal fibers. The hollow fibers can be aligned parallel to the reinforcement fibers so that these have minimal influence on the mechanical properties of the host composite.[110] However, it is not possible to reconnect the hollow fibers to design interconnected networks that make their refilling difficult.

6.19.1.1 Sacrificial Fibers/Scaffolds

Sacrificial fibers/scaffolds are the 3D structures formed by materials that can easily be removed, dissolved, or degraded. After integration inside the polymeric host material, such a structure is expected to remain unchanged during the curing process of the host material. After complete curing, the sacrificial fibers/scaffolds are removed by either of the processes, such as physical methods, temperature increase, or pH change of the

environment.[111] The drawbacks of this strategy are the possibility of a 1D structure only and the creation of invisible damage during the processes adopted after curing of the matrix.

6.19.1.2 3D Printing of Sacrificial Scaffolds

Design of a 3D network system can be done by using 3D printing technology. The drawback lies in the complete removal of the sacrificial system. Poly (lactic acid) (PLA) acts as a very good sacrificial material as it is converted into gas during thermal depolymerization at a high temperature. The sacrificial structures are constituted by 3D-printed PLA.[112] Further modifications can be done to make it more convenient.

6.19.1.3 Melt-Spinning and Electrospinning

After making the fibers by any of the processes, these are embedded in a matrix and cured. Later by adopting the dissolution method, the sacrificial materials are removed to make hollow fibers. As an example, sugar fibers made by these methods are embedded in the PDMS matrix and cured.[113] The matrix after curing is immersed in water to remove the sugar fibers by dissolution.

6.19.1.4 Replication of Existing Patterns

The pattern of a vascular network is replicated by first building it and then building a "negative" version of the structure on the substrate made of glass or pre-oxidized PDMS.[114] The cavity in the substrate is filled with gelatin, which is allowed to solidify and placed inside hydrogel and the system is heat-treated. After the gelatin has melted, it is automatically removed to leave a microvascular network inside the hydrogel.

6.19.2 Electrostatic Discharge

A microvascular network with natural designs can rapidly be fabricated by applying an electric discharge. As an example, irradiation of PMMA with an electron beam results in the accumulation of electrical charges inside the material. By connecting this specimen to the ground, a rapid discharge similar to lightning occurs that results in the formation of a tree-shaped, branched microvascular network inside the specimen block (Figure 6.9).[115]

6.19.3 Laser Direct-Write

The technique, known as laser direct-write, can be adopted to fabricate channels in micro-fluidic devices. These channels perform the function of vessels in vascular systems.[116] The microchannels on a PDMS block are manufactured by employing a high-brightness diode-pumped Nd:YAG slab laser. Also, very complex two-dimensional (2D) patterns are fabricated by the lasers directly on polymeric materials. The vascular network architectures do not significantly alter the tensile behavior of fiber-reinforced composites when the fiber alignment remains unaltered and there is no distortion of the reinforcement fiber architecture.[117] The healing efficiency of the vascular self-healing composites is mostly below 80%, which is lower than the capsule-based system. Healing performance is significantly

FIGURE 6.9 Schematic 3D vascular pattern produced by electrostatic discharge.

affected by other structural and dynamic factors.[106] Vessel size and roughness determine the extent of the liquid flow during damage, which determines the efficiency of the healing. Vessel size and roughness also have a role to play in healing as a high amount of the hollow vascular structure inside the composite allows sufficient flow of the liquid for healing, while it decreases the mechanical properties. It is established that high healing efficiency is exhibited at suitably high temperatures and has a long healing time. The amount of outflow of the healing agents and the mixing quality of the multiple parts can be manipulated by the hydraulic pressure applied inside the vascular network. Moisture and oxygen have also a role to play during the healing.

6.19.4 Intrinsic Self-Healing Materials

The inherent reversibility of bonding of the matrix polymer has been used during repair via intrinsic self-healing. Various physical parameters, such as hydrogen bonding, ionic interaction, thermally reversible reactions, a dispersed meltable thermoplastic phase, and molecular diffusion are commonly used for intrinsic self-healing.[118] For intrinsic systems, the problems associated with the integration of an isolated healing agent and its compatibility are avoided as the matrix is inherently self-healing. However, it has to be remembered that the desired mechanical, chemical, and optical properties required for particular applications are maintained. The components of a self-healing system, that undergo reversible reaction, can be reversibly transformed from the monomeric state to the cross-linked polymeric state through the addition of external energy. After damage, the enhanced mobility in the damaged region for bond reformation, and self-healing can be achieved by application of heat or intense photo-illumination. The most common strategy is the use of the Diels-Alder (DA) and retro-Diels-Alder (rDA) reactions. For thermoset materials, self-healing can be achieved by incorporating a meltable thermoplastic polymer, which has the ability to be redispersed into the

crack plane after melting and becoming mechanically interlocked with the surrounding matrix after cooling. Ionic clusters, made by ionic segments of ionomeric self-healing copolymers, can act as reversible cross-links, and can be activated by external stimuli such as temperature or ultraviolet (UV) irradiation. The self-healing supramolecular assembly is formed via multiple complementary, reversible hydrogen bonds available with polymers with end group and/or side group. Time- and temperature-dependent void closure, surface interaction, and molecular entanglement are involved in molecular diffusion-based self-healing. Additionally, the molecular diffusion can inhibit corrosion. A $CeCl_3$ inhibitor solution and a polymer-coated zinc cathode exhibit self-healing barrier properties.[119]

6.20 Self-Healing Polymer Nanocomposites

Polymer nanocomposites, comprising polymer and nanomaterials, form an important class of system having excellent strength and reasonable elongation and other mechanical properties, together with designed functional properties. However, in case of damage, there are serious problems with delamination, cracks, etc. This deficiency can be fixed by introducing self-healing characteristics.[120] Computer simulation is used for the theoretical study, which establishes that by adding nanoparticles to the polymers, the nanoparticles can be confined to the nanoscale cracks that practically form "patches" to repair the damage.[121] The optimal conditions for the nanoparticles to act as responsive, self-assembled "band-aids" for composite materials can be predicted from the properties of the undamaged, damaged, and healed state systems through micromechanics simulations. By using this approach, it is found that the mechanical properties of the repaired composites can be restored to 75–100% of the undamaged material. By using molecular dynamics to study CNTs as nanoreservoirs for self-healing applications, their actions as both self-healing containers as well as the reinforcing component have been established.[122] It is also possible that the size of the crack decides the amount of organic molecules (healing agent) released during self-healing. By employing self-consistent field theory, it is established that in thermally reversible reactions, various components can be assembled into hierarchical structures having nano-sized mineral bridges.[123] The neighboring platelets are connected by the bridges and a certain quantity of nanoparticles migrates to the platelet surfaces, resulting in an interphase layer between the organic and inorganic platelets phases.

6.20.1 Nanocomposite Healing by the Diels-Alder Reaction

Along with other techniques, the properties of polysiloxanes can be improved by incorporating inorganic nanoparticles. The nanoparticles can be suitably designed to act as a cross-linker and constitute self-healing composites. Modified inorganic nanoparticles together with polymers of low viscosity generally increase the DA/rDA reaction rates. The hydrosilylation of the substituted spherosilicate with furfuryl allyl ether results in the formation of an inorganic nano-building-block, which can be incorporated into various reversible Diels-Alder reaction-based self-healing hybrid materials. By modifying the spherosilicates with furan, the healing can be completed within minutes.[124] An elastomeric material results in the substitution of this type of cross-linker with an oligomeric siloxane substituted bismaleimide.

Synthesis of nanocomposites can be done by cross-linking poly (butyl methacrylates) and various polysiloxane structures with surface modified silica nanoparticles via a Diels-Alder reaction.[125] Terminally modified polymers, with high mobility and enough space around the Diel-Alder moieties, are preferred in nanocomposites. As the mobility of the non-end functional groups is controlled by the diffusion characteristics of the polymeric chains in block and homopolymers, they are less suitable for self-healing composite preparation. The healing of the nanocomposites does not occur even at high temperature due to less proximity of the DA groups to react. By adding chloroform repeatedly during heating, the self-healing can be performed quite satisfactorily as the addition of chloroform increases the mobility of the polymer chains.

6.20.2 Bromobutyl Rubber/CNT Nanocomposites

Introduction of nanomaterial can also enhance the ionic interaction. As an example, a chemically attached imidazolium group on BIIR can be mixed with CNT to enhance properties and performance.[126] The ionic networks formed by the ionic imidazolium group exhibit definite self-healing behavior. The improved rubber-filler interaction, originating from the cation-π bondings between the modified BIIR and the CNT surface, enhances the mechanical performance and electrical conductivity of the composites. Clusters formed by the diffusion of the ionic groups act as physical cross-linking points that are responsible for the self-healing of the ionically modified BIIR. Heating above ambient temperature is the requisite for diffusion of the ionic groups through the polymer matrix. Heating across a damaged surface is provided by an electrical current. The healing process is an accelerated Joule heating effect due to the presence of CNT.

6.20.3 Nanocomposite Hydrogels

The main idea of designing nanocomposite hydrogels is to make them tough along with the ability of self-healing. Production of the nanocomposites is performed by introducing nanofillers during polymerization. For example, hydrogels are synthesized from the cationic polyacrylamide and graphene oxide (GO) by free-radical polymerization of AM, N,N′-methylene-bisacrylamide, and 2-(dimethylamino) ethyl acrylate methochloride (DAC) in the presence of GO, which acts as a micro cross-linker.[127] The GO content and the mass ratio of AM and DAC are varied to optimize the mechanical properties as well as the self-healing ability. Dissipation of energy and rebuilding the networks are dependent on the hydrogen bonds between AM and GO and the ionic bonds between DAC and GO. Recovery of toughness and stiffness varies in the range of 98–69% and 98–91% respectively up to 20 cycles. Healing efficiency increases to a large extent when water is incorporated, which originates from the reformation of ionic and hydrogen bonds via water assistance. The reformation of the

loose and weak contacts between polymer chains and oxygen groups on GO is facilitated by the swelling process with water in the composite hydrogels. The diffused polymer chains are driven to closely interact with the oxygen group on the GO sheets during subsequent drying that results in the formation of additional cross-links in the final hydrogels. The high mechanical strength of the hydrogels after self-healing is due to the re-established electrostatic interaction and hydrogen bonding.

6.20.4 Conducting Nanocomposites Hydrogels

By adopting a strategy of insertion of hard nanosheets into the matrix, both the mechanical strength and retention of the mobility of a polymer chain can be improved. 2D boron nitride (BN) nanosheets, having interesting electronic thermal and mechanical properties, excellent chemical stability, and biocompatibility, can be functionalized and their nanocomposite is made with poly(N-isopropylacrylamide) (PNIPAM) and clay to produce hydrogels with conducting and self-healing properties.[128] The hydrogel forms a large number of hydrogen bonds among various functional groups including -OH,–CONH, and many others, so that the mechanical energy dissipation is quite high and the self-healing abilities are retained by maintaining the mobility of the polymer chains. Further, a range of pores is developed that are connected by micropores, which contribute to maintaining high mechanical performance.

6.20.5 Polyurethane Nanocomposites Based on the Diels-Alder Adduct

Polyurethane based on the Diels-Alder chemistry (PU-DA) can be employed to make a composite material. The reaction product of furfuryl alcohol and bismaleimide is further reacted with isocyanate terminated PU, along with covalent linking with functionalized graphene nanosheets (FGNS). The resultant composite is mechanically robust and has self-healing ability under ambient conditions with the help of infrared (IR) laser irradiation.[129] At a higher temperature, the DA adducts undergo retro-DA reaction by disconnecting the furan and maleimide moieties so that they have enough mobility to heal the fractured place of the sample via reconnection of the network during the natural cooling down to room temperature. Graphene has good IR absorbing capacity and, therefore, the presence of FGNS enables the nanocomposites to show strong IR absorption that makes FGNS-PU-DA nanocomposites self-healing in nature, together with photo-thermal conversion ability. In a similar way, self-healing PU based on a DA reaction can be synthesized from isophorone diisocyanate and polypropylene glycol-based prepolymer with furfuryl alcohol and bismaleimide.[130] However, this kind of structure exhibits r-DA reaction at very high temperature, 120°C.

DA cycloaddition of a bisanthracene ester with C60-fullerene in the solid state by the mechanochemical route and the r-DA reaction in the temperature range 40–65°C are possible.[131] With this information, several routes have been explored. The copolymer of lauryl methacrylate and anthracene containing monomer, anthracen-9-ylmethyl methacrylate, and others copolymers undergo reversible DA reaction.[132] The r-DA reaction depends on the chemical structure of monomers reaching as low as 40°C.

6.20.6 Self-Healing Polycarbonate Containing PU/CNT Nanocomposites

The nanocomposites of polycarbonate-based PU and CNT can exhibit high mechanical properties along with enhanced autonomous self-healing characteristics.[133] The polymer segments are so flexible that CNT moves in association with the segments. Even then the tensile strength varies in the range of 1500–3000 psi and reversible extension of 500% with 0.2–1% CNT loading. Fully cured surfaces do not adhere together, while the broken surfaces exhibit a strong affinity for one another. The self-healing is very quick in this type of thermoplastic elastomer.

6.20.7 Elastic Nanocomposites with Graphene Oxide

Elastic self-healing materials being rare, amorphous polymers with hydrogen bonding network can be used, which have the ability to reversibly associate or dissociate at room temperature.[134] However, these polymers suffer from inferior mechanical properties. Design of polymers with superior properties requires a high density of hydrogen bonds and covalent cross-linking sites. Recoverable extensibility up to several hundred percents and negligible creep under load are possible for the molecules that associate together via hydrogen bonds, to form both chains and cross-links.[40] The mechanism of self-healing at room temperature can be given as the creation of new surfaces via damage, that results in a broken network mostly by the occurrence of separation of associated moieties. The damaged surfaces have a large number of free moieties in the vicinity, which can re-associate and heal after the damaged surfaces are brought back into contact. The polymers with high T_g do not self-heal spontaneously. Addition of plasticizer lowers the T_g as well as the mechanical strength. The strength can be augmented by adding metal powders. However, a high dose of fillers reduces the hydrogen-bond density that slows down the healing speed. Elastic nanocomposites can be made by combining the properties of a polymer with the hydrogen bonding ability with graphene oxide (GO) acting as a macro-cross-linker, which has the ability to rapidly self-heal at room temperature.[135] GO is used in low doses (<2%) so that density is kept low, while the mechanical property is enhanced and the self-healing speed remains optimum. Normally, the self-healing efficiency reduces after the damaged surfaces are kept apart for several hours. However, the presence of GO, maintaining a low covalent cross-linking density, slows down the decay in self-healing efficiency.

6.20.8 Supramolecular Polymer Composites

An amine-terminated hyperbranched polymer and 1–2 wt% isocyanate modified GO are mixed to make nanocomposites, which do not show stress concentration due to the uniform dispersion of the latter.[136] The presence of carbamate esters and multiple reactive sites ensures a sufficient density of hydrogen bonding sites in the modified GO that enhances the mechanical strength due to the formation of supramolecular assembly. The healing rate decreases by increasing the modified –GO content, which occurs due to a disturbance in the polymer chains movement.

6.21 Self-Healing in Electrochemical Systems

Flexible, lightweight, and miniaturized supercapacitors are being developed to cater to the rapidly growing advances in the area of portable electronic devices. However, under actual working conditions, most of the electrode materials of these supercapacitors become vulnerable to mechanical damage during bending or during charge/discharge processes. Even in the case of polymeric flexible substrates, mechanical damage resulting from deformation over time is possible. Thus, the reliability and lifetime of the supercapacitors remain a concern that causes several inconveniences which result in the breakdown and generation of electronic waste and also create safety hazards.[137] An ideal supercapacitor is expected to retain high capacitance, portability, a prevention mechanism for structural fractures of the electrode materials or the mechanism to restore the devices' configuration integrity and electrical properties post mechanical damage.

6.21.1 Self-Healable Supercapacitors

The supercapacitor electrodes with mechanical and electrical self-healing ability can be fabricated by spreading SWCNT films onto self-healing substrates.[138] A supramolecular network with T_g below room temperature and hierarchical flower-like TiO_2 nanostructures is ideally suited for making self-healing substrates. The rutile flower-like TiO_2 nanospheres are first uniformly dispersed in the as-synthesized oligomers, which are then thermally cross-linked with urea. The shape of the supramolecular networks is kept intact by the interaction between them and the TiO_2 nanospheres. The composite can be thermally compressed and self-adhered onto various hydrophilic substrates due to the presence of reversible hydrogen bonds. The substrate is chosen from flexible polyethylene terephthalate (PET), oxygen plasma-treated poly(dimethylsiloxane) (PDMS) sheets, and others. The presence of a large amount of carboxylic and the hydroxyl group grafted onto the SWCNTs' surface, due to the acid treatment, allows it to be firmly adhered with uniform distribution onto the self-healing substrates. After mechanical damage, the separated areas of the SWCNT layer are brought back into contact by the lateral movement of the self-healing composite layer, and thus the device's configuration and conductivity are recovered. The supercapacitors, thus prepared, are electrochemically excellent with the ability to self-heal.

A biochar-based composite electrode and a polyampholyte hydrogel electrolyte are combined to design a flexible and self-healing supercapacitor with high energy density in low-temperature operations.[139] Having a self-healing ability and mechanical flexibility and toughness, the polyampholytes hydrogel can be used at low temperature in aqueous electrolytes. Biochar is basically a carbon-based material and its synthesis is based on the low-temperature pyrolysis of biological wastes. The mechanical integrity and electrical conductivity of this material are improved by incorporating reduced GO. The reversible nature of ionic cross-linking in polyampholyte is responsible for its self-healing ability. The polyampholyte-based supercapacitor has potential application in the wearable and flexible electronics area due to its bendable and self-healing nature.

A stretchable matrix is the general requirement for wearable electronics, biomedical devices, electronic paper displays, artificial skin incorporating sensors, etc. More specifically, large levels of strain need to be sustained without much degradation in electronic properties. Mechanically strong stretchable supercapacitors are made using various approaches, the most popular being the application of layers of electrode materials onto a stretchable substrate. The disadvantage associated with this approach is that extra weight and volume are added, which lowers the performance of the device. Also, there is a restriction on stretching and complex shape change imposed by the nature of the deformation of the supporting substrate. Although a couple of successful measures, such as a graphene-based fiber-incorporated supercapacitor are available to design a stretchable supercapacitor, there is no provision for mending the damage, if it occurs. Spring-like reduced graphene oxide (RGO)-based thick composite fibers can be designed, which can have 300% stretchability with the ability to reconnect the tiny broken fibers.[140] Stretchable carboxylated polyurethane can be coated onto spring-like fiber electrodes to act as a self-healing material that maintains stretchability and the self-healing properties of the supercapacitor. Fiber electrodes comprise polypyrrole (PPy)-decorated RGO/MWCNTs that can be twisted into springs. Interfacial hydrogen bonding of carboxylated PU fashions the self-healing of the electrode. This design of supercapacitor has about 82% capacitance retention after 100% stretching, which reduces with the repetition of healing . Even after a 3000 galvanostatic charge/discharge cycle of the three-times healed sample, it retains about 84% capacitance without degradation.

6.21.2 Self-Healable Battery Electrodes

Normally, degradation and damage to the battery material occur due to repeated electrochemical reactions leading to structural changes, and it ends up non-functional. Lithium-ion batteries suffer from this kind of failure mechanism. Silicon anode in a lithium ion battery undergoes volumetric expansion up to three times followed by contraction during lithium insertion and extraction, which causes fast capacity decay and a short cycle-life. In fact, the cracking and pulverization in the electrode occur due to these extreme volumetric changes and, finally, a loss of electrical contact and excessive solid–electrolyte interphase growth occurs.[141] The promising but damage-prone battery materials can be stabilized by combining them with self-healing polymers. Silicon can be combined with polymers to make a battery's negative electrode. It maintain its self-healing property that heals the cracks formed during use of the battery and can be recharged while maintaining much higher energy-storing capacity and a higher number of recharges. Conductive carbon particles are mixed with the polymer to ensure the flow of electricity through the battery, which is then converted to anode by combining with silicon microparticles.[142] During charging and discharging of the battery, there is continuous expansion and contraction of the

silicon, leading to fractures. The polymer is there to self-heal the fracture by pulling everything back together.

In another approach, a coating of soft self-healing polymer is applied to the silicon electrode, which imparts stretchability and spontaneous repairability to the mechanical damage and cracks in the electrode and thus stable mechanical and electrical connections are maintained among the silicon particles.[143] Branched polymer with dynamic hydrogen-bonding ability, low T_g, and an amorphous nature, is combined with conducting carbon nanoparticles to make electrodes with a self-healing agent and modest conductivity. There is a dynamic re-association of hydrogen bonds at room temperature after the damage occurs that leads to self-healing of the electrodes spontaneously. In a composite self-healing coating, the stretchability and spontaneous self-healing capabilities are responsible for bringing the shrunken silicon particles back into contact with the polymer binder and also avoid the appearance of large non-healable cracks in the polymer binder.

Aligned carbon nanotube sheets loaded with $LiMn_2O_4$ and $LiTi_2(PO_4)_3$ nanoparticles placed on a self-healing polymer substrate are used as electrodes, and aqueous lithium sulfate/sodium carboxymethylcellulose (Li_2SO_4/CMC) is used as both gel electrolyte and separator to make all-solid flexible self-healing aqueous lithium-ion batteries.[144] High flexibility and good recovery of electrochemical performance have become possible by combining self-healing polymer with aligned CNT sheets electrodes with reasonable electrical and healing properties and Li_2SO_4/CMC gel electrolyte. The polymer is supramolecular in nature with a high density of hydrogen bonds that take part in the reconstruction after breaking. The aligned CNT composite electrodes exhibit better relinking of the broken aligned CNTs via Van der Waals forces.

Dynamic ionic bonding can be introduced between active Si nanoparticles and a polymer binder via modification of the silicon with an amine group, which finds use in silicon composite electrodes.[145] The attachment of amine groups to Si nanoparticles via covalent bond can be performed by using a coupling reaction of the amino silane compound. The amine-functionalized Si nanoparticles are combined with a poly(acrylic acid) binder to make a composite electrode. The self-healing ability is realized by the formation of ionic bonds between the amine and carboxyl groups that increase the cycle lifetime and reliability of the Si composite anodes.

6.22 Miscellaneous Self-Healing

6.22.1 Shape Memory Polyurethanes in Structural Health Monitoring

To avoid catastrophic failures, the critical infrastructures in services, civil engineering, and the automotive industries must have a proper mechanism of monitoring and maintenance. The present system depends on monitoring the health of these structures, which is reported to the maintenance authority. However, the time lag and other dependent mechanisms make the situation worse. Future implementation of structural health monitoring (SHM) needs to be coupled with materials that can respond automatically to the damage by healing themselves so that the service lifetime is prolonged and inspection frequency is reduced. Among intrinsic and extrinsic mechanisms, the latter is concerned with the release of healing agents on the occurrence of cracks in the polymeric matrix. Although it exhibits good healing performance, accommodation of external healing agents in the complex fabrication remains a serious challenge. SMPs associated with a stable polymeric network have the ability to tackle several limitations by returning to their original shape after temporary deformation. It is obvious that the need for an external stimulus is not required to close the crack. Thermosetting SMPs, such as polyurethanes, remaining lightly cross-linked, exhibit considerable ductility above T_g that makes the shape recovery possible after a temporary shape change.[146] Self-repairing syntactic foam based on SMP also exhibits self-recovery repeatedly after the damage caused at the micro and macro scales. In this type of foam, a cored sandwich composite, with tolerance for low-velocity impact damage, can be recovered completely after self-repair.[147]

6.22.2 Autonomous Healing Systems

A fully autonomous healing system can be designed by embedding fiber optics in SMPs so that they can detect the damage and also are able to deliver thermal energy. The crack is closed by the SMP and a feedback control loop heats the crack tip while increasing the toughness of the cracked area.[148] However, the broken bonds cannot be reformed even though the strength of the recovery is very good. By using a reversible DA reaction in a polyurethane (PU) network, active damage detection and repeatable autonomous healing can be integrated.[149] The PU network takes care of autonomous crack closure using its shape memory characteristic and the reversible DA reaction takes care of self-healing. During crack propagation in a polymer with embedded fiber optic system, the optical fiber undergoes fracture, resulting in a drop in the laser power emitted through it. This power reduction activates a controller so that the power of the inflow light is increased in a feedback loop, prompting both shape recovery and healing by delivering the thermal energy at the crack area via photo-thermal heating through the transmission loss of the optical fiber.

6.22.3 Polyelectrolyte Multilayer as a Super Oxygen Barrier

An ionically bonded complex, having the ability to rapidly self-heal, is extremely efficient in various applications: gas barrier, energy storage, flame resistance, antifouling surfaces, drug delivery, etc. Mostly, the low T_g polymers, with large free volume as well as mobile chains, have a self-healing ability. On the other hand, high T_g and small free volume are normally suitable for barrier properties. Ionic strength, pH, etc. act as stimuli for ionically bonded polymers. PAA and poly (allylamine hydrochloride) forming an ionically bonded complex exhibits rapid

and efficient self-healing behavior in the presence of sodium chloride as a promoter.[150] High elastic modulus and an ultrahigh oxygen barrier property and humidity-induced self-healing at room temperature are exhibited by the polyelectrolyte multilayer nanocoating obtained by alternate deposition of PEI and PAA.[151] In the presence of high humidity, swelling of the hydrophilic PEI/PAA multilayer thin films reduce the modulus, increase the chain mobility and seal the crack via plasticization. The 8-bilayer PEI/PAA film, exhibiting a super oxygen barrier, is completely recovered after the healing in minimum time and at high humidity (10 min at 97% RH). Protection of food, pharmaceuticals, and electronics, etc. can be done by using a film formed by a simple self-healing nanocoating.

6.23 Understanding Molecular Self-Healing

Various techniques, such as rheology, a tensile test, and scratch healing have been exploited to characterize morphologically the healing process in self-healing polymers. In fact, the morphological approaches are applied to determine the material's properties before and after the self-healing process, so that the efficiency of the self-healing is quantified. However, the chemistry of the molecular self-healing process remains unanswered or only partially answered by these techniques. The characterization of self-healing materials on a molecular level is done by vibrational spectroscopy such as infrared absorption or Raman spectroscopy.[152] The practical difficulty arises in the characterization of healing in gels and soft rubbery surfaces. The fracture processes of rubbers and gels involve large and highly delocalized deformations and also the magnitude of the mechanical properties before and after fracture differs, that creates several practical difficulties during the experimental determination of healing. It becomes difficult to separate the surface properties related effects from the effects due to changes in the mechanical properties of the surrounding bulk matrix. Various techniques have been developed to quantify the surface adhesion characteristics. However, the creation of a controlled damaged surface is difficult. The cohesive strength between fracture surfaces is much higher than that of melt-pressed surfaces. Thus, the self-healing property of these rubbers has a direct relation with the damaging processes.[153]

6.24 Conclusion

This chapter has been divided into various sections, such as thermoplastic polymers, polyurethane, epoxy, elastomer, hydrogel, coatings, textile coatings, etc. Self-healing polymers have been discussed in terms of their self-healing ability, mechanical properties, and other important properties, such as electrical, optical, and so on. A variety of healing actions with mechanisms have been incorporated in the discussion. Healing due to hydrogen bonding, ionic association, supramolecular assembly, the reversible Diels-Alder reaction, the Schiff-base interaction, solvents, heat, light, magnets, the release of healing agents due to mechanical damage of microcapsules and microvascular networks, etc. have been discussed, along with a typical polymer matrix.

Blends, composites, and nanocomposites were introduced, due to their participation in enhancing strength, assisting self-healing and sometimes the response to light. Various polymer structures have been discussed separately, according to their self-healing strategy and applications. Miscellaneous self-healing processes have been incorporated for their specific applications. Structural health monitoring has been discussed with a possible strategy. Finally, attempts have been made to understand the various aspects of self-healing.

REFERENCES

1. M.D. Hager, P. Greil, C. Leyens, S. van der Zwaag, and U.S. Schubert, "Self-healing materials." *Adv. Mater.* 22(47) (2010): 5424–5430.
2. S.R. White, N.R. Sottos, P.H. Geubelle, J.S. Moore, M.R. Kessler, S.R. Sriram, E.N. Brown, and S. Viswanathan. "Autonomic healing of polymer composites." *Nature* 409(6822) (2001): 794–797.
3. R.S. Trask, G.J. Williams, and I.P. Bond. "Bioinspired self-healing of advanced composite structures using hollow glass fibers." *J. R. Soc. Interface* 4(13) (2007): 363–371.
4. C.J. Hansen, W, Wu, K.S. Toohey, N.R. Sottos, S.R. White, and J.A. Lewis. "Self-healing materials with interpenetrating microvascular networks." *Adv. Mater.* 21(41) (2009): 4143–4147.
5. K.S. Toohey, C.J. Hansen, J.A. Lewis, S.R. White, and N.R. Sottos. "Delivery of two-part self-healing chemistry via microvascular networks." *Adv. Funct. Mater.* 19(9) (2009): 1399–1405.
6. S.D. Mookhoek, H.R. Fischer, and S. van Der Zwaag. "A numerical study into the effects of elongated capsules on the healing efficiency of liquid-based systems." *Comput. Mater. Sci.* 47(2) (2009): 506–511.
7. C.B. Lin, S. Lee, and K.S. Liu. "Methanol-induced crack healing in poly(methyl methacrylate)." *Polym. Eng. Sci.* 30(1) (1990): 1399–1406.
8. K. Takeda, H. Unno, and M. Zhang. "Polymer reaction in polycarbonate with Na2CO3." *J. Appl. Polym. Sci.* 93(2) (2004): 920–926.
9. E.B. Murphy, E. Bolanos, C. Schaffner-Hamann, F. Wudl, S.R. Nutt, and M.L. Auad. "Synthesis and characterization of a single-component thermally remendable polymer network: Staudinger and Stille revisited." *Macromolecules*, 41(14) (2008): 5203–5209.
10. J-L. Wietor, A. Dimopoulos, L.E. Govaert, R.A.T.M. van Benthem, G. de With, and R.P. Sijbesma. "Preemptive healing through supramolecular cross-links." *Macromolecules* 42(17) (2009): 6640–6646.
11. J. Canadell, H. Goossens, and B. Klumperman. "Self-healing materials based on disulfide links." *Macromolecules* 44(8) (2011): 2536–2541.
12. D. Montarnal, P. Cordier, C. Soulié-Ziakovic, F. Tournilhac, and L. Leibler. "Synthesis of self-healing supramolecular rubbers from fatty acid derivatives, diethylenetriamine, and urea." *J. Polym. Sci. Part A: Polym. Chem.* 46(24) (2008): 7925–7936.
13. Y. Yang, X. Ding, and M.W. Urban. "Chemical and physical aspects of self-healing materials." *Prog. Polym. Sci.* 49–50 (2015): 34–59.
14. R. Shunmugam and G.N. Tew. "Terpyridine–lanthanide complexes respond to fluorophosphate containing nerve gas G-agent surrogates." *Chem. Eur. J.* 14(18) (2008): 5409–5412.
15. S. Burattini, H.M. Colquhoun, B.W. Greenland, and W. Hayes. "A novel self-healing supramolecular polymer system." *Faraday Discuss.* 143 (2009): 251–264.
16. M.M. Caruso, D.A. Davis, Q. Shen, S.A. Odom, N.R. Sottos, S.R. White, and J.S. Moore. "Mechanically-induced chemical changes in polymeric materials." *Chem. Rev.* 109(11) (2009): 5755–5798.
17. J. Zhang, M. Huo, M. Li, T. Li, N. Li, J. Zhou, and J. Jiang. "Shape memory and self-healing materials from supramolecular block polymers." *Polymer* 134 (2018): 35–43.
18. C-M. Chung, Y-S. Roh, S-Y. Cho, and J-G. Kim. "Crack healing in polymeric materials via photochemical [2+2] cycloaddition." *Chem. Mater.* 16(21) (2004): 3982–3984.
19. H. Ying and J. Cheng. "Hydrolyzable polyureas bearing hindered urea bonds." *J. Am. Chem. Soc.* 136(49) (2014): 16974–16977.
20. S. Zechel, R. Geitner, M. Abend, M. Siegmann, M. Enke, N. Kuhl, N. Klein, Moritz, J. Vitz, … and M.D. Hager. "Intrinsic self-healing polymers with a high E-modulus based on dynamic reversible urea bonds." *NPG Asia Mater.* 9 (2017): e420.
21. F. Ruiperez, M. Galdeano, E. Gimenez, and J.M. Matxain. "Sulfenamides as building blocks for efficient disulfide-based self-healing materials. A quantum chemical study." *Chem. Open* 7 (2018): 248–255.
22. J.L. Wardell. "Preparation of thiols." In S. Patai (Ed.), *The Chemistry of the Thiol Group*, vol. 1. London: Wiley,1974, pp. 163–269.
23. N.V. Tsarevsky and K. Matyjaszewski. "Reversible redox cleavage/coupling of polystyrene with disulfide or thiol groups prepared by atom transfer radical polymerization." *Macromolecules* 35(24) (2002): 9009–9014.
24. N.B. Pramanik, P. Mondal, R. Mukherjee, and N.K. Singha. "A new class of self-healable hydrophobic materials based on ABA triblock copolymer via RAFT polymerization and Diels-Alder 'click chemistry'." *Polymer* 119 (2017): 195–205.
25. M. Nakahata, S. Mori, Y. Takashima, H. Yamaguchi, and H. Harada. "Self-healing materials formed by cross-linked polyrotaxanes with reversible bonds." *Chem* 1 (2016): 766–775.
26. T. Sakakura, J.C. Choi, and H. Yasuda. "Transformation of carbon dioxide." *Chem. Rev.* 107(6) (2007): 2365–2387.
27. Y. Yang and M.W. Urban. "Self-repairable polyurethane networks by atmospheric carbon dioxide and water." *Angew. Chem. Int. Ed.* 53(45) (2014): 12142–12147.
28. R.J. Sarma, S. Otto, and J.R. Nitschke. "Disulfides, imines, and metal coordination within a single system: interplay between three dynamic equilibria." *Chem. Eur. J.* 13(34) (2007): 9542–9546.
29. A.M. Belenguer, T. Friscic, G.M. Day, and J.K.M. Sanders. "Solid-state dynamic combinatorial chemistry: Reversibility and thermodynamic product selection in covalent mechanosynthesis." *Chem. Sci.* 2(4) (2011): 696–700.
30. J.M. Matxain, J.M. Asua, and F. Ruiperez. "Design of new disulfide-based organic compounds for the improvement of self-healing materials." *Phys. Chem. Chem. Phys.* 18(3) (2016): 1758–1770.
31. S. Nevejans, N. Ballard, J.I. Miranda, B. Reck, and J.M. Asua. "The underlying mechanisms for self-healing of poly(disulfide)s." *Phys. Chem. Chem. Phys.* 18(39) (2016): 27577–27583.

32. X. Jian, Y. Hu, W. Zhou, and L. Xiao. "Self-healing polyurethane based on disulfide bond and hydrogen bond." *Polym. Adv. Technol.* 29(1) (2018): 463–469.

33. R. Liu, X. Yang, Y. Yuan, J., Liu, and X. Liu. "Synthesis and properties of UV-curable self-healing oligomer." *Prog. Org. Coat.* 101(2016): 122–129.

34. Y-K. Guo, H. Li, P-X. Zhao, X-F. Wang, D. Astruc, and M.B. Shuai. "Thermo-reversible MWCNTs/epoxy polymer for use in self-healing and recyclable epoxy adhesive." *Chinese J. Polym. Sci.* 35(6) (2017): 728–738.

35. P. Chakma, L.H.R. Possarle, Z.A. Digby, B. Zhang, J.L. Sparks, and D. Konkolewicz. "Dual stimuli-responsive self-healing and malleable materials based on dynamic thiol-Michael chemistry." *Polym. Chem.* 8(42) (2017): 6534–6543.

36. B. Zhang, Z.A. Digby, J.A. Flum, P. Chakma, J.M. Saul, J.L. Sparks, and D. Konkolewicz. "Dynamic thiol-Michael chemistry for thermoresponsive rehealable and malleable networks." *Macromolecules* 49(18) (2016): 6871–6878.

37. U.S. Chung, J.H. Min, P-C. Lee, and W-G. Koh. "Polyurethane matrix incorporating PDMS-based self-healing microcapsules with enhanced mechanical and thermal stability." *Colloids and Surfaces A: Physicochem. Eng. Aspects* 518 (2017): 173–180.

38. R.W. Style, R. Boltyanskiy, B. Allen, K.E. Jensen, H.P. Foote, J.S. Wettlaufer, and E.R. Dufresne. "Stiffening solids with liquid inclusions (preprint)." *Nat. Phys.* 11 (2014): 82–87.

39. A. Chipara, P. S. Owuor, S. Bhowmick, G. Brunetto, S.A.S. Asif, M. Chipara, R. Vajtai, J. Lou, … and P.M. Ajayan. "Structural reinforcement through liquid encapsulation." *Adv. Mater. Interfaces* 4(2) (2017): 1600781.

40. P. Cordier, F. Tournilhac, C. Soulié-Ziakovic, and L. Leibler. "Self-healing and thermoreversible rubber from supramolecular assembly." *Nature.* 451(7181) (2008): 977–980.

41. M. Montarnal, P. Cordier, C. Soulé-Ziakovic, F. Tournilhac, and L. Leibler. "Synthesis of self-healing supramolecular rubbers from fatty acid derivatives, diethylene triamine, and urea." *J. Polym. Sci., Part A: Polym. Chem.* 46(24) (2008): 7925–7936.

42. D. Montarnal, F. Tournilak, M. Hidalgo, J.L. Couturier, and L. Leibler. "Versatile one-pot synthesis of supramolecular plastics and self-healing rubbers." *J. Am. Chem. Soc.* 131(23) (2009): 7966–7967.

43. Y. Chen, A.M. Kushner., G.A. Williams, and Z. Guan. "Multiphase design of autonomic self-healing thermoplastic elastomers." *Nat. Chem.* 4(6) (2012): 467–472.

44. Y. Chen and Z. Guan. "Multivalent hydrogen bonding block copolymers self-assemble into strong and tough self-healing materials." *Chem. Commun.* 50(4) (2014): 10868–10870.

45. Y. Chen and Z. Guan. "Self-healing thermoplastic elastomer brush copolymers having a glassy polymethylmethacrylate backbone and rubbery polyacrylate-amide brushes." *Polymer* 69 (2015): 249–254.

46. Y. Wang, X.X. Liu, S. Li, T. Li, Y. Song, Z. Li, W. Zhang, and J. Sun. "Transparent, healable elastomers with high mechanical strength and elasticity derived from hydrogen-bonded polymer complexes." *ACS Appl. Mater. Interfaces* 9(34) (2017): 29120–29129.

47. S.J. Kalista Jr., T.C. Ward, and J. Oyetunji. "Self-healing of poly (ethylene-co-methacrylic acid) copolymers following projectile puncture ionic interaction and electric heating." *Mech. Adv. Mater. Struct.* 14(2007): 391–397.

48. R.J. Varley and S. van der Zwaag. "Towards an understanding of thermally activated self-healing of an ionomer system during ballistic penetration." *Acta Materialia* 56(19) (2008): 5737–5750.

49. P. Zheng and T.J. McCarthy. "A surprise from 1954: Siloxane equilibration is a simple, robust, and obvious polymer self-healing mechanism." *J. Am. Chem. Soc.* 134(4) (2012): 2024–2027.

50. J.S. Parent, A.M.J. Porter, M.R. Kleczek, and R.A. Whitney. "Imidazolium bromide derivatives of poly (isobutylene-co-isoprene): A new class of elastomeric ionomers." *Polymer* 52 (2011): 5410–5418.

51. A. Das, A. Sallat, F. Böhme, M. Suckow, D. Basu, S. Wiessner, K.W. Stöckelhuber, B. Voit, and G. Heinrich. "Ionic modification turns commercial rubber into a self-healing material." *ACS Appl. Mater. Interfaces* 7(37) (2015): 20623–20630.

52. C. Xu, L. Cao, B. Lin, X. Liang, and Y. Chen. "Design of self-healing supramolecular rubbers by introducing ionic cross-links into natural rubber via a controlled vulcanization." *ACS Appl. Mater. Interfaces* 8(27) (2016): 17728–17737.

53. T. Boontheekul, H. Kong, and D.J. Mooney. "Controlling alginate gel degradation utilizing partial oxidation and bimodal molecular weight distribution." *Biomaterials* 26(15) (2005): 2455–2465.

54. K.L. Tan and E.N. Jacobsen. "Indium-mediated asymmetric allylation of acylhydrazones using a chiral urea catalyst." *Angew. Chem., Int. Ed.* 46(8) (2007): 1315–1317.

55. Z. Wei, J.H. Yang, Z.Q. Liu, F. Xu, Z.X. Zhou, M. Zrínyi, Y. Osada, and Y.M. Chen "Novel biocompatible polysaccharide-based self-healing hydrogel." *Adv. Funct. Mater.* 25(9) (2015): 1352–1359.

56. R. Wang, Q. Li, B. Chi, X. Wang, Z. Xu, Z. Xu, S. Chen, and H. Xu. "Enzyme-induced dual-network ε-poly-L-lysine-based hydrogels with robust self-healing and antibacterial performance." *Chem. Commun.* 53(35) (2017): 4803–4806.

57. T.J. Ji, Y. Zhao, Y. Ding, and G. Nie. "Using functional nanomaterials to target and regulate the tumor microenvironment: Diagnostic and therapeutic applications." *Adv. Mater.* 25(26) (2013): 3508–3525.

58. J. Qu, X. Zhao, P.X. Ma, and B. Guo. "pH-responsive self-healing injectable hydrogel based on N-carboxyethyl chitosan for hepatocellular carcinoma therapy." *Acta Biomaterialia* 58 (2017): 168–180.

59. S. Maity, A. Datta, S. Lahiri, and J. Ganguly. "A dynamic chitosan-based self-healing hydrogel with tunable morphology and its application as an isolating agent." *RSC Adv.* 6(84) (2016): 81060–81068.

60. Y. Zhang, B. Yang, X. Zhang, L. Xu, L. Tao, S. Li, and Y. Wei. "A magnetic self-healing hydrogel." *Chem. Commun.* 48(74) (2012): 9305–9307.

61. Y. Zhang, L. Tao, S. Li, and Y. Wei. "Synthesis of multiresponsive and dynamic chitosan-based hydrogels for controlled release of bioactive molecules." *Biomacromolecules* 12(8) (2011): 2894–2901.

62. Z.Wei, J. He, T. Liang, H. Oh, J. Athas, Z. Tong, C. Wang, and Z. Nie. "Autonomous self-healing of poly(acrylic acid) hydrogels induced by the migration of ferric ions." *Polym. Chem.* 4(17) (2013): 4601–4605.

63. H. Zhou, G. Xu, J. Li, S. Zeng, X. Zhang, Z. Zheng, X. Ding, W. Chen, Q. Wang, and W. Zhang. "Preparation and self-healing behaviors of poly(acrylic acid)/cerium

ions double-network hydrogels." *Macromol. Res.* 23(12) (2015): 1098–1102.

64. H. Zhou, M. Zhang, J. Cao, B. Yan, W. Yang, X. Jin, A. Ma, W. Chen, … and C. Luo. "Highly flexible, tough, and self-healable hydrogels enabled by dual cross-linking of triblock copolymer micelles and ionic interactions." *Macromol. Mater. Eng.* 302(2) (2016): 1600352.

65. X. Li, Q. Yang, Y. Zhao, S. Longa and J. Zheng. "Dual physically crosslinked double network hydrogels with high toughness and self-healing properties." *Soft Matter* 13(5) (2017): 911—920.

66. Y. Gao, Q. Luo, S. Qiao, L. Wang, Z. Dong, J. Xu and J. Liu. "Enzymatically regulating the self-healing of protein hydrogels with high healing efficiency." *Angew. Chem. Int. Ed.* 53(35) (2014): 9343–9346.

67. C. Xu, Q. Li, X-T. Hu, C.F. Wang and S. Chen. "Dually crosslinked self-healing hydrogels originated from cell-enhanced effect." *J. Mater. Chem. B* 5(21) (2017): 3816–3822.

68. Y-L. Liu and Y-W. Chen. "Thermally reversible cross-linked polyamides with high toughness and self-repairing ability from maleimide- and Furan-functionalized aromatic polyamides." *Macromol. Chem. Phys.* 208(2) (2007): 224–232.

69. Y. Zhang, A.A. Broekhuis, and F. Picchioni. "Thermally self-healing polymeric materials: The next step to recycling thermoset polymers?" *Macromolecules* 42(6) (2009): 1906–1912.

70. N.C. Sharma, H.J. Galustians, J. Qaquish, A. Galustians, K.N. Rustogi, M.E. Petrone, P. Chalnis, L. Garcia, A.R. Volpe and H.M. Proskin. "The clinical effectiveness of a dentifrice containing triclosan and a copolymer for controlling breath odor measured organoleptically twelve hours after toothbrushing." *J. Clin. Dent.* 10(4) (1999): 131–134.

71. X. Ma, N. Zhou, T. Zhang, W. Hu, and N. Gu. "Self-healing pH-sensitive poly[(methyl vinyl ether)-alt-(maleic acid)]-based supramolecular hydrogels formed by inclusion complexation between cyclodextrin and adamantine." *Mater. Sci. Eng. C* 73 (2017): 357–365.

72. M. Nadgorny, J. Collins, Z. Xiao, P.J. Scales, and L.A. Connal. "3D-printing of dynamic self-healing cryogels with tunable properties." *Polym. Chem.* 9(13) (2018): 1684–1692.

73. N. Selvakumar, K. Jeyasubramanian, and R. Sharmila. "Smart coating for corrosion protection by adopting nanoparticles." *Prog. Org. Coat.* 74(3) (2012): 461–469.

74. T. Szabóa, J. Telegdia, and L. Nyikosa. "Linseed oil-filled microcapsules containing drier and corrosion inhibitor – Their effects on self-healing capability of paints." *Prog. Org. Coat.* 84(2015): 136–142.

75. Z. Yang, Z. Wei, L. Le-Ping, W. Hong-Mei, and L. Wu-Jun. "The self-healing composite anticorrosive coatings." *Physics Procedia* 18 (2011): 216–221.

76. S. Park and E. Koh. "The synthesis of polyurethane microcapsules and evaluation of self-healing paint protection properties." Paper presented at 21st International Conference on Composite Materials, Xi'an, 2017.

77. S.H. Cho. "Self-healing coatings for an anti-corrosion barrier in damaged parts." *Corr. Sci. Technol.* 8(6) (2009): 223–226.

78. S.H. Cho, S.R. White, and P.V. Braun. "Self-healing polymer coatings." *Adv. Mater.* 21(6) (2009): 645–649.

79. S.H. Cho, S.R. White, and P.V. Braun. "Room-temperature polydimethylsiloxane-based self-healing polymers." *Chem. Mater.* 24(21) (2012): 4209–4214.

80. S.J. García, H.R. Fischer, P.A. White, J. Mardel, Y. González-García, J.M.C. Mol, and A.E. Hughes. "Self-healing anticorrosive organic coating based on an encapsulated water reactive silyl ester: Synthesis and proof of concept." *Prog. Org. Coat.* 70(2–3) (2011): 142–149.

81. J.D.D. Melo, A.P.C. Barbosa, M.C.B. Costa, and G.N. de Melo. "Encapsulation of solvent into halloysite nanotubes to promote self-healing ability in polymers." *Adv. Compos. Mater.* 23(5–6) (2014): 507–519.

82. A.A. Nazeer and M. Madkour. "Potential use of smart coatings for corrosion protection of metals and alloys: A review." *J. Mol. Liq.* 253 (2018): 11–22.

83. D.V. Andreeva, D. Fix, H. Möhwald, and D.G. Shchukin. "Self-healing nanotechnology anticorrosion coatings as an alternative to toxic chromium." *Adv. Mater.* 20(14) (2008): 2789–2794.

84. A.G. Skirtach, A.A. Antipov, D.G. Shchukin, and G.B. Sukhorukov. "Remote activation of capsules containing Ag nanoparticles and IR dye by laser light." *Langmuir* 20(17) (2004): 6988–6992.

85. Z. Lu, M.D. Prouty, Z. Guo, V.O. Golub, C.S.S.R. Kumar, and Y.M. Lvov. "Magnetic switch of permeability for polyelectrolyte microcapsules embedded with Co@Au nanoparticles." *Langmuir* 21(5) (2005): 2042–2050.

86. D.G. Shchukin, K. Kohler, and H. Mohwald. "Microcontainers with electrochemically reversible permeability." *J. Am. Chem. Soc.* 128(14) (2006): 4560–4561.

87. J-L. Wietor, A. Dimopoulos, L.E. Govaert, R.A.T.M. van Benthem, G. de With, and R.P. Sijbesma. "Preemptive healing through supramolecular cross-links." *Macromolecules* 42(17) (2009): 6640–6646.

88. Y. Gonzalez-Garcia, J.M.C. Mol, T. Muselle, I. De Graeve, G. Van Assche, G. Scheltjens, B. Van Mele and H. Terryn. "A combined mechanical, microscopic and local electrochemical evaluation of self-healing properties of shape-memory polyurethane coatings." *Electrochim. Acta* 56(26) (2011): 9619–9626.

89. E.D. Rodriguez, X. Luo, and P.T. Mather. "Linear/network poly(ε-caprolactone) blends exhibiting shape memory assisted self-healing (SMASH)." *ACS Appl. Mater. Interfaces* 3(2) (2011): 152–161.

90. X. Luo and P.T. Mather. "Shape memory assisted self-healing coating." *ACS Macro Lett.* 2(2) (2013): 152–156.

91. K.S. Toohey, N.R. Sottos, J.A. Lewis, J.S. Moore, and S.R. White. "Self-healing materials with microvascular networks." *Nat. Mater.* 6(8) (2007): 581–585.

92. S.K. Rath, J.G. Chavan, S. Sasane, M. Patri, A.B. Samui, and B.C. Chakraborty. "Two-component silicone modified epoxy foul release coatings: Effect of modulus, surface energy and surface restructuring on pseudobarnacle and macrofouling behavior." *Appl. Surf. Sci.* 256(6) (2010): 2440–2446.

93. S. Sommer, A. Ekin, D.C. Webster, S.J. Stafslien, J. Daniels, L.J. Van der Wal, S.E.M. Thompson, M.E. Callow, and J.A. Callow. "A preliminary study on the properties and fouling-release performance of siloxane-polyurethane coatings prepared from poly(dimethylsiloxane) (PDMS) macromers." *Biofouling* 26(8) (2010): 961–972.

94. C. Liu, C. Ma, Q. Xie, and G. Zhang. "Self-repairing silicone coating for marine anti-biofouling." *J. Mater. Chem. A* 5(3) (2017): 15855–15861.

95. X-Y. Liang, L. Wang, Y-M. Wang, L-S. Ding, B-J. Li and S. Zhang. "UV-blocking coating with self-healing capacity." *Macromol. Chem. Phy.* 218(19) (2017): 1700213

96. J-H. Park and P.V. Braun. "Coaxial electrospinning of self-healing coatings." *Adv. Mater.* 22(4) (2010): 496–499.

97. A. Greiner and J.H. Wendorff. "Electrospinning: A fascinating method for the preparation of ultrathin fibers." *Angew. Chem, Int. Ed.* 46(30) (2007): 5670–5703.

98. D. Gaddes, H. Jung, A. Pena-Francesch, G. Dion, T. Srinivas, W.J. Dressick, and M.C. Demirel. "Self-healing textile: Enzyme encapsulated layer-by-layer structural proteins." *ACS Appl. Mater. Interfaces* 8(31) (2016): 20371–20378.

99. N. García-Huete, W. Post, J.M. Laza, J.L. Vilas, L.M. León, and S.J. García. "Effect of the blend ratio on the shape memory and self-healing behavior of ionomer-polycyclooctene crosslinked polymer blends." *Eur. Polym. J.* 98 (2018): 154–161.

100. S.J. Benight, C. Wang, J.B.H. Tok, and Z. Bao. "Stretchable and self-healing polymers and devices for electronic skin." *Prog. Polym. Sci.* 38(12) (2013): 1961–1977.

101. T. Wang, Y. Zhang, Q. Liu, W. Cheng, X. Wang, L. Pan, B. Xu, and X. Xu. "A self-healable, highly stretchable, and solution processable conductive polymer composite for ultrasensitive strain and pressure sensing." *Adv. Funct. Mater.* 28(7) (2018): 1705551.

102. J. Ahner, M. Micheel, R. Geitner, M. Schmitt, J. Popp, B. Dietzek, and M.D. Hager. "Self-healing functional polymers: Optical property recovery of conjugated polymer films by uncatalyzed imine metathesis." *Macromolecules* 50(10) (2017): 3789–3795.

103. J.R. Davidson, G.A. Appuhamillage, C.M. Thompson, W. Voit and R.A. Smaldone. "Design paradigm utilizing reversible Diels-Alder reactions to enhance the mechanical properties of 3D printed materials." *ACS Appl. Mater. Interfaces* 8(26) (2016): 16961–16966.

104. R.C. Boutelle and B.H. Northrop. "Substituent effects on the reversibility of furan–maleimide cycloadditions." *J. Org. Chem.* 76(19) (2011): 7994–8002.

105. G.A. Appuhamillage, J.C. Reagan, S. Khorsandi, J.R. Davidson, W. Voit, and R.A. Smaldone. "3D printed remendable polylactic acid blends with uniform mechanical strength enabled by a dynamic Diels-Alder reaction." *Polym. Chem.* 8(13) (2017): 2087–2092.

106. Y. Wang, D.T. Pham, and C. Ji. "Self-healing composites: A review." *Cogent Engineering* 2(1) (2015): 1075686.

107. W.T. Li, Z.W. Jiang, Z. H. Yang, N. Zhao, and W.Z. Yuan. "Self-healing efficiency of cementitious materials containing microcapsules filled with healing adhesive: mechanical restoration and healing process monitored by water absorption." *PLoS One* 8 (2013): e81616.

108. C.M. Dry. "Three designs for the internal release of sealants, adhesives, and waterproofing chemicals into concrete to reduce permeability." *Cem. Concr. Res.* 30(12) (2000): 1969–1977.

109. F. Xing, Z. Ni, N.X. Han, B.Q. Dong, X.X. Du, Z. Huang, and M. Zhang. "Self-healing mechanism of a novel cementitious composite using microcapsules, Advances in concrete structural durability." *Proc. ICDCS* 2008 1(2) (2008): 195–204.

110. A. Kousourakis and A.P. Mouritz. "The effect of self-healing hollow fibers on the mechanical properties of polymer composites." *Smart Mater. Struct.* 19(8) (2010): 085021.

111. C.J. Norris, I.P. Bond, and R.S. Trask. "Healing of low-velocity impact damage in vascularised composites." *Composites Part A: Appl. Sci. Manuf.* 44(1) (2013): 78–85.

112. L. Liu, M.R. Zachariah, S.I. Stoliarov, and J. Li. "Enhanced thermal decomposition and kinetics of poly (lactic acid) sacrificial polymer catalyzed by metal oxide nanoparticles." *RSC Adv.* 5(123): 101745–101750.

113. L.M. Bellan, S.P. Singh, P.W. Henderson, T.J. Porri, H.G. Craighead, and J.A. Spector. "Fabrication of an artificial 3-dimensional vascular network using sacrificial sugar structures." *Soft Matter* 5(7) (2009): 1354–1357.

114. A.P. Golden and J. Tien. "Fabrication of microfluidic hydrogels using molded gelatin as a sacrificial element." *Lab on a Chip* 7(6) (2007): 720–725.

115. J-H. Huang, J. Kim, N. Agrawal, A.P. Sudarsan, J.E. Maxim, A. Jayaraman, and V.M. Ugaz. "Rapid fabrication of bio-inspired 3D microfluidic vascular networks." *Adv. Mater.* 21(35) (2009): 3567–3571.

116. D. Lim, Y. Kamotani, B. Cho, J. Mazumder, and S. Takayama. "Fabrication of microfluidic mixers and artificial vasculatures using a high-brightness diode-pumped Nd: YAG laser direct write method." *Lab on a Chip* 3(4) (2003): 318–323.

117. A.M. Coppola, P.R. Thakre, N.R. Sottos, and S.R. White. "Tensile properties and damage evolution in vascular 3D woven glass/epoxy composites." *Composites Part A: Appl. Sci. Manuf.* 59 (2014): 9–17.

118. B.J. Blaiszik, S.L.B. Kramer, S.C. Olugebefola, J.S. Moore, N.R. Sottos, and S.R. White. "Self-healing polymers and composites." *Annu. Rev. Mater. Res.* 40 (2010): 179–211.

119. K. Aramaki. "Preparation of chromate-free, self-healing polymer films containing sodium silicate on zinc pretreated in cerium (III) nitrate solution for preventing zinc corrosion at scratches in 0.5 M NaCl." *Corros. Sci.* 44(6) (2002): 1375–1389.

120. V.K. Thakur and M.R. Kessler. "Self-healing polymer nanocomposite materials: A review." *Polymer* 69 (2015): 369–383.

121. J.Y. Lee, G.A. Buxton, and A.C. Balazs. "Using nanoparticles to create self-healing composites." *J. Chem. Phys.* 121 (2004): 5531–5540.

122. G. Lanzara, Y. Yoon, H. Liu, S. Peng, and W-I. Lee. "Carbon nanotube reservoirs for self-healing materials." *Nanotechnology* 20(33) (2009): 335704.

123. G-K. Xu, G. Lu, X-Q. Feng, and S-W. Yu. "Self-assembly of organic-inorganic nanocomposites with nacre-like hierarchical structures." *Soft Matter* 7(10) (2011): 4828–4832.

124. T. Engel and G. Kickelbick. "Furan-modified spherosilicates as building blocks for self-healing materials." *Eur. J. Inorg. Chem.* 2015(7) (2014): 1226–1232.

125. S. Schäfer and G. Kickelbick. "Self-healing polymer nanocomposites based on Diels-Alder reactions with silica nanoparticles: The role of the polymer matrix." *Polymer* 69 (2015): 357–368.

126. H.H. Le, S. Hait, A. Das, S. Wießner, K.W. Stöckelhuber, F. Böhme, U. Reuter, K. Naskar, G. Heinrich, and H.-J. Radusch. "Self-healing properties of carbon nanotube filled natural rubber/bromobutyl rubber blends". *eXPRESS Polym. Lett.* 11(3) (2017): 230–242.

127. C. Pan, L. Liu, Q. Chen, Q. Zhang, and G. Guo. "Tough, stretchable, compressive novel polymer/graphene oxide nanocomposite hydrogels with excellent self-healing performance." *ACS Appl. Mater. Interfaces* 9(43) (2017): 38052–38061.

128. X. Tong, L. Du, and Q. Xu. "Tough, adhesive and self-healing conductive 3D network hydrogel of physically

linked functionalized-boron nitride/clay/poly(N-iso-propyl acrylamide)." *J. Mater. Chem. A*, 6(7) (2018): 3091–3099.

129. S. Wu, J. Li, G. Zhang, Y. Yao, G. Li, R. Sun, and C. Wong. "Ultrafastly self-healing nanocomposites via infrared laser and its application in flexible electronics." *ACS Appl. Mater. Interfaces*, 9(3) (2017): 3040–3049.

130. L. Irusta, M.J. Fernandez-Berridi, and J. Aizpurua. "Polyurethanes based on isophorone diisocyanate trimer and polypropylene glycol crosslinked by thermal revers-ible Diels-Alder reactions." *J. Appl. Polym. Sci.* 134(9) (2016): 44543.

131. G.W. Wang, Z.X. Chen, Y. Murata, and K. Komatsu. "Fullerene adducts with 9-substituted anthracenes: mechanochemical preparation and retro Diels-Alder reac-tion." *Tetrahedron* 61(20) (2005): 4851–4856.

132. J. Kötteritzsch, R. Geitner, J. Ahner, M. Abend, S. Zechel, J. Vitz, S. Hoeppener, B. Dietzek, … and M.D. Hager. "Remendable polymers via reversible Diels-Alder cyclo-addition of anthracene-containing copolymers with fullerenes." *J. Appl. Polym. Sci.* 135(10) (2017): 45916.

133. J.P. Harmon and R. Bass. "Self-healing polycarbonate containing polyurethane nanotube composite." US Patent No. 8,846,801 B1–2014 (2014).

134. L. Brunsveld, B.J. Folmer, E.W. Meijer, and R.P. Sijbesma. "Supramolecular polymers." *Chem. Rev.* 101(12) (2001): 4071–4098.

135. C. Wang, N. Liu, R. Allen, J.B.H. Tok, Y. Wu, F. Zhang, Y. Chen, and Z. Bao. "A rapid and efficient self-healing thermo-reversible elastomer crosslinked with graphene oxide." *Adv. Mater.* 25(40) (2013): 5785–5790.

136. Y.G. Luan, X.A. Zhang, S.L. Jiang, J.H. Chen, and Y.F. Lyu. "Self-healing supramolecular polymer composites by hydrogen bonding interactions between hyperbranched polymer and graphene oxide." *Chinese J. Polym. Sci.* 36(5) (2018): 584–591.

137. L. Huang, N. Yi, Y. Wu, Y. Zhang, Q. Zhang, Y. Huang, Y. Ma, and Y. Chen. "Multichannel and repeatable self-healing of mechanical enhanced graphene-thermoplastic polyurethane composites." *Adv. Mater.* 25(15) (2013): 2224–2228.

138. H. Wang, B. Zhu, W. Jiang, Y. Yang, W.R. Leow, H. Wang, and X. Chen. "Mechanically and electrically self-healing supercapacitor." *Adv. Mater.* 26(22) (2014): 3638–3643.

139. X. Li, L. Liu, X. Wang, Y.S. Ok, J.A.W. Elliott, S.X. Chang, and H-J. Chung. "Flexible and self-healing aqueous supercapacitors for low-temperature applications: polyampholyte gel electrolytes with biochar electrodes." *Sci. Rep.* 7(1) (2017): 1685.

140. S. Wang, N. Liu, J. Su, L. Li, F. Long, Z. Zou, X. Jiang, and Y. Gao. "Highly stretchable and self-healable supercapacitor with reduced graphene oxide-based fiber springs." *ACS Nano* 11(2) (2017): 2066–2074.

141. T.D. Hatchard and J.R. Dahn. "In situ XRD and electro-chemical study of the reaction of lithium with amorphous silicon." *J. Electrochem. Soc.* 151(6) (2004): A838–A842.

142. K. Bourzac. "A gooey cure for crack-prone high-capacity batteries." *MIT Tech. Rev.* Nov. 17, 2013.

143. C. Wang, H. Wu, Z. Chen, M.T. McDowell, Y. Cui, and Z. Bao. "Self-healing chemistry enables the stable oper-ation of silicon microparticle anodes for high-energy lithium-ion batteries." *Nat. Chem.* 5 (2013): 1042–1048.

144. Y. Zhao, Y. Zhang, H. Sun, X. Dong, J. Cao, L. Wang, Y. Xu, J. Ren, Y. Hwang, … and H. Peng. "A self-healing aqueous lithium-ion battery." *Angew. Chem. Int. Ed.* 55 (2016): 1–6.

145. S. Kang, K. Yang, S.R. White, and N. R. Sottos. "Silicon composite electrodes with dynamic ionic bonding." *Adv. Energy Mater.* 7(1) (2017): 1700045.

146. G. Li. *Self-Healing Composites: Shape Memory Polymer Based Structures*. New York: Wiley, 2014.

147. G. Li and D. Nettles. "Thermomechanical characterization of a shape memory polymer based self-repairing syntactic foam." *Polymer* 51(2010): 755–762.

148. M.E. Garcia, Y. Lin, and H.A. Sodano. "Autonomous materials with controlled toughening and healing." *J. Appl. Phys.* 108 (2010): 093512.

149. A. Kazemi-Lari, M.H. Malakooti, and H.A. Sodano. "Active photo-thermal self-healing of shape memory polyurethanes." *Smart Mater. Struct.* 26 (2017): 055003.

150. A. Reisch, E. Roger, T. Phoeung, C. Antheaume, C. Orthlieb, F. Boulmedais, P. Lavalle, J.B. Schlenoff, B. Frisch, and P. Schaaf. "On the benefits of rubbing salt in the cut: Self-healing of saloplastic PAA/PAH compact polyelectrolyte complexes." *Adv. Mater.* 26(16) (2014): 2547–2551.

151. Y. Song, K.P. Meyers, J. Gerringer, R.K. Ramakrishnan, M. Humood, S. Qin, A. A. Polycarpou, S. Nazarenko, and J.C. Grunlan. "Fast self-healing of polyelectrolyte multi-layer nanocoating and restoration of super oxygen barrier." *Macromol. Rapid Commun.* 38(10) (2017): 1700064.

152. R. Geitner, F-B. Legesse, N. Kuhl, T.W. Bocklitz, S. Zechel, J. Vitz, M. Hager, U.S. Schubert, … and J. Popp. "Do you get what you see? Understanding molecular self-healing." *Chemistry* 24(10) (2018): 2493–2502.

153. F. Maes, D. Montarnal, S. Cantournet, F. Tournilhac, L. Corté, and L. Leibler. "Activation and deactivation of self-healing in supramolecular rubbers." *Soft Matter* 8(5) (2012): 1681–1687.

7

Drug Delivery and Biotechnological Applications of Polymers

Vinayak Kamble
Institute of Chemical Technology, Mumbai, India

Prakash Mahanwar
Institute of Chemical Technology, Mumbai, India

CONTENTS

Abbreviations

PNIPAM	poly(N-isopropylacrylamide)
LbL	layer-by-layer
PAA	poly (acrylic acid)
PEO-b-PCL	poly (ethylene oxide)-block-poly(caprolactone)
PEG-b-(PDEAEMA-bPHEMA-g-FA)2]	poly (ethylene glycol)-b-(poly(2-(diethylamino) ethyl methacrylate)-b-poly (hydroxyethyl methacrylate)-g-folic acid)
MPEG-b-PNLG	methoxy poly (ethylene glycol)-b-poly[N-[N-(2-aminoethyl)-2-aminoethyl]-L-glutamate]
TPA	terephthalaldehyde
LCST	lower critical solution temperature
UCST	upper critical solution temperature
PIPAAm-PBMA	poly(N-isopropylacrylamide-b-butyl methacrylate)
PBMA	poly (butyl methacrylate)
PSt	polystyrene
MDR	multidrug resistance
b-CD	b-cyclodextrin
Ad-PEG	adamantyl-containing poly(ethylene glycol)
PEG-b-PNIPAM	poly (ethylene glycol)-b-poly(N-isopropylacrylamide)
PEG-g- PNIPAM	poly(ethylene glycol) g-poly(N-isopropylacrylamide)
CyA	cyclosporin A
IND	Indomethacin
PNI-UDPy	pendant U-DPy groups
ADR	adriamycin
Biotin-P(NIPAAm-co-HMAAm)-b-PMMA)	biotin-poly (N-isopropylacrylamide-co-N-hydroxymethylacrylamide)-block-poly(methyl methacrylate)
PTNs	polytyrosine nanoparticles
DOX	doxorubicin
PG	polyglycerol

7.1 Introduction

Aa well as the development of new drugs, current research is focused on the development of new drug delivery systems to improve their biological activity and specificity. A number of approaches have been studied in the past decades in smart sensitive drug delivery systems. This can be achieved by incorporating or embedding special capabilities in the macromolecular structure to release the drug on the application of external stimuli. This class of polymers is generally known as "stimuli-responsive polymers." "Responsive polymer" or "smart polymers" are the attractive class of materials that undergo conformational changes and respond with small changes in their environment. A wide variety of responsive polymers have been reported that react to various external stimulus, such as temperature, pH, electric or magnetic fields, light intensity, biological molecules, etc., that

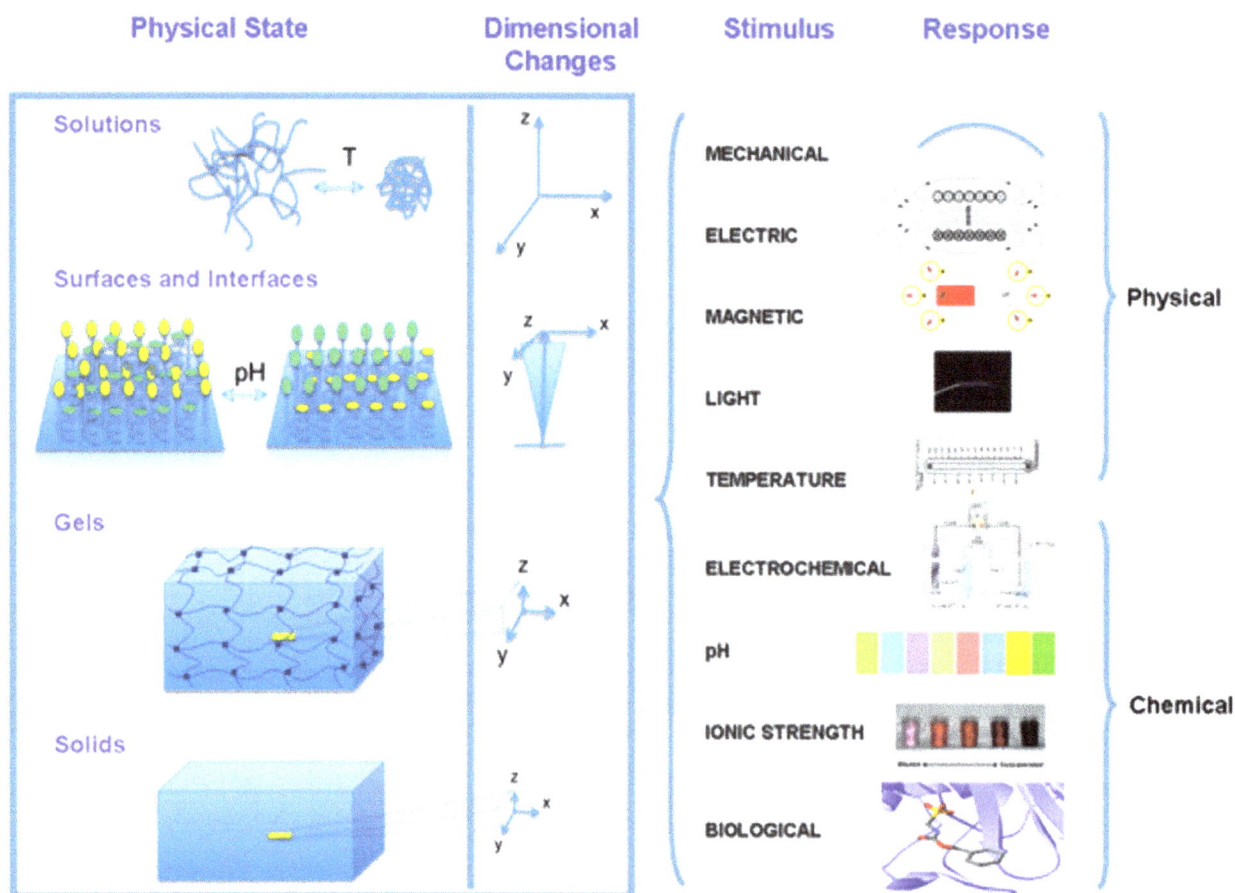

FIGURE 7.1 Schematic representation of dimensional changes in polymeric solutions, at surfaces and interfaces, in polymeric gels, and polymer solids resulting from physical or chemical stimuli. Responses are classified into physical and chemical categories, where multiple stimuli may result in one or more responses, or one stimulus may result in more than one response.

Source: [9].

induce macroscopic responses in the material, such as swelling, collapse, or sol-to-gel transitions, depending on the physical state of the chains. Some polymer systems may also react to a multiple trigger.[1–4] Example responses are presented in Figure 7.1.

For example, pH-sensitive polymers have pendant ionizable groups in their structure, which can be protonated or deprotonated in response to the environmental pH. This results in the alteration of the hydrodynamic volume, chain conformation, and solubility of the polymer. These conformational/property changes of a polymer can lead to the pH-induced release of the drug from a pH-sensitive carrier.[5,6] Where shape memory polymers are morphologically responsive, materials exist in a "memorized" macroscopic shape, temporarily exist in another shape, and then revert back to their original shape upon exposure to a stimulus.[7]

Nowadays tremendous opportunities have been created for stimuli-responsive polymers in biomedical applications. Various reactions were observed for stimuli-responsive changes in shape, surface characteristics, solubility, the formation of intricate molecular self-assembly and a sol-gel transition, which have enabled several novel applications in the delivery of therapeutics, tissue engineering, cell cultures, bio-separation, and

sensor or actuator systems.[8,9] Responsive polymers have widespread applications in diverse fields and a number of reviews have summarized the research progress in biomimetic actuators, immobilized biocatalysts, drug delivery, thermoresponsive surfaces, bioseparation and bioconjugates, controlled drug release, and water-borne coatings[10–13] (Figure 7.2).

The stimuli can be classified accordingly as physical, chemical, or biological, based on the type of stimuli required to produce the desired effect on the polymer.[14] They can alter their structure or properties in response to light, pH, electricity, magnetic fields, ion concentration, chemical and enzyme degradation, and temperature. Smart polymers can be dissolved in or phase-separated out of aqueous solutions or may be adsorbed on or grafted onto surfaces or be used as physically or chemically cross-linked hydrogels, in addition to graft copolymers, and molecular brushes and statistical/block copolymers.[15–17] The stimuli-responsive polymers may be applied in combination with the bioactive molecules by chemical conjugation, complexation or mixed physically with the bioactives, such as nucleic acids, carbohydrates, organic molecules, and proteins and peptides.[15] The majority of such systems are based on poly(N-isopropylacrylamide) (PNIPAM) or related polymers.[18]

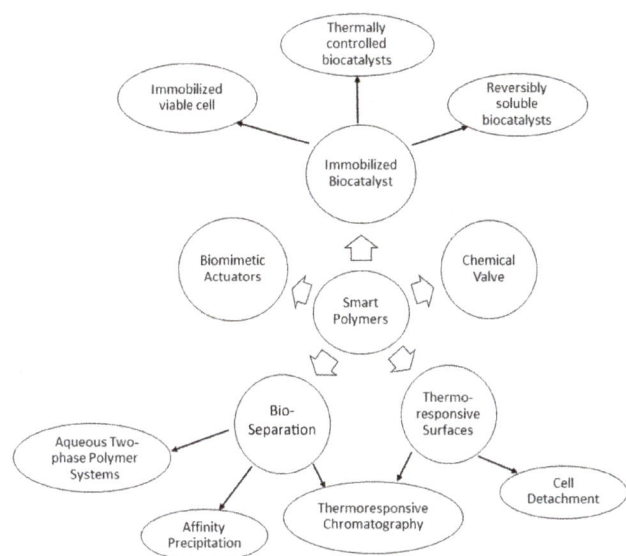

FIGURE 7.2 Applications of smart polymers in biotechnology and medicine.

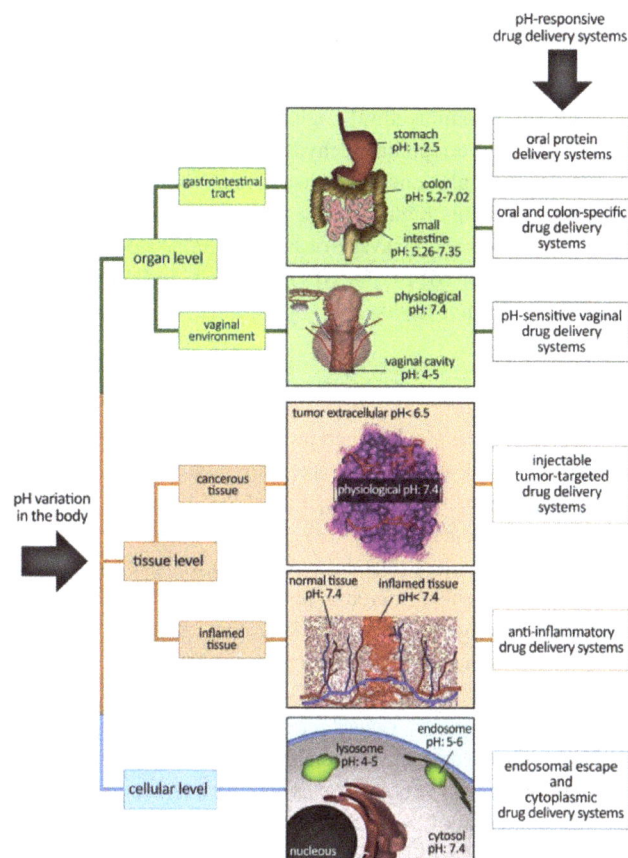

FIGURE 7.3 pH variation in the human body, at the organ, tissue, and cellular levels.

Source: [19].

7.2 Applications of Smart Polymers

7.2.1 pH-Responsive Delivery Systems

One of the most important stimuli used for the design of responsive drug delivery system is pH. Considering the varying pH in the human body, it has been extensively used for the design of various drug delivery systems by selecting the smart polymer to match the desired pH range.[14–19] pH-responsive polymers are systems whose solubility, volume, and chain conformation can be manipulated by changes in pH, co-solvent, and electrolytes.[20] pH-responsive polymers contain an ionizable functional group that undergoes reversible (e.g., hydration) or irreversible (e.g., hydrolysis) changes in response to a change in the environmental pH. The reversible pH-sensitive polymers are typically polyelectrolytes that contain multiple weakly acidic or basic functional groups, whose ionization status can change in response to the environmental pH, while irreversible pH-responsive polymers include copolymers and polymers with linkers susceptible to hydrolysis that is susceptible to acid or base catalysis. Such polymers lose structural integrity upon a change in pH. The pH responsiveness of the polymer is generally triggered by the protonation and deprotonation cycle of the weak polyacid or weak polybase or a combination of both in the block polymers at different pH or pH-induced conformation changes of the copolymer.[14,21]

In drug delivery applications, the drug release can be controlled by two different strategies. In the first strategy, the drug is released from the polymer due to swelling/de-swelling behavior or degradation of the micelle structure with the pH change. In the second strategy, polymeric materials release drugs with the cleavage of the covalent bonds between the drug and the polymer by pH changes. The pH-labile bonds used in polymer structures are hydrazine,[22,23] acetal/ketal,[24,25] cis-acotinyl,[26,27] imine,[28] substituted trityl,[29] orthoester,[30] and others.[31]

In the human body, there exists a pH gradient which spans all the way from very acidic to physiological pH value (7.4). Figure 7.3 shows the variation in pH that is found in the human body at the organ, tissue, and cellular levels.[19]

Polymeric micelles are a supramolecular assembly with characteristic properties, such as a core-protecting double-layer structure that is tens of nanometers in diameter, has low toxicity in the human body, and has a prolonged circulation in the blood, owing to its high water-solubility, thus avoiding phagocytic and renal clearance.[32–34] Polymeric micelles can be incorporated into layer-by-layer (LbL) films, based on hydrogen bonding, without the use of charged polyelectrolytes. LbL films are constructed by employing the hydrogen bonding between biologically compatible poly (acrylic acid) (PAA) and a biodegradable block copolymer micelle of poly (ethylene oxide)-block-poly(caprolactone) (PEO-b-PCL) as the drug delivery vehicles from surfaces. These polymeric micelles have been extensively used to provide a highly versatile nanometer-sized delivery platform for drugs, proteins, DNA, and personal care products.[33,35] Folic acid grafted and tertiary amino-based pH-responsive pentablock Poly (ethylene glycol)-b-(poly(2-(diethylamino) ethyl methacrylate)-b-poly (hydroxyethyl methacrylate)-g-folic acid) (PEG-b-(PDEAEMA-bPHEMA-g-FA)2) polymeric micelles can be prepared for targeted anticancer drug delivery.[36] Chitosan-based pH-sensitive polymeric micelles containing curcumin are prepared for colon-targeted drug delivery.[37]

A pH-responsive polymeric nanogel has been developed for tumor therapy.[38] The pH-responsive nanogels are constructed using hydrophilic methoxy poly (ethylene glycol)-b-poly[N-[N-(2-aminoethyl)-2-aminoethyl]-L-glutamate] (MPEG-b-PNLG) and hydrophobic terephthalaldehyde (TPA) as a cross-linker by the formation of pH-sensitive benzoic imine bonds. At pH 7.4, the nanogel exhibits high stability, while in the acidic environments (pH ~6.4), the cleavage of benzoic imine bonds induces the destruction of MPEG-b-PNLG nanogels and leads to the rapid release of their payloads.

7.2.2 Thermoresponsive Delivery Systems

The most widely employed smart polymers are thermoresponsive polymers, and one of their characteristic properties exhibits the volume phase transition, resulting in a sudden change in the solvation state.[39,40] Generally, thermoresponsive polymers exhibit a temperature-dependent miscibility or solubility gap in aqueous solutions that can be observed in phase diagrams of temperature versus polymer fraction volume. This phase separation occurs due to hydrophobic interactions between the polymer chains, and this enables the self-assembly and aggregation of the polymers in aqueous solutions. Lower critical solution temperatures (LCSTs) or upper critical solution temperatures (UCSTs) appear for thermoresponsive polymers when a polymer solution is phase separated above or below a specific temperature. In the phase diagram, the minimum and maximum temperatures are indicated as being the LCST and UCST, respectively.[41,42] The most commonly used thermoresponsive polymer is poly(*N*-isopropylacrylamide) (PNIPAAm) because of its lower critical solution temperature, its LCST of around 32°C, a very useful temperature for biomedical applications since it is close to the body temperature (37°C).[43,44]

Preparation of the temperature-responsive polymeric micelle is done by using block copolymers of (poly(N-isopropylacrylamide-b-butyl methacrylate) (PIPAAm-PBMA)) for the release of the adriamycin drug.[45] The micelle inner core is formed by self-aggregates of PBMA segments and the outer shell of PIPAAm chains play a role in stabilizing and initiating the micellar thermo-response. The outer shell hydrophilicity prevents drug interaction with biocomponents, and other micelles can suddenly be switched to hydrophobic at a specific site by a local temperature increase beyond the LCST (32.58°C). PIPAAm is hydrophilic and water-soluble with extended chain conformation below its LCST and undergoes a phase transition to an insoluble and hydrophobic aggregate above its LCST.[46] Modification of the thermoresponsive behavior of the polymer micelle is demonstrated by preparing the block copolymer of PIPAAm (poly(N-isopropylacrylamide)) with either PBMA (poly (butyl methacrylate)) or PSt (polystyrene). This polymeric micellar structure comprises a hydrophilic outer shell of hydrated PIPAAm segments and a hydrophobic inner core. The inner core can be loaded with hydrophobic drugs, while the PIPAAm outer shell plays the role of aqueous solubilization and temperature-responsiveness. Also, the changes in the inner core structure result in a change in response to the PIPAAm outer shell.

X. Song et al. developed a thermoresponsive micellar drug delivery system formed by a non-covalently connected supramolecular block polymer to overcome the limitation of chemotherapy, such as low water solubility of anticancer drugs and multidrug resistance (MDR) in cancer cells.[47] The system is based on the host-guest interaction between a well-defined b-cyclodextrin (b-CD) based poly(N-isopropylacrylamide) star host polymer and an adamantyl-containing poly(ethylene glycol) (Ad-PEG) guest polymer.

Of all the hydrophilic-hydrophilic block and graft copolymers, poly (ethylene glycol)-b-poly(N-isopropylacrylamide) (PEG-b-PNIPAM) or poly(ethylene glycol)g-poly(N-isopropylacrylamide) (PEG-g- PNIPAM) is an interesting one, where the PNIPAM is one of the most studied responsive polymers that exhibits a LCST in water around 32°C [48–50] and the PEG is normally used for the stabilization of dispersions and emulsions.[51,52] The thermally responsive cholesteryl end-capped poly(N-isopropylacrylamide-co-N, N-dimethylacrylamide) and cholesteryl grafted poly[N-isopropylacrylamide-co-N-(hydroxymethyl) acrylamide] amphiphilic polymers are synthesized to encapsulate cyclosporin A (CyA) and indomethacin (IND) within core-shell nanoparticles by the membrane dialysis method.[53] The LCST values were discovered to be 33.4°C and 38.3°C, respectively. C.-C. Cheng et al. have investigated pendant U-DPy groups (PNI-UDPy) in an aqueous solution, with extremely high micellar stability and drug-loading capacity (up to 16%), excellent thermoresponsive behavior and a rapid drug release rate due to the U-DPy-induced physical cross-linking[54] (Figure 7.4).

In the case of pristine PNIPAM, the chains adopt a hydrophilic coil conformation below LCST. The PNIPAM in water is enabled by the formation of intermolecular hydrogen bonding interactions between the amide groups of PNIPAM and water molecules. The introduction of hydrogen bond functionality into the PNIPAM backbone exerts a beneficial effect on their self-assembly process and results in a "hydrophilic globule conformation," due to the occurrence of supramolecular interactions contributed by the strongly associating U-DPy or DAP functional groups. Adriamycin (ADR)-loaded thermally responsive polymeric micelles composed of poly(N-isopropylacrylamide-coN,N-dimethylacrylamide)-b-poly(D,L-lactide) are prepared by the dialysis method.[55] ADR-loaded micelles did not show much cytotoxic activity against bovine aorta endothelial cells at 37°C, while they show high cytotoxicity at 42.5°C. Thus, thermally responsive polymeric micelles show distinct control of ADR cytotoxic activity by temperature, while free ADR does not have any control. Cross-linked hybrid micelles can be prepared from the P(NIPAAm-co-MPMA)-b-PMMA via a two-step process: self-assembly of P(NIPAAm-co-MPMA)-b-PMMA to form polymer micelles in an aqueous solution and then cross-linking of the peripheral P(NIPAAm-co-MPMA) block through an inorganic "silica-based" cross-linking strategy.[56] The drastic changes are observed in the drug release profile with an increase in temperature above the LCST. Below LCST (at 25°C) about 60% of the drug is released from the micelle within 213 hrs. but at a temperature above the LCST, the drug release is accelerated and reached 80% for a period of 213 hrs. Thermosensitive

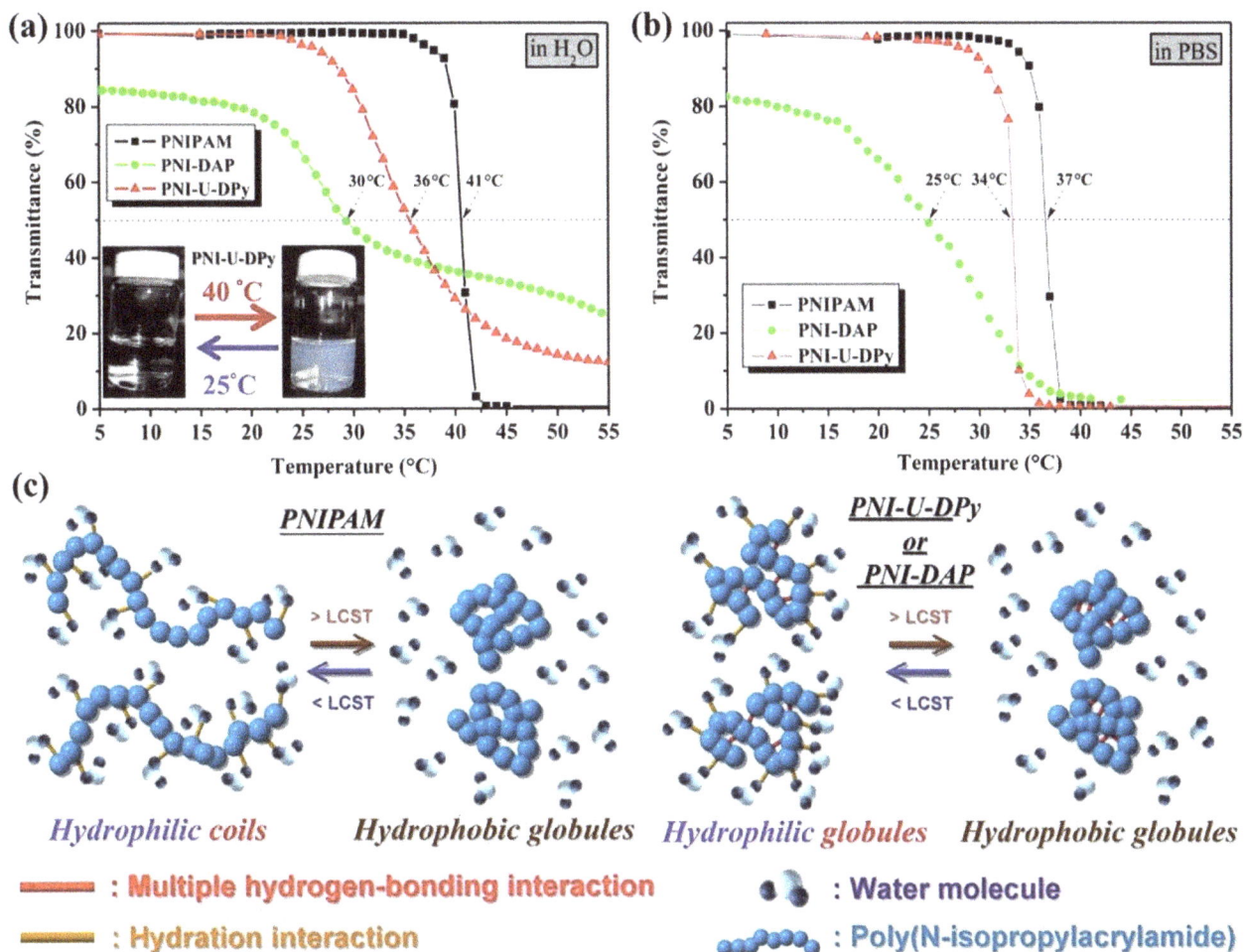

FIGURE 7.4 Temperature-responsive behavior of supramolecular polymeric micelles. Transmittance curves of SPN-functionalized PNIPAMs in (a) water and (b) PBS buffer. (a) The inset photographs show the phase-transition behavior of PNI-U-DPy micelles in aqueous solution at 25 C and 40 C. (c) Suggested mechanisms for the formation of micellar particles in aqueous solution.

Source: [54].

and biotinylated biotin-poly (N-isopropylacrylamide-co-N-hydroxymethylacrylamide)-block-poly(methyl methacrylate) (biotin-P(NIPAAm-co-HMAAm)-b-PMMA) block copolymers are prepared by radical polymerization.[54]

7.2.3 Enzymatic Responsive Delivery Systems

Enzymes are key biological catalysts as they are highly selective and specific biomolecules. They generally require milder conditions compared to other conventional chemical reactions, i.e., enzyme reactions can occur in aqueous environments, at relatively low temperatures, and typically at neutral or slightly acidic or alkaline pH. These defining advantages make enzyme-catalyzed reactions suitable for biomedical applications. The synthesis of polytyrosine nanoparticles (PTNs) self-assembled from poly(ethylene glycol)-b-poly(L-tyrosine) block copolymer enables the ultra-high loading and rapid enzyme-responsive release of doxorubicin (DOX). The DOX-loaded PTNs (PTN-DOX) exhibit small sizes, narrow polydispersity, good colloidal stability, and ultrahigh drug loading content up to 63.1 wt%.

The fast release of the DOX from the nanoparticles results in an enhanced antiproliferative effect as compared with the clinically used LP-DOX in both RAW 264.7 cells and HCT-116 human colorectal cancer cells.[57]

Synthetic polymers consisting of enzyme-cleavable phosphate functionalities can exhibit triggered self-assembling and aggregation characteristics. Enzymatically sensitive polymers are also widely used in the case of polymeric micelles that physically encapsulate their cargo and release it upon enzymatic degradation of the polymer of which they are composed.[58] They are a synthesized natural and simple core–shell-structured microcapsule, which releases its cargo only when exposed to lipase. In the process, the cargo is trapped inside a gel matrix, which is surrounded by a double-layer shell containing an inner solid lipid layer and an outer polymer layer. The polymer layer consists of casein and poloxamer 338. The lipase-degradable lipid layer of the microcapsule shell can reduce the inherent limitations of slow and incomplete dissolution of poorly water-soluble drugs and facilitate the formation of solubilized fragments from which absorption may occur.

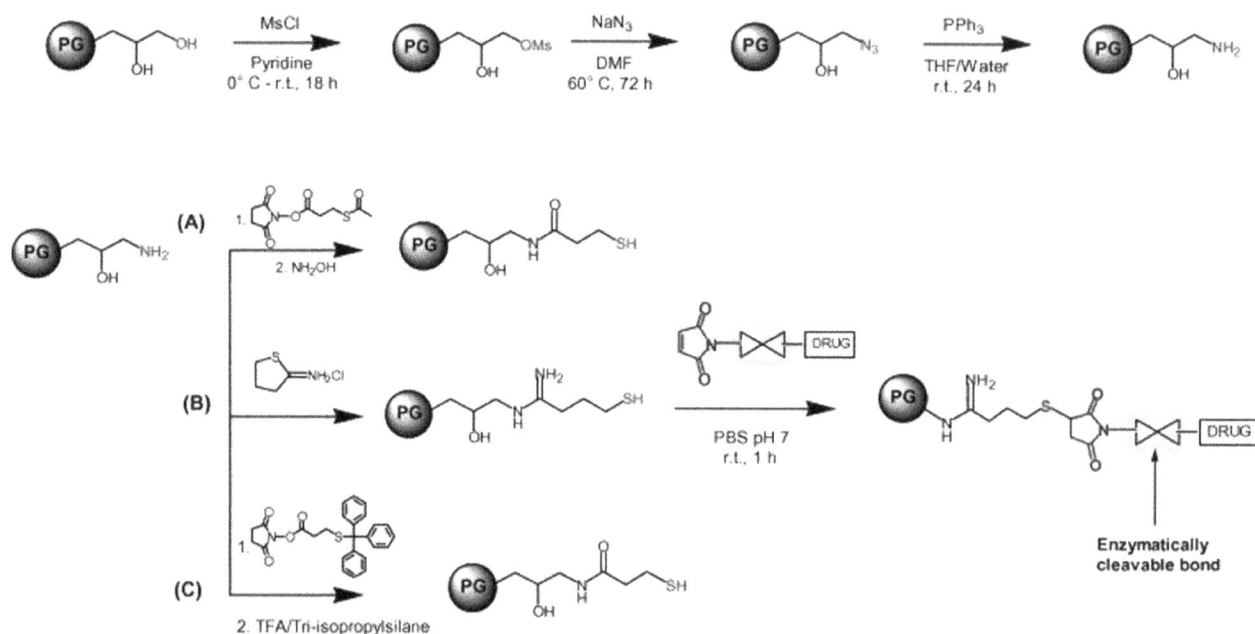

FIGURE 7.5 Studied pathways for synthesis of thiolated polyglycerol and prodrug coupling. (A) 3-(tritylthio)propionic acid based thiolation; (B) 2-iminothiolane based thiolation; (C) acetyl-thiopropionic acid based thiolation.

The synthesis of poly (ethylene glycol) (PEG) hydrogel cross-linked using thiol-ene photopolymerization is done for the selective release of protein therapeutics upon exposure to HNE.[59] The work demonstrates that thiol-ene photopolymerization can be used as a method for fabricating enzyme-responsive PEG hydrogels for the application of controlled protein release.

The synthetic protocol consists of four steps. The first three steps are shown in Figure 7.5, For the synthesis of the thiolated derivatives, three different pathways were studied using 3-(tritylthio) propionic acid (A), 2-iminothiolane (B), or acetyl-thiopropionic acid (C).

Another approach is based on the synthesis of a scaffold derived from hyperbranched polyglycerol (PG) and its subsequent conjugation with enzymatically cleavable prodrugs of doxorubicin and methotrexate[60] (Figure 7.5).

7.3 Conclusion

In conclusion, smart polymers play an important role in the development of a variety of medical devices, including biosensors, actuators, and microfluidics-based systems. Smart polymers exhibit stimuli-responsive behavior that results in the small change of atmosphere triggered by light, pH, electricity, magnetic fields, ion concentration, chemical and enzyme degradation, and temperature. They offer a wide range of applications and their properties can be tuned by several mechanisms. These smart polymers not only enhance the properties of drugs, such as solubility, bioavailability, and prolonged circulation times but also can be made to selectively release their payload at the desired site of action. However, a few critical parameters should be taken into consideration, such as biocompatibility and toxicology of

these multi-component polymer-based systems, their ability to provide the required levels of the drug, and addressing the necessary formulation issues in dosage design (e.g., shelf life, sterilization, reproducibility). In the future, the smart polymer-based stimuli-sensitive drug delivery systems are expected to cure a large number of diseases.

REFERENCES

1. S. Guragain, B.P. Bastakoti, V. Malgras, K. Nakashima, and Y. Yamauchi. "Multi-stimuli-responsive polymeric materials." *Chemistry* 21(38) (2015): 13164–13174.
2. V. Aina, G. Malavasi, C. Magistris, G. Cerrato, G. Martra, G. Viscardi, L. Menabue, and G. Lusvardi. "Conjugation of amino-bioactive glasses with 5-aminofluorescein as probe molecule for the development of PH sensitive stimuli-responsive biomaterials." *Journal of Materials Science: Materials in Medicine* 25(10) (2014): 2243–2253.
3. K.P.Y. Qiu. "Environment-sensitive hydrogels for drug delivery." *Advanced Drug Delivery* 53 (2001): 321–339.
4. D. Roy, J.N. Cambre, and B.S. Sumerlin. "Future perspectives and recent advances in stimuli-responsive materials." *Progress in Polymer Science (Oxford)* 35(1–2) (2010): 278–301.
5. H. Almeida, M.H. Amaral, and P. Lobão. "Temperature and PH stimuli-responsive polymers and their applications in controlled and self-regulated drug delivery." *Journal of Applied Pharmaceutical Science* 2(6) (2012): 01–10.
6. A.E. Felber, M.H. Dufresne, and J.C. Leroux. "pH-sensitive vesicles, polymeric micelles, and nanospheres prepared with polycarboxylates." *Advanced Drug Delivery Reviews* 64(11) (2012): 979–992.
7. J.G. Hardy, M. Palma, J. W. Shalom, and M.J. Biggs. "Responsive biomaterials: advances in materials based on shape-memory polymers." *Advanced Materials* 28(27) (2016): 5717–5724.

8. B. Jeong and A. Gutowska. "Lessons from nature: Stimuli-responsive polymers and their biomedical applications." *Trends in Biotechnology* 20(7) (2002): 305–311.

9. F. Liu and M.W. Urban. "Recent advances and challenges in designing stimuli-responsive polymers." *Progress in Polymer Science (Oxford)* 35(1–2) (2010): 3–23.

10. I.Y. Galaev and B. Mattiasson. "'Smart' polymers and what they could do in biotechnology and medicine." *Trends in Biotechnology* 7799 (1999): 335–340.

11. A.S. Hoffman, P.S. Stayton, V. Bulmus, G. Chen, J. Chen, C. Cheung, A. Chilkoti, Z. Ding, L. Dong, …, and T. Miyata. "Really smart bioconjugates of smart polymers and receptor proteins." *Journal of Biomedical Materials Research* 52(4) (2000): 577–586.

12. Z. Ge, D. Xie, D. Chen, X. Jiang, Y. Zhang, H. Liu, and S. Liu. "Stimuli-responsive double hydrophilic block copolymer micelles with switchable catalytic activity." *Macromolecules* 40(10) (2007): 3538–3546.

13. S.K. Li and A.D. Emanuele. "On-off transport through a thermoresponsive hydrogel composite membrane." *Journal of Controlled Release* 75(1–2) (2001): 55–67.

14. A.A. Moghanjoughi, D. Khoshnevis, and A. Zarrabi. "A concise review on smart polymers for controlled drug release." *Drug Delivery and Translational Research* 6(3) (2016): 333–340.

15. A.S. Hoffman. "Stimuli-responsive polymers: Biomedical applications and challenges for clinical translation." *Advanced Drug Delivery Reviews* 65(1) (2013): 10–16.

16. J.K. Chen and C.J. Chang. "Fabrications and applications of stimulus-responsive polymer films and patterns on surfaces: A review." *Materials* 7(2) (2014): 805–875.

17. N. Rapoport. "Physical stimuli-responsive polymeric micelles for anti-cancer drug delivery." *Progress in Polymer Science (Oxford)* 32(8–9) (2007): 962–990.

18. M. Ballauff and Y. Lu. "'Smart' nanoparticles: Preparation, characterization, and applications." *Polymer* 48(7) (2007): 1815–1823.

19. S. Bazban-Shotorbani, M.M. Hasani-Sadrabadi, A. Karkhaneh, V. Serpooshan, K.I. Jacob, A. Moshaverinia, and M. Mahmoudi. "Revisiting structure-property relationship of PH-responsive polymers for drug delivery applications." *Journal of Controlled Release* 253 (2017): 46–63.

20. S. Dai, P. Ravi, and K.C. Tam. "PH-responsive polymers: synthesis, properties, and applications." *Soft Matter* 4(3) (2008): 435–449.

21. J. Du and R.K. O'Reilly. "Advances and challenges in smart and functional polymer vesicles." *Soft Matter* 5(19) (2009): 3544–3561.

22. Y. Gu, Y. Zhong, F. Meng, R. Cheng, C. Deng, and Z. Zhong. "Acetal-linked paclitaxel prodrug micellar nanoparticles as a versatile and potent platform for cancer therapy." *Biomacromolecules* 14(8) (2013): 2772–2780.

23. J. Cui, Y. Yan, Y. Wang, and F. Caruso. "Templated assembly of pH-labile polymer-drug particles for intracellular drug delivery." *Advanced Functional Materials* 22(22) (2012): 4718–4723.

24. E.R. Gillies, A.P. Goodwin, and J.M.J. Fréchet. "Acetals as pH-sensitive linkages for drug delivery." *Bioconjugate Chemistry* 15(6) (2004): 1254–1263.

25. Y. Chan, T. Wong, F. Byrne, M. Kavallaris, and V. Bulmus. "Acid-labile core cross-linked micelles for pH-triggered release of antitumor drugs." *Biomacromolecules* 9(7) (2008): 1826–1836.

26. J.H. Park, Y.W. Cho, Y.J. Son, K. Kim, H. Chung, S.Y. Jeong, K. Choi, C.R. Park, … and I.C. Kwon. "Preparation and characterization of self-assembled nanoparticles based on glycol chitosan bearing adriamycin." *Colloid and Polymer Science* 284(7) (2006): 763–770.

27. H.S. Yoo, E.A. Lee, and T.G. Park. "Doxorubicin-conjugated biodegradable polymeric micelles having acid-cleavable linkages." *Journal of Controlled Release* 82(1) (2002): 17–27.

28. C. Ding, J. Gu, X. Qu, and Z. Yang. "Preparation of multi-functional drug carrier for tumor-specific uptake and enhanced intracellular delivery through the conjugation of weak acid labile linker." *Bioconjugate Chemistry* 20(6) (2009): 1163–1170.

29. J. Zou, F. Zhang, S. Zhang, S.F. Pollack, M. Elsabahy, J. Fan, and K.L. Wooley. "Poly(ethylene oxide)-block-polyphosphoester-graft-paclitaxel conjugates with acid-labile linkages as a PH-sensitive and functional nanoscopic platform for paclitaxel delivery." *Advanced Healthcare Materials* 3(3) (2014): 441–448.

30. C.C. Song, R. Ji, F.S. Du, D.H. Liang, and Z.C. Li. "Oxidation-accelerated hydrolysis of the ortho ester-containing acid-labile polymers." *ACS Macro Letters* 2(3) (2013): 273–277.

31. W. Chen, F. Meng, F. Li, S.J. Ji, and Z. Zhong. "PH-responsive biodegradable micelles based on acid-labile polycarbonate hydrophobe: synthesis and triggered drug release." *Biomacromolecules* 10 (7) (2009): 1727–1735.

32. L. Kang, Z. Gao, W. Huang, M. Jin, and Q. Wang. "Nanocarrier-mediated co-delivery of chemotherapeutic drugs and gene agents for cancer treatment." *Acta Pharmaceutica Sinica B* 5(3) (2015): 169–175.

33. K. Kataoka, A. Harada, and Y. Nagasaki. "Block copolymer micelles for drug delivery: design, characterization and biological significance." *Advanced Drug Delivery Reviews* 47(1) (2001): 113–131.

34. U. Kedar, P. Phutane, S. Shidhaye, and V. Kadam. "Advances in polymeric micelles for drug delivery and tumor targeting." *Nanomedicine: Nanotechnology, Biology, and Medicine* 6(6) (2010): 714–729.

35. R. Savić, L. Luo, A. Eisenberg, and D. Maysinger. "Micellar nanocontainers distribute to defined cytoplasmic organelles." *Science* 300(5619) (2003): 615–618.

36. Q. Chen, J. Zheng, X. Yuan, J. Wang, and L. Zhang. "Folic acid grafted and tertiary amino based PH-responsive pentablock polymeric micelles for targeting anticancer drug delivery." *Materials Science and Engineering C* 82 (2018): 1–9.

37. T. Woraphatphadung, W. Sajomsang, T. Rojanarata, T. Ngawhirunpat, P. Tonglairoum, and P. Opanasopit. "Development of chitosan-based PH-sensitive polymeric micelles containing curcumin for colon-targeted drug delivery." *AAPS PharmSciTech* (2017).

38. Y. Li, Q.N. Bui, L.T.M. Duy, H.Y. Yang, and D.S. Lee. "One-step preparation of PH-responsive polymeric nanogels as intelligent drug delivery systems for tumor therapy." *Biomacromolecules* 19(6) (2018): 2062–2071.

39. X. Yin, A.S. Hoffman, and P.S. Stayton. "Poly(N - isopropylacrylamideco -propylacrylic acid) copolymers that respond sharply to temperature and Ph." *Biomacromolecules* 7 (2006):1381–1385.

40. Y.J. Kim and Y.K. Matsunaga. "Thermo-responsive polymers and their application as smart biomaterials." *Journal of Materials Chemistry B* 5(23) (2017): 4307–4321.

41. Y. Zhu, R. Batchelor, A.B. Lowe, and P.J. Roth. "Design of thermoresponsive polymers with aqueous LCST, UCST, or both: modification of a reactive poly(2-vinyl-4,4-dimethylazlactone) scaffold." *Macromolecules* 49(2) (2016): 672–680.

42. C. Liu, H. Qin, and P.T. Mather. "Review of progress in shape-memory polymers." *Journal of Materials Chemistry* 17(16) (2007): 1543–1558.

43. B.R. Twaites, A.C. De Las Heras, M. Lavigne, A. Saulnier, S.S. Pennadam, D. Cunliffe, D.C. Górecki, and C. Alexander. "Thermoresponsive polymers as gene delivery vectors: cell viability, DNA transport, and transfection studies." *Journal of Controlled Release* 108(2–3) (2005): 472–483.

44. K.B. Doorty, T.A. Golubeva, A.V. Gorelov, Y.A. Rochev, L.T. Allen, K.A. Dawson, W.M. Gallagher, and A.K. Keenan. "Poly(N-isopropylacrylamide) co-polymer films as potential vehicles for delivery of an antimitotic agent to vascular smooth muscle cells." *Cardiovascular Pathology* 12(2) (2003): 105–110.

45. J.E. Chung, M. Yokoyama, M. Yamato, T. Aoyagi, Y. Sakurai, and T. Okano. "Thermo-responsive drug delivery from polymeric micelles constructed using block copolymers of Poly(N-isopropylacrylamide) and poly(butylmethacrylate)." *Journal of Controlled Release* 62(1–2) (1999): 115–127.

46. J.E. Chung, M. Yokoyama, and T. Okano. "Inner core segment design for drug delivery control of thermo-responsive polymeric micelles." *Journal of Controlled Release* 65(1–2) (2000): 93–103.

47. X. Song, J.L. Zhu, Y. Wen, F. Zhao, Z.X. Zhang, and J. Li. "Thermoresponsive supramolecular micellar drug delivery system based on star-linear pseudo-block polymer consisting of β-cyclodextrin-poly(N-isopropylacrylamide) and adamantyl-poly(ethylene glycol)." *Journal of Colloid and Interface Science* 490 (2017): 372–379.

48. K. Kubota, S. Fujishige, and I. Ando. "Single-chain transition of poly(N-isopropylacrylamide) in water." *Journal of Physical Chemistry* 94(12) (1990): 5154–5158.

49. S. Fujishige, K. Kubota, and I. Ando. "Phase transition of aqueous solutions of poly(N-isopropylacrylamide) and poly(N-isopropylmethacrylamide)." *Journal of Physical Chemistry* 93(8) (1989): 3311–3313.

50. C. Wu and S. Zhou. "Thermodynamically stable globule state of a single poly(N-isopropylacrylamide) chain in water." *Macromolecules* 28(15) (1995): 5388–5390.

51. W. Zhang, L. Shi, K. Wu, and Y. An. "Thermoresponsive micellization of poly(ethylene glycol)-B-poly(N-isopropylacrylamide) in water." *Macromolecules* 38(13) (2005): 5743–5747.

52. K.E. Uhrich, S.M. Cannizzaro, R.S. Langer, and K.M. Shakesheff. "Polymeric systems for controlled drug release." *Chemical Reviews* 99(11) (1999): 3181–3198.

53. C.S. Chaw, K.W. Choi, X.M. Liu, C.W. Tan, L. Wang, and Y.Y. Yang. "Thermally responsive core-shell nanoparticles self-assembled from cholesteryl end-capped and grafted polyacrylamides: drug incorporation and in vitro release." *Biomaterials* 25(18) (2004): 4297–4308.

54. C-C. Cheng, F-C. Chang, W-Y. Kao, S-M. Hwang, L-C. Liao, Y-J. Chang, M-C. Liang, J-K. Chen, and D-J. Lee. "Highly efficient drug delivery systems based on functional supramolecular polymers: In vitro evaluation." *Acta Biomaterialia* 33 (2016): 194–202.

55. F. Kohori, K. Sakai, T. Aoyagi, M. Yokoyama, M. Yamato, Y. Sakurai, and T. Okano. "Control of adriamycin cytotoxic activity using thermally responsive polymeric micelles composed of poly(N-isopropylacrylamide-co-N,N-dimethylacrylamide)-B-poly(-lactide)." *Colloids and Surfaces B: Biointerfaces* 16(1–4) (1999): 195–205.

56. H. Wei, C.C. Chang, W.Q. Chen, S.X. Cheng, X.Z. Zhang, and R.X. Zhuo. "Synthesis and applications of shell cross-linked thermoresponsive hybrid micelles based on poly(N-isopropylacrylamide-co-3-(trimethoxysilyl)propyl methacrylate)-B-poly(methyl methacrylate)." *Langmuir* 24(9) (2008): 4564–4570.

57. X. Gu, M. Qiu, H. Sun, J. Zhang, L. Cheng, C. Deng, and Z. Zhong. "Polytyrosine nanoparticles enable ultra-high loading of doxorubicin and rapid enzyme-responsive drug release." *Biomaterials Science* 6(6) (2018): 1526–1534.

58. R. Ravanfar, G.B. Celli, and A. Abbaspourrad. "Controlling the release from enzyme-responsive microcapsules with a smart natural shell." *ACS Applied Materials and Interfaces* 10(6) (2018): 6046–6053.

59. A.A. Aimetti, A.J. Machen, and K.S. Anseth. "Poly(ethylene glycol) hydrogels formed by thiol-ene photopolymerization for enzyme-responsive protein delivery." *Biomaterials* 30(30) (2009): 6048–6054.

60. M. Calderón, R. Graeser, F. Kratz, and R. Haag. "Development of enzymatically cleavable prodrugs derived from dendritic polyglycerol." *Bioorganic and Medicinal Chemistry Letters* 19(14) (2009): 3725–3728.

8

Technical Textiles

A.K. Sidharth
Institute of Chemical Technology, Mumbai, India

Junaid Parkar
Institute of Chemical Technology, Mumbai, India

Ravindra Kale
Institute of Chemical Technology, Mumbai, India

Ramanand Jagtap
Institute of Chemical Technology, Mumbai, India

CONTENTS

DOI: 10.1201/9781003037880-8

Abbreviations

PCM	phase change material
PEG	polyethylene glycol
SMPs	shape memory polymers
PU	polyurethane
DMF	dimethylformamide
PVC	polyvinyl chloride
PLA	poly(DL-lactic acid).
WVP	water vapor permeability
PTFE	poly (tetrafluoro ethylene)
PTGPU	polytetramethylene ether/ester copolymer-type polyurethane
WVTR	water vapor transmission rate
WCA	water contact angle
CMC	critical micelle concentration
CAC	critical aggregation concentration
PVG	poly-N-vinylguanidine
PHA	poly-hydroxamic acid
PAN	poly (acrylo nitrile)
PUD	polyurethane dispersion
HDI-T	hexamethylene diisocyanate-trimer
DHPDMS	di-hydroxy-butyl terminated poly-di-methyl-siloxane
APTES	amino propyl triethoxy silane
TEM	transmission electron microscopy
DMF	dimethylformamide
DMPA	dimethyl propionic acid
WBFs	waterproof breathable fabrics
WVT	water vapor transmission
WVR	water vapor resistance
RH	relative humidity
WBFs	waterproof breathable fabrics
PVDF	polyvinylidine fluoride

8.1 Introduction

In the advanced socio-economic structure of our society, textiles are no longer only limited for use as apparel clothing but appear as aids in every walk of life. With respect to the priorities of scientists around the world, who are innovating to enhance the quality of human life through protection against various hazards, as well as protection of the environment, textiles play a crucial role. The fastest-growing area of textile consumption in the world is "technical textiles." In these circumstances, textiles are playing a major role through their diversified applications and India has great potential for the production, consumption and export of technical textiles. The major applications of technical textiles are described here.

8.2 Agrotextiles

With the growing global population, the land formerly used for agriculture is being developed for housing and infrastructures. This is leading to a reduction in the area of actual cultivable land. Thus it is important to use the land as well as agricultural produce wisely. In order to store grains, vegetables, and flowers, it is necessary to improve the yield and quality of agro-products. However, it is not advisable to totally rely on pesticides and herbicides. Additionally, these methods are expensive and have an enduring ecological impact on the soil as well as on the final product.

Today, agriculture, horticulture, and other allied areas are opting for various technologies to obtain higher overall yield and quality for agro products. Adopting new farming techniques, where textile structures may be used; could thereby enhance the quality and thus the overall yield of agro products. Textiles in various forms can be used for shade or for greenhouses and also in the open fields so as to control environmental factors, such as water, temperature, and humidity. This thus protects the agro

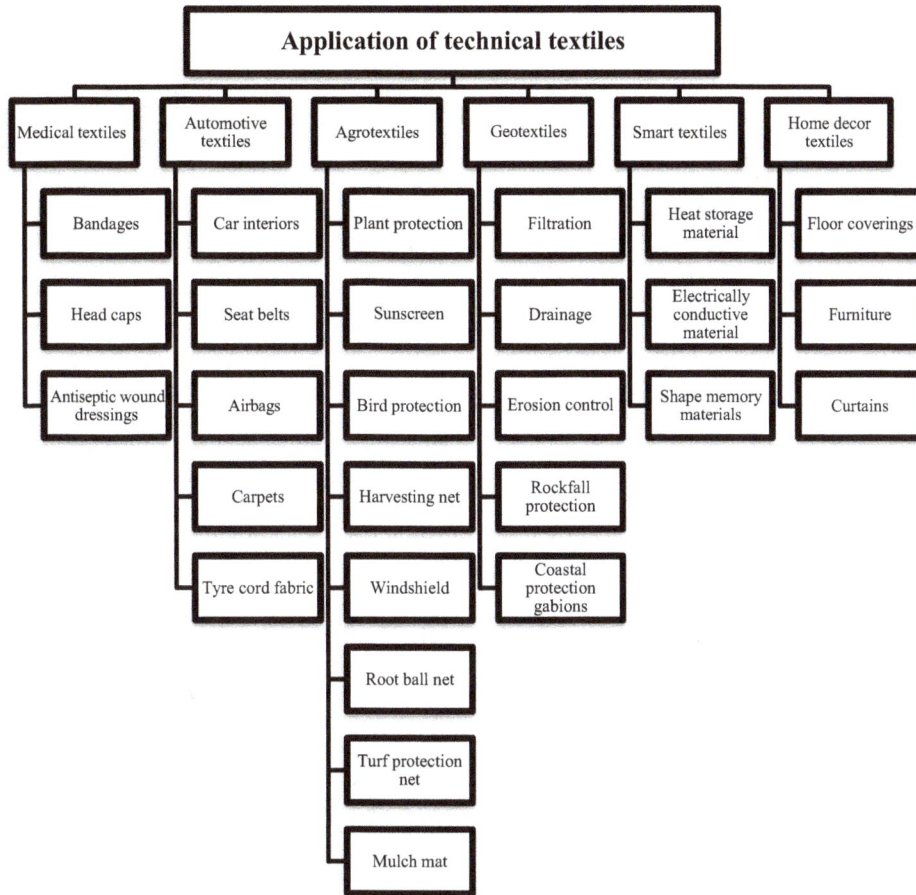

products from being damaged by rain, wind or, in some cases, birds. Agro textiles, such as mulch mats, sunscreens, hail protection nets, harvesting nets, bird nets, windshields, etc. can be used for these purposes.

A wide variety of agro textile products are available and the selection of the suitable type of product depends on the protection that the crop needs. The agro textile selection is greatly influenced by its use in a specific geographical location. Therefore, the selection of agro textiles is done as per the location and the desired protection from the external agencies.

8.2.1 Sunscreen

In order to protect fields from intense solar radiation for healthy plant growth and a good harvest, sunscreen nets with open mesh construction are used to manage the sunshine and thereby the amount of shade required. Sunscreen fabrics are available with different shade coefficients, such as 35%, 50%, 65%, and 95%. These net fabrics allow the airflow freely, so the excess heat does not build up under the sunscreen. They are made of nylon or polyester (Figure 8.1). A UV-blocking dye is used to color the fabrics.

8.2.2 Bird Protection Net

Net fabrics of open-mesh type are used to protect fruit, crops, and vegetables from birds. The plantation area is covered from the top and at the side to stop the birds from entering the fields

FIGURE 8.1   Sunscreen.

(Figure 8.2). The advantage of using nets over other means of protection is that the nets only have to be erected once in the season. Generally, nylon is employed for the construction of these nets.

8.2.3 Root Ball Net

For the safe and speedy growth of young plants, it is important that the root system is intact when they are dug up, transported,

FIGURE 8.2   Bird protection net.

FIGURE 8.4   Turf protection net.

FIGURE 8.3   Root ball net.

FIGURE 8.5   Mulch films.

need for herbicides. Biodegradable and non-biodegradable types of mulch mats are available.

8.2.6 Nets for Covering Pallets

For safe transportation of fruits and vegetables, individual boxes are placed in larger units which are covered with wide, large mesh nets on pallets to stop the boxes being turned upside down or squashing each other. This helps to keep the products safe and intact during transport to the market.

8.2.7 Packing Materials for Agricultural Products

Nets can be used in the form of sacks, bags, and tubes for packaging farm products for many end applications, including packing sacks for vegetables and tubular packing nets for fruits. Net structures are preferred because of their high strength, low weight, transparency, air permeability, and low cost. The flat tape yarns exert minimal pressure and hence are used to carry soft fruits.

8.3 Automotive Textiles

Over the past few decades, automotive textiles have increasingly been developed and used because of the following factors:

or replanted. Traditionally, the root balls are wrapped in cloth but nowadays elastic netting tubes are an effective alternative. The elastic yarns apply gentle pressure on the soil to keep the roots intact and undamaged, so that the root balls can be transported safely (Figure 8.3). When the plants have been transported, the nets on the outside do not have to be removed since the roots can protrude through the nets.

8.2.4 Turf Protection Net

Nets are laid over the grassy areas on riverbanks so that lumps of earth are not removed while animals are grazing there (Figure 8.4). This will help to minimize soil erosion loss and thus improve conservation of the fields or riverbanks.

8.2.5 Mulch Films

In horticulture applications, mulch mats (needle-punched non-woven and black plastic sheets) are used to suppress weed growth (Figure 8.5). The mulch mat covers the soil, blocking out the light and preventing the competitive weed growth around the seedlings. This also helps in weed control and thus reduces the

FIGURE 8.6 Car seats.

FIGURE 8.7 Car seat belt.

1. Improved standards of living mean increased private vehicle use.
2. As a part of a luxury car, the car interior is very important.
3. For improved fuel economy, lightweight vehicles are in demand, replacing metal applications by fiber composites.
4. Stringent legislations demand safety devices in vehicles such as airbags and seat belts.
5. Ecological improvements demand textiles in an automobile for easy recycling of used cars.
6. Textiles industries have also come up with the solution to automobile engineering problems, such as acoustics protection, tire reinforcement, petrol filters, and air filtration.

8.3.1 Car Interiors

People are spending more time in their cars, due to increased traffic density, greater mobility, and long-distance traveling. Car interiors are an important feature of the raised consumer expectations of car buyers and the car interior's comfort is a priority reflected in their cost. Today, over 90% of car seat covers are made of polyester filament yarns (Figure 8.6). Apart from abrasion resistance and UV resistance, polyester also offers good tearing strength and ease of cleaning.

8.3.2 Seat Belts

Wearing a seat belt reduces fatalities and serious injuries by 50%. All new cars have diagonal seat belts made of 250 grams of fabrics (Figure 8.7). Seat belts are an energy-absorbing device which controls the forward movement of the wearer in the event of a sudden deceleration of the vehicle. The seat belt is designed to keep the load imposed on the victim's body during a collision, reduced to survivable limits and deliver non-recoverable extension in the event of a crash. This non-recoverable extension prevents the occupant from being thrown back into the seat and sustaining impact injuries soon after the impact. Today in modern cars, seat belts are designed to hold the passengers in the sitting position to fall into the airbag when it is inflated. Due to non-recoverable stretch, the used seat belts have to be replaced after major accidents. For seat belts, polyester is the preferred fiber as it satisfies the requirements of maximum extension up to 24–30%, offers heat and light resistance, good abrasion resistance, and is light in weight. Basically, seat belts are narrow fabrics (mostly 46 mm) made of polyester filaments.

8.3.3 Airbags

Due to stringent legislation in most countries, airbags are mandatory for all passenger cars. In 2002 alone, airbag systems contributed to a 20% reduction in fatalities resulting from front impact. How an airbag works is a precision application. In just 0.03 seconds airbags should begin and by 0.06 seconds the bag should be fully inflated after the collision. Mostly, airbags contain nitrogen gas. The fabric used for airbags must withstand the force of hot gases and at the same time the fabric must not be penetrated or rip and must contain the gases. Typically, airbags are typically woven from multifilament nylon 6,6 or are two pieces sewn together with suitable thread. But recently the one-piece weaving system has been used directly on the loom to produce airbags. Also, research is going on to use non-woven textiles for this end use. Airbags can be coated or uncoated.

8.3.4 Carpets

Each car normally contains about 3.5–4.5 sq. meters of carpet (Figure 8.8). It is not only used for sensual comfort, but it also plays a significant role in acoustic and vibration control. The popularity of multipurpose vehicles and headliner carpets have also increased the demand for the same. In a few countries, road noise is considered environmental pollution. One of the pressures on automobile manufacturers is to reduce external noise about 50%, so carpets also contribute to solving this problem.

8.3.5 Tire Cord Fabrics

Tire cord fabric is a skeleton structure for reinforcement to give dimensional stability to the tire and to hold the uniform rubber

FIGURE 8.8   Car carpet.

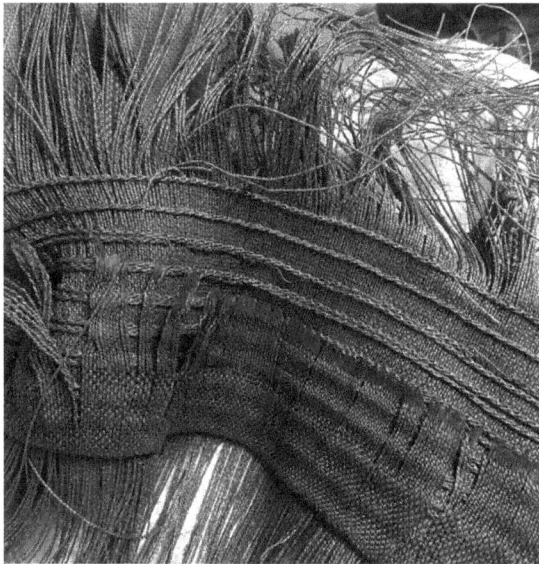

FIGURE 8.9   Tire cord.

mass of the tire (Figure 8.9). A tire is a pressure vessel, so the cord fabric helps to keep it dimensionally stable. Cord fabric also is the basis for the skeleton stage, offers load-carrying capacity, and reduces fatigue and resistance. Mainly, polyester, nylon, and rayon are used in tire cords.

8.4 Geotextiles

Geotextiles have been defined as woven or non-woven fabrics having applications in civil engineering, such as interfacing of fabrics with soil to give a reinforced structure. These fabrics are permeable and ,when used in soil, have the ability to separate and filter it along with providing reinforcement, protection, or drainage.

8.4.1 Filtration

Geotextiles find a wide application in filtration. They help to retain the soil particles while permitting water to pass via the plane of the geotextile from the soil (Figure 8.10).

Geonet made of polypropylene is capable of withstanding high temperatures and is suitable for asphalt reinforcement. Geonets are non-corrosive, inert to chemical attack, non-biodegradable, and thermally stable. They are available in a width up to 10 meters and a suitable roll length. The aggregate in the overlay is interlocked in the mesh openings and a stiff layer is formed at the interface of the base layer and overlay. This system prevents further propagation of cracks. Advantages include control of reflection cracking in overlays, increased fatigue life, and improved load distribution.

8.4.2 Drainage

The drainage function of geotextiles involves the transmission of liquid in the plane of the fabric without soil loss (Figure 8.11). The difference between filtration and a drainage system is the direction of flow which thereby makes the in-plane permeability critical for the drainage function. Non-woven geotextiles help in the collection and also the removal of this damaging water. Geotextiles used at the aggregate interface protect the drainage system from contamination that can occur due to soil fines, thus allowing the free flow of water.

8.4.3 Erosion Control

Waves and tides near coastlines and riverbanks can undermine and damage the soil. The following elements require protection: soil, the filter unit protecting the soil, the armor protecting the filter. The filtration properties of geotextiles allow them to be used instead of traditional filter layers (Figure 8.12). For example, a single layer of geotextile fabric can replace a set of stone filter layers.

8.4.4 Rockfall Protection

Hill slopes composed of rocks are prone to generate rock falls. Rock falls are highly dangerous to life and property because of the large momentum they pick up during the rolling and bouncing motion. Hence the use of a high strength net very close to the surface profile, with an anchor behind the crest of the cutting and at the toe rock face, is necessary. The vertical section will also have anchoring as far as possible with split bolts of capacity 3000 kg. The net is thus under reasonable tension and exerts a compression or inward force holding the loose mass together and prevents the rocks from getting easily disturbed due to surface movement of water, movement of fines and any untoward disturbance caused by vibrations (Figure 8.13).

8.4.5 Coastal Protection Gabions

Currents and wave energy are generally dissipated using angular stones. However, based on the water depth and the wave surge, the stone sizes that have to be used may be too large to handle. Hence gabions are used to hold the small boulders together and thus offer a solution for coastal protection. Gabions are made of polypropylene twisted ropes and are woven by a process to fabricate it in different sizes (Figure 8.14). They are available in a

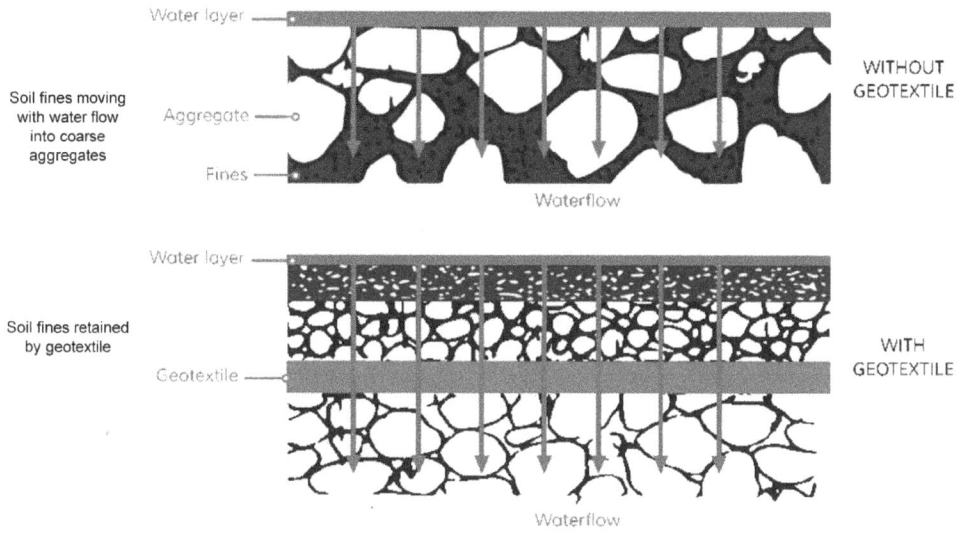

FIGURE 8.10 Schematic showing the use of geotextile for filtration purposes.

FIGURE 8.11 Drainage protection.

FIGURE 8.12 Erosion control.

pre-fabricated but collapsible form with the sides held together with a flip-open top lid.

The advantages of these polymeric gabions include resistance to acidic as well as alkaline environments. In addition to this, they are immune to rotting, safe from marine organisms, and are flexible to handle the riverbed contours with no rusting and no damage to fish. They also do not affect the water and are non-biodegradable with high tensile strength, high abrasion resistance as well as sufficiently high thermal stability.

FIGURE 8.13 Rockfall protection.

FIGURE 8.14 Coastal protection gabions.

FIGURE 8.15 Smart textile developed by the National Physical Laboratory. Source: Wills (n.d.).

8.5 Smart Textiles

Some active smart fibers contain thermally and electrically conductive materials or phase change materials (PCM) that result in a specific behavior of the textile, as per the intended application (Figure 8.15).

As the material is heated, it expands and thus reduces conductivity between particles. These materials help to regulate the on/off control of the electricity and thus keep the temperature stable. The US space travel agencies succeeded in developing phase change material microcapsules. In the process of production, phase change material is introduced into the fiber matrix in the form of microcapsules. These capsules thus prevent temperature variations that may occur by discontinuing the temperature increase when the phase change temperature is attained. The energy that is fed from the environment or from the human body to the phase change material is stored in it, increasing its thermal capacity. Inversely, phase change material materials give out the stored heat back to the environment as it cools down.

8.5.1 Heat Storage Textiles

The main aim of heat storage or thermo-regulatory textiles is to keep the wearer in a state of thermo-physiological comfort even under tough workloads as well as ambient conditions. They are novel textiles that can be used to absorb, redistribute, and also release heat by the process of phase change in low melting point materials, which is in accordance with the change in the surrounding temperature. The National Aeronautics and Space Administration (NASA) planned to put the phase change materials into the gloves worn by pilots, to keep their hands warm. NASA has also developed textiles that aim to enhance the protection of instruments as well as the astronauts against any kind of extreme fluctuations in temperature in space on the basis of the heat-absorbing and temperature-regulating technologies. It is quite challenging to process the composite fibers by a melt spinning process, only using phase change materials, because the melt viscosity of the textile grade phase change materials remains lower than required. Thus they do not possess the required spin-ability. Temperature-regulating fibers can be spun after mixing the phase change materials with PEG of a sufficient molecular weight.

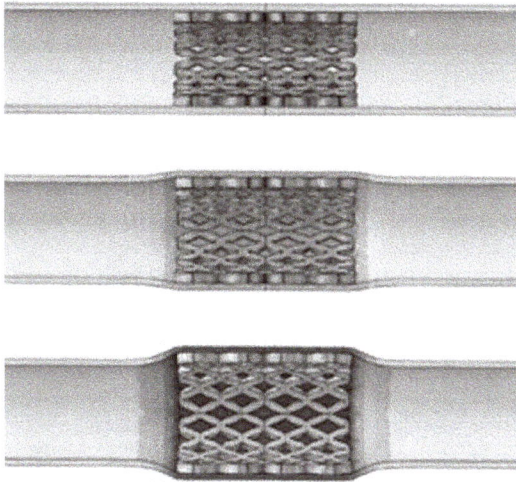

FIGURE 8.16 Smart textile being incorporated for medical applications.

Source: Barbur (n.d.).

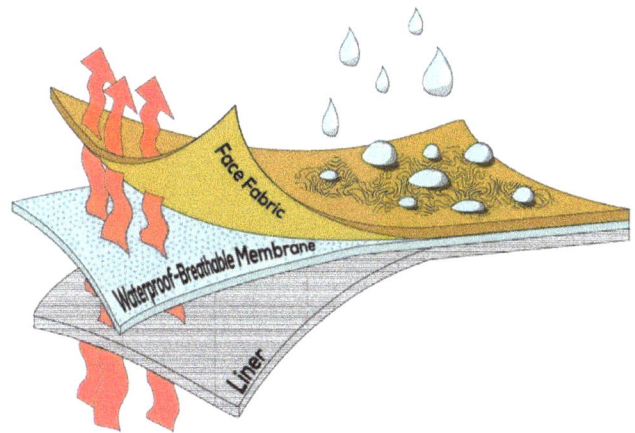

FIGURE 8.17 Illustration explaining breathability of the fabric.

Source: Showers pass (n.d.).

Mitsubishi Rayon developed a "thermo catch heat insulation system" using acrylic fibers that contains the fiber core with ceramic particles inside which convert light to heat and include, within the fiber sheath, an antimony-stannic oxide component. A mobile-thermal snow jacket has been incorporated with such bi-component yarn along with an integral heating system, assuring precise temperature control.

8.5.2 Shape Memory Materials

Shape memory materials (SMPs) can simply be explained as materials that are chemically stable at two or more temperatures. With the change in temperature, these materials have the potential to attain or deform to different shapes. Shape memory polymers are those thermo-sensitive materials that have some hard segments as well as soft segments, e.g., polyurethane, polyester, ether, styrene-butadiene copolymer (Figure 8.16). The SMPs tend to exhibit some novel behavior, such as damping, actuation, sensitivity, and adaptive response to external conditions, such as temperature, which can thereby be employed in various ways for smart systems.

Military clothing and textiles can be designed by incorporating garments with SMPs to protect soldiers against heat and cold. A garment, if incorporated with various different components of the hard and soft segments, can help achieve variable insulating properties that have great versatility and thus can be used as an inter-liner. This technique can also be incorporated into multiple layer fabric systems that can change shape within a specific temperature range and therefore can be used to change the density between the individual layers of the multiple layer fabric systems (Figure 8.17). A similar type of smart material is one that can be used in place of an inner liner in a fabric which can allow the movement of water vapor and air through it but at the same time restricts the penetration of water droplets through the coated fabric. Polyether-based polyurethanes have been used for a long time for this purpose.

Recently researchers have started to discover some more polyurethanes (PUs) based on polyester polyols (Fuensanta et al. 2017; Jassal et al. 2004; Liu et al. 2015; Tang et al. 2013) which help them to enhance the properties of the final fabric that is being manufactured. PU can thereby help researchers to enhance textiles when used as a waterproof coating. PU can alter fabrics to make materials that have properties suitable for shoes, bags, and fashion garments (Davis n.d.).

Polyurethane that is being synthesized by established and industrially accepted methods requires a solvent, e.g., dimethylformamide (DMF), to be incorporated during the process, which leads to conditions that create occupational health risks. The addition also leads to environmental pollution. Production of polyurethanes involves a three-step process:

1. Creating a base layer, which helps to absorb the heat and water vapor from the skin.
2. The middle layer acts as a membrane to transfer the water vapor from the inner layer to the outer layer.
3. In the end, preparing a skin layer by using a transfer coating process, which then laminates the two layers.

It should also be noted that this process requires large amounts of water during the washing stage and thus also requires high levels of heat requirement to dry the substrate and recover the DMF. It is, therefore, an energy-intensive and resource-intensive process.

Hence a revolutionary idea was developed to incorporate polyurethane (PU) in the form of a dispersion, i.e., to cut down on solvent and use water instead has promoted this new process further in the textile industry. Waterborne PU dispersion offers three principal benefits:

1. It results in a safer working environment for workers.
2. Pollution risk is eliminated.
3. The process becomes more effective and efficient.

Engineered polyurethane dispersions removed the DMF from the process and thus removed the post-treatment requirements, such as washing, thereby leading to a significant reduction in the total drying time, making the process friendly to the environment and hence less energy-intensive. As per a study, the production of waterborne PU uses 95% less process water than with conventional PU. It also reduced energy costs by 50% (Covestro 2017).

8.6 Polyurethane: An Overview of the Textile Industry

Polyurethane dispersions are not only applicable for full penetration of the textile but can also be used for the surface treatment of the fabric in many applications (Figure 8.18). The latter application can be achieved by setting the required viscosity of the dispersion. With the inclusion of additives and adhesion promoters, polyurethanes can be tailored so that they adhere to the non-polar fibers (Kyosev 2016). It is possible to generate an abrasion-resistant fabric by combining PU with other additives, e.g., silicones, waxes, or oils.

Repellent properties with respect to oil and water are crucial on fabrics that are being used for leisure and outdoor garments. The repellent can also be incorporated into military textiles, as discussed earlier, along with umbrellas. "Waterproof" as a term is normally applied to textile materials that can help prevent the penetration and absorption of water into the fibers of the fabric. Hence, waterproof material is thereby expected to provide a good barrier for water in all end-use conditions. The most common method of manufacturing a waterproof fabric is by coating the fabric with a polymeric coating (e.g., polyurethanes, neoprene, PVC) (Pritchard et al. 2000).

In fabrics that are used in apparel, the movement of air and moisture is required to pass through the fabric so that sufficient levels of thermo-physical and thermo-physiological conditions

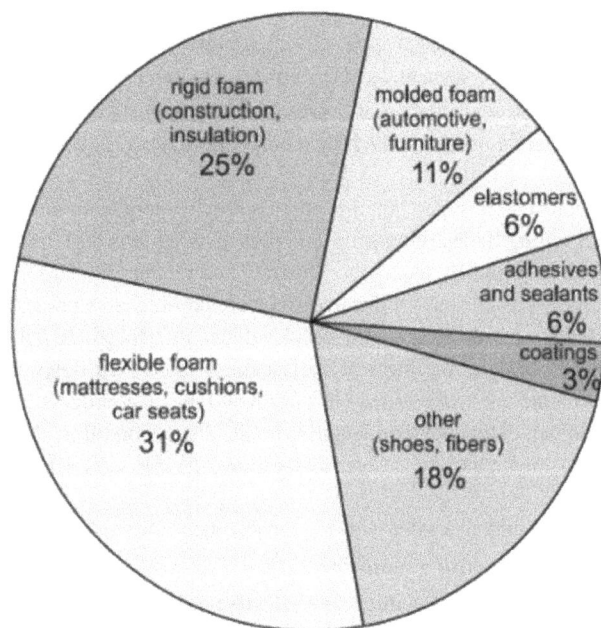

FIGURE 8.18 Applications of polyurethanes.

are maintained. Water is a hydrogen-bonded liquid with high surface tension (72.75mN/m at 20°C) (Dorsey 2015), whereas hydrocarbon-based oils have a much lower surface tension (20–31mN/m at 20°C) (ibid.).

It should be understood that water-repellent finishes, if applied, can be adequate to repel water but might be inadequate to perform similarly with oils. Silicones are used for finishes to acquire water repellency, but to achieve water and oil repellency, fluorocarbon-based finishes have been widely used. However, fluorocarbon finishes are currently not favored because they are difficult to degrade and hence can accumulate in the environment (Pritchard et al. 2000). Various water-repellent finishes have been used for technical textiles with selective properties and performance for a particular end application. Extenders can be used to increase the repellency of silicones and fluorocarbons, which are generally cross-linking agents. Using these, maximum efficiencies can be achieved (Unkelhäßuer et al. 2014).

8.6.1 Bio-Based and Biodegradable Polyurethanes

Biodegradable polyurethanes were synthesized using bio-sources by Chuanpit Khaokong et al. (Khaokong et al. 2013). They made use of natural rubber and poly(DL-lactic acid). The molecular weight so obtained was sufficient enough to produce polyurethane dispersion for textile coatings. Increasing the PLA content increased the tensile strength of the material, thereby making it a perfect raw material for textile coatings where durability and flexibility both play a vital role.

In addition to this, the polyester polyols that are used for polyurethanes can be derived from bio-sources, such as 1,3-butanediol, succinic acid, sebacic acid, etc. The discovery of pentamethylene diisocyanate (Meng 2017) from a bio-source, such as corn, has helped the manufacturers proudly be involved in the bio-sourced polyurethane synthesis (Mannari 2011), thereby reducing the carbon footprint of a polyurethane-based product tremendously. This will lead to easy acceptance in the market and thus will limit the fossil fuel used for polyurethane synthesis.

8.7 The Need for Coatings

The coating process includes the application of a polymeric layer directly onto one or both surfaces of the fabric. To make the polymer coating adhere to the textile, a blade or similar aperture is used to control the thickness of the coating layer and then the coated fabric is heated to cure the polymer. The major factors in fabric coating include the chemical formulation of the coating, the form of the technical textile, the number of layers, the coating thickness and weight, and the nature of any pre-treatment. The choice of application method of the coating formulation and machinery must be versatile to eliminate problems in knitted fabrics, such as curling selvage and to minimize the tension on the fabric that may lead to distortion or stretch. Weft straightening, scroll opening equipment, selvage uncurlers, or tension bars may thus be used.

8.8 Different Binders Used in Coatings

A binder is required in the formulation to hold all the ingredients together and form the film. It is a mandatory component; regardless of the presence or absence of other components, it is essential because it produces the desired properties of the final paint film. The binder gives protection to the substrate as well as the components within the film. It also influences vital properties such as flexibility, durability, gloss, and toughness. The binder consists of one or more polymer systems or basic resin (synthetic or natural resin). Typical types of binders include polyurethanes, polyesters, acrylics, melamine resins, silicates, oils, epoxies, depending upon the substrate and film characteristics required. The binders are categorized as per curing mechanism or drying.

8.8.1 Simple Solvent Evaporation

Coatings dried by the simple solvent evaporation method result in a solid film as the solvent evaporates over time. Depending upon the film thickness, the solvent type, and the resin construction, the time required for drying can be quick or can take hours, days, or weeks. A solid film can re-dissolve in a solvent and hence is not applicable to chemical resistance-based properties.

8.8.2 Single Component Catalyzed/Cross-Linked Polymerization

A single component material comprises all the necessary ingredients required to produce a film. It uses external factors, such as the presence of ultraviolet light, heat, moisture in the air or heat to speed up, initiate, or complete the curing process. Epoxies, polyurethanes, or polyimides are used as single component systems that are cured by heat and typically consist of two pre-mixed components which eliminates the need for metering and mixing. Heat excites the polymer chains with subsequent chemical reactions to cross-link with the material. The type of polymer or resin system determines the strength of the cross-linked structure and affects the performance of the final film.

8.8.3 Two-Component Catalyzed/Cross-Linked Polymerization

Two components polymerization consists of part A, normally the resin, and a second part B, the hardener or catalyst (in this case, the former is pre-mixed with the resin). Depending upon the resin or polymer type, the mixing ratios can differ from 1:1 to 10:1 as needed to mix the two components. Mostly a two-part material formulation can cure at room temperature.

8.9 Characteristics of Waterproof Breathable Fabrics

Moisture vapor is transported through a clothing system and waterproof breathable fabric, depending on the humidity of the clothing, the microclimate, the temperature gradient across the waterproof breathable layer, and the interaction between the water vapors and the clothing layers. When designing a breathable fabric, there are 15 basic fundamental considerations:

1. Waterproof ability.
2. Comfort level.
3. Mass of the fabric.
4. Durability/flexibility of the coating/lamination.
5. Flex and abrasion resistance.
6. Aesthetic properties.
7. Good hydrostatic resistance.
8. Water vapor transmission.
9. Good hydrostatic resistance.
10. Effectiveness of clothing against wind chill factor.
11. Repellent to insects.
12. Durability: tear tensile and peel strength.
13. Good washability or dry cleanability.
14. Strength of coating.
15. Good laundering ability.

8.10 Preparation of Breathable Waterproof Fabrics

Breathable waterproof fabric can be prepared by various methods such as microporous polyurethanes, laminates, and hydrophilic polyurethanes (Figure 8.19) The advantages of hydrophilic polyurethane coatings are good adhesion on a textile substrate, resistance to water and solvents, high moisture permeable properties, high gloss, and less expensive than microporous and laminated polyurethane film (Chinta and Satish, 2014).

Hydrophilic polyurethanes are manufactured by the calculated combination of hydrophilic and hydrophobic urethane components to give optimum moisture vapor transmission properties without

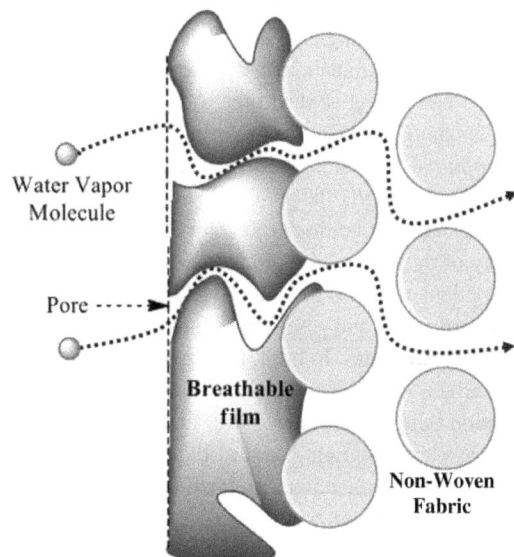

FIGURE 8.19 Breathable film for medical applications.

compromising their physical properties. A monolithic hydrophilic coating method can be applied by conventional coating methods and hence can be less expensive than microporous or laminated films. On the contrary, hydrophilic films have low water vapor permeability (WVP) (De et al. 2005). Further, it was found that the waterproofness hydrostatic head is 180 cm, 2130 cm, and 1380 cm for cotton, PTFE, and PU respectively. Also, the water permeability, tested by an upright cup method at 35°C ,was found to be 4625, 5200, and 3475 $g/m^2/day$ respectively (ibid.).

The sportswear textile industries have seen market segmentation and diversification for fibrous materials which have further contributed to the need for the advance of textile science and technology. The sports goods requirements often demand widely different properties from their fabrics and constituent fibers, such as providing a barrier to snow, rain, heat, cold, and offering strength. At the same time, they must also fulfill the consumers' requirements of comfort, drape, fit, and ease of movement.

Most medical protective clothing is made from non-woven, one-time use materials. Generally, the outer surface is treated with fluorocarbon to repel liquids, whereas if a total barrier is required, the non-woven fabric is laminated onto a waterproof film or in some cases waterproof breathable films are also used (Fung 2002).

The non-woven composite fabric acts as a barrier to blood vessels and infection challenges and also provides breathability for ease of healing. Also, it is particularly suited for use as a disposable surgical gown. The fabric is made, first, of a microporous ply, consisting of a structural microporous resin that has been extrusion-coated onto a non-woven fabric base and then stretched to form micro-porosity and at least one additional ply is positioned next to the first ply (Langley et al. 2004). A comparison of the physiological functions of water-vapor impermeable construction and breathable material was made by Bartels and Umbach. In such a fabric, the ratio of evaporation was high and moisture accumulation was low for breathable construction. And the water vapor resistance of the hydrophilic membrane laminates was not dependent on the temperature. It was especially tested for water vapor transport characteristics. They also found the physiological function of a water vapor permeable protection garment is superior to non-permeable clothing and, due to evaporation, there is variation in the water accumulation on the garment (Umbach 2002).

Yiwei Du and Zhang studied the coating of a water- and oil-repellent finishing agent on non-woven polyester fabrics. The technical optimal conditions used are: finishing agent (40 g/l), the bridging agent (6 g/l), the pre-curing condition (100°C for 2–3 min) and the curing condition (160°C for 2.5 min). The thickness and density of the fabrics also affect the evaluation of the water- and oil-repellent effect (Yiwei Du and Zhang 2009).

Diffusion of vapor and gases through a non-porous membrane occurs by various routes. In brief, the gas is first absorbed in the exposed surface of the polymer thus proportionally there is a concentration build-up on the surface. Due to the concentration gradient, gas starts migrating toward the opposite surface. This migration can be visualized as a sequential step where molecules pass through a potential barrier separating one position from the next. For a successful migration, a sufficient size of passage (a pore) should be available and it also depends on the thermal motion of the polymer chains (Ashley 1985).

The hydrophilic framework depends on a synthetic chain response between non-porous film and the moisture molecule. The membrane comprises carbon, hydrogen, and oxygen atoms connected together in long molecular chains. Positive and negative charges emerge at the different points, forcing the atoms to make weak bonds with water molecules, which remain stable until the point when dislodged by another water molecule, making all the water molecules travel through the film. This property is possessed by the hydrophilic polymer backbone in a coating material which builds the affinity of the polymer toward the water and gives breathability. The proximity of the water-loving hydrogen molecules on the polymer chain helps in the transport of the water atoms from a higher to a lower phase of humidity (Shekar et al. 2003).

The basis of moisture vapor transmission in a group of films is the chemical adsorption of moisture vapor under the correct temperature/humidity gradient conditions through "soft" segments (e.g., the polyether chains) of the film while water droplets are repelled by "hard" segments (e.g., the urethane chains). The chemical groups possessing "hydrophilic" actions that can be added to the film's chemistry are polyethylene oxide groups.(Ashley 1985) The amorphous region (where the soft segment is) acts like intermolecular pores, permitting passage to the water vapor molecules, at the same time, due to the solid nature of the membrane, it also prevents penetration of liquid. Figure 8.19 shows the schematic diagram of the mechanism of the vapor transport through the hydrophilic polymer.

Diffusion of water vapor is introduced by the integration of hydrophilic functional groups such as $-NH_2$, $-O-$, $-OH$ or $CO-$ into the polymer or in a block copolymer. The hydrophilic group forms reversible hydrogen bonds with the water molecules, which then diffuse through the film along the molecular chains in a stepwise action (Lomax 1990). However, many of the hydrophilic polymers, such as polyvinyl alcohol and polyethylene oxide, are highly sensitive to water and would either dissolve completely or swell so much in contact with rainwater that the flex and abrasion resistance property would be poor. Hence, hydrophilic polymer coating should, therefore, have a sufficient swelling index not only to transmit water vapor but also to retain suitable film strength. To obtain the desired functional property of the film, optimization of the hydrophilic-hydrophobic ratio is required (Scott 1995). Incorporating polyethylene oxide into a hydrophobic polymer chain and the use of segmented copolymers have been found to be successful (Lomax 1985).

In isothermal tests, the transport properties of clothing systems made up of hydrophilic polymers, especially those with low transmission rates, are improved considerably more than those clothes with microporous polymers in a temperature gradient test (Gretton et al. 1998).

The advantages of hydrophilic films include strength, their windproof nature, excellent odor barrier, and a wide range of chemical and solvent resistance. Thinner film laminates are placed nearer to the first single layer material and, in addition ,they can be built to give higher breathability than microporous material with a safer window of leakage through the laminates.

Thermoplastic elastomers, such as co-polyether-esters, polyether-polyester block amides have alternative hard and soft polyester segments used as breathable films for protective clothing (Bajaj 2001). "Sympatex" is one of the hydrophilic polyester membranes which includes the polyether groups to impart hydrophilicity to the membrane. Commercial films 10–25 mm thick are manufactured by an extrusion process. The membranes are colorless and opaque in appearance and provide water vapor permeability when accompanied by ~5% swelling (Krishnan 1991). Generally, thinner films are more breathable (Painter 1996). A "Sympatex" membrane with only 1/100 mm of thickness has over 584 billion hydrophilic polyester units per square centimeter which allow moisture vapor to be transported from the body to the outside of the fabric.

Based on shape memory polymer, textile material can be manufactured as intelligent, waterproof, breathable fabrics (Hayashi, Ishikawa, and Giordano 1993; Ding, Hu and Tao 2004). The fabric limits the loss of body heat by interrupting the transfer of heat and vapor at low and high temperatures and it transfers more heat and water vapor from inside clothing to the outside than breathable, waterproof fabrics (Hayashi 1992). In shape memory polymers, polyurethane is one of the widely used polymers (Crowson 1996; Ding, Hu and Tao 2004). These fabrics are hence called smart breathable fabrics, which control the transmission of water vapor. At a temperature higher than the transition temperature, due to the high proportion of hydrophobic interactions, the coating exists in a collapsed state, resulting in the opening of the microcracks, whereas at a lower temperature, due to the absorption of water from the surroundings, the coating substance exists in a swollen state on the fabric, resulting in the closure of the microcracks. Apart from temperature, there are other factors, such as the change in diffusion flux governed by changes in both the diffusion path of the water molecules through the swollen and collapsed coating and the diffusion coefficient. Because of these two factors, at a lower temperature, the diffusion flux is likely to decrease, compared with a higher temperature.

Transfer of water vapor of breathable fabrics depends on the atmospheric conditions. Apart from breathable fabrics, the garment fabric layers underneath affect the condensation and alter the transport properties of the clothing. Hence in breathable fabrics, minimizing the condensation of the water inside the fabric is one of the most stimulating aspects in developing strategies. The condensation problem may be resolved by changing the physical properties of a three-layer waterproof breathable fabric. The formation of condensation can be reduced and the water vapor transfer out of the fabric can be improved by increasing the average diffusion coefficient of the outer layer and membrane or decreasing the thickness of the waterproof membrane and the outer layer fabric. As a matter of fact, decreasing the thickness of the lining can increase the water vapor transfer from the hot side to the interface of the dry-wet regions, but it also triggers condensation. Water vapor transfer from the hot side and condensation can also be triggered by increasing the diffusion coefficient of the lining (Ren and Ruckman 2004).

Using the wet coagulation process as an alternative development strategy, blending of various ratios of polytetramethylene ether/ester copolymer-type polyurethane (PTGPU) and PU at a different molecular weight was performed (Enomoto et al. 1997), in this case, the fabrics consist of three layers:

1. the inner surface (in contact with the body) has hydrophobic micro-networks resulting from the concentration gradient of PTGPU;
2. the intermediate porous layer of PU is hydrophilic when compared with PTGPU and is useful for moisture transportation;
3. the outer surface is composed of a hydrophobic nylon fabric.

Fabrics for sportswear are waterproof, breathable, and elastic (Herliky 1995), consisting of three layers:

1. a shell fabric consisting of polypropylene and polyurethane elastomer fibers;
2. a waterproof membrane laminated to the backing of the shell fabric;
3. a mesh lining attached to the back of the membrane.

Toray has developed a high performance waterproof, breathable fabric consisting of a three-layer polyurethane coating membrane including two different microporous layers (Toray Inc. 1993):

- the top layer is the foam layer with a microporous structure which improves waterproofness and durability;
- the middle layer which is regularly configured microporous increases moisture permeability;
- the bottom layer of pores in the resin surface enhance waterproofing and inhibit dew condensation.

Kanebo developed Isofix Super as a coating which incorporates structural technologies and new processing by fusing the anti-sublimation/antimigration coating technique to a special ceramic with countless number of pores. The fabric exhibits a dry feel and a soft touch along with the anti-condensation property (Kanebo Ltd. 1994).

Fabric manufacturers falsely claim the superiority of breathable characteristics of their fabrics, often without following testing procedures and also simulating practical situations. In a physiological trial, a person is tested on instruments while wearing identical garments made up of a range of materials. The study showed that there is a much smaller difference in vapor permeability between materials when worn in garment form than in the laboratory test (Gray and Millard, n.d.). In a comparative evaluation of Ventile, PTFE film laminates, and PU coating-based breathable fabrics, ventilates have shown low waterproofness but higher permeability.

To compare a series of breathable fabrics, Salz has developed a laboratory-heated cup method in combination with an artificial rain installation (Salz 1988). The test of a microporous PU-coated fabric exhibited a water vapor transmission rate (WVTR) of 142 g/m^2.hr in a dry atmosphere, whereas when exposed to rain-like conditions, it decreased to 34 g/m^2.hr. Two-layered PTFE laminated fabric showed a WVTR of 205 g/m^2.hr and 269 g/m^2.hr in dry and rain-like conditions respectively. Three-layered PTFE laminate showed WVTR of 174 g/m^2.hr and 141 g/m^2.hr in dry and rain-like conditions respectively. These values differed by 119 g/m^2.hr and 23 g/m^2.hr respectively for hydrophilic and

PU laminate. A microfiber fabric was found to have a WVTR of 190 g/m².hr and 50 g/m².hr in dry and rain-like conditions respectively. Thus the performance of microporous coating and laminates is found to be more efficient than hydrophilic laminate (ibid.).

In most cases, except for two-layer PTFE-laminated fabric, there is a significant reduction in water vapor transmission in rain. Laminated fabrics were found to perform better than coated fabrics under cold ambient conditions, preventing condensation from forming, transmitting more water vapor for a longer period of time (Holmes 2000a). Thus coating has advantages over laminates at a lower price of base materials and flexibility of the process (Kannekens 1994).

To clarify the mechanisms of diffusion-based water vapor transfer in waterproof breathable fabrics for clothing, simple experiments using a glass dish were carried out under steady-state conditions with and without a temperature gradient in the climatic chamber. It was found that vapor pressure, as well as natural convection within the air gap, affects the level of water vapor transfer. For different fabrics, rates of water vapor transfer are ranked differently depending on the presence of temperature gradient. The ranking of fabric is as follows:

1. micro-fiber fabrics
2. PTFE-laminated fabrics
3. cotton Ventile
4. hydrophilic-laminated fabrics
5. poromeric polyurethane-laminated fabrics
6. polyurethane-coated fabrics.

In the presence of a temperature gradient, condensation also play a crucial role, especially if the temperature is below 0°C. The amount of condensation is least on the inner surface of PTFE-laminated fabrics, followed by cotton Ventile, microfiber fabrics, hydrophilic-laminated fabrics, polymeric polyurethane laminated fabrics, and polyurethane-coated fabrics.

For garment comfort, the movement of water vapor through a textile is an important factor. The human body maintains a core body temperature around 37⁰C, where the balance between perspiration and heat produced and lost by the body is the comfort factor. The body would be in a state of comfort when there is no moisture on the skin and the body temperature is about 35⁰C (Ruckman 1997). So to maintain this comfort, the primary functions of breathable textiles are as follows:

1. To provide protection against the cold.
2. To keep the wearer dry (from external water and from internally generated condensation).
3. To protect against the wind chill factor, which is perhaps the bonus feature that the waterproof breathable textiles have, due to their wind-resistant nature.

The total heat transfer through the clothing on the body to the environment, considering the thermal and evaporation resistance of the clothing, has been given by Woodcock (Sen 2001) as Equation (8.1):

$$H = \frac{Ts - Ta}{I} + \frac{Ps - Pa}{E} \qquad (8.1)$$

where

H = total heat transfer
Ts - Ta = temperature difference between skin and ambient temperature
Ps - Pa = water vapor pressure difference between skin and ambient temperature
I = insulation of the clothing
E = evaporation resistance of the clothing.

Perspiring helps to maintain the body's temperature at a comfortable balance, hence the body perspires to cool down during and after periods of energetic activities (Holmes 2000b). When the environmental temperature is greater than the skin temperature, heat loss by evaporation is the only way to dissipate heat from the body. On the skin surface, the liquid sweat is transformed into vapor and passes into the environment to cool. Hence, evaporation of moisture from the skin surface is tremendously effective in disposing of the body heat. The greater the rate of moisture evaporation, the greater will be the comfort. The heat-flux through fabrics and the loss of heat through perspiration are crucial factors where part of this heat is dissipated by moisture evaporation. Mass diffusion also results from a temperature gradient in a system and is called thermal diffusion. Similarly, the concentration gradient gives rise to a temperature gradient which consequently results in heat transfer. An impermeable fabric is windproof and waterproof but also not permeable to water vapor. On the other hand, a breathable fabric is water vapor permeable and can cope with the features of foul weather (Sen 2001). Approximate work and perspiration rates associated with various activities can be measured to aid in constructing a breathable fabric (Holmes 2000a), such as, for sleeping, the work rate is 60 W while its perspiration rate is 2280 g/day. Compared to this, in gently walking, 200 W of energy is used, resulting in a perspiration rate of 7600 g/day. Active walking takes around 500 W of energy perspiring at the rate of 19,000 g/day. In climbing, trekkers perspire up to 38,400 g/day, using up 800 W of energy. Workers in an industrial area with high work rate uses 1200 W of energy so their clothing should have the capacity to perspire 45,000 g/day of water vapor.

8.11 Water Contact Angle: A Vital Concept for Hydrophobicity

The water contact angle (WCA) depends on two factors: the chemical composition and the surface roughness. The latter plays a more important role in determining the WCA value when the fluorine/carbon (F/C) ratios have no evident difference. To enhance the WCA using the chemistry of molecules, fluorine was used in a study (J. Zhao et al. 2017), as shown in Figure 8.20 (a). Higher concentrations of fluorine help to provide more waterproof behavior for the fabric. But it must be noted that just like surfactants that possess the critical micelle concentration (CMC),

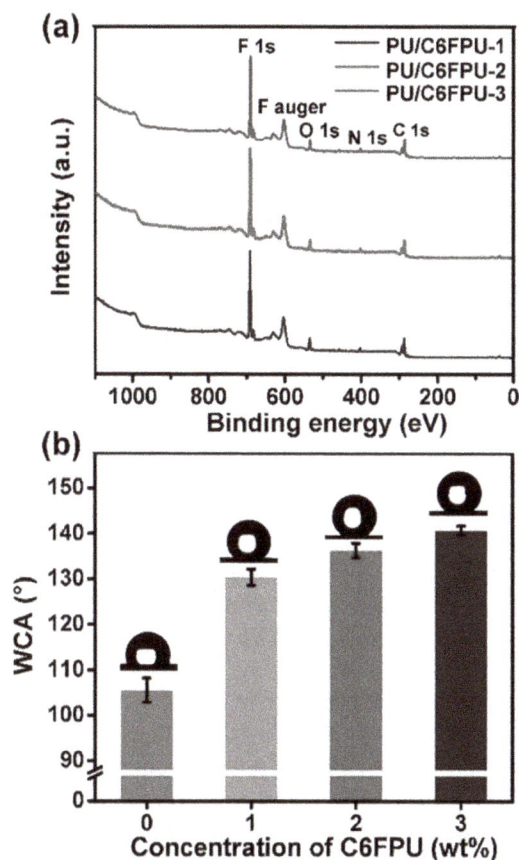

FIGURE 8.20 (a) XPS spectra and (b) WCA of PU/C6FPU fibrous membranes with different C6FPU concentrations.

Source: J. Zhao et al. (2017). Reprinted (adapted) with permission. © 2017 American Chemical Society.

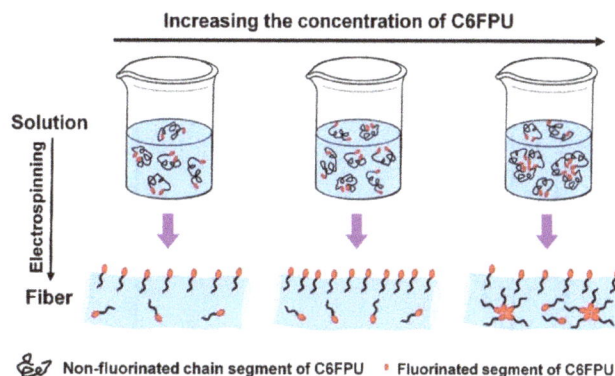

FIGURE 8.21 Representation of the perfluoroalkyl chains behavior with the gradually increased C6FPU concentration before and after the electrospinning process.

Source: J. Zhao et al. (2017). Reprinted (adapted) with permission. © 2017 American Chemical Society.

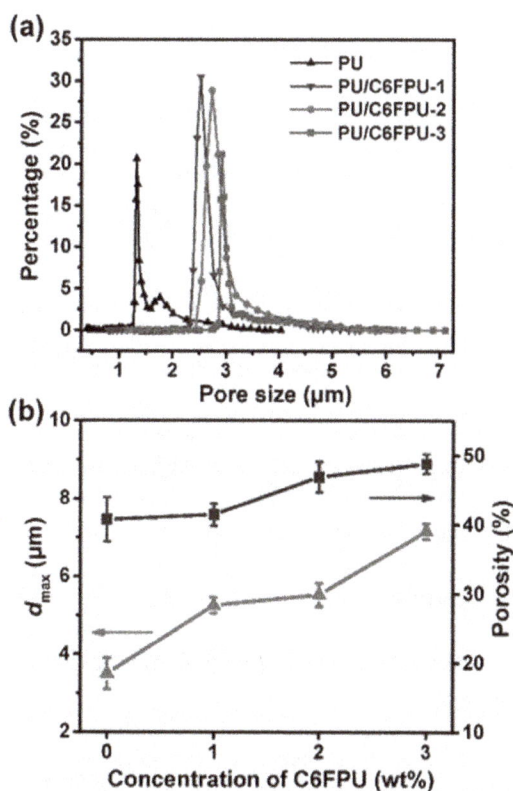

FIGURE 8.22 (a) Pore size distribution, (b) d_{max} and porosity of PU/C6FPU fibrous membranes with different C6FPU concentrations.

Source: J. Zhao et al. (2017). Reprinted (adapted) with permission. © 2017 American Chemical Society.

as the C6FPU concentration reaches a certain critical value, as shown in Figure 8.21 of the study, there comes a point when the perfluoroalkyl chains start to aggregate, which can be called the critical aggregation concentration (CAC). Once above CAC, further addition of C6FPU only leads to an increase in the number of aggregates in solution. When the C6FPU concentration increased to 1 wt.%, the contact angle increased to 135° (Figure 8.20 (b)), which can be ascribed to the sharply increased F/C ratio.

As presented in Figure 8.22, there was an increased tendency for the average pore size with the increasing of C6FPU concentration. As the C6FPU concentration increased, d_{max} and porosity of the fibrous membranes increased. These results can be ascribed to the decreased bonding points and the fluffy accumulated fibers with the increasing content of C6FPU.

The waterproof ability of PU/C6FPU membranes with 0–3 wt% C6FPU was 4.8–20.3 kPa, respectively, which was the result of the combined effect of surface wettability and d_{max}. Meanwhile, as the concentration of C6FPU increased, the moisture vapor permeability increased, as shown in Figure 8.23. It is a result of the increscent porosity induced by the decreased adhesion structure, which brought about more interconnected channels for the transmission of the water vapor, thus leading to better moisture vapor permeability.

8.12 Non-Porous and Porous Films for Breathable Laminates

Waterproof and breathable textiles are constructed by laminating or coating a fabric with microporous film. Water penetration is primarily affected via the micropores of the film which are

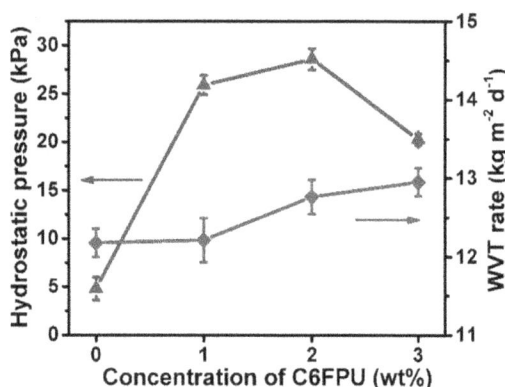

FIGURE 8.23 Hydrostatic pressure and WVT rate.

Source: J. Zhao et al. (2017). Reprinted (adapted) with permission. © 2017 American Chemical Society.

produced via a special process during the production of the film or while it is being coated. The coating is so designed that perspiration can easily escape from the surface of the body into the environment very quickly but at the same time, it will not permit small drops of rain or spray to penetrate this breathable system, despite its porous structure. This is because the diameter of the water vapor molecules is 0.35 nm and that of water droplets is about 1 mm (Lomax 2001b). Besides regarding the widely used PTFE membranes, in which the pores are made during the production due to stretching of the membranes, there are also quite a few varieties of microporous hydrophobic polyurethane membranes (Kubin 2001).

The advantages of microporous membranes are (Kramar 1998):

1. *Superior breathability*: hydrophilic lamination/coatings are influenced by the thickness of the laminates/coatings and also by the number of hydrophilic groups present in the structure of the film.
2. *Superior handling*: the hydrophilic coating is supposed to have stiffer handling.

Despite superior breathability, these films also have several disadvantages:

1. During use, pore sealing of the micropore of the coating or laminates occurs and thereby affects breathability. Micropores can be contaminated by various interventions, including pesticide residues, insect repellent, detergents and surfactants used for laundering or dry cleaning, particulate and air-borne dirt, sunscreen lotions, salts, and skin exudates All of these contaminants are expected to lower the breathability of the film (Mooney and Schwartz 1985).
2. Distortion of the microporous film during use causes the films to break or an increase in the pore size so that the waterproofness becomes insufficient for practical use. Small degraded segments that have detached from the fabric can easily mix with human sweat, thus becoming an ideal medium for bacteria and mold to grow.

3. Compared to the solid structure of a non-porous film, the microporous film has inferior tearing strength.
4. Also, it can be noted that a hydrophilic coating is prone to wrinkling in wet conditions.

Advantages of microporous films over non-porous breathable films are as follows:

1. The process of film-making is simple, thus rendering higher production speed.
2. As they have a solid structure and of course no holes, the non-porous films are thereby less sensitive to possible degradation, as mentioned above.

Due to the presence of solid structures, non-porous films possess distinct advantages compared to microporous films. A certain amount of WVP, i.e., water vapor permeability, is required so as to give the wearer a comfortable feel. Improving the permeability of the non-porous membranes is undoubtedly an immense challenge for polymer chemists. The permeability of non-porous membranes may be enhanced by the introduction of hydrophilic groups into the backbone of the polymer, as the permeability prevails through the non-porous films via a molecular mechanism, i.e., absorption–diffusion–desorption. But too many hydrophilic groups cause weight loss and swelling of the film during washing, due to increased solubility, due to which, after each wash, the waterproofness is found to decrease continuously (Kramar 1998).

The choice can be established in the form of shape memory polymer films (SMPs). In these films, the large change in mechanical and thermo-mechanical properties generally occurs across the glass transition temperature (T_g) or at the soft segment crystal melting point temperature (T_{ms}) (Hayashi et al. 1993). In addition to these properties of SMPs, it is found that SMPs also experience a large alteration in moisture permeability below and above the T_g/T_{ms}. Based on the T_g/T_{ms} of the SPMs set at room temperature, the SMPs exhibit low moisture permeability below the T_g/T_{ms} or during the glassy state and have high moisture permeability above T_g/T_{ms}. This behavior can be used with SMP-laminated textiles that could contribute as thermal insulation in cold temperatures and, on the other hand, as the highly permeable membrane at room temperature or above.

The features of non-porous films are (Johnson and Samms 1997):

- windproof;
- waterproof and liquid-proof;
- selective permeability;
- high water entry pressure;
- good tearing strength;
- low water vapor transmission.

The features of microporous films are (ibid.):

- waterproof and liquid-resistant;
- non-selective permeability;
- low water entry pressure;

- low tearing strength;
- high water vapor transmission.

It should be noted that the non-porous films are pinhole-free and quite dense polymer membranes. These polymer films are also generally hydrophilic so they are able to absorb water very readily. This is a significant property that develops a "wicking" action which helps to actively attract water vapor molecules. These films allow the conveyance of water vapor through a process known as active diffusion. The permeant, in this case, being the water vapor, dissolves on the surface of the non-porous film on one side that has the highest concentration and then further diffuses through the film. When the water vapor emerges on the opposite surface of the non-porous film, the permeant, i.e., the water vapor, desorbs and typically becomes part of the surrounding airspace as gas or vapor.

There is a mass transfer linked with convection in which water vapor is transported from one side of the film to the other in the flow system (see Figure 8.19). This pattern of channeling occurs on a macroscopic level and is usually understood through the concepts of fluid mechanics. In other words, mass transfer is the consequence of diffusion from the zone of high concentration, i.e., the inner side of the fabric, to regions of low concentration, i.e., the outer side of the fabric, due to the concentration gradient. Higher concentration means there are more molecules (of water vapor) per unit volume. Mass diffusion may also result due to the temperature gradient between the inner and outer zones and is thus termed "thermal diffusion." On the contrary, a concentration gradient may lead to a rise in temperature, thus creating a temperature gradient and thereby a region of heat transfer.

The exchange of water vapor because of concentration or temperature through the film similar to the non-permeable layer can be aggregately termed "pervasion" and this is an aggregate procedure of dispersion and assimilation. In straightforward terms, the porousness of the mass atoms of the water in the polymer will depend on both the diffusive and solvency association of the non-permeable film with the water vapor (Cussler 1997). We can thus infer that the concoction structure of the non-permeable film and its thickness are the primary outline criteria for assembling a penetrable nonporous layer (Baker 2012).

Fiber mats were functionalized using the layer-by-layer method by using poly-N-vinylguanidine (PVG) and poly-hydroxamic acid (PHA) (Chen et al. 2010). The water vapor diffusion resistance of these fiber mats was estimated, as shown in Figure 8.24 and contrasted with standard reference material. The investigation recommends that the functional coatings involve the pores of electrospun fiber mats to some extent, which enhances the barrier performance against the convective wind current when contrasted with the untreated PAN fiber mats, however, such pore filling does not significantly trade off the water vapor porousness.

8.13 Factors Influencing Film Permeability

8.13.1 Cohesive Energy of the Polymer

Not only the polymer-penetrant interaction but also the first structure of the polymer itself is vital for an understanding of film

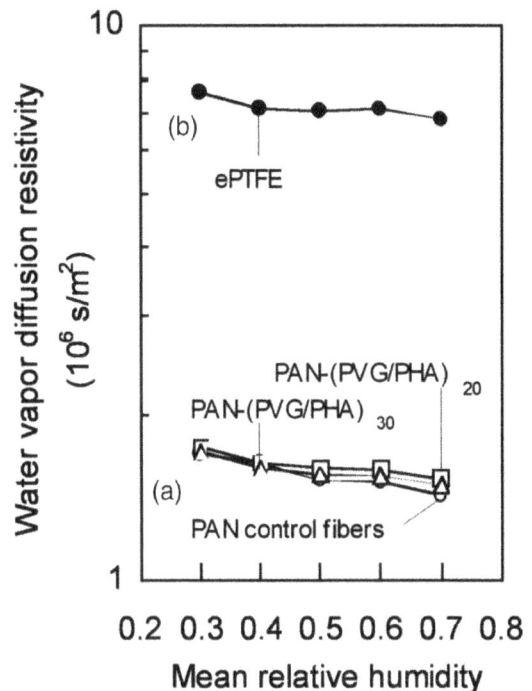

FIGURE 8.24 Water vapor diffusion resistivities of (a) control and functionalized fiber mats and (b) an expanded polytetrafluoroethylene (ePTFE) membrane with pores in the range of 200 nm, standard reference material for breathable protective clothing.

Note: T = 30°C.

Source: Chen et al. (2010). Reprinted (adapted) with permission. © 2010 American Chemical Society.

functions such as sorption, diffusivity, and porousness of tiny molecules (Porter and Porter 2009). The cohesive energy of the polymer membranes is set by such factors as chain flexibility (internal rotation of repetition unit), coulombic interaction, van der Waals interaction, H bonding, and so on. A high porousness constant is mostly earned once every factor except flexibility is low and is not suffering from the penetrant, so the polymer that contains a monomer unit with a high charge density, dipole moment, and capability for H bonding, can provide a low porousness constant. Cipriano et al. (1991) stated that the permeating fluxes increase with the increasing value of the pre-polymers' molecular weight. This suggests that longer molecular chains create larger polymer network holes and thus higher water vapor fluxes.

8.13.2 Hydrophilic and Hydrophobic Content

The water vapor permeability (WVP) values increase with a rise in the hydrophilic component whereas the water penetration resistance will increase with a rise in the hydrophobic element. By varying the relative proportion of hydrophilic and hydrophobic parts, the WVP and water penetration resistance for various applications are optimized (Jassal et al. 2004).

The higher NCO/OH molar ratios provide polyurethane dispersion (PUD) films with a substantial percentage of urea and amide structures, thus higher tensile strength, better solvent resistance, and modulus. The mobility of molecular chains due to the temperature gradient is decreased as the NCO/OH molar ratio is increased. PUD films with higher NCO/OH molar ratio tend to be comparatively stiffer (Lin and Lee 2017).

Crystallization temperature T_c and the melting temperature T_m progress to a lower point due to the use of hexamethylene diisocyanate-trimer (HDI-T). Coated films tend to exhibit an increase in hydrophobicity or, in other words, decreased water adsorption. Meanwhile, the tensile strength and Young's modulus, along with the pencil hardness of the films, exhibit improvement with increasing HDI-T content, but, on the contrary, their elongation at break tends to decrease. The thermal stability thus becomes inferior with the inclusion of more C-N bond which is a consequence due to the increase in the HDI-T content. The water absorption in this investigation after 7 days for polyurethane with 0%, 5%, 10%, 15%, and 20% excess HDI was found to be 27, 18, 11, 6, and 5.5 respectively (H. Zhao et al. 2017).

Researchers have developed a fluorine-free textile coating by incorporating silica nanoparticles (NPs). (Sheng et al. 2017) Figure 8.25 shows the variation of thermal conductivity, water vapor transmission rate, and air permeability with the different concentrations of SiO_2 NPs. When increasing the volume of the nanoparticles, it was reported that the porosity decreased. NPs of SiO_2 would block the spaces between the adjacent nanofibers so as to reduce the porosity of the membranes. Reduced porosity resulted in the still air decreasing in the membranes, as a consequence of which, the heat primarily is transferred via the nanofibers rather than the air. These investigations indicated that there is a negative linear relationship between the porosity and TC, which is exhibited in Figure 8.25 (b).

It can be established from Figure 8.25 (c) that with increasing SiO_2 NPs concentration, the WVTR of the modified membrane decreases and air permeability reduces the same way. These decreasing trends can be explained by Fick's diffusion model, which proposes that the diffusion flux is positively related to the diffusion coefficient, which is well determined by the porosity. Therefore, the relationship between air permeability, WVTR, and porosity is significantly in positive correlation (Figure 8.25 (d)). On the basis of the hydrostatic pressure, thermal conductivity, WVTR, and air permeability results, the waterproofness and breathability can be flexibly regulated by varying the SiO_2 NPs amount, which is an easy way to supply different levels of barrier and comfort performance to meet diverse end uses and consumer needs.

8.13.3 Film Thickness

Fick's law can be used to elaborate on the rate of transport of the permeant through the non-porous film under the existing temperature and concentration gradient. In cases in which the

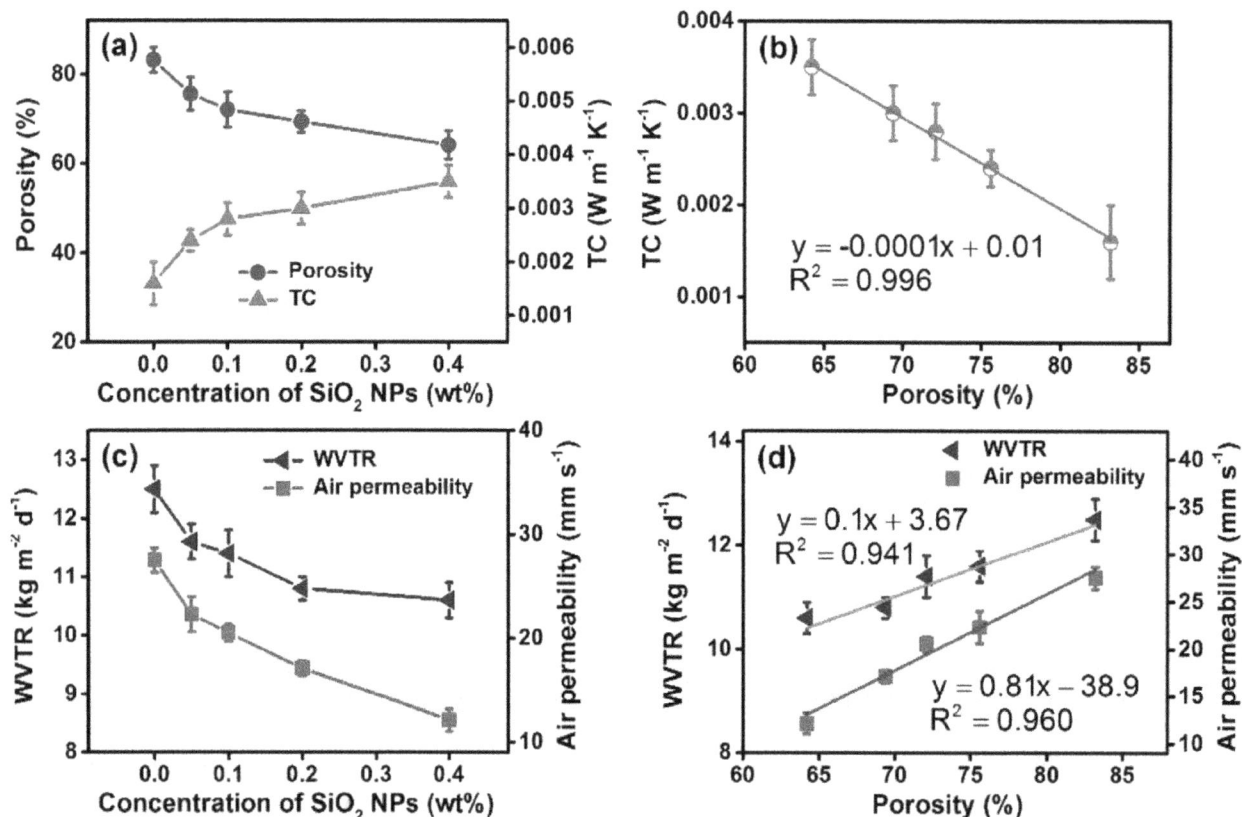

FIGURE 8.25 The heat, moisture and air transfer performance of PAN@ASO-1/SiO_2 modified membranes with various concentrations of SiO_2 NPs: (a) porosity and TC; (b) the linear relationship between TC and porosity; (c) WVTR and air permeability; and (d) the linear relationship between WVTR and porosity, air permeability and porosity, respectively.

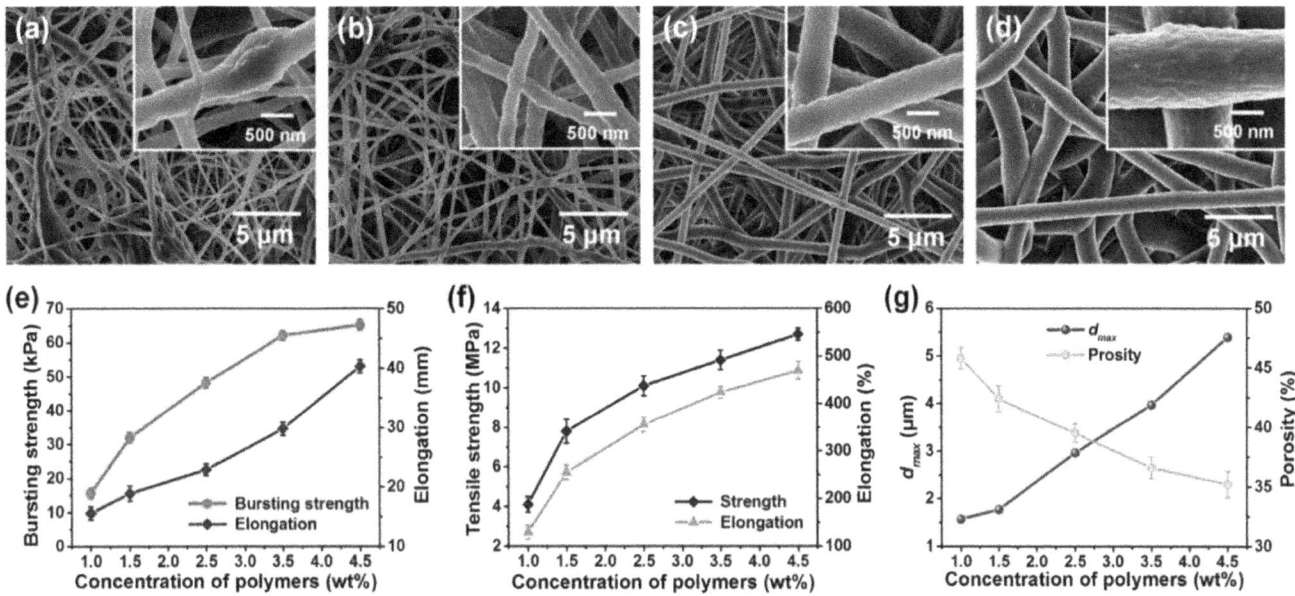

FIGURE 8.26 Morphology, mechanical properties and pore structure of FPU/PU fibrous membranes. FE-SEM images of FPU/PU fibrous membranes fabricated from polymers solutions with different concentrations: (a) 1.0 wt%, (b) 1.5 wt%, (c) 2.5 wt%, and (d) 3.5 wt%, respectively. (e) Bursting strength and elongation, (f) tensile strength and elongation at break, (g) d_{max} and porosity of FPU/PU fibrous membranes fabricated from polymers solutions with different concentrations.

Note: The FPU/PU weight ratio was kept 1:8.

Source: Li et al. (2015). Reprinted (adapted) with permission. © 2015 American Chemical Society.

diffusion coefficient is not dependent on concentration, then the integration across the film thickness "l" yields (Cussler 1997):

$$J = \frac{D(c_1 - c_2)}{l}$$

where:

c_1 = concentrations of permeant at the high-pressure face of the film surface

c_2 = concentrations of permeant at the low-pressure face of the film surface

l = thickness of the membrane.

A linear relationship between the concentration of water vapor in equilibrium with the film and actual concentration of water vapor dissolved in the film is assumed by Henry's law, as given below, which holds for many polymers:

$$c = Sp$$

Substituting, values of c_1 and c_2, the equation becomes:

$$J = \frac{DS(p1 - p2)}{l} \quad (8.2)$$

where:

p_1 = the external partial pressure of the vapor on the high-pressure side of the membrane

p_2 = the external partial pressure of the vapor on the low-pressure side of the membrane

DS = the permeability (P).

From Equation (8.2), it is clear that for ideal systems, the permeation rate of permeant is directly proportional to the pressure gradient and inversely proportional to the membrane thickness.

On the contrary, macro-porous membranes made of fluorinated polyurethane and their performance can be enhanced by carbon nanotubes (Li et al. 2015). The randomly arranged carbon nanotubes, as shown in Figure 8.26 (a)-(d), form micro-wrinkles and thus entrap air in them. This entrapped air functions as a cushion for water droplets, thus making them hydrophobic but at the same time allowing for the flow of water vapor from the human body into the atmosphere. Hierarchical surface structures with micro/nanoscale wrinkles are clearly visible on close observation of all samples (insets in Figures 8.26 (a)–(d)).

Figures 8.26 (e) and 8.26 (f) demonstrate that the mechanical properties could be improved by increasing the concentrations of polymer solutions, which would be attributed to the increase in the fiber diameter that provides stronger individual fibers, as well as the existence of adhesion structures that prevent the randomly oriented fibers from pulling apart. With the polymers' concentrations increased, the d_{max} (maximum pore size) of the fibrous membranes also increased, as shown in Figure 8.26 (g), which could be attributed to the accumulation of thicker fibers that formed a larger space between the adjacent fibers. This transmission of porosity could be related to the thicker fibers and the higher adhesion rate that took up more space between the fibers.

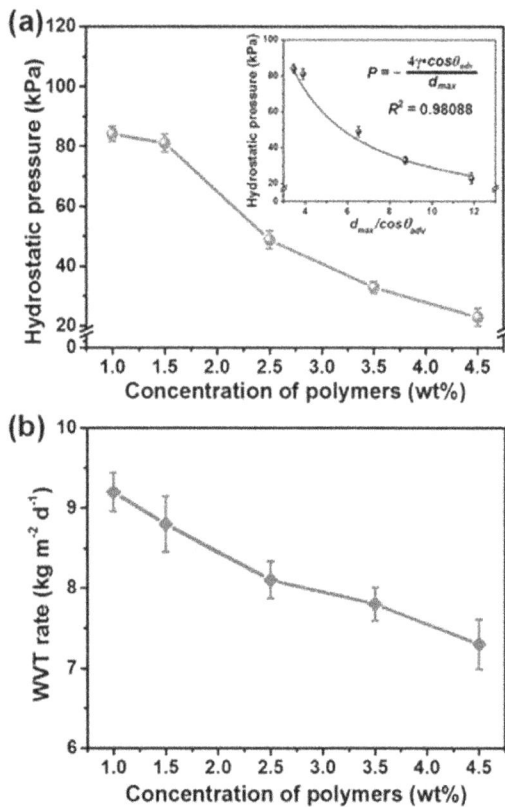

FIGURE 8.27 Waterproof and breathable properties of FPU/PU fibrous membranes. (a) Hydrostatic pressure and (b) WVT rate of FPU/PU fibrous membranes fabricated from polymers solutions with various concentrations. The inset in (a) shows the relationship among hydrostatic pressure, d_{max}, and θ_{adv}.

Source: Li et al. (2015). Reprinted (adapted) with permission. © 2015 American Chemical Society.

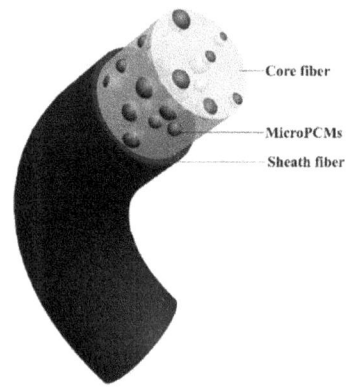

FIGURE 8.28 Schematic diagram of thermo-regulated composite fiber with core-sheath structure.

Source: Li et al. (2014). Reprinted (adapted) with permission. © 2014 American Chemical Society.

With the increase in polymer concentration from 0% to 1.5 wt%, the hydrostatic pressure goes down slightly from 90 kPa to 81.1 kPa. And the further increase of polymer concentrations, as shown in Figure 8.27 (a) resulted in a dramatic decrease of hydrostatic pressure. Since all the membranes exhibit similar wettability, the decreased hydrostatic pressure of the membranes obtained from higher polymer concentrations should be mainly attributed to the enormous increased d_{max}, which would be penetrated under lower external water pressure. In Figure 8.27 (b), the WVT rate of the fibrous membranes decreased from 9.2 to 7.3 kg/m²/day gradually with the increase of polymer concentrations, revealing impaired breathability. This is attributed to the fact that membranes with lower porosity possess less interconnected passages for water vapor transmission.

8.13.4 The Glass Transition Temperature as a Transition Point for WVP

Decreasing the glass transition temperature leads to increased free volume in the polymer, and the micro-Brownian motion increases due to the soft segment and thus makes the intermolecular gap wide enough to allow water vapor molecules to be transported through the film. In other words, the diffusivity of the water vapor molecules in the film increases with increasing temperature. Hence, significant changes in moisture vapor permeability below and above the T_g of the soft segment are observed. On the other hand, the glassy state of the soft segment at lower temperatures acts as the water vapor barrier, thereby decreasing the water vapor permeability and thus is a waterproof barrier at lower temperatures (Hu et al. 2003).

A core-sheath thermo-regulated fiber was produced by coating a polyurethane sheath (Yeh et al. 2008) on alginate composite fibers, as shown in Figure 8.28. When the copolymer shell was heated sufficiently to form a highly cross-linked structure and permitted the continuous polymerization of the residual shell-forming monomers to occur, as a result, the thermal stability of high pressure and high temperature-treated micro-phase change materials (microPCMs) was enhanced. This can be shown in Figure 8.29.

When increasing the amount of microPCMs to 10 and 15 wt%, the average enthalpy of exothermic and endothermic enthalpy rose to 2.2 and 7.2 J/g respectively. Also, it must be noted that with the microPCM's loading increasing, the peak temperature of composite fibers increased from 29.6 to 31.2°C steadily, as shown in Figure 8.30 due to the enhanced thermal conductivity of the composite fibers with increasing microPCMs loading.

8.13.5 Crystal Melting of Soft Segment Temperature as a Transition Point for WVP

Theoretically, in the thermogram of a polymer, the phase transition from the crystalline to the amorphous phase results in an increased amorphous area leading to an increase in free volume. As per the free volume theory for dense samples, the water vapor transition takes place by diffusion through the film which is driven by the vapor concentration difference. In the case of films, when the temperature reaches the crystal melting point, there is an increase in free volume as the relative amorphous area increases. Therefore, the film provides more paths for water vapor permeation, resulting in significant increase in WVP (Ding, Hu, and Tao 2004).

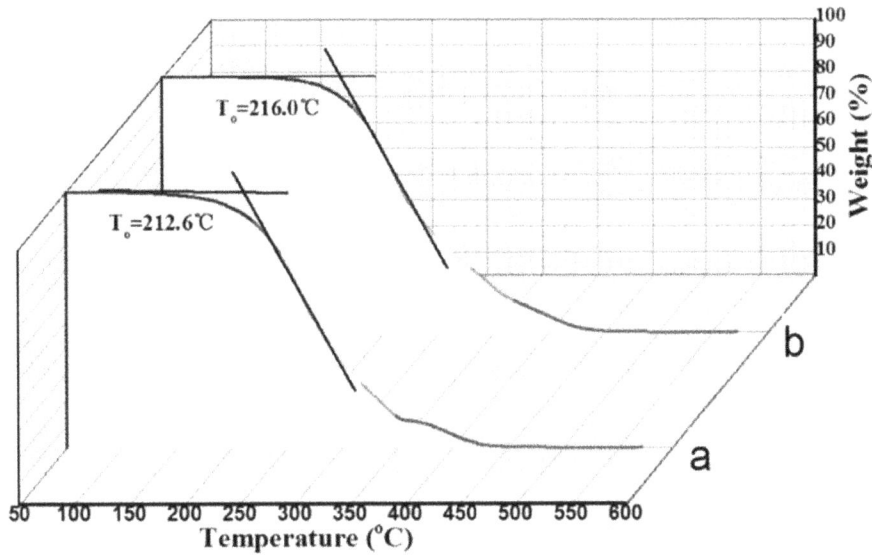

FIGURE 8.29   TGA thermograms of (a) microPCMs and (b) HPHT treated microPCMs. To: onset temperature of the TGA curve.

Source: Li et al. (2014). Reprinted (adapted) with permission. © 2014 American Chemical Society.

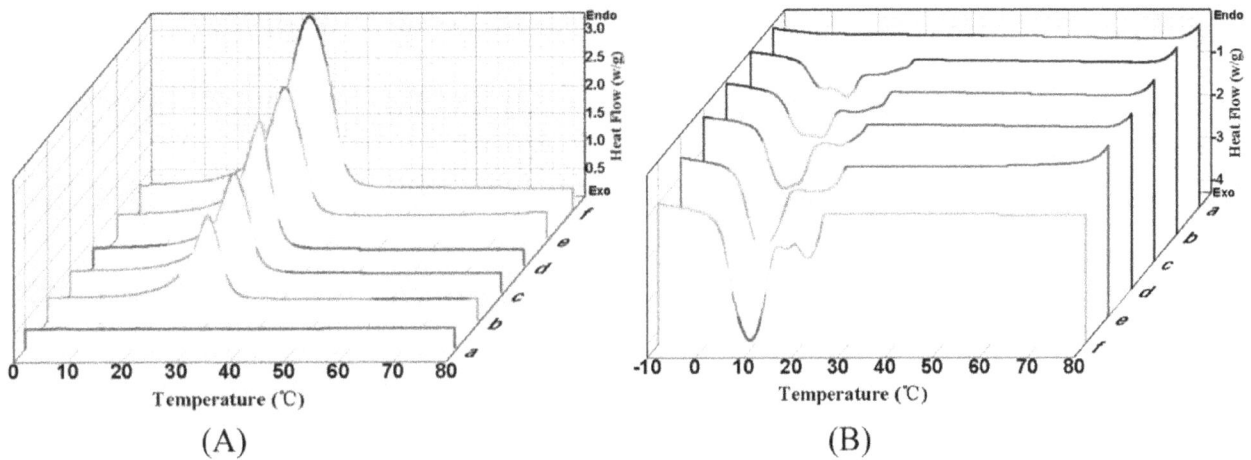

FIGURE 8.30   DSC thermograms of composite fibers containing various micro PCM contents and micro PCM: (a) 0 wt.%; (b) 60 wt.%; (c) 75 wt.%; (d) 82 wt.%; (e) 87 wt.%; (f) micro PCMs. (A) DSC heating curve; (B) DSC cooling curve.

Source: Li et al. (2014). Reprinted (adapted) with permission. © 2014 American Chemical Society.

8.14 Effect of Solid Content on Properties of Polyurethane Dispersion

Increasing the solid content in polyurethanes leads to a decrease in the particle size distribution of the particles and therefore to the polydispersity of the polyurethane dispersions. In a study, polyurethane with 37% solid content was found to contain particles of 358 nm and 1919 nm. On the other hand, polyurethane with 44% solid content was found to have a particle size of 1748 nm. It can also be noted that the same investigation discovered that, as the solid content increased, the Brookfield viscosity of the PUDs tends to increase and thereby led to the decreased mobility of the particles and consequently decreased the probability of particle-particle interactions (García-Pacios et al. 2011).

This was confirmed by the relative intensity of several IR bands of the polyurethanes, which exhibited differences in the hard segment content. The increase in the solid content even led to an increased loss in weight, which can easily be related to the degradation of the soft segments. In other words, the thermal stability increased as the solid content of the polyurethanes increased. The increase in the solid content decreased the glass transition temperature, pointing to a higher degree of phase separation. Increase in the solid content did not significantly affect the adhesive strength under the presence of peel stresses. However, it was observed that the shear strength and a cohesive failure in the polyurethane decreased, which can occur due to the decrease in the hard segments (García-Pacios et al. 2011),

8.15 Effect of -Si- Bonds on Polyurethane Dispersion

Organosilane molecules have the unique ability to conform to covalent bonding between organic and inorganic compounds. Siloxanes have inherent stability and flexibility, hence their molecules can provide multiple benefits in a broad range of coating systems. They are an effective adhesion characteristic, especially if used in additives/primers for paints, sealants, inks, adhesives, and coatings.

When used as additives, they emigrate to the interface between the substrate and adhesive layer, thus performing effectively. When used as a primer, the silane, which acts as a coupling agent, is applied to an inorganic substrate. After this, the product to adhere is applied.

Organofunctional silanes can bond with organic polymers, thus resulting in a reaction of the silyl group onto the polymer chain. This mechanism can be used to cross-link polyethylene, urethanes, and acrylics to impart water resistance, durability, along with heat resistance to adhesives, coatings, and paints. The organic group on the silane can be either reactive or non-reactive. Non-reactive silanes are often used as dispersing agents and are listed below:

- methyl-methoxy-silane
- methyl-tri-ethoxy-silane
- dimethyl-di-methoxy-silane
- propyl-tri-methoxy-silane
- iso-butyl-tri-methoxy-silane.

They attribute a non-reactive group (X), which is compatible with the silyl group, and an alkoxy group that reacts with the substrate on which it is being applied. Other reactive silanes include a couple of amino functional silanes as listed below:

- amino-propyl-tri-ethoxy-silane
- amino-ethyl-amino-propyl-tri-methoxy-silane.

Other reactive organosilanes are:

- γ-methacryloxy-propyl-tri-methoxy-silane
- γ-glycidoxy-propyl-tri-methoxy-silane
- γ-chloropropyl-tri-methoxy-silane
- vinyl-tri-methoxy-silane
- vinyl-tri-acetoxy-silane
- mercapto-propyl-tri-ethoxy-silane.

Organosilanes for their application require the silane molecule to engage in hydrolysis and condensation reactions. Silicon hydride (Si-H) reacts with water to form silanol (Si-OH) species, which is capable of reacting further. The impact of a functional group on silanol reactivity depends exclusively upon the distance between them (Materne et al. 2012).

- *Acid-catalyzed hydrolysis*: Acid-catalyzed hydrolysis involves the protonation of the leaving OR group (Figure 8.31).

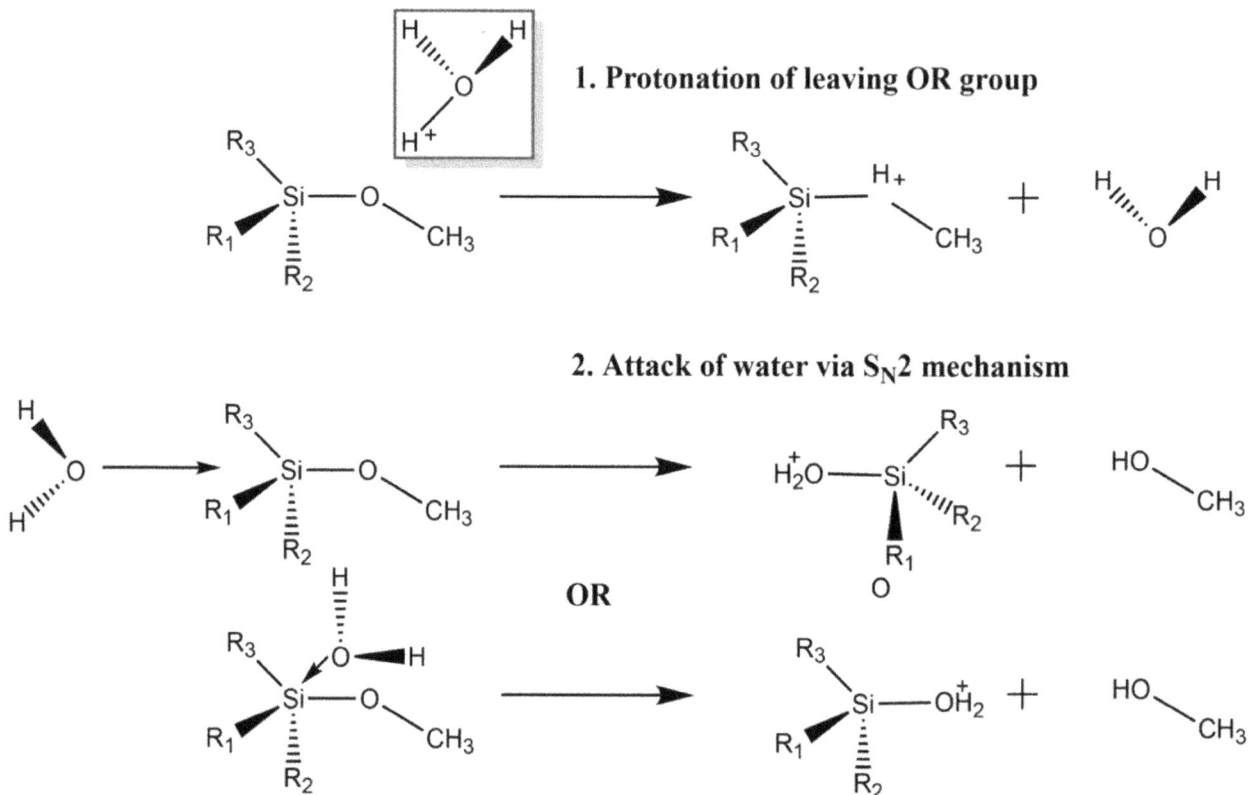

FIGURE 8.31 Acid catalyzed hydrolysis.

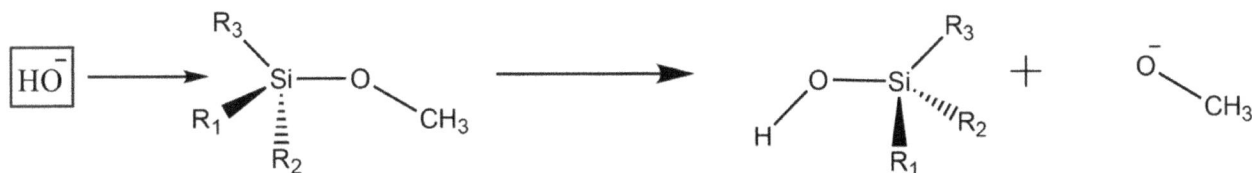

FIGURE 8.32 Base catalyzed hydrolysis.

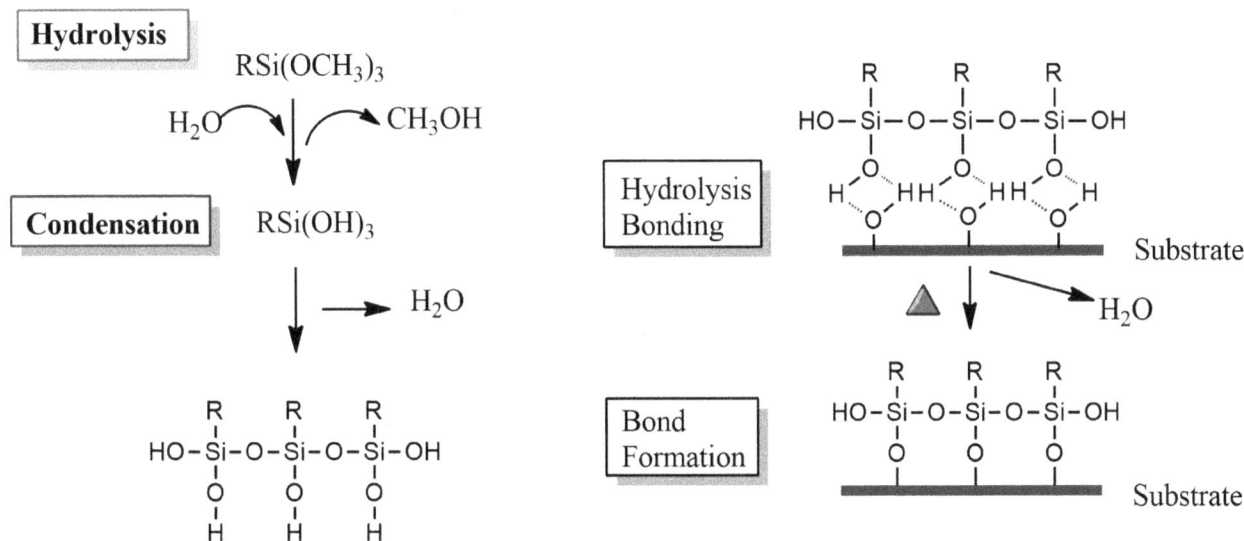

FIGURE 8.33 Mechanism of hydrolysis and condensation.

- *Base-catalyzed hydrolysis*: This comprises an attack on silicon by the hydroxyl ion to form a pentacoordinate intermediate. This is succeeded by hydroxyl displacing an alkoxy. High-electron acceptors group, if present next to the silico atom, will hereby increase hydrolysis under the influence of basic pH (Figure 8.32).

- *Hydrolysis rate*: The nature of the organic group on silicon and the displaced alkoxy group influence the rate of hydrolysis by both mechanisms (Figure 8.33). As pH tends to change, the rate of hydrolysis also changes. Figure 8.34 denotes the optimum pH for the hydrolysis and condensation reactions.

A series of siloxane-modified PUDs with different di-hydroxy-butyl terminated poly-di-methyl-siloxane (DHPDMS) content was compared with the conventional WPU by Ge and Luo (Ge and Luo 2013). They discovered that there was a great improvement in the thermal stability of the siloxane-modified copolymers. After siloxane-modification, the surface energy and the water absorption of the PUD's films decreased significantly. As per the study, the surface energy decreased from 34 mN/m to 23 mN/m when the PDMS content was increased from 0% to 9%; resulting in decreased water absorption from 14% to 2%. This investigation thus concludes that there was mobility of siloxane chains to the surface of siloxane-modified films.

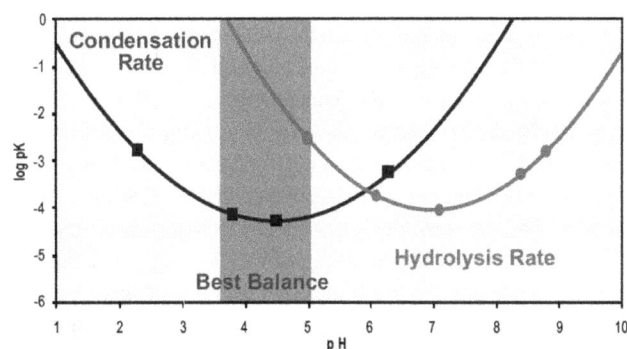

FIGURE 8.34 The hydrolysis/ condensation balance.

The investigation further suggests that the siloxane-modified films possess good adhesion, flexibility, chemical resistance, and hardness. An increase in the DHPDMS content led to a decrease in the gloss of the coatings of siloxane-modified PUDs, but the presence of the higher DHPDMS promotes a shortened drying time for the siloxane-modified PUD and thus improves the impact resistance of the films.

Amino propyl triethoxy silane (APTES) was used to prepare inorganic-organic water-based polyurethane through the sol-gel approach (Florian et al. 2010). The inorganic precursor (APTES), which helps in the formation of thermally stable Si-O-Si linkages, led to notable improvement in the thermal properties of the

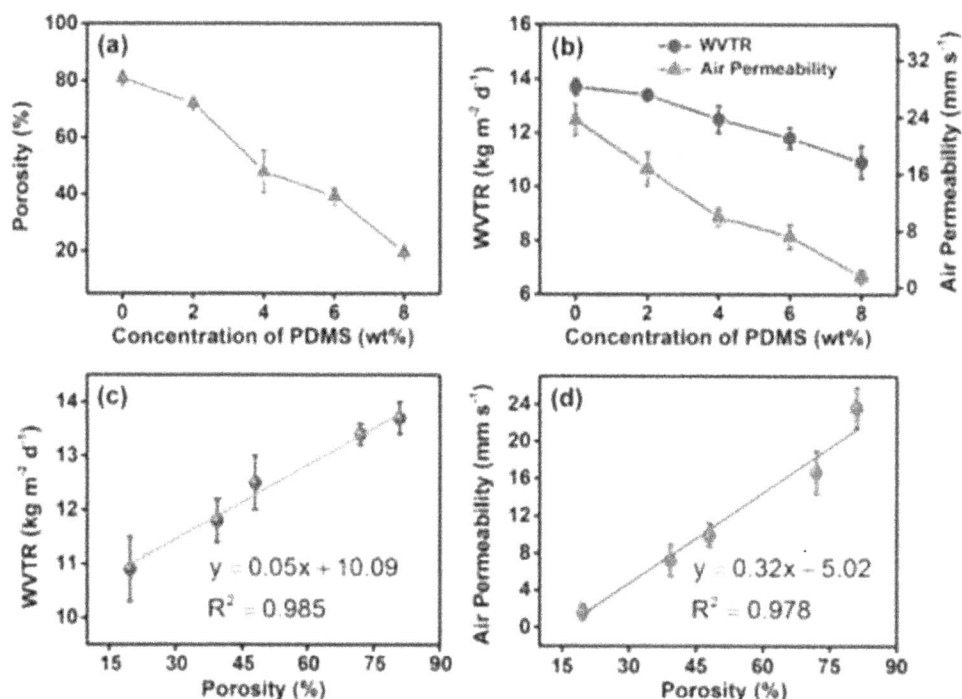

FIGURE 8.35 Breathable property of water vapor and air of relevant PAN@PDMS NFM with different PDMS concentrations. (a) The porosity of PAN@PDMS NFM. (b) WVTR and air permeability of PAN@PDMS NFM. (c) The linear relationship between WVTR and porosity. (d) The linear relationship between air permeability and porosity.

Source: Sheng et al. (2016). Reprinted (adapted) with permission. © 2016 American Chemical Society.

resulting compound. Glass transition temperatures increased in the resulting hybrid compound compared with that of diamine and the diol system. Water-based polyurethane dispersions studied in this investigation render a good balance between required properties.

With an increase in the concentration of PDMS (Sheng et al. 2016), the structural transformation from fluffy to adhesive resulted in the porosity gradually decreasing, as displayed in Figure 8.35 (a). Figure 8.35 (b) displayed that WVTR and air permeability had the same decline tendency as that of porosity. The average pore size PDMS modification was far bigger than the size of water vapor and air molecules. On account of this phenomenon and according to Darcy's law, porosity played an important role in WVTR and air permeability. Therefore, the relationship between porosity and breathability (WVTR and air permeability) was investigated, as shown in Figures 8.35 (c, d). WVTR and air permeability demonstrated a linear relationship with porosity. In other words, every single porosity provided a consistent amount of water vapor and air permeability.

Yet another cotton fabric coated with silica NPs is known to exhibit super-hydrophobicity (Liu et al. 2018). As shown in Figure 8.36 (a), when a drop of water falls onto the untreated cotton fabric, the water wets the fabric immediately with a water contact angle approaching 0° because of complete infiltration. By contrast, after coating with a silica NP/PDMS layer, the water contact angle of the textile dramatically increases to 161°, while the sliding angle is less than 5° (Figure 8.36 (d)), indicating an extraordinary waterproofness. In addition, the treated super-hydrophobic textile also exhibits an anti-fouling function in dirty

water, as indicated in the case with red chalk powders dissolved in water. As shown in Figures 8.36 (b, c), the original cotton cloth is dyed red by immersing in dirty water within a short time, whereas the super-hydrophobic cotton fabric shows a remarkable anti-fouling ability in dirty water (Figures 8.36 (e, f)). Furthermore, the washing durability was evaluated by ultrasonication in a mixed solution of water and ethanol. The change of the water contact angle is less than 9° after 66 h of ultrasonication, indicating that the cloth might be durable to washing (Figures 8.36 (g, h)).

8.16 Effect of the Curing Agent Concentration

The effect of the curing agent concentration on the particle size and morphology was investigated by Sardon et al. (Sardon et al. 2010). The final particle size of the dispersions increased with the aminopropyl triethoxy-silane (APTES) concentration, especially for high concentrations. As per the study, the increase in APTES concentration from 0% to 20% resulted in an increase of the particle size from 30 nm to 140 nm. The effect of pH on the zeta potential of the polyurethane with different percentages of APTES was studied. A very clear difference can be found when the pH of the polyurethane dispersion is 11. The dispersion with no APTES was found to be the most stable of all of the studied dispersion. The zeta potential tends to be more negative, to a maximum value of -80 mV as the concentration of APTES was increased from 0% to 20%

In addition to this, no silica domains could be detected at low APTES concentrations, by TEM. But, when the APTES

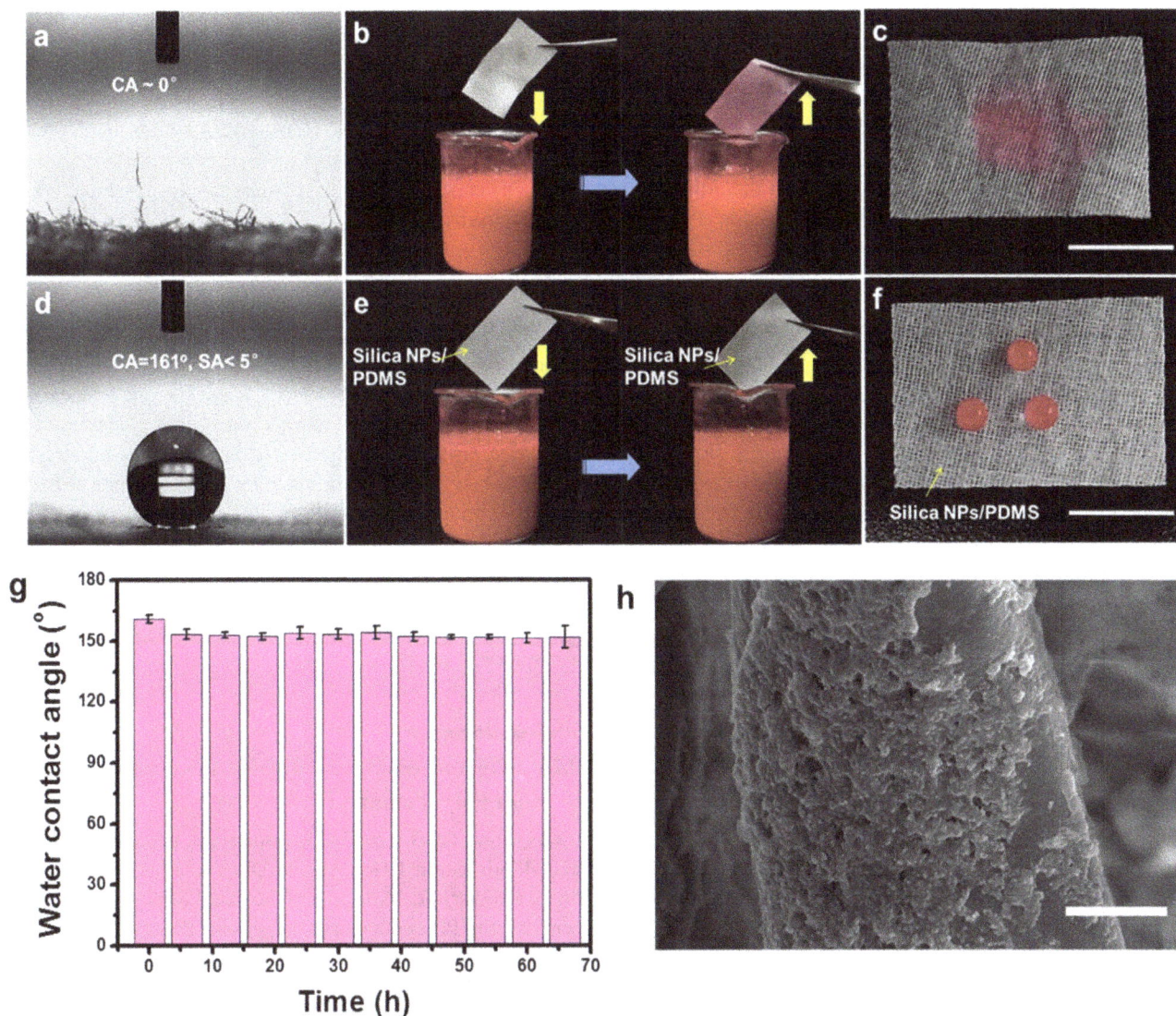

FIGURE 8.36 Wettability of the cotton textile before and after coating silica NPs. (a) Water drop (2 μL) on the untreated cotton textile, showing a contact angle close to 0°. (b) Untreated normal cotton cloth is polluted when immersed in dirty water. (c) Untreated cotton textile is completely infiltrated with dirty water drops. (d) Water drop (2 μL) on the cotton textile with silica NPs, showing a contact angle of ~161° and a sliding angle of less than 5°. (e) After hydrophobic treatment, the cotton fabric becomes anti-fouling to dirty water. (f) Dirty water drops on the treated fabric, showing a remarkable anti-fouling property. (g) Change of water contact angle with ultrasonic cleaning duration. (h) SEM image of the coating surface after 66 h of ultrasonication.

Note: Scale bars are 1 cm in (c), (f) and 5 μm in (h).

Source: Liu et al. (2018). Reprinted (adapted) with permission. © 2018 American Chemical Society.

concentrations were higher than 9.7 wt%, the polyurethane contained inorganic-rich domains. This confirms that there was partial condensation of alkoxy groups during the process of emulsification. This is held responsible, as per the investigation, for promoting their aggregation. The measurements of zeta potential indicate that the higher silanol groups on the surface of the particle were obtained when using 9.7 wt% of APTES (ibid.).

A study of cross-linking of polyurethanes has testified that the polymers are soluble in DMF below 12 mol% of the triol (Petrović et al. 1991). The triol, if used in higher amounts, led to gelation and thereby network formation. An increase in strain at break and tensile strength, with increasing degree of cross-linking, was observed up to 56 mol% of the triol. The rate of stress-relaxation was found to be contingent on both: the degree of cross-linking and the extension. This idea can be studied and extrapolated with proper investigations to develop quick drying coating for textiles.

8.17 Effect of the Catalyst on the Synthesis of Polyester Polyols

In self-catalyzed and acid-catalyzed polyesterification of adipic acid with ethylene glycol, l,4-butanediol, and l,6-hexanediol under constant reaction temperatures, the conversion versus reaction time data were studied and fitted into the rate equations. Chin-Tsou (Chin-Tsou 1989) calculated the activation energies which tend to decrease with increasing the chain length of the alkyl group. Also, a significant drop in the activation energies of the acid-catalyzed reactions was observed, as compared to self-catalyzed reactions. The study has been summarized in Table 8.1.

Jacquel et al. (Jacquel et al. 2011) studied the efficiency of different catalysts for the transesterification reaction step, which is of prime importance for the purpose of building the desired molecular weight for the synthesis of polyester polyol used to synthesize polyurethane dispersions. The study reveals that out of all the possible catalysts that were put to the test, the titanium-based catalyst proved to be the most efficient. This was followed by zinc and tin-based catalysts. It also emphasizes the concentration of the titanium-based catalyst compared to others. Titanium, being proved the best, was tested further to determine the best concentration that would lead to an optimized reaction rate: 100 ppm of this catalyst proved really insufficient; 200, 300, and 400 ppm had a good effect on the reaction rate, with the reaction rate reducing to half as the concentration was increased from 200 ppm to 400 ppm to reach a ΔC of 16 N.m. In the same way Zr and Sn catalysts were tested and were evaluated for the same time of 350 minutes. To reach a competitive stage where the rate of reaction provided by the Ti, Zr, and Sn catalysts was same (i.e., to reach $\Delta C = 16$ N.m), 200 ppm, 1200 ppm, and 900 ppm of Ti, Zr ,and Sn catalysts were incorporated in the reaction mixture in the polycondensation step.

This study also made it clear that the addition of the catalyst before the transesterification step led to the delayed completion of the reaction, unlike the case where it was added during the transesterification step. This behavior of the catalyst is due to the poisoning of the catalyst during the esterification process wherein it is poisoned by water which is generated as a by-product of the esterification.

Considering the previous recommendation of adding the catalyst after the water is removed, along with the suggestion of using it throughout the time of the reaction, we can conclude that excess catalyst can be added so that even if a few moles of it are poisoned by water, still a significant amount of it would be available to speed the reaction.

8.18 Effect of the Material Composition

Polyurethane–silica hybrid materials in the form of freestanding film and the effects of the material composition on the optical clarity of WPU and surface wettability, were studied by Yeh et al.(Yeh et al. 2008). The freestanding film of such materials at low silica loading (5 wt.%) compared to WPU shows about 5% growth in the water contact angle. The contact angle studies reveal that the contact angle increased from 67.2^0 to 78.4^0 when the concentration of TEOS was increased from 0% to 9%. At low silica contents of polyurethane–silica hybrid materials, optical clarity remains high.

8.19 Effect of the Thickness of the Original Fabric

It can be concluded from the study conducted by Aditi Bakshi (Bakshi 2015) that when the percentage of silicone oil increases, breathability decreases and also the waterproof properties of the fabric decrease. The study heavily emphasizes the thickness of the fabric being used. According to the author, in spite of having high waterproof properties, the highest waterproof ratings could not be achieved and the main reason for this is the loose weave of the fabric. Lowering the waterproof capabilities may also contribute to the lower thickness of the original fabric. Hence, tighter weave and thicker fabric have to be used to obtain higher waterproofing (ibid.).

8.20 Effect of Di-Methyl-Propionic-Acid (DMPA) on Polyurethane Dispersion

It was shown that for stable PU dispersions, a minimum amount of DMPA is required. However, the DMPA concentration depends on many other variables, such as the initial phase-inversion temperature, the PU content, and the solvent affinity to water. The study reveals that initial PU content greatly affects the requirement of DMPA to obtain a stable dispersion. The requirement of DMPA reduces from 0.6 mmol/g_{pol} to 0.1 mmol/g_{pol} when the initial PU content is increased from 50% to 75%. Moreover, it is important to take into account that the evaporation conditions at phase inversion can create instability in the PU dispersions, hence, mild temperatures (25–30^0C) have to be maintained to avoid the generation of large aggregates (Sardon 2010). The study reveals that, for 60% solid content, the particle size was 240 nm, 160 nm, 130 nm, 35 nm, and 30 nm for dispersions made at 50, 40, 35, 30, and 25^0C respectively.

8.21 Effect of Solvent Affinity to Water

According to the study when using 0.30 mmol of DMPA/g of polymer and working at 25^0C, 60 wt% of PU is needed to form stable water dispersions. However, if the DMPA concentration increases, both the PU affinity to water and the ionic strength increase with the DMPA content. However, when the temperature is increased, the PU solubility in the solvents changes (Sardon 2010).

The mechanism of use of acetone concentration promotes two different particle formation mechanisms. When the acetone concentration exceeds a critical value, emulsification takes place by spinodal decomposition, producing coarse emulsions which

TABLE 8.1

Decrease in the Activation Energy with An Increasing Chain Length of the Alkyl Group

System	Self-Catalyzed (kcal/mol)	Acid-Catalyzed (kcal/mol)
AA/EG	14.23	9.95
AA/BDO	11.75	9.00
AA/HDO	11.11	8.71

separate rapidly. In contrast, when the acetone concentration is lower than a critical value, the particles are stabilized by means of spontaneous emulsification, generating small stable particles (ibid.).

8.22 Testing of Breathable Fabrics and Their Associated Problems

Breathability testing of waterproof breathable fabrics (WBFs) is met with confusion and controversy. This applies mostly to the evaluation of commercial fabrics in the laboratory and extends to wearer trials of garments in climatic chambers or in the field. The basic problem originates from the difference in water vapor transport mechanisms between hydrophilic layers and microporous layers, which is further deteriorated by the swelling effects in the former. However, WBFs are designed as per the specifications of the clothing manufacturers or the end-users, to find a universal standard test method.

8.22.1 Fundamentals

Diffusion of a substance through a layer is described by Fick's 1st Law for steady state and 2nd Law which relates to the non-steady state condition (Ashley 1985). Fick's main hypothesis states that the rate of transfer through the unit area of a section, F(x), is directly proportional to the concentration gradient measured as normal to the section, i.e.

$$F(x) = -D\left(\frac{\partial c}{\partial x}\right) \qquad (8.3)$$

where:

x = space coordinate normal to the section
c = concentration of diffusing substance
D = diffusion coefficient

The water vapor transmission (WVT) properties of WBFs can be derived from Equation (8.3) by the different proportionality constant (k) to relate WVT with the pressure difference across the surfaces (p_1 - p_2) and its thickness d (Lomax 2001a):

$$m/At = k(p_1 - p_2)/d \qquad (8.4)$$

Many recognized breathability tests are used to measure the amount of water vapor (m) passing through a fixed area of material (A) over a time period (t) by gravimetry. The WVT rate (m/At) obtained under steady-state conditions has an SI unit of kg m^{-2} s^{-1}, although the trivial unit g m^{-2} day^{-1} is most widely recognized by the textile industry.

One major concern with WVT testing is that results quoted do not take the magnitude of the driving force (p_1 - p_2) into consideration. More often, values for p_1, p_2 at each surface of the fabric are more guess-estimated than accurately measured. A hydrophilic membrane shows unusual and strong dependence on applied relative temperature and humidity, which determines the p. These two factors alone make comparing data obtained from different sources unreliable. For instance, it is possible to generate an apparent 40-fold increase in WVT of WBFs merely by changing the set conditions of the "standard" test method (Lomax 1990).

An alternative method is to visualize the fabric coating as a barrier to water vapor diffusion rather than as a transmitter by the reciprocal quantity of water vapor resistance (WVR). The limiting WVR to zero indicates the infinitely breathability of the fabric coating, but, practically, due to the barrier layer of finite thickness, zero value can never be achieved. Measuring the WVR's value has two major advantages:

1. The pressure gradient is considered.
2. The data are additive, which is useful for assessing the overall resistance of hypothetical and real clothing assemblies if the R values of individual layers are already known.

For air-permeable membranes and base fabrics, WVR data are reliable but less reliable for hydrophilic polymers which are strongly dependent on the surrounding water content or water vapor pressure. In the case of true microporous components having low WVRs, they barely change with relative humidity (RH) at a given temperature (Hu 2008). Conversely, hydrophilic WBFs are not especially breathable at low humidity, but their WVRs decline as the RH rises. As a result, swelling in hydrophilic membranes occurs as a response to a build-up of water vapor inside a clothing assembly, thereby becoming more breathable. This "sense and react" mechanism can be classified as an intelligent textile response (Leitch and Tassinari 2000). Hence, as shown, hybrid fabrics containing a mixture of both hydrophilic and microporous layers show intermediate properties.

Body thermo-regulation involves an interactive set of physiological responses to balance metabolic heat production (M) under workload (expended energy, P_w) with the heat losses which are necessary to maintain body temperature (T_C) at approximately 37^{0}C (Havenith 2002). Heat loss through clothing occurs mostly by combined dry processes (conduction, convection, and radiation) and wet processes (insensible perspiration and sweat), and this process can be equated as Equation (8.5):

$$M_{output} - P_w = \left(H_{cond} + H_{conv.\ dry\ loss} + H_{rad}\right)$$
$$+ \left(H_{insens} + H_{sweat\ wet\ loss} + H_{resp}\right) \pm \Delta S_{debt} \qquad (8.5)$$

The heat loss component depends on the ambient atmosphere and workload, the person's metabolism rate, and the clothes worn. A sedentary person has a heat output, which is about 100 W, a respiratory heat loss (H_{resp}) of 25 W, and water vapor loss through the skin (H_{insens}) is 250–300 gm^{-2} day^{-1}, which are equivalent to cooling of 12–15 W. Dry heat loss accounts for the remainder (ca. 60 W) and the wearer usually remains comfortable in light clothing. In contrast, heat output in heavy workloads can be as high as 1000–1250 W, but these levels of exertion are sustained for short periods of time. In such cases, sweating is a cooling mechanism because water has a very high latent heat of evaporation (0.672 Wh g^{-1} at 35^{0}C). This infers that a sweat rate of 1 kg h^{-1} provides 670 W cooling, provided all the sweat

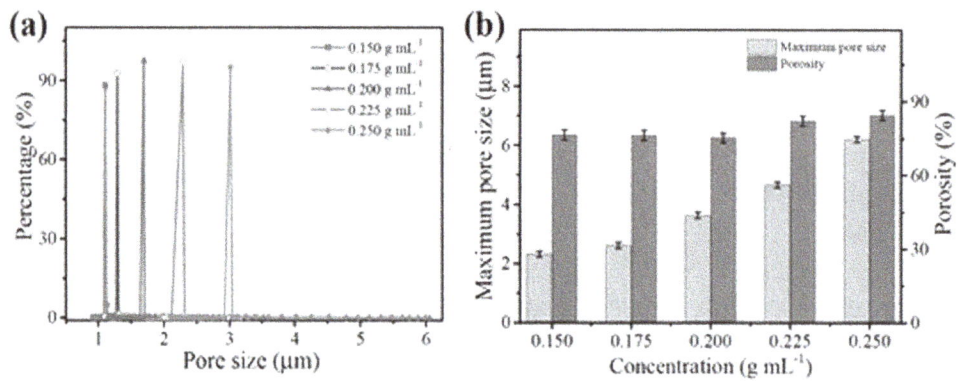

FIGURE 8.37 Microporous structure and waterproofness of PVDF membranes prepared with various PVDF concentrations. (a) Pore size distribution. (b) Maximum pore size and porosity.

Source: Jiang et al. (2018). Reprinted (adapted) with permission. © 2018 American Chemical Society.

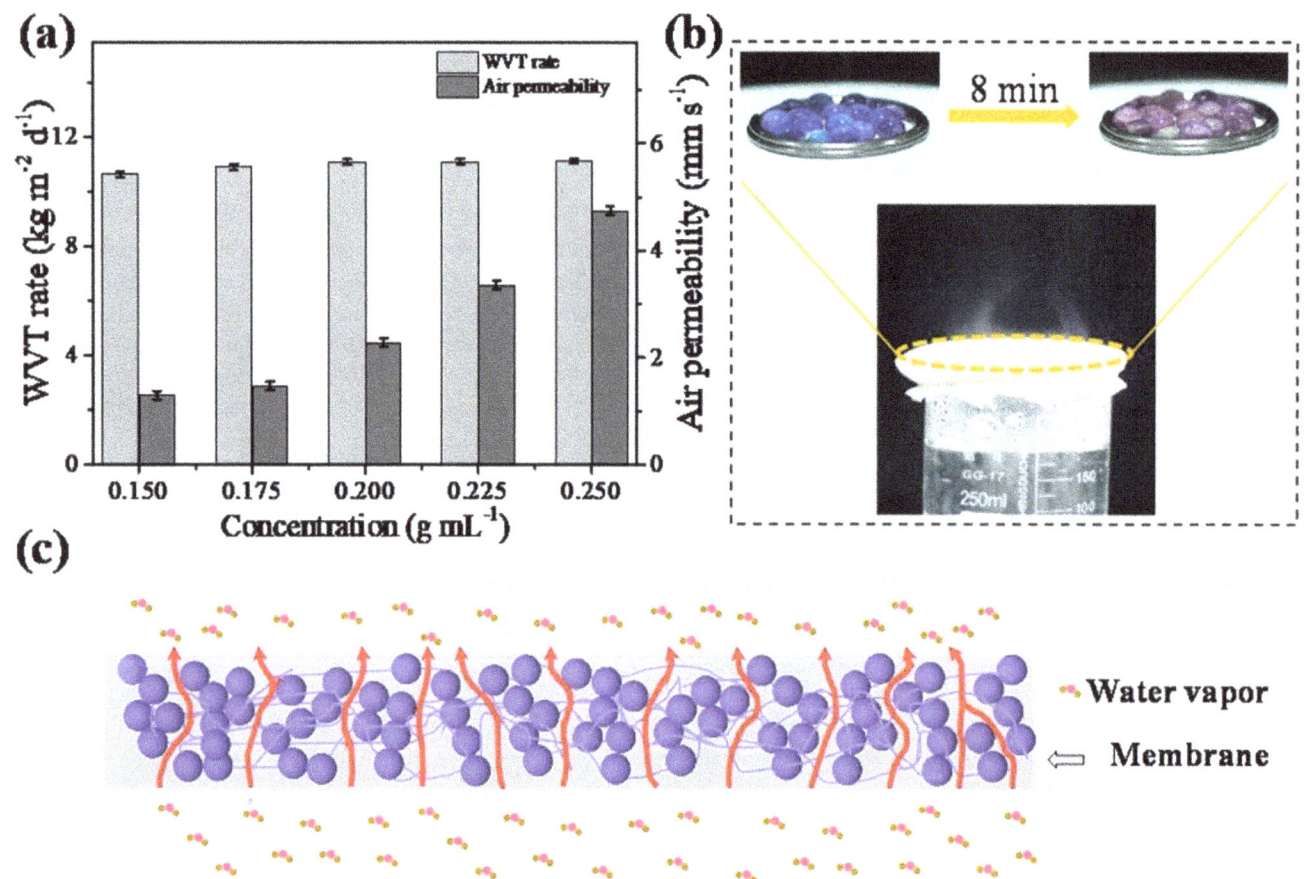

FIGURE 8.38 (a) WVT rate and air permeability of PVDF membranes prepared with various PVDF concentrations. (b) A typical test demonstrating the breathable performance. (c) Schematic illustrating the process of breathable behavior through the membrane.

Source: Jiang et al. (2018). Reprinted (adapted) with permission. © 2018 American Chemical Society.

evaporates from the body. Clothing interferes with this process by trapping moist air, resulting in condensation within the textile layers (Holmé 2006). When heat output and heat loss are in dynamic balance, i.e., the subject is in a state of thermo-physiological comfort, the heated debt (ΔS) is zero. In the case where sustained heat output exceeds heat losses (+ve ΔS) or vice versa

(-ve ΔS), it can lead to hyperthermia or hypothermia, respectively (ibid.).

It is also observed (Jiang et al. 2018) that when the membrane porosity increases to more than 80%, the WVT rate increases greatly. A higher porosity leads to a higher WVT rate. The proportional relationship between WVT rate and membrane porosity can be explained by the Fickian diffusion model in which the expounded diffusion flux is positively related to the diffusion coefficient that is determined by porosity (Figure 8.37).

A study was carried out (ibid.) in which the breathability of a membrane was determined by a typical experiment. As a humidity indicator to confirm the water vapor transmitting through the membrane, allochroic silica gel particles are used. As presented in Figure 8.38, a polyvinylidene fluoride (PVDF) membrane is put as a cover on a beaker filled with boiling water. At 100°C, large quantities of steam are visibly seen passing through the membrane. The allochroic silica gel particles change from blue to a pink color within 8 minutes.

REFERENCES

Ashley, R.J. (1985). Permeability and Plastic Packaging. In J. Comyn (Ed.), *Polymer Permeability*. London: Elsevier, pp. 269–308.

Bajaj, P. (2001). Eco-Friendly Finishes for Textiles. *Indian Journal of Fibre and Textile Research* 26(1): 162–186.

Baker, R.W. (2012). *Membrane Technology and Applications*. Hoboken, NJ: Wiley.

Bakshi, A. (2015). Development and Study of Waterproof Breathable Fabric Using Silicone Oil and Polyurethane Binder. Thesis.

Barbur, V. (n.d.). Medical Textiles: How Smart Do They Have to Be? Available at: www.textileworld.com/textile-world/nonwovens-technical-textiles/2012/12/medical-textiles-how-smart-do-they-have-to-be/ (accessed May 31, 2018).

Bartels, V.T. and Umbach, K.H. (2002). Water Vapour Transport through Protective Textiles at Low Temperatures. *Textiles Research Journal* 72, 899–905.

Chen, L., Bromberg, L., Lee, J.A., … Rutledge, G.C. (2010). Multifunctional Electrospun Fabrics via Layer-by-Layer Electrostatic Assembly for Chemical and Biological Protection. *Chemistry of Materials* 22(4), 1429–1436.

Chin-Tsou, K. (1989). Kinetics of Polyesterification: Adipic Acid. *Journal of Polymer Science: Part A: Polymer Chemistry* 27, 2793–2803.

Chinta, S.K. and Satish, D. (2014). Studies in Waterproof Breathable Textiles. *International Journal of Recent Development in Engineering and Technology* 3(2), 2347–2435.

Cipriano, M.M., Diogo, A., and de Pinho, M.N. (1991). Polyurethane Structure Design for Pervaporation Membranes. *Journal of Membrane Science* 61, 65–72.

Covestro. (2017). Annual Report. Available at: www.google.com/url?sa=tandrct=jandq=andesrc=sandsource=webandcd=2andved=0ahUKEwiMi7203rzbAhUJtY8KHczLBj8QFggsMAEandurl=https%3A%2F%2Finvestor.covestro.com%2Fsecured1%2F14297andusg=AOvVaw32yWsVOuglK595h1q1T8GJ

Crowson, A. (1996). Smart Materials Based on Polymeric Systems. Paper presented at Smart Structures and Materials, Smart Materials Technologies and Biomimetics.

Cussler, E.L. (1997). Mass Transfer in Fluid Systems. In *Diffusion: Mass Transfer in Fluid Systems*, 1st edn. Cambridge: Cambridge University Press.

Davis, R. (n.d.). Polyurethane Coatings Go Green. *Textile World*. Retrieved from www.textileworld.com/textile-world/features/2016/02/polyurethane-coatings-go-green/ (accessed June 5, 2018).

De, P., Sankhe, M.D., Chaudhari, S. S., and Mathur, M.R. (2005). UV-resist, Water-repellent Breathable Fabric as Protective Textiles. *Journal of Industrial Textiles* 34(4), 209–222.

Ding, X.M., Hu, J.L., and Tao, X.M. (2004). Effect of Crystal Melting on Water Vapour Permeability of Shape-Memory Polyurethane Film. *Textile Research Journal* 74(1), 39–43.

Dorsey, N.E. (2015). *The Surface Tension of Water and of Certain Dilute Aqueous Solutions, Determined by the Method of Ripples*. Berlin: Nabu Press.

Enomoto, M., Suehiro, K. and Muraoka, Y. (1997). Effect of Composition and Coagulation Structure of Coating Polymers Made from Hydrophobic and Hydrophilic Polyurethane Resin on Moisture Transporting Properties in Waterproof/Moisture Permeable Fabrics. *Journal of Textile Machinery Society of Japan* 50(9), 77–85.

Florian, P., Jena, K.K., Allauddin, S., Narayan, R., and Raju, K.V.S.N. (2010). Preparation and Characterization of Waterborne Hyperbranched Polyurethane-Urea and Their Hybrid Coatings. *Industrial and Engineering Chemistry Research* 49(10), 4517–4527.

Fuensanta, M., Jofre-Reche, J.A., Rodríguez-Llansola, F., Costa, V., Iglesias, J.I., and Martín-Martínez, J.M. (2017). Structural Characterization of Polyurethane Ureas and Waterborne Polyurethane Urea Dispersions Made with Mixtures of Polyester Polyol and Polycarbonate Diol. *Progress in Organic Coatings*, 112(April), 141–152.

Fung, W. (2002). Products from Coated and Laminated Fabrics. In *Coated and Laminated Textiles*. Oxford: Woodhead, pp. 149–249.

García-Pacios, V., Iwata, Y., Colera, M., and Martín-Martínez, J.M. (2011). Influence of the Solids Content on the Properties of Waterborne Polyurethane Dispersions Obtained with Polycarbonate of Hexanediol. *International Journal of Adhesion and Adhesives* 31(8), 787–794.

Ge, Z. and Luo, Y. (2013). Synthesis and Characterization of Siloxane-Modified Two-Component Waterborne Polyurethane. *Progress in Organic Coatings* 76(11), 1522–1526.

Gray, N.C. and Millard, C.E. (n.d.). *Moisture Vapour Permeable Garments: A Physiological Assessment*. Farnborough: DERA.

Gretton, J.C., Brook, D.B., Dyson, H.M., and Harlock, S.C. (1998). Moisture Vapour Transport Through Waterproof Breathable Fabrics and Clothing Systems Under a Temperature Gradient. *Textile Research Journal* 68(12), 936–941.

Havenith, G. (2002). Interaction of Clothing and Thermoregulation. *Exogenous Dermatology* 1(5), 221–230.

Hayashi, S. (1992). Makeup Material for Human Use. Google Patents.

Hayashi, S., Ishikawa, N., and Giordano, C. (1993). High Moisture Permeability Polyurethane for Textile Application, *Journal of Coated Fabrics* 23(7), 74–83.

Herliky, D.J. (1995). Waterproof Breathable Fabric for Outdoor Athletic Apparel. Google Patents.

Holmé, R, I. (2006). Protective Clothing in Hot Environments. *Industrial Health* 44(3), 404–413.

Holmes, D.A. (2000a). Performance Characteristics of Waterproof Breathable Fabrics. *Journal of Coated Fabrics* 29(4), 306–316.

Holmes, D.A. (2000b). Waterproof Breathable Fabrics. In A.R. Horrocks and S.C. Annan (Eds.), *Handbook of Technical Textiles*. Cambridge: Woodhead, pp. 282–315.

Hu, J. (2008). *Fabric Testing*. Oxford: Elsevier Science.

Hu, J.L., Zeng, Y.M., and Yan, H.J. (2003). Influence of Processing Conditions on the Microstructure and Properties of Shape Memory Polyurethane Membranes. *Textile Research Journal* 73(2), 172–178.

Jacquel, N., Freyermouth, F., Fenouillot, F., ... Saint-Loup, R. (2011). Synthesis and Properties of Poly(Butylene Succinate): Efficiency of Different Transesterification Catalysts. *Journal of Polymer Science, Part A: Polymer Chemistry* 49(24), 5301–5312.

Jassal, M., Khungar, A., Bajaj, P., and Sinha, T.J.M. (2004). Waterproof Breathable Polymeric Coatings Based on Polyurethanes. *Journal of Industrial Textiles* 33(4), 269–280.

Jiang, G., Luo, L., Tan, L., ... Jin, J. (2018). Microsphere-Fiber Interpenetrated Superhydrophobic PVDF Microporous Membranes with Improved Waterproof and Breathable Performance. *ACS Applied Materials and Interfaces* 10(33), 28210–28218.

Johnson, L. and Samms, J. (1997). Thermoplastic Polyurethane Technologies for the Textile Industry. *Journal of Coated Fabrics* 27, 48.

Kanebo Ltd. (1994). Isofix Super. 407.

Kannekens, A. (1994). Breathable Coatings and Laminates. *Journal of Coated Fabrics* 24(1), 51–59.

Khaokong, C., Suwanmanee, J., Pilard, J.F., Pasetto, P., and Tanrattanakul, V. (2013). Synthesis and Properties of New Biodegradable Polyurethane Containing Natural Rubber and Poly(lactic Acid): Effect of NR and PLA Ratio. *Materials Science and Nanotechnology I* 531–532(December), 317–320.

Kramar, L. (1998). Recent and Future Trends for High Performance Fabrics Providing Breathability and Waterproofness. *Journal of Coated Fabrics* 28(2), 106–115.

Krishnan, S. (1991). Technology of Breathable Coatings. *Journal of Coated Fabrics* 21(1), 71–74.

Kubin, I. (2001). Functional and Fashion Coating for Apparel. *Melliand International* 7, 134–138.

Kyosev, Y. (2016). *Advances in Braiding Technology: Specialized Techniques and Applications*. Oxford: Elsevier Science.

Langley, J.D., Hinkle, B.S., Carroll, T.R. and Vencill, C.T. (2004). Breathable Blood and Viral Barrier Fabric. Google Patent.

Leitch, P. and Tassinari, T.H. (2000). Interactive Textiles: New Materials in the New Millennium. Part 1. *Journal of Coated Fabrics* 29(3), 173–190.

Li, W., Ma, Y., Tang, X., Jiang, N. Zhang, R., Han, N., and Zhang, X. (2014). Composition and Characterization of Thermoregulated Fiber Containing Acrylic-Based Copolymer Microencapsulated Phase-Change Materials (MicroPCMs), , *Industrial and Engineering Chemistry Research* 53(13), 5413–5420.

Li, Y., Zhu, Z., Yu, J., and Ding, B. (2015). Carbon Nanotubes Enhanced Fluorinated Polyurethane Macroporous Membranes for Waterproof and Breathable Application. *ACS Applied Materials and Interfaces* 7(24), 13538–13546.

Lin, W.T. and Lee, W.J. (2017). Effects of the NCO/OH Molar Ratio and the Silica Contained on the Properties of Waterborne Polyurethane Resins. *Colloids and Surfaces A: Physicochemical and Engineering Aspects* 522, 453–460.

Liu, N., Zhao, Y., Kang, M., ... Li, Q. (2015). The Effects of the Molecular Weight and Structure of Polycarbonatediols on the Properties of Waterborne Polyurethanes. *Progress in Organic Coatings* 82, 46–56.

Liu, Q., Huang, J., Zhang, J., ... Guo, C.F. (2018). Thermal, Waterproof, Breathable, and Antibacterial Cloth with a Nanoporous Structure. *ACS Applied Materials and Interfaces* 10(2), 2026–2032.

Lomax, G.R. (1985). The Design of Waterproof, Water Vapour-Permeable Fabrics. *Journal of Coated Fabrics* 15(1), 40–66.

Lomax, G.R. (1990). Hydrophilic Polyurethane Coatings. *Journal of Coated Fabrics* 20(10), 88–107.

Lomax, G.R. (2001a). Paper presented at 9th International Techtextil-Symposium for Technical Textiles, Nonwovens and Textile-Reinforced Materials, Frankfurt.

Lomax, G.R. (2001b). "Intelligent" Polyurethane for Interactive Clothing. Textile Asia, pp. 39–50.

Mannari, V. (2011). Bio-Based Polyurethane Dispersion Compositions and Methods. US Patent Office.

Materne, T., de Buyl, F., and Witucki, G.L. (2012). Organosilane Technology in Coating Applications: Review and Perspectives. *Dow Corning*, pp. 1–16.

Meng, G.K. (2017). Method for Synthesizing Biobased 1,5-pentadiisocyanate. China Patent Office.

Mooney, C.L. and Schwartz, P. (1985). Effect of Salt Spray on the Rate of Water Vapour Transmission n Microporous Fabric. *Textile Research Journal* 55(8), 449–452.

Painter, C.J. (1996). Waterproof, Breathable Fabric Laminates: A Perspective from Film to Market Place. *Journal of Coated Fabrics* 26(2), 107–130.

Petrović, Z.S., Ilavský, M., Dušek, K., Vidaković, M., Javni, I., and Banjanin, B. (1991). The Effect of Crosslinking on Properties of Polyurethane Elastomers. *Journal of Applied Polymer Science* 42(2), 391–398.

Porter, M.C. and Porter, J. (2009). *Handbook of Industrial Membrane Technology*. Birmingham, AL: Crest Publishing House.

Pritchard, M., Sarsby, R.W., and Anand, S.C. (2000). Textiles in Civil Engineering. In A.R. Horrocks and S.C. Annand (Eds.), *Handbook of Technical Textiles*. Cambridge: Woodhead, pp. 372–406.

Ren, Y.J. and Ruckman, J E. (2004). Condensation in Three-Layer Waterproof Breathable Fabrics for Clothing. *International Journal of Clothing Science and Technology* 16(3), 335–347.

Ruckman, J.E. (1997). Water Vapour Transfer in Waterproof Breathable Fabrics: I Under Steady State Conditions. *International Journal of Clothing Science and Technology* 9(1), 10–22.

Salz, P. (1988). Performance of Protective Clothing: Second Symposium. In *Testing the Quality of Breathable Textiles*. Pennsylvania: ASTM International, pp. 296–304.

Sardon, H. (2010). Waterborne Polyurethane Dispersions Obtained by the Acetone Process: A Study of Colloidal Features. *Journal of Applied Polymer Science* 120(4), 2054–2062.

Sardon, H., Irusta, L., Fernández-Berridi, M. J., Lansalot, M., and Bourgeat-Lami, E. (2010). Synthesis of Room Temperature Self-Curable Waterborne Hybrid Polyurethanes Functionalized With (3-Aminopropyl)Triethoxysilane (APTES). *Polymer* 51(22), 5051–5057.

Scott, R.A. (1995). Coated and Laminated Fabrics. In *Chemistry of the Textiles Industry*. Berlin: Springer, pp. 210–248.

Sen, A.K. (2001). Principles and Applications. In J, Damewood (Ed.), *Coated Textiles* New York: Technomic Publishing Co., pp. 133–154.

Shekar, R.I., Yadav, A.K., Kumar, K., and Tripathi, V.S. (2003). Breathable Apparel Fabrics for Defence Applications. *Man-Made Textiles in India* 46(12), 9–16.

Sheng, J., Xu, Y., Yu, J., and Ding, B. (2017). Robust Fluorine-Free Superhydrophobic Amino-Silicone Oil/SiO_2 Modification of Electrospun Polyacrylonitrile Membranes for Waterproof-Breathable Application. *ACS Applied Materials and Interfaces* 9(17), 15139–15147.

Sheng, J., Zhang, M., Xu, Y., Yu, J., and Ding, B. (2016). Tailoring Water-Resistant and Breathable Performance of Polyacrylonitrile Nanofibrous Membranes Modified by Polydimethylsiloxane. *ACS Applied Materials and Interfaces* 8(40), 27218–27226.

Showers pass. (n.d.). Waterproof Breathable Technology. Available at: www.showerspass.com/pages/waterproof-breathable-technology (accessed May 31, 2018).

Tang, D., Thiyagarajan, S., Noordover, B.A.J., Koning, C.E., van Es, D.S., and van Haveren, J. (2013). Fully Renewable Thermoplastic Poly(ester Urethane Urea)s from Bio-based Diisocyanates. *Journal of Renewable Materials* 1(3), 222–229.

Toray Inc. (1993). High Performance Waterproof/Breathable Fabric '"Entrant G II."' *Japan Textile News* 459(39).

Unkelhäußer, T., Hallack, M., Gorlitzer, H., and Lomölder, R. (2014). The Best of Two Worlds. *European Coatings Journal* 78, 21–25.

Wills, K. (n.d.). Smart Textiles: The Future of Remote Healthcare: National Physical Laboratory. Available at: www.npl.co.uk/news/smart-textiles-the-future-of-remote-healthcare (accessed May 31, 2018).

Yeh, J.M., Yao, C.T., Hsieh, C.F., Yang, H.C., and Wu, C.P. (2008). Preparation and Properties of Amino-Terminated Anionic Waterborne-Polyurethane-Silica Hybrid Materials Through a Sol-Gel Process in tthe Absence of an External Catalyst. *European Polymer Journal* 44(9), 2777–2783.

Yiwei Du, K. L. and Zhang, J. (2009). Application of the Water and Oil Repellent Finishing Agent in Polyester Nonwovens. *Asian Social Science* 5(7), 168–173.

Zhao, H., Hao, T.H., Hu, G.H., Shi D., Huang, D., Jiang, T., and Zhang, Q.C. (2017). Preparation and Characterization of Polyurethanes with Cross-Linked Siloxane in the Side Chain by Sol-Gel Reactions. *Materials* 10(3). doi:10.3390/ma10030247.

Zhao, J., Li, Y., Sheng, J., Wang, X., Liu, L., Yu, J., and Ding, B. (2017). Environmentally Friendly and Breathable Fluorinated Polyurethane Fibrous Membranes Exhibiting Robust Waterproof Performance. *ACS Applied Materials and Interfaces* 9(34), 29302–29310.

9

Polymeric Materials as a Holographic Recording Medium

Asit Baran Samui
Institute of Chemical Technology, Mumbai, India

Alips Srivastava
Naval Materials Research Laboratory, Ambernath, India

CONTENTS

DOI: 10.1201/9781003037880-9

Abbreviations

BER	bit error rate
LC	liquid crystal
LCD	liquid crystal display
MW	molecular weight
HDS	holographic data storage
LED	light emitting diode
CTE	coefficient of thermal expansion
PVA	polyvinyl alcohol
TEA	triethanol amine
LCP	liquid crystalline polymer
I	isotropic
N	nematic
MWCNT	multi-walled carbon nanotube
PVK	polyvinyl carbazole
SN	substituted naphthalenediimide
S	singlet state
QD	quantum dot
QDDLC	quantum dot doped liquid crystal
PDLC	polymer dispersed liquid crystals
PSLC	polymer stabilized liquid crystals
PCBM	fullerene phenyl-C61 butyric acid methyl ester
WORM	write once read many times
AA	acrylamide
DA	diacetone acrylamide
PS	polystyrene
BEA	zeolite beta

Symbols with Units

Symbol	Description	Unit
n	Average refractive index of the recording medium	Dimensionless
λ	Recording wavelength	nm, nanometer
θ	Half of the angle between the two recording beams	Degree
Λ	Fringe spacing	cm
S	Sensitivity	cm/J
E	Total exposure energy	Joule
Δn	Value of refractive index modulation	Dimensionless
S_η	Sensitivity	Dimensionless
I	Total intensity of	W/m^2
L, d	Medium thickness	μm
t	Exposure time	sec
η	Diffraction efficiency	%
$M\#$	Dynamic range	Per cm
θ_1	Initial slant angles	Degree
θ_0	Final slant angles	Degree
L	Optical path length	μm

9.1 Introduction

Holography is a very common term these days, and we usually know this terminology in the context of certain companies affixing their holograms on their products to distinguish the original products (with a hologram affixed on them) from duplicate/counterfeit/spurious products, and also to promote their brands. Apart from security and aesthetic purposes, holography has applications in other fields as well, such as optical elements, as a data storage medium, to provide crucial heat-transfer data that aids in the safe design of containers for the transport/storage of nuclear materials, in bar-code readout systems, in telephone credit cards, in TVs, in recording of valuables or fragile museum objects, in surgical planning, in forensic science investigations, in advertising, portraitures, for vibration analysis,[1] etc.

Holography is essentially a method of recording a three-dimensional image (which exhibits parallax) of an object by recording the optical interference pattern formed by the interaction of the coherent optical object and the reference beams.[2,3] The vast applicability area for holography is attributed to its unique and inherent feature of recording and reconstructing light waves. However, holography is not restricted to just light waves; rather, a hologram can be made by using other electromagnetic waves, even by using sound waves. Depending on the kind of radiation that is being employed to construct a hologram, we can have X-ray holography, UV holography, microwave holography, acoustic holography, so on. X-ray holography or UV holography can be used to record images of atoms/molecules (particles of a size less than the wavelength range for visible radiation). Microwave holography can be used to gather information about space, by studying the radio waves emitted by objects in space. Likewise, acoustic holography can be used to study the internal features of solid objects. A remarkable feature of holograms is that each portion of a hologram can reproduce an entire image of the object.[2]

9.2 Background

A British-Hungarian scientist named Dennis Gabor developed the theory of holography in 1948. This work, however, was basically an experiment in serendipity, as described by Dennis Gabor himself. The term holography was coined by Gabor, and is of Greek origin. Holography is a combination of two Greek words, "holos" meaning "whole" and "graphe" meaning "writing/record." A hologram is, thus, a record of the entire information about an object. Denis Gabor was awarded a Nobel Prize for this work in 1971. In the early years of the development of this technology, laser technology was not known and a mercury arc lamp was used as the light source for recorded holograms. However, distortions and peripheral twin images are inherent in these holograms due to the low coherence of the recording light. All this hampered any significant development in this field until 1960 when the first laser was developed by Theodore Maiman. Since then, this field has witnessed an exponential increase both technology-wise and application-wise.[3,4]

Holography, like photography, is an image-recording process. However, there are certain fundamental differences between holography and photography. In photography, each point on the photograph contains information about only the intensity (i.e., the square of the amplitude) of light that illuminates that particular point. However, a light wave has a phase component as well, apart from the amplitude and wavelength components, and information about the phase component is completely absent in a photograph. In a hologram, both the intensity and phase information about the illuminating light are stored by ensuring coherent illumination and the introduction of a reference beam. The interference pattern created by this reference beam and the object beam is recorded on photographic film, and thus the three-dimensional information of the illuminated object is stored in a hologram.[3]

There are numerous photosensitive materials known today that can record amplitude variations in light used to view an object; however, none of these materials can record the corresponding phase variations in light. To record phase variations, the concept of a reference beam was introduced by Gabor in 1948. Interference of the reference and the object beams converts information about the phase to amplitude variation that is recorded on photosensitive materials. These photosensitive materials undergo physical and/or chemical changes because of the recording of an interference pattern on them. Such changes are manifested as a change in the absorption, refractive index or thickness of the storage media, and thus, the illuminating interference pattern is copied. If this recording is now illuminated with the original reference beam, it reconstructs the object beam because of its diffraction by the stored interference pattern.[5] Since holography requires interference between reference and object beams, the two beams must be coherent, and this condition necessitates the requirement of a laser to record a hologram (Figure 9.1).

The laser light is thus split into two beams to generate two coherent beams, one of which forms the reference beam and the other carries the information of the object and is called the object beam. The photosensitive material is placed in the region where the two beams meet to generate an interference pattern.[6]

If the hologram material is thin, the readout beam can be used that has a wavelength different from the reference beam. Similarly, the readout beam angle for a hologram during reconstruction can also be different from the reference beam angle used during the hologram recording. However, for thicker recording media, the reconstructed object will only appear if the readout and

the reference beams are identical in angle and wavelength. This feature is made use of to store multiple holograms throughout the volume of a thick hologram recording medium. Each hologram can be selectively accessed by using the recording reference beam. However, this technique has a disadvantage in that there is variation of diffraction efficiency of the nth recorded hologram inversely with n^2.

9.3 Types of Holograms

Holograms are, basically, of two types: transmission holograms and reflection holograms. When reference and object beams approach the recording medium from the same side, the hologram obtained is called a transmission hologram, whereas if these two beams approach the recording medium from opposite sides, then a reflection hologram is formed. In case of transmission holograms, the fringes formed in the medium, as a result of interference between the two beams, are perpendicular to the surface of the medium (Figure 9.2). Further, during reconstruction, these fringes refract transmitted light to reconstruct the image. As against this, in reflection holograms, the interference fringes are constructed parallel to the plane of the recording medium, and these fringes refract incident light during reconstruction to form the image.[3]

A collimated laser beam (beam dia: 1 cm) can be split in two in the ratio of 1:1 (w.r.t. beam power density) and these two beams are allowed to interfere at the recording plate (photopolymer film) at a specific angle and produce transmission grating. Interference of the two beams produces a grating in the film. The grating period/fringe spacing is given by Λ:

$$\Lambda = \lambda / 2n \sin\theta \qquad (9.1)$$

where n is the average refractive index of the recording medium, λ is the recording wavelength, and θ is half of the angle between the two recording beams.[6, 7] Resolution of the recorded hologram is defined by the fringe spacing. Grating resolution is calculated from the fringe spacing as 1000/fringe spacing, and is stated as lines per mm. As evident from the above formula, the recording angle influences the fringe spacing, and thus this also influences the grating resolution.

After formation of a grating/hologram in the photopolymer film, the grating is illuminated with a readout laser beam. The beam undergoes diffraction after passing through the grating. The ratio $I_d/(I_0+I_d)$ gives the diffraction efficiency of the grating (expressed

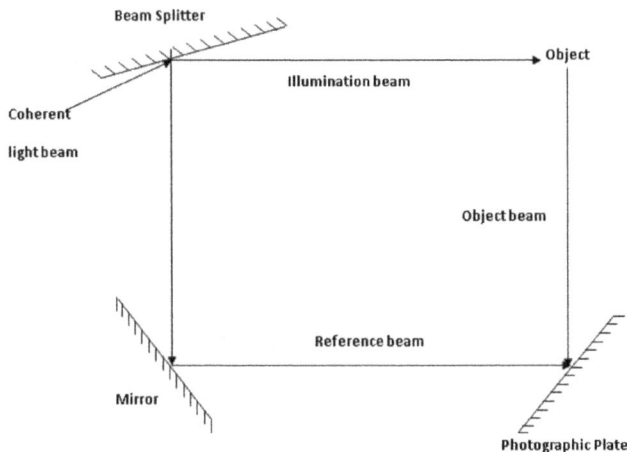

FIGURE 9.1 Schematics of holographic recording set-up.

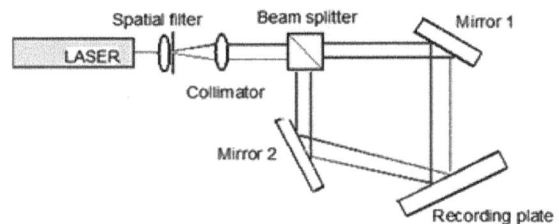

FIGURE 9.2 Schematics of transmission grating recording set-up.

in %), where I_d is the first-order diffracted beam intensity and I_0 is the intensity of the transmitted beam. Thus, the losses, such as linear absorption and light scattering are accounted for in the calculation.[8–10] Some studies report the calculation of diffraction efficiency as $I_{diffracted}/I_{incident}$, where I indicates light intensity.[11] The grating resolution is measured from the recording angle.[6,12]

There is another way of classifying holograms as phase holograms and amplitude holograms. In phase holograms, the refractive index is modulated (or, sometimes thickness) in the recording medium, whereas in amplitude holograms, a variation in absorption is recorded. Amplitude holograms have, in general, lower efficiency than phase holograms. There are also full-view holograms and rainbow holograms.[3]

9.4 Holographic Recording Media

The sensitivity and dynamic range are used to characterize a storage medium in terms of its holographic recording properties. The sensitivity can be expressed as:

$$\text{Sensitivity, S (cm}^2\text{/J)} = \Delta n/E \qquad (9.2)$$

where E is the total exposure energy and Δn is the value of refractive index modulation.[13] Diffraction efficiency is directly proportional to the square of the index modulation times the thickness. Thus, recording sensitivity is commonly expressed in terms of the square root of diffraction efficiency, η:

$$S_\eta = (\eta^{1/2}) / (ILt) \qquad (9.3)$$

where I is the total intensity, L is the medium thickness, and t is the exposure time; this form of sensitivity is usually given in units of cm/J. Because of variation of thickness in the recording materials, a more useful comparison can be made by using modified sensitivity defined as follows:

$$S'_\eta = S_\eta \times 1 \qquad (9.4)$$

This quantity has units of cm²/J. It appears as the inverse of the exposure time required to produce a standard signal level. S_η, can be used to assess the recording ability of a storage material, with the condition that the sample is extremely thin. In contrast, S'_η signifies the ability of a specific sample to respond to a recording exposure in quantitative manner.

The term dynamic range refers to the total response of the medium in which holograms are multiplexed in a common volume of material. It is usually denoted as M#, where

$$M\# = \Sigma \, \eta^{1/2} \qquad (9.5)$$

and the sum is over all the M holograms in one location. M# also describes the scaling of diffraction efficiency as M is increased, i.e.,

$$\eta = (M\# / M)^2 \qquad (9.6)$$

Dynamic range is very important as it has great influence on the data storage density that can be achieved.[5] The basic criteria for selection of a material for holographic recording are: optical quality, diffraction efficiency, dynamic range, recording properties and stability, since these properties influence the maximum achievable data density and capacity, the data input and output rates, the BER (bit error rate), and recording-erasure times.[14,15,5]

To describe these criteria more vividly, an ideal material should satisfy the following requirements:[16]

1. High resolution and a flat spatial frequency response are the prerequisites of the material that allows storage of the complete interference pattern without any loss of any fine fringe detail.

2. The exposure and the amplitude of the reconstructed wave must have a linear relationship to ensure conformity of the image during replay.

3. The material must have a large dynamic range to enable it to have sufficient modulation during recording so that a good signal-to-noise ratio is maintained.

4. High optical efficiencies (bright images) can be obtained by using lossless material of high optical quality.

5. The material should not be susceptible to environmental condition variations and the hologram should have stability for a long duration.

6. The material used as a recording medium should have sufficient sensitivity so that recording can be done with low energy exposure.

But there is no single material that meets all these criteria. So, keeping these prerequisites in mind, several materials have been used as recording media,[3,17,18] the most important of them being silver halide emulsion,[18–20] dichromated gelatin,[19,21–23] photorefractive polymers,[5,24–26] thermoplastic polymers,[27,28] photoresists,[2,3,5] photochromic polymers,[2,5] photoactive LC polymers, polyelectrolytes, and photopolymers.[3,5] But all these materials have their share of merits and demerits.

9.4.1 Silver Halide Emulsion

Holographic plates based on silver halide emulsions consist of plates coated with grains of silver halide crystals suspended in an emulsion in gelatin. The silver halides normally used are chloride and bromide salts.[29] The photochemistry of this holographic medium is depicted in Figure 9.3.

These materials are inexpensive, have good resolution, wide spectral sensitivity, and high photosensitivity, and most importantly, they are suitable for both reflection and transmission mode hologram recording, and are also suitable for both amplitude and phase holograms. The limitations with this material include a requirement for wet processing, resulting in shrinkage or swelling of the material and hence distortion of the hologram, and random scatter due to the particulate nature of the dispersed phase, that reduces the dynamic range and produces noise gratings.[2]

Among the three types of silver-halides, silver chloride makes low sensitivity emulsions, the chloride/bromide combination makes light sensitivity emulsion, while the bromide/

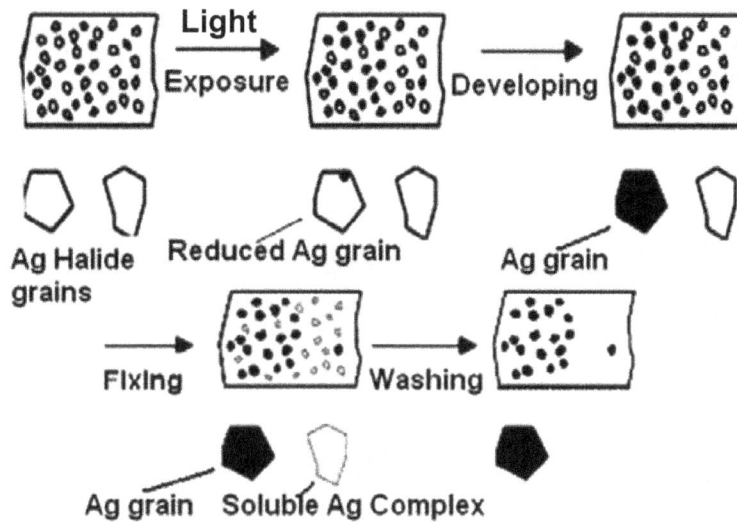

FIGURE 9.3 Holographic photochemistry for silver halide emulsion.

Source: [2].

iodide emulsion produces an even higher sensitivity formulation. Conventionally, silver iodide is used in a mixture with silver bromide, which normally is added around 5% or less, that exhibits a higher sensitivity and contrast than pure silver bromide emulsions having a similar grain size.[29]

9.4.2 Dichromated Gelatin

High diffraction efficiency, good resolution, and high optical quality are the characteristics of the dichromated gelatin medium that lead to a high intensity signal as compared to noise. The difference in swelling characteristics of the exposed and unexposed gelatin has been used as the criterion for recording. In the exposure condition, the exposed gelatin is cross-linked via the photo-induced decomposition products of the dichromate sensitizer (Figure 9.4). The main drawbacks of this material are: poor shelf life, the requirement of wet processing, it is environmentally unstable and is relatively less photosensitive than silver halide emulsion.[2]

9.4.3 Photorefractive Polymers

To be photorefractive, the polymer must have charge-generating, charge-transporting and electro-optic properties. When a spatially varying light pattern is beamed onto the photorefractive polymer, mobile charge carriers are generated in the "illuminated" regions. Under the influence of an external field or due to the concentration gradient, the mobile charge carrier generated moves to the nearby dark area,[30,31] and gets trapped there. Thus, the charge redistribution, so created, constitutes the space charge field in the material. This space-charge pattern produces a spatially varying electric field that modulates the refractive index through the electro-optic effect, which creates the equivalent of a phase grating.[32] In a photorefractive polymer, the mobility of one charge is much greater than its counter-charge, and in most of the organic materials, it's the hole that is more labile than

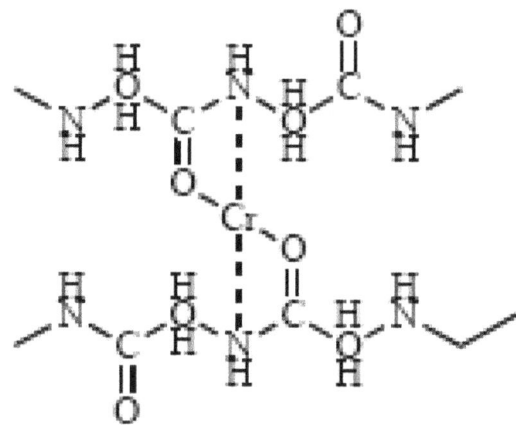

FIGURE 9.4 Cross-linked structure of dichromated gelatin.

Source: [2]. **Reprinted with permission from Bentham Science.**

the electron. Thus, the electron is considered localized near the generation site while the hole migrates and gets trapped in the darker regions.[31] To erase the space charge pattern, so created, uniform illumination of the polymer is conducted, and after that a new recording can be done.[3] Photorefractive polymers exhibit diffraction efficiencies of 100% and a refractive modulation as high as 7×10^{-3}. However, poling under a high external electric field is required to align the chromophores within the film. This has limited the use of the materials.[33]

9.4.4 Photochromic Polymers

Photochromic polymers, like photorefractive polymers, exhibit high diffraction efficiency. However, mostly non-reversible holograms result in the case of photochromic organic materials. Also, unlike photorefractives, they don't require any inherent order.[30,33] In photochromic polymers, the photo-oxidation mechanism is the predominant mechanism responsible for the refractive index modulation via a change of absorption that leads to hologram formation.[33]

9.4.5 Thermoplastic Polymers

A thermoplastic polymer is a multilayer structure comprising a thin, transparent, conducting layer of indium oxide, a photoconductor, and a thermoplastic on a glass or Mylar substrate. Initial sensitization of the film is done in the dark by applying a uniform electric charge to the top surface. On exposure and recharging, a spatially varying electrostatic field is created. The thermoplastic is then heated for a short period to make it soft enough to be deformed by this field, and then cooled to fix the variations in thickness.[3] These recording media alternatives offer advantages of high photosensitivity, yield thin phase holograms with good diffraction efficiency, and are erasable (by heating the substrate) and thus rewritable. They also have the advantage of fast processing in situ, but the main drawback of this material is that it is difficult to synthesize a photosensitive layer in a thermoplastic material.

9.4.6 Photopolymers

Photopolymers offer several significant advantages over the other holographic recording media described above and have, therefore, become the material of choice for recording holograms. Inherent advantages of photopolymer systems are: high refractive index contrast and, therefore, very high diffraction efficiency, real-time dry processing, wide spectral sensitivity, good photospeed, long shelf-life, wide process latitude, large dynamic range, good optical quality, i.e., high resolution, good photosensitivity, easy tailorability, and potentially low production cost.[13,34]

9.5 Photopolymers for Holography

Photopolymers first came to be used for holography at the end of the 1960s, following which extensive research on the development of these photopolymers became one of the branches of holography making the fastest progress.[35] Photopolymers have been explored from their early days for their applicability as a holographic recording medium, LCD displays, helmet-mounted displays, holographic diffusers, security holograms, and many more.[36]

9.5.1 The Hologram Recording Mechanism

An interference pattern generated by incident laser beams on articles creates spatial modulations of refractive index, which are stored in the photopolymer. The photoreactions of the material produce a different refractive index of irradiated areas than that of dark areas. A higher difference of refractive index between the bright and dark regions permits the material to have a higher data storage capacity.[36]

Photopolymer is made by combining a photo-inactive polymeric binder, monomer/s having polymerizable ability, and a photo-initiator system. The spatial variations in incident light intensity via irreversible changes in the refractive index are recorded in the photopolymer recording medium. The processes, viz. monomer polymerization and monomer diffusion, occur to form the hologram. The polymerization proceeds via a free radical mechanism. In the first step, the initiation occurs via generation of free radicals that activate the monomer to start a chain. After initiation, the propagation p starts by adding other monomers to make long polymer radicals. Finally, one polymer radical may react with another polymer radical or free radical to terminate the chain, creating dead polymers.

There is the occurrence of "spreading" of photopolymer during polymerization, which is due to a non-local response originating from the growth of the chain away from the initiation location. Thus, during illumination, the hologram formation is considered to be the result of the interplay between the processes of polymerization and the diffusion of the monomer (from the non-exposed to the exposed zones).[37,12] The monomer diffusion process is influenced both by the mobility of the monomer molecules and by the chain flexibility of the photopolymer (a mixture of the polymer matrix and the photopolymerizable components).[11]

In the "bright" regions of the incident interference pattern, a change in density as well as a change in the structure of the medium occurs due to photopolymerization that results in a change in the refractive index in that region. As a result of polymerization, the monomer concentration gradient is created between the "bright" and the "dark" zones, which leads to migration of the monomer molecules from the "dark" to the "bright" zones. This also helps in enhancement in the refractive index contrast (Figure 9.5). The phase information of the incident light is thus encoded by the refractive index modulation between the "bright" and the "dark" regions.

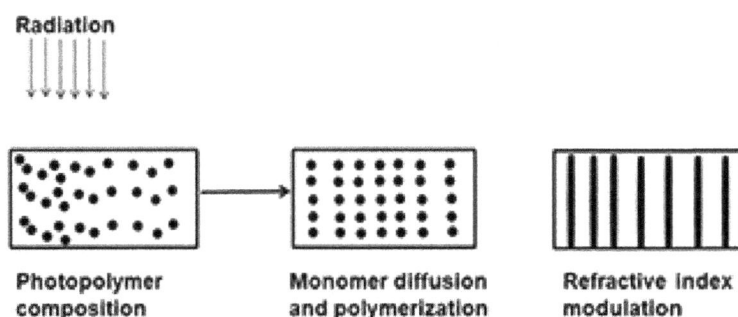

FIGURE 9.5 Holographic recording mechanism in a photopolymer.

The polymerization reaction continues until it is terminated either by a combination of two propagating polymer radicals or by a combination of a propagating radical with the initiator radical. The concentration of radicals (initiator), the viscosity of the medium, and the percentage conversion of the monomer govern the termination mechanism. After recording, the hologram can be permanently fixed by either UV cure or by thermal treatment. During this step, polymerization of the remaining monomer occurs, and thus the refractive index modulation is further amplified. Dye bleaching also occurs during this step.

A note of caution is that the presence of oxygen should be taken care of in case the polymerization proceeds by the free radical mechanism, since oxygen is known to be a retarder of free-radical polymerization, by acting as a quencher of the initiator radical, and also as a scavenger on initiating and growing polymer radicals as it forms low reactive peroxide radicals. This results in an inhibition period during which there is no polymerization. This leads to longer exposure times for hologram recording, and also lower index contrast, and hence lower diffraction efficiency due to shorter polymer chains being formed.[38,39] The retarding effect of oxygen is very strongly observed in photopolymer systems containing acrylate. To reduce the effect of oxygen in these systems, holographic recording is best carried out by irradiating with green light with thiol incorporation in the photopolymer film.

9.5.2 Factors Affecting the Storage Capacity of a Photopolymer

The refractive index modulation magnitude enhances the storage capacity of the photopolymer. To completely utilize the refractive index modulation of the recording material, it is imperative that the material should be able to store the maximum possible holograms, i.e., multiplexing of holograms in the material should be feasible. Storage capacity also increases with the medium thickness (1–5 mm) because of their very narrow angular responses, thereby accommodating many holograms in a given volume of material without overlapping.[36,38] The diffraction efficiency of the phase grating (modulated index) is thus improved.[36] To achieve a commercially viable (~100 bits/μm^2) holographic data storage capacity, the index contrast should be very high in thick photopolymer materials. In spite of a better performance of the thicker samples than thinner materials, their holograms are affected even with minor dimensional changes in the sample. This adversely influences their recording capacity since the shrinkage resulting from photopolymerization shifts the hologram positions significantly during encoding and also distorts the holograms. Thus, working with thick photopolymers has several disadvantages as well, including material shrinkage, hologram destabilization, drying-induced material deformations, and so on.

As stated above, high refractive index modulation leads to enhanced storage capacity. However, purely photopolymer-based holographic material cannot produce large index modulation. To develop a large index contrast in such materials, different high and low index monomers are required to be polymerized. However, the resulting copolymer possesses an averaged index along with diminished index contrast. The resonance stabilization and polarization factors for vinyl radicals, along with calculations by the Alfrey-Price equations, confirm the occurrence of copolymerization, indicating limited storage capacity in such photopolymer materials. Anyway, thick films with high optical clarity can easily be formed with photopolymeric storage materials. If the low and high index components of a material are not allowed to form simultaneously, following the same chemistry, a much better index contrast can be achieved. This can be achieved by mixing the photopolymerizable monomers with an inert, preformed polymer and modulating the index between these two components. Commercially available photopolymer-based holographic materials are formed by following this technique.[36] The preformed polymers in these materials do not serve to produce refractive index contrast. They are basically used as a binder for their physical characteristics that help the unexposed materials to be handled and stored as dry films prior to hologram formation. It is interesting to know that the binder can also play a role in hologram formation, which remained an unknown fact until a report by Smothers et al. in 1990.[40] Since then, investigation on the preformed polymers began, to maximize the index contrast. This strategy, however, is not ideal for practical purposes. For a solid polymer binder, complicated processing is required. Generally, the solution of polymer and monomers is evaporated to deposit the film. This process suffers from practical difficulties and only good film with a low thickness of around 100 μm can be achieved that limits the storage capacity. On the other hand, solvent-less processing is much simpler by which a fluid polymer is used as a binder and photopolymerizable monomers are simply mixed. However, the initial index modulation recorded is erased by mass transport in the fluid. The transport ceases after sufficient setting-up of the material occurs. For the purpose of thickening the mixture, photo exposure can be applied before hologram recording. This has the inherent disadvantage in terms of reduced index contrast as the initially recorded portion becomes a part of the binder that reduces the overall contrast between the binder and the unreacted photopolymer. These problems can be tackled by introducing a room temperature curing reaction to form a solid binder (or matrix). Thick films, to the tune of several millimeters, can easily be prepared by applying this method, for example, polyamine-hardened epoxy resins containing vinyl monomers constitute thick films. The ratio of amine to epoxy can be varied to control the cross-link density of the matrix and thereby adjust the mechanical properties so that the holographic performance can be comfortably optimized.

9.6 Components of a Photopolymer System

The general composition of a photopolymer system to be used for hologram recording includes a binder, monomer(s), and a photoinitiation system comprising of an initiator and a photosensitizer. Other components, viz., plasticizers, surfactants,[41] chain transfer agents, cross-linkers, etc. can be added to control the parameters such as viscosity, the degree of polymerization, mechanical stability, etc. The choice and design of each component are crucial to the performance of the system as a whole in recording a good hologram. The selection of each component is guided by the intended application of that photopolymer system.

9.6.1 Photo-Initiator

Selection of the initiation system is based upon the recording wavelength for the hologram. The sensitizing dye determines the absorption wavelength of the photopolymer film, while the photo-initiator has very little or no effect on the energy absorption. The photo-initiator's role, a radical generator, is to accept and transfer the energy produced by the dye to the free monomer to initiate polymerization.[41] The sensitizer molecules absorb the energy of irradiation and are raised to the triplet state, that is followed by its de-excitation and transfer of energy released to the initiator molecule. The energized initiator breaks into initiator radicals and then reacts with the monomer to record the grating. Thus, the lower the sensitizer concentration below an optimum value, the lower will be a reduction in diffraction efficiency because of the lower number of monomer radicals produced during the exposure duration. If the monomer radical concentration is kept low, there will be a high degree of polymerization. However, there will be an increase in the time required for complete polymerization to reach a high degree of polymerization. Thus, within the given exposure time with a low concentration of sensitizer, the polymerization reaction cannot be completed. This lower extent of conversion of monomer to polymer produces a low refractive index contrast that leads to low diffraction efficiency. Also, the "dark" reactions are likely to occur that adversely affect the grating quality. The diffraction efficiency also decreases for the sensitizer concentration higher than the optimum value, as it produces a higher concentration of initiator radicals that, in turn, generate a higher concentration of monomer radicals. At a high concentration of monomer and initiator radicals, chain termination competes with the polymerization process, resulting in the formation of short polymer molecules. Two opposing effects take place: an increase in the degree of polymerization leads to a significant increase in the "bright" region compared to the "dark" region that increases the refractive index contrast, while the short polymer molecules formed will lead to lower refractive index contrast. Therefore, sensitizer concentration should be such, during the stipulated exposure time, to enable completion of the polymerization process together with the high degree of polymerization, during the exposure. Thus, a stable grating with a higher diffraction efficiency is possible by choosing the sensitizer concentration at such a level that a sufficient number of initiator/monomer radicals is generated and the high degree of polymerization occurs, and is completed during the pre-decided exposure time.

There is an optimum even for the initiator concentration. As the initiator concentration will increase, more initiator radicals and thus more monomer radicals will be generated. This leads to the disappearance of the monomer molecules quite rapidly from the "bright" regions that create a drag for monomer molecules from "dark" to "bright" regions. Thus, with an increase in initiator concentration, there would be an increase in photospeed and a decrease in recording energy. However, beyond a certain concentration, the number of initiator radicals and thus monomer radicals will become too high so that the chain termination reaction becomes prominent and competes with the polymerization reaction, which may be the reason for reduced diffraction efficiency. The availability of a large number of well-studied photo-initiators makes it possible to use radical polymerization as the preferred choice for writing holograms. Ring-opening polymerization chemistry is also employed for hologram writing. Recently the macro-photo-initiator has been considered for use in holographic writing. This has the advantage of undergoing slower unwanted diffusion into unexposed regions and has less probability of cage recombination reactions that would produce unwanted volatile products.[42]

9.6.2 Monomers and Binders

Monomers and binders influence the physical and optical properties, and hence the final performance, profoundly. Thus, the selection of the monomer and the binder should be made to create the maximum possible refractive index modulation upon illumination of the photopolymer film by the interference pattern.

9.6.2.1 Binders

The binder is basically a support matrix for the other components of the photopolymer system. However, it is this support matrix that dictates the physical and mechanical properties of the photopolymer film, including rigidity, environmental stability, dimensional variation during exposure to light, and obtainable thickness of the film. It also influences the diffusion behavior of the monomer, thereby influencing the photo-induced refractive index modulation. This point can be explained by considering two polymeric binders which differ only in their molecular weights. Incorporation of higher molecular weight polymers maintains a higher matrix viscosity that adversely affects the mobility, which restricts the diffusion of molecules during recording. This, in turn, will require higher exposure energy to achieve the targeted diffraction efficiency. Thus there will be decrease in film speed. A higher binder molecular weight also decreases the refractive index contrast, and hence the diffraction efficiency. Further, the polymerization rate being slower in a viscous medium with a high molecular weight, there will be "dark" reactions long after the recording is over. This has an adverse effect on the grating stability of a photopolymer film. However, in certain photopolymers, using a high molecular weight (MW) binder improves the exposure sensitivity, compared to when a low MW binder is used. This higher diffusion coefficient in a higher MW binder might be due to the increased effect of the higher MW binder on the conformation of the photopolymerized polymer, ultimately leading to polymer phase separation.[43] Thus, proper selection of a binder with regards to its chain flexibility (dictated by the binder structure as well as its viscosity), which affects both the monomer diffusibility and its mobility, should be made.

9.6.2.2 Monomers

Along with the binder, the monomer plays a great role in enhancing the refractive index contrast. A proper monomer-polymeric binder combination, such as an aliphatic binder and an aromatic monomer, or vice versa, can be chosen for its enhanced properties. Other important considerations with regards to the monomer is that it (and the polymer obtained from the polymerization of this monomer) should have good solubility in the binder to impart good optical quality to the photopolymer film and render it usable for hologram recording purposes, and it should also have a high rate of polymerization that would lead to high photospeed with lower recording energy.

The volume fraction of a writing monomer must be properly optimized, as by increasing its concentration, there will be an increase in noise along with the increase of modulation of recorded features.[44] Further, the distortion of recorded features, especially in thick samples, is observed because of the volume shrinkage resulting from recording. The index-contrast benefit achieved per bond converted can be maximized, and the shrinkage and scatter penalties accumulated per bond converted can be minimized to improve the signal-to-noise in the recording. Monomers with a large number of index-contrasting moieties per reactive group can be employed to maximize the index-contrast benefit per bond converted. Highly dendronized macromonomers exhibit reduced volume shrinkage while maintaining a moderate index contrast.[45] To minimize the scatter penalties per bond converted, the development time needs to be longer than the exposure time and that can be achieved by introducing a polymerization retarder. The result is that most of the polymerization occurs in the dark post-exposure, so that this feedback loop is suppressed. Cationic ring-opening writing chemistries can be adopted to minimize the shrinkage penalty due to the volume increase from ring opening.[46] Moreover, the cationic polymerization is not inhibited by oxygen, and a linear response is exhibited even at high exposure intensities.

9.6.3 Chain-Transfer Agent

This is another important component of most of the photopolymer systems used as storage media for holograms. To have a high resolution, it is required that the polymerization reaction should not proceed away from its point of initiation, i.e., a nonlocal material response should be prevented. This can be achieved by keeping the average polymer chain length low (achieved by adding a suitable chain transfer agent) and/or by increasing the monomer mobility (achieved by the addition of a plasticizer to the system). Addition of a chain transfer agent has also an adverse effect, as it is likely to decrease the rate of monomer polymerization. So, there is an optimum concentration even for the chain transfer agent.

9.6.4 Cross-Linker

Sometimes, the modulation in the refractive index is increased by adding a cross-linker to the photopolymer system (due to the formation of an extended chain polymer), to improve the energy sensitivity (due to the increase in polymerization rate), and to impart greater stability to the stored grating.[38] The concentration of the cross-linker in the final composition should neither be too low nor too high, as too low a cross-linker concentration will not be

sufficient to achieve a high and stabilized grating efficiency and too high a cross-linker concentration will increase the cross-link density to a considerable extent that retards the monomer mobility during hologram recording. If the cross-linker has limited solubility in the binder, the recorded films may appear opaque when its concentration is very high. Therefore, an optimum concentration of cross-linker will be ideal to ensure an enhanced refractive index contrast leading to high diffraction efficiency, and also long-term storage stability of the recorded grating. Grating stability results from the formation of a cross-linked network structure as it prevents the movement of migrating species from the "bright" regions to the "dark" zones. It also maintains the maximum utilization of the monomer during the irradiation stage so that the "dark" reactions are either eliminated or minimized. The stability of the recorded grating is ensured by both factors.

At times, the addition of a cross-linker is done even to cross-link the binder so that the storage stability of the photopolymer film is increased and the polymerization-induced volume shrinkage is minimized.[47] In certain cases, where the cross-linker for the binder has preferentially more affinity for the growing polymer chain, its addition further enhances the diffraction efficiency.[43] However, a high chain cross-link density will place a limitation on holographic performance, since a high cross-link density will lead to an increase in matrix rigidity, resulting in hindrance to monomer diffusion during photopolymerization, and thereby an insufficient energy sensitivity.

9.6.5 Plasticizer

Sometimes, a plasticizer is also added to the photopolymer system.[41] The plasticizer serves to provide better compatibility among the component molecules.

9.6.6 Design of Photopolymers

Commercial development of photopolymers is targeted at two different types of applications, such as thick (mm range) film, including holographic data storage (HDS), and the thin film (μm range), including display and security.[48] Because of the exploits of HDS being volumetric rather than surface patterning, thick media layers are required to attain high storage densities.[49] The diffusers for LED backlighting, transparent heads-up displays, concentrators for solar panels, printed display holograms, and security holograms are based on thin films, which require high index modulations, and rigorous spatial resolution. Severe environmental situations, such as abrasion or exposure to ultraviolet in sunlight can be sustained by the holograms.

While considering the active optical properties, the passive/bulk optical properties including optical, transparency, and phase uniformity must be considered, and more so for thick media layers. These passive optical properties are achieved by adopting two general strategies, viz., "single-chemistry" or "two chemistry." In the single-chemistry strategy, a monomer, photo-initiator, and high molecular weight binder are considered. Single photopolymerization chemistry is applied for both recordings and cross-linking the entire media layer. In the two-chemistry approach, the solid host matrix is formed first, and then orthogonal chemistry is applied to initiate photopolymerization that is mostly radical in nature. Thus, this two-chemistry approach has additional design freedom, such as the matrix polymer controlling the passive properties can be planned free from any interference from the writing polymer that controls the recording properties. Superior passive optical properties, particularly during the thermal curing of the matrix via a step-growth process are realizable because of this additional design freedom. Importantly, this approach is commercially preferred. Thick holographic elements are highly sensitive to polymerization-induced volume shrinkage that causes distortion of the holographic fringes and is typically anisotropic, due to the mechanical constraints enforced by the packaging of the media layer.

On the other hand, the thick holographic elements, having high sensitivity to thermal expansion, lead to design compromise. As is well known, a polymeric matrix with high T_g has a low coefficient of thermal expansion (CTE). A low T_g, on the other hand, allows recording via fast diffusion. Also, greater index contrast is fashioned due to a lower matrix refractive index.[50] Elastic modulus and scratch hardness also become important for the cases where the system is likely to be exposed directly to the environment without any protective packaging. Thus, the properties of materials used and the final hologram produced need thorough consideration, as demanded by the application criteria, and a design trade-off is made for that purpose.

As discussed above, materials development for holographic photopolymers falls into two categories. An example of single-chemistry development can be named as the DuPont's Omnidex films, comprising high index acrylate writing monomers and cellulose-based binders that produce good index modulation via thermal post-processing which facilitates monomer diffusion. Polaroid's DMP-128 films comprise acrylate writing monomers and a polyethyleneimine binder. After recording, the application of solvent processing results in the formation of micropores in the exposed regions.[51] High index modulation with good sensitivity can be achieved by filling the pores either with air or a high index contrasting solvent. In the two-chemistry strategy, the binder is replaced by a mm-thick cross-linked host matrix layer while maintaining excellent scatter and shrinkage resistance. A nanoporous glass, infiltrated with a photopolymer resin, can be used in place of the polymeric host matrix. The system maintains excellent bulk mechanical rigidity, and negligible volume shrinkage, together with fast diffusion of resin through the porous network, make it an ideal system.[52] Although the fabrication cost of nanoporous glass remains prohibitive, the hybrid inorganic-organic sol-gel glasses can be used as an alternative.[53] High refractive index modulation, low scattering, and negligible shrinkage, and very high diffraction efficiencies (close to the theoretical maxima) are inherent in photopolymerizable glass.

9.7 Photopolymer Systems

The photopolymer systems, developed for hologram recording, belong to polyacrylates, polymethacrylates, polystyrene, polyesters, polysulphones, polyamide, and azopolymers.

However, only limited photopolymer systems can be successfully used for holographic recording purposes. This may be because a compatible pair of polymers with a large refractive index difference is not common.[11] Of the various photopolymer systems that have been investigated until now for the purpose of hologram recording, the majority are based on polyacrylates, polymethacrylates, polystyrene, polyesters, polysulphones, polyamide, and azopolymers. For example, the commercial DMP-128 of the Polaroid Corporation contains a mixture of monomers, such as methylene bisacrylamide and lithium acrylate and poly-N-vinylpyrrolidone as a binder. The exact components of DuPont Omnidex photopolymer systems are proprietary, but the various binder materials reported to have been researched included cellulose acetate butyrate, polymethyl methacrylate, etc.[16] The Canon system was based on poly(N-vinyl carbazole).[54] Several other combinations of mono-functional and multi-functional acrylates in suitable polymeric binders have also been successfully researched for holographic recording purposes.

9.7.1 Photochemical Processes

In a typical free radical photopolymerization process five main reactions such as initiation, propagation, termination, inhibition, and chain transfer occur.[55] The main chemical reactions associated with each process can be given as:[56]

- initiation;
- propagation;
- termination.

9.7.1.1 Initiation

During laser irradiation, the reaction between the photosensitizer and the co-initiator leads to the production of initiator radicals, R•, which are likely to react with the monomers to initiate a chain:[57]

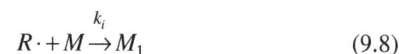

$$I \xrightarrow{h\nu} 2R \qquad (9.7)$$

$$R\cdot + M \xrightarrow{k_i} M_1 \qquad (9.8)$$

where I, hν, k_i and M represent the initiator, the energy absorbed and the rate constant for initiation, and the monomer respectively.

9.7.1.2 Propagation

The chain initiator, M_1•, will react with another monomer and so on via opening of the monomer double bond, resulting in a polymer radical with a live front. Thus, the growth of a polymer radical takes place by multiple propagation steps.[16]

$$M_n^{\cdot} + M \xrightarrow{k_p} M_{n+1}^{\cdot} \qquad (9.9)$$

where k_p, $M\bullet_n$, $M\bullet_{n+1}$ represent the propagation rate constant, growing macroradical chains of n and (n + 1) monomeric units (n≥1).

9.7.1.3 Termination

Termination takes place primarily via two mechanisms. When growing polymer radicals interact with each other and form two polymers, it is called termination via disproportionation. When the same interaction results in only one polymer, it is called termination via combination. The termination modes can be shown as:

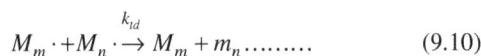

$$M_m\cdot + M_n\cdot \xrightarrow{k_{td}} M_m + m_n\ldots\ldots \quad (9.10)$$

$$M_m\cdot + M_n\cdot \xrightarrow{k_{tc}} M_{m+n} \quad (9.11)$$

where k_{td} and k_{tc} represent termination rate constants via disproportionation and combination, respectively. M_m, M_n and M_{m+n} represent terminated chains, a dead polymer without any radical center. Other possibilities, such as chain transfer, primary radical termination, etc. also exist, depending on the reaction parameters and monomer molecules.

One photopolymer system that is widely investigated has acrylamide as the monomer, probably because of the good compatibility between acrylamide and poly(vinyl alcohol) (PVA) (the binder used in these systems), high diffraction efficiency, high energy sensitivity (arising out of high rate of polymerization of the acrylamide), high signal-to-noise ratio and processing under dry conditions. These photopolymers contain triethanolamine (TEA) as the photo-initiator, which is highly efficient in transferring the product from photoreaction of the sensitizing dyes to the monomers. High diffraction efficiency and high energy sensitivity can be achieved with these photopolymers. Also, this system is a commercially viable system. The hologram writing in these systems involves a set of reactions (Figure 9.6).[58,59]

However, acrylamide suffers from the drawback of low photosensitivity. Also, these photopolymers tend to absorb water from the surrounding atmosphere when the humidity level exceeds 60%, resulting in excessive light scattering which deteriorates the product quality. The addition of a secondary photo-initiator like N-phenylglycine,[41] or the addition of a cross-linking agent, such as glutaraldehyde, for the binder, serves to overcome this shortcoming. Hologram recording on polyvinyl alcohol (PVA)/acrylamide-based photopolymer compositions has been widely studied.[34,6] These holograms have been used even as humidity and temperature sensors.

9.7.2 Liquid Crystalline Polymers

A new class of material, which has recently generated great attention for hologram recording, is called "liquid crystals." Liquid crystals (LC) are considered promising materials for holography because of their ability to undergo change in molecular alignment to produce a large change in the refractive index. The refractive index changes of LCs are dependent mostly on π-electron conjugation, molecular shape, and order parameter. Since the refractive index change directly transforms into diffraction efficiency achievable with that material, the LC molecules enable achievement of enhanced properties with respect to refractive index modulation and diffraction efficiency in holography. The spatial resolution of photorefractive liquid crystals can be enhanced either by replacing the low-molecular-mass liquid crystals with polymeric liquid crystals,[60] or by producing phase-separated morphology in a mixture of low-molecular-mass liquid crystals and a polymer,[61] (polymer-dispersed liquid crystals). The latter can be designed as miscible but not mesogenic (polymer/nematogen composites), or miscible and mesogenic (polymer-dissolved liquid-crystal composites) types. Further, unlike the low molecular weight LCs, the polymer-based systems do not require any glass cells. By using a simple spin coating technique, the polymer solutions can be used for direct coating onto a glass plate, which is not the case with low-molecular-weight LCs.

To enhance the nonlinearity of an LC-based holographic recording medium, dopants are used. Optimization of the charge generation, charge transport, and electro-optic characteristics of the doped polymers can be performed to achieve a higher net photorefractive gain than the inorganic single crystal counterparts. The orientational alignment of the birefringent, nonlinear optical chromophores within the viscous polymer is possible by a space-charge field via reduction of T_g. The total change in the index of refraction is highly influenced by the nonlinear electro-optic effect.[62] In fact, the contribution of the orientational effect to the photorefractive gains in many polymers is higher than that of the traditional linear electro-optic effect.[63]

9.7.2.1 Dye-Doped LC Polymers

Among the various dopants used, the most frequently employed dopants are the dye molecules, particularly azo-dye molecules, which trigger changes in the alignment of LC molecules via their photo-induced isomerization.[64] The reversible nature of the isomerization of azo dyes ensures dynamic control of grating formation.[65] A significant change in molecular alignment can be

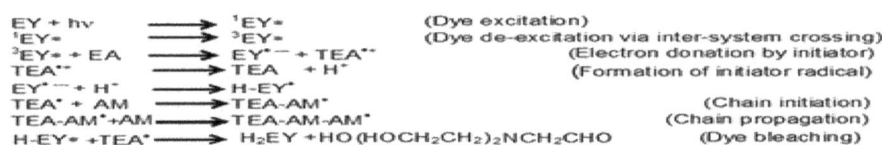

$$
\begin{array}{lll}
EY + h\nu & \longrightarrow {}^1EY* & \text{(Dye excitation)}\\
{}^1EY* & \longrightarrow {}^3EY* & \text{(Dye de-excitation via inter-system crossing)}\\
{}^3EY* + EA & \longrightarrow EY^{*-} + TEA^{*\cdot} & \text{(Electron donation by initiator)}\\
TEA^{*\cdot} & \longrightarrow TEA^* + H^* & \text{(Formation of initiator radical)}\\
EY^{*-} + H^* & \longrightarrow H\text{-}EY^* & \\
TEA^* + AM & \longrightarrow TEA\text{-}AM^* & \text{(Chain initiation)}\\
TEA\text{-}AM^* + AM & \longrightarrow TEA\text{-}AM\text{-}AM^* & \text{(Chain propagation)}\\
H\text{-}EY^* + TEA^* & \longrightarrow H_2EY + HO(HOCH_2CH_2)_2NCH_2CHO & \text{(Dye bleaching)}\\
\end{array}
$$

FIGURE 9.6 Reactions occurring during hologram recording.

produced in an azo dye-doped LC/amorphous polymer film when irradiated with linearly polarized light. The production of excited states of molecules by absorption of light fulfills a few exclusive requirements, such as the coincidence of the electric field vector of light with the transition moment direction of the molecule. As an example, the transition moments of trans-azobenzenes are nearly parallel to the molecular long axis, while the linearly polarized light absorption by cis-azobenzenes is a function of the angle. Thus, the trans-azobenzene molecules can be activated very effectively to reach their excited state if their π–π* transition moments are parallel to the polarization direction of linearly polarized light, which is followed by trans-cis isomerization, whereas the molecules having perpendicular transition moments are not responsive to isomerization. After repetitive trans-cis-trans isomerization cycles, a certain proportion of azobenzene molecules have their alignment perpendicular to the light polarization. This is well known as the "Weigert effect."[66] This creates the basis for creating anisotropic structures in holography.

The azo moiety can also be chemically linked to a polymer chain by placing the moiety in the main chain or in the side chain. The photochemical phase transition of the side chain azobenzene derivative can be used to form a dynamic grating on PLC. Donor-acceptor derivatives of azobenzene photosensitive groups undergo fast cis-trans thermal back-isomerization, while that for non-donor-acceptor derivatives is quite slow.[67] The thermal back isomerization can be called the thermal Isotropic(I) - Nematic (N) phase transformation. The phase structure of the PLC is decided by the trans-cis isomerization of the azobenzene moiety. Thus, the photochemical N-I and thermal I-N phase transition causes a reversible change of refractive index, which can be seen as the formation and erasure of the grating.[68]

9.7.2.2 Polymeric Azobenzene with the Dual Characteristics of LC and Chromophore

Polymeric azobenzene can function as both an LC and a chromophore. The rod-like structure gives it an LC nature and the trans-cis photo-isomerization gives it chromophoreic properties. As it is a polymer, it has a wide temperature range for the nematic LC phase and the recording at room temperature can be stored in the cis-phase, particularly below T_g. The photo-isomerization of azobenzenes occurs mainly through an inversion mechanism.[69] A much smaller sweep volume is required by this mechanism for isomerization so that the isomerization of azobenzene derivatives can take place in relatively rigid matrices, such as polymer matrices below T_g, as the movement of side chains is still possible in this condition. A rapid optical response of 200 microseconds is possible for the nematic-to-isotropic phase transition due to the trans-cis isomerization of azobenzene with a laser pulse.[70]

9.7.2.3 Nanocarbons and Other Dopants

In addition to the already established dye molecules, the dopants such as fullerene C60 and carbon nanotubes,[71] electron-donor and electron-acceptor groups[72] and quantum dots,[73] being photoconductors, are also found to be potential materials for enhancement of the optical nonlinearities of liquid crystals. The

effective underlying mechanism for liquid crystals doped with those photoconductive dopants is the photorefractive effect, as mentioned earlier. The refractive index changes in these doped polymers can be maximized by varying the dopants and applied fields so that high diffraction efficiencies can be achieved for holography.

LC polymers (LCP), such as polymethacrylate with four-cyanophenyl benzoate side groups, together with low-molar-mass LC molecules, doped with a photoconductive sensitizer such as fullerene, exhibit orientational photorefractive Bragg diffraction.[74] The composition, exhibiting no phase separation, is in the nematic phase. Because the long rod-shaped molecules in nematic LCs produce greater bulk birefringence than those with a non-LC polymer, doped with a birefringent chromophore, these are considered very attractive for use in photorefractive materials. Also, for a given space-charge field, the directional order of low viscosity LCs permits greater orientational displacement. Large electric fields are required to pole the photorefractive polymeric materials so that a bulk electro-optic effect is obtained, whereas only a weak electric field is sufficient for LCs to induce directional charge transport and to enhance the quadratic electro-optic effect.[27] Unwanted absorption from the birefringent chromophore is kept low as most of the LCs have very little absorption in the visible as well as in near-infrared regions.

The optical nonlinearity resulting from doping with a multi-walled carbon nanotube (MWCNT) can lead to a dramatic enhancement in the self-diffraction efficiency.[75] The applied DC field influences this phenomenon, and the grating constant influences the memory effect in a nematic film. Liquid crystals, doped with MWCNT and sandwiched between two polyvinyl carbazole (PVK) layers, can be used for recording hologram gratings.[76] The nematic LCs are aligned in the DC electric field if the external voltage applied is higher than the Fréedericksz threshold. The interference pattern from two intersecting coherent beams generates the photocharges that are transported to polymeric films along photocurrent paths across the sample. Then there is movement of charges by diffusion and drift to the zones of the photoconducting films where the construction takes place optically. During this movement, there is the possibility that they will be trapped on the way. In fact, the entrapment occurs at the defect sites, and a space-charge field is established, whose evanescent component penetrates the liquid crystal layer, resulting in a reoriented profile of the nematic rods. A modulated liquid crystal refractive index is thus created.

Large orientational photorefractive effects in nematic LCs can also be realized through the systematic addition of both electron donors and electron acceptors that are easily oxidized and reduced, respectively, to generate stable radical ions. In one typical composition, perylene, having high solubility and extended visible range absorption up to 514 nm, acts as a donor, and substituted naphthalenediimide (SN) acts as an acceptor due to its high solubility and no visible absorption near 514 nm. Thus perylene generates the charge by absorbing light and then transfers it (as electron) to SN.[72] Perylene and SN undergo reversible oxidation and reduction.[73] The redox potentials for perylene and SN are conducive to an efficient charge generation in a cyanobiphenyl-based LC mixture that occurs by the transfer of perylene to its lowest excited singlet state (S), and then transferring electrons from S to SN, resulting in perylene$^+$ and NI$^-$ ions. The directors of the LCs, and the long axes of perylene and SN align along the same direction, which make it possible to increase the concentrations of perylene and SN to large extent, without adversely affecting the LC phase of the LC mixture.

9.7.2.3.1 Quantum Dot Dopant

Quantum dots (QDs) are another class of dopants for nematic LCs used for holographic recording purposes. QDs are semiconductor materials by nature that have electrons or holes confined in all three dimensions. Quantum confinement for the quantum dot is possible due to their smaller size compared to Exciton Bohr Radius.[77] Excitation of the bandgap via irradiation results in a charge separation:[78]

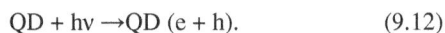

$$QD + h\nu \rightarrow QD (e + h). \qquad (9.12)$$

where e is the electron and h is the hole

When the electrons and holes recombine, a photon $h\nu'$ is emitted:

$$QD (e + h) \rightarrow QD + h\nu' \qquad (9.13)$$

The size of the quantum dot influences the emitted photon $h\nu'$.[79]

By doping an aligned nematic liquid crystal with QD, it is possible to change the refractive index of the liquid crystal by applying either an external electrical field or irradiation with alternative light. The reorientation of the liquid crystal by the local space charge field effects causes changes in the refractive index and is akin to the photorefractive effect. It has to be remembered that the refractive index grating designed in the QDDLC is temporary in nature. When the recording beams are removed, the photo-induced space charge undergoes recombination, leading to the disappearance of the index grating and both processes are very fast.

By applying an electric field of 1 V/μm to QD-doped LCs, a real-time holographic display can be built up within a few milliseconds and with a diffraction efficiency higher than 25%.[80] By using these combinations, it is possible to build holographic videos of RGB colors at a reasonable refresh rate.

9.7.2.3.2 Organic/Inorganic Semiconducting Material Composites

To enhance the diffraction efficiency, higher photorefractive sensitivity is required. Composites of organic/inorganic semiconducting materials produce larger refractive index modulations and faster transport speeds compared to pure QDDLCs. Such composites mostly are based on ZnSe, which is a semiconducting material well known for its large charge carrier mobility, photoconductivity, and high quality.

A composite of the photoconducting agent, chromophore, and sensitizer has been additionally found to have a substantial orientational photorefractive effect. By using carbazole and cyanobiphenyl-based polymers as a photoconductive agent and a nonlinear chromophore, respectively, and incorporating

fullerene-60 as a sensitizer, good stability and performance are possible.[81] In addition, the composite possesses low light scattering, high speed, and high gain properties. In fact, the need for a conventional plasticizing agent can be avoided by using a room temperature liquid crystal at high concentration. The orientational photorefractive effect observed in this material is quite high.

9.7.2.4 Polymer Dispersed Liquid Crystals (PDLCs)

By dispersing liquid crystal droplets in a polymer matrix, polymer dispersed liquid crystals (PDLCs) can be made. By applying an electric field to the composite films, switching can be done between the opaque and the transparent state. Holographic polymer-dispersed liquid crystal films, containing liquid crystals and a photoreactive prepolymer, can be exposed to a coherent interference pattern generated by a laser.[82] Upon holographic exposure (irradiation with a laser beam), there is the onset of a counter-diffusion process which results in the movement of the liquid crystals to the "dark" regions and the monomer to the "light" regions of the interference pattern. This is followed by polymerization of the monomer so that the periodic structure of alternating liquid crystal-rich and polymer-rich zones is set. The main advantage of this technique is the realization of large area grating structures by a fast and single-step process. The liquid crystal-rich zones contain randomly oriented sub-micrometer droplets, whose size depends on the exposure time, laser beam intensity, concentration of liquid crystal, and the polymer, respectively. The diffraction efficiency of the gratings can be modulated by applying an electric field. The index modulation can be erased by applying a sufficiently strong electrical field so that the liquid crystal molecules and the material will be optically homogeneous, which is possible by matching the ordinary index of the liquid crystal with the polymer,

9.7.2.5 Polymer Stabilized Liquid Crystals

Polymer stabilized liquid crystals (PSLC) are either made by dispersing polymer in low molar-mass LC molecules or by dissolving both components in a solvent. In both cases, the LC phase separates out during the solidification of the polymer. In the first case, the polymer forces the LC phase to separate out during its solidification, while, in the second case, solvent-induced phase separation occurs. Usually, a monomer is dissolved in low-molar-mass LC molecules along with a photo-initiator and polymerization is effected by irradiation, while keeping the LC in preferred alignment to get PSLC. After the polymer network attains a critical molecular weight and/or branch content, the primary particles separate out. These primary particles compact into small nodular beads that aggregate into a thin network by a diffusion-controlled cluster aggregation mechanism.[83] The schematics of the process are given in Figure 9.7.

During the progress of polymerization, there is continuous change in the shape of the beads and their overall morphology because of the expulsion of solvent from the region with an abundance of polymer. The orientation-based coupling between the polymer network and the low-molar-mass liquid crystal solvent remains strong. The orientational order, polymer morphology,

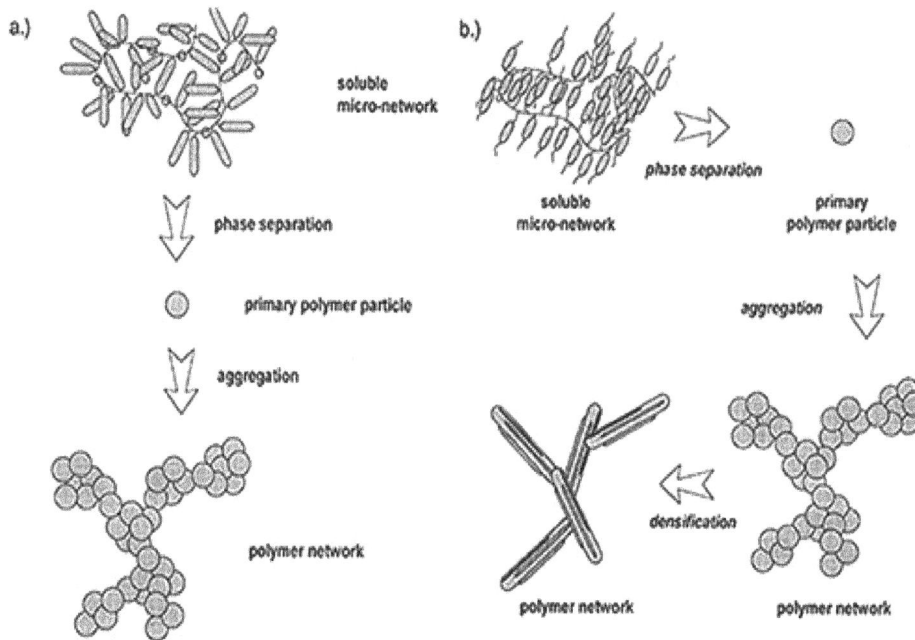

FIGURE 9.7 Schematic representation of a proposed model for the evolution of polymer network morphology of (a) p(BAB) and (b) p(BAB6).

Note: Some mesogens are drawn foreshortened, to indicate out-of-plane orientation.

Source: Reprinted with permission from[83]. Copyright 1995 American Chemical Society.

and the orientational coupling can be controlled by flexible spacer length and the polymerization conditions.

A typical PSLC is made up of polymethylmethacrylate, nematic mixture E7 and a fullerene derivative (PCBM), which exhibits higher diffraction efficiency at higher PCBM loading.[84] The high extent of charge generation and trapping, because of the presence of polymer network, enhances the diffraction efficiency quite considerably. With the help of an external electric field, the grating can be recorded by charge trapping in the bulk with a polymer network and the same can be read without an external applied field.

9.8 Limitations of Photopolymers

However, as with other materials, photopolymers also have certain drawbacks, viz., polymerization-induced shrinkage of the photopolymer system, which results in distortion of the recorded hologram,[39] thickness limitation,[38] which leads to lower dynamic range and greater angular bandwidth compared to thicker materials; non-rewritable (also called WORM (write once read many times) type materials); and non-local material response, which is due to the formation of a smeared profile of the recorded pattern originating from the polymer chains growing away from their point of initiation, This causes its inability to record high spatial frequency gratings.

9.8.1 Shrinkage

During the polymerization of monomers, shrinkage is due to the decreased distance between the monomers in the chain, the close packing of the polymer chains, leading to a decrease of volume. Shrinkage is dependent mostly on monomers and also on the interaction between the chains.[85] The shrinkage effect has a direct influence on the performance of holograms. For reflection gratings, the volume change of the polymer matrix results in a blue shift of the reflection peak because the pitch of the stratified dispersion is lower than the original interference pattern. Scattering losses also occur, mostly with thick samples. The squeezing of the droplets resulting from shrinkage leads to high shape anisotropy. Although no optical effect is observed for reflection gratings, the polarization dependence observed in the transmission gratings is quite high.[86] Also, the shrinkage is likely to affect the optimum electrical field and the response time because of the influence of the shape anisotropy of the droplet on the electro-optic performance equations.[87] The droplet shape is an important factor for relaxation time.

There are many models to investigate the shrinkage during holographic recording. Volume shrinkage and volume relaxation during polymerization of acrylate monomers have been described with the help of a free-volume model.[88] The diffraction efficiency in dichromated gelatin films and Polaroid photopolymer is explained by modeling the vacuoles and pores or voids.[89] As the shrinkage of the polymer matrix is relatively small, the diffusion of voids is ignored. From the calculation based on the number of voids formed, the shrinkage of the H-PDLCs can be calculated.[82]

Shrinkage in a DuPont photopolymer is determined by using the Bragg curve shift method.[90] The change in the sample thickness is Δd and d is the original thickness. The fractional change in sample thickness can be given as:

$$\frac{\Delta d}{d} = \frac{\tan\theta_1}{\tan\theta_0} - 1 \qquad (9.14)$$

where θ_1 and θ_0 are initial and final slant angles, and are calculated from the position of the Bragg curve peak before and after the occurrence of shrinkage resulting from exposure.[91] It is assumed that the layer undergoes a dimension change in the direction perpendicular to its surface. Any change in the layer's plane is mostly ignored.

To study objects with precision and sensitivity, holographic interferometry is commonly used for surfaces that are not so optically smooth. A hologram can be recorded in a photopolymer layer at a wavelength of 633 nm in which negligible absorbance occurs.[92] By recording a simple grating in that photopolymer at 532 nm, shrinkage is induced in the layer, which, in turn, changes the optical path length of the 633 nm object beam. The changing object beam in real time interferes with the reconstructed beam from the unperturbed object hologram. Interference fringes are produced on the reconstructed image from the hologram. The interference fringes, captured in real time, using an appropriate camera, can be used to calculate the optical path length (L) of the beam from the object by getting the number of fringes n from the equation:

$$L = n\lambda \qquad (9.15)$$

where λ is the laser wavelength used.

The length of the optical path can be equated with shrinkage, Δd of the polymer layer:

$$L = \Delta d\left(\cos\theta_1 + \cos\theta_2\right) \qquad (9.16)$$

where θ_1 is the angle during illumination and θ_2 is the angle during reconstruction. Equations (9.15) and (9.16) can be combined as:

$$\Delta d = \frac{n\lambda}{\left(\cos\theta_1 + \cos\theta_2\right)} \qquad (9.17)$$

9.8.1.1 Shrinkage Reduction

A compensation method can be adopted to correct the holographic recording/baking-induced shrinkage effect. An angular deviation of 85' with surface normal leads to 5.25% variation of thickness with a DuPont photopolymer (HRF series). The surface-normal configuration is employed to correct the Bragg diffraction angle shift by 1°21' corresponding to 5.25% film shrinkage.[93] A diffraction efficiency of 80% is possible with a volume hologram after shrinkage correction. Permanent holographic storage with high diffraction efficiency in the holographic film, based on PVA/AA photopolymer with TiO_2-nanoparticle dispersion, can be used to make high efficiency permanent holographic storage that will have holographic writing-related shrinkage.[94] The shrinkage exhibits a minimum around a TiO_2 concentration of 1.5 wt%.

The replacement of the monomer from acrylamide (AA) to a larger one, diacetone acrylamide (DA), results in an average shrinkage reduction of 10–15%, compared to the AA-based system.[95] The monomer molecule sizes and the corresponding diffusion rates are responsible for the difference; the higher size DA is slower to diffuse. The production of polymer chains of longer length depends on the above two parameters. It is also apparent that with a sample thickness there is an increase in shrinkage, while with recording intensities there is a decrease. Another method is also found to be suitable to reduce shrinkage during hologram formation. As an example, 2.5 wt% BEA-type zeolite nanoparticles incorporated in an AA-containing photopolymer produce a 13% reduction in relative shrinkage compared to that without nanoparticles, due to the stability enhancement of the grating structure.

9.9 Nanoparticle Assemblies in Photopolymers

The dry photopolymers for hologram recording are practical viable alternatives for recording, only if the refractive index modulation as well as the dimensional stability of the holograms formed are quite high. However, achieving these objectives with all-organic photopolymers is difficult due to the limitation in the refractive index range. Inorganic materials, on the other hand, have the ability to provide a high refractive index. Therefore, optically mobile inorganic nanoparticles can be added to all-organic photopolymer syrup to obtain large Δn, which is higher than the all-organic photopolymers.[96] The polymerization-related shrinkage is also substantially suppressed by the incorporation of nanoparticles that enhance the dimensional stability. The holographic manipulation technique of photopolymer syrup with nanoparticle assemblies and its prospective use in photonic applications is described in (Figure 9.8).

When a nanoparticle-filled photopolymer system is illuminated with an interference pattern, then along with the occurrence of monomer polymerization in "bright" regions and the resultant diffusion of monomers present in the "dark" regions to "bright" regions; there also occurs a simultaneous counter-diffusion of photo-insensitive nanoparticles (which do not get consumed during the hologram recording process) from the "illuminated" region to the "dark" regions. This happens as a result of an

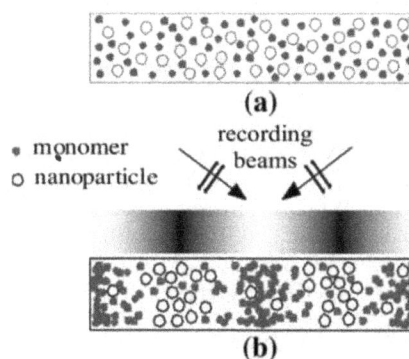

FIGURE 9.8 Distribution of constituents (monomer molecules and nanoparticles) both (a) before and (b) during holographic exposure.

Source: [96].

increase in the chemical potential of these nanoparticles in the "bright" regions because of the depletion of the monomer. These two opposing diffusion processes continue till the completion of photopolymerization. Thus, there is a redistribution of nanoparticles that results in a higher compositional and density difference between the two regions.

Holographic photopolymerization can be used to direct the arrangement of particles, including nanoparticles of gold, polystyrene (PS) latex spheres, and powders of layered silicate.[97] Parallel valleys (dark regions), separated by dense polymer planes, cover the entire thickness of the films containing the gold and layered silicate particles and a sharp modulation in the surface topography is exhibited by the PS spheres. High diffraction efficiencies and extrapolation to much smaller dimensions and other geometries are possible by using this strategy.

REFERENCES

1. R.L. Powel and K.A. Stetson. "Interferometric vibration analysis by wavefront reconstruction." *J. Opt. Soc. Am.* 55(12) (1965): 1593–1598.
2. A.B. Samui. "Holographic recording medium." *Recent Patents on Materials Science* 1(1) (2008): 74–94.
3. P. Hariharan. *Basics of Holography*. Cambridge: Cambridge University Press, 2002.
4. Wikipedia. "Holography.". Date updated: July 17, 2018. Available at: https://en.wikipedia.org/w/index.php?title= Holography&oldid=850700500. (accessed August 12, 2018).
5. J. Ashley, M.P. Bernal, G.W. Burr, H. Coufal, H. Guenther, J.A. Hoffnagle, C.M. Jefferson, B. Marcus, … and G.T. Sincerbox. "Holographic data storage technology." *IBM J. Res. Develop.* 44(3) (2000): 341–368.
6. I. Naydenova, R. Jallapuram, V. Toal, and S. Martin. "Characterisation of the humidity and temperature responses of a reflection hologram recorded in acrylamide-based photopolymer." *Sens. Act. B* 139(1) (2009): 35–38.
7. D.Y. Kim, L. Li, X.L. Jiang, V. Shivshankar, J. Kumar, and S.K. Tripathy. "Polarized laser induced holographic surface relief gratings on polymer films." *Macromolecules* 28(26) (1995): 8835–8839.
8. M. Feuillade, C. Croutxe´-Barghorn, L. Mager, C. Carre, and A. Fort. "Photopatterning of hybrid sol–gel glasses: Generation of volume phase gratings under visible light." *Chem. Phys. Lett.* 398(1–3) (2004): 151–156.
9. F. Ling, L. Dan, H. Zhou, and J. Yao. "Influence of thiol on the holographic properties of silica nanoparticle-embedded acrylate photopolymer films." *Opt. Mater.* 31(2) (2008): 206–208.
10. E. Tolstik, A. Winkler, V, Matusevich, R. Kowarschik, U.V. Mahilny, D.N. Marmysh, Y.I. Matusevich, and L.P. Krul. "PMMA-PQ photopolymers for head-up-displays." *IEEE Photon. Technol. Lett.* 21(12) (2009): 784–786.
11. W.S. Kim, Y.C. Jeong, J.K. Park, C.W. Shin, and N. Kim. "Diffraction efficiency behavior of photopolymer based on P(MMA-co-MAA) copolymer matrix." *Opt. Mater.* 29(12) (2007): 1736–1740.
12. C. Neipp, S. Gallego, M. Ortuno, A. Marquez, A. Belendez, and I. Pascual. "Characterization of a PVA/acrylamide photopolymer: Influence of a cross-linking monomer in the final characteristics of the hologram." *Opt. Commun.* 224(1–3) (2003): 27–34.
13. Y.C. Jeong, S. Lee, and J.K. Park. "Holographic diffraction gratings with enhanced sensitivity based on epoxy-resin photopolymers." *Opt. Exp.* 15(4) (2007): 1497–1504.
14. I.G. Marino, D. Bersani, and P.P. Lottici. "Holographic gratings in DR1-doped sol–gel silica and ORMOSILs thin films." *Opt. Mater.* 15 (4) (2001): 279–284.
15. M. Ortuno, A. Marquez, E. Fernandez, S. Gallego, A. Belendez, and I. Pascual, *Opt. Commun.* 281(6) (2008): 1354–1357.
16. J.R. Lawrence, F.T. O'Neill, and J.T. Sheridan. "Photopolymer holographic recording material." *Optik* 112(10) (2001): 449–463.
17. H.M. Smith. "Holographic recording materials." In *Basic Holographic Principles*. Berlin: Springer-Verlag, 1977, pp. 1–20.
18. P. Hariharan. "Holographic recording materials: Recent developments." *Opt. Engg.* 19(5) (1980): 636.
19. R.A.A. Syms. *Practical Volume Holography*. Oxford: Oxford University Press. 1990.
20. Kodak Professional "Special Holographic Glass Plate 120–01, 120–02." (2000). Technical Information Data Sheet, Eastman Kodak Company. Available at: www.kodak.com/global/en/professional/support/techPubs/ti2526/ti2526.pdf].
21. V.W. Krongauz and A.D. Trifunac. *Processes in Photoreactive Polymers*. New York: Chapman & Hall. 1994.
22. H.I. Bjelkhagen and D. Vukicevic. "Investigation of silver-halide emulsions for holography." *Holographic Imaging and Materials*, *Proc. SPIE 2043*, 12 January, 1994.
23. B.J. Chang. "Dichromated gelatin as a holographic storage medium." In K.G. Leib (Ed.), *Optical Data Storage*, *Proc. SPIE* 177, 71–81 (10 July 1979).
24. G. Li, L. Liu, B. Liu, and Z. Xu. "High-efficiency volume hologram recording with a pulsed signal beam." *Opt. Lett.* 23(16) (1998): 1307–1309.
25. D.D. Nolte. *Photorefractive Effects and Materials*. Dordrecht: Kluwer, 1995.
26. S. Ducharme, J.C. Scott, R.J. Twieg, and W.E. Moerner. "Observation of the photorefractive effect in a polymer." *Phys. Rev. Lett.* 66(14) (1991): 1846.
27. W.E. Moerner and S.M. Silence. "Polymeric photorefractive materials." *Chem. Rev.* 94(1) (1994): 127–155.
28. H.Y.S. Li and D. Psaltis. "Three-dimensional holographic disks." *Appl. Opt.* 33(17) (1994): 3764–3774.
29. H.I. Bjelkhagen. *Silver Halide Materials for Holography and Their Processing*. Berlin: Springer-Verlag, 1993.
30. S.J. Zilker. "Materials design and physics of organic photorefractive systems." *Chem. Phys. Chem.* 1(2 (2000): 72–87.
31. O. Ostroverkhova, D. Wright, U. Gubler, W.E. Moerner, M. He, A. Sastre-Santos, and R.J. Twieg. "Recent advances in understanding and development of photorefractive polymers and glasses." *Adv. Funct. Mater.* 12(9) (2002): 621–629.
32. O. Ostroverkhova and W. E. Moerner. "Organic photorefractives: mechanisms, materials, and applications." *Chem. Rev.* 104(7) (2004): 3267–3314.
33. O. Levi, S. Shalom, I. Benjamin, G. Perepelitsa, A.J. Agranat, R. Neumann, Y. Avny and D. Davidov. "Conjugated polymeric composites for holographic storage." *Synth. Met.* 102(1–3) (1999): 1178–1181.
34. E. Fernandez, M. Ortuno, S. Gallego, A. Marquez, C. Garcia, R. Fuentes, A. Belendez, and I. Pascual. "Optimization of a holographic memory set-up using an LCD and a PVA-based photopolymer." *Optik – Int. J. Light and Electron Opt.* 121(2) (2010): 151–158.

35. I. Banyasz. "Hologram build-up in a near infrared sensitive photopolymer." *Opt. Commun.* 181(4–6) (2000): 215–221.

36. T.J. Trentler, J.E. Boyd, and V.L. Colvin. "Epoxy resin–photopolymer composites for volume holography." *Chem. Mater.* 12(5) (2000): 1431–1438.

37. S. Gallego, C. Neipp, M. Ortuno, A. Belendez, E. Fernandez, and I. Pascual. "Analysis of monomer diffusion in depth in photopolymer materials." *Opt. Commun.* 274(1) (2007): 43–49.

38. S. Gallego, C. Neipp, M. Ortuno, E. Fernandez, A. Belendez, and I. Pascual. "Analysis of multiplexed holograms stored in a thick PVA/AA photopolymer." *Opt. Commun.* 281(6) (2008): 1480–1485.

39. S.H. Lin, P.L. Chen, Y.N. Hsiao, and W.T. Whang. "Fabrication and characterization of poly(methyl methacrylate) photopolymer doped with 9,10-phenanthrenequinone (PQ) based derivatives for volume holographic data storage." *Opt. Commun.* 281(4) (2008): 559–566.

40. W.K. Smothers, K.C. Doraiswamy, M.L. Armstrong, T.J. Trout, "Dry film process for altering wavelength response of holograms." US Patent No. US 4,959,283-1990 (1990).

41. C. Sun, H. Lu, R. Li, Y. Xiao, D. Tang, and M. Huang. "Study of holographic characteristics of photopolymers with two photoinitiator systems." *Optik* 120(4) (2009): 183–187.

42. V. Castelvetro, M. Molesti, and P. Rolla. "UV-curing of acrylic formulations by means of polymeric photoinitiators with the active 2,6-dimethylbenzoylphosphine oxide moieties pendant from a tetramethylene side chain." *Macromol. Chem. Phys.* 203(10–11) (2002): 1486–1496.

43. V. Weiss, E. Millul, and A.A. Friesem. "Photopolymeric holographic recording media: in-situ and real-time characterization." *Holographic Materials II. Proc. SPIE* 2688 (1996): 11–21

44. L. Dhar, A. Hale, H.E. Katz, M.L. Schilling, and M.E. Schnoes. "Recording media that exhibit high dynamic range for digital holographic data storage." *Opt. Lett.* 24(7) (1999): 487–489.

45. A. Khan, A.E. Daugaard, A. Bayles, S. Koga, Y. Miki, K. Sato, J. Enda. S. Hvilsted, G.D. Stucky, and C.J. Hawker. "Dendronized macromonomers for three-dimensional data storage." *Chem. Commun.* 4 (2009): 425–427.

46. D.A. Waldman, R.T. Ingwall, P.K. Dhal, M.G. Horner, E.S. Kolb, H.Y.S. Li, R.A. Minns, and R.A. Schild. "Cationic ring-opening photopolymerization methods for volume hologram recording." In *Diffractive and Holographic Optics Technology III: Proc. SPIE* 2689 (1996).

47. P.W. Labuschagne, W.A. Germishuizen, S.M.C. Verryn, and F.S. Moolman. "Improved oxygen barrier performance of poly(vinyl alcohol) films through hydrogen bond complex with poly(methyl vinyl ether-co-maleic acid)." *Eur. Polym. J.* 44(7) (2008): 2146–2152.

48. B.A. Kowalski, and R.R. McLeod. "Design concepts for diffusive holographic photopolymers." *J. Polym. Sci. Part B: Polym. Phys.* 54(11) (2016): 1021–1035.

49. H.J. Coufal, G.T. Sincerbox, and D. Psaltis. *Holographic Data Storage.* New York: Springer-Verlag, 2000.

50. P. Wang, B. Ihas, M. Schnoes, S. Quirin, D. Beal, and S. Setthachayanon.; "Optical data storage" *Proc. SPIE* 5380 (2004): 283–288.

51. R.T. Ingwall and M. Troll. "The mechanism of hologram formation in dmp-128 photopolymer." *Opt. Eng.* 28(4) (1989): 586–591.

52. M.G. Schnoes, L. Dhar, M.L. Schilling, S.S. Patel, and P. Wiltzius. "Photopolymer-filled nanoporous glass as a dimensionally stable holographic recording medium." *Opt. Lett.* 24(10) (1999): 658–660.

53. M.L. Calvo and P. Cheben. "Photopolymerizable sol–gel nanocomposites for holographic recording." *J. Opt. A: Pure Appl. Opt.* 11(2) (2009): 024009.

54. T.J. Trout, J.J. Schmieg, W.J. Gambogi, and A.M. Weber. "Optical photopolymers: Design and applications." *Adv. Mater.* 10(15) (1998): 1219–1224.

55. G. Odian. "Radical chain polymerization" In *Principles of Polymerization.* 4th edn. New York: Wiley, 1991, pp. 198–349.

56. F.K. Bruder and F. Thomas. "Materials in optical data storage." *Int. J. Mater. Res.* 101(2) (2010): 199–215.

57. M.R. Gleeson and J.T. Sheridan. "Nonlocal photopolymerization kinetics including multiple termination mechanisms and dark reactions. Part I. Modeling." *J. Opt. Soc. Am. B* 26(6) (2009): 1736–1745.

58. S. Martin, P.E.L.G. Leclere, V. Toal, and Y.F. Lion. "Characterization of an acrylamide-based dry photopolymer holographic recording material." *Opt. Eng.* 33(1994): 3942–3946.

59. R. Jallapuram, I. Naydenova, S. Martin, R. Howard, V. Toal, S. Frohmann, S. Orlic, and H.J. Eichler. "Acrylamide-based photopolymer for microholographic data storage." *Opt. Mater.* 28(12) (2006): 1329–1333.

60. T. Sasaki, S. Hamada, Y. Ishikawa, and T. Yoshimi. "Photorefractive effect of liquid crystal polymer poly-4-methacryloyloxyhexyloxy(4'-nitrobenzylidene)aniline." *Chem. Lett.* 26(11) (1997): 1183–1184.

61. H. Ono and N. Kawatsuki. "Orientational photorefractive effects observed in polymer-dispersed liquid crystals." *Opt. Lett.* 22(15) (1997): 1144–1146.

62. W.E. Moerner, S.M. Silence, F. Hache, and G.C. Biorklund. " Orientationally enhanced photorefractive effect in polymers." *J. Opt. Soc. Am: B* 11(2) (1994): 320–330.

63. K. Meerhoz, B.L. Voodin, Sandaphon, B. Kippeen. N. Peyghambarian. "A photorefractive polymer with high optical gain and diffraction efficiency near 100%." *Nature* 371(6497) (1994): 497–500.

64. L. Lucchetti, M. Di Fabrizio, O. Francescangeli, and F. Simoni. "Colossal optical non-linearity in dye-doped liquid crystals." *Opt. Commun.* 233 (2004): 417–424.

65. X. Li, C.P. Chen, H.Y. Gao, Z.H. He, Y. Xiong, H.J. Li, W. Hu, Z.C. Ye, G.F. He, J.G. Lu, and Y.K. Su. "Video-rate holographic display using azo-dye-doped liquid crystal." *J. Disp. Technol.* 10 (2014): 438–443.

66. F. Weigert. "Uber einen neuen Effekt der Strahlung in lichtempfindlichen Schichten." *Verh. Dtsch. Phys. Ges.* 21(1919): 479–491.

67. S. Xie, A. Natansohn, and P. Rochon. "recent developments in aromatic azo polymers research." *Chem. Mater.* 5(4) (1993): 403–411.

68. M. Hasegawa, T. Yamamoto, A. Kanazawa, T. Shiono, and T. Ikeda. "A dynamic grating using a photochemical phase transition of polymer liquid crystals containing azobenzene derivatives." *Adv. Mater.* 11(8) (1999): 675–677.

69. H. Rau and E.J. Luddecke. "On the rotation-inversion controversy on photoisomerization of azobenzenes. Experimental proof of inversion." *J. Am. Chem. Soc.* 104(6) (1982): 1616–1620.

70. T. Ikeda and O. Tsutsumi. "Optical switching and image storage by means of azobenzene liquid-crystal films." *Science* 268(5219) (1995): 1873–1875.

71. I.C. Khoo, J. Ding, Y. Zhang, K. Chen, and A. Diaz. "Supra-nonlinear photorefractive response of single-walled carbon nanotube- and c-60-doped nematic liquid crystal." *Appl. Phys. Lett.* 82 (2003): 3587–3589.

72. G.P. Wiederrecht, B.A. Yoon, and M.R. Wasielewski. "High photorefractive gain in nematic liquid crystals doped with electron donor and acceptor molecules." *Science* 270(5243) (1995): 1794–1797.

73. X. Li, C.P. Chen, Y. Li, P. Zhou, X. Jiang, X. Rong, S. Liu, G. He, J. Lu, and Y. Su. "High-efficiency video-rate holographic display using quantum dot doped liquid crystal." *J. Disp. Technol.* 12(4) (2016): 362–367.

74. H. Ono and I. Saito. "Photorefractive Bragg diffraction in high- and low-molar-mass liquid crystal mixtures." *Appl. Phys. Lett.* 72(16) (1998): 1942–1944.

75. W. Lee and C.S. Chiu. "Observation of self-diffraction by gratings in nematic liquid crystals doped with carbon nanotubes." *Opt. Lett.* 26(8) (2001): 521–523.

76. W. Lee and C.C. Lee. "Holographic grating recording in a hybrid cell of nanotube-doped nematic sandwich." In T.R. Wolinski, M. Warenghem, and S-T. Wu (Eds.), *Liquid Crystals: Optics and Applications. Proc. SPIE* (2005): 59470F.

77. P. Zhou, Y. Li, X. Li, S. Liu, and Y. Su. "holographic display and storage based on photoresponsive liquid crystals." *Liq. Cryst. Rev.* 4(2) (2016): 83–100.

78. P.V. Kamat. "Quantum dot solar cells. semiconductor nanocrystals as light harvesters." *J. Phys. Chem. C.* 112(48) (2008): 18737–18753.

79. L. Brus. "Electronic wave functions in semiconductor clusters: experiment and theory." *J. Phys. Chem.* 90(12) (1986): 2555–2560.

80. X. Li, Y. Li, Y. Xiang, N. Rong, P. Zhou, S. Liu, J. Lu, and Y. Su. "Highly photorefractive hybrid liquid crystal device for a video-rate holographic display." *Opt. Express* 24(8) (2016): 8824–8831.

81. J. Zhang and K.D. Singer. "Homogeneous photorefractive polymer/ nematogen composite." *Appl. Phys. Lett.* 72(23) (1998): 2948.

82. J. Qi, M. DeSarkar, G.T. Warren, and G.P. Crawford. "In-situ shrinkage measurement of holographic polymer dispersed liquid crystals." *J. Appl. Phys.* 91(2002): 4795–4800.

83. CV. Rajaram, S.D. Hudson, and L.C. Chien. "Morphology of polymer-stabilized liquid crystals." *Chem. Mater.* 7(12) (1995): 2300–2308.

84. A. Denisov and J-L.D.V. de La Tocnaye. "Soluble fullerene derivative in liquid crystal: polymer composites and their impact on photorefractive grating efficiency and resolution." *Appl. Opt.* 48(10) (2009): 1926–1931.

85. J-P. Fouassier. *Photoinitiation, Photopolymerization, and Photocuring: Fundamentals and Application.* Munich: Hanser, 1995.

86. R.L. Sutherland, L.V. Natarajan, V.P. Tondiglia, and T.J. Bunning. "Bragg gratings in an acrylate polymer consisting of periodic polymer-dispersed liquid-crystal planes." *Chem. Mater.* 5(10) (1993): 1533–1538.

87. B-G. Wu, J.H. Erdmann, and J.W. Doane. "Response times and voltages for PDLC light shutters." *Liq. Crys.* 5(5) (1989): 1453–1465.

88. C.N. Bowman and N.A. Peppas. "Coupling of kinetics and volume relaxation during polymerizations of multiacrylates and multimethacrylates." *Macromolecules* 24(8) (1991): 1914–1920.

89. R.T. Ingwall and M. Troll. "The mechanism of hologram formation in DMP-128 photopolymer." In *Holographic Optics: Design and Applications. Proc. SPIE* (1988): 0883.

90. C. Zhao, J. Liu, Z. Fu, and R.T. Chen. "Shrinkage correction of volume phase holograms for optical interconnects." *Optoelectronic Interconnects and Packaging IV. Proc. SPIE* 3005 (1997), 224–229.

91. J.T. Gallo and C.M. Verber. "Model for the effects of material shrinkage on volume holograms." *Appl. Opt.* 33(29) (1994): 6797–6806.

92. M. Moothanchery, V. Bavigadda, V. Toal, and I. Naydenova. "Shrinkage during holographic recording in photopolymer films determined by holographic interferometry." *Appl. Opt.* 52(35) (2013): 8519–8527.

93. C. Zhao, J. Liu, Z. Fu, and R.T. Chen. "Shrinkage-corrected volume holograms based on photopolymeric phase media for surface-normal optical interconnects." *Appl. Phys. Lett.* 71(11) (1997): 1464–1466.

94. L. Zhao, J-H. Han, R-P. Li, L-G. Wang, and M-J. Huang. "Resisting shrinkage properties of volume holograms recorded in TiO$_2$ nanoparticle-dispersed acrylamide-based photopolymer." *Chin. Phys. B* 22(12) (2013): 124207.

95. D. Cody, M. Moothanchery, E. Mihaylova, V. Toal, S. Mintova, and I. Naydenova. "Compositional changes for reduction of polymerisation-induced shrinkage in holographic photopolymers." *Adv. Mater. Sci. Eng.* 2016 (2016): 8020754.

96. Y. Tomita. "Holographic assembly of nanoparticles in photopolymers for photonic applications." *SPIE Newsroom* 12 December 2006. doi: 10.1117/2.1200612.0475. Available at: http://talbot.ee.uec.ac.jp/

97. R.A. Vaia, C.L. Dennis, L.V. Natarajan, V.P. Tondiglia, D.W. Tomlin, and T.J. Bunning. "One-step, micrometer-scale organization of nano- and mesoparticles using holographic photopolymerization: A generic technique." *Adv. Mater.* 13(20) (2001): 1570–1574.

10

Shape Memory Polymers

Asit Baran Samui
Institute of Chemical Technology, Mumbai, India

CONTENTS

DOI: 10.1201/9781003037880-10

Abbreviations

SMP	shape memory polymer
CNT	carbon nanotube
PVAc	poly(vinyl acetate)
PLA	poly(lactic acid)
PCl	polycaprolactone
LCEs	liquid crystalline elastomers
POSS	polyhedral oligomeric silsesquioxane
PPEGMA	poly (oligomer polyethylene glycol monomethyl ether methacrylate)
PDL	ω-pentadecalactone
CB	carbon black
SCF	short carbon fiber
MWCNT	multi-walled carbon nanotube
SMASH	shape memory-assisted self-healing
CTH	close-then-heal
SRPs	stimuli responsive polymers
SMPU	shape memory polyurethane
TRPGs	thermo-responsive polymer hydrogels
PCMs	phase change materials
SMHs	shape memory hydrogels
PNIPAm	poly(N-isopropylacrylamide)
2D	two-dimensional
4D	four-dimensional
SMECs	shape memory elastomeric composites
BSMPs	biodegradable shape memory polymers

Symbols with Units

Symbol	Description	Unit
E_g	Glassy modulus	Pascals
E_r	Rubbery modulus	Pascals
f_{IR}	Viscous flow strain	Pascal seconds
f_a	Viscous strain for $t \gg t_r$	Pascal seconds
R_f	Strain fixixity rate	Percentage
R_r	Strain recovery rate	Percentage
ε_u	Strain at application of force	Percentage
T_d	Deformation temperature	°C
$\varepsilon_p(N\text{-}1)$	Permanent strain at deformation temperature T_d	Dimensionless
$\varepsilon_u(N)$	Strain recorded after unloading.	Dimensionless
σ	Deformation stress at T_d	Pascals
$\varepsilon_l^d(N)$	Recorded deformed sample length at T_d	Centimeter
T_f	Lower fixing temperature	°C
$\varepsilon_l(N)$.	Fixed strain observed at T_f under loading	Dimensionless
σ_c	Specific lower constrain stress	Pascals
$\varepsilon_u(N)$	Strain recorded after unloading	Dimensionless
$\varepsilon_p(N)$	Recorded final recovered strain	Dimensionless
ε_m	Maximum strain imposed on the material	Dimensionless
ε_p	Strain in the sample at the end of the same cycle before yield stress is applied	Dimensionless

10.1 Introduction

The polymers which can deform by the application of a certain external stimulus and maintain the deformation even after removal of the stimulus are called shape memory polymers

(SMPs). By applying an appropriate stimulus again, the polymer can be brought back to its initial state. Chemical or physical cross-linking memorizes the permanent shape, and the glass transition or melting transition functions to fix the temporary shape. The high strength-by-weight ratio and high rigidity of SMPs and their composites are being studied for use in various important areas. Two or sometimes three shapes are remembered by SMPs, and the inter-shape transition is triggered by a variation in temperature, electric/magnetic field, light, or a solution. A wide range of polymer properties are covered by the SMPs. For example, SMPs can recover strains up to 800% and higher.[1] The SMPs are made from either thermoplastic or thermoset polymeric materials.

A new area of research and applications for shape-changing materials will cater to a wide range of disciplines. For example, architectural structures responding to the environment and medical devices that are required to undergo shape changes inside the human body, robotics, etc. will be the areas of application. Artificial shape-changing materials are in their early stages of development, while the biological morphing materials are already showing shape change potential efficiently in response to external stimuli.

Two important quantities, namely, strain recovery rate (R_r) and strain fixity rate (R_f) are normally used to measure the shape memory effects. The ability of the material to memorize its permanent shape is characterized by the strain recovery rate, and the switching segments' ability to fix the mechanical deformation is characterized by the strain fixity rate.[2]

The mathematical model used to describe the shape memory effect is given as:[3]

$$R_f(N) = 1 - \frac{E_r}{E_g} \qquad (10.1)$$

$$R_r(N) = 1 - \frac{f_{IR}}{f_\alpha (1 - E_f / E_g)} \qquad (10.2)$$

where E_g and E_r represent the glassy and rubbery modulus, respectively; f_{IR} represents viscous flow strain, and f_α is the strain for $t \gg t_r$.

The SMPs are mostly thermally stimulated. The temperature which enables the polymer to return to its permanent shape is called the switching or transformation temperature (T_{trans}), and is related either to the glass transition (T_g) or to the melting temperature (T_m). Thus, the switching types are responsible for classification to T_g-based or T_m-based SMPs. Considering the reversibility of switches, other mechanisms such as liquid crystal-related transitions, supramolecular assembly formation/disassembly, reversible network formation via irradiation, percolation network formation and its disorder are also likely to function as SMPs.[4] The physical or chemical network linkages, responsible for permanent shape, are designated as net points. Above the T_{trans}, mechanical deformation is applied to create the temporary shape. Below the T_{trans} values (i.e., T_g- and T_m-linked temperatures), deformation is also possible.[5] By selecting a proper programming temperature, the optimized performance can be realized by altering a number of properties, namely, recovery temperature, shape fixity ratio, and maximum as well as final recovery stresses.

The macroscopic deformation applied causes the molecular chains and molecular segments of the networks to undergo conformational changes or macroscopic deformation of the switching phase.[6] By maintaining the deformation via cooling, the temporary shape is fixed, leading to vitrification/crystallization in T_g-/T_m-based SMPs. The deformation energy is thus stored that is released by heating it above T_{trans}. Heating induces the reversal of the conformational changes that occurred during the formation of the temporary shape. Thus, the molecular chains and segments of the networks go through rearrangements. By melting the crystals, the release of the constraints on macroscopically deformed phases help in the restoration of the permanent shape. The depiction as furnished above is concerned with one-way SMPs, which means that the transformation from temporary to permanent shape is only activated by external stimulus or, in the case of multi-shape SMPs, from one temporary shape to another one. Two-way SMPs (2W-SMPs), undergoing a reversible shape change between the permanent and temporary shapes upon application of external stimulus, are also possible. If the reversible shape memory is two-way, a variety of applications are offered. For example, it will add smart ability to actuators, artificial muscles, self-locomotion robotics, etc. Movement between two distinct shapes occurs with such polymers by exposing them to stimuli, generating the opposing conditions, such as heating and cooling. The 2W-SMPs with unique properties are one of the most desired shape memory materials, although they are hard to achieve.

The working principle of a T_g-based system can be seen in Figure 10.1.[7] Changing from a permanent to a temporary shape can be done by heating above T_g and thereafter, by maintaining the deformation stress, the cooling is performed below T_g. Recovery to a permanent shape is done by raising the temperature to the transition temperature, which is above T_g. The stabilization of the network and its net points is done by introducing cross-links. Also, high recovery stress and rate are maintained by the cross-links, compared to the linear counterparts. By applying a similar methodology, the cross-linked networks can be given temporary shapes and the permanent shape can be recovered thereafter. Highly unsaturated natural oils are copolymerized with styrene and divinylbenzene to form random cross-linked copolymer networks, whose glass transitions and rubbery properties can be tuned by varying the monomer ratio.[8]

10.1.1 SMP with High Recovery Stress

The energy stored during its deformation decides the recovery stress of an SMP. It is to be remembered that because of the deformations taking place in the rubbery state, the elastic moduli are in the order of several megapascals (MPa). Thus, the amount of energy that could be stored has some limitations, which indicates the presence of relatively low shape recovery stress. Incomplete shape recovery within a spatially constrained environment results from low shape recovery stress. Examples of this may be a collapsed vertebral disc or a narrowed blood vessel. The elastic modulus in the rubbery state can be improved by several means, such as increasing the thermoset cross-link density or increasing the thermoplastic physical cross-linking extent in a SMP. By incorporating various organic fillers, such as carbon

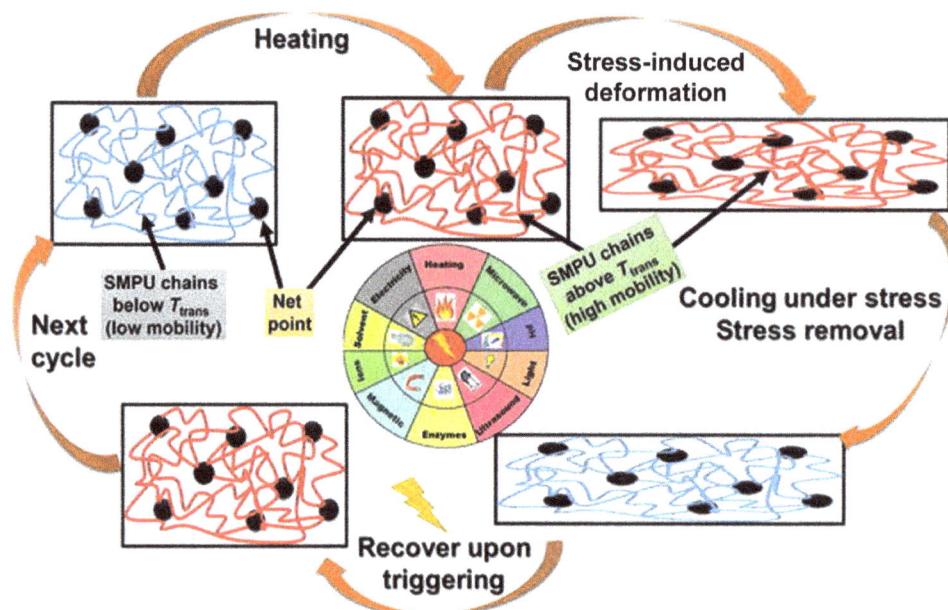

FIGURE 10.1 The molecular mechanism of the shape memory effect under different stimuli.

Note: Black dots: net points; blue lines: SMPU chains below T_{trans} (low mobility); red lines: SMPU chains above T_{trans} (high mobility).

Source: [7].

powders, carbon nanotubes (CNTs), and inorganic fillers, such as glass fibers, Kevlar fibers, SIC particles, etc., within the SMP network, the elastic modulus can be enhanced.

10.1.2 Ultrahigh Molecular Weight Polymers as SMPs

Due to the lack of flow above T_g and the good shape fixing by the vitrification of the polymers with T_g above room temperature, ultra-high molecular weight polymers, such as 10^6 g mol^{-1}, may function as SMPs. A substantial number of entanglements per chain that function as physical cross-links on the time scale of typical deformations (<10 s) qualify a polymer to act as an SMP. Excellent elasticity above the glass transition is maintained by the three-dimensional networks formed by such physical cross-links. However, because of difficult thermal processing, it is necessary to adopt a solvent-based processing.

10.2 Architecture of Shape Memory Polymers

10.2.1 Physically Cross-Linked Glassy Copolymers as Shape Memory Polymers

Technological advances in shape memory polymers will only be possible if processing is easy. The melt-miscible blend of poly(vinyl acetate) (PVAc) and poly(lactic acid) (PLA) meets this requirement because of its simple processing technique. The physical cross-links are formed by the crystalline or rigid amorphous domains in this type of thermoplastics SMP, which afford the super-T_g elasticity required for the development of a shape memory. The phase-separated block copolymers exhibit this type of behavior. The material will flow by raising

its temperature (T_{high}) above T_m or T_g of these distinct physical domains and in that condition it becomes able to be processed and reshaping can be done at this stage. Another existing continuous phase, with lower T_m or T_g, exhibits a rubbery phase at $T_{low} < T < T_{high}$ and this phase is fixed by cooling it to $T < T_{low}$.

Another SMP class in the same group includes polymers with low crystallinity, semi-crystalline homopolymers, or melt-miscible polymer blends that have at least one semi-crystalline compatible component and are compatible in the molten and amorphous states.[9] In such a system, the crystals act as physical cross-links and the T_g of the amorphous region is selected as the transition temperature. For these types of systems, the blend composition can be varied to adjust the glass-transition temperature of the amorphous phase and the work involved during shape recovery.

Hydrogen-bonding ionic clusters can also be physical cross-links to set the network. These interactions strengthen the hard domains by decreasing the chain slippage during deformation, which allows an increased extent of shape recovery.

10.2.2 Semi-Crystalline Cross-Linked Polymer Networks

Semi-crystalline networks can also be used to generate shape recovery. In the semi-crystalline networks, it is preferable to fix the secondary shapes by crystallization instead of using vitrification, as with amorphous polymers. Chemical cross-linking takes care of the permanent shape that cannot be reshaped after processing. Recovery is done by heating above T_m, followed by cooling below T_c (the crystallization temperature). Below the critical temperature this class of material is more compliant,

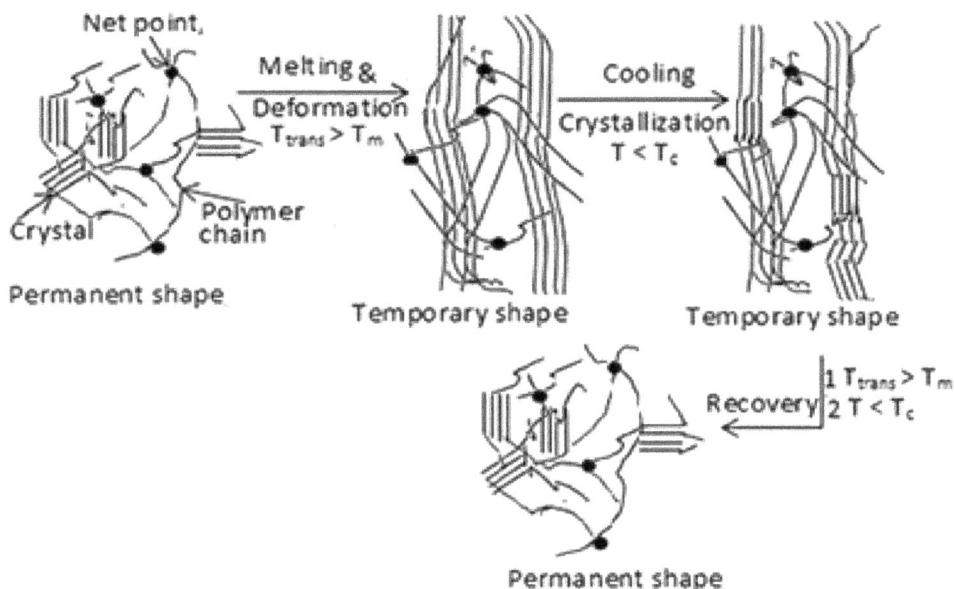

FIGURE 10.2 Schematics of molecular mechanisms of thermally-induced shape memory effect in semi-crystalline polymers.

Note: The temporary shape is created by tensile deformation.

with a stiffness that is dependent on the degree of crystallinity, and the extent of cross-linking. Faster shape-recovery speeds are observed for this first-order transition and sharper transition zones are also observed at times. Bulk polymers, such as semi-crystalline rubbers, liquid-crystal elastomers (LCEs), along with the hydrogels with phase-separated crystalline microdomains, belong to this class of materials.[10] The molecular mechanism of the shape memory function in T_m-activated SMPs is presented in Figure 10.2. The polymer is deformed with stress by heating above its T_m. The fixing of a deformation-induced temporary shape is done by cooling below the crystallization temperature (T_c) during which crystallizations takes place. T_m is never the same as T_c, mostly due to the subcooling effect. Recovery is done by heating above T_m, followed by cooling below T_c. It is accepted that the initial crystalline structure may not be completely restored after completion of the shape memory cycle, which may lead to a reduction in the R_r value.

10.2.3 Physically Cross-Linked Semi-Crystalline Block Copolymers as Shape Memory Polymers

Because of the crystallization of the soft domain in some block copolymers, their T_m values act as shape memory transition temperatures instead of T_g, and thus the crystallization of the soft domains fixes the secondary shapes.

Conventionally, polyurethanes (with PCL diol) with an alternate arrangement of oligomeric hard and soft segments constitute multiblock copolymers; the former undergoes physical cross-linking via polar interaction, hydrogen bonding, or crystallization. Further, such types of cross-links remain mostly unaffected at moderately high temperatures. The thermally reversible phase is formed by the soft segments (e.g., oligocaprolactone) and the secondary shape is determined by the crystallization of these soft

segments. The advantages associated with polyurethanes are their ability to easily adjust the room temperature stiffness, the transition temperature, and also the work output ability can be adjusted by varying their compositions.

The POSS containing polyurethane can also function similarly via its segregation from the soft domain, followed by crystallization with a moderately low melting temperature (~110°C). These polyurethanes can be thermally processed. The rubbery properties are adjusted by varying the hard/soft domain ratio, while the transition temperatures are manipulated by varying the melting or the glass transition of the soft domains.

10.3 SMPs with Special Properties

10.3.1 SMPs with Multi-Shape Memory Effect

Generally, the number of discrete reversible phase transitions possible within the network of an SMS is closely related to the number of shapes it can memorize. Only one permanent shape is remembered by conventional SMPs, which is the most relaxed state of the switching segments exhibiting the dual-shape memory effect. Discrete transition temperatures within a cross-linked network can be incorporated by combining two types of polymer chains, that leads to the triple shape memory effect.[11] For example, PCL segments and poly (cyclohexyl methacrylate) segments, exhibiting T_m and T_g around 50°C and 140°C respectively, can be used for triple shape memory characteristics. The PCL segments and the poly (oligomer polyethylene glycol monomethyl ether methacrylate) (PPEGMA), exhibiting two T_ms above 50°C and around 17–39°C (dependent on the PPEGMA content), make the second network. The programming of the shape memory can be done by deforming at a particular temperature followed by its fixing. For example, the shape C of a sample is deformed to shape B at a temperature (T_{high}) that is above the higher one of the two T_{trans}s, and then it is fixed by cooling the sample to a temperature (T_{mid}) between the two T_{trans}s, while maintaining the stress. The sample is then deformed to shape A at T_{mid} (temperature in-between high and low transition temperature) and fixed by cooling below the lower T_{trans}. The sample is heated above T_{high} so that the shape B and shape C are recovered sequentially. The deformed shapes B and A may not be unidirectional in nature, which exactly fits the complex and multi-directional shape recovery applications. Fine adjustments of the ratio of the two discrete phases are the prerequisite of these systems to exhibit the triple-shape memory effect. To introduce a triple shape memory effect in a cross-linked macroscopic polymer, bilayers with two well-separated phase transitions are required.[12] An epoxy polymer layer is cured to have a lower T_g (38°C) on top of another pre-formed epoxy polymer layer with a higher T_g (78°C) to form a bilayer SMP.

10.3.2 SMPs with Multiple Functional Properties

Smart applications of polymers require the polymers to have multiple functionalities. For example, multiple factors governing the efficiency and their clinical safety are considered for in vivo applications of SMP. For biomedical SMPs, there are requirements of a combination of mechanical properties, shape memory effect together with biofunctionality and degradability, which is essential as well as challenging.[13] While fabricating smart SMP bone substitute, consideration of a patient's defect configuration for matching is essential. To facilitate the surgical insertion of this SMP scaffold, it is deformed into a minimally invasive temporary shape. The scaffold is thermally triggered after it reaches the designated location to attain the permanent shape to fit into the defect location while providing structural and/or mechanical functions. The function of the scaffold ends after the ingrowth of new bone and it should then ideally be resorbable. The requirements of an SMP network are that it should have adequate mechanical strength at body temperature, both before and after the thermal deployment, and the ability to completely recover at a physiologically safe triggering temperature. Controlled degradation rates almost equal to the new bone growth rate are expected to be exhibited by such SMPs.

10.3.3 Two-Way Reversible Shape Memory Polymers

By applying an external stimulus, conventional SMPs are allowed to recover their permanent shape from a temporary shape. Conventionally, the one-way shape memory effect is exhibited by SMPs as one programming step can accommodate only one shape memory cycle. It needs an extra programming step to exhibit an additional shape memory cycle. The requirement of a two-way reversible shape memory is found for various applications, such as actuators, artificial muscles, and self-locomotion robotics.[14]

Two types of two-way reversible shape memory effects (2W-SMEs) are: 2W-SMEs under stress and stress-free conditions. After heating it to a temperature (T_H), which is above the

melting temperatures (T_m), a stress is applied in both cases. After deforming the sample and holding it at deformed state, it is cooled to a low temperature (T_L), which is below the crystallization temperature (T_c). The stress is removed at this point for the stress-free configuration, while the stress is maintained in the stress condition. The stress-free condition is suitable for the 2W-SMEs. The crystallization-induced elongation upon cooling and melting-induced contraction upon heating with or without tensile stress remain the sole mechanism in both the 2W-SMEs and are normally found in liquid crystalline elastomers (LCEs) or semi-crystalline polymer networks.[15]

Although the 2W-SMPs in the stress-free condition have limitations in certain applications, they have attracted a lot of attention. To design a thermally induced 2W-SMP in the stress-free condition, the methods normally used are:

1. Structures with polymer laminates: Reversible strain change is negligible.

2. Semi-crystalline polymer networks with chemical cross-linking having one broad or two T_m: By heating above both T_ms. it is possible to erase the remembered shapes and reprogramming to other shapes is possible.

3. Two-stage cross-linking process adopted for building polymer networks (liquid crystalline elastomers or semi-crystalline polymers): During the preparation process of the polymer network the shapes are fixed that are not erasable by heating.

4. By choosing the deformation temperature in the T_m range, it is possible for the thermoplastic semi-crystalline polymers to exhibit 2W-SME under the stress-free condition: Compared to the chemically cross-linked polymer networks, the reversible strain change is inferior.

Generally, the best 2W-SME in the stress-free condition is exhibited by the chemically cross-linked semi-crystalline polymer networks because of their larger reversible strain change and reprogrammable shapes. Another successful strategy is the use of cocrystallizable comonomers in a copolymer system, such as random prepolymers of CL and ω-pentadecalactone (PDL) to tune the actuation temperature.[16] Adjustment of T_m is done by varying the comonomer ratio. In addition, the cross-linking is done by using a thio-ene reaction. The ability is inherent in the materials to undergo reversible bending–unbending and coiling–uncoiling motions. The microstructure evaluation is given in Figure 10.3.

After being exposed to the heating and cooling cycle, the lamellae of the unstretched sample are expected to be exfoliated and dispersed randomly in the polymer matrix (ε_0). The melting of the crystals and the oriented polymer chains occurs during heating above the T_ms of both segments and then deforms to a high temperature strain (ε_H) under 0.3 MPa. After cooling the sample below the T_cs of both segments, when the external stress is removed, an elongated strain (ε_L) results that is due to the crystallization of the polymer chains along the stretching direction. When the sample is heated to a temperature between

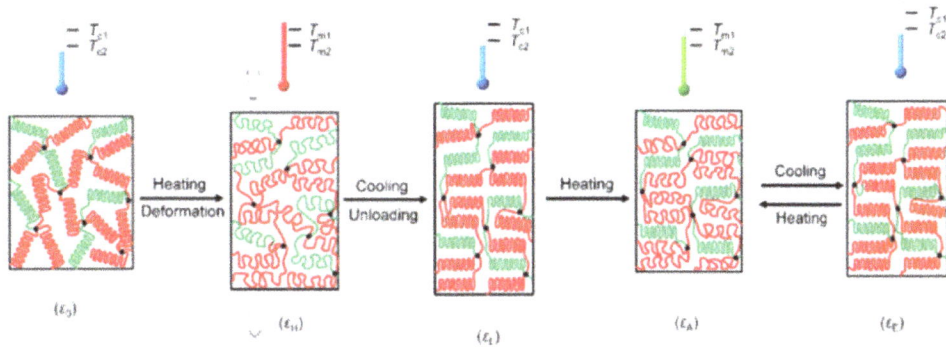

FIGURE 10.3 Microstructure evolution of 2W-SME under stress-free condition.

Source: [16]. Reprinted (adapted) with permission. Copyright (2017) American Chemical Society.

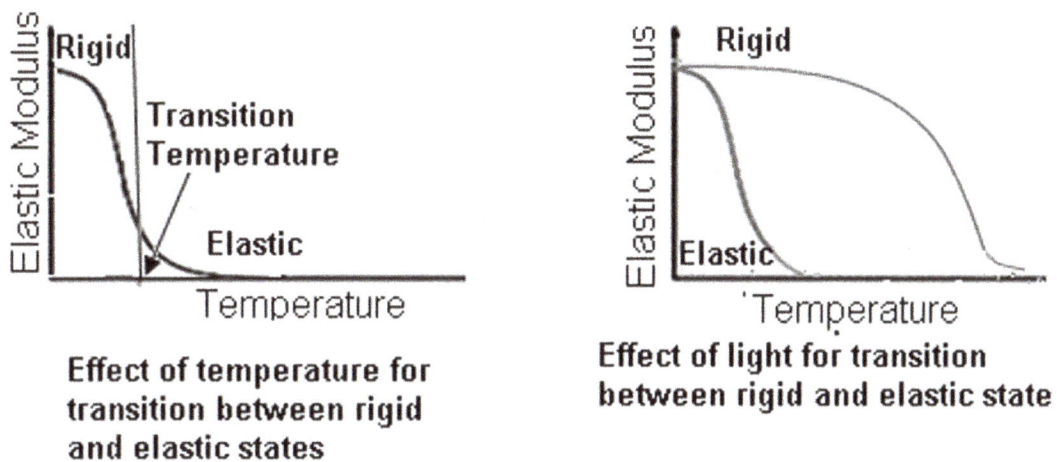

FIGURE 10.4 Difference between SMP and LASMP.

Source: [17]: Figure 4.

the distinct T_ms of two segments, the PCL crystalline phase melts while for the P(CL-co-PDL) 1:2, its crystalline phase is maintained in the polymer network. During this process, the elastic energy stored in the PCL crystalline phase is released, resulting in the contraction of the sample, leading to an intermediate shape (ε_C). Now, the P(CL-co-PDL)1:2 crystalline phase is compressed during the process, generating an internal tensile force. Thus, when the sample is cooled in the PCL segment, the PCL segment undergoes crystallization-induced elongation when cooled below its T_c. This is caused by the internal tensile force of the P(CL-co-PDL)1:2 crystalline phase to produce an elongated shape (ε_E).

10.4 Mechanisms of Shape Memory Effect

10.4.1 Light-Activated Shape Memory Polymers (LASMPs)

Faster response times of products developed for various technologies using SMPs were required. In light-activated shape

memory polymers (LASMPs), the T_g is shifted by irradiating the SMP with specific wavelengths of light. The design approach and photo-responsive chemistry are treated as basic parameters for this class of materials.[17] The basic requirement of the SMPs is to behave as a high-performance polymer below the T_g, and as a high-performance elastomer above the T_g, The difference between the SMP and the LASMP is elaborated in Figure 10.4.

The cross-linking density of the LASMP is modified with photo-cross-linking and photo-cleaving, which indicates that the softening temperature can be affected by light. It can be explained thus: in stage 1, the SMP is fully cured with low T_g, In stage 2, the cross-linking is done with a particular wavelength of light so that the T_g, is increased. In the reverse direction, irradiation is done by another wavelength of light to reversibly break the cross-links (photo-cleaving).Thus, the operational temperature range is between the stage 1 product and the stage 2 product. The switching between the elastomeric material and the rigid polymer can thus be performed. The $[2\pi+2\pi]$ cycloaddition induced by light irradiation and the corresponding photo-cleavage of the cyclobutane derivatives constitutes the mechanism of photo-switching. The

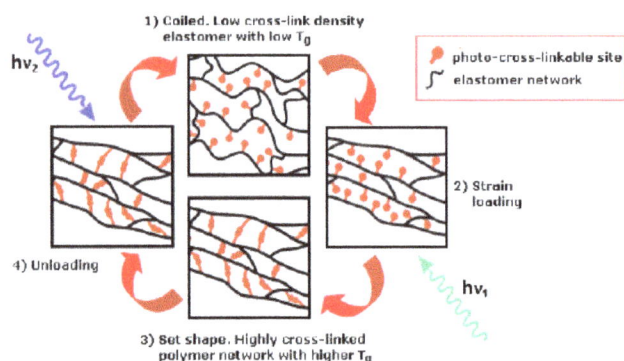

FIGURE 10.5 Detailed pictorial description of LASMP.

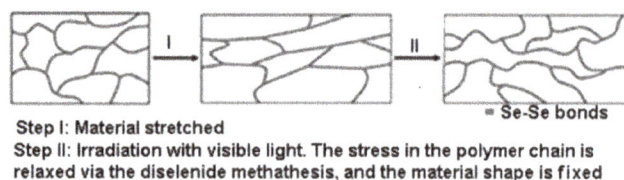

FIGURE 10.6 Brief illustration of the light-induced plasticity in diselenide bond-containing materials.

schematics of LASMP functioning with the help of radiation is presented in Figure 10.5.

The application of the LASMP technology has the advantages of dynamic modulus transition without heating, a quicker response time, and significantly lower energy requirements for the transition. LASMPs function in exactly the similar manner as SMPs except for the transition energy. By using the advantage of SMP technology, the morphing function has already started in both aerospace and space industries.[18]

Depending on the type of doping, different wavelengths can be employed for light-driven reactions. In fact, stimulation can be done by light of all wavelengths with the proper selection of materials. As it has good penetration ability, infrared light and microwaves can be suitably used in shape memory polymer-based biomedical instruments. Because of the heat-producing nature of electromagnetic waves, wave-absorbing material is used during the employment of infrared light and microwaves. The functioning of wave-absorbing materials depends on the presence of resistance, an electric medium or a magnetic medium.

10.4.2 Visible Light-Induced Shape Memory Polymers

While considering the plasticity of thermosets, the idea of a vitrimer has appeared. Vitrimers, derived from thermosets, are able to reform their covalent polymer networks while specific conditions are maintained. Bond exchange reactions, such as transesterification reactions or dynamic covalent bond formation and breaking, are the routes to restrict the network.[19] Most of the vitrimers or vitrimer-like materials are generally reprocessed via heating to a high temperature or by light irradiation. The reprocessing of the vitrimer-like materials is mainly based on the photochemical reactions or disulfide bond exchanges that require ultraviolet (UV) light.[20] The UV light is likely to damage the polymer chains, resulting in deterioration of their mechanical properties and lifetimes. The penetration power of UV light is limited, for example, below 300 nm. Therefore, the UV light becomes partially effective for polymers, such as for some PUs, even those with 1 mm thickness. Spatial and temporal control and cost-effectiveness are the advantages of using visible light, together with its non-damaging nature on the polymer chains. The exchange reactions via dynamic covalent diselenide bonds are triggered by visible light.[21] The introduction of diselenide bonds into the polymeric architecture, such as PUs, enables it to undergo polymer chain reformation under visible light, leading to self-healing. The loaded stress can be relaxed during the chain reformation through the whole material that indicates that the material can be reprocessed (Figure 10.6).

Diselenide bonds can be introduced by reacting polyol ends, while preserving the unreacted hydroxyl group, which will react with isocyanate to form polyurethane. Altering the permanent shape is easier due to the orthogonality of light and heat, as a mold or manual fix is not required for the shape during this process.

10.4.3 Water-Driven Shape Memory Polymers

Water-triggered recovery and/or gradient glass transition temperature in the polyurethane SMPs can be realized because of the effects of moisture in reducing the T_g of PU as well as of many others. The field of applications of SMPs is thus enlarged with the application of the above technique. Polyurethane wires (dia:1.5 mm) (SMP MM3520 from Mitsubishi Heavy Industries, Japan), with a T_g of 36°C, are used to test the water-driven recovery.[22] In the process, the straight SMP wire is given a circular shape at 40°C that is fixed while cooling back to room temperatures (~22°C). When the wire is maintained in a cabinet with humidity around 30 RH, no apparent recovery is observed even after one week at ambient temperature. However, there is gradual recovery when it is kept immersed in water at room temperature for about 30 min. The recovery by the influence of water is further confirmed by the evaluation of the glass transition temperature gradient. The wire is further divided into three segments of equal length and each segment is immersed in water at room temperature for three different periods (0 min, 30 min, 5 h). Three different T_g s of 36°C, 28°C and 15°C, respectively are observed, indicating the plasticizing effect. Similarly, the wire is deformed into an "m" shape at 40°C and then cooled to 10°C to fix the shape. The bottom segment recovers in about 30 s by exposing it in the air, while the middle segment straightens in about 1 min when heated to about 30°C. When heated to 40°C for about 1 min, the top segment regains its original shape. Thus, the SMPs can be used by tuning for programmable recovery via actuation by water. The mechanism with the PU is depicted as: water is adsorbed by the PU SMP that becomes bound with the polymer, resulting in the weakening of the hydrogen bond between N–H and C=O groups and a significant decrease of T_g occurs.

The water-induced shape memory from the fixation and recovery of the temporary deformation of the hydrophilic polymer results from the formation and dissolution of poly(ethylene glycol) (PEG) crystal.[23] As over 75% of the water-induced shape recovery is achieved even at 0°C, it can be concluded that its sensitivity to temperature is quite low. Generally, the hydrophobic segment, such as poly(ε-caprolactone) (PCL) or poly(L-lactide) (PLA), and the hydrophilic segment, such as PEG, constitute such types of materials. Another advantage of PLA or PCL segments is that the materials are biodegradable under physiological conditions. PEG crystals, formed in dried materials, are

destroyed during swelling in water, and rubber-like materials are formed (Figure 10.7).

Different shapes are derived from the deformation of these swollen samples. Following the swelling and deforming, it is dried so that the PEG crystalline structure appears again and the predetermined shapes are fixed. The stability of these shapes is assured for a long time at room temperature, and the original shapes are recovered via the disappearance of the PEG crystals by immersing in water. The temperature limitation during fixation and recovery with water uptake is eliminated by the ability of polymers to form different predetermined temporary shapes and the recovery of their initial shapes within a wide temperature range. These materials are used widely in biomedical fields, such as a self-tightening suture, artificial lens, smart stents, and an artificial scaffold with complex geometries because of the above characteristics.

The PU polymers, having both a thermal and moisture-driven shape and their intrinsic bio-compatibility, can be applied in bio-related fields, particularly minimally invasive surgery and cell surgery.

Porous polymers, with scope for important applications in tissue engineering, can be used as a scaffold so that there are cellular attachments and tissue development.[24] Commonly used organic solvents may harm the tissue. Due to the presence of a trace amount of solvents along with the implant, this can harm the surrounding tissues. Therefore, high moisture-absorbing PU can be employed to make porous thin films by using water as a non-toxic component.[25] By varying the duration of immersion in water and the subsequent heating temperature, bubbles can be formed of various sizes and density (Figure 10.8). By applying heat, there is shrinkage in the bubbles and, finally, they disappear completely. During heating or immersion in water, medicine can be pumped out due to the shape recovery (shrinkage) of these bubbles.[26]

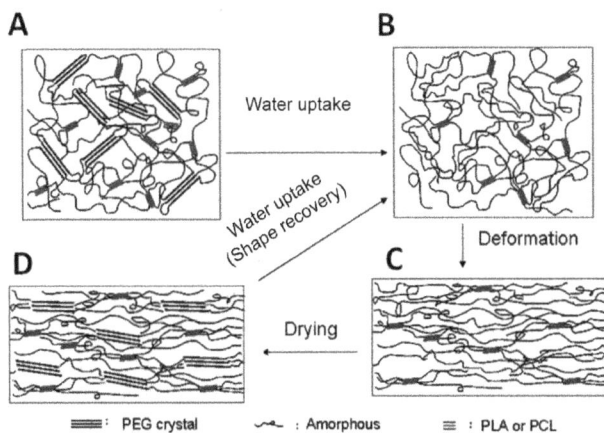

FIGURE 10.7 Schematic of water-triggered shape memory and restoration. With the uptake of water, the PEG crystals in material (A) transform into amorphous state (B); and the wet material is easily elongated or deformed (C); after drying, the PEG crystals are restored and the deformed shape is fixed or memorized (D); with water uptake. the reformed material returns back its initial shape.

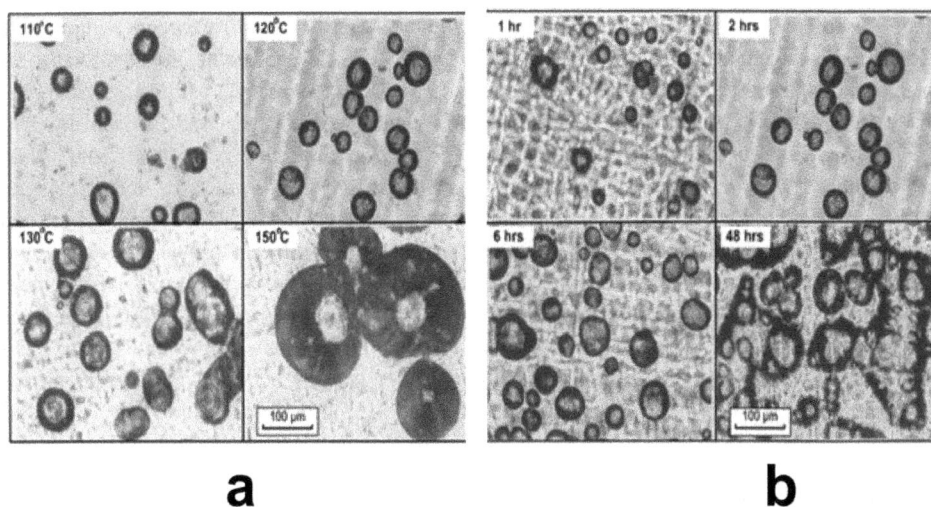

FIGURE 10.8 Porous SMP films using water as a non-toxic agent. (a) Heating at different temperature after immersing in room temperature water for two hours; (b) immersing in water for a different period of time and heated at 120°C.

Source: [21].

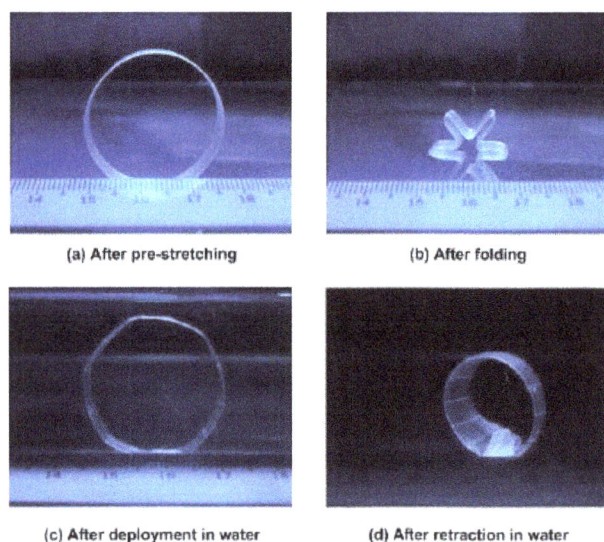

(a) After pre-stretching

(b) After folding

(c) After deployment in water

(d) After retraction in water

FIGURE 10.9 Retraction of a pre-deformed polyurethane SMP STENT in water. (a) Deformed polyurethane SMP stent; (b) second deformation of stent via multiple folding; (c) mechanical expansion of the stent; (d) shrinkage of the stent in water.

Source: [24]. Reprinted with permission from SPIE.

Currently, self-expandable stents of various types are being used. Stainless steel stents were normally used in the early days. During further development, an alternative has been found in NiTi shape memory alloy. Currently, the stents made of various polymers (including biodegradables and/or SMP) or stents coated with polymers have become the materials of choice.[27] A catheter delivers a stent to the exact location that is expanded in the next stage by adopting a mechanical (elastic) expansion method or the shape memory effect by using shape memory materials. It is mostly a difficult task to remove the first category of stent after placing it in the desired location. On the other hand, a foldable/unfoldable stent can easily be removed. A thermo-moisture responsive SMP can be used for this purpose. The total process can be explained as: a piece of thin film PU SMP (0.5 mm thick) is first fabricated and pre-stretched by 50%, and then bent into a round tube shape (Figure 10.9 (a)). By using mechanical force, a star-shaped tube is designed by folding (Figure 10.9 (b)). This folded tube with a reduced diameter can now be placed in the required location by using a catheter. By using mechanical means the stent is expanded (Figure 10.9 (c)). After some time, when the SMP is immersed in water, it absorbs the water and the shape recovery takes place. The immersion in water is akin to the environment inside a human body. The process induces shrinkage of the tube and there is a substantial reduction in diameter (Figure 10.9 (d)). At this stage, the tube is ready for removal.

10.4.4 Solvent-Driven Shape Memory Polymers

Solvent-driven functionally controllable SMPs can be designed using the principle of reduction of the T_g with polar solvents that increase the application areas of the SMPs. To check the possibilities, the shape memory behavior has to be established as a function of the solubility parameter. As an example, when a

constrained shape memory natural rubber is exposed to solvent vapor, a mechanical stress is generated.[28] As soon as the solvent vapor is removed from the chamber, the material regains its original condition. The reversibility of the material is confirmed and the stress remains proportional to the amount of solvent vapor. Further, the stress is specific to the solvent.

10.5 Special Types of Shape Memory Polymers

10.5.1 High- and Low-Temperature-Resistant Shape Memory Polymers

Space structures have to endure both high as well as low temperature thermal cycling. It is known that many polymers become brittle at a low temperature and also many others degrade at a high temperature. For the polymers to work in harsh environments, such as space, for a long period of time, they are required to perform normally as well as sustain the intensity of thermal cycling.[29] For use in aerospace and spacecraft, the aromatic polyimides (PI) are the materials of choice because of their excellent thermal, mechanical, and radiation-shielding capabilities, along with low creep.[30] Shape memory polyimide, enjoying the combined effects of shape memory polymer and the PI with uncommon properties, has potential applications in various fields. The primary requirement is that the T_g of the SMP is higher than the surrounding temperature. However, there is a problem if the shape is triggered with too high a T_g. This demands the combination of different T_gs to satisfy the temperature requirements under different environmental conditions. A typical shape memory polyimide, exhibiting a T_g between 229 and 243°C, can be synthesized from a dianhydride derivative of bis phenol A (BPADA) and aromatic diamine derivative of diphenyl ether (ODA).[31] The aromatic polyimide chains are thermally stable and the flexible portions along the backbone help in the shape memory process.

The shape recovery process is shown in Figure 10.10 for the PI sample (B2: 3% star-shaped tri aromatic amine used as a crosslinker). Recovery at 240°C occurs within 1–6 seconds.

It is generally believed that the shape memory effects are affected by the matrix parameters, such as chain flexibility, molecular weight, and cross-link density. Thermoplastic shape memory polyimide must have sufficiently high M_n to form physical cross-links effectively. Low M_n polymers, having short chains, cannot undergo chain entanglement. On the contrary, the high M_n polymers, with long chains, undergo chain entanglements. In fact, the formation of the physical cross-links of the shape memory process is dependent mainly on chain entanglements. Furthermore, the aryl interactions can also be used for this purpose. It is also apparent that the better shape memory properties are exhibited by the thermoset shape memory polyimides as compared to the thermoplastic polyamides. The applicability of this shape memory polyimide is confirmed by a negligible change in properties during thermal cycling from −50°C to +150°C for 200 h.

10.5.2 Porous Shape Memory Polymers

Porous SMPs can be formed from foams, meshes, etc. which have three-dimensional porous macrostructures. In addition to the

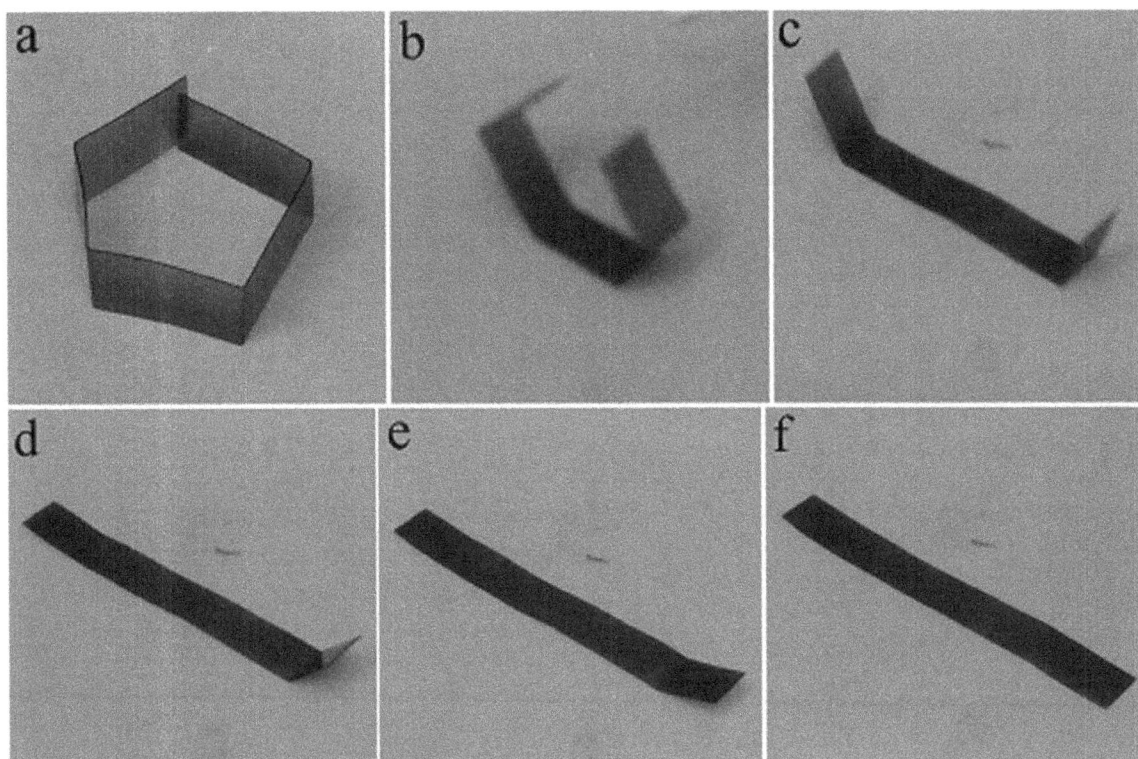

FIGURE 10.10 Shape recovery process of the shape memory polyimide: (a) shows the deformed shape of the sample B2, and (b–f) show the shape recovery process on 240°C hot-stage at 1, 2, 3, 4 and 6 s.

Source: [31].

conventional properties exhibited by neat SMPs, additional properties, such as a greater volumetric expansion capability as well as stimuli-responsive dynamic permeability upon actuation, are exhibited by the porous shape memory polymeric materials. The porosity-dependent environmental responses are exhibited by the porous SMPs during actuation, which qualifies them to function as a material of choice for biotechnological applications, e.g., as scaffolds, in which physiological reactions are influenced by porosity and pore size.[32] Required designs with controlled pore sizes and geometries can be achieved along with uniformity in a wide range of polymeric systems. The architectural framework and also the chemical constitution decide the behavior of the material.[33]

Either open-celled or closed-celled isotropic foams with pore size variation, covering the range from sub-micron to mm, can be generated by using both thermoplastic and thermoset polymers. Physical mixing of hollow particles, such as hollow glass microbeads, with a liquid prepolymer resin and then locking the microbeads via curing result in the formation of syntactic foams. By using salt leaching, porous structures can be designed. A salt such as sodium chloride can be added to a solution of monomer/cross-linker or thermoplastic polymer in an organic solvent to cast a film, which allows the leaching of the salt particles when immersed in water to form porous polymer substrates. The fabrication techniques and the process parameters in the cell membrane reticulation method are so selected that porous materials will exhibit all three types, such as an open, closed, and mixed cellular morphology.

Regarding the application of SMP foam based on polyurethane (FLEGMAT, made by Ranwal Ltd, Great Britain), above the material's T_g, it is compliant enough to grip an object with irregular geometries or delicate objects that are prone to damage by stiffer grips. The foam's compliance is reduced to a large extent by lowering the temperature below its T_g. In that state, it attains a firm state that allows the transportation of the object.[34]

Low mass, low launch volume, deployable structures for aerospace applications can ideally be designed with SMP foams. When reverting back to its original structure from the compressed secondary geometry, the SMP foam undergoes large volumetric expansion by the application of minimal energy. This makes it a potential material in space technology to replace the conventional technologies which have various deficiencies, such as a complicated design that requires heavy inflation systems which may be damaged by debris and micrometeorite strikes. The SMP foam structures are self-deployable and reliable, and can be made at low cost. Also, the SMP foam will be able to absorb energy from any impact by an outside object, thus preventing damage, unlike non-foam structures.[35] Further, no effect on shape recovery properties is observed for the SMP foams during storage for long periods of time. The SMP foams with micron-sized cells and a thin structure have the potential to act as alternatives to the complicated actuation mechanisms and large support structures. Solar sails can be made by integrating the thin-film electronics, sensors, actuators and power sources with the SMP foam.[36]

The traditional method of gravel packing, used as a sand management tool in oil well operations, can be replaced by SMP foams. The foam is compressed radially for installation in the well, which on actuation generates pressure on the sandface of the well to maintain an open wellbore while the foam is held in place.[37] The shape of the memory pillow, developed by Bayer, is adjusted to the contour of the neck and shoulder at around body temperature. Similarly, memory mattresses were designed by using SMP foams that can be used to provide body comfort and support and also as insoles for improved shoe fitting.

10.5.3 Shape Memory Coatings

There are several disadvantages in conventional shape memory textiles, hair shaping methods/coatings, and protective coatings for metal, glass, and plastic, such as solvent-based coatings are not desirable for substrates, such as textiles and hair. As a permanent strategy, the shape of the hair is fixed by first using reducing agents for the cleavage of disulfide bonds in the hair and then the disulfide bonds are reconnected by using an oxidation process. Although this permanent method is effective for a long time, the chemical treatment of the hair has an adverse effect on it and also the process is troublesome. The SMP chemical compounds/compositions can be used to coat certain substrates in the form of an aqueous dispersion. The design can be such that the ionic groups imparts the ability to disperse in water, and a cross-linkable moiety for shape memory constitute the SMPs. Water-sensitive SMPs are selected for this purpose. When such materials are exposed to water vapor or liquid water, a trigger is generated for a change from a temporary to a permanent shape. In household products, such water-triggered actuation may be used to dispense material on contact with water, and so on.[38] An example of a smart application of the SMPs is dispensing a drug when it comes into contact with body fluids. An autonomous control system, having the desired shape change characteristics upon exposure to water, is another example of an industrial product that replaces the complex water sensor and control electronics. The aqueous solution-based SMP coating is responsible for the shape memory properties of the monofilament wires with a coating thick enough to maintain comparable stiffness with the monofilament core.

10.5.4 Shape Memory Polymer Surfaces

Complex surface interactions, with physical and chemical properties, are behind the way the biological or natural nanomaterials behave and also devices and processes depend on this. This makes the development of shape memory polymer surfaces interesting. Therefore, the shape memory polymer surfaces have a great role to play in practical applications.[39]

The specific surface of a material can be expanded to a considerable extent by the surface microstructures. In fact, by going to very low dimensions, the special mechanical and optical properties of materials exhibit geometric relation, i.e., the properties can be changed by inducing any geometric deformation. Due to the basic nature of variable geometry, the SMPs can generate a large number of surface microstructure materials that exhibit variable properties. The types of geometric deformation-induced variable properties are called affine and non-affine. The former leads to the tuning of properties via tuning tension or compression, while the latter exhibits properties in terms of their existence or disappearance via the creation or erasure of the surface microstructure.[40]

Superficial hydrophobicity, fluid drag reduction, transparency, and also one- or two-dimensional photonic structures are generated by the surface microstructure created as waves, bumps, tree protrusions, and column arrays.

Three basic methods, such as top-down, bottom-up, and their combination create polymer surface microstructures. In the top-down approach, the microstructures are created by macroscopically treating the polymer surface, such as using lithography or imprinting. In the bottom-up approach, the shape memory effect is employed. The films or lines on the polymer substrate are buckled to generate the surface microstructures.

10.6 Shape Memory Polymer Composites

For shape memory polymer composites, basically reinforcement and effective stimulation methods are the two prominent requirements. The stimulation can be made more precise by using shape memory polymer composites and highly selective shape memory behaviors can be maintained.[41] With the aim of supplying heat via non-contact mode by light, electricity, and magnetic forces, various multifunctional and smart shape memory polymer materials can be designed. By using SMP as a matrix, the reinforcement can be done by doping with carbon materials, polymers, metal, and ceramic particles used in their zero, one, or two dimensions. Moreover, a multifunctional composite material can be made by using a shape memory polymer as the functional filler in the form of fibers and similar design. A composite structure can be made with an appropriate design that may exhibit two-way recovery and other interesting and useful functions. The proportion of filler is decided by the need to design a particular type. The thermal and mechanical properties are reduced by using excessive fillers and also the transition temperature is lowered. In fact, the enhancement of the shape memory effect can be done by adding only a small amount of filler. Generally, most of the fillers adversely affect the shape memory effect of a polymer. The driving force for the recovery process and better carrying capacity corresponding to a high elastic modulus are increased by the reinforcement of the SMP. Good compatibility with polymer-based materials, considerable reinforcement, together with superior physical and chemical properties, make the carbon material widely used for reinforcement. Glass fiber, carbon fiber, spandex fiber, Kevlar fiber, silica, etc. are also used for this purpose. The reinforcing ability in SMP follows the order: long fibers > short fibers > particles. Dual functions are played by several fillers in the form of providing reinforcement and increasing the shape memory effect. The formation of supramolecular systems is the necessary requirement for such behavior. As an example, the reinforcement, as well as the interaction with the polymer network through hydrogen bonds, are maintained by cellulose whiskers and graphene oxide in a reversible supramolecular effect, that fixes the programmed shape.[42] The presence of carbon black in the polymer also increases the fixation rate and recovery rate of the material because of its self-organizing nature.[43]

10.6.1 Electroactive Shape Memory Polymer Composites

Among all kinds of stimulation, a convenient and precise one is electric stimulation and its use has widened the application potential of shape memory polymers.

The polymers mixed with conductors enjoy the advantages of ease of design and superior properties, such as high efficiency and stability. The passage of a current through the composite generates heat, which enables the matrix to initiate the process of shape recovery. Doping of the polymers for this purpose is done by using nanocarbons as well as nanoscale granules of metal. For example, a mixture of carbon black (CB) nanoparticle and short carbon fiber (SCF) is blended with the SMP to improve its electrical property. The basic requirement is the conductive three-dimensional network structure and the uniform distribution of dopants. Thus, in a styrene-based thermoset shape memory resin, the fibrous component, the SCF, transports the charge over a long distance by forming local conductive networks and particulate CB is dispersed homogeneously and acts as interconnect among the fibers.[44] The mechanical properties and electrical conductivities are enhanced by using this strategy. By applying an electric current, the SMP can be heated efficiently, thus qualifying it to realize its shape recovery induced by electricity.

To impart electrical conductivity as well as enhance the mechanical properties of SMPs, various high performance materials have been used:

- *SMPs filled with carbon nanotubes*: Both surface-modified or unmodified multi-walled carbon nanotubes (MWCNTs) exist; the latter maintains higher electrical conductivity. The surface-modified MWCNTs can be used to make composites with improved mechanical properties.
- *SMPs filled with electromagnetic filler*: The shape memory polymer matrices with surface-modified super-paramagnetic nanoparticles undergo complex shape transitions by applying electromagnetic fields from a remote location.
- *SMPs filled with Ni chain*: During the curing process, the magnetic particles in the SMP can form chains under an applied magnetic field that reduces the electrical resistance to a considerable extent. By applying a voltage of 6 V, the chained sample with a high volume percentage of Ni (10) can be heated from a temperature of 20°C to 55°C.[45]
- *SMPs filled with hybrid fibers*: Both particulate and fibrous fillers are used, which form a conducting pathway in the matrix. The magnitude of the applied voltage and the electrical resistivity of the SMP determine the rate of the shape recovery. The potential of these composite SMPS in a wide range of technological applications originates from their electric triggering.

10.6.2 Self-Healing Shape Memory Polymer Composites

Similar to the injuries of plant and animals and their self-healing ability, materials face the threats of fatigue and accidental damage in the form of bending, stretching, electric shock, or chemical corrosion. Usually, normal materials undergo irreversible damage that leads to increased cost along with other problems. Therefore, the interest in self-healing materials has grown over time. Polymer systems can easily be designed with self-healing abilities, i.e., the long segments present in the system can be moved and modified. Therefore, for a self-healing polymer, two concepts are important: (1) the re-attachment of the polymers through their own physical and chemical properties; and (2) doping with a healing agent to repair the damage. Polymer nanocomposites can also function by either providing healing agents contained via microencapsulation, or by introducing a framework that can bring the fracture surface back into contact with the help of an external force.

Slow self-healing is observed because the interaction between the surfaces of the broken parts is very weak, along with the low expansion speed of the chain parts. Spontaneous self-healing can be made possible by using a shape memory polymer matrix or by doping with a shape memory polymer fiber. A thermoplastic polymer, such as PCL, can be introduced for self-repair in general thermosetting shape memory polymers. For example, a PCL-based shape memory polymer fiber can be blended with epoxy resin to obtain a self-healing shape memory polymer composite.[46]

In self-healing polymers, there are two kinds of automatic merging: (1) shape memory-assisted self-healing (SMASH); and (2) close-then-heal (CTH). SMASH can be done more easily compared to CTH, but a wider range of applicability is possible with CTH. In the absence of any ambient load, the shape memory polymer is effective and, therefore, the SMASH method can be used, while with an external load, the shape memory polymer is not able to complete the recovery process. The CTH method is very effective in this case.[47] By removing the external force with the shape memory polymer matrix, recovery to the original shape is possible. Therefore, an external constraint upon recovery is required for the preprogrammed shape memory polymer base. A fabric or grid can function as an external compression constraint.

10.6.3 Magnetic Nanocomposites for the Shape Memory Effect

The magnetically induced shape memory effect can be produced from a remote location via noncontact triggering. The general method of preparing the magnetically induced shape memory composites is by incorporating magnetic nanoparticles in thermoplastic shape memory polymers. The magnetic field-induced inductive heating of magnetic particles, acting as mini-antennas, converts the electromagnetic energy to heat energy. The shape recovery of SMPs in this approach enjoying the advantages of remote heating, provides a fast heating rate, and has applications in medical imaging techniques. For example, in fluoroscopy or computed tomography scans, the implanted device is detected without additional surgery required for the proper device placement. An alternating magnetic field is applied for inductive heating so that the shape memory effect of the composite is activated.[48]

The shape recovery behaviors of magnetic particle-reinforced SMP composites can be studied by using finite element simulations.[49] As an example, the experimental data from an acrylate-based SMP is used to determine the required model

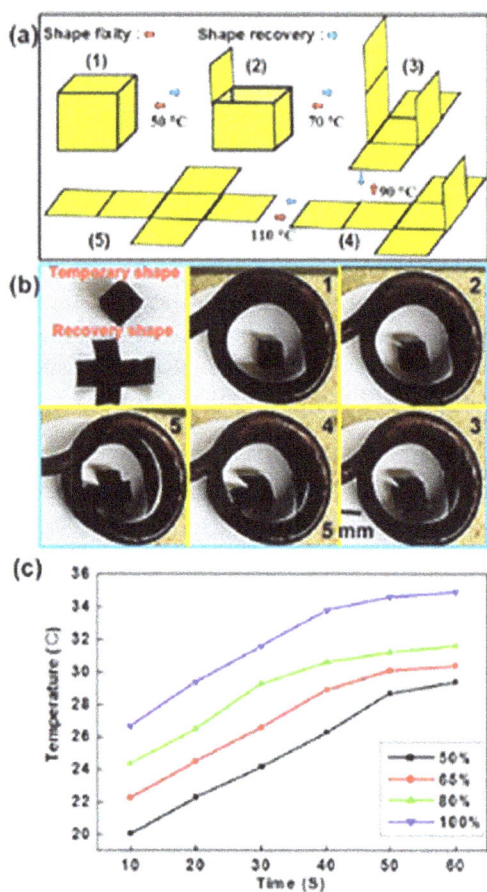

FIGURE 10.11 3D shape recovery performance: (A) the designed shape recovery process of a box; (B) magnetic field actuated shape recovery; (C) temperature vs time at different output powers.

parameters. Generally, by reducing the particle sizes or increasing the volume fractions, the recovery rate can be enhanced. However, there exists a critical particle size suitable for a fixed target recovery temperature and a particle volume fraction. The recovery time is controlled by the material intrinsic recovery time below this critical size and is therefore independent of the particle size. Similarly, above the critical volume fraction, the heating efficiency cannot be improved just by increasing the volume fraction.

More than five shapes can be programmed with nonwoven fiber composite Nafion/Fe_3O_4 films to induce several sequential fast transformations by varying the intensities of the same stimulus.[50] All the transformations can be introduced in a sequence of < 18 seconds by varying the power settings. The different parts of the glass transition are activated by the nanoparticles due to the generation of heat by an internal mechanism. In fact, by heating above 110°C, there is sufficient heat diffusion to maintain the surface temperature of the bulk sample at only 40°C or less that is very near to body temperature. The scheme can be seen in Figure 10.11.

10.7 SMP Textiles

Novel functions such as textile displays, motion-sensing dresses, self-cleaning textiles, temperature-regulated textiles, and so on are possible with smart functional textiles, with embedded digital components, and electronics, etc. Stimuli-responsive polymers (SRPs), which can sense and respond to environmental signals, have motivated many researchers to design self-regulated structures via the development of smart textiles.

10.7.1 Thermally Responsive Textiles

The switches based on the glass or melting transitions of SMPs have a physical cross-linking structure, a crystalline/ amorphous hard phase, or a chemical cross-linking structure and a low temperature transition to a crystalline or amorphous phase. For practical applications, the SMPs should have a thermal transition temperature that is required to be higher than the temperature of the environment in which they are applied. The switching of SMPs can be set at a temperature around the body temperature.

10.7.2 Shape Memory Finishing of Fabrics

The thermo-responsive properties of SMPs are imparted to textiles by introducing thermo-responsive SMPs into the textile via garment finishing and fiber spinning. Due to the shape recovery effect of the SMPs, a wrinkle-free effect is observed in cotton by its treatment with SMP.[51] The de-bonding and slippage of hydrogen bonds cause wrinkles in cotton fabric easily under low stress during wearing or storage.[52] The original flat shape of SM polyurethane (SMPU) treated cotton fabric is recovered within a minute upon blowing steam over it. The effect of SMPU finishing of cotton is evident in the increase in its mechanical strength and also its retention of a wrinkle-free effect after repeated washing (about 100 times). By treating with SMP emulsion, the increased dimensional stability and reduced felting effect of the wool garment are seen, as compared to untreated fabric. The thickness of the coating is optimized to achieve the appropriate effect.

10.7.3 Shape Memory Fibers

As compared to SMPU films, a lower shape fixity, higher shape recovery as well as recovery stress are induced by the molecular orientation of spun SMPU fibers. Electro-responsive SMP fibers can be designed by incorporating CNT into the fiber and the resulting composite exhibits shape recovery by electrical stimulation. The drawback in the material is that it requires a high voltage for shape recovery. The SMP fibers have better compatibility with the human body as compared to SM alloys, while the look and feel are similar to that of conventional clothing fabrics and are also much cheaper.

10.7.4 Fabrics Made of Shape Memory Polymer Fibers

Shape memory fabrics, with in-built SMP fibers in textiles and clothing, are used to design self-adaptable textiles with self-regulated structures. They respond to any changes in environmental temperature. Biomedical areas such as wound dressing, scaffolding materials, and orthodontics are considering the use of fiber/fabric made of SMPs because of the very good compatibility of the shape memory textiles with human skin. SMP fabric, used as wound dressing material, changes modulus in response to a change in body temperature. A low pressure can be applied to the wound by tuning the pressure on the wound. The structure of the fabric can be manipulated to adjust the water vapor permeability. The roughness of the SMP fibers is tackled by combining SMPU filaments with other fibers, such as cotton or wool.

10.7.5 Two-Way Shape Memory Fabrics

Two-way shape memory alloy springs can be used in a firefighter's coat by placing them between the two cloth layers. Having a switching temperature of around 50°C, the fabric extends when the temperature near the firefighter goes above 50°C. This way the two cloth layers separate from each other to make an increased air gap in between them that acts as a barrier to the surrounding heat. The length of the spring decreases below 50°C, while the two layers of cloth come back to their original position. Similar mechanisms can be adopted to design a smart shirt that has sleeves that roll up and roll down during a change in the surrounding temperature.

10.7.6 Damping Fabrics

It is known that good damping properties are possible around the glass transition temperature (switching temperature). SMPs can also be designed to absorb impact energy.[53] Automotive seat belt fabric can be designed to absorb kinetic energy by using the damping effect of the SMP fibers to improve a passenger's safety.[54] As an example, poly(ethylene terephthalate)–poly(caprolactone) block copolymers are melt spun to make SMP fibers.

10.7.7 Reversible Super-Hydrophilic/Super-Hydrophobic Fabrics

Designing self-cleaning surfaces, microfluidic tools, tunable optical lenses, etc. for industrial applications is done by using smart surfaces that can switch between super-hydrophobicity and super-hydrophilicity.[55]

Lotus leaves have a combination of micro and nano pillars covered with wax to make them super-hydrophobic in nature, due to the hydrophobic nature of wax, along with their unique microstructure. Similarly, the surface chemical compositions of thermo-responsive polymer hydrogels (TRPGs) can be changed with a variation of temperature that, in turn, changes the surface wettability. Below the lower critical solution temperature (LCST), the hydrophilicity exhibited by poly N-isopropylacrylamide (PNIPAAm) is due to its extensive intermolecular H bonding with water molecules. On the other hand, above the LCST, a compact and collapsed conformation results because of intramolecular H bonding in the PNIPAAm chains that leads to hydrophobicity.[56] Thus, the TRPG treated substrate can maintain the surface roughness that helps in increasing the hydrophobicity and hydrophilicity to a very high level. High roughness of the textile surfaces substrate is necessary to impart reversible switching behavior between super-hydrophilicity and super-hydrophobicity.

10.7.8 Chiral Two-Way Shape Memory Polymer Fibers

Chirality, the helical structure, is very common in natural materials. Unusual mechanical, optical, and magnetic properties can be imparted by chirality. Tensile actuation can be improved to large extent by using artificial muscle made of 2W-SMP fiber with a chiral design. The twist insertion and coiling are the two consecutive steps followed to fabricate artificial muscles.[57] The axial actuation of the coiled artificial muscle can be improved at lower actuation temperatures due to the two-way shape memory effect of the precursor and twisted (chiral) fibers.

10.8 Phase Change Materials

Phase change materials (PCMs) are known to take up and store heat energy at a high temperature, and at a lower temperature the stored heat energy is released via a phase change at an almost constant temperature. Coating or an encapsulation process is applied to incorporate PCMs in textiles to make thermo-regulated textiles. A solid-solid PCM can be prepared by using PEG as the soft segment. For the dual requirement of the shape memory effect as well as the maintenance of constant temperature, such materials are employed. Thus, the PCMs possessing shape memory effect can be integrated into the temperature-controlling textiles, houses, and packaging.

10.9 Moisture-Responsive Shape Memory Polymers

The rear vents in a smart shirt can respond to the sweating of the wearer and open up to allow perspiration and heat to escape. Under dry condition, the vents close automatically. Thus, water/moisture-responsive textiles can effectively control the moisture and heat management between the skin and the fabric via a change of macroshape or microstructure.

10.10 Shape Memory Hydrogels

The volumetric expansion of a polymer network with hydrophilic chain segments results from its strong interaction with water that leads to the formation of a hydrogel. Smart hydrogels can be realized by tuning the polymer chain segments so that they undergo reversible hydration under fluctuating environmental conditions that allow it to shrink/swell on demand. On the other hand, a special approach is needed for the switching of shape of a hydrogel rather than volume. In fact, a spatially directed movement of shape memory hydrogels (SMHs) without much change in volume is conceptually and mechanistically different from the normal approaches. For hydrophobic polymers, the shape memory effect allows the corresponding materials to revert back to a distinct previous shape in response to an external stimulus and that occurs for macroscopic as well as for microscopic matrices.[58]

The challenges associated with the application of concepts to SMHs are many, such as the hydrophilic chain segments of a hydrogel connecting the covalent net points are in the extended state, while the same is not true for hydrophobic, non-swollen materials. Therefore, the possible deformation involved in programming SMHs to their temporary shape is not satisfactory. The structural elements capable of forming a temporary cross-link in a water-rich environment are required to fix the temporary shape. The hydrogel properties, such as mechanical strength or ability to swell, change due to the temporary cross-link formation and cleavage, which, in turn, interfere with the preplanned macroscopic effect.

By way of anisotropic swelling, the hydrogels can be designed for directed movement. As an example, hydrogel layers with different ability to swell can be combined to form a gradient polymer network structure that will behave anisotropically. An LCST with sol-gel/gel-sol transition is exhibited by a poly(N-isopropylacrylamide) (PNIPAm) network in which small-scale variation in swelling is able to induce a reversible internal mechanical stress, leading to reversible movement.[59] A periodic assembly of stripes of different PNIPAm compositions undergoes shape transition from a planar sheet into a helical shape with the variation of the ionic strength or the pH of the medium.[60]

Further modification can be made by loading with gold nanoparticles, graphene, or CNT to harness their inherent properties for shape memory activities. The swelling-induced movements in DNA-based hydrogels can be effected by using self-assembly. After actuation of the SMH, the programming deformation can be reversed to control the directional movement. The pending moieties, such as short crystallizable side chains, oligomeric crystallizable side chains, groups capable of undergoing host-guest interactions, or complex-formation, form molecular switches used to fix the temporary shape for SMHs. The persistent phase segregation of oligomeric chains of oligo(tetrahydrofuran) and others into hydrophobic domains even above their T_{trans} prompts them to exhibit almost temperature-independent swelling behavior. The oligomeric switching segments are also capable of undergoing entanglements. The oligomeric side chains form the domains that are crystalline in the swollen polymer network. Isotropic X-ray scattering is exhibited by them in the permanent shape, while an anisotropic alignment of crystallites is maintained in the elongated temporary shape.[61]

By way of physical incorporation of the crystallizable units as molecular switches, double network architectures can be produced. By forming interpenetrating poly(vinyl alcohol) chains in a chemically cross-linked poly(ethylene glycol) hydrogel, crystalline domains can be formed that are assisted by freezing/thawing cycles for stabilization of the temporary shape.[62]

10.10.1 Network Architectures and Concepts for Temporary Cross-Links

Rapid diffusion of small molecules is possible by the hydrogel condition that acts as a trigger for the SME. When hydrogen bonding, dipole-dipole interactions, or ion complexation are used to fix a temporary shape instead of crystallizable domains, the cleaving of the molecular switches can be done by using complexing agents, changes of pH, or redox reactions. By introducing two different types of crystallizable side chains in a hydrophilic network structure, the triple-shape hydrogels can be realized that have the ability to produce two steps of shape switching,. For example, a copolymer network

can be designed by using ethylene glycol oligomers-cross-linked N-vinylpyrrolidone main chains and oligo(ω-pentadecalactone) and oligo(tetrahydrofuran) as side chains, that exhibit recovery to the intermediate and permanent shape by heating because of the presence of isolated thermal transitions of the crystalline domains of each combination of side chains.[63]

10.10.2 Hydrogel Microstructures

The swelling/de-swelling of SMHs during cleaving/forming of temporary cross-links changes the material dimension that interferes with the directed movement and the recovery performance. By integrating the large interconnected pores created by leaching techniques, cross-linking reactions in emulsions, or gas-triggered foaming, the volume change by swelling/de-swelling can be minimized and in this way the material functionality of a superstructure is improved. The diffusion processes of nutrients, ions, oxygen, and other molecules, taking place from outside to inside of the material, are facilitated by the microscale superstructure.[64] The ultrasound waves penetrate through the interconnected pores of the porous structure. The collapse of the cavitation bubbles inside the macropores is also induced by the porous structure that improves the efficiency of energy generation from ultrasound to the polymer.

In situations where the insertion dimensions are critical, the shape memory property of the hydrogel provides treatment options. Minimally invasive spinal surgery uses these properties well wherein the hydrogel is configured to a particular shape via collapsing and then after implantation it transforms into a larger and different, functional shape. The configuring of implants to a minimized shape facilitates insertion with little tissue damage and the implant functions in a fully hydrated state as a nucleus augmentation implant.

10.11 Shape Memory Polymer-Based Sealants

The expansion joints in small-span bridge decks and concrete pavements, etc. are sealed by using a variety of sealants. Along with aging, adhesive failure and cohesive failure are very common for these sealants. The expansion of a concrete wall at high temperatures poses serious problems to compression-sealed sealants. The programming in a two-dimensional (2D) stress condition, i.e., compression in the horizontal direction and tension in the vertical direction, can be done with SMP-based syntactic foam, while the transition temperature is controlled below the highest temperature to be experienced during use.[65] The accumulated compressive stress can be significantly released due to a drop in the stiffness of the foam at a temperature above the transition temperature. As a result, the crushing of the concrete and the sealant is avoided. As the foam shrinks in the vertical direction during recovery of the foam, the squeezing-out problem is eliminated. The compression programming in the horizontal direction of the 2D-programmed smart sealant traffic ensures the recovery of plastic deformation at high temperatures while the contact with the concrete wall is maintained and the leaking problem is minimized.

10.12 Biodegradable Shape Memory Polymers

Along with their response to several environmental conditions, the biodegradability of SMPs makes them a potential candidate for many medical applications. The nature of SMPs allows the conversion of bulky implants into a compressed shape that is inserted into the body through a small incision, and then it takes its normal designed shape. An intelligent suture for wound closure can be made from biodegradable shape memory polymer. Stress is applied to the wound lips by the material via its shape memory effect. The implant should degrade within the designed time interval so that follow-up surgery is not required for its removal. Introduction of weak, hydrolyzable bonds that break under physiological conditions is required to maintain the biodegradability of the SMP.

10.12.1 Biocompatible Electrically Conductive Nanofibers

For nerve tissue engineering applications, a porous shape memory scaffold with both biomimetic structures and electrical conductivity has an important role. For example, inorganic polydimethylsiloxane segments and organic poly (ε-caprolactone) (PCL) segments (switching segment) containing SMPU, with carbon black incorporation, can be electrospun to obtain electrically conductive nanofibers.[66] The shape memory properties of these nanofibers are not influenced much by the carbon black addition. However, even after five thermo-mechanical cycles, the shape recovery ratios and shape fixity ratios are maintained above 90% and 82% respectively. By promoting cell-cell interactions, good biocompatibility is maintained for PC12 cells cultured on the shape memory nanofibers.

10.13 3D Printing of Shape Memory Hydrogels

The design ability of stimuli-responsive smart polymers with a 3D printer enhances the visual effects, along with their applicability for engineering purposes. The limitations associated with design and cost during traditional mold processing are eliminated in this technology due to its geometric independence, economic viability, and easy and rapid fabrication of complex design. Shape deformable polymers and gels can be fabricated by 3D printing to design actuator valves, self-foldable, and healable materials for engineering and biomedical applications. By using a 3D optical printing method, commonly known as a stereolithographic process, shape memory hydrogels can be made using a new type of material based on acrylamide and acrylate containing copolymers.[67] Their physical properties are tuned by both hydrophobic and hydrophilic components. The shape fixity ratio and the recovery ratio are above 99% for almost all the cycles. They can be adjusted by simply varying the molar ratio of the monomers in the copolymer matrix to adjust the toughness and flexibility. At elevated temperatures, extraordinary strain percentage is exhibited by this hydrogel.

10.14 Design Strategies for Shape Memory Polymers

A proper design strategy is required to have sufficient control over material properties, functions, and possible applications. Several strategies are now discussed.[68]

10.14.1 Chemical Cross-Linking of a High MW Thermoplastic Polymer

Every polymer has at least a minimum of one thermal transition available which can be used as a switching mechanism for shape memory. The network structure resulting from chemical cross-linking defines the permanent shape or memory. However, less susceptibility of saturated polymers to free radical cross-linking, compared to unsaturated polymers, leads to incomplete cross-linking. This adversely affects the shape memory performance, as both recovery and permanent deformation remain quite low. By physically blending the polymer with an unsaturated species, acting as a sensitizer, the gel fraction can be increased via gamma irradiation-induced cross-linking. In general, the sensitizer decreases the gelation extent and maintains good shape memory properties. Also, the base polymer is able to be processed using conventional processing techniques.

10.14.2 One-Step Polymerization of Monomers/ Pre-Polymers/Cross-Linking Agents

One-step polymerization is done by reacting together the selected monomers, reactive pre-polymers, and cross-linking agents so that polymerization and cross-linking are completed in one step. The shape memory and other material properties are controlled in a much better way by this method as the system undergoes a sol-gel transition similar to any thermosetting resins. As an example, T_m-based SMPs are synthesized by using a reactive pre-polymer equivalent to the crystallizable switching segment along with other monomers and/or cross-linking agents.

10.14.3 One-Step Synthesis of Phase-Segregated Block Copolymers

Although this process is similar to that described above, it is targeted to yield a thermoplastic polymer so that it will be possible to process it by using more conventional plastics processing techniques. The SMPs obtained are usually SMPs with two blocks having different transition temperatures results due to phase segregation of the block copolymers. Generally, the shape memory, as well as other material properties of the SMPs, obtained by this method, is nicely controlled in this system. However, sometimes these SMPs exhibit various properties that are inferior to thermoset SMP due to the absence of cross-links.

10.14.4 Direct Blending of Different Polymers

Another major strategy to make SMPs is the physical blending of thermoplastic polymers having nil or limited shape memory. Two approaches are mostly followed to acquire memory from one polymer and a switching mechanism from the other. Keeping in mind the absence of network structures, one thermoplastic polymer is blended with a second thermoplastic polymer having a higher transition temperature, in the first approach, so that it functions as a physical cross-link for temperatures between the two transitions. In the second approach, blending of a polymer of the desired T_g or T_m is done with an elastomer. Blending of an elastomeric ionomer (e.g., sulfonated) is also done with various low-MW fatty acids and fatty acid salts. The transitions for shape memory are maintained by the melting of these fatty acids and salts. Although the processes appear simple, there are various complications in the blending strategy that make it challenging in practice. Further, changing the property of one component can influence the property of the second one.

10.14.5 Other Strategies

The other strategies incorporate shape memory elastomeric composites (SMECs). For example, semi-crystalline PCL existing as non-woven fabrics of microfibers, and evenly distributed in a continuous silicone rubber make a two-phase morphology. Useful shape memory properties are not possible with either of the components. However, in combination, the silicone rubber provides an entropic network (the memory) and the PCL functions as a T_m-based switching segment, and it exhibits excellent shape memory performance.[69]

10.15 Characteristics of Shape Memory Polymers

Various methods of characterization of SMPs can be followed to determine the relevant parameters to gain an understanding of the shape memory effect. Some tests and their requirements are mentioned to provide a basic understanding.[70]

10.15.1 Mechanical Properties

- *Static tensile test*: This is used to investigate the elastic moduli and strains of SMP samples at different temperatures.
- *Dynamic mechanical analysis test*: The T_g of SMPs are determined by using a dynamic mechanical analyzer.
- *Thermo-mechanical cycle test*: The shape fixity and recovery ability of SMP samples are found by using the thermo-mechanical cycle test.

10.15.2 Percent Recovery Testing

The shape memory polymer under test is clamped to one platform and the heat gun is mounted on another horizontal and adjustable platform to test the percent recovery. The heat gun can be turned on to heat the SMP for one minute. The percent recovery is calculated by using the formula

% recovery = ((stored angle) − (recovered angle))/stored angle

The SMP is always heated up before deforming with the heat gun while placing it at a distance of 7 inches. The heat gun is placed at varied distances away during recovery so that different recovery temperatures are induced.

10.15.3 Force Testing

Force testing uses the same apparatus as was used in percent recovery testing. The drift of the force sensor's result originating from the change in temperature is minimized by employing a longer probe on the force sensor. Before deforming for force testing, the SMP is always heated up with the heat gun at a distance of 7 inches. Different recovery temperatures are maintained during the recovery process by placing the heat gun at various distances.

10.15.4 Static Friction Testing

In static friction testing, the static coefficient of friction between the SMP samples and aluminum, ABS plastic's rough as well as smooth side are determined. In the test, each sample is placed on the surface being tested and then the surface is tilted to the point that the SMP slips down the surface. The coefficient of friction is calculated as the tangent of the angle at which the slip occurred.

10.15.5 Cyclic Thermo-Mechanical Test

A mechanical testing instrument, equipped with a temperature control unit, is used for quantitative analysis of shape memory performance.[71] This test is applied for both stress-controlled and strain-controlled mode. A predefined stress and temperature increase rate are applied to the SMP during a stress-controlled cyclic thermo-mechanical test, and the strain over time is obtained as output. Similarly, a strain-controlled cyclic thermo-mechanical test is conducted by applying a predefined strain and temperature rise rate while the stress over time is obtained as output. A typical cyclic thermo-mechanical test with stress-control in the Nth cycle constitutes four distinctive steps (Figure 10.12).

The steps can be given as:

Step 1. The SMP is equilibrated with a recorded permanent strain $\varepsilon_p(N\text{-}1)$ at deformation temperature T_d. Predefined deformation stress σ at T_d is applied to the SMP. ε_l^d (N) is the recorded deformed sample length at T_d.

Step 2. The SMP is cooled to a lower fixing temperature (T_f) by applying a constant load σ so that the temporary length ε_l^d (N) is fixed. Under loading, the fixed strain observed at T_f is ε_l (N).

Step 3. The stress is unloaded to zero or a specific lower constrain stress (σ_c). The strain is recorded as $\varepsilon_u(N)$ after unloading.

Step 4. The shape is recovered with either zero stress or a specific constrain stress σ_c at T_r.

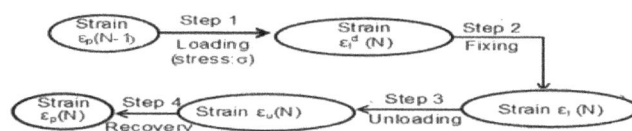

FIGURE 10.12 Programming steps in a stress-controlled cyclic thermo-mechanical test of an SMP.

Notes: p = permanent, l = with loading, d = deformation temperature T_d, u = unloading.

$\varepsilon_p(N)$ is recorded as the final recovered strain. The SMP recovers in the absence of external constraints in a free-strain recovery mode, either variable with temperature during a transient heating or with time during an isothermal process. Alternatively, in the fixed-strain recovery mode, stress within the SMP is generated with full deformation constraint, either variable with temperature during a transient heating or with time during an isothermal process. E_r, representing the shape recovery stress, which is generated by a SMP during recovery, is recorded during this mode.

The status of the SMP held in a strained temporary shape is assessed by using the shape fixing ratio (R_f). The ratio of the deformation after unloading and that at T_f under the loading σ give the value of R_f, and this is calculated by using Equation (10.3):

$$R_f = \frac{\varepsilon_u(N) - \varepsilon_p(N-1)}{\varepsilon_l(N) - \varepsilon_p(N-1)} \qquad (10.3)$$

The completeness of recovery of an SMP to its memorized state is understood from the shape recovery ratio (R_r). The ratio of the recovered deformation at T_r and the fixed deformation under stress at T_d refers to the R_r value, and this is calculated by using Equation (10.4):

$$R_r = \frac{\varepsilon_u(N) - \varepsilon_p(N)}{\varepsilon_u(N) - \varepsilon_p(N-1)} \qquad (10.4)$$

10.15.6 Quantitative Determination of Shape Memory Properties

The amount of the applied strain that is recovered in the same cycle following heating above the TS is considered the percentage strain recovery:

$$\% \text{ strain recovery} = \frac{\varepsilon_m - \varepsilon_p}{\varepsilon_m} \times 100\% \qquad (10.5)$$

where ε_m is the maximum strain imposed on the material. ε_p is the strain in the sample at the end of the same cycle before the yield stress is applied.

The ability of the switching segment to hold shape after the sample has been stretched is expressed as the percentage of strain fixity.

$$\% \text{ strain fixity} = \frac{\varepsilon_p}{\varepsilon_m} \times 100\% \qquad (10.6)$$

10.16 Applications of SMPCs

During the 1960s, shape memory polymers were used for their first large-scale application as PE thermal contraction tubes because of their good flame-retardant property, insulation, and temperature-tolerance properties. In insulated joint protection, erosion prevention, as well as in electrical work, these materials are widely used. Since then, the studies of shape memory polymers have undergone rapid development.

The simple shape memory polymers have a limitation in their performance because of their lower thermodynamic properties, along with the single stimulation mode. This suggests that shape memory polymer composites should be developed for further study and their use. SMPCs can be used to replace shape memory alloys as structural materials for aerospace applications because of their light weight, high deformability, and high fixation rate. While maintaining these advantages, the mechanical, shape recovery, and various thermodynamic properties of composite fiber materials can be enhanced to a considerable extent. Appliances, brackets, and occluders are the main biomedical instruments in which SMPs are employed. SMP smart structures applied inside the body for minimally invasive surgery require methods of stimulation without direct contact. Indirect heating of doping materials is used for this purpose. Sensors, energy devices, and electronic accessories, etc. can also be made by using SMPs. Their applications in the fields of smart fabrics, 4D printing, etc. are considered a possibility.

10.16.1 Applications in Aerospace and Aviation

For applications in all kinds of aerospace expansion structures and in driving machine-based locking-release structures with limited deformation times, the stable single-deformation ability of SMPs is ideally suited. The polymer-based shape memory composites, having low mass, high toughness, and large deformation, can be used to design matrices of large expansion structures[72] in which a combination of driving devices and structural materials is provided.

The environment in space is severe compared to our normal environment. More severe erosion of polymers and other organic material than metal or ceramic materials by atomic oxygen and ultraviolet radiation, together with extreme temperature differences, lead to loss of mass of polymers, deterioration in their dynamic functions, and even complete loss of performance. Good tolerance of the outer-space environment is observed with epoxy resin-based and cyanate ester-based SMPs. In both high and low temperatures, the chemical stability of the polyimides makes them suitable for future applications in outer space.

10.16.2 Biomedical Devices Based on SMPCs

Suitable recovery force as well as good biological adaptability of polymers can be exploited by using them in biomedical devices. Polyurethane, polycaprolactone, polyether, polyurethane, and polylactic acid are suitable as SMPs and they are also degradable in the body. Also, the driving methods of SMPs can be varied by making composites. In fact, controlling from a remote location is possible with both light driving (e.g., microwave) and magnetic driving techniques. Excellent potential exists for SMPCs with these properties to be integrated into medical instruments used inside the body. Like SMPs, the SMPCs have found use in many biomedical instruments, such as thrombus cleaners, surgical sutures, intravascular stents, aneurysm occluders, and others. SMP is used to make aneurysm coils as a replacement of traditional platinum coils that eliminates the aneurysm re-opening caused by biological inertia.[73] By blending the tantalum filler with an SMP, X-ray impermeability can be increased, which means an X-ray image of the tantalum-filled SMP coil can be obtained and the response of the coil does not affect the blood flow.

10.16.3 Smart Textiles

For complicated structures, the SMPs play an important role in eliminating the system difficulties. Greater deformation ability is possible with shape memory polymers and their composites, which make it easier to produce more complex structures, compared to those with SMA. Bottle-shaped and S-shaped air duct structures are easily produced with SMPs, whose dabbers can be used repeatedly.[74] Material deformation caused by external forces can recover to its original shape by using stimulations like heating. Textiles can be made from SMPs, which, on heat application, allow the wrinkles to disappear automatically.

10.16.4 Smart Electronics

To design soft electronics, which has application possibilities in the biological electronic techniques, SMPs can be used. To overcome the difficulty in adjusting to complex and soft biological surfaces with traditional hard electronic devices, the SMP-based soft electronic devices exhibit better properties together with higher comfort in use.[75] The benefits of using SMPs can be given as:

- Variable hardness.
- The SMPs are in a soft and programmable state at higher temperature, which allows the scope for self-adjustment of their properties or self-finishing.
- A tough and stable state is attained by cooling it to room temperature so that its properties are stabilized and structural supports are possible.

Excellent electric properties and reliability are achieved with SMPs used in crystal tubes with organic thin membranes via photolithography and their function is retained even after 100 cycles of softening and bending.[76] The resistors, made from carbon nanotubes and shape memory polyurethane composite material, have a sensitivity to water that makes them suitable for making moisture sensors.[77] Soft temperature sensors are made of composite materials via a combination of surface layers of shape memory polyurethane with silver nanowire that show sensitivity to temperature.[78]

10.16.5 4D Printing

3D printing can be improved to 4D printing by incorporating time into the printed matter and the change involves shape, and other properties in design directions. The direct-write printing of UV cross-linkable poly(lactic acid)-based inks with excellent shape memory ability is employed to make 3D printing along with doping with ferrite to achieve remote control of the shape change.[79] The global deformation is used to design a 4D print that is based on swelling equilibrium, which uses variable exposure time for UV light. This permits the hydrogels to achieve different cross-linking conditions so that the swell equilibrium is changed. This technology is based on the exposure of the plane simultaneously and it no longer relies on the one-dimensional print head.[80]

10.16.6 A Boon for the Disabled

The handles of spoons made from SMPs can be deformed to a predetermined extent by placing them in hot water. Subsequently, the spoon is immersed in cold water to fix the shape that can be used by the disabled person .The original shape is recovered by heating the spoon in boiling water.

10.16.7 Applications in Breathable Textiles

Garments coated with shape memory polymer can be used to enhance the comfort level. Garments can be coated with shape memory polyurethanes to add a permeability controlling feature to a large extent. At a higher temperature the water vapor permeability is realized, while at a lower temperature the permeability is quite low. This feature allows it to be employed in breathable textiles.

10.16.8 Self-Tightening Sutures

Self-tightening sutures can be made by using SMPs. To prevent damage to the cells of the skin, the sutures are stitched very carefully, even for very tight stitching. On the other hand, the required functions are not done properly if the threads are kept very loose. By using SMP sutures, the threads can be stitched loosely, which contract to tighten by coming into contact with the body temperature. Shape memory polymers can also be used in the removal of clots. The SMP straight rod is inserted into a vessel. This is the secondary shape fixed; the primary shape being a tapered cork screw. When it passes through the clot, a laser is beamed on it to set it in the shape recovery that allows the initial corkscrew shape to return to hold the clot firmly. For normal blood flow, both of them are removed.

10.16.9 Biodegradable Shape Memory Polymers in Medicine

The scope and clinical utility of SMPs in medical applications have been multiplied by designing biodegradable SMPs (BSMPs).[81] In general, minimally invasive procedures are also facilitated by using SMPs. The biodegradability in these types of applications has an added benefit in terms of the requirement of the procedure to remove implanted non-degradable materials no longer exists, which makes it simple and cost-effective. Two common features of sensitive bonds in biodegradable polymers are: (1) the presence of hydrolytically sensitive functional groups like esters, amides, urea, etc.; and (2) the presence of enzymatically sensitive polymers like polysaccharides, or their derivatives. Because of the well-established biodegradability and use of aliphatic polyesters, such as PCL, poly(lactide) (PLA), and their copolymers, BSMPs find widespread use in medical applications.

10.16.10 Poly(ε-caprolactone)

This material suffers from low fixity, low recovery parameters together with crystallinity-related issues. To eliminate the problem, cross-linking the material via acrylate end-groups is commonly employed.[82] The derived materials exhibit high R_f and R_r values. Alternatively, it can be blended with a polymer with a much higher T_m to ensure that the physical cross-links are maintained over the temperature range of the temporary shape transformation.[83]

10.16.11 Poly(lactide)- and Lactide-Based Copolymers

Uncross-linked PLLA exhibits shape memory behavior because of its semi-crystalline nature; the crystalline regions act as physical cross-links that prohibit chain slippage during programming.[84] The uncross-linked high molecular weight copolymers of D- and L-lactide (PDLLA) exhibit good shape memory properties because of the formation of network structure by chain entanglements.[85] The T_g of PLLA is quite high compared to physiological temperatures, which can be used for in vivo applications only by arranging for heating from an external source to activate the shape changes. This is tackled by making copolymers. Of course, the hydrolytic degradation of PLLA is quite common. In addition, the use of BSMPs in dynamic surfaces allows the control of cell orientation, growth, and differentiation. Although the patterned microgrooves control the cell orientation and morphology, the triggering of shape recovery induces the aligned cells cultured on microgrooves to lose their alignment.[86]

REFERENCES

1. M. Voit, T. Ware, R.R. Dasari, P. Smith, L. Danz, D. Simon, S. Barlow, S.R. Marder, and K. Gall. "High-strain." *Shape-Memory Polymers* 20(1) (2010): 162–171.
2. Wikipedia. "Shape-memory polymer." Available at: ,https://en.wikipedia.org/w/index.php?title=Shape-memory_polymer &oldid=853402770 (accessed November 8, 2018).
3. B.K. Kim, S.Y. Lee, and M. Xu. "Polyurethanes having shape memory effects." *Polymer* 37(26) (1996): 5781–5793.
4. J. Hu, H. Meng, G. Li, and S.L. Ibekwe. "A review of stimuli-responsive polymers for smart textile applications." *Smart Mater. Struct.* 21 (2012): 053001.
5. L. Sun, W.M. Huang, C.C. Wang, Y. Zhao, Z. Ding, and H. Purnawali. "Optimization of the shape memory effect in shape memory polymers." *J. Polym. Sci. Part A: Polym. Chem.* 49(16) (2011): 3574–3581.
6. J. Karger-Kocsis and S. Kéki. "Biodegradable polyester-based shape memory polymers: Concepts of (supra) molecular architecturing." *eXPRESS Polym. Lett.* 8(6) (2014): 397–412.
7. S. Thakur and J. Hu. "Polyurethane: A shape memory polymer (SMP)." In F. Yilmaz (Ed.), *Aspects of Polyurethanes.* Intech Open, 2017. doi: 10.5772/intechopen.69992.
8. F. Li, A. Perrenoud, and R.C. Larock. "Thermophysical and mechanical properties of novel polymers prepared by the cationic copolymerization of fish oils, styrene and divinylbenzene." *Polymer*, 42(26) (2001): 10133–10145.
9. C. Liu and P.T. Mather. *Proceedings of the Annual Technical Conference – Society of Plastics Engineers.* Brookfield, CT, 2 (2003) 1962–1966.
10. C. Liu, H. Qin, and P.T. Mather. "Review of progress in shape-memory polymers." *J. Mater. Chem.* 17(16) (2007): 1543–1558.
11. I. Bellin, S. Kelch, R. Langer, and A. Lendlein. "Polymeric triple-shape materials." *Proc. Nat. Acad. Sci. of U.S.A.* 103(48) (2006): 18043–18047.
12. T. Xie, X.C. Xiao, and Y.T. Cheng. "Revealing triple-shape memory effect by polymer bilayers." *Macromol. Rapid Commun.* 30(21) (2009): 1823–1827.
13. J. Xu and J. Song. (2011). "Thermal responsive shape memory polymers for biomedical applications, biomedical engineering." In R. Fazel-Rezai (Ed.), *Biomedical Engineering: Frontiers and Challenges.* Intech Open. doi: 10.5772/19256. Available at: www.intechopen.com/books/biomedical-engineering-frontiers-and-challenges/thermal-responsive-shape-memory-polymers-for-biomedical-applications
14. Q. Peng, H. Wei, Y. Qin, Z. Lin, X. Zhao, F. Xu, J. Leng, X. He, A. Cao, and Y. Li. "Shape-memory polymer nanocomposites with a 3D conductive network for bidirectional actuation and locomotion application." *Nanoscale* 8(42) (2016): 18042–18049.
15. A.H. Torbati and P.T. Mather. "A hydrogel-forming liquid crystalline elastomer exhibiting soft shape memory." *J. Polym. Sci., Part B: Polym. Phys.* 54(1) (2016): 38–52.
16. K. Wang, Y-G. Jia, and X.X. Zhu. "Two-way reversible shape memory polymers made of cross-linked cocrystallizable random copolymers with tunable actuation temperatures." *Macromolecules*, 50(21) (2017): 8570–8579.
17. E. Havens, E.A. Snyder, and T.H. Tong. "Light-activated shape memory polymers and associated applications." In E.V. White (Ed.), *SPIE Smart Structures and Materials + Nondestructive Evaluation and Health Monitoring*, vol. 5762, 7–10 March 2005/
18. D.A. Perkins, H. Reed, and E. Havens. "Adaptive wing structures," In E.H. Anderson (Ed.), *Proceedings of SPIE Conference*, vol. 5388, 14–18 March, 2004.
19. H. Zhou, C. Xue, P. Weis, Y. Suzuki, S. Huang, K. Koynov, G.K. Auernhammer, R. Berger, H.J. Butt, and S. Wu. "Photoswitching of glass transition temperatures of azobenzene-containing polymers induces reversible solid-to-liquid transitions." *Nat. Chem.* 9 (2017): 145–151.
20. I. Azcune and I. Odriozola. "Aromatic disulfide crosslinks in polymer systems: self-healing, reprocessability, recyclability and more." *Eur. Polym. J.* 84 (2016): 147–160.
21. S. Ji, J. Xia, and H. Xu. "Dynamic chemistry of selenium: Se–N and Se–Se dynamic covalent bonds in polymeric systems." *ACS Macro Lett.* 5 (2016): 78–82.
22. W.M. Huang, B. Yang, N. Liu, and S.J. Phee. "Water-responsive programmable shape memory polymer devices." *Proceedings of SPIE* 6423(2007): 64231S1–7.
23. G. Niu and D. Cohn. "Water triggered shape memory materials." *Science Insights.* 3(1) (2013): 49–50.
24. R.C. Thompson, A.K. Shung, M.J. Yaszemski, and A.G. Mikos. *The Principles of Tissue Engineering.* San Diego, CA: Academic Press, 2000.
25. W.M. Huang, B. Yang, L.H. Wooi, S. Mukherjee, J. Su, and Z.M. Tai. "Formation and the adjustment of bubbles in a polyurethane shape memory polymer." In H.P. Glick (Ed.), *Materials Science Research Horizons.* New York: Nova Science Publishers, 2007, pp. 235–50,
26. W.M. Huang. "Thermo-moisture responsive polyurethane shape memory polymer for biomedical devices." *The Open Med. Devices J.* 2 (2010): 11–19.

27. H.M. Wache, D.J. Tartakowska, A. Hentrich, and M.H. Wagner. "Development of a polymer stent with shape memory effect as a drug delivery system." *J. Mater. Sci: Mater. Med.* 14 (2003): 109–112.

28. D. Quitmann, N. Gushterov, G. Sadowski, F. Katzenberg and J.C. Tiller. "Solvent-sensitive reversible stress-response of shape memory natural rubber." *ACS Appl. Mater. Interfaces* 5(9) (2013): 3504–3507.

29. A. Paillous and C. Pailler. "Degradation of multiply polymer-matrix composites induced by space environment." *Composites* 25 (1994): 287–295.

30. M. Lebron-Colon, M.A. Meador, J.R. Gaier, F. Solá, D.A. Scheiman, and L.S. McCorkle. "Reinforced thermoplastic polyimide with dispersed functionalized single wall carbon nanotubes." *ACS Appl. Mater. Interfaces* 2(3) (2010): 669–676.

31. X. Xiao, D. Kong, X. Qiu, W. Zhang, Y. Liu, S. Zhang, F. Zhang, Y. Hu, and J. Leng. "Shape memory polymers with high and low temperature resistant properties." *Sci. Rep.* 5 (2015): 14137.

32. L. Sun, W.M. Huang, Z. Ding, Y. Zhao, C.C. Wang, H. Purnawali, and C. Tang. "Stimulus-responsive shape memory materials." *Rev. Mater. Des.* 33 (2012): 577–640.

33. K. Hearon, P. Singhal, J. Horn, W. Small, C. Olsovsky, K.C. Maitland, T.S. Wilson, and D.J. Maitland. "Porous shape memory polymers." *Polym. Rev.* 53(1) (2013): 41–75.

34. G.J. Monkman and P.M. Taylor. "Memory foams for robot grippers." *Proc. 5th Int. Conf. Adv. Robotics.* Pisa, Italy, 1(1991): 339–342.

35. W.M. Sokolowski and S.C. Tan. "Advanced self-deployable structures for space applications." *J. Spacecraft and Rockets* 44(2007): 750–754.

36. W. Sokolowski, S. Tan, P. Willis, and M. Pryor. "Shape memory self-deployable structures for solar sails." *Proceedings of SPIE – The International Society for Optical Engineering* (2008): 7267.

37. C. Ozan, W. Van der Zee, W. Brudy, and J. Vinson. "Mechanical modeling of shape memory polyurethane foam for application as a sand management solution." *SPE Annual Technical Conference and Exhibition* 3 (2011): 2210–2219.

38. P. Mather, K. Ishida, and P. Wilson. "Waterborne shape memory polymer coatings," WO 2014/055985-. 2014.

39. S. Sharafi and G. Li. "Multiscale modeling of vibration damping response of shape memory polymer fibers." *Composites Part B: Engineering* 91 (2016): 306–314.

40. S. Bauer, S. Bauer-Gogonea, I. Graz, M. Kaltenbrunner, C. Keplinger, and R. Schwodiauer. "25th anniversary article: A soft future: from robots and sensor skin to energy harvesters." *Adv. Mater.* 26(1) (2014): 149–162.

41. T. Mu, L. Liu, X. Lan, Y. Liu, and J. Len. "Shape memory polymers for composites." *Compos. Sci. Technol.* 160(26) (2018): 169–198.

42. X. Qi, X Yao, S. Deng, T. Zhou, and Q. Fu. "Water-induced shape memory effect of graphene oxide reinforced polyvinyl alcohol nanocomposites." *J. Mater. Chem. A* 2(7) (2014): 2240–2249.

43. X. Qi, H. Xiu, and W. Yuan. "Enhanced shape memory property of polylactide/thermoplastic poly (ether) urethane composites via carbon black self-networking induced co-continuous structure." *Comp. Sci. Tech.* 139 (2017): 8–16.

44. J. Leng, H. Lv, Y. Liu, and S. Du. "Synergic effect of carbon black and short carbon fiber on shape memory polymer actuation by electricity." *J. Appl. Phys.* 104(2008): 104917.

45. J.S. Leng, X. Lan, Y.J. Liu, S.Y. Du, W.M. Huang, N. Liu, S.J. Phee, and Q. Yuan. "Electrical conductivity of thermoresponsive shape-memory polymer with embedded micron sized Ni powder chains." *Appl. Phys. Lett.* 92(2008): 014104.

46. D. R. Erika, X. Luo, and P.T. Mather. "Linear/network poly (ε-caprolactone) blends exhibiting shape memory assisted self-healing (SMASH)." *ACS Appl. Mater. Interfac.* 3(2) (2011): 152–161.

47. P. Zhang and G. Li. "Advances in healing-on-demand polymers and polymer composites." *Prog. Polym. Sci.* 57 (2016): 32–63.

48. R. Mohr, K. Kratz, T. Weigel, M. Lucka-Gabor, M. Moneke, and A. Lendlein. "Initiation of shape-memory effect by inductive heating of magnetic nanoparticles in thermoplastic polymers." *PNAS* 103 (10) (2006): 3540–3545.

49. K. Yu, K.K. Westbrook, P.H. Kao, J. Leng, and H.J. Qi. "Design considerations for shape memory polymer composites with magnetic particles." *J. Compos. Mater.* 47(1) (2013): 51–63.

50. F.H. Zhang, Z.C. Zhang, C.J. Luo, I-T. Lin, Y. Liu, J. Leng, and S.K. Smoukov. "Remote, fast actuation of programmable multiple shape memory composites by magnetic fields." *J. Mater. Chem. C* 3(43) (2015): 11290–11293.

51. Y. Li, S. Chung, L. Chan, and J.L. Hu. "Characterization of shape memory fabrics". *Textile Asia* 35(2004): 32–37.

52. J. Hu, Y. Zhu, H. Huang, and J. Lu. "Recent advances in shape–memory polymers: Structure, mechanism, functionality, modeling and applications." *Prog. Polym. Sci.* 37(12) (2012): 1720–1763.

53. J.H. Yang, B.C. Chun, Y.C. Chung, J.W. Cho, and B.G. Cho. "Vibration control ability of multi-layered composite material made of epoxy beam and polyurethane copolymer with shape memory effect." *J. Appl. Polym. Sci.* 94 (2004): 302–307.

54. K.Y. Lim, B.C. Kim, and K.J. Yoon. "Effect of structural characteristic on physical properties of copolyesters from poly(ethylene terephthalate) oligomer and polycaprolactone." *J. Polym. Sci. Part B* 40 (2002): 2552–2560.

55. B. Xin and J. Hao. "Reversibly switchable wettability." *Chem. Soc. Rev.* 39(2010): 769–782.

56. X. Feng and L. Jiang. "Design and creation of superwetting/antiwetting surfaces." *Adv. Mater.* 18(23) (2006): 3063–3078.

57. O. Yang, J. Fan, and G. Li. "Artificial muscles made of chiral two-way shape memory polymer fibers." *Appl. Phys. Lett.* 109 (2016): 183701.

58. C. Wischke and A. Lendlein. "Method for preparation, programming and characterization of miniaturized particulate shape-memory polymer matrices." *Langmuir* 30(2014): 2820–2827.

59. Y. Zhang and L. Ionov. "Reversibly cross-linkable thermoresponsive self-folding hydrogel films." *Langmuir* 31(2015): 4552–4557.

60. Z.L. Wu, M. Moshe, J. Greener, H. Thérien-Aubin, Z.H. Nie, E. Sharon, and E. Kumacheva. "Three-dimensional shape transformations of hydrogel sheets induced by small-scale modulation of internal stresses." *Nat. Commun.* 4 (2013): 1586.

61. M. Balk, M. Behl, U. Nöchel, and A. Lendlein. "Shape-memory hydrogels with crystallizable oligotetrahydrofuran side chains." *Macromol. Symp.* 345(2014): 8–13.

62. G. Li, H.J. Zhang, D. Fortin, H.S. Xia, and Y. Zhao. "Poly(vinyl alcohol)-poly(ethylene glycol) double-network hydrogel: A general approach to shape memory and self-healing functionalities." *Langmuir* 31 (2015): 11709–11716.

63. U. Nöchel, M. Behl, M. Balk, and A. Lendlein. "Thermally-induced triple-shape hydrogels: soft materials enabling complex movements." *ACS Appl. Mater. Interfaces* 8 (2016): 28068–28076.

64. C. Löwenberg, M. Balk, C. Wischke, M. Behl, and A. Lendlein. "Shape-memory hydrogels: Evolution of structural principles to enable shape switching of hydrophilic polymer networks." *Acc. Chem. Res.* 50 (2017): 723–732.

65. G. Li, A. King, T. Xu, and X. Huang. "Behavior of thermoset shape memory polymer-based syntactic foam sealant trained by hybrid two-stage programming." *J. Mater. Civil Eng.* 25(3) (2013): 393–402.

66. D. Kai, M.J. Tan, M.P. Prabhakaran, B.Q.Y. Chan, S.S. Liow, S. Ramakrishna, and X.J. Loh. "Biocompatible electrically conductive nanofibers from inorganic-organic shape memory polymers." *Colloids Surf. B: Biointerfaces* 148 (2016): 557–565.

67. M.D.N.I. Shiblee, K. Ahmed, A. Khosla, M. Kawakami, and H. Furukawa. "3D printing of shape memory hydrogels with tunable mechanical properties." *Soft Matter*, 14(38) (2018): 7809–7817.

68. X. Luo and P.T. Mather. "Design strategies for shape memory polymers." *Curr. Opin. Chem. Eng.* 2 (2013):103–111.

69. X. Luo and P.T. Mather. "Preparation and characterization of shape memory elastomeric composites." *Macromolecules* 42 (2009):7251–7253.

70. J.L. Smith. "Material testing of shape memory polymers for modular robotics applications and development of a prototype SMO gripper for mini-Pr2 robot." Summer Undergraduate Fellowship in Sensor Technologies (2010).

71. W. Wagermaier, K. Kratz, M. Heuchel, and A. Lendlein. "Characterization methods for shape-memory polymers." *Shape-Memory Polym.* 226 (2010): 97–145.

72. P. Keller, M. Lake, D. Codell, R. Barrett, and R. Taylor. "Development of elastic memory composite stiffeners for a flexible precision reflector." Paper presented at 47th AIAA/ASME/ASCE/AHS/ASC Structures, Structural Dynamics & Materials Conference, 2006.

73. J. Hampikian, B. Heaton, and F. Tong. "Mechanical and radiographic properties of a shape memory polymer composite for intracranial aneurysm coils." *Mater. Sci. Eng. C* 26(8) (2006): 1373–1379.

74. M. Everhart and S. Jaime. "Reusable shape memory polymer mandrels." *Proceedings of SPIE* 5762 (2005): 27–34.

75. M. Amjadi, K. Kyung, I. Park, and M. Sitti. "Stretchable, skin-mountable, and wearable strain sensors and their potential applications: a review." *Adv. Funct. Mater.* 26 (2016): 1678–1698.

76. A. Adrian, W. Taylor, and A. David. "Mechanical cycling stability of organic thin film transistors on shape memory polymers." *Adv. Mater.* 25(22) (2013): 3095–3099.

77. H. Luo, Y. Ma, and W. Li. "Shape memory-enhanced water sensing of conductive polymer composites." *Mater. Lett.* 161 (2015): 189–192.

78. H. Luo, Z. Li, and G. Yi. "Temperature sensing of conductive shape memory polymer composites." *Mater. Lett.* 140 (2015): 71–74.

79. H. Wei, Q. Zhang, Y. Yao, L. Liu, Y. Liu, and J. Leng. "Direct-write fabrication of 4D active shape-changing structures based on a shape memory polymer and its nanocomposite." *ACS Appl. Mater. Interfac.* 9(1) (2017): 976–883.

80. L. Huang, R. Jiang, J. Wu, J. Song, H. Bai, B. Li, Q. Zhao, and T. Xie. "Ultrafast digital printing toward 4D shape changing materials." *Adv. Mater.* 29(7) (2017): 1605390.

81. G.I. Peterson, A.V. Dobrynin, and M.L. Becker. "Biodegradable shape memory polymers in medicine." *Adv. Healthcare Mater.* 6(21) (2017): 1700694.

82. M. Zarek, N. Mansour, S. Shapira, and D. Cohn. "4D printing of shape memory-based personalized endoluminal medical devices." *Macromol. Rapid Commun.* 38(2) (2017): 1600628.

83. X. Jing, H-Y. Mi, H-X. Huang, and L-S. Turng. "Shape memory thermoplastic polyurethane (TPU)/poly(ε-caprolactone) (PCL) blends as self-knotting sutures." *J. Mech. Behav. Biomed. Mater.* 64 (2016): 94–103.

84. Y.S. Wong, Y. Xiong, S.S. Venkatraman, and F.Y.C. Boey. "Shape memory in un-cross-linked biodegradable polymers." *J. Biomater. Sci., Polym. Ed.* 19(2) (2008): 175–191.

85. S. Petisco-Ferrero, J. Fernández, M.M. Fernández San Martín, P.A. Santamaría Ibarburu, and J.R. Sarasua Oiz. "The relevance of molecular weight in the design of amorphous biodegradable polymers with optimized shape memory effect." *J. Mech. Behav. Biomed. Mater.* 61(2016): 541–553.

86. M. Ebara, K. Uto, N. Idota, J.M. Hoffman, and T. Aoyagi. "Shape-memory surface with dynamically tunable nanogeometry activated by body heat." *Adv. Mater.* 24(2) (2012): 273–278.

87. Ji, S., F. Fan, C. Sun, Y. Yu, and H. Xu. "Visible light-induced plasticity of shape memory polymers." *ACS Appl. Mater. Interfaces* 9(38) (2017): 33169–33175.

11

Hydrophobic and Super-Hydrophobic Polymer Coatings

Kirti Thakur
Nano Surface Texturing Lab, Defence Institute of Advanced Technology (DU), Ministry of Defence, Girinagar, Pune, India

Swaroop Gharde
Nano Surface Texturing Lab, Defence Institute of Advanced Technology (DU), Ministry of Defence, Girinagar, Pune, India

Sarang Jamdade
Indian Coast Guard, Ministry of Defence, Government of India, India

Balasubramanian Kandasubramanian
Nano Surface Texturing Lab, Defence Institute of Advanced Technology (DU), Ministry of Defence, Girinagar, Pune, India

CONTENTS

Abbreviations

LCD	liquid crystal display	CNF	carbon nanofiber
CA	contact angle	SEM	scanning electron microscope
WCA	water contact angle	FDTS	perfluorodecyltrichlorosilane
PVDF	polyvinylidene fluoride	PDMS	polydimethyl siloxane
SAMS	self-assembly of monolayers	ZP	zinc phosphate
CNT	carbon nanotube	GMA	glycidyl methacrylate
		AIBN	azobis-isobutyronitrile

DOI: 10.1201/9781003037880-11

pDVB	poly-divinyl benzene
pPFDA	poly-perfluorodecylacrylate
CVD	chemical vapour desposition
RMS	root mean square
UHMWPE	ultra-high molecular weight polyethylene
FEP	perfluoroethylene propylene
PC	polycarbonate
PLA	polylactic acid
Silane-f-rGONR	silane functionalized reduced Graphene oxide nanoribbons
PU	polyurethane
APCMPs	aminopyridine imbibed conjugated microporous polymers
FAS	fluoroalkylsilane
PTFE	polytetrafluoroetheylene
ESO	epoxidized soybean oil
SA	sebacic acid
HCP	hexagonal close packing
AA	acrylic acid
MMA	methyl methacrylate
OPSZ	organopolysilazane
F-SiO$_2$ NPs	fluorinated silica nanoparticles
LLDPE	linear low-density polyethylene
μTAS	micro total analysis systems
PCU	polycarbonate urethane
MEMS	micro electro mechanical systems

11.1 Introduction

Biomimetics is a field combining nature and technology to obtain surfaces whose functions are governed by the dimensions and designs based on the naturally occurring phenomena that possess unique properties (Sarikaya et al. 2003; Vincent et al. 2006). The inspiration obtained from nature in terms of adaptation has led to the invention and enhancement of a multitude of technologies, such as bird-inspired aircraft, butterfly-inspired liquid crystal displays (LCDs), mussel-inspired underwater anti-adherence coatings, beetle-inspired water harvesting from fog, leaves-inspired solar cells, etc. (Bar-Cohen 2006; Parker and Townley 2007; Bhushan 2009). One of the major singularities that has made an impact in the research field as well as in industry is the advent of hydrophobic and super-hydrophobic surfaces following biomimicry off various plants and animals, such as rice leaves, mosquito eyes, cicada wings, gecko feet, spider silk, etc. but the first reported literature was on the lotus leaves and its self-cleaning properties (Wagner, Neinhuis, and Barthlott 1996; Parker and Lawrence 2001; Gao and Jiang 2004; Byun et al. 2009). Hydrophobic/super-hydrophobic surfaces have an aversion to water to such an extent that water forms a sphere when in contact with the surface of the substrate and the extent of its hydrophobicity is categorized by the contact angle measurements at the solid-liquid interface (Vincent et al. 2006). The extraordinary anti-sticking, self-cleaning, corrosion-resistant, anti-biofouling qualities that the super-hydrophobic structures offer make it an in-demand technology, which has the potential to change the way

we perceive the nano-dimensions, and the naturally occurring structures provide new insights for the design of biomimicry of architectural marvels (Barthelat 2007; Bhushan 2009).

The rapid rise in the number of studies on hydrophobicity began in the work published by Onda et al. on fractal surfaces synthesized with alkylketene dimer that displayed a super-hydrophobic water contact angle (WCA) of 174° (Onda et al. 1996). The basic theory proposed by Wenzel (Wenzel 1936) and Cassie-Baxter (Cassie and Baxter 1944) and cases based on the same theory existed decades ago, but they were not able to have the same impact as the studies by Onda et al. Reinhoudt compiled the techniques that helped in fabricating hydrophobic/super-hydrophobic surfaces. Highly water-repellent surface were biologically inspired by a lotus leaf that gave the process the nomenclature 'the lotus effect' (Von Baeyer 2000) and became the pivot for all the super-hydrophobic research and was followed by the explanation of the 'petal' (Bhushan and Nosonovsky 2010), which strictly followed the Cassie state of wetting theory rather than the Cassie impregnated wetting state (Barthlott and Neinhuis 1997). These surface-related phenomena established the dependence of super-hydrophobicity on the surface roughness and it was evident that to enhance the hydrophobicity and produce highly non-wetting surfaces, it was essential to alter the morphology of the substrate's topmost stratum (Kwon et al. 2009). Although hydrophobicity is beneficial in many applications, the exact underlying principles for the high contact angle are still unclear (Padhi et al. 2018). The attempt to understand the exact phenomena of wetting and develop methods to fabricate super-hydrophobic materials in bulk is a challenge for the scientific as well as the industrial community, but it has little effect on the progress of these surfaces as new development methods are being reported for their fabrication on a large scale (Sahoo and Kandasubramanian 2014a). This provides space for experimentation in the variation of materials as well as in the fabrication techniques to develop newer surface patterns and textures, that in turn enhance hydrophobicity (ibid.). Properties such as optical transparency, conductivity, strength, flexibility, durability and the ability to be developed on any substrate are the key to large-scale development of such morphologically heterogeneous structures (Badhe et al. 2015). Taking these factors into account, polymers have made their mark in the fabrication of hydrophobic coatings that can be applied to any surface to enhance its hydrophobicity (Bangar et al. 2018). Polymers are the best choice when tuning of a material affects the properties it displays and their ability to incorporate any other material as fillers also adds to the positive attitude towards these smart materials (Sahoo, Balasubramanian and Sucheendran 2015; Gupta and Balasubramanian 2016). A large number of polymer combinations with other polymers or nanofillers have been reported or simple surface modification techniques have been developed to apply these network-forming materials, which makes them highly sought after for applications in such fields as anti-corrosion (Hendry and Pilliar 2001), anti-fogging (Fürstner et al. 2005), anti-icing (Cao et al. 2009) and transparent systems (Yadav, Zachariah and Balasubramanian 2016), heat-sinking applications (Gore et al. 2016), self-cleaning, and drug delivery (Yadav and Balasubramanian 2015).

In this chapter, we will discuss the theory and various models that affect the hydrophobicity of materials, followed by explanations of various fabrication techniques, such as electrospinning, spin coating, plasma and laser technology, phase separation, sol-gel, etc. Finally, an overview of applications with case studies is included to better understand the choice of materials, fabrication techniques and wide applicability of hydrophobic/super-hydrophobic materials.

11.1.1 Wettability

The wetting properties of surfaces when in contact with liquids is of interest in the research on material sciences and surface chemistry. The formation of a droplet or complete wetting of the solid surface are termed hydrophobic or hydrophilic respectively (Durand et al. 2011) and are generally measured by measuring the contact angle when the solid and the liquid are in contact with each other (Yuan and Lee 2013).

The application of coatings to form hydrophobic and super-hydrophobic surfaces has been a topic of attention in the past few decades and their wide applicability in research as well as in industry covers environmental as well as manufacturing-related issues. The first known applicability of super-hydrophobic coatings was observed in cameras installed on the highway in Germany to detect speed limit violations (Mohammadi, Wassink and Amirfazli 2004; Kavale et al. 2011).

The hydrophobicity can be inculcated on a surface by factors that are responsible for controlling wettability. Surface free energy, surface roughness and homogeneity play a key role but the effect of the first two controlling parameters is paramount. Hence, it is important to understand how we can manipulate the surface to tune the surface for increasing hydrophobicity(Onda et al. 1996; Ma and Hill 2006).

11.1.1 Wettability Theories

Laplace related the adhesion of liquids on solid surfaces to the central fields of force of the fluid considered, while investigating capillarity, which also formed the basis of a network of correlations between surface tension, surface energy and internal pressure and laid the foundation for the fundamental equation of capillarity (Cain et al. 1983). As science progressed and the knowledge of the structure of liquids and solids increased, the concept of a central field of force was not substantiated and was replaced by electromagnetic and wave mechanics concepts. Moreover, the parameters required by the Laplace explanation were not experimentally possible and it failed to explain the most common phenomenon of adhesion.

The contemplation of the contact angle of the fluid droplet on the topmost stratum as a result of mechanical equilibrium, due to three surface tensions, was proposed by Thomas Young, in 1805, and is still theoretically relevant (Young 1805). The three surface tensions corresponded to the interactions of the liquid-vapour boundary, the solid-liquid boundary and the solid-vapour boundary and were the major parameters introduced by Young in the form of an equation giving a relation for wettability widely accepted by the scientific community (ibid.). It is therefore important to understand how a droplet is formed by the action of surface tension of the liquid (Snoeijer and Andreotti 2008). When similar liquid molecules in bulk surround a single liquid molecule, the net force exerted on the molecule is zero (Crick and Parkin 2013). When the molecules on a surface are considered, the force is not balanced and the similar neighbouring molecules exert a similar force inwards, pulling the surface molecules along with them and this in turn creates an internal pressure (Winandy and Shupe 2010). The liquid tries to limit its surface area and surface free energy to a minimum. This intermolecular force that results in the contracting to form a spherical drop is known as surface tension (Snoeijer and Andreotti 2008). The contact angle (θ_c) is therefore determined by the surface tension as well as forces such as gravity and is characteristic of a given solid-liquid pair under specific conditions (Kwok and Neumann 1999; Lam, Lu and Neumann 2002; Ruiz-Cabello, Rodríguez-Valverde and Cabrerizo-Vílchez 2014).

Therefore, considering the three interfaces (solid-liquid, solid-gas, liquid-gas) and their mechanical equilibrium on an ideal solid surface, Young gave the relation, which is now commonly known as Young's equation:

$$\gamma_{lv} \cos \theta c = \gamma_{sv} - \gamma_{sl} \tag{11.1}$$

where γ_{lv}, γ_{sv} and γ_{sl} represent the liquid-vapour, solid-liquid, and solid-vapour surface tensions, respectively, and θ_c is the contact angle (Equation 11.1) and the physical representation is depicted in Figure 11.1.

The spreading of a drop of fluid was categorized by the angle that the droplet formed with the substrate surface on which it was placed (Good 1992; Padday 1992). If the fluid spreads completely, depending on the viscosity of the fluid and roughness of the solid, the contact angle (θ_c) is recorded as zero (Princen 1969). When the fluid forms a droplet on the surface, we categorize the surface on the extent of the angle formed between the solid and the fluid droplet, and when the fluid is water, the term

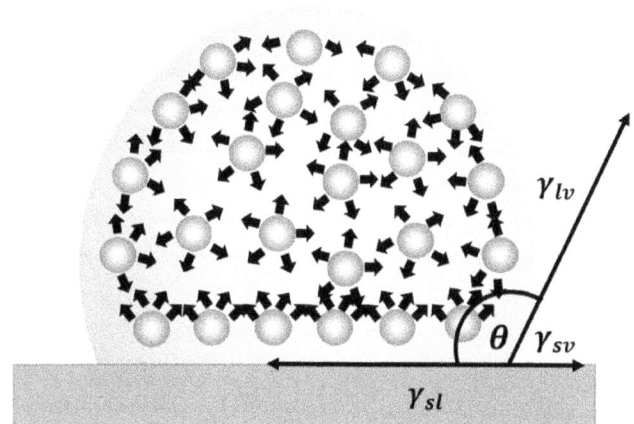

FIGURE 11.1 Unbalanced forces of liquids molecules generating surface tension and depiction of Young's contact angle.

'hydro' is generally prefixed, this categorization is presented in Table 11.1 and visually in Figure 11.2.

The possibility of $\theta_c = 180°$ is highly improbable as there is always some adhesion of fluid when in contact with a dense substrate. When we consider the homogeneous surface, thus the contact angle is independent of the volume of the spherical liquid drop and is a useful tool to measure wettability (Barati Darband et al. 2018).

11.1.2 Contact Angle Hysteresis

The contact angle by Young's equation is valid for ideal cases where the thermodynamic parameters related to the interfaces complement a unique contact angle. It has been observed that the contact angle for real cases does not match the theoretical value and this can be attributed to metastable states of the liquid drop on the solid's surface, as the wetting phenomena is dynamic, i.e., the contact line of the three phases is in motion to obtain a dynamic contact angle (as depicted in Figure 11.3) (Johnson and Dettre 1964; Joanny and de Gennes 1984; Gao and McCarthy 2006; Krumpfer and McCarthy 2010). McCarthy and co-workers worked on the expansion and contraction of the liquids' contact angle and termed it an advancing (θ_a) and receding (θ_r) contact angle respectively, where at a low-speed rate of recording the contact angle, the observed value is very close to the static contact angle recorded (Schwartz and Garoff 1985). These angles are responsible for pinning forces, which in turn induce hydrophobicity (Gore et al. 2016; Gupta and Balasubramanian 2016). The difference between θ_a and θ_r is known as hysteresis (H) and is mathematically formulated as:

$$H = \theta_a - \theta_r \tag{11.2}$$

Further investigation of contact angle hysteresis led to the generalization that the homogeneity or surface roughness was responsible for it. For non-homogeneous surfaces, there exist barriers that refrain the liquid from being in complete contact with the solid substrate. The variations in the slope of the surface create obstacles and alter macroscopically the contact angles by expanding or contracting the motion of the water. The use of Young's contact angle might match the observed contact angle in these ideal surface conditions (Kwok et al. 1998; Kwok and Neumann 1999). For a morphologically homogeneous but chemically heterogeneous surface, the resultant sorption or swelling might be a basis for the contact angle (CA) to be different than the theoretically calculated Young's CA but it will be a good approximation when advancing a CA (Kwok et al. 1998; Kwok and Neumann 1999). Although when rough solid surfaces are concerned, no correlation can be drawn between θ_a and θ_c and thus explaining wettability in terms of contact angle seems meaningless (Sedev, Petrov and Neumann 1996). Therefore, a thermodynamic equilibrium contact angle has been derived by various scientists considering various rough and heterogeneous surfaces and was very different from Young's contact angles, and this is widely accepted for real cases. These angles were introduced by Wenzel (Wenzel 1936) and Cassie and Baxter (Baxter and Cassie 1945), considering different approaches to heterogeneous surfaces and thus are called after them.

In 1936, Wenzel established the relationship between roughness and contact angle and stated that the roughness and wettability are directly proportional to each other (Wenzel 1936). This led to the fact that if hydrophobicity needs to be induced or enhanced, this could easily be achieved by augmenting the roughness of the surface. The Wenzel equation is used to mathematically state this as:

$$\cos\theta_m = r\cos\theta_Y \tag{11.3}$$

where θ_m is the Wenzel contact angle, θ_c is Young's contact angle and r is a term for the roughness ratio, which is a ratio between the real and the probable area of the surface under consideration and is estimated from a 3D roughness parameter. The value of r is 1 for a smooth surface and is greater than 1 for a morphologically heterogeneous surface. The roughness parameter is essentially defined as surface roughness and is a combination of a set of parameters (Marmur 2006a; Moutinho, Figueiredo and Ferreira 2007). The 2D parameters are denoted by R while the 3D parameters that help to include the topography of the surface are

TABLE 11.1
Young's Contact Angle and Wettability

Contact Angle (θ_c)	Wettability
$\theta_c < 30°$	Super-hydrophillic
30°–90°	Hydrophilic
90°–150°	Hydrophobic
$\theta_c > 150°$	Super-hydrophobic

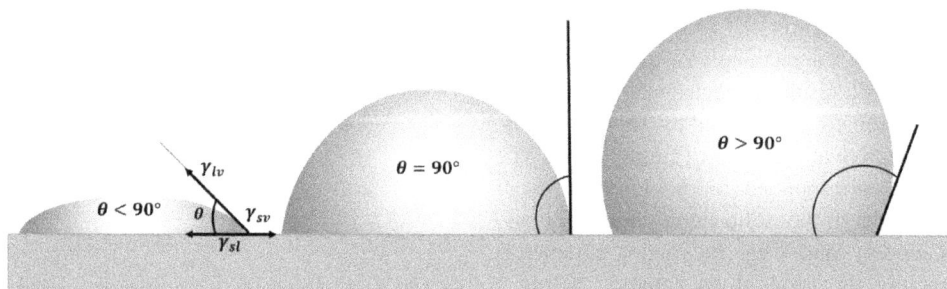

FIGURE 11.2 Sessile drop method for measuring contact angle on homogeneous solid strata.

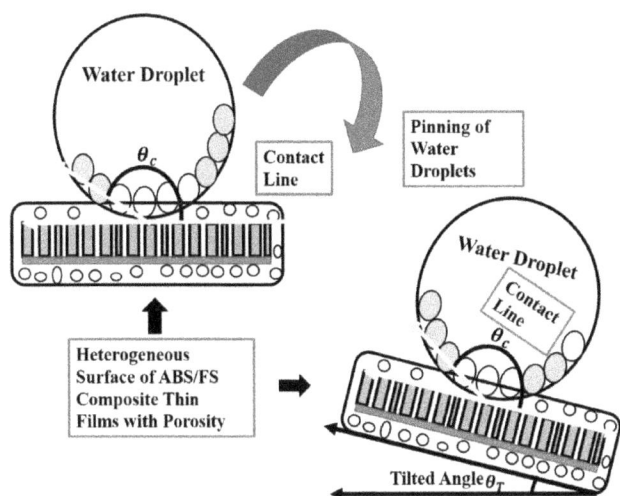

FIGURE 11.3 Schematic for contact angle hysteresis.

denoted by the letter S and a summary of these combinations can be obtained from ISO 25178 (Blunt and Jiang 2003). The surface and the projected area differ in value and their ratio can be used to determine the roughness ratio, as stated in Equation (11.3). The Wenzel equation is valid only when the liquid infiltrates the coarse terrain and the droplet is two to three times bigger than the grooves on the surface and this model is categorized as a homogeneous wetting regime (He, Lee and Patankar 2004).

The disadvantage of this regime is that it cannot explain the modification of hydrophilic materials to super-hydrophobic materials as it does not consider any other factor than roughness and thus the composition of the surface under consideration is completely neglected. Therefore, considering the Wenzel contact angle for non-homogeneous surfaces is not advised, instead the Cassie equation was established for these surfaces (Koch, Amirfazli and Elliott 2014; Ba et al. 2016).

The Cassie equation was determined to establish a wettability relation for heterogeneous surfaces with two different material compositions on the surface and is mathematically represented by:

$$\cos \theta_m = f_1 \cos \theta_1 + f_2 \cos \theta_2 \qquad (11.4)$$

where f is the fraction area covered by the two materials (1 and 2 are used to differentiate between the two materials). If one of the material compositions is air, then the equation is represented as:

$$\cos \theta_m = x_1 \left(\cos \theta_Y + 1 \right)^{-1} \qquad (11.5)$$

As the sum of the fraction area of the two materials is unity ($f_2 = 1 - f_1$), the contact angle with respect to air is 180° ($\cos \theta_2$ is -1). This is because of the assumption that the fluid droplet and the solid's topmost strata are in contact only at the pinning position or the roughness tips and air pockets are trapped in the grooves and are not wetted by the liquid (Hey and Kingston 2007).

The above equation was developed by both Cassie and Baxter and is commonly called the Cassie-Baxter equation of wettability.

To obtain the Cassie-Baxter angle, the surface of the material must be manufactured precisely. As the surface roughness reduces, the Cassie-Baxter angle comes into close approximation of the Young's contact angle. In real cases, Wenzel and Cassie and Baxter have been in good agreement with the data, but in some cases, a combination of these models has also yielded good results (Cassie and Baxter 1944; Giacomello et al. 2012; Liu, Chen, and Kim 2015; Hao and Wang 2016). These models are depicted pictorially in Figure 11.4.

11.2 Fabrication of Super-Hydrophobic Coatings

Many technologies have been used to fabricate super-hydrophobic structures and, based on the pros and cons of each method, one can choose between various parameters, such as high accuracy, cost-effectiveness, etc. Some of the methods are briefly discussed below and depicted in Figure 11.5.

11.2.1 The Sol-Gel Process

The sol-gel method comprises the formation of a colloidal solution of monomers that convert into an integrated network of polymers, i.e., gels when the solvent evaporates (Sahoo and Balasubramanian 2014b). The advantages of using this technique are its low thermal energy requirement, it is economical, and offers excellent control of the process (Hench and West 1990) and it can be applied to coat a wide variety of substrates, such as silica, alumina, titania, etc. (Mahltig and Böttcher 2003; Pilotek and Schmidt 2003; Hikita et al. 2005; Wu, Zheng and Wu 2005; Yu et al. 2007; Jitianu et al. 2010). Factors such as variation of precursor solutions, polycondensation, hydrolysis and the rate of gel formation aid in the tunability of the surface energy and its architecture. The possibility of transparent super-hydrophobic materials is offered by the use of glass or other transparent material as substrates (Tadanaga, Morinaga and Minami 2000; Shang et al. 2005; Venkateswara Rao et al. 2009). Synthesis of nano cauliflower and nano broccoli-shaped composite coating with the help of PVDF/carbon soot particles was carried out via the gelation technique (Sahoo and Balasubramanian 2014b).

11.2.2 Electrospinning

When a jet of polymeric solution flows from a drip at the tip of a needle, it follows a straight path, which then curves owing to the electrostatic forces. The curving of the polymeric strand with respect to electric forces applied results in thinning and the evaporation of the solvent leaves nano-strands on the collecting plate. The electrostatic formation of fibres using electric force to form polymeric strands can produce fibres with diameters ranging up to micrometres, and if the parameters are well controlled, they can be generated in the nanometre regime, and this is known as electrospinning. Both natural and synthetic polymers can be used to produce bulk nano-dimensional fibres and thus they find great applicability in research as well as commercially (Ahn et al. 2006; Lannutti et al. 2007; Hunley and Long 2008; Reneker and Yarin 2008). Viscosity, conductivity, surface tension,

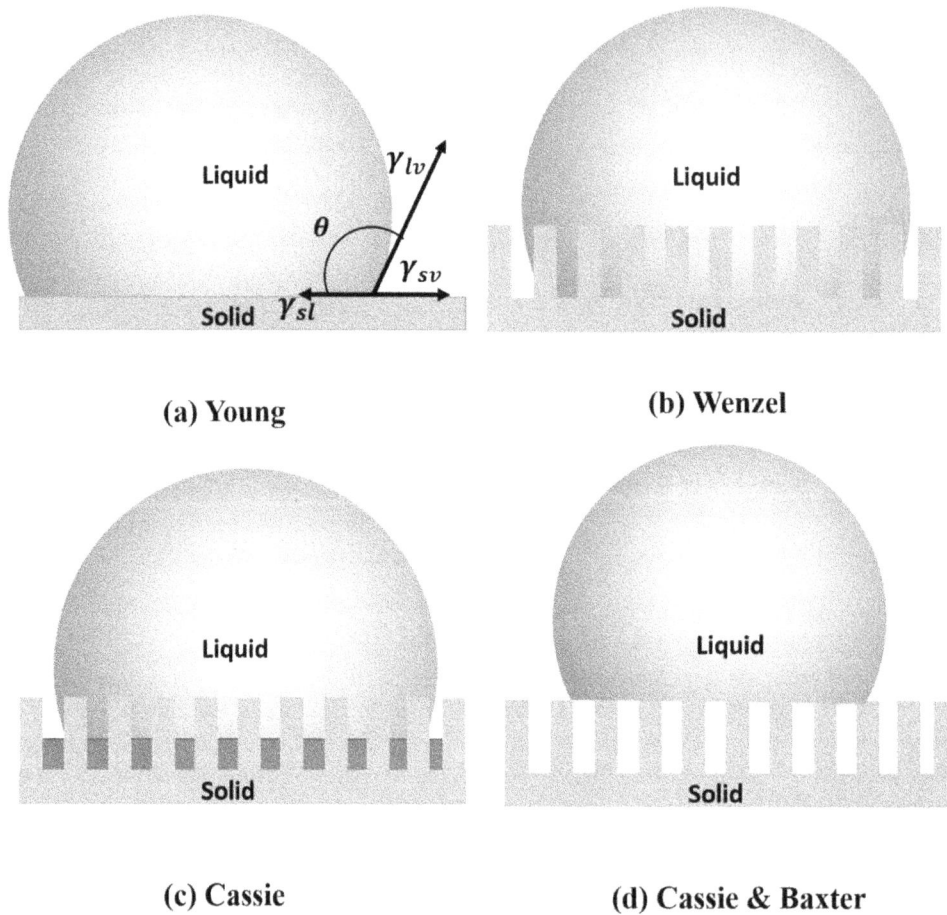

(a) Young   **(b) Wenzel**

(c) Cassie   **(d) Cassie & Baxter**

FIGURE 11.4   Different wetting models, (a) Young model; (b) Wenzel model; (c) Cassie model; and (d) Cassie & Baxter model.

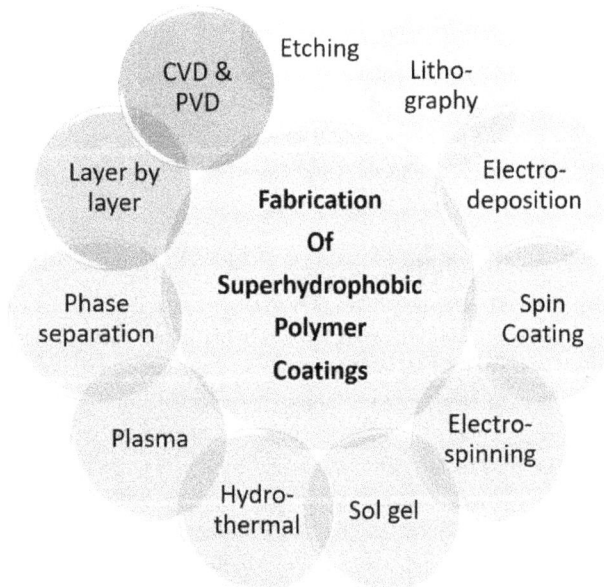

FIGURE 11.5   Fabrication techniques for development of super-hydrophobic polymer coatings.

polymer molecular weight are some of the properties essential for electrospinning (Ma et al. 2005).

11.2.3 Electrochemical Methods

Electrochemical methods to form hydrophobic and super-hydrophobic coatings are generally used for morphologically heterogeneous surfaces (Shi, Wang and Zhang 2005; Zhao et al. 2005). The hydrophobic nature of polymers makes them a good candidate for water-repellent coatings and they can easily be deposited electrochemically (Shi et al. 2007; Ahmad 2015). This technique also ensures complete coverage for the coating of objects and the ease of control of growth kinetics helps to create large coated areas with different morphologies, such as needles, dendrites, sheets, fibres, etc. (Youngblood and McCarthy 1999; Wang et al. 2005). Electrochemical deposition, anodic oxidation, polymerization reactions can be used to manufacture super-hydrophobic surfaces.

11.2.4 Layer-by-Layer Assembly

The variation in surface charge of the substrate multi-layer films can be fabricated in the nano to micrometre dimensions (Han et al. 2005; Zhao et al. 2005). The wettability can be tuned by using nanoparticles that upsurge the surface roughness of the substrate. The layer-by-layer deposition ensures complete control over the thickness and is favoured to produce transparent layers. Another major advantage is that the surface of the substrate may or may not be morphologically homogeneous for this hydrophobic-inducing technique (F. Zhang et al. 2008).

11.2.5 Treatment with Plasma and Laser Technology

Deposition of plasma, i.e., an electrically charged cloud of materials, and subsequent treatment by polymers lead to the generation of morphological heterogeneity and reduction in surface energy, which can be accomplished by using oxygen plasma. However, the ageing of the plasma-treated surface is disadvantageous as the surface roughness offered by the plasma treatment, which is easy to adjust, makes it a good option for the fabrication of coatings imparting excellent optical properties (Morra, Occhiello and Garbassi 1989; Youngblood and McCarthy 1999; Coulson et al. 2000; F. Zhang et al. 2008). A coating of fluorine is widely reported to be achieved by using this technique.

The laser treatment re-deposits the vaporized material as clusters on the treated surface. The specification offered by the laser can be used to fabricate different morphological variations on the same substrate, this holds true even for the plasma technique (Lau et al. 2003; Woodward et al. 2003).

11.2.6 Spray Coating and Spin Coating

This method involves the physical application of a hydrophobic coating on the substrate's surface with the help of a spray gun or a spin coater. For a small surface, spin coating can be used but for larger surfaces spray coating acts as a good alternative although the precision of the spin-coated substrate is better than that of the spray-coated substrate (Magisetty, Shukla and Kandasubramanian 2018). The main characteristics of this fabrication technique are the use of inexpensive materials and easy manufacturing. The thickness of the coating depends on several parameters such as the viscosity of the polymeric solution, and the method of coating (Thanawala and Chaudhury 2000; Chakrabarty, Nisenholt and Wynne 2012).

11.2.7 Lithography and Chemical Etching

Lithography is a technique that employs stencils or optical activity of the substrate to form patterns that make the surface morphologically heterogeneous and in turn hydrophobic. The dimensions of the patterns formed on the surface can range from the nano dimension to micrometres (Öner and McCarthy 2000; Shiu et al. 2004; Li, Breedveld and Hess 2012). This includes various subtypes like soft lithography or stencil lithography that give different resolutions to the preparation of the hydrophobic surfaces.

Chemical etching is the reaction of the substrate with chemicals that etch or remove the areas they come into contact with, resulting in an increase in the surface roughness and therefore enhancing hydrophobicity (Boisier et al. 2009).

11.2.8 Other Fabrication Methods

Fabrication methods, such as self-assembly of monolayers (SAMS), physical and chemical vapour deposition methods, and phase separation methods, can also be employed to generate

surface roughness with specific conditions. One of the main fabrication technologies is the hydrothermal process and it is highly efficient in generating a dimensionally similar roughness on the surface of the substrate. However, this is disadvantageous for bigger structures due to the high pressure and closed systems required for fabrication to occur (Xu et al. 2009; Akram Raza et al. 2010; Du and He 2011). Thus, these methods can be used to devise newer polymeric coatings and some examples of these coatings are mentioned in Section 11.3.

11.3 Application-Related Case Studies

The water-repelling properties of substrates are highly sought after by industry for various applications where properties of other materials are hampered and cause an economic loss. Environmentally-friendly materials are more favoured. In this section, some case studies are presented to show the importance of hydrophobic and super-hydrophobic coatings, sub-divided according to the applications in research or industry.

11.3.1 Anti-Corrosive Coatings

Corrosion is a term used to define the degradation of a surface, generally metallic, when environmental factors such as humidity and the pH of the surrounding atmosphere act on it (Cicek 2014). Metallic surfaces are very common in our daily lives, as using metals in the home as well as in industry is inevitable. From the tip of pens to large hulls of ships, it is necessary to avoid corrosion as it does irreparable damage, rendering the metals useless. Metals form the foundation of most structural marvels and thus the sanctity of architectural constructions especially require anti-corrosive coatings (Kritzer 2004). Every year a lot of damage is done economically due to corrosion and thus researchers as well as industrialists are always looking for newer, cost-effective and highly efficient technologies (Walker 1993).

Jamdade et al. (Jamdade, Nimje and Balasubramanian 2015) worked on anti-corrosive coatings for ship hulls. It is imperative for ships to be cocooned with anti-corrosive coatings due to the high salt content and humidity in their working environment. This is achieved by coating the ship with specialized anti-corrosive paint. Epoxy resin is used as a binder because of its low shrinkage, better bonding capability, dimensional stability, chemical and corrosion-resistance capabilities. It provides surface hardness and flexibility to the paint coating. The epoxy resin also has some disadvantages that need to be addressed. The epoxy resin absorbs water of a volume equivalent to 2% mass. Although it is tough, the surface abrasion and wear are still an issue and the initiation and propagation of cracks also make it unsuitable without any modifications. These defects lead to the ingress of water and aggressive species such as oxygen which lead to localized corrosion and delamination of the coating. These limitations can be addressed by incorporating a second phase that is miscible with epoxy polymer to decrease the porosity and minimize the paths for aggressive species. This second phase is known as fillers. Conventional coatings used 25–40% micron fillers and this high concentration leads to the deterioration of the epoxy properties, due to the large interparticle distance, which allows the water molecules to diffuse and initiate corrosion. Nanotechnology solved this by reinforcing epoxy with nano-dimensional organic and inorganic materials based on carbon. Carbon nanotubes (CNTs), carbon nanofibres (CNFs), nanoclays, metal oxide nanoparticles, etc. showed promising results but carbon-based fillers were preferred due to their availability and a history of use of coal tar in the paint which was deemed unsuitable as it harmed the environment. Camphor soot particles are a cost-effective way to fix this, shown in Figure 11.6.

This has also been described in detail by Sahoo and Balasubramanian and studies show the stability and super-hydrophobicity of these soot particles (Sahoo and Balasubramanian 2014a; Padhi et al. 2018). The soot particles are primarily composed of carbon and consist of agglomerated particles with a diameter of about 20–60 nm. The soot particles are rich in hydrophobic functional groups and impart super-hydrophobic features. Sahoo et al. confirmed the super-hydrophobic nature of camphor soot particles with WCA ~170° and enhanced surface roughness. Studies. have shown the presence of γ and π bands in camphor soot, indicating the presence of crystalline graphitic carbon. The SEM image of the camphor-generated soot particles is given in Figure 11.7.

The contact angle of the epoxy resin without any fillers was recorded as 50°. After the addition of fillers, it was recorded to be >90° which makes the epoxy coating hydrophobic. Several compositions of the fillers were made which ranged from 0.5% soot to 5% soot and a gradual rise in contact angle was detected. The use of the surfactant Nonidet P-40 improved the dispersal of the camphor soot entities in the epoxy matrix and its spherical shape made the particles spread easily, compared to CNT and CNF. Other properties like abrasion resistance were also analysed and are presented in Table 11.2.

The increase in nanoparticle composition also improved the quality of the cured epoxy coating, reduced the porosity of the matrix and inhibited the penetration of aggressive corrosion-causing entities. It was observed that 1.5 and 2 wt% compositions showed overall improvement and this composition is much less than what the industry conventionally uses. It was also proven that thermal stability increases with an increase in concentration of carbon filler.

Super-hydrophobic polysiloxane coatings offer another solution to the problem of corrosion. Arukalam et al. (Arukalam,

FIGURE 11.6   Formation of camphor soot particles.

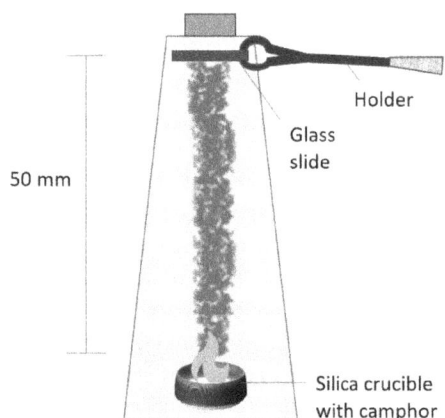

FIGURE 11.7   FESEM of carbon soot particles obtained via controlled incineration of camphor.

TABLE 11.2
Different Tests Performed on Coatings with Varying Amounts of Filler

Coating with % of Filler Tests	Pristine Epoxy	0.5%	1%	1.5%	2%
Tensile Strength (MPa) ASTM D -882	60.62	56.96	54.2	69.68	59.04
Abrasion Resistance [Weight Loss (in mg)/500 cycles] ASTM D 4060	26.9	50.5	44.4	44.2	29.7
Flexibility [Conical Mandrel Bend Test] ASTM D522	Pass up to 1'	Pass up to 1'	Pass up to 1'	Pass up to 1'	Pass up to 1'
Impact Resistance ASTM D 2794	Pass	Pass	Pass	Pass	Pass
Cross-hatch Adhesion ASTM D 3589	5	5	5	5	5
Scratch Resistance for Tolerance Weight 03 kg BS 3900 Part E2	Pass	Pass	Pass	Pass	Pass

Oguzie and Li 2018) have reported super-hydrophobic perfluorodecyltrichlorosilane (FDTS) with the incorporation of different nano-ZnO particles, with variation in size, and poly(dimethylsiloxane) to obtain different ratios of (PDMS):nano-ZnO composite materials. Properties such as contact angle, thermal and mechanical responses, or surface morphology were assessed and it was found that the variation in size of nano-ZnO particles influenced the fluid-repellence, surface topographies and tensile properties. The 30 nm ZnO:PDMS(1:1) composite showed the best results for barrier and spray corrosion tests and showed highest anti-corrosion performance as the impedance modulus |Z| was recorded to be >109 Ω.cm^2 with respect to frequency (Hz).

Jadhav et al. (Jadhav, Holkar and Pinjari 2018) have reported another hydrophobic filler incorporated epoxy resin which shows a self-healing mechanism (Figure 11.8). This is done by synthesis of super-hydrophobic imidazole, encapsulated in hollow zinc phosphate (ZP) nanoparticles, that is added to the epoxy resin and exists as a dispersion and is further treated with octanol which ensures a hydrophobic nature. When a defect or a crack appears in the coating, the ZP releases the imidazole which covers the cracks, thus prohibiting contact with the corrosion-causing agents and the coating self-repairs. The mesoporous ZP synthesized by ultrasonication and suitable surfactant acts as a concealer for the imidazole and assures the prohibition of undesired leakage when it interacts with water and its release when the coating is hampered. This phenomenon is supported by the super-hydrophobicity that increases the synergistic effect. The spherical shape of the hollow ZP ensures better spreading of the epoxy resin and grafting with octanol preserves the wetting properties. The coating was tested on mild steel and EIS and the salt spray tests provide excellent correlation with the anti-corrosive behaviour. The super-hydrophobic Im-ZP-EPo coating showed 5.34 ×10^{-5}mm/year, 0.0012 mm/year, 0.0042 mm/year that determines the anti-corrosion activity in presence of conc. HCl, NaOH and NaCl medium respectively.

11.3.2 Anti-Icing

Icing or freezing of water on surfaces is a menace as this causes many problems such as icy roads, trees falling, and electricity powerlines breaking, stalling of the air-foil in aircraft and can result in heavy economic losses (Cober, Isaac and Strapp 2001; Lasher-Trapp et al. 2008). This icing is caused by super-cool water and is also termed freezing rain, or impact ice. A lot of data has been reported on super-hydrophobic surfaces but with super-cooled water these surfaces do not show the same properties. The speculation that the super-hydrophobic surfaces involve anti-icing properties has been a topic of debate among researchers for many years. But some studies do exist that show the possibility of having anti-icing coatings with the help of polymeric coatings (Saito, Takai and Yamauchi 1997; Nakajima, Hashimoto and Watanabe 2001; Farhadi, Farzaneh and Kulinich 2011).

Cao et al. (Cao et al. 2009) carried out free radical polymerization of styrene, butyl methacrylate(BMA), and glycidyl methacrylate(GMA) in toluene and initiated this reaction with the

help of azodiisobutyronitrile (AIBN) to form the resin which was turned into a binder by incorporating it with silicone. The binder was found to have a hydrophobic angle of 107°. Silica particles of various dimensions, ranging from 20 nm to 20 µm were mixed with the binder and spray painted on aluminium plates at 30 psi pressure and with a total curing time of 12 hours. The nanocomposites were observed to possess a super-hydrophobic character of more than 150° advancing angle and a receding angle with < 2° hysteresis, except for the 20 µm nanoparticle-doped nanocomposite, which had a hydrophobic contact angle with 4° hysteresis. Testing of anti-icing capability was done by arranging two aluminium plates where one was coated with the nanocomposite. Super-cooled water (T = -20°C) was poured on both the aluminium plates from a distance of about 5 cm and the ice formation was inspected. The formation of ice on the non-coated surface and with no effect on the uncoated surface (excluding the build-up of ice on the untreated surface) proved the efficacy of these materials as anti-icing polymeric coatings. It was also noted that anti-icing properties were 100% efficient with 20 and 50 nm coating but reduce in efficiency for nanoparticle dimension of more than 50 nm diameter. This experiment was also performed in natural environment conditions where only 50 nm Si doped composite was used to cover surfaces of dish antennas, etc. and was compared to the non-coated surfaces. The results can be visualized in Figures 11.9.

Sojoudi et al. (Sojoudi et al. 2018) developed a bilayer of poly-divinyl benzene and poly-perfluorodecylacrylate (pDVB/pPFDA) coatings which were durable and robust for anti-icing properties (Figure 11.10). The coatings were synthesized by initiated CVD (chemical vapour deposition) that eliminates the need for solvent to decrease the adhesion of ice to the silicon and steel substrates. The thickness of the pDVB was much greater than that of the layer of pPFDA which gives it a rough texture to be deposited on morphologically heterogeneous steel. The pDVB/pPFDA composite which has 40:1 ratio thickness has a hydrophobic nature and low surface energy owing to the fluorine groups present in its

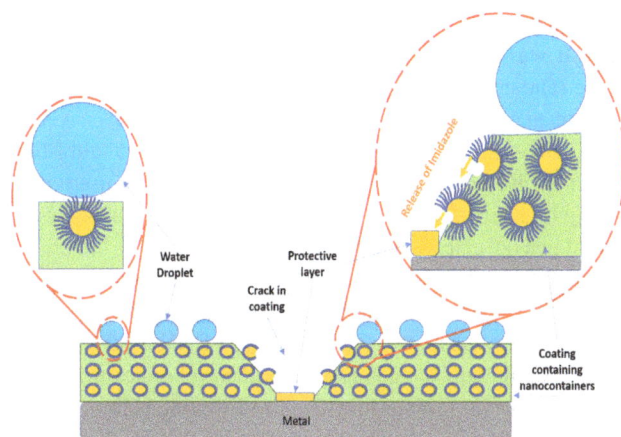

FIGURE 11.8 Anti-corrosion by self-healing activity of Su-Im-HZPn-Epo coated substrate.

Source: Jadhav, Holkar, and Pinjari (2018).

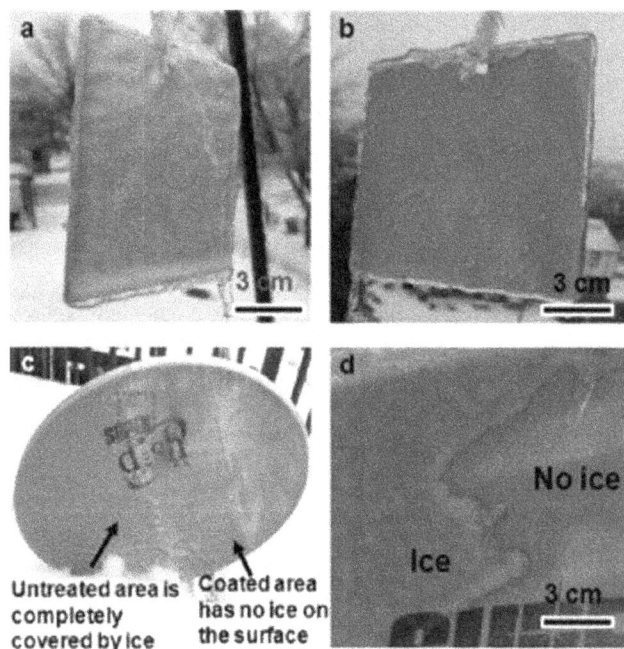

FIGURE 11.9 Anti-icing properties on coated and non-coated surfaces on different substrates, (a) untreated side of an aluminum plate after the natural occurrence of "freezing rain"; (b) treated side of the aluminum plate coated with a superhydrophobic composite after the "freezing rain"; (c) satellite dish antenna after the freezing rain; (d) close-up view of the area labeled by a red square in caption (c), showing the boundary between the coated (no ice) and uncoated area (ice) on the satellite dish antenna.

Source: Cao et al. (2009).

structure and when exposed to water is restricted from orienting because of the fluorine interaction with the pDVB. The ascending and receding CA were recorded to be 153.9 $\pm$ 2.21 and 157.0 $\pm$ 4.51 and the receding CAs were determined to be 139.0 $\pm$ 3.51 and 148.1 $\pm$ 2.91 on silicon and steel, respectively. RMS (Root mean square) roughness values of Rq = 178.0 $\pm$ 17.5 nm and Rq = 312.7 $\pm$ 23.5 nm were recorded and are shown in Figure 11.9. A reduction in ice adhesion from 1010.95 kPA to 180.85 kPA was observed when the steel surfaces were coated with these bilayer films. It was also determined that these coatings were durable by the sand erosion test as well as examination of numerous ice adhesion and de-adhesion rotations.

A flame-sprayed mixture of polyethylene and ultra-high molecular weight polyethylene (UHMWPE) was used as a coating by Koivuluoto et al. (2017) and anti-icing behaviour was studied in a novel icing wind tunnel that replicated the conditions in nature. Fluoropolymer perfluoroethylene propylene (FEP) was also mixed with polyethylene and the coating was applied to steel and aluminium substrates. Flame-sprayed PE and PE/FEP layers were tested for anti-icing behaviour and corroborated the theory of dependence on surface-roughness and wetting behaviour. The polished surfaces had decreased ice adhesion values when coated with as-sprayed coatings which shows that polished surfaces had a hydrophobic nature and as-sprayed surfaces had a hydrophilic wetting nature on icing behaviour. When compared with polyurethane coatings, the flame-sprayed PE and PE+FEP coatings

showed exemplary performance in a high-velocity impact test. The energy absorption abilities of the coatings were temperature-dependent and had a directly proportional relationship with the high velocity impact tests (ibid.).

Coatings of PE were studied by Koivukuoto et al. (ibid.) which were used in gas pipeline coatings. UHMWPE coatings possessed protective properties (Petrovicova and Schadler 2002). They were used in the fabrication of corrosion resistance coatings and were also used in decreasing friction, generating non-stick surfaces, and for ornamental purposes. These coatings have multiple applications and this can be used for multi-utility coatings in the future (Leivo et al. 2004). Fluoro-based materials also have anti-icing properties (Yang et al. 2011).

11.3.3 Oil-Water Separation

Water pollution is affecting the marine environment to such an extent that the marine ecosystem is suffering. As water is essential for all beings on planet Earth, it is imperative that we need to keep the existing water resources clean and safe as our very existence will otherwise be challenged. Separation of oil and water is also essential in various industrial methods. Various polymeric coatings haves been developed to separate oil and water. Oil and water do not mix but sometimes form a suspension of tiny droplets, which are not easily separable. For the coating to enhance the oil-water separation process, we need oleophilic-hydrophobic or oleophobic-hydrophilic coatings. Here we will discuss research based on oil-loving/water-hating (oleophilic-hydrophobic) materials (Nordvik et al. 1996; Sahoo and Kandasubramanian 2014a; Gore, Dhanshetty and Balasubramanian 2016; Gore et al. 2017; Gore et al. 2018).

Kavalenka et al. (Kavalenka et al. 2017) produced a polycarbonate (PC) and polylactic acid (PLA) nano-fur surface by applying the hot pulling method, resulting in the fabrication of nano and micro hairs during the separation of hot sand-blasted steel plate. The topology of the surface makes the material superhydrophobic and super-oleophilic to produce an efficient oil-water separation system that can be recycled. The short PC nanofur oil absorption capacity was noted to be 127 g/m^2 while that for the PLA nanofur was 479 g/m^2 when the hydraulic oil-water mixture was used to determine the separation efficiency. The WCA of the PC was recorded to be ~ 165°, making the surface superhydrophobic. The study also mentions that the material switches to being super-oleophobic when treated with argon plasma. This study confirms the tuning of surface morphology can help in developing super-hydrophobic surfaces that help in the separation of the oil-water mixtures and their efficiency in helping control environmental pollution caused by oil spill.

Simon et al. produced an oleophilic and hydrophobic low-cost camphor soot composite (Simon et al. 2018), by an easy and cost-effective approach and using a phase separation method. The composite is highly porous, owing to the solvent-non-solvent interaction, when the non-solvent evaporated, the system caused the pores to develop. This gives superior absorption and handling capacity and absorbs oil up to 25 times its own weight, without contaminating the water resources. The oil absorption capacity is increased compared to its predecessors, PVDF foam and nascent carbon soot. The carbon soot used also has high recyclability

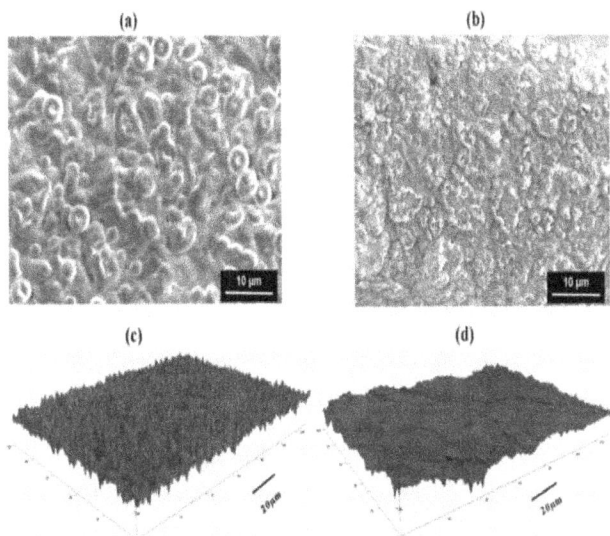

FIGURE 11.10 SEM images (a and b) and 3D optical profile; (c and d) of linker grafted bilayer DVB/pPFDAon silicon (left) and steel (right) substrates.

Source: Sojoudi et al. (2018).

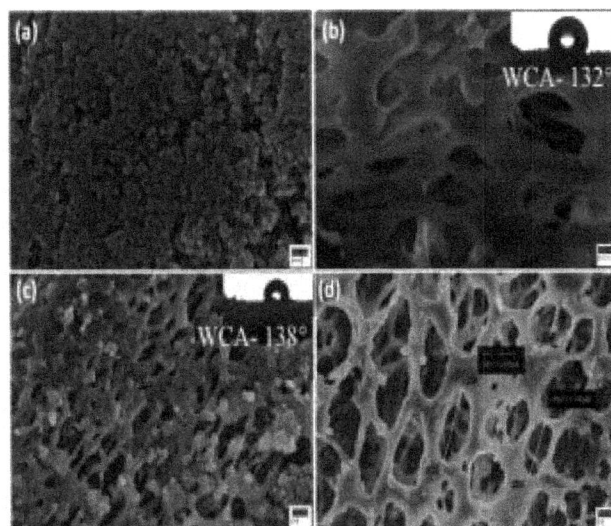

FIGURE 11.11 (a) Interconnected network of carbon nanospheres inside camphor soot; (b) pure PVDF foam; (c) PVDF-CS foam imprinted with CS particles; (d) microporous nature of composite.

Source: Mishra and Balasubramanian (2014).

and a reasonable ability to absorb oil and also has a super-hydrophobic surface. This composite can be used as a coating on any rough surface and thus provides an economic, easy-to-handle, easy-to-synthesize composite with an enhanced hydrophobic nature, and thus it has an edge over the existing techniques (Mishra and Balasubramanian 2014) (Figure 11.11).

Gore et al. produced a cotton-based super-hydrophobic and super-oleophilic fabric coated with PLA and PLA/nanoclay. The synthesis method used is electrospinning as it produces nanofibres that coat the surface of the cotton fabric with good adhesion to the substrate. The sample obtained has a WCA of ~ 152° and 140° and an OCA of 0° for both the PLA/nanoclay and the PLA, respectively, as shown in Figure 11.12. It was also found that the material possesses low ice adhesion properties with respect to a highly heterogeneous surface morphology and an oil-water separation efficiency of 99.16% and it has the ability to be reused for 30 wash cycles. The maximum permeation was recorded at around 65000 Lm^{-2} h^{-1} for *n*-hexane from a mixture of water and solvent. The environmental impact of polymers has been a topic of interest, thus, this polymer was designed to be biodegradable, and 78% loss in weight was observed in a span of 28 days. For biodegradation studies, the fabric was subjected to a biodegradation set-up, that was developed in-house, made of cow dung and compost in equal proportions in a litre of water and the samples were maintained in an oven at 37°C. The maximum deterioration was caused by *Bacillus polymyxa* bacteria. The uniqueness of this fabric was that it retained its properties in harsh conditions, such as an acidic to mildly basic condition (pH ~ 1–10), hyper-saline conditions, a mild detergent solution, ultraviolet radiation exposure as well as sub-zero temperatures (up to -20°C). This proved the efficacy of the Janus fabric (possessing a dual nature) and it has great potential for oil-water separation efficiency in many applications (Gore and Kandasubramanian 2018).

Similarly, other composites such as silane functionalized reduced graphene oxide nanoribbons (silane-f-rGONR)-coated polyurethane (PU) sponges (Qiang et al. 2018), cellulose nanocrystal-coated super-hydrophobic cotton fabric (Cheng, Guan et al. 2018), poly-*p*-phenylenes solution on filter paper (Bai et al. 2018), aminopyridine imbibed conjugated microporous polymers (APCMPs) (Jiao et al. 2018), fluoroalkylsilane-modified natural rubber-encapsulated silica latex (FAS-modified NR/SiO$_2$) (Saengkaew et al. 2018), porous metal sintered nanofiber felt with (PTFE) (Y. Chen et al. 2018), epoxidized soybean oil (ESO), sebacic acid (SA) and nano-ZnO sprayed cotton fabric (Cheng, Liu et al. 2018), super-hydrophobic nanocomposites of fluoroelastomer (Simon et al. 2018), etc. are some examples that can be an interesting read to increase knowledge of oil-water emulsion separation techniques and materials used as coatings.

11.3.4 Anti-Fogging

Fogging is a common phenomenon that decreases the functioning of reflective surfaces, such as mirrors, glass, etc. and also in various industry-related functions. Fogging can be eliminated or controlled by studying the interaction between the substrate, i.e., where fogging appears and the liquid which causes this phenomenon. Super-hydrophobicity enhanced anti-fogging glasses have been installed by most automobile industries in various vehicles available on the market and these coated glasses have proven their efficacy by improving drivers' visibility. There are two approaches by which any substrate can have an anti-fogging surface: (1) make the substrate super-hydrophilic; or (2) make the substrate super-hydrophobic which ensures longevity of the coatings (Das et al. 2018).

One of the most explicit examples was presented by Gao et al. (Gao et al. 2007) by creating a super-hydrophobic anti-fog coating for materials. This was done by mimicking mosquito

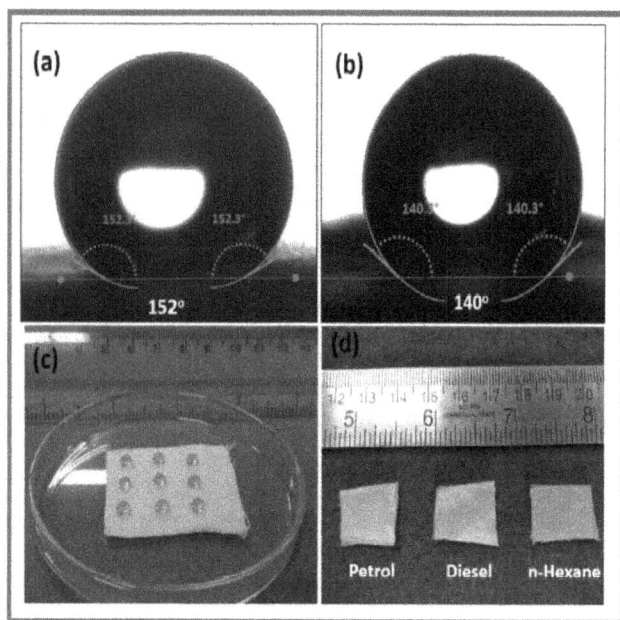

FIGURE 11.12 Static WCA of (a) cotton- PLA/NC fabric; (b) cotton- PLA fabric; (c) water droplets resting on Janus fabric; (d) superoleophilic nature of coated substrates.

Source: Gore and Kandasubramanian (2018).

compound eyes, inspired by Lee et al., who proposed a 3D optical method to synthesize artificial compound eyes, made up of microscale ommatidia placed on a hemisphere-shaped dome-like structure made of polymers (Lee and Szema 2005; Jeong, Kim and Lee 2006). The synthesis included several stages of patterning by soft lithography. First, the glass surface was covered with a photoresist coating to obtain a pattern that had the dimensions of the eyes of the mosquito, i.e., 20 μm with a 5 μm spacing. This was then heated to 120°C to obtain dome-like hemispheres made of the photoresist. The next step involved transferring this pattern onto PDMS (poly-dimethyl siloxane) which acted as a mould to replicate the photoresist structure which was also constructed by PDMS. This final PDMS structure was then pressed onto the substrate with SiO_2 nanospheres at a temperature of 100°C for 3 hours and the nanospheres were transferred on the PDMS which was peeled off after cooling. The hemispheres had a 22 μm diameter and an HCP (hexagonal closed packing) structure. The static WCA and tilted angle were 155° and 15° and showed an excellent super-hydrophobic nature and also anti-fogging properties.

11.3.5 Anti-Fouling and Anti-Scaling

Biofouling is a process which mainly comprises the substitution of a solid substrate-WCA by a solid substrate-biological material contact (Callow and Callow 2002; Marmur 2006b). There is a close relationship between the wetting phenomena and the setting of biological matter on the solid surface (Bers and Wahl 2004; Hoipkemeier-Wilson et al. 2004). Anti-fouling coatings are generally required for surfaces that remain under water for a long time, providing ambient conditions for the microbes to grow. One of the

major applications of anti-fouling is in marine applications as the hulls of ships are prone to fouling (Ng, Gani and Dam-Johansen 2007). The marine coatings are divided into two types, namely, micro-fouling and macro-fouling. First, micro-fouling takes place, i.e., a film of bacteria adheres to the surface of the ship's hull. This provides a base for the larger organisms to adhere to the surface. The resulting effect is that the fuel consumption of the ships is increased as the friction increases when the ship sails (Buskens et al. 2013).

Various research groups have reported on anti-fouling properties provided by hydrophobic coatings. Zhang et al. focused on hydrophobic coating made of fumed silica, alkyl trialkoxysilane, and polysiloxane with varying surface chemistry and roughness, compared to the uncoated substrates. These coatings showed excellent anti-fouling activity but the tests could not be carried out as the coatings did not last for a long time. An interesting thing to note was the fish attack marks on the coating which led to the conclusion that fish considered the coating to be food, which was not the desired effect (Zhang, Lamb and Lewis 2005).

Wouters and co-workers (Wouters, Rentrop and Willemsen 2010) modified sepiolite that changed the hydrophobicity of clay. This was done by the addition of aliphatic or polysiloxane tails either separately or in a blend which affects the super-hydrophobicity as well as their interaction with the formulation of the coating polymer. The obtained coating was produced by the sol-gel method and the roughness was specified. They also determined that surface roughness played a significant role in the formation and release of bionically formed films.

Brady and Singer (2000) determined that elastic modulus, critical surface, and thickness of the polymeric coating are linearly related to each other and affect the strength of the interface. The fracture mechanics on the solid-bionic interface validate that dense coatings have low elastic modulus and surface energy and do not restrain biofouling as effectively. Holland et al. (Holland et al. 2004) experimented on PDMSE and acidified glass and the growth of *Amphora coffeaeformis var. perpusilla*, *Craspedostaurosaustralis* and *Naviculaperminuta* was studied. It was observed that the PDMSE allowed the biological growth but the adherence of the bio-film was less compared to the glass and thus could easily be removed.

Chaudhury et al. (Chaudhury et al. 2005) investigated the release of adhered spores and young of green alga and reported on the effect of modulus and depth of film for the favourability of their growth. PDMS elastomers were used where the thickness was maintained but the elasticity was varied by varying the cross-linking with differing degrees of polymerization using silicone oligomers. They discussed the deformation and adherence of the alga on PDMS and also applied fracture mechanic models for hard fouling and proposed models for soft fouling by the considered algal species.

Schmidt et al. (Schmidt et al. 2004) studied the biofouling release by coatings that had perfluoroalkyl groups on the surface. The synthesis of the coatings was a cross-linking reaction of the copolymer of 1H, 1H,2H,2H-heptadecafluorodecyl acrylate and acrylic acid (AA) with varying concentrations of poly (2-isopropenyl-2-oxazoline) and methyl methacrylate (MMA). The results indicated that marine biofouling was inhibited by surfaces which displayed low WCA hysteresis and high receding contact angle.

There are three main drawbacks of hydrophobic coatings with foul release capabilities:

1. The substrate should be in motion for effective removal of fouling agents.
2. Soft coatings are prone to damage.
3. Leachable organisms might be responsible for biofouling.

This implies that anti-fouling and anti-scaling properties are only valid when the object with the coating is in motion, and their efficiency changes with respect to time. After considering the drawbacks, hydrophobic coatings are still considered the best in their category for ship hull coatings (Das et al. 2018).

11.3.6 Self-Cleaning Coatings

Some plants have super-hydrophobic leaves that exhibit self-cleaning property after a rain shower. Water dewdrops form on the surface of the leaf and roll off even at a slight inclination. Super-hydrophobic surfaces on leaves and other parts of plants possess a roughness of micro and nano dimensions (Barthlott and Neinhuis 1997). Dust particles settle on the tips of this terrain and have low adhesion forces. When a water droplet rolls, it carries this dust particle along with it and this results in the cleaning of the surface. This phenomenon has been widely studied and has been applied to develop self-cleaning coatings (Fürstner et al. 2005).

Organopolysilazane (OPSZ) and fluorinated silica nanoparticles (F-SiO$_2$ NPs) were synthesized to create super-hydrophobic nanocomposites which exhibited mechanical durability, a rare property shown by super-hydrophobic surfaces. The materials were alternatively coated in layers of OPSZ and F-SiO$_2$ NPs obtained by spray coating on glass substrates. The roughness and growth in bilayer deposits were directly proportional to each other. A five-bilayer coating exhibited super-hydrophobicity with a WCA of 158.3° and sliding CA of 3°. Y. Chen et al. (Y. Chen et al. 2018) while working on this research also determined the adhesive force to be 13 μN and proved the self-cleaning efficacy of these bilayers. Mechanical durability was determined by immersing the sample in strong acid and alkaline solutions and the uniqueness of this material was due to the super-hydrophobicity maintained, even after being subjected to such harsh conditions. This proves the efficacy of polysilazane for super-hydrophobicity for self-cleaning applications in severe conditions.

Satapathy et al. (Satapathy et al. 2018) coated a glass substrate with linear low-density polyethylene (LLDPE) embedded with SiO$_2$ nanoparticles, which was fabricated by the dip-coating method and displayed exceptional super-hydrophobic properties. The material was synthesized in two variants: porous and non-porous. For the porous material, the non-solvent used was ethanol and the phase separation method helped to induce porosity by the evaporation of the non-solvent. Chemical, thermal and mechanical durability was also established by subjecting the samples to extreme conditions to test the respective parameters. The porous coating helped in achieving a WCA of 170° and a sliding angle of 3.8° and also showed exceptional self-cleaning properties even after the temperatures were increased up to 120°. The water jet impact test was also done to determine the water-repellent properties of the coating. The non-porous material exhibited more stability for varying pH conditions (2 to 9) than non-porous coatings but the abrasion test revealed that the mechanical durability was enhanced for porous coatings.

TiO$_2$ affected the surface morphology of micro/nano porous hydrophobic fluoropolymer composites which showed an excellent super-hydrophobic nature (Sahoo, Kandasubramanian and Thomas 2013). Super-hydrophobic (fluoroalkyl)silane coated TiO$_2$ photocatalytic films (Akira Nakajima et al. 2000), P25 (nano TiO$_2$)/ATP (attapulgite)/ER (epoxy resin)/PDMS (polydimethylsiloxane) coating (Yu et al. 2018), a polyvinylidene fluoride matrix (PVDF) with SiO$_2$ as fillers (Kumar, Li and Chen 2016), TEMPO-oxidized nanopaper made of cellulose nanofibrils (TOCNF) and poly-siloxanes, are some of the notable mentions which have recently been developed and exhibit exceptional self-cleaning properties (Y. Chen et al. 2018).

11.3.7 Medical Applications and Microfluidics

Polymer coatings are widely used in the medical field for drug delivery and dental applications. One of the major advances in the medical domain is the advent of microfluidic devices. Microfluidic devices are instruments that use small amounts of fluid on a microchip for lab-scale testing and are often used for biological assays (Fiorini and Chiu 2005). These are termed 'lab-on-a chip' devices because of their ability to perform many tests which otherwise would be time-inefficient or costly. The cost of these tests is further reduced by using polymer materials that provide simple and disposable alternatives and easily devised micro total analysis systems (μTAS) (Becker and Gärtner 2000) PDMS. A range of thermoplastics (Tsao and DeVoe 2009) are generally used in microfluidics due to their ability to be moulded, their functional feasibility and other properties that make them indispensable for the production of microfluidic devices (Tsao 2016).

PCU (polycarbonate urethane) (Zdrahala and Zdrahala 1999) has biocompatible properties and long-term bio-stability and, when combined with 20% fluorinated alkyl side chains, decreases the surface energy and enhances the properties of synthesized polymers and this makes it advantageous for biomedical applications. Carbon nanotubes (CNT) templates were dip-coated and dried to obtain a thin coating of the polymer on their surface. The wettability of the nanostructured thin films increased drastically and a CA of 163.6 ± 1.18° was observed which proved the super-hydrophobicity of the coating. The adherence of platelets was also negligible for these surfaces which was established by immunofluorescence tests. This property helps in medical technologies related to artificial organ, blood vessel transplants, pacemakers, etc. (Peppas and Langer 1994; Sun et al. 2005).

Pacetti and Hossainy (Pacetti and Hossainy 1999) patented an implantable substrate that was coated with polycondensation polymers made up of diketene acetal and a diol. A large number of variations were provided and the fundamental application for which this technology was developed is implantable drug eluting vascular stents.

Photo-cross-linked PU (polyurethane) fluoro-functionalized silica nanoparticles (F-SiO$_2$ NPs) have been shown to prevent microleakage restoration of dental composites. The nanoparticles imparted low surface energy to the composite while the PU enforced it with excellent mechanical stability. The studies were performed on various PU substrates incorporated with different ratios of F-SiO$_2$ nanoparticles and it was observed that the ratio of 1:3 for PU/F-SiO$_2$ composite displayed an ordered papillae architecture wherein air was found to be trapped in the grooves. A high CA of ~160.1° with a low sliding CA of ~<1° and an additional property of transparency was also observed. An MTT assay performed resulted in certifying the cell viability and biocompatibility while super-hydrophobicity prohibited the water infiltration in the assessment of the resin composite restoration (Cao et al. 2017).

Bio-MEMS (micro-electro-mechanical systems) are attracting great interest owing to their highly efficient capabilities for various applications. The epoxy-based polymer SU-8 which possesses UV sensitivity has been used by various research groups to fabricate microchannels as it possesses biocompatibility, transparency and a chemically inert nature. This polymer was used to introduce hydrophobicity into the channel walls of the microfluidic chips, which are made of glass or PDMS to aid the segmented flow of fluid droplets that are in motion with the non-polar carrier liquid, by coating the surfaces with octadecyl trichloroslane (Schumacher et al. 2008).

The use of polymer coatings in these applications provides a glimpse into the wide potential of generated hydrophobic structures and shows there is still a huge possibility to develop various polymeric coatings and apply them in other fields such as drug delivery (Yadav and Balasubramanian 2015), architecture and construction, smart textiles, defence applications, heat sinking applications (Katiyar and Balasubramanian 2015), cosmetics, etc.

11.4 Conclusion

The naturally bestowed phenomenon of hydrophobicity in biological entities such as the eyes of a mosquito and rose petals has aroused academic and industrial interest and thus super-hydrophobicity is recognized as one of the top-notch technologies. The need to address serious problems such as corrosion, biofouling, biocompatibility, etc. that use hydrophobic/super-hydrophobic surfaces for applications in aircraft, automobiles, or drug delivery has increased exponentially in the last decade. There has been a significant change in the fabrication technologies from a multi-step complex method to a single-step simplistic method. Several techniques such as spin coating, SAMS, sol-gel, layer-by-layer deposition, etc. have attracted interest in the development of these polymeric coatings owing to their feasibility and reproducibility. This chapter has presented a detailed overview of wettability theories and their applicability in solving various engineering problems. Although polymer coatings have proven efficacy, there

are still some challenges that need to be addressed for large-scale industrial usage. The need for environmentally-friendly polymer coatings with great economic value, longer shelf life and excellent adhesion strength is the primary requirement for the current era. Even the Wenzel and Cassie-Baxter theories require improvements for a better understanding of the mechanism of hydrophobicity. Thus, the aim of this chapter is to impart knowledge about the basics of hydrophobicity, the fabrication technology and various case studies for different applications which can act as a basis for a better understanding of this subject and prove to be an essential tool for further research and industrial activities.

REFERENCES

Ahmad, Zaki (ed.) 2015. *New Trends in Alloy Development: Characterization and Application*. InTech. https://doi.org/10.5772/59221.

Ahn, Y.C., S.K. Park, G.T. Kim, Y.J. Hwang, C.G. Lee, H.S. Shin, and J.K. Lee. 2006. 'Development of High Efficiency Nanofilters Made of Nanofibers'. *Current Applied Physics* 6 (6): 1030–35. https://doi.org/10.1016/j.cap.2005.07.013.

Akira, Nakajima, Kazuhito Hashimoto, Toshiya Watanabe, Kennichi Takai, Goro Yamauchi, and Akira Fujishima. 2000. 'Transparent Superhydrophobic Thin Films with Self-Cleaning Properties'. https://doi.org/10.1021/LA000155K.

Akram Raza, Muhammad, E. Stefan Kooij, Arend van Silfhout, and Bene Poelsema. 2010. 'Superhydrophobic Surfaces by Anomalous Fluoroalkylsilane Self-Assembly on Silica Nanosphere Arrays'. *Langmuir* 26(15): 12962–72. https://doi.org/10.1021/la101867z.

Arukalam, Innocent O., Emeka E. Oguzie, and Ying Li. 2018. 'Nanostructured Superhydrophobic Polysiloxane Coating for High Barrier and Anticorrosion Applications in Marine Environment'. *Journal of Colloid and Interface Science* 512 (February): 674–85. https://doi.org/10.1016/j.jcis.2017.10.089.

Ba, Yan, Qinjun Kang, Haihu Liu, Jinju Sun, and Chao Wang. 2016. 'Three-Dimensional Lattice Boltzmann Simulations of Microdroplets Including Contact Angle Hysteresis on Topologically Structured Surfaces'. *Journal of Computational Science* 17: 418–30. https://doi.org/10.1016/j.jocs.2016.03.015.

Badhe, Y., K. Balasubramanian, M. Singh, and A. Aswathy. 2015. 'Nano-Engineered Hybrid Hydroxyapatite-Grafted Biocomposites for Euspria Pulchella Mimicking through Chaotic Flow Regimes'. *RSC Advances*. https://doi.org/10.1039/c4ra14792h.

Bai, Weibin, Meiqin Guan, Ningshan Lai, Rijin Yao, Yanlian Xu, and Jinhuo Lin 2018. 'Superhydrophobic Paper from Conjugated Poly(p-Phenylene)s: Self-Assembly and Separation of Oil/Water Mixture'. *Materials Chemistry and Physics* 216 (September): 230–6. https://doi.org/10.1016/J.MATCHEMPHYS.2018.06.014.

Bangar, G.Y., D. Ghule, R.K.P. Singh, and Balasubramanian Kandasubramanian. 2018. 'Thermally Triggered Transition of Fluid Atomized Micro- and Nanotextured Multiscale Rough Surfaces'. *Colloids and Surfaces A: Physicochemical and Engineering Aspects*. https://doi.org/10.1016/j.colsurfa.2018.03.044.

Barati Darban, G., M. Aliofkhazraei, S. Khorsand, S. Sokhanvar, and A. Kaboli. 2018. 'Science and Engineering of Superhydrophobic Surfaces: Review of Corrosion Resistance, Chemical and Mechanical Stability'. *Arabian Journal of Chemistry*, 2018. https://doi.org/10.1016/j.arabjc.2018.01.013.

Bar-Cohen, Yoseph. 2006. 'Biomimetics: Using Nature to Inspire Human Innovation'. *Bioinspiration and Biomimetics*. https://doi.org/10.1088/1748-3182/1/1/P01.

Barthelat, François. 2007. 'Biomimetics for Next Generation Materials'. *Philosophical Transactions of the Royal Society A: Mathematical, Physical and Engineering Sciences*. https://doi.org/10.1098/rsta.2007.0006.

Barthlott, W. and C. Neinhuis. 1997. 'Purity of the Sacred Lotus, or Escape from Contamination in Biological Surfaces'. *Planta* 202 (1): 1–8. https://doi.org/10.1007/s004250050096.

Baxter, S. and A.B.D. Cassie. 1945. '8—The Water Repellency of Fabrics and a New Water Repellency Test'. *Journal of the Textile Institute Transactions* 36 (4): 67–90. https://doi.org/10.1080/19447024508659707.

Becker, Holger and Claudia Gärtner. 2000. 'Polymer Microfabrication Methods for Microfluidic Analytical Applications'. *Electrophoresis*. https://doi.org/10.1002/(SICI)1522-2683(20000101)21:1<12::AID-ELPS12>3.0.CO;2-7.

Bers, A. Valeria and Martin Wahl. 2004. 'The Influence of Natural Surface Microtopographies on Fouling'. *Biofouling* 20 (1): 43–51. https://doi.org/10.1080/08927010410001655533.

Bhushan, B. 2009. 'Biomimetics: Lessons from Nature – an Overview'. Philosophical Transactions of the Royal Society A: Mathematical, Physical and Engineering Sciences. https://doi.org/10.1098/rsta.2009.0011.

Bhushan, Bharat and Michael Nosonovsky. 2010. 'The Rose Petal Effect and the Modes of Superhydrophobicity'. *Philosophical Transactions of the Royal Society A: Mathematical, Physical and Engineering Sciences*. https://doi.org/10.1098/rsta.2010.0203.

Blunt, L. and X. Jiang. 2003. *Advanced Techniques for Assessment Surface Topography: Development of a Basis for 3D Surface Texture Standards 'Surfstand'*. Oxford: Elsevier. https://doi.org/10.1016/B978-1-903996-11-9.X5000-2.

Boisier, Grégory, Alain Lamure, Nadine Pébère, Nicolas Portail, and Martine Villatte. 2009. 'Corrosion Protection of AA2024 Sealed Anodic Layers Using the Hydrophobic Properties of Carboxylic Acids'. *Surface and Coatings Technology* 203 (22): 3420–26. https://doi.org/10.1016/j.surfcoat.2009.05.008.

Brady, Robert F. and Irwin L. Singer. 2000. 'Mechanical Factors Favoring Release from Fouling Release Coatings'. *Biofouling* 15 (1–3): 73–81. https://doi.org/10.1080/08927010009386299.

Buskens, Pascal, Mariëlle Wouters, Corné Rentrop, and Zeger Vroon. 2013. 'A Brief Review of Environmentally Benign Antifouling and Foul-Release Coatings for Marine Applications'. *Journal of Coatings Technology and Research* 10 (1): 29–36. https://doi.org/10.1007/s11998-012-9456-0.

Byun, Doyoung, Jongin Hong, Saputra Jin, Hwan Ko, Young Jong Lee, Hoon Cheol Park, Bong-Kyu Byun, and Jennifer R. Lukes. 2009. 'Wetting Characteristics of Insect Wing Surfaces'. *Journal of Bionic Engineering* 6 (1): 63–70. https://doi.org/10.1016/S1672-6529(08)60092-X.

Cain, J.B, D.W. Francis, R.D. Venter, and A.W. Neumann. 1983. 'Dynamic Contact Angles on Smooth and Rough Surfaces'. *Journal of Colloid and Interface Science* 94 (1): 123–30. https://doi.org/10.1016/0021-9797(83)90241-2.

Callow, Maureen E. and James E. Callow. 2002. 'Marine Biofouling: A Sticky Problem'. *Biologist (London, England)* 49 (1): 10–14. www.ncbi.nlm.nih.gov/pubmed/11852279.

Cao, Danfeng, Yingchao Zhang, Yao Li, Xiaoyu Shi, Haihuan Gong, Dan Feng, Xiaowei Guo, Zuosen Shi, Song Zhu, and Zhanchen Cui. 2017. 'Fabrication of Superhydrophobic Coating for Preventing Microleakage in a Dental Composite Restoration'. *Materials Science and Engineering: C* 78 (September): 333–40. https://doi.org/10.1016/j.msec.2017.04.054.

Cao, Liangliang, Andrew K. Jones, Vinod K. Sikka, Jianzhong Wu, and Di Gao. 2009. 'Anti-Icing Superhydrophobic Coatings'. *Langmuir* 25 (21): 12444–8. https://doi.org/10.1021/la902882b.

Cassie, A.B.D. and S. Baxter. 1944. 'Wettability of Porous Surfaces'. *Transactions of the Faraday Soc.* 40: 546–51.

Chakrabarty, Souvik, Mark Nisenholt, and Kenneth J. Wynne. 2012. 'PDMS–Fluorous Polyoxetane–PDMS Triblock Hybrid Elastomers: Tough and Transparent with Novel Bulk Morphologies'. *Macromolecules* 45 (19): 7900–13. https://doi.org/10.1021/ma301447f.

Chaudhury, Manoj K, John A. Finlay, Jun Young Chung, Maureen E. Callow, and James A. Callow. 2005. 'The Influence of Elastic Modulus and Thickness on the Release of the Soft-Fouling Green Alga *Ulva Linza* (Syn. *Enteromorpha Linza*) from Poly(Dimethylsiloxane) (PDMS) Model Networks'. *Biofouling* 21 (1): 41–8. https://doi.org/10.1080/08927010500044377.

Chen, Yannan, Gangqiang Zhu, Mirabbos Hojamberdiev, Jianzhi Gao, Runliang Zhu, Chenghui Wang, Xiumei Wei, and Peng Liu. 2018. 'Three-Dimensional Ag_2O/Bi_5O7I P–n Heterojunction Photocatalyst Harnessing UV–vis–NIR Broad Spectrum for Photodegradation of Organic Pollutants'. *Journal of Hazardous Materials* 344: 42–54. https://doi.org/10.1016/j.jhazmat.2017.10.015.

Chen, Zhifeng, Guangji Li, Liying Wang, Yinlei Lin, and Wei Zhou. 2018. 'A Strategy for Constructing Superhydrophobic Multilayer Coatings with Self-Cleaning Properties and Mechanical Durability Based on the Anchoring Effect of Organopolysilazane'. *Materials & Design* 141 (March): 37–47. https://doi.org/10.1016/J.MATDES.2017.12.034.

Cheng, Quan-Yong, Cheng-Shu Guan, Meng Wang, Yi-Dong Li, and Jian-Bing Zeng. 2018. 'Cellulose Nanocrystal Coated Cotton Fabric with Superhydrophobicity for Efficient Oil/Water Separation'. *Carbohydrate Polymers* 199 (November): 390–6. https://doi.org/10.1016/j.carbpol.2018.07.046.

Cheng, Quan-Yong, Mei-Chen Liu, Yi-Dong Li, Jiang Zhu, An-Ke Du, and Jian-Bing Zeng. 2018. 'Biobased Super-Hydrophobic Coating on Cotton Fabric Fabricated by Spray-Coating for Efficient Oil/Water Separation'. *Polymer Testing* 66 (April): 41–7. https://doi.org/10.1016/j.polymertesting.2018.01.005.

Cicek, Volkan. 2014. 'Factors Influencing Corrosion.' *Corrosion Engineering*. https://doi.org/10.1002/9781118720837.

Cober, Stewart G., George A. Isaac, and J. Walter Strapp. 2001. 'Characterizations of Aircraft Icing Environments that Include Supercooled Large Drops' (accessed 29 August 2018). http://citeseerx.ist.psu.edu/viewdoc/download?doi=10.1.1.555.2932&rep=rep1&type=pdf.

Coulson, S.R., I. Woodward, J.P.S. Badyal, S.A. Brewer, and C. Willis. 2000. 'Super-Repellent Composite Fluoropolymer Surfaces'. *The Journal of Physical Chemistry B* 104 (37): 8836–40. https://doi.org/10.1021/jp0000174.

Crick, Colin R. and Ivan P. Parkin. 2013. 'Relationship between Surface Hydrophobicity and Water Bounces: A Dynamic Method for Accessing Surface Hydrophobicity'. *Journal of Materials Chemistry A* 1 (3): 799–804. https://doi.org/10.1039/c2ta00880g.

Das, Sonalee, Sudheer Kumar, Sushanta K. Samal, Smita Mohanty, and Sanjay K. Nayak. 2018. 'A Review on Superhydrophobic Polymer Nanocoatings: Recent Development and Applications'. *Industrial & Engineering Chemistry Research* 57 (8): 2727–45. https://doi.org/10.1021/acs.iecr.7b04887.

Du, Xin and Junhui He. 2011. 'A Self-Templated Etching Route to Surface-Rough Silica Nanoparticles for Superhydrophobic Coatings'. *ACS Applied Materials & Interfaces* 3 (4): 1269–76. https://doi.org/10.1021/am200079w.

Durand, Nelly, David Mariot, Bruno Améduri, Bernard Boutevin, and François Ganachaud. 2011. 'Tailored Covalent Grafting of Hexafluoropropylene Oxide Oligomers onto Silica Nanoparticles: Toward Thermally Stable, Hydrophobic, and Oleophobic Nanocomposites'. *Langmuir* 27 (7): 4057–67. https://doi.org/10.1021/la1048826.

Farhadi, S., M. Farzaneh, and S.A. Kulinich. 2011. 'Anti-Icing Performance of Superhydrophobic Surfaces'. *Applied Surface Science* 257 (14): 6264–9. https://doi.org/10.1016/j.apsusc.2011.02.057.

Fiorini, Gina S. and Daniel T. Chiu. 2005. 'Disposable Microfluidic Devices: Fabrication, Function, and Application'. *BioTechniques*. https://doi.org/10.2144/05383RV02.

Fürstner, Reiner, Wilhelm Barthlott, Christoph Neinhuis, and Peter Walzel. 2005. 'Wetting and Self-Cleaning Properties of Artificial Superhydrophobic Surfaces'. *Langmuir*. https://doi.org/10.1021/la0401011.

Gao, Lichao and Thomas J. McCarthy. 2006. 'Contact Angle Hysteresis Explained'. *Langmuir* 22 (14): 6234–7. https://doi.org/10.1021/la060254j.

Gao, X., X. Yan, X. Yao, L. Xu, K. Zhang, J. Zhang, B. Yang, and L. Jiang. 2007. 'The Dry-Style Antifogging Properties of Mosquito Compound Eyes and Artificial Analogues Prepared by Soft Lithography'. *Advanced Materials* 19 (17): 2213–17. https://doi.org/10.1002/adma.200601946.

Gao, Xuefeng and Lei Jiang. 2004. 'Water-Repellent Legs of Water Striders'. *Nature* 432 (7013): 36. https://doi.org/10.1038/432036a.

Giacomello, Alberto, Simone Meloni, Mauro Chinappi, and Carlo Massimo Casciola. 2012. 'Cassie-Baxter and Wenzel States on a Nanostructured Surface: Phase Diagram, Metastabilities, and Transition Mechanism by Atomistic Free Energy Calculations'. *Langmuir* 28 (29): 10764–72. https://doi.org/10.1021/la3018453.

Good, Robert J. 1992. 'Contact Angle, Wetting, and Adhesion: A Critical Review'. *Journal of Adhesion Science and Technology* 6 (12): 1269–302.

Gore, Prakash M.M., Mamta Dhanshetty, and Balasubramanian Kandasubramanian 2016. 'Bionic Creation of Nano-Engineered Janus Fabric for Selective Oil/Organic Solvent Absorption'. *RSC Advances* 6 (112): 111250–60. https://doi.org/10.1039/C6RA24106A.

Gore, Prakash M., and Balasubramanian Kandasubramanian. 2018. 'Heterogeneous Wettable Cotton Based Superhydrophobic Janus Biofabric Engineered with PLA/Functionalized-Organoclay Microfibers for Efficient Oil–water Separation'. *Journal of Materials Chemistry A* 6 (17): 7457–79. https://doi.org/10.1039/C7TA11260B.

Gore, Prakash M., Latika Khurana, Rohit Dixit, and B. Kandasubramanian. 2017. 'Keratin-Nylon 6 Engineered Microbeads for Adsorption of Th (IV) Ions from Liquid Effluents'. *Journal of Environmental Chemical Engineering* 5 (6): 5655–67. https://doi.org/10.1016/j.jece.2017.10.048.

Gore, Prakash Macchindra, Latika Khurana, Suhail Siddique, Anjana Panicker, and K, Balasubramanian. 2018. 'Ion-Imprinted Electrospun Nanofibers of Chitosan/1-Butyl-3-Methylimidazolium Tetrafluoroborate for the Dynamic Expulsion of Thorium (IV) Ions from Mimicked Effluents'. *Environmental Science and Pollution Research* 25 (4): 3320–34. https://doi.org/10.1007/s11356-017-0618-6.

Gore, Prakash M., Susan Zachariah, Prashant Gupta, and K, Balasubramanian. 2016. 'Multifunctional Nano-Engineered and Bio-Mimicking Smart Superhydrophobic Reticulated

ABS/Fumed Silica Composite Thin Films with Heat-Sinking Applications'. *RSC Advances* 6 (107): 105180–91. https://doi.org/10.1039/c6ra16781k.

Gupta, Prashant, and K. Balasubramanian. 2016. 'Numerical Investigation of Heat Loss in Nano-Fumed Silica Reinforced Styrene Acrylonitrile Hydrophobic Thermo-Sheath for Heat Inhibition in Hydronic Boiler'. *Materials Focus* 5 (6): 556–64. https://doi.org/10.1166/mat.2016.1362.

Han, Joong Tark, Yanli Zheng, Jeong Ho Cho, Xurong Xu, and Kilwon Cho. 2005. 'Stable Superhydrophobic Organic-Inorganic Hybrid Films by Electrostatic Self-Assembly'. *The Journal of Physical Chemistry B* 109 (44): 20773–8. https://doi.org/10.1021/jp052691x.

Hao, Jun Hua, and Zheng Jia Wang. 2016. 'Modeling Cassie-Baxter State on Superhydrophobic Surfaces'. *Journal of Dispersion Science and Technology* 37 (8): 1208–13. https://doi.org/10.1080/01932691.2015.1089407.

He, Bo, Junghoon Lee, and Neelesh A. Patankar. 2004. 'Contact Angle Hysteresis on Rough Hydrophobic Surfaces'. *Colloids and Surfaces A: Physicochemical and Engineering Aspects* 248 (1–3): 101–4. https://doi.org/10.1016/j.colsurfa.2004.09.006.

Hench, Larry L. and Jon K. West. 1990. 'The Sol-Gel Process'. *Chemical Reviews* 90 (1): 33–72. https://doi.org/10.1021/cr00099a003.

Hendry, Jason A. and Robert M. Pilliar. 2001. 'The Fretting Corrosion Resistance of PVD Surface-Modified Orthopedic Implant Alloys'. *Journal of Biomedical Materials Research* 58 (2): 156–66. https://doi.org/10.1002/1097-4636(2001)58:2<156::AID-JBM1002>3.0.CO;2-H.

Hey, Michael J. and John G. Kingston. 2007. 'The Apparent Contact Angle for a Nearly Spherical Drop on a Heterogeneous Surface'. *Chemical Physics Letters* 447 (1–3): 44–8. https://doi.org/10.1016/j.cplett.2007.08.080.

Hikita, Masaya, Keiji Tanaka, Tetsuya Nakamura, Tisato Kajiyama, and Atsushi Takahara. 2005. 'Super-Liquid-Repellent Surfaces Prepared by Colloidal Silica Nanoparticles Covered with Fluoroalkyl Groups'. *Langmuir* 21 (16): 7299–302. https://doi.org/10.1021/la050901r.

Hoipkemeier-Wilson, Leslie, James F. Schumacher, Michelle L. Carman, Amy L. Gibson, Adam W. Feinberg, Maureen E. Callow, John A. Finlay, James A. Callow, and Anthony B. Brennan. 2004. 'Antifouling Potential of Lubricious, Micro-Engineered, PDMS Elastomers against Zoospores of the Green Fouling Alga *Ulva (Enteromorpha)*'. *Biofouling* 20 (1): 53–63. https://doi.org/10.1080/08927010410001662689.

Holland, R., T.M. Dugdale, R. Wetherbee, A.B. Brennan, J.A. Finlay, J.A. Callow, and Maureen E. Callow. 2004. 'Adhesion and Motility of Fouling Diatoms on a Silicone Elastomer'. *Biofouling* 20 (6): 323–9. https://doi.org/10.1080/08927010400029031.

Honkanen, Anna and Victor Benno Meyer-Rochow. 2009. 'The Eye of the Parthenogenetic and Minute Moth Ectoedemia Argyropeza (Lepidoptera: Nepticulidae)'. *European Journal of Entomology*. https://doi.org/10.14411/eje.2009.078.

Hunley, Matthew T. and Timothy E. Long. 2008. 'Electrospinning Functional Nanoscale Fibers: A Perspective for the Future'. *Polymer International* 57 (3): 385–9. https://doi.org/10.1002/pi.2320.

Jadhav, Ananda J., Chandrakant R. Holkar, and Dipak V. Pinjari. 2018. 'Anticorrosive Performance of Super-Hydrophobic Imidazole Encapsulated Hollow Zinc Phosphate Nanoparticles on Mild Steel'. *Progress in Organic Coatings* 114 (January): 33–9. https://doi.org/10.1016/j.porgcoat.2017.09.017.

Jamdade, Sarang, S.V. Nimje, and Krishnamoorthi Balasubramanian. 2015. 'Hydrophobic Hierarchical Structures – Anti Corrosive Coating'. *Journal of Material Science and Mechanical Engineering (JMSME)* 2 (13): 45–9.

Jeong, K.-H., Jaeyoun Kim, and Luke P. Lee. 2006. 'Biologically Inspired Artificial Compound Eyes'. *Science* 312 (5773): 557–61. https://doi.org/10.1126/science.1123053.

Jiao, Rui, Lulu Bao, Wanli Zhang, Hanxue Sun, Zhaoqi Zhu, Chaohu Xiao, Lihua Chen, and Li An. 2018. 'Synthesis of Aminopyridine-Containing Conjugated Microporous Polymers with Excellent Superhydrophobicity for Oil/Water Separation'. *New Journal of Chemistry*. https://doi.org/10.1039/C8NJ02500B.

Jitianu, Andrei, John Doyle, Glenn Amatucci, and Lisa C. Klein. 2010. 'Methyl Modified Siloxane Melting Gels for Hydrophobic Films'. *Journal of Sol-Gel Science and Technology* 53 (2): 272–9. https://doi.org/10.1007/s10971-009-2087-y.

Joanny, J.F. and P.G. de Gennes. 1984. 'A Model for Contact Angle Hysteresis'. *The Journal of Chemical Physics* 81 (1): 552–62. https://doi.org/10.1063/1.447337.

Johnson, Rulon E. and Robert H. Dettre. 1964. 'Contact Angle Hysteresis. III. Study of an Idealized Heterogeneous Surface'. *The Journal of Physical Chemistry* 68 (7): 1744–50. https://doi.org/10.1021/j100789a012.

Katiyar, Neha and K. Balaasubramanian. 2015. 'Nano-Heat-Sink Thin Film Composite of PC/Three-Dimensional Networked Nano-Fumed Silica with Exquisite Hydrophobicity'. *RSC Advances* 5 (6): 4376–84. https://doi.org/10.1039/C4RA11597J.

Kavale, Mahendra S., D.B. Mahadik, V.G. Parale, P.B. Wagh, Satish C. Gupta, A. Venkateswara Rao, and Harish C. Barshilia. 2011. 'Optically Transparent, Superhydrophobic Methyltrimethoxysilane Based Silica Coatings without Silylating Reagent'. *Applied Surface Science* 258 (1): 158–62. https://doi.org/10.1016/j.apsusc.2011.08.023.

Kavalenka, Maryna N., Felix Vüllers, Jana Kumberg, Claudia Zeiger, Vanessa Trouillet, Sebastian Stein, Tanzila T. Ava, Chunyan Li, Matthias Worgull, and Hendrik Hölscher. 2017. 'Adaptable Bioinspired Special Wetting Surface for Multifunctional Oil/Water Separation'. *Scientific Reports* 7 (1): 39970. https://doi.org/10.1038/srep39970.

Koch, Brendan M.L., A. Amirfazli, and Janet A.W. Elliott. 2014. 'Wetting of Rough Surfaces by a Low Surface Tension Liquid'. *Journal of Physical Chemistry C* 118 (41): 23777–82. https://doi.org/10.1021/jp5071117.

Koivuluoto, Heli, Christian Stenroos, Mikko Kylmälahti, Marian Apostol, Jarkko Kiilakoski, and Petri Vuoristo. 2017. 'Anti-Icing Behavior of Thermally Sprayed Polymer Coatings'. *Journal of Thermal Spray Technology* 26 (1–2): 150–60. https://doi.org/10.1007/s11666-016-0501-x.

Kritzer, Peter. 2004. 'Corrosion in High-Temperature and Supercritical Water and Aqueous Solutions: A Review'. *Journal of Supercritical Fluids*. https://doi.org/10.1016/S0896-8446(03)00031-7.

Krumpfer, Joseph W. and Thomas J. McCarthy. 2010. 'Contact Angle Hysteresis: A Different View and a Trivial Recipe for Low Hysteresis Hydrophobic Surfaces'. *Faraday Discussions* 146: 103. https://doi.org/10.1039/b925045j.

Kumar, Divya, Lin Li, and Zhong Chen. 2016. 'Mechanically Robust Polyvinylidene Fluoride (PVDF) Based Superhydrophobic Coatings for Self-Cleaning Applications'. *Progress in Organic Coatings*. https://doi.org/10.1016/j.porgcoat.2016.09.003.

Kwok, D.Y., C.N.C. Lam, A. Li, K. Zhu, R. Wu, and A.W. Neumann. 1998. 'Low-Rate Dynamic Contact Angles on Polystyrene and the Determination of Solid Surface Tensions'. *Polymer Engineering & Science* 38 (10): 1675–84. https://doi.org/10.1002/pen.10338.

Kwok, D.Y. and A.W. Neumann. 1999. 'Contact Angle Measurement and Contact Angle Interpretation'. *Advances in Colloid and Interface Science*. 81. https://doi.org/10.1016/S0001-8686(98)00087-6.

Kwon, Yongjoo, Neelesh Patankar, Junkyu Choi, and Junghoon Lee. 2009. 'Design of Surface Hierarchy for Extreme Hydrophobicity'. *Langmuir*. https://doi.org/10.1021/la803249t.

Lam, C.N. Catherine, J. James Lu, and A. Wilhelm Neumann. 2002. 'Measuring Contact Angle'. In K. Holmberg, D.O. Shah, and M.J. Schwuger (eds), *Handbook of Applied Surface and Colloid Chemistry*, vol. 2. Hoboken, NJ: John Wiley & Sons, pp. 251–80. https://doi.org/10.1604/9780471490838.

Lannutti, J., D. Reneker, T. Ma, D. Tomasko, and D. Farson. 2007. 'Electrospinning for Tissue Engineering Scaffolds'. *Materials Science and Engineering C* 27 (3): 504–9. https://doi.org/10.1016/j.msec.2006.05.019.

Lasher-Trapp, Sonia, Sarah Anderson-Bereznicki, Ashley Shackelford, Cynthia H. Twohy, and James G. Hudson. 2008. 'An Investigation of the Influence of Droplet Number Concentration and Giant Aerosol Particles upon Supercooled Large Drop Formation in Wintertime Stratiform Clouds'. *Journal of Applied Meteorology and Climatology* 47 (10): 2659–78. https://doi.org/10.1175/2008JAMC1807.1.

Lau, Kenneth K.S., José Bico, Kenneth B.K. Teo, Manish Chhowalla, Gehan A.J. Amaratunga, William I. Milne, Gareth H. McKinley, and Karen K. Gleason. 2003. 'Superhydrophobic Carbon Nanotube Forests'. *Nano Letters* 3 (12): 1701–5. https://doi.org/10.1021/nl034704t.

Lee, L.P. and Robert Szema. 2005. 'Inspirations from Biological Optics for Advanced Photonic Systems'. *Science* 310 (5751): 1148–50. https://doi.org/10.1126/science.1115248.

Leivo, E., T. Wilenius, T. Kinos, P. Vuoristo, and T. Mäntylä. 2004. 'Properties of Thermally Sprayed Fluoropolymer PVDF, ECTFE, PFA and FEP Coatings'. *Progress in Organic Coatings*. https://doi.org/10.1016/j.porgcoat.2003.08.011.

Li, Lester, Victor Breedveld, and Dennis W. Hess. 2012. 'Creation of Superhydrophobic Stainless Steel Surfaces by Acid Treatments and Hydrophobic Film Deposition'. *ACS Applied Materials & Interfaces* 4 (9): 4549–56. https://doi.org/10.1021/am301666c.

Liu, Tingyi Leo, Zhiyu Chen, and Chang Jin Kim. 2015. 'A Dynamic Cassie-Baxter Model'. *Soft Matter* 11 (8): 1589–96. https://doi.org/10.1039/c4sm02651a.

Ma, Minglin and Randal M. Hill. 2006. 'Superhydrophobic Surfaces'. *Current Opinion in Colloid & Interface Science* 11 (4): 193–202. https://doi.org/10.1016/j.cocis.2006.06.002.

Ma, Minglin, Yu Mao, Malancha Gupta, Karen K. Gleason, and Gregory C. Rutledge. 2005. 'Superhydrophobic Fabrics Produced by Electrospinning and Chemical Vapor Deposition'. *Macromolecules* 38 (23): 9742–8. https://doi.org/10.1021/ma0511189.

Magisetty, Ravi Prakash, Anuj Shukla, and Balasubramanian Kandasubramanian. 2018. 'Dielectric, Hydrophobic Investigation of ABS/NiFe2O4 Nanocomposites Fabricated by Atomized Spray Assisted and Solution Casted Techniques for Miniaturized Electronic Applications'. *Journal of Electronic Materials* 47 (9): 5640–56. https://doi.org/10.1007/s11664-018-6452-x.

Mahltig, B. and H. Böttcher. 2003. 'Modified Silica Sol Coatings for Water-Repellent Textiles'. *Journal of Sol-Gel Science and Technology* 27 (1): 43–52. https://doi.org/10.1023/A:102262 7926243.

Marmur, Abraham. 2006a. 'Soft Contact: Measurement and Interpretation of Contact Angles'. *Soft Matter* 2 (1): 12–17. https://doi.org/10.1039/B514811C.

Marmur, Abraham. 2006b. 'Super-Hydrophobicity Funda-mentals: Implications to Biofouling Prevention'. *Biofouling* 22 (2): 107–15. https://doi.org/10.1080/08927010600562328.

Mishra, Paridhi and K. Balasubramanian. 2014. 'Nanostructured Microporous Polymer Composite Imprinted with Superhydrophobic Camphor Soot, for Emphatic Oil–water Separation'. *RSC Adv.* 4 (95): 53291–6. https://doi.org/10.1039/C4RA07410F.

Mohammadi, R., J. Wassink, and A. Amirfazli. 2004. 'Effect of Surfactants on Wetting of Super-Hydrophobic Surfaces'. *Langmuir* 20 (22): 9657–62. https://doi.org/10.1021/la049268k.

Morra, M., E. Occhiello, and F. Garbassi. 1989. 'Contact Angle Hysteresis in Oxygen Plasma Treated Poly(Tetrafluoroethylene)'. *Langmuir* 5 (3): 872–6. https://doi.org/10.1021/la00087a050.

Moutinho, Isabel, Margarida Figueiredo, and Paulo Ferreira. 2007. 'Evaluating the Surface Energy of Laboratory-Made Paper Sheets by Contact Angle Measurements'. *Tappi Journal* 6 (6): 26–32.

Nakajima, Akira, Kazuhito Hashimoto, and Toshiya Watanabe. 2001. 'Recent Studies on Super-Hydrophobic Films'. *Monatshefte für Chemie/Chemical Monthly* 132 (1): 31–41. https://doi.org/10.1007/s007060170142.

Ng, K.M., R. Gani, and K. Dam-Johansen. 2007. *Chemical Product Design: Toward a Perspective through Case Studies*. Oxford: Elsevier.

Nordvik, Atle B., James L. Simmons, Kenneth R. Bitting, Alun Lewis, and Tove Strøm-Kristiansen. 1996. 'Oil and Water Separation in Marine Oil Spill Clean-up Operations'. *Spill Science & Technology Bulletin* 3 (3): 107–22. https://doi.org/10.1016/S1353-2561(96)00021-7.

Onda, T., S. Shibuichi, N. Satoh, and K. Tsujii. 1996. 'Super-Water-Repellent Fractal Surfaces'. *Langmuir* 12 (9): 2125–7. https://doi.org/10.1021/la950418o.

Öner, Didem and Thomas J. McCarthy. 2000. 'Ultrahydrophobic Surfaces: Effects of Topography Length Scales on Wettability'. *Langmuir* 16 (20): 7777–82. https://doi.org/10.1021/la000598o.

Pacetti, Stephen and Hossainy, Syed F.A. 1999. 'Hydrophobic Biologically Absorbable Coatings for Drug Delivery Devices and Methods for Fabricating the Same'. US Patent No. 7056591B1, issued May 24, 1999. https://patents.google.com/patent/US7056591B1/en.

Padday, J.F. 1992. 'Spreading, Wetting, and Contact Angles'. *Journal of Adhesion Science and Technology* 6 (12): 1347–58. https://doi.org/10.1163/156856192X00665.

Padhi, Sandhya, Suresh Gosavi, Ramdayal Yadav, and Balasubramanian Kandasubramanian. 2018. 'Quantitative Evolution of Wetting Phenomena for Super Hydrophobic Surfaces'. *Materials Focus*. https://doi.org/10.1166/mat.2018.1509.

Parker, Andrew R. and Chris R. Lawrence. 2001. 'Water Capture by a Desert Beetle'. *Nature* 414 (6859): 33–4. https://doi.org/10.1038/35102108.

Parker, Andrew R. and Helen E. Townley. 2007. 'Biomimetics of Photonic Nanostructures'. *Nature Nanotechnology*. https://doi.org/10.1038/nnano.2007.152.

Peppas, N.A. and R. Langer. 1994. 'New Challenges in Biomaterials'. *Science (New York, N.Y.)* 263 (5154): 1715–20. www.ncbi.nlm.nih.gov/pubmed/8134835.

Petrovicova, E. and L.S. Schadler. 2002. 'Thermal Spraying of Polymers'. *International Materials Reviews* 47 (4): 169–90. https://doi.org/10.1179/095066002225006566.

Pilotek, S. and H.K. Schmidt. 2003. 'Wettability of Microstructured Hydrophobic Sol-Gel Coatings'. *Journal of Sol-Gel Science and Technology* 26 (1/3): 789–92. https://doi.org/10.1023/A:1020779011844.

Princen, H.M. 1969. 'Capillary Phenomena in Assemblies of Parallel Cylinders'. *Journal of Colloid and Interface Science* 30.

Qiang, Fei, Li-Li Hu, Li-Xiu Gong, Li Zhao, Shi-Neng Li, and Long-Cheng Tang. 2018. 'Facile Synthesis of Super-Hydrophobic, Electrically Conductive and Mechanically Flexible Functionalized Graphene Nanoribbon/Polyurethane Sponge for Efficient Oil/Water Separation at Static and Dynamic States'. *Chemical Engineering Journal* 334 (February): 2154–66. https://doi.org/10.1016/j.cej.2017.11.054.

Reneker, Darrell H. and Alexander L. Yarin. 2008. 'Electrospinning Jets and Polymer Nanofibers'. *Polymer* 49 (10): 2387–425. https://doi.org/10.1016/j.polymer.2008.02.002.

Ruiz-Cabello, F., J. Montes, M.A. Rodríguez-Valverde, and M.A. Cabrerizo-Vílchez. 2014. 'Equilibrium Contact Angle or the Most-Stable Contact Angle?' *Advances in Colloid and Interface Science* 206: 320–7. https://doi.org/10.1016/j.cis.2013.09.003.

Saengkaew, Jittraporn, Duy Le, Chanatip Samart, Hideo Sawada, Masakazu Nishida, Narong, Chanlek, Suwadee Kongparakul, and Suda Kiatkamjornwong. 2018. 'Superhydrophobic Coating from Fluoroalkylsilane Modified Natural Rubber Encapsulated SiO2 Composites for Self-Driven Oil/Water Separation'. *Applied Surface Science* 462 (December): 164–74. https://doi.org/10.1016/J.APSUSC.2018.08.059.

Sahoo, Bichitra Nanda and K. Balasubramanian. 2014a. 'Controlled Fabrication of Superhydrophobic Hierarchical Structures Using Epf/Carbon Based Composites'. Paper presented at The Eighth International Conference in Materials Technology and Modelling, MMT. Available at: www.ariel.ac.il/sites/conf/mmt/mmt-2014/service files/papers/4-122-130.pdf.

Sahoo, Bichitra Nanda and K. Balasubramanian. 2014b. 'Facile Synthesis of Nano Cauliflower and Nano Broccoli like Hierarchical Superhydrophobic Composite Coating Using PVDF/Carbon Soot Particles via Gelation Technique'. *Journal of Colloid and Interface Science* 436 (December): 111–21. https://doi.org/10.1016/j.jcis.2014.08.031.

Sahoo, Bichitra Nanda, K. Balasubramanian, and Mahesh Sucheendran. 2015. 'Thermally Triggered Transition of Superhydrophobic Characteristics of Micro- and Nanotextured Multiscale Rough Surfaces'. *The Journal of Physical Chemistry C*, June, 150610094322003. https://doi.org/10.1021/acs.jpcc.5b02917.

Sahoo, Bichitra Nanda, K. Balasubramanian, and Amrutha Thomas. 2013. 'Effect of TiO$_2$ Powder on the Surface Morphology of Micro/Nanoporous Structured Hydrophobic Fluoropolymer Based Composite Material'. *Journal of Polymers* 2013 (June): 1–4. https://doi.org/10.1155/2013/615045.

Saito, H., K. Takai, and G. Yamauchi. 1997. 'Water- and Ice-Repellent Coatings'. *Surface Coatings International* 80 (4): 168–71. https://doi.org/10.1007/BF02692637.

Sarikaya, Mehmet, Candan Tamerler, Alex K.Y. Jen, Klaus Schulten, and François Baneyx. 2003. 'Molecular Biomimetics: Nanotechnology through Biology'. *Nature Materials*. https://doi.org/10.1038/nmat964.

Satapathy, M., P. Varshney, D. Nanda, S.S. Mohapatra, A. Behera, and A. Kumar. 2018. 'Fabrication of Durable Porous and Non-Porous Superhydrophobic LLDPE/SiO$_2$ Nanoparticles Coatings with Excellent Self-Cleaning Property'. *Surface and Coatings Technology* 341 (May): 31–9. https://doi.org/10.1016/J.SURFCOAT.2017.07.025.

Schmidt, Donald L., Jr., Robert F. Brady, Karen Lam, Dale C. Schmidt, and Manoj K. Chaudhury. 2004. 'Contact Angle Hysteresis, Adhesion, and Marine Biofouling'. https://doi.org/10.1021/LA035385O

Schumacher, J.T., A. Grodrian, C. Kremin, M. Hoffmann, and J. Metze. 2008. 'Hydrophobic Coating of Microfluidic Chips Structured by SU-8 Polymer for Segmented Flow Operation'. *Journal of Micromechanics and Microengineering* 18 (5): 055019. https://doi.org/10.1088/0960-1317/18/5/055019.

Schwartz, Leonard W. and Stephen Garoff. 1985. 'Contact Angle Hysteresis on Heterogeneous Surfaces''. *Langmuir* 1 (2): 219–30. https://doi.org/10.1021/la00062a007.

Sedev, R.V., J.G. Petrov, and A.W. Neumann. 1996. 'Effect of Swelling of a Polymer Surface on Advancing and Receding Contact Angles'. *Journal of Colloid and Interface Science* 180 (1): 36–42. https://doi.org/10.1006/jcis.1996.0271.

Shang, H.M., Y. Wang, S.J. Limmer, T.P. Chou, K. Takahashi, and G.Z. Cao. 2005. 'Optically Transparent Superhydrophobic Silica-Based Films'. *Thin Solid Films* 472 (1–2): 37–43. https://doi.org/10.1016/j.tsf.2004.06.087.

Shi, F., J. Niu, J. Liu, F. Liu, Z. Wang, X.-Q. Feng, and X. Zhang. 2007. 'Towards Understanding Why a Superhydrophobic Coating Is Needed by Water Striders'. *Advanced Materials* 19 (17): 2257–61. https://doi.org/10.1002/adma.200700752.

Shi, F., Z. Wang, and X. Zhang. 2005. 'Combining a Layer-by-Layer Assembling Technique with Electrochemical Deposition of Gold Aggregates to Mimic the Legs of Water Striders'. *Advanced Materials* 17 (8): 1005–9. https://doi.org/10.1002/adma.200402090.

Shiu, Jau-Ye, Chun-Wen Kuo, Peilin Chen, and Chung-Yuan Mou. 2004. 'Fabrication of Tunable Superhydrophobic Surfaces by Nanosphere Lithography'. *Chemistry of Materials* 16 (4): 561–4. https://doi.org/10.1021/cm034696h.

Simon, Shilpa and K. Balaasubramanian, 2018. 'Facile Immobilization of Camphor Soot on Electrospun Hydrophobic Membrane for Oil-Water Separation'. *Materials Focus* 7 (2): 295–303. https://doi.org/10.1166/mat.2018.1511.

Simon, Shilpa, Ankit Malik, and Balasubramanian Kandasubramanian. 2018. 'Hierarchical Electrospun Super-Hydrophobic Nanocomposites of Fluoroelastomer'. *Materials Focus* 7. https://doi.org/10.1166/mat.2018.1499.

Snoeijer, Jacco H. and Bruno Andreotti. 2008. 'A Microscopic View on Contact Angle Selection'. *Physics of Fluids* 20 (5): 057101. https://doi.org/10.1063/1.2913675.

Sojoudi, Hossein, Hadi Arabnejad, Asif Raiyan, Siamack A. Shirazi, Gareth H. McKinley, and Karen K. Gleason. 2018. 'Scalable and Durable Polymeric Icephobic and Hydrate-Phobic Coatings'. *Soft Matter* 14 (18): 3443–54. https://doi.org/10.1039/C8SM00225H.

Sun, Taolei, Hong Tan, Dong Han, Qiang Fu, and Lei Jiang. 2005. 'No Platelet Can Adhere—Largely Improved Blood Compatibility on Nanostructured Superhydrophobic Surfaces'. *Small* 1 (10): 959–63. https://doi.org/10.1002/smll.200500095.

Tadanaga, Kiyoharu, Junichi Morinaga, and Tsutomu Minami. 2000. 'Formation of Superhydrophobic-Superhydrophilic Pattern on Flowerlike Alumina Thin Film by the Sol-Gel Method'. *Journal of Sol-Gel Science and Technology* 19 (1/3): 211–14. https://doi.org/10.1023/A:1008732204421.

Thanawala, Shilpa K. and Manoj K. Chaudhury. 2000. 'Surface Modification of Silicone Elastomer Using Perfluorinated Ether'. *Langmuir* 16 (3): 1256–60. https://doi.org/10.1021/la9906626.

Tsao, Chia Wen. 2016. 'Polymer Microfluidics: Simple, Low-Cost Fabrication Process Bridging Academic Lab Research to Commercialized Production'. *Micromachines*. https://doi.org/10.3390/mi7120225.

Tsao, Chia Wen and Don L. DeVoe. 2009. 'Bonding of Thermoplastic Polymer Microfluidics'. *Microfluidics and Nanofluidics*. https://doi.org/10.1007/s10404-008-0361-x.

Venkateswara Rao, A., Sanjay S. Latthe, Digambar Y. Nadargi, H. Hirashima, and V. Ganesan. 2009. 'Preparation of MTMS Based Transparent Superhydrophobic Silica Films by Sol-gel Method'. *Journal of Colloid and Interface Science* 332 (2): 484–90. https://doi.org/10.1016/j.jcis.2009.01.012.

Vincent, Julian F.V., Olga A. Bogatyreva, Nikolaj R. Bogatyrev, Adrian Bowyer, and Anja Karina Pahl. 2006. 'Biomimetics: Its Practice and Theory'. *Journal of the Royal Society Interface*. https://doi.org/10.1098/rsif.2006.0127.

Von Baeyer, H.C. 2000. 'The Lotus Effect'. *The Sciences*. https://doi.org/10.1002/j.2326-1951.2000.tb03461.x.

Wagner, Thomas, Christoph Neinhuis, and Wilhelm Barthlott. 1996. 'Wettability and Contaminability of Insect Wings as a Function of Their Surface Sculptures'. *Acta Zoologica* 77 (3): 213–25. https://doi.org/10.1111/j.1463-6395.1996.tb01265.x

Walker, Robert. 1993. 'Principles and Prevention of Corrosion'. *Materials & Design*. https://doi.org/10.1016/0261-3069(93)90066-5.

Wang, Shutao, Lin Feng, Huan Liu, Taolei Sun, Xi Zhang, Lei Jiang, and Daoben Zhu. 2005. 'Manipulation of Surface Wettability between Superhydrophobicity and Superhydrophilicity on Copper Films'. *ChemPhysChem* 6 (8): 1475–8. https://doi.org/10.1002/cphc.200500204.

Wenzel, Robert N. 1936. 'Resistance of Solid Surfaces to Wetting by Water'. *Industrial & Engineering Chemistry* 28 (8): 988–94. https://doi.org/10.1021/ie50320a024.

Winandy, Jerrold E. and Todd F. Shupe. 2010. 'From Hydrophilicity to Hydrophobicity: A Critical Review: Part I. Wettability and Surface Behavior Cheng Piao'. *Wood and Fiber* 42 (4): 490–510.

Woodward, I., W.C.E. Schofield, V. Roucoules, and J.P.S. Badyal. 2003. 'Super-Hydrophobic Surfaces Produced by Plasma Fluorination of Polybutadiene Films'. *Langmuir* 19 (8): 3432–8. https://doi.org/10.1021/la020427e.

Wouters, Mariëlle, Corné Rentrop, and Peter Willemsen. 2010. 'Surface Structuring and Coating Performance: Novel Biocidefree Nanocomposite Coatings with Anti-Fouling and Fouling-Release Properties'. *Progress in Organic Coatings* 68 (1–2): 4–11. https://doi.org/10.1016/J.PORGCOAT.2009.10.005.

Wu, Xuedong, Lijun Zheng, and Dan Wu. 2005. 'Fabrication of Superhydrophobic Surfaces from Microstructured ZnO-Based Surfaces via a Wet-Chemical Route'. *Langmuir* 21 (7): 2665–7. https://doi.org/10.1021/la050275y.

Xu, Yuxi, Wenjing Hong, Hua Bai, Chun Li, and Gaoquan Shi. 2009. 'Strong and Ductile Poly(Vinyl Alcohol)/Graphene Oxide Composite Films with a Layered Structure'. *Carbon* 47 (15): 3538–43. https://doi.org/10.1016/j.carbon.2009.08.022.

Yadav, Ramdayal and K. Balaasubramanian. 2015. 'Polyacrylonitrile/Syzygium Aromaticum Hierarchical Hydrophilic Nanocomposite as a Carrier for Antibacterial Drug Delivery Systems'. *RSC Advances* 5 (5): 3291–8. https://doi.org/10.1039/C4RA12755B.

Yadav, Rohit, Susan Zachariah, and K. Balaasubramanian. 2016. 'Thermally Stable Transparent Hydrophobic Bio-Mimetic Dual Scale Spherulites Coating by Spray Deposition'. *Advanced Science, Engineering and Medicine* 8 (3): 181–7. https://doi.org/10.1166/asem.2016.1842.

Yang, Shuqing, Qiang Xia, Lin Zhu, Jian Xue, Qingjun Wang, and Qing Min Chen. 2011. 'Research on the Icephobic Properties of Fluoropolymer-Based Materials'. *Applied Surface Science*. https://doi.org/10.1016/j.apsusc.2011.01.003.

Young, T. 1805. 'An Essay on the Cohesion of Fluids'. *Philosophical Transactions of the Royal Society of London* 95 (January): 65–87. https://doi.org/10.1098/rstl.1805.0005.

Youngblood, Jeffrey P. and Thomas J. McCarthy. 1999. 'Ultrahydrophobic Polymer Surfaces Prepared by Simultaneous Ablation of Polypropylene and Sputtering of Poly(Tetrafluoroethylene) Using Radio Frequency Plasma'. *Macromolecules* 32 (20): 6800–6. https://doi.org/10.1021/ma9903456.

Yu, Minghua, Guotuan Gu, Wei-Dong Meng, and Feng-Ling Qing. 2007. 'Superhydrophobic Cotton Fabric Coating Based on a Complex Layer of Silica Nanoparticles and Perfluorooctylated Quaternary Ammonium Silane Coupling Agent'. *Applied Surface Science* 253 (7): 3669–73. https://doi.org/10.1016/j.apsusc.2006.07.086.

Yu, Nanlin, Xinyan Xiao, Zhihao Ye, and Guangming Pan. 2018. 'Facile Preparation of Durable Superhydrophobic Coating with Self-Cleaning Property'. *Surface and Coatings Technology* 347 (August): 199–208. https://doi.org/10.1016/J.SURFCOAT.2018.04.088.

Yuan, Yuehua and T. Randall Lee. 2013. 'Contact Angle and Wetting Properties'. In G. Bracco and B. Holst (eds), *Surface Science Techniques*. Berlin: Springer, pp. 3–34. https://doi.org/10.1007/978-3-642-34243-1_1.

Zdrahala, Richard J. and Ivanka J. Zdrahala. 1999. 'Biomedical Applications of Polyurethanes: A Review of Past Promises, Present Realities, and a Vibrant Future'. *Journal of Biomaterials Applications* 14 (1): 67–90. https://doi.org/10.1177/0885328 29901400104.

Zhang, Fazhi, Lili Zhao, Hongyun Chen, Sailong Xu, David G. Evans, and Xue Duan. 2008. 'Corrosion Resistance of Superhydrophobic Layered Double Hydroxide Films on Aluminum'. *Angewandte Chemie* 120 (13): 2500–3. https://doi.org/10.1002/ange.200704694.

Zhang, H., R. Lamb, and J. Lewis. 2005. 'Engineering Nanoscale Roughness on Hydrophobic Surface—preliminary Assessment of Fouling Behaviour'. *Science and Technology of Advanced Materials* 6 (3–4): 236–9. https://doi.org/10.1016/j.stam.2005.03.003.

Zhao, Nan, Feng Shi, Zhiqiang Wang, and Xi Zhang. 2005. 'Combining Layer-by-Layer Assembly with Electrodeposition of Silver Aggregates for Fabricating Superhydrophobic Surfaces'. *Langmuir* 21 (10): 4713–16. https://doi.org/10.1021/la0469194.

12

Molecularly Imprinted and Ion Imprinted Polymers for Selective Recognition and Sensing of Organics and Ions

Pankaj E. Hande
Indian Institute of Technology Bombay, Powai, Mumbai, India

Asit Baran Samui
Institute of Chemical Technology, Mumbai, India

CONTENTS

Abbreviations

MIP	molecularly imprinted polymer
IIP	ion imprinted polymer
FTIR	Fourier Transform infrared spectroscopy
NMR	nuclear magnetic resonance
FE-SEM	field emission scanning electron microscopy
HR-TEM	high resolution transmission electron microscopy
AFM	atomic force microscopy
XPS	X-ray photoelectron spectroscopy
BET	Brunauer-Emmett-Teller
TGA	thermogravimetric analyzer
CNT	carbon nanotube
IPNs	interpenetrating polymer networks
RAM	restricted accessed material
APTES	3-aminopropyltriethoxysilane

DOI: 10.1201/9781003037880-12

TEOS	tetraethoxy silane
Lys	lysozyme
[AAPIM]Cl	1-(α-allyl acetate)-3-N-(3-aminopropyl)-imidazolium chloride
MS	mesoporous silica
EGDMA	ethylene glycol dimethacrylate
AAPTS	3-(2-aminoethyl amino) propyltrimethoxysilane
RhB	rhodamine B
MWCNT	multi-walled carbon nanotube
CPE	carbon paste electrode
GTFX	gatifloxacin
SPE	solid phase extraction
IPTS	3-isocyanatopropyltriethoxysilane
CTAB	cetyltrimethylammonium bromide
DMG	dimethylglyoxime
APDC	ammonium pyrrolidine dithiocarbamate
DET	N,N'-diethylthiourea
HPLC	high-performance liquid chromatography
RAFT	reversible addition-fragmentation chain-transfer
MAA	methacrylic acid
4-VP	4-vinyl pyridine
PAEs	phthalic acid esters
EDCs	endocrine disrupting chemicals
DnPP	di-N-pentyl phthalate
CA	clofibric acid
TDS	total dissolved solids
2,6-DIPP	2, 6–diisopropyl phenol
DNT	dinitrotoluene
RDX	cyclotrimethylenetrinitramine
CWA	chemical warfare agents
SERS	surface-enhanced raman scattering
IMS	ion mobility spectrometry
SPR	surface plasmon resonance
QDs	quantum dots
QCM	quartz crystal microbalance
SAW	surface acoustic wave
CEC	capillary electrochromatography
CE	capillary electrophoresis
TNT	trinitrotoluene
DPA	diphenylamine

Symbols with Units

Symbol	Description	Unit
K_d	Distribution coefficient	ml/g
C	Concentration	g/dl
m	Mass	G
k_d	Selectivity co-efficient	Unitless
q	Adsorption capacity	mg/g
ΔG	Gibbs free energy	Kilojoules
Ka	Association constant	l/mole
a_a, b_a	Concentration	moles/l
δ	Chemical shift in NMR	ppm
f_r	Mole fraction	Unitless

12.1 Introduction

A molecularly imprinted polymer (MIP) is a polymer that has the memory of the shape and the functional groups of a template molecule. The advantage of this material is that it can recognize selectively the template molecule used in the imprinting process, in the presence of compounds with structure and functionality similar to those of the template. Normally, the MIPs are economical, can be quickly produced by conventional methods, and are robust and stable during storage. They can perform under difficult conditions, such as at elevated temperatures, in organic solvents, and at extreme pH values. A higher sample load capacity is exhibited for small molecules than is typical for immuno-affinity-based sorbents that result in higher recoveries for analytical applications. In trace analysis applications, the efficient washing procedures are effective in leaching out the extremely low level of compounds from the MIP.

12.2 Introduction to Molecular Imprinting/Ion Imprinting

Molecular imprinting or ion imprinting of polymers is an artificial method of creating selective recognition sites in a cross-linked polymer matrix. In simple terms, the imprinting is akin to welding of various parts of a lock with the key inside during welding. After welding, the key is removed and the keyhole will allow only this particular key to enter and open the lock. The process involved in the development of MIP/IIP is shown in Figure 12.1.

The imprinted polymer is synthesized by polymerization in the presence of a template using a monomer, a cross-linker and a radical initiator, respectively. It is essential that the template does not participate in the polymerization. The monomers and cross-linkers used to design imprinted polymers are listed in Figure 12.2.

After the formation of the polymer matrix, the template is extracted using a suitable leaching agent/solvent and cavities are created with the size of the template. The extraction of the template is carried out by different techniques, such as physically assisted, supercritical, and Soxhlet extraction. After extraction of the template, the polymer is called an imprinted polymer. An IIP is different from MIP in terms of the template. In an IIP, the ion (cation or anion) is a template while in MIP, the molecule is a template. There are two important approaches in imprinting: (1) a covalent;[1] and (2) a noncovalent or self-assembly approach.[2] The two approaches are based on template-monomer interactions. In the case of the MIP, a noncovalent approach is more effective and widely used while the covalent approach is used only in special cases. In the case of an IIP, the template-monomer interaction is based on the coordination approach. The MIP/IIP is stable, easy to synthesize, inexpensive, and reusable, making it more suitable for many applications over other methods. Because of high selectivity, MIPs/IIPs are found to be most effective man-made mimics for various applications, such as solid phase extraction, fluorescent sensors, electrochemical sensors, etc.

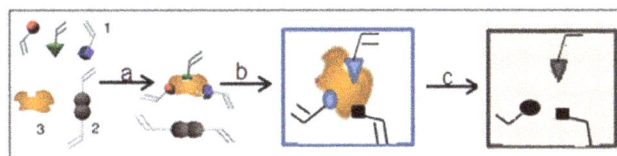

FIGURE 12.1 Process involved in the development of the MIP/IIP: (1) monomer; (2) cross-linker; (3) template; (a) monomer-template interaction; (b) cross-linking; and (c) leaching.

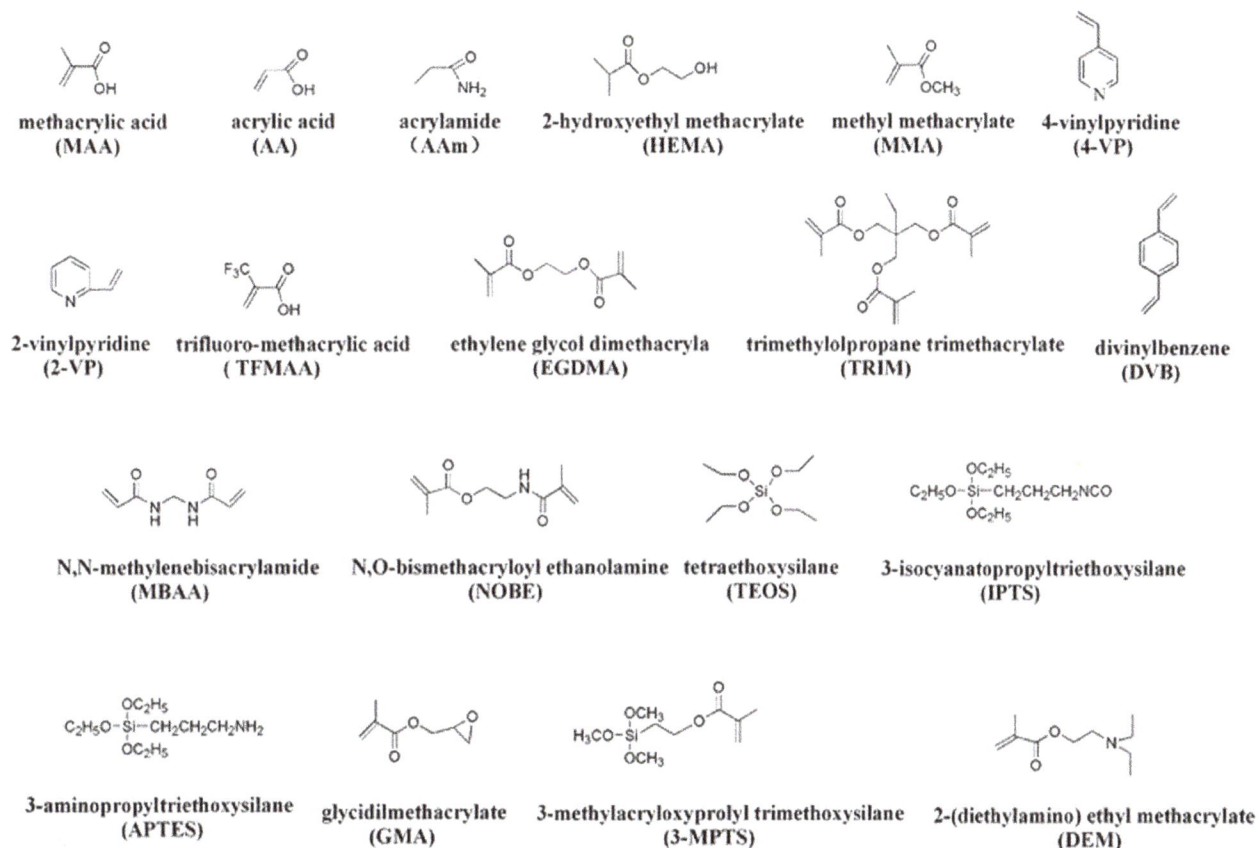

FIGURE 12.2 List of commonly used monomers and cross-linkers for developing MIP/IIP.

Several preparation methods have been reported for the development of MIPs/IIPs, such as bulk polymerization, suspension polymerization, emulsion polymerization, dispersion polymerization, and precipitation polymerization. In bulk polymerization, the initiator, a template-monomer, and a cross-linker are added together in one phase where the initiator is completely soluble in the template-monomer and cross-linker. Suspension polymerization is a heterogeneous radical polymerization where the template, monomer, cross-linker, and initiator are soluble in each other and mixed together by high-speed mechanical stirring in a liquid phase such as water to get polymeric beads. In emulsion polymerization, the template, monomer, and cross-linker are polymerized using an oil-in-water emulsion in the presence of a surfactant where the initiator is water-soluble. Precipitation polymerization is the type of radical polymerization where the template, monomer, cross-linker, and initiator are dissolved in a solvent but during the process of reaction (polymerization), precipitation takes place.

There are various instrumental techniques used for the characterization of MIPs/IIPs. For structural and functional group confirmation, Fourier Transfer infrared spectroscopy (FTIR) and nuclear magnetic resonance spectroscopy (NMR) are used. The surface morphology of the polymers can be studied using field emission scanning electron microscopy (FE-SEM), high resolution transmission electron microscopy (HR-TEM), atomic force microscopy (AFM), X-ray photoelectron spectroscopy (XPS), and Brunauer-Emmett-Teller (BET) surface area analysis. Thermal stability studies of the polymers can be achieved by using thermogravimetric analysis (TGA),

The analytical performance of the designed MIP/IIP is one of the most important studies and hence, a relative selectivity coefficient (k), and adsorption capacity (q) have been used to study the performance of MIPs/IIPs. The distribution coefficient of the adsorbate can be given by Equation (12.1):

$$K_d = \frac{C_i - C_f}{C_f} \frac{V}{m} \tag{12.1}$$

where K_d is the distribution coefficient; C_i and C_f are the initial and final concentrations of adsorbate, respectively. V is the volume of the solution and m is the mass of the adsorbent. The selectivity coefficient (k) for the binding of adsorbate in the presence of competitor species can be obtained from Equation (12.2):

$$k = \frac{K_{d(template)}}{K_{d(interferent)}} \tag{12.2}$$

A relative selectivity coefficient k' is given by Equation (12.3):

$$k' = \frac{k_{imprinted}}{k_{non\ imprinted}} \tag{12.3}$$

Adsorption capacity (q) of IIP adsorbent is obtained from Equation (12.4):

$$q = \frac{(C_i - C_f)V}{m} \tag{12.4}$$

where q represents the maximum adsorption capacity and C_i and C_f are initial and final concentrations of adsorbate, respectively; V represents the volume of the adsorbate solution; m is the weight of the adsorbent.

12.2.1 Thermodynamics of the Template-Monomer and Template-MIP/IIP Recognition

The physical approach of the template-monomer interaction is very important in the development of MIPs/IIPs.[3] In a thermodynamic study, a change in Gibbs free energy plays a significant role in the extent of the template complexation in the template-monomer recognition.[4,5] A change in Gibbs free energy of the monomer-template binding is given by Equation (12.5):[6]

$$\Delta G_{bind} = \Delta G_r + \Delta G_h + \Delta G_{t+r} + \Delta G_{vib} + \Delta G_{conf} + \Delta G_{vdw} + \sum \Delta G_P \tag{12.5}$$

where ΔG_{bind}, ΔG_r, ΔG_h, ΔG_{t+r}, ΔG_{vib}, ΔG_{conf}, ΔG_{vdw}, and $\Sigma \Delta G_r$ are the change in Gibbs free energy for complex formation (*bind*), restriction of rotors upon complexation (*r*), translational (*t*) and rotational (*r*), hydrophobic interactions (*h*), residual soft vibrational modes (*vib*), adverse conformational changes (*conf*), unfavorable van der Waals interactions (*vdw*), and the sum of the interacting polar group contributions (*p*), respectively. The ΔG_{bind} is the non-covalent monomer-template interaction. The ΔG_r is the rigidity of the template and, thus, for better template-monomer recognition, a template with less rigidity is favored.

In the development of MIPs, non-polar organic solvents are used mostly, but sometimes water is also used as a polymerization solvent. ΔG_h is a change in the Gibbs free energy, when there is an interaction between the hydrophobic template and the monomer. The ΔG_{t+r} describes the changes in translational and rotational modes related to the different entities in a complex. The ΔG_{vib} is related to the vibrational modes of the solution and, thus, depends upon the polymerization temperature. The ΔG_{conf} and ΔG_{vdw} are the template conformation and van der Waals interaction for effective solvation during polymerization. The term $\Sigma \Delta G_P$ is the number and relative strength of the template-monomer interactions. In a thermodynamic study of the binding of the template to the MIP, ΔG_{conf} and ΔG_{vdw} are not considered because the formation of a recognition site in the MIP using the monomer-template interaction is controlled thermodynamically. Therefore, the population of the template shows minimum conformational strain with maximum complementarity in the MIP recognition sites. Thus, in the binding of a template to an MIP, the conformational strain and van der Waals interactions are not considered. The thermodynamic parameters related to template-MIP recognition are given by Equation (12.6):

$$\Delta G_{bind} = \Delta G_{t+r} + \Delta G_r + \Delta G_h + \Delta G_{vib} + \sum \Delta G_P \tag{12.6}$$

The key points from the thermodynamics equations are; (1) the stability of the complex is enhanced by increasing the interaction between the monomer and the template (increased $\Sigma \Delta G_P$), which gives better side fidelity; (2) multivalent monomers with stoichiometric composition offer an energetic advantage due to less reduction in degrees of freedom on complex formation (reduction of ΔG_{t+r}); and (3) in an aqueous medium, the hydrophobic effect can be used to reduce the weak electrostatic interactions by maximizing ΔG_h, while, during rebinding of the template, the medium is chosen to attenuate electrostatic and hydrophobic interaction.

12.2.2 Determination of the Template-Monomer Interaction

The extent of the monomer-template interaction is important when designing a polymer with good molecular recognition properties. NMR spectroscopy is used to investigate the strength of interaction between the template and the monomer. To establish the strength of association of the template and the functional monomer, the ^{1}H NMR spectra of template/monomer mixture in a suitable solvent are recorded at various monomer concentrations. The association constant K_a can be determined by Equation (12.7):[7]

$$\frac{\Delta A}{b_0^n} = -K_a \Delta A + K_a \Delta_\in a_0 \tag{12.7}$$

where n indicates the number of compositions of the complex and K_a refers to the association constant, a_o is the template concentration, and b_o is the monomer concentration. The plot of $\Delta A/b_o$ versus ΔA gives a straight line of slope K_a.

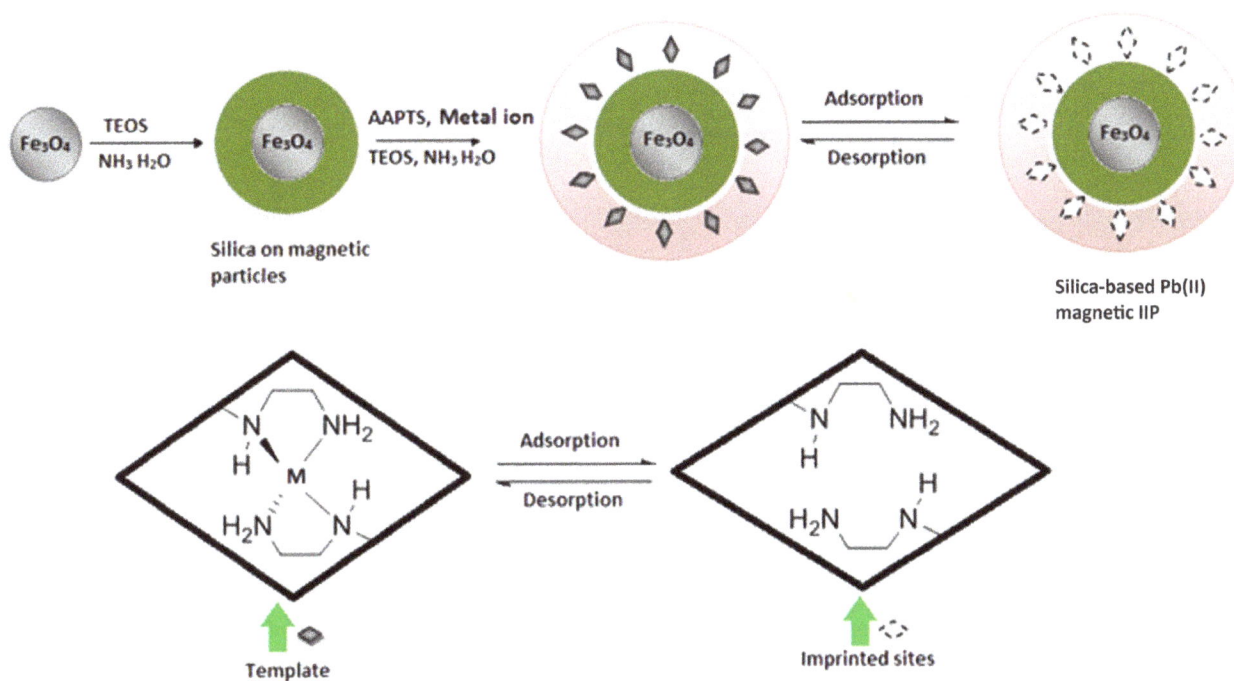

FIGURE 12.3 Schematic expression of core-shell silica-based magnetic metal ion imprinted polymer.

The change in a chemical shift of ^{1}H NMR spectra gives an indication of the molecular interaction leading to complex formation. The stoichiometry of the complexation is investigated by using Job's plot (Figure 12.3).[8] The complex concentration is calculated by using the product $(\delta_{obs} - \delta_T)f_T$, where $(\delta_{obs} - \delta_T)$ is the change in chemical shift of the complexing proton of the template in the presence of the monomer and f_T is the mole fraction of the template.

12.3 Methods of Imprinting

MIPs are modified to enhance the adsorption rate, adsorption capacity, selectivity, thermal stability, and mechanical strength, etc. Some of the important types in the development of MIPs are magnetic, silica, fiber, carbon nanotubes (CNTs), interpenetrating polymer network (IPNs), and restricted accessed material (RAM).

12.3.1 Magnetic Imprinted Polymers

Magnetic imprinting is one of the important methodologies in the development of MIPs/IIPs, as the handling of MIP/IIP has become very convenient and the total operation time has drastically reduced. As an example, the magnetic IIP ($Fe_3O_4@SiO_4$) is synthesized and used for the determination of Pb(II) at trace level.[9] A typical core-shell magnetic ion-imprinted polymer used for the removal of Pb(II) is shown in Figure 12.4.

The synthesis of IIP involves preparation of Fe_3O_4 nanoparticles and then reacting the active surface of the nanoparticles with a chelating monomer such as 3-aminopropyltriethoxysilane (APTES), which is further allowed to form the chelate with the template,

and finally cross-linking with tetraethoxysilane (TEOS) gives the polymer. To stabilize the surface-active nanoparticles, the copolymerization strategy is adopted in which methacryloylated APTES is first reacted with magnetic nanoparticles, leaving behind a reactive double bond which is copolymerized with 4-vinyl pyridine.[10] Proteins have complex conformation, a flexible structure, and a large size. The enzymes' activities are susceptible to the immobilization process, operation temperature, pH, and humidity.[11] Mild experimental conditions are therefore required during the imprinting of enzymes. Another challenge faced by bio-macromolecules for imprinting applications is diffusion limitation, as removing and rebinding the template are difficult. Thus, the preparation of MIPs for bio-macromolecules is difficult.[12] Lysozyme (Lys) is considered an important index in the diagnosis of various diseases involving bronchopulmonary dysplasia in newborns, kidney problems, conjunctivitis, and leukemia.[13] Highly selective extraction of lysozyme (Lys) can be performed by using a core-shell ionic liquid (IL) MIP that is synthesized by using double bond-functionalized Fe_3O_4 nanoparticles.[14] The Fe_3O_4 nanoparticles are protected from oxidation and conglomeration in the detection medium by the coating made by the polymers during imprinting. 1-(a-allyl acetate)-3-N-(3-aminopropyl)-imidazolium chloride ([AAPIM] Cl), used as the functional monomer of the MIPs, can form strong interactions with a template that has improved imprinting effect.

12.3.2 Silica-Based Imprinted Polymers

Increased surface area plays a key role in improving the MIP/IIP performance in terms of the adsorption capacity. Mesoporous silica with uniform, tailorable pore size and a high surface area

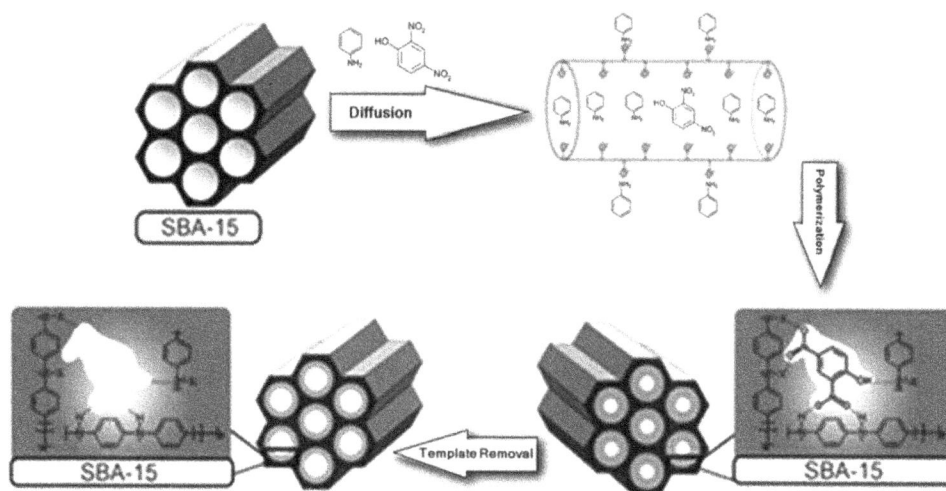

FIGURE 12.4 Process involved in surface imprinting of SBA-15/polyaniline nanocomposite.

can be used in a number of applications, such as indoor air cleaning, bio-catalysis, wastewater treatment, CO_2 capture, bioanalytical sample preparation, drug delivery, catalysis, and so on. Thus, synthesis of the MIP on silica has been successfully performed to enhance efficiency for detection and separation. A large surface area (up to 1000 cm² g⁻¹) and pore volumes with easy regeneration qualify the ordered MS as sorptive for metals ions. In fact, the functionalization of mesoporous silica with various functionalities allows the selective removal of metal ions from aqueous systems. For example, mesoporous silica (MCM-41) is modified by incorporating a dipyridyl ligand into an EGDMA matrix in the presence of Au(II) for selective extraction of gold at trace level (detection limit of < 0.2 ng mL⁻¹ and recovery > 99%).[16] Similarly, surface imprinting on silica gel is done by using As(V) as a template, 3-(2-aminoethyl amino) propyltrimethoxysilane (AAPTS) as a monomer and an epichlorohydrin (cross-linker).[17] The removal rate of the imprinted sorbent for As(V) from the multi-competitive synthetic wastewater is about 75%. Silica-based imprinting is also reported for the selective recognition of organic moieties. Rhodamine B (RhB)-imprinted polymer with SiO_2 nanoparticles can be prepared by a combination of silica gel modification and surface imprinting to determine the RhB content in red wine and beverages.[18] The 2,4-dinitrophenol imprinted SBA-15/polyaniline nanocomposite is synthesized for the selective recognition of the explosive 2,4-dinitrophenol (Figure 12.5).[19]

12.3.3 Carbon Nanotube-Based Imprinted Polymers

Carbon nanotubes (CNTs) have wide applications in chemical, physical, and material sciences due to their excellent overall properties, such as high surface area, easy functionalization, and good conductivity. This has prompted modification of the electrode substrate with multi-walled carbon nanotubes (MWCNTs) for use in analytical application for low-level detection of the targets.[20] The Cu(II)-IIP is prepared by using 1-(2-pyridylazo)-2-naphthol and methacrylic acid.[21] In a chemically modified

carbon paste electrode (CPE), Cu(II)-IIP, graphite powder and CNTs are used. The variation of CNT content affects the selectivity and the best selectivity is obtained with 5wt% of CNTs in the matrix. Similarly, Hg(II)-IIP is modified using MWCNTs for the detection of toxic Hg(II).[22] Because of the wide range of applications, MWCNTs are also used to design MIPs. For example, an MWCNTs-based MIP has been developed to sense uric acid in the human body.[23] For speedy separation, magnetic CNTs are used to design the MIP to detect gatifloxacin (GTFX) in blood serum.[24]

12.3.4 Imprinted Polymer Fibers

Recently, fibers have been used as a grafting substrate for MIPs/IIPs, which have some advantageous properties, such as an increased active surface area and less hindered approach.[25] The Cu(II) imprinted fibers are used for the selective removal of Cu(II).[26] The typical mode of synthesis of metal ion imprinted fibers is shown in Figure 12.6. Cellulosic cotton fibers have advantages, such as strong mechanical properties, easy availability, and biodegradability. This makes functionalization of cellulose a preferred route and acts as a very good metal complexing agent. Cellulosic cotton fibers modified with thiourea were used to synthesize Hg(II) ion-imprinted polymeric fibers.[27] Molecularly imprinted nanowires can be synthesized using nano-porous alumina membrane. The advantage of using theophylline-immobilized nanopores in imprinted nanowires is the recognition sites close to the surface, which make them more accessible to analytes to diffuse into.[28] Being nanowires, they can be well dispersed in solution, and their applications become compatible with other techniques.

12.3.5 Interpenetrating Polymer Network (IPN)-Based Imprinted Polymers

Interpenetrating polymer network (IPN) gel is known for its high sorption capacity and excellent mechanical strength, thus, it is used in solid phase extraction (SPE) columns as ion/molecule

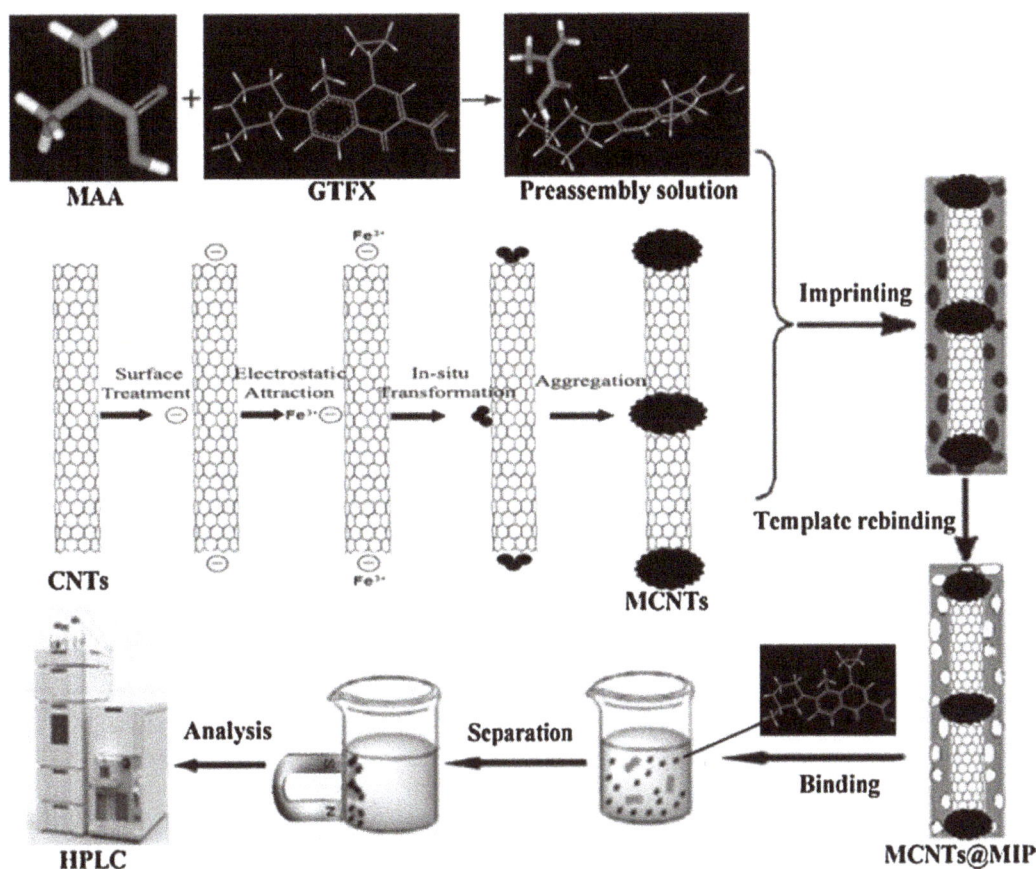

FIGURE 12.5 Scheme for the synthetic route of MWCNTs-MIP.

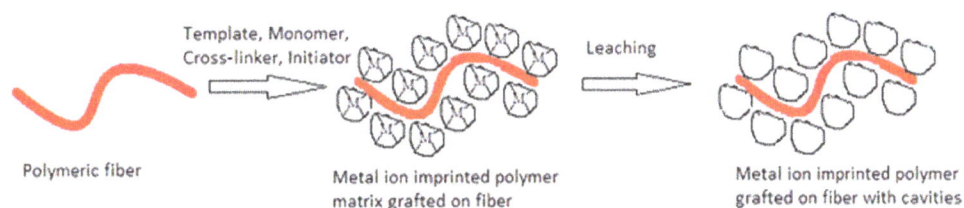

FIGURE 12.6 Schematic for the synthesis of metal ion imprinted fiber.

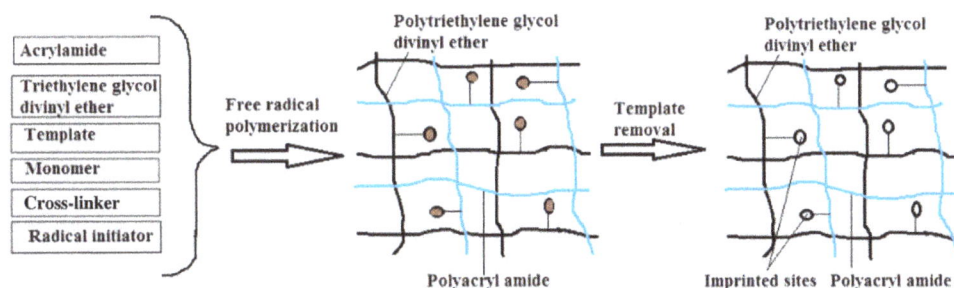

FIGURE 12.7 Schematic for synthesis of interpenetrating polymer network (IPN)-based IIP.

imprinted sorbents. For example, Cd(II)-imprinted IPN gel is prepared for the selective separation of Cd(II) from an aqueous solution.[29] Similarly, thermosensitive Zn(II)-imprinted IPN hydrogel is synthesized for the selective extraction of Zn(II) from wastewater.[30] The process involved in the synthesis of IPN-based IIP is shown in Figure 12.7.[15] Similarly, a semi IPN-MIP membrane is synthesized for the selective separation of atrazine.[31]

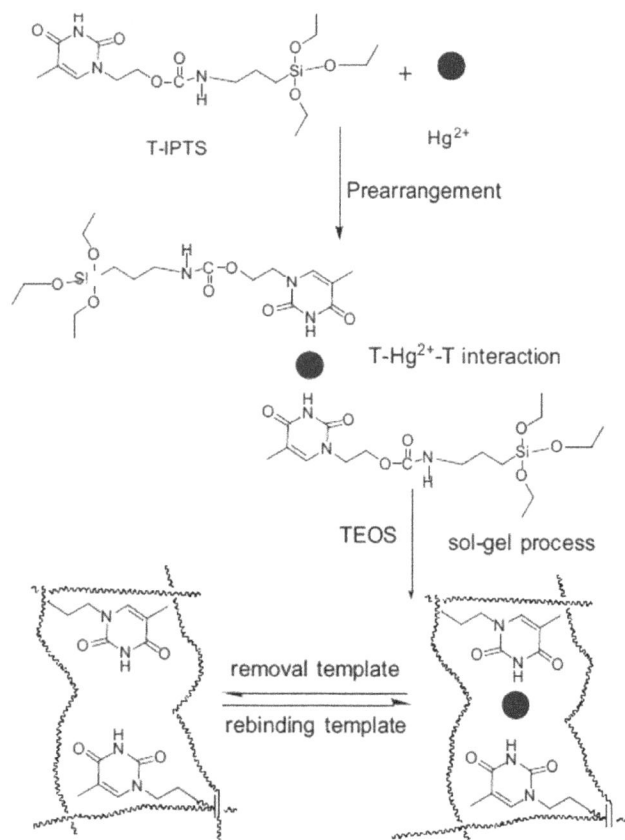

FIGURE 12.8 Schematic for the preparation of Hg(II)-IIP.

12.4 Applications

Because of its high selectivity, IIPs/MIPs are widely used for different applications, as discussed below.

12.4.1 Removal of Toxic Metals from Contaminated Water and Aqueous Solutions

These days, there is great concern over water contamination with toxic metals. Most people in rural areas suffer from this contamination. There are various methods available for the adsorption of heavy metal ions from contaminated water. However, selective adsorption and preconcentration are very challenging tasks. IIPs have been used for the selective removal of metals from an aqueous solution.

Mercury is one of the most studied toxic metals and there are various sources of mercury through which it is released into the ecosystem. [32] The binding behavior of the metal depends upon the chelating agent used in developing the IIP. Hg(II)-imprinted diazoaminobenzene-vinyl pyridine copolymer packed-bed columns have been developed to remove Hg(II) from an aqueous solution.[33] Adsorption of Hg (II) onto IIP depends on the pH. The adsorption increases with the increase in pH, while at a low pH, the H+ ions are adsorbed rather than adsorbing Hg(II). At a higher pH value, the H+ ions concentration decreases and hence, adsorption of Hg(II) is enhanced. However, above pH 10, the OH- concentration increases and interferes with the template

ion resulting in a decrease in the adsorption capacity. Sulfur-containing monomers are used for the development of Hg(II) imprinted polymer as they have a very high affinity to Hg(II) ions. Considering Hg(II)-thymine (T) interaction, a functional monomer, 3-isocyanatopropyltriethoxysilane (IPTS) bearing thymine unit (T-IPTS) is synthesized and used for the imprinting of Hg(II) (Figure 12.8).[34]

Lead is one of the most toxic metals, causing adverse effects on the human body.[35] So far, many Pb(II)-IIPs have been developed for selective removal of Pb(II). For example, a core-shell magnetic lead imprinted polymer is used for the selective removal of lead from an aqueous solution.[36] These magnetic sorbents possess a large surface area and a short diffusion route that increases the extraction frequency. Silica gel is one of the ideal supporting materials because of its high thermal stability, high mass exchange property, and is a non-swelling inorganic material in nature. Pb(II)-imprinted amino-functionalized silica gel sorbents have been used for the adsorption of lead from plants and water solutions.[37] A crown ether monomer with a good chelating ability is used for synthesizing Pb(II)-IIP.[38] These Pb(II)-IIPs have been successfully applied for the selective extraction of Pb(II) from sea water, environmental samples, lake water, saline water, etc. Highly selective chitosan-MAA-EGDMA-based Pb(II) II-IPNs are synthesized by simultaneous polymerization.[39] Pb(II) II-IPNs have unique interpenetration among both kinds of polymer chains. Chitosan with hydroxyl and MAA with carboxyl functionality form complexes with Pb(II).

Some 15,000 t of cadmium is produced annually worldwide for pigments, batteries, stabilizers, coatings, and alloys.[40] It accumulates in the body with a low excretion rate. Thus, selective removal from aqueous systems and preconcentration of cadmium is very important. The WHO limit for cadmium in drinking water is 3 μg L^{-1}. Cd(II)-IIPs have been used for the selective removal of Cd(II) from an aqueous solution. The monomer, (2Z)-N,N-bis(2-aminoethylic) but-2-enediamide, is used to synthesize IIP for the extraction of Cd(II) from wastewater with a detection limit of 0.14 μg.L^{-1}, the column packed with Cd(II)-IIPs works well for Cd(II) extraction from a mixture of Hg(II), Cu(II), and Zn(II), respectively.[41] To enhance the adsorption capacity and selectivity, nannochloropsis sp biomass is used with silica as a support.[42] Algae biomass from several algae species binds metal ions from the aqueous medium as it is associated with several functional groups, which can act as ligands to metal ions.[43]

It's very difficult to remove arsenic from an aqueous solution because of its complex chemistry in water. Arsenic exists in two forms: pentavalent and trivalent. Pentavalent arsenate (H$_2$AsO$_4^-$ or HAsO$_4^{2-}$) is easy to remove from wastewater because of its oxyanion as compared to trivalent (H$_3$AsO$_3$). The WHO limit for arsenic is 10 μg L^{-1} in drinking water. IIPs can detect levels lower than the WHO guidelines. Surface imprinting enhances the mass transfer rate; thus, a surface ion imprinted amine-functionalized silica gel sorbent is employed for the selective removal of arsenic (V) from an aqueous solution.[17]

According to European Council Directive 98/83/EC, the drinking water limit for chromium is 50 μg L^{-1}. Cr exists in Cr(III) and Cr(VI) oxidation states. Cr(VI) is highly toxic to humans compared to Cr(III). IIP techniques have been used for the selective removal of Cr in both oxidation states. For example, Cr(III)-imprinted ethylene glycol dimethacrylate-methacryloyl histidine polymer beads have been synthesized and can be used in harsh conditions such as concentrated acid and high temperatures respectively to remove Cr(III) from an aqueous solution.[44] Similarly, AAPTS functionalized Cr(III)-imprinted polymer is synthesized and exhibits an excellent selectivity coefficient (700) in the presence of Mn (II). [45] Even, Cr(VI) has an affinity toward its IIP in an extremely acidic condition. Cr(III)-IIP with pyrrolidine dithiocarbamate (PDC) as a chelating agent possesses higher selectivity toward Cr(III) in the presence of Cr(VI) ions, because of the formation of the complementary recognition sites in Cr(III)–PDC for Cr(III) ions.[46] 4-vinyl pyridine as a complexing monomer is used in the synthesis of IIPs for selective removal of Cr(VI) from a contaminated solution.[47] The copolymerization of a Cr(VI) chelated monomer with another monomer by radical polymerization induces maximum matrix stability with less interference in terms of the selectivity. Nanoparticles exhibit high surface area and thus possess a high adsorption capacity. Therefore, nano Cr(VI)-IIP has been synthesized for selective removal of Cr(VI) from wastewater and it shows a good adsorption capacity.[48]

Copper is used for many industrial applications and it is released into the environment during activities like mining, metal cleaning, metal finishing, etc. Therefore, many Cu(II)-IIPs have been prepared for various applications. Multi-column chromatography is used for large-scale continuous separation, while batch column chromatography is restricted in its use in large-scale separation. Thus, for continuous separation, a three-zone[49] and four-zone carousel process[50] is used by packing with Cu(II)-IIP. Therefore, these methods have attracted attention from metal-related industries where large-scale separation is demanded. There are two ways of imprinting, 3D and 2D. 3D imprinting involves high cross-linking during polymerization and it is widely used but it has a disadvantage as a slow mass transfer, while the 2D method is simple and quick in mass transfer where the ligands are self-assembled at the surface of the polymer matrix.[51]

Iron plays a significant role in many biological activities but all the same it is toxic at higher concentrations.[52] IIPs have been used for the selective removal of Fe(III). For example, organic/inorganic hybrid material-based imprinting using a methacryloylamidoantipyrine-Fe(III) complex monomer and smectite organoclay modified with quartamine has been done successfully for Fe(III).[53] Double template imprinting is the combination of two imprinting methods, micelle templating synthesis and molecular imprinting, and this approach can be precisely controlled at the pore structures and adsorption sites.[54] Double template imprinted polymers are found to be superior to organic polymer-based imprinted materials in terms of easy preparation, high thermal and chemical stability, and fast mass transfer.[55] For the preparation of double imprinted polymers of Fe(III), AAPTS, cetyltrimethylammonium bromide (CTAB), and tetraethoxysilane (TEOS) are used.

Innumerable sources of contamination are the reason for the removal of toxic metals. Most of the components for IIP synthesis are concentrated around N and S containing chelating agents/monomers.

12.4.2 Preconcentration of Valuable Metal Ions for Subsequent Use

Actinides, U(VI) and Th(IV), are mostly used in a nuclear power plant for electricity production. U(VI) is water-soluble and can easily contaminate the environment. Uranium is hazardous to the ecosystem. Inhalation of uranium causes serious hazards to human beings, such as kidney failure, lung disease, etc.[56] The WHO limit for uranium is 0.6 μg kg^{-1} of body weight/day and in drinking water it should be less than 9 μg L^{-1}. Therefore, the removal and preconcentration of uranium are challenging. There are many methods for the extraction of uranium such as cation exchange, liquid-liquid extraction, chemically modified silica, adsorption on activated carbon, etc. but selective removal is challenging. For the extraction of uranium from the waste of the nuclear plant, salicylaldoxime-based IIPs have been used.[57] Similarly, chloroacrylic acid-based U(VI)-IIPs have been prepared for extraction of U(VI), and the performance was checked in the presence of +2, +3 and +4 oxidation state potential interfering metal ions.[58]

Selective extraction of Th(IV) is challenging because of its complex nature. When designing Th(IV)-IIPs, different chelating agents such as N-(o-carboxyphenyl)maleamic acid,[59] 1-phenyl-3-methylthio-4-cyano-5-acrylicacidcarbamoyl-pyrazole,[60] acryloyl-b-cyclodextrin,[61] N,N'-bis(3-allyl salicylidene)o-phenylenediamine,[62] etc. have been used. Similarly, the IIP technique has become a promising tool for the extraction of erbium from the mixture of lanthanides.[63] The effect of γ-irradiation on the extraction of Dy(III) is very interesting with 35–180-fold increase in its selectivity coefficient.[64]

Furthermore, for the preconcentration of Ag(I) from an aqueous solution, an IIP is used.[65] Similarly, an IIP modified with porous silica is used for the extraction of gold.[16] Pd(II) is one of the most costly metals extensively used in metallurgy, chemical synthesis (catalysis), jewelry, etc. Various chelating agents, such as dimethylglyoxime (DMG), ammonium pyrrolidine dithiocarbamate (APDC), and N,N'-diethylthiourea (DET), along with vinyl pyridine, form a complex with palladium and are used in the formulation to design IIPs for preconcentration of Pd(II).[66] This method is also used for the analysis of trace Pd(II) in geological samples. Platinum metal preconcentration from an aqueous solution is achieved using a thiosemicarbazone-based IIP.[67] Tl(III) is used for many industrial and medicinal applications. Because of the toxicity of Tl(III), its continuous separation using an IIP-based mini-column has been carried out.[68]

12.4.3 Removal of Toxic Organics

There are various products, including medicine, food, and many other materials, in which small contamination affects the health and other processes involved. Using MIPs for the detection and removal of these toxic chemicals has gained widespread acceptability. For example, MIPs modified with silica nanoparticles synthesized using surface imprinting are used for the selective removal of rhodamine B. The MIP, having a high affinity, is dependent on the electrostatic interactions and hydrogen bonding involved between the rhodamine B and the imprinted sites.

TABLE 12.1

Analytical Performances of the QDs Fluorescent MIP Sensors for Various Applications

QDs MIP	Target molecule	Limit of Detection	Application	Ref.
Carbon	α-amanitin	15 ng mL^{-1}	Biomedical	[74]
	Dopamine	1.7 nM	Biomedical	[75]
	Paranitrophenol	9.00 ng mL^{-1}	Energetic	[76]
	Nifedipine	0.076 μM	Biomedical	[77]
CdS	Guanosine	–	Biomedical	[78]
CdSe	Ractopamine	–	Biomedical	[79]
	Pyrethroids	3.6 μg^3 L^{-1}	Biomedical	[80]
	Bisphenol A	7.57×10^{-10} M	Biodegradation	[81]
ZnS	Cocaine	–	Biomedical	[82]
	4-nitrophenol	76 nM	Energetic	[83]
	Cyphenothrin	9.0 nM	Environmental	[84]
	Chlorpyrifos	17 nM	Environmental	[85]
CdSe/ZnS	Mycotoxin zearalenone	0.002 mmol L^{-1}	Biomedical	[86]
	Tocopherol	0.023 mmol L^{-1}	Biomedical	[87]
	Sesamol	$4.8 \times 10{-4}$ mol L^{-1}	Sesame oil	[88]
CdTe	Cysteine	0.85 μM	Biomedical	[89]
	Deltamethrin	0.16 μg mL^{-1}	Environmental	[90]
	Bovine hemoglobin	9.4 nM	Biomedical	[91]
	Bisphenol A	4 nM	Environmental	[92]
CdTe/ZnS	Ractopamine	1.47×10^{-10} M	Biomedical	[93]
carboxylated QDs	Protein	-	Biomedical	[93]

Further, MIPs in combination with HPLC are used for the determination of rhodamine B in spiked red wine and beverages, with recovery rare at around 92%.[18] The molecular recognition of a poorly functionalized template such as phenol and its derivatives is required as its presence in food or ground and surface water poses a significant danger to health. To detect m-nitrophenol, acrylamide is used as a monomer and bis-acrylamide as a cross-linker.[69] The rebinding study established that by using UV-Vis spectroscopy, effective binding in the template cavity of the MIP and the selectivity coefficient is detected that reaches as high as 12 (Table 12.1).

The MIPs were prepared by the aqueous RAFT precipitation polymerization of acrylic acid and a cross-linker in the presence of aristolochic acid I (AAI) as the template.[70] The MIPs display excellent selective extraction of aristolochic acid I and play a great role in the purification of natural products. Heptachlor, chlorinated dicyclopentadiene is used to control termites. For its selective recognition, MIP was designed using functional monomers such as MAA, 4-VP or styrene and cross-linkers like EGDMA or DVB.[71] Phthalic acid esters (PAEs) is the endocrine disrupting the chemicals (EDCs) added to plastic products to increase the plasticity and to cosmetics to provide versatility. Magnetic MIPs (Fe@SiO$_2$@MIP) have been developed using N-isopropylacrylamide, MAA and EGDMA, for the selective removal of di-n-pentyl phthalate (DnPP) from wastewater.[72] The MIP possesses good adsorptive capacity for DnPP (194 mg g^{-1}). Adsorption is optimum at neutral pH and a lower temperature (25°C). The extent adsorption of DnPP is more than 4.8 times higher than other phthalates. Clofibric acid (CA)-imprinted polymer is prepared using 2-vinyl pyridine as a monomer and

divenyl benzene as a cross-linker, for the removal of CA from an environmental water sample.[73] The MIP exhibits excellent selectivity in the presence of total dissolved solids (TDS) in water and can be reused up to 12 times effectively without loss of performance.

Recognition of explosives can also be done by imprinting with a structural analog to avoid explosion hazard during MIP synthesis. For example, 2,6-diisopropyl phenol (2,6-DIPP) can be used as a template as an analog of dinitrotoluene (DNT) while polymerizing the MIP. This MIP behaves excellently during sensing of both 2,6-DIPP as well as DNT. Figure 12.9 details the imprinting scheme.[74]

12.4.4 MIPs/IIPs as a Sensor

Chemical sensors are increasingly attracting interest in the field of modern-day analytical chemistry for clinical diagnostics, food analysis, environmental analysis, etc. In forensic investigation, the only requirement is the presence of a sufficient amount of DNA molecules with distinguishable 0.01% DNA sequence.[75] Thus, for the enrichment of DNA fragments from biological fluid samples, an MIP is used.[76] Nitrobenzene, dinitrotoluene (DNT), trinitrotoluene (TNT) and cyclotrimethylenetrinitramine (RDX) have been investigated by using MIPs. Chemical warfare agents (CWA) are required to be analyzed for defense and national security requirements. Pinacolyl methyl phosphonate, a hydrolyzed product of soman, is detected at ppt level within seconds by using an MIP coupled with a fluorinated polymer.[77] MIPs have been successfully used in combination with different transducers and methods.[78]

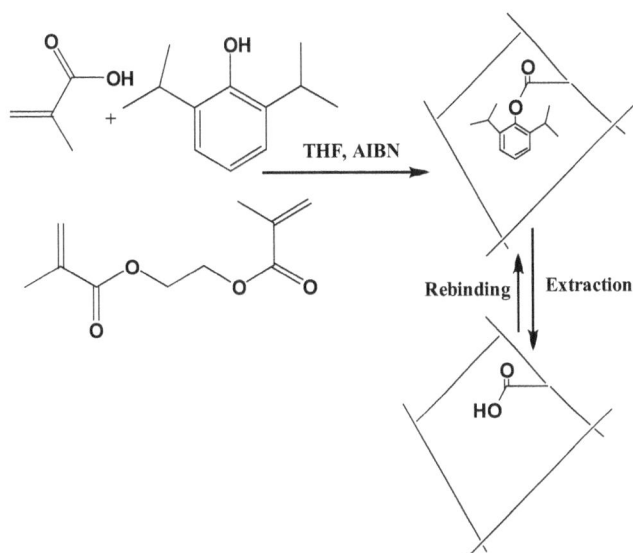

FIGURE 12.9 Synthesis of 2,6-DIPP imprinted polymer for detection of DNT.

Source: [74].

12.4.4.1 Spectroscopy

The integration of MIPs into the commercial system has several advantages in terms of design and performance:

- *Surface-enhanced Raman scattering (SERS)*: The instrument is based on molecular vibration. By integrating SERS with MIPs by depositing imprinted xerogels, the selectivity can be enhanced to a great extent. SERS possess high porosity and large surface areas which make the system more sensitive to substances such as explosives.[79]
- *Ion mobility spectrometry (IMS)*: Mobility of gas phase ions in a weak electric field is the principle behind this method and therefore it is used to detect trace amounts of vapors. By incorporating MIPs in IMS, the selectivity and detection range can be improved.[80]

12.4.4.2 Optical Sensors

- *Surface plasmon resonance (SPR)*: The principle behind this method is the optical excitation of surface electrons of metal on a metal-dielectric interface. Molecular imprinting-based SPR sensors can be employed for biological species like antibodies, proteins, hormones, viruses, cells, aptamers, etc.[81] Similarly, they can be used to sense drugs, explosives, toxic gases, etc.
- *Colorimetric sensing*: The periodic nanostructures of photonic crystals affect the motion of photons which makes a label-free sensing agent because of its structural coloration ability.[82] By combining photonic crystals with an MIP, maintaining its porous nature, sensors can be designed.[83]
  - *Chemiluminescence sensing* is very sensitive colorimetric sensing technique. Sensing of morphine and methamphetamine hydrochloride (MA) can be

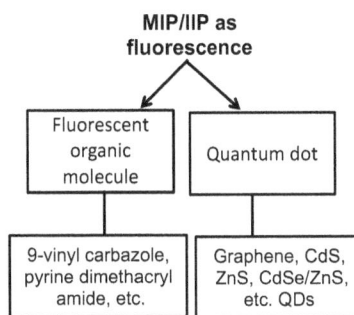

FIGURE 12.10 Classification of the fluorescence-based MIP/IIP sensor.

done by combining imprinted sol-gel polymers with a light-emitting material and a multi-walled carbon nanotube composite with a detection limit of 4.0 x 10^{-15} M.[84]

- *Fluorescence sensing* is the ultrasensitive detection technique. Therefore, an MIP can be combined with this technique to perform low-level detection. Fluorescence emission in MIPs can be introduced by a variety of fluorophore dyes or different-sized quantum dots. Details of the fluorophores are given in (Figure 12.10). Fluorescent probes with MIPs have gained attention in the area of sensing explosive vapors, poisons, and for fire debris analysis.[85]

12.4.4.3 Quantum Dots (QDS)-Based Fluorescent MIPs

There are various fluorescent-based MIPs based on QDs. These include carbon QDs, CdS, CdSe, ZnS, CdSe/ZnS, CdTe, and CdTe/ZnS-MIPs. Carbon/graphene QDs-MIP attracted attention due to their fascinating features such as high fluorescence activity and photostability. For bio-imaging applications, carbon/graphene QDs are more efficient due to their eco-friendly nature, low cytotoxicity, and impressive biocompatibility. The QDs-MIPs have mostly been used in the sensing of organic compounds. The QDs-based fluorescent MIP sensors have mostly been used in biomedical applications.

12.4.4.4 Organic Monomer/Complexing Agent-Based Fluorescent Imprinting

In this technique, the fluorescent monomer is developed in such a way that the fluorescence is quenched or enhanced after interacting with the template. These types of fluorescent MIP sensors are mostly reported in the sensing of organic moieties. Fluorescent monomers for the design of fluorescent MIP are shown in Table 12.2. Fluorescence sensors can be designed by merging MIPs with an optic fiber. To sense Al(III), Al(III)-IIP is synthesized based on fluorescence.[102] Further, Hg(II)-IIP is synthesized to detect Hg(II) and the sensor shows enhancement in the fluorescence after interacting with mercury with detection up to 50 μM.[103] Similarly, fluorescence quenching sensors, Cu(II)-IIP, have been developed for Cu(II) sensing.[104] Fluorescent functional monomer 4-[(E)-2-(4′-methyl-2,2′-bipyridine-4-yl)vinyl] phenyl methacrylate is used to design the Cu(II)-IIP. Fluorescent IIPs have not yet been fully developed.

TABLE 12.2

List of Various Fluorescent Organic Monomers Used to Design MIPs

Fluorescent Agent/Monomer for MIP	Target Molecule	Sensing Applications	Ref.
1,2-diphenyl- 6-vinyl-1H-pyrazole-[3,4-b]-quinoline	Cyclic guanosine 3,5-monophosphate	Biomedical	97
Trans-4-[p-(N,N-dimethylamino)styryl]-N-vinyl benzylpyridinium chloride	Cyclic adenosine 3,5-monophosphate sodium salt	Biomedical	98
Mono [6-O-(8-quinolyl)]-β-cyclodextrin	Spermidine	Biomedical	99
N-(1-pyrenyl)maleimide	Homocysteine	Biomedical	100
Rhodamine B	L-cysteine	Biomedical	101
Pyrene dimethacrylamide	Diquat (DQ), paraquat (PQ)	Environmental	102

12.4.4.5 Mass Detection

Piezoelectric transduction-based sensors can be used for gas phase detection and, therefore, have wide applications. The most commonly used mass sensitive sensor systems are quartz crystal microbalances, surface acoustic wave-guides, micro-electro-mechanical and microcantilevers.

- *A quartz crystal microbalance (QCM)* is based on the principle that change in mass per unit area is measured with the help of change in the frequency of a quartz crystal resonator. The frequency change is carried out either by the addition or by removal of a small mass at the surface of the acoustic resonator. If the MIP is integrated with QCM, the capture will be mostly the target molecule, which can be sensed at a very low level. For example, an MIP with QCM is used to detect microcystin-LR, which is a cyanobacteria toxin.[110]
- *Surface acoustic wave (SAW) and bulk acoustic wave (BAW) sensors* are based on the principle that change in the velocity of the acoustic waves is observed due to the changed mass at the transducer. An MIP modified SAW sensor has been used to sense an anabolic steroid, nandrolone. To design the sensor, a SAW device is coated with an MIP and integrated with a frequency-determining component. For example, a mono[6-deoxy-6-[(mercaptodecamethylene)thio]]-β-CD and O-ethyl-S-2-diisopropylaminoethyl methylphosphonothiolate-based MIP, coated with a SAW device, exhibits a rapid and high-frequency response at extremely low concentrations.[111]

12.4.4.6 Chromatography

MIPs in combination with chromatography are found to be an effective technique in the area of forensic science such as the determination of gunshot residues from hands and clothes, similarly, to find out the metabolites produced during the biodegradation of various drugs, polycyclic aromatic compounds, etc.

- *High-performance liquid chromatography (HPLC)*: HPLC can be extensively used for chiral separation by combining with an MIP. For example, MIPs in combination with LC are used for the separation of amino acid derivatives.[112] Later on, MIPs have become a popular stationary phase in HPLC.
- *Capillary electrochromatography (CEC)* This is widely used in forensic chemistry for analysis of drugs, inks, dyes, gunshot/explosive residues, etc. For example, a combination of the MIP with capillary electrophoresis (CE) has high separation efficiency.[113] Column efficiency is higher when separations are launched in CEC mode as compared to that in HPLC mode.

12.4.4.7 IIP-/MIP-Based Electrochemical Sensors

In electrochemical sensors like potentiometry and voltammetry, IIPs/MIPs are used as a selective recognition material. Thus, IIPs/MIPs have been used to sense metal ions as well as organic compounds. The designing of IIP-/MIP-based sensors is simple where the IIP/MIP is used to fabricate an electrode sensor. Different types of IIP-/MIP-based electrode sensors are a glassy carbon electrode, a carbon paste electrode (CPE), and a carbon paper electrode or thin film electrode, etc.[15]

For example, a modified glassy carbon electrode sensor has been developed using 5,10,15,20-tetrakis(3-hydroxyphenyl) porphyrin-based Hg(II)-imprinted polymeric nanobeads for mercury sensing.[22] Electrode substrate modification with MWCNTs maintains high sensitivities, low detection limits, reduction of over-potential, and resistance to surface fouling. When developing sensors, simplicity and selectivity are important. Thus, an Hg(II)-IIP-based electrode sensor has been developed using 4-vinyl pyridine as a monomer.[114] In designing the process, Hg(II)-IIP with graphite powder and melted n-eicosane is used to fabricate a carbon paste electrode sensor. Similarly, for the other metal ions these electrode sensors are designed. For example, for the detection of Pb(II) using (Pb(II)-IIP-CPE) which is developed with the help of chelating groups, bismuth, organic clay nanoparticles, etc.[115] The Pb(II) imprinted modified gold electrode as an electrochemical sensor for sensing Pb(II) has been developed. Its lower detection limit is higher than Pb(II)-IIP-CPE, however, it possesses good stability in a dilute lead solution.[16] Further, magnetic ion-imprinted polymer ($Fe_3O_4@SiO_2@IIPs$) modified CPE has been designed for the fast detection of Pb(II).[9] For Cd(II) detection, a CPE sensor has been developed using Cd(II) IIP (developed using quinaldic acid and 4-VP), carbon powder, and melted n-eicosane.[117] Low-level Cd(II) detection in the presence of cations and anions shows good analytical characteristics of the sensor with 100% recovery of Cd(II) and can be used even after 5 months without a decrease in efficiency.

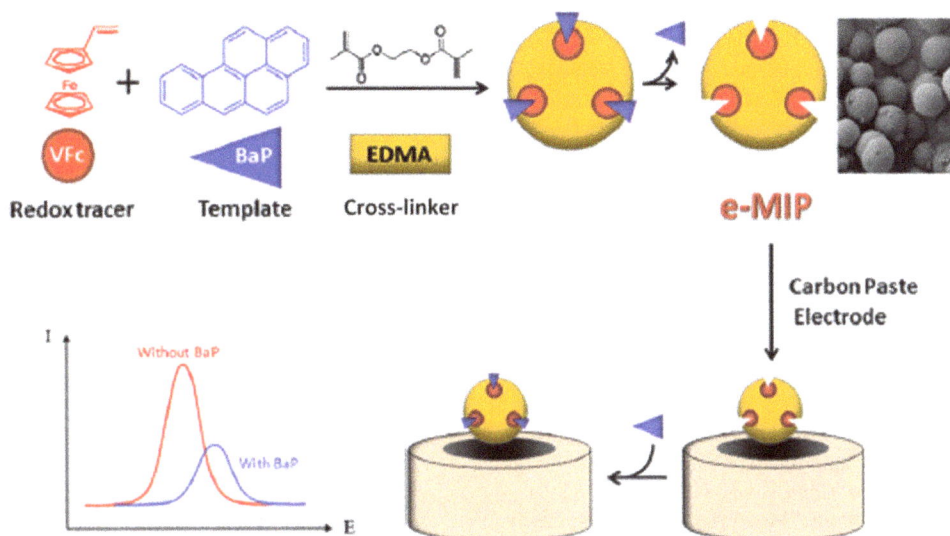

FIGURE 12.11 Schematic for designing of MIP-CPE electrode sensor for benzo[a]pyrene.

Source: [119]. Reprinted with permission from Royal Society of Chemistry.

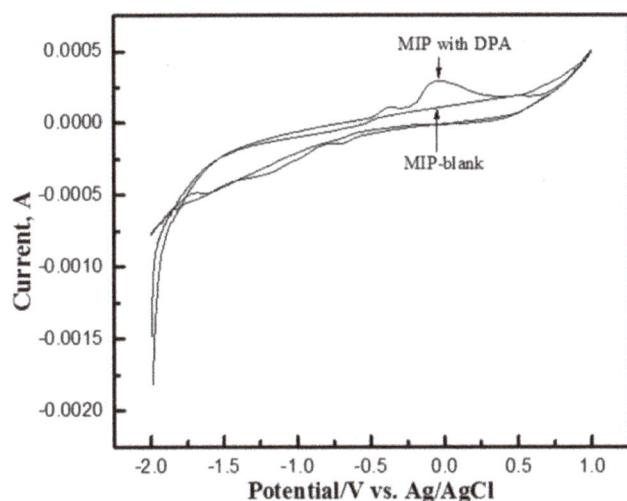

FIGURE 12.12 Electrochemical behavior of DPA-MIP and blank-MIP carbon paste electrode (potential scan: 2 to 1.0 V; potential scan rate 0.1 Vs1; DPA concentration 2 mM).

Because of their excellent performance and efficiency, MIP modified electrochemical sensors have been designed for the detection of many organic compounds. For example, in defense applications, MIP modified electrode (MIP-CPE) sensors have been developed to sense nitro explosives like trinitrotoluene (TNT).[118] In the same way, an MIP-CPE probe has been designed for the detection of toxic and carcinogenic hydrocarbon benzo[a]pyrene, as shown in Figure 12.11.[119]

Similarly, an MIP-gold nanoparticles-based electrochemical sensor has been designed for the determination of TNT where detection is based on π-donor-acceptor interactions.[120] To enhance the efficiency of MIP-based electrode sensors in terms of sensitivity and selectivity, various modifications have been performed. In this respect, an MIP-polypyrrole-graphene-gold nanoparticle modified electrode has been developed for the sensing of levofloxacin (antibiotic).[121] Similarly, for the detection of the antibiotic trimethoprim, an MIP graphene modified CPE has been developed.[122] Further, MIP-CPE is also used for the detection of diphenylamine (DPA) stabilizer in propellant.[123] In MIP-CPE, the peak at 0 V corresponds to oxidation of DPA to DPAH+• radical cation and is further dimerized to phenyl benzidine (Figure 12.12). IIP/MIP-modified electrochemical sensors have huge scope because of their low detection limit, good selectivity, high stability, ease of operation, and low cost.

12.5 Conclusion

Overall, the MIPs/IIPs are found to be excellent man-made mimics with wide applications in diverse fields, such as biomedicinal, sensing, selective recognition, wastewater treatment, enrichment of precious metals, etc. These man-made mimics can be designed according to the relevant applications. For example, in sensing applications such as a fluorescence sensor or an electrochemical sensor, the MIPS/IIPs are designed with excellent selectivity, while for selective separation, the MIPS/IIPs are designed with excellent adsorption capacity.

REFERENCES

1.   G. Wulff and A. Sarhan. "The use of polymers with enzyme-analogous structures for the resolution of racemates." *Angew. Chemie Int. Ed. English* 11(4) (1972): 341.

2.   R. Arshady and K. Mosbach. "Synthesis of substrate-selective polymers by host-guest polymerization." *Die Makromol. Chemie.* 182(2) (1981): 687–692.

3.   I.A. Nicholls, K. Adbo, H.S. Andersson, P.O. Andersson, J. Ankarloo, J. Hedin-Dahlström, P. Jokela, J.G. Karlsson, L. Olofsson, … S. Wikman. "Can we rationally design molecularly imprinted polymers?" *Anal. Chim. Acta.* 435(1) (2001): 9–18.

4. E.E. Bittar, B. Danielsson, L. Bülow, I.A. Nicholls. "An approach toward the semiquantitation of molecular recognition phenomena in noncovalent molecularly imprinted polymer systems: consequences for molecularly imprinted polymer design." *Adv. Mol. Cell Biol.* 15 (1996): 671–679.

5. I.A. Nicholls. "Towards the rational design of molecularly imprinted polymers." *J. Mol. Recognit.* 11(1–6) (1998): 79–82.

6. D.H. Williams, J.P.L Cox, A.J. Doig, M. Gardner, U. Gerhard, P.T. Kaye, A.R. Lal, I.A. Nicholls, C.J. Salter, R.C. Mitchell. "Toward the semiquantitative estimation of binding constants: Guides for peptide-peptide binding in aqueous solution." *J. Am. Chem. Soc.* 113(18) (1991): 7020–7030.

7. Z. Jie and H. Xiwen. "Study of the nature of recognition in molecularly imprinted polymer selective for 2-aminopyridine." *Anal. Chim. Acta* 381(1) (1999): 85–91.

8. G. Wulff and G. Kirstein. "Measuring the optical activity of chiral imprints in insoluble highly cross-linked polymers." *Angew. Chem. Int. Ed. Engl.* 29(6) (1990): 684–686.

9. Y. Cui, J-Q. Liu, Z-J. Hu, X-W. Xu, H-W. Gao. "Well-defined surface ion-imprinted magnetic microspheres for facile onsite monitoring of lead ions at trace level in water." *Anal. Methods* 4(10) (2012): 3095–3097.

10. N.T. Tavengwa, E. Cukrowska, L. Chimuka. "Synthesis, adsorption and selectivity studies of N-propyl quaternized magnetic poly(4-vinyl pyridine) for hexavalent chromium." *Talanta* 116 (2013): 670–677.

11. S. J. Cho, H.B. Noh, M.S. Won, C.H. Cho, K.B. Kim, Y.B. Shim. "A selective glucose sensor based on direct oxidation on a bimetal catalyst with a molecularly imprinted polymer." *Biosens. Bioelectron.* 99 (2018): 471–478.

12. E. Shoghi, S.Z. Mirahmadi-Zare, R. Ghasemi, M. Asghari, M. Poorebrahim, M.H. Nasr-Esfahani. "Nanosized aptameric cavities imprinted on the surface of magnetic nanoparticles for high-throughput protein recognition." *Microchim. Acta.* 185(4) (2018): 241.

13. C. Ocana, A. Hayat, R. Mishra, A. Vasilescu, M.D. Valle, J.L. Marty. "A novel electrochemical aptamer–antibody sandwich assay for lysozyme detection." *Analyst* 140(12) (2015): 4148–4153.

14. W. Xu, Q. Dai, Y. Wang, X. Hu, P. Xu, R. Nia, J. Meng. "Creating magnetic ionic liquid-molecularly imprinted polymers for selective extraction of lysozyme." *RSC Adv.* 8(39) (2018): 21850–21856.

15. P.E. Hande, A.B. Samui, P.S. Kulkarni. "Highly selective monitoring of metals by using ion-imprinted polymers." *Environ. Sci. Pollut. Res.* 22(10) (2015): 7375–7404.

16. H. Ebrahimzadeh, E. Moazzen, M.M. Amini, O. Sadeghi. "Novel ion-imprinted polymer coated on nanoporous silica as a highly selective sorbent for the extraction of ultra-trace quantities of gold ions from mine stone samples." *Microchim. Acta.* 180(5–6) (2013): 445–451.

17. H-T. Fan, X. Fan, J. Li, M. Guo, D. Zhang, F. Yan, T. Sun. "Selective removal of arsenic (V) from aqueous solution using a surface-ion-imprinted amine-functionalized silica gel sorbent." *Ind. Eng. Chem. Res.* 51(14) (2012): 5216–5223.

18. Z. Long, W. Xu, Y. Lu, H. Qiu. "Nanosilica-based molecularly imprinted polymer nanoshell for specific recognition and determination of rhodamine B in red wine and beverages." *J. Chromatogr.* 1029–1030 (2016): 230–238.

19. A. Mehdinia, M. Ahmadifar, M.O. Aziz-Zanjani, A. Jabbari, M.S. Hashtroudi. "Selective adsorption of 2,4-dinitrophenol on molecularly imprinted nanocomposites of mesoporous silica SBA-15/polyaniline." *Analyst* 137(18) (2012): 4368–4374.

20. H. Beitollahi and I. Sheikhshoaie. "Electrocatalytic and simultaneous determination of isoproterenol, uric acid and folic acid at molybdenum (VI) complex-carbon nanotube paste electrode." *Electrochim. Acta* 56(27) (2011): 10259–10263.

21. H Ashkenani and M.A. Taher. "Selective voltammetric determination of Cu(II) based on multiwalled carbon nanotube and nano-porous Cu-ion imprinted polymer." *J. Electroanal. Chem.* 683 (2012): 80–87.

22. H.R. Rajabi, M. Roushani, M. Shamsipur. "Development of a highly selective voltammetric sensor for nanomolar detection of mercury ions using glassy carbon electrode modified with a novel ion imprinted polymeric nanobeads and multi-wall carbon nanotubes." *J. Electroanal. Chem.* 693 (2013): 16–22.

23. P.Y. Chen, P.C. Nien, C.W. Hu, K.C. Ho. "Detection of uric acid based on multi-walled carbon nanotubes polymerized with a layer of molecularly imprinted PMAA." *Sensors Actuators, B Chem.* 146(2) (2010): 466–471.

24. D. Xiao, P. Dramou, N. Xiong, H. He, H. Li, D. Yuan, H. Dai. "Development of novel molecularly imprinted magnetic solid-phase extraction materials based on magnetic carbon nanotubes and their application for the determination of gatifloxacin in serum samples coupled with high-performance liquid chromatography." *J. Chromatogr. A* 1274 (2013): 44–53.

25. E.H.M. Koster, C. Crescenzi, W. den Hoedt, K. Ensing, G. de Jong. "Fibers coated with molecularly imprinted polymers for solid-phase microextraction." *J. Anal. Chem.* 73(13) (2001): 3140–3145.

26. T. Li, S. Chen, H. Li, Q. Li, L. Wu. "Preparation of an ion-imprinted fiber for the selective removal of Cu2+." *Langmuir* 27(11) (2011): 6753–6758.

27. M. Monier, I.M. Kenawy, M.A. Hashem. "Synthesis and characterization of selective thiourea modified Hg(II) ion-imprinted cellulosic cotton fibers." *Carbohydr. Polym.* 106 (2014): 49–59.

28. Y. Li, X-F. Yin, F-R. Chen, H-H. Yang, Z-X. Zhuang, X-R. Wang. "Synthesis of magnetic molecularly imprinted polymer nanowires using a nanoporous alumina template." *Macromolecules* 39(13) (2006): 4497–4499.

29. J. Wang and F. Liu. "Synthesis and application of ion-imprinted interpenetrating polymer network gel for selective solid phase extraction of Cd2+." *Chem. Eng. J.* 242 (2014): 117–126.

30. J. Wang, J. Li, H. Li, L. Ding. "Modulated ion recognition by thermosensitive ion-imprinted hydrogels with IPN structure." *Mater. Lett.* 131 (2014): 9–11.

31. T.A. Sergeyeva, O.V. Piletska, S.A. Piletsky, L.M. Sergeeva, O.O. Brovko, G.V. El'ska. "Data on the structure and recognition properties of the template-selective binding sites in semi-IPN-based molecularly imprinted polymer membranes." *Mater. Sci. Eng. C* 28(8) (2008): 1472–1479.

32. Q. Wang, D. Kim, D.D. Dionysiou, G.A. Sorial, D. Timberlake. "Sources and remediation for mercury contamination in aquatic systems-a literature review." *Environ. Pollut.* 131(2) (2004): 323–336.

33. Y. Liu, X. Chang, D. Yang, Y. Guo, S. Meng. "Highly selective determination of inorganic mercury(II)

after preconcentration with Hg(II)-imprinted diazoaminobenzene–vinyl pyridine copolymers." *Anal. Chim. Acta* 538(1–2) (2005): 85–91.

34. S. Xu, L. Chen, J. Li, Y. Guan, H. Lu. "Novel Hg^{2+}-imprinted polymers based on thymine-Hg^{2+}-thymine interaction for highly selective preconcentration of Hg^{2+} in water samples." *J. Hazard. Mater.* 237–238 (2012): 347–354.

35. P.E. Body, G. Inglis, P.R. Dolan, D.E. Mulcahy. "Environmental lead: A review." *Crit. Rev. Environ. Control* 20(5–6) (1991): 299–310.

36. M. Zhang, Z. Zhang, Y. Liu, X. Yang, L. Luo, J. Chen, S. Yao. "Preparation of core shell magnetic ion imprinted polymer for selective extraction of Pb(II) from environmental samples." *Chem. Eng. J.* 178 (2011): 443–450.

37. X.B. Zhu, Y.M. Cui, X.J. Chang, X.J. Zou, Z.H. Li. "Selective solid-phase extraction of lead(ii) from biological and natural water samples using surface-grafted lead(ii)-imprinted polymers." *Microchim Acta* 164(1–2) (2009): 125–132.

38. X. Luo, L. Liu, F. Deng, S. Luo. "Novel ion-imprinted polymer using crown ether as a functional monomer for selective removal of Pb(II) ions in real environmental water samples." *J. Mater. Chem. A* 1(28) (2013): 8280–8286.

39. P.E. Hande, S. Kamble, A.B. Samui, P.S. Kulkarni. "Chitosan-based lead ion-imprinted interpenetrating polymer network by simultaneous polymerization for selective extraction of lead(II)." *Ind. Eng. Chem. Res.* 55(12) (2016): 3668–3678.

40. R.B. Hayes. "The carcinogenicity of metals in humans." *Cancer Causes Control* 8(3) (1997): 371–385.

41. Y. Zhai, Y. Liu, X. Chang, S. Chen, X. Huang. "Selective solid-phase extraction of trace ca mium(II) with an ionic imprinted polymer prepared from a dual-ligand monomer." *Anal. Chim. Acta* 593(1) (2007): 123–128.

42. Buhani, Suharso, Sumadi. "Adsorption kinetics and isotherm of Cd(II) ion on *Nannochloropsis* sp biomass imprinted ionic polymer." *Desalination* 259(1–3) (2010): 140–146.

43. R. Patel and S. Suresh. "Kinetic and equilibrium studies on the biosorption of reactive black 5 dye by Aspergillus foetidus." *Bioresour. Technol.* 99(1) (2008): 51–58.

44. E. Birlik, A. Ersöz, E. Açıkkalp, A. Denizli, R. Say. "Cr(III)-imprinted polymeric beads: Sorption and preconcentration studies." *J. Hazard. Mater.* 140(1–2) (2007): 110–116.

45. N. Zhang, J.S. Suleiman, M. He, B. Hu. "Chromium(III)-imprinted silica gel for speciation analysis of chromium in environmental water samples with ICP-MS detection." *Talanta* 75(2) (2008): 536–543.

46. B. Leśniewska, B. Godlewska-żyłkiewicz, A.Z. Wilczewska. "Separation and preconcentration of trace amounts of Cr(III) ions on ion-imprinted polymer for atomic absorption determinations in surface water and sewage samples." *Microchem. J.* 105 (2012): 88–93.

47. G. Bayramoglu, M.Y. Arica. "Synthesis of Cr(VI)-imprinted poly(4-vinyl pyridine-co-hydroxyethyl methacrylate) particles: Its adsorption propensity to Cr(VI)." *J. Hazard. Mater.* 187(1–3) (2011): 213–221.

48. M. Uygun, E. Feyzioğlu, E. Özçalışkan, M. Caka, A. Ergen, S. Akgöl, A. Denizli. "New generation ion-imprinted nanocarrier for removal of Cr(VI) from wastewater." *J. Nanoparticle Res.* 15(8) (2013): 1833.

49. S-H. Jo, S-Y. Lee, K-M. Park, S.C. Yi, D. Kim, S. Mun. "Continuous separation of copper ions from a mixture of heavy metal ions using a three-zone carousel process packed with metal ion-imprinted polymer." *J. Chromatogr. A* 1217(45) (2010): 7100–7108.

50. S-H. Jo, C. Park, S.C. Yi, D. Kim, S. Mun. "Development of a four-zone carousel process packed with metal ion-imprinted polymer for continuous separation of copper ions from manganese ions, cobalt ions, and the constituent metal ions of the buffer solution used as eluent." *J. Chromatogr. A* 1218(33) (2011): 5664–5674.

51. N.W. Turner, C.W. Jeans, K.R. Brain, C.J. Allender, V. Hlady, D.W. Britt. "From 3D to 2D: A review of the molecular imprinting of proteins." *Biotechnol. Prog.* 22(6) (2006): 1474–1489.

52. R. Crichton. *Iron Metabolism: From Molecular Mechanisms to Clinical Consequences.* Chichester: John Wiley & Sons. Ltd, 2009.

53. M. Karabörk, A. Ersöz, A. Denizli, R. Say. "Polymer–clay nanocomposite iron traps based on intersurface ion-imprinting." *Ind. Eng. Chem. Res.* 47(7) (2008): 2258–2264.

54. S. Dai, M.C. Burleigh, Y.H. Ju, H.J. Gao, J.S. Lin, S.J. Pennycook, C.E. Barnes, Z.L. Xue. "Hierarchically imprinted sorbents for the separation of metal ions." *J Am. Chem. Soc.* 122(5) (2000): 992–993.

55. M.E. Diaz-Garcia, R.B. Laínño. "Molecular imprinting in sol-gel materials: recent developments and applications." *Microchim. Acta* 149(1–2) (2004): 19–36.

56. J.L. Domingo. "Reproductive and developmental toxicity of natural and depleted uranium: A review." *Reprod. Toxicol.* 15(6) (2001): 603–609.

57. C.R. Preetha, J.M. Gladis, T.P. Rao, G. Venkateswaran. "Removal of toxic uranium from synthetic nuclear power reactor effluents using uranyl ion imprinted polymer particles." *Environ. Sci. Technol.* 40(9) (2006): 3070–3074.

58. G.D. Saunders, S.P. Foxon, P.H. Walton, M.J. Joyce, S.N. Port. "A selective uranium extraction agent prepared by polymer imprinting." *Chem. Commun.* 4 (2000): 273–274.

59. Q. He, X. Chang, Q. Wu, X. Huang, Z. Hu, Y. Zhai. "Synthesis and applications of surface-grafted Th(IV)-imprinted polymers for selective solid-phase extraction of thorium(IV)." *Anal. Chim. Acta.* 605(2) (2007): 192–197.

60. C. Lin, H. Wang, Y. Wang, Z. Cheng. "Selective solid-phase extraction of trace thorium(IV) using surface-grafted Th(IV)-imprinted polymers with pyrazole derivative." *Talanta* 81(1–2) (2010): 30–36.

61. X.Z. Ji, H.J. Liu, L.L. Wang, Y.K. Sun, Y.W. Wu. "Study on adsorption of Th(IV) using surface modified dibenzoylmethane molecular imprinted polymer." *J. Radioanal. Nucl. Chem.* 295(1) (2013): 265–270.

62. F.\F. He, H.Q. Wang, Y.Y. Wang, X.F. Wang, H.S. Zhang, H.L. Li, J.H. Tang. "Magnetic Th(IV)-ion imprinted polymers with salophen Schiff base for separation and recognition of Th(IV)." *J. Radioanal. Nucl. Chem.* 295(1) (2013): 167–177.

63. R. Kala, J.M. Gladis, T.P. Rao. "Preconcentration separation of erbium from Y, Dy, Ho, Tb, and Tm by using ion imprinted polymer particles via solid phase extraction." *Anal. Chim. Acta* 518(1–2) (2004): 143–150.

64. V.M. Biju, J.M. Gladis, T.P. Rao. "Effect of gamma-irradiation of ion-imprinted polymer (IIP) particles for the preconcentration separation of dysprosium from other selected lanthanides." *Talanta* 60(4) (2003): 747–754.

65. M. Ahamed, X.Y. Mbianda, A.F. Mulaba-Bafubiandi, L. Marjanovic. "Ion-imprinted polymers for the selective extraction of silver(I) ions in aqueous media: Kinetic modeling and isotherm studies." *React. Funct. Polym.* 73(3) (2013): 474–483.

66. B. Godlewska-Żyłkiewicz, B. Leśniewska, I. Wawreniuk. "Assessment of ion-imprinted polymers based on Pd(II) chelate complexes for preconcentration and FAAS determination of palladium." *Talanta.* 83(2) (2010): 596–604.

67. B. Leśniewska, M. Kosińska, B. Godlewska-Żyłkiewicz, E. Zambrzycka, A.Z. Wilczewska. "Selective solid phase extraction of platinum on an ion-imprinted polymer for its electrothermal atomic absorption spectrometric determination in environmental samples." *Microchim Acta* 175(3–4) (2011): 273–282.

68. A. Darroudi, A.M.H. Zavar, M. Chamsaz, G. Zohuri, N. Ashraf. "Ion-imprinted polymer mini-column for on-line preconcentration of thallium(III) and its determination by flame atomic absorption spectrometry." *Anal. Methods.* 4(11) (2012): 3798–3803.

69. T. Kanai, C. Sanskriti, P. Vislawath, A.B. Samui. "Acrylamide based molecularly imprinted polymer for detection of m-nitrophenol." *J. Nanosci. Nanotech.* 13 (2013): 3054–3061.

70. Y. Xiao, R. Xiao, J. Tang, Q. Zhu, X. Li, Y. Xiong, X. Wu. "Preparation and adsorption properties of molecularly imprinted polymer via RAFT precipitation polymerization for selective removal of aristolochic acid I." *Talanta,* 162 (2017): 415–422.

71. K. Singh, A. Pasha, B. Amitha Rani. "Preparation of molecularly imprinted polymers for heptachlor: An organochlorine pesticide." *Chronicles Young Sci.* 4(1) (2013): 46–50.

72. J. Li, Q. Zhou, Y. Yuan, Y. Wu. "Iron-based magnetic molecular imprinted polymers and their application in removal and determination of di-n-pentyl phthalate in aqueous media." *Royal Society Open Science,* 4(8) (2017): 170672.

73. C. Dai, J. Zhang, Y. Zhang, X. Zhou, S. Liu. "Application of molecularly imprinted polymers to selective removal of clofibric acid from water." *PLoS ONE* 8(10) (2013): e78167.

74. A.B. Samui, T. Kanai, P. Vislawath. "Development of molecularly imprinted polymer for detection of explosives." (2012) Unpublished results.

75. L. Chen, X. Wang, W. Lu, X. Wu, J. Li. "Molecular imprinting: Perspectives and applications." *Chem. Soc. Rev.* 45 (2016): 2137–2211.

76. R. Uzek, L. Uzun, S. Senel, A. Denizli. "Nanospines incorporation into the structure of the hydrophobic cryogels via novel cryogelation method: An alternative sorbent for plasmid DNA purification." *Coll. Surf. B Biointerface.* 102 (2013): 243–250.

77. J.W. Boyd, G.P. Cobb, G.E. Southard, G.M. Murray. "Development of molecularly imprinted polymer sensors for chemical warfare agents." *Johns Hopkins APL Tech. Dig.* 25 (2004): 44–49.

78. S.A. Piletsky, N.W. Turner, P. Laitenberger. "Molecularly imprinted polymers in clinical diagnostics: Future potential and existing problems." *Med. Eng. Phys.* 28 (2006): 971–977.

79. E.L. Holthoff, D.N. Stratis-Cullum, M.E. Hankus. "A nanosensor for TNT detection based on molecularly imprinted polymers and surface-enhanced Raman scattering." *Sensors* 11(3) (2011): 2700–2714.

80. W. Lu, H.Y. Li, Z.H. Meng, X.X. Liang, M. Xue, Q.H. Wang, X. Dong. "Detection of nitrobenzene compounds in surface water by ion mobility spectrometry coupled with molecularly imprinted polymers." *J. Hazard. Mater.* 280 (2014): 588–594.

81. E. Yilmaz, D. Majidi, E. Ozgur, A. Denizli. "Whole cell imprinting based Escherichia coli sensors: A study for SPR and QCM." *Sens. Actuator B Chem.* 209 (2015): 714–721.

82. E. Yılmaz, B. Garipcan, H.K. Patra, L. Uzun. "Molecular imprinting applications in forensic science." *Sensors* 17(4) (2017): 691.

83. X. Hu, G. Li, J. Huang, D. Zhang, Y. Qiu. "Construction of self-reporting specific chemical sensors with high sensitivity." *Adv. Mater.* 19(24) (2007): 4327–4332.

84. C. Han, Z. Shang, H. Zhang, Q. Song. "Detection of hidden drugs with a molecularly imprinted electrochemiluminescence sensor." *Anal. Method* 5(21) (2013): 6064–6070.

85. R.C. Stringer, S. Gangopadhay, S.A. Grant. "Detection of nitroaromatic explosives using a fluorescent labelled imprinted polymer." *Anal. Chem.* 82(10) (2010): 1773–1779.

86. Y. Mao, Y. Bao, D. Han, F. Li, L. Niu. "Efficient one-pot synthesis of molecularly imprinted silica nanospheres embedded carbon dots for fluorescent dopamine optosensing." *Biosens. Bioelectron.* 38(1) (2012): 55–60.

87. Y. Zhou, Z. Qu, Y. Zeng, T. Zhou, G. Shi. "A novel composite of graphene quantum dots and molecularly imprinted polymer for fluorescent detection of para-nitrophenol." *Biosens. Bioelectron.* 52 (2014): 317–323.

88. R. Jalili, M. Amjadi. "Surface molecular imprinting on silane-functionalized carbon dots for selective recognition of nifedipine." *RSC Adv.* 5(90) (2015): 74084–74090.

89. S.E. Diltemiz, R. Say, S. Büyüktiryaki, D. Hür, A. Denizli, A. Ersöz. "Quantum dot nanocrystals having guanosine imprinted nanoshell for DNA recognition." *Talanta* 75(4) (2008): 890–896.

90. H. Li, Y. Li, J. Cheng. "Molecularly imprinted silica nanospheres embedded CdSe quantum dots for highly selective and sensitive optosensing of pyrethroids." *Chem. Mater.* 22(8) (2010): 2451–2457.

91. J. Liu, H. Chen, Z. Lin, J-M. Lin. "Preparation of surface imprinting polymer capped Mn-doped ZnS quantum dots and their application for chemiluminescence detection of 4-nitrophenol in tap water." *Anal. Chem.* 82(17) (2010): 7380–7386.

92. X. Ren and L. Chen. "Quantum dots coated with molecularly imprinted polymer as fluorescence probe for detection of cyphenothrin." *Biosens. Bioelectron.* 64 (2015): 182–188.

93. X. Ren, H. Liu, L. Chen. "Fluorescent detection of chlorpyrifos using Mn(II)-doped ZnS quantum dots coated with a molecularly imprinted polymer." *Microchim. Acta* 182(1–2) (2014): 193–200.

94. G. Fang, C. Fan, H. Liu, M. Pan, H. Zhu, S. Wang. "A novel molecularly imprinted polymer on CdSe/ZnS quantum dots for highly selective optosensing of mycotoxin zearalenone in cereal samples." *RSC Adv.* 4(6) (2014): 2764–2771.

95. H. Liu, G. Fang, C. Li, M. Pan, C. Liu, C. Fan, S. Wang. "Molecularly imprinted polymer on ionic liquid-modified CdSe/ZnS quantum dots for the highly selective and sensitive optosensing of tocopherol." *J. Mater. Chem.* 22(37) (2012): 19882–19887.

96. H. Liu, D. Wu, Y. Liu, H. Zhang, T. Ma, A. Aidaerhan, J. Wang, B. Sun. "Application of an optosensing chip based on molecularly imprinted polymer coated quantum dots for the highly selective and sensitive determination of sesamol in sesame oils." *J. Agric. Food Chem.* 63(9) (2015): 2545–2549.

97. M. Chao and C. Hu. "Fluorescent turn-on detection of cysteine using a molecularly imprinted polyacrylate linked to alkyl thiol-capped CdTe quantum dots." *J. Microchem. Acta.* 181(9–10) (2014): 1085–1091.

98. S. Ge, J. Lu, L. Ge, M. Yan, J. Yu. "Development of a novel deltamethrin sensor based on molecularly imprinted silica nanospheres embedded CdTe quantum dots." *J. Spectrochim. Acta - Part A Mol. Biomol. Spectrosc.* 79(5) (2011): 1704–1709.

99. D.Y. Li, X.W. He, Y. Chen, W.Y. Li, Y.K. Zhang. "Novel hybrid structure silica/CdTe/molecularly imprinted polymer: synthesis, specific recognition, and quantitative fluorescence detection of bovine hemoglobin." *ACS Appl. Mater. Interfaces* 5(23) (2013): 12609–12616.

100. A. Lourenço, R. Viveiros, A. Mouro, J.C. Lima, V.D.B. Bonifácio, T. Casimiro. "Supercritical CO₂-assisted synthesis of an ultrasensitive amphibious quantum dot-molecularly imprinted sensor." *RSC Adv.* 4(108) (2014): 63338–63341.

101. H. Liu, D. Liu, G. Fang, F. Liu, C. Liu, Y. Yang, S. Wang. "A novel dual-function molecularly imprinted polymer on CdTe/ZnS quantum dots for highly selective and sensitive determination of ractopamine." *Anal. Chim. Acta* 762 (2013): 76–82.

102. S.M. Ng, R. Narayanaswamy. "Fluorescence sensor using a molecularly imprinted polymer as a recognition receptor for the detection of aluminum ions in aqueous media." *Anal. Bioanal. Chem.* 386(5) (2006): 1235–1244.

103. F. Karagöz, O. Güney. "Development and characterization of ion-imprinted sol-gel-derived fluorescent film for selective recognition of mercury(II) ion." *J. Sol-Gel Sci. Technol.* 76(2) (2015): 349–357.

104. S.C.L. Pinheiro, A.B. Descalzo, I.M. Raimundo, G. Orellana, M.C. Moreno-Bondi. "Fluorescent ion-imprinted polymers for selective Cu(II) optosensing." *Anal. Bioanal. Chem.* 402(10) (2012): 3253–3260.

105. L-Y. Yu, H-M. Shen, Z-L. Xu. "PVDF–TiO₂ composite hollow fiber ultrafiltration membranes prepared by TiO₂ sol-gel method and blending method." *J. Appl. Phys.* 113(3) (2009): 1763–1772.

106. B. Wandelt, A. Mielniczak, P. Cywinski. "Monitoring of cAMP-imprinted polymer by fluorescence spectroscopy." *Biosens. Bioelectron.* 20(6) (2004): 1031–1039.

107. Y. Cheng, P. Jiang, X. Dong. "Molecularly imprinted fluorescent chemosensor synthesized using quinoline-modified-β-cyclodextrin as monomer for spermidine recognition." *RSC Adv.* 5(68) (2015): 55066–55074.

108. C.F. Chow, M.H.W. Lam, M.K.P. Leung. "Fluorescent sensing of homocysteine by molecular imprinting.' *Anal. Chim. Acta* 466(1) (2002): 17–30.

109. X. Cai, J. Li, Z. Zhang, G. Wang, X. Song, J. You, L. Chen. "Chemodosimeter-based fluorescent detection of L-cysteine after extracted by molecularly imprinted polymers." *Talanta* 120 (2014): 297–303.

110. I. Chianella, S.A. Piletsky, I.E. Tothill, B. Chen, A.P.F. Turner. "MIP-based solid phase extraction cartridges combined with MIP-based sensors for the detection of microcystin-LR." *Biosens. Bioelectron.* 18(2–3) (2003): 119–127.

111. Y. Pan, L. Yang, N. Mu, S. Shao, W. Wang, X. Xie, S. He. "A SAW-based chemical sensor for detecting sulfur-containing organophosphorus compounds using a two-step self-assembly and molecular imprinting technology." *Sensors* 14(5) (2014): 8810–8820.

112. B. Sellergren, B. Ekberg, K. Mosbach. "Molecular imprinting of amino acid derivatives in macroporous polymers: Demonstration of substrate-and enantio-selectivity by chromatographic resolution of racemic mixtures of amino acid derivatives." *J. Chromatogr. A* 347 (1985): 1–10.

113. Y.P. Huang, C. Zheng, Z.S. Liu. "Molecularly imprinted polymers for the separation of organic compounds in capillary electrochromatography." *Curr. Org. Chem.* 15(11) (2011): 1863–1870.

114. T. Alizadeh, M.R. Ganjali, M. Zare. "Application of an Hg2+ selective imprinted polymer as a new modifying agent for the preparation of a novel highly selective and sensitive electrochemical sensor for the determination of ultra-trace mercury ions." *Anal. Chim. Acta* 689(1) (2011): 52–59.

115. T. Alizadeh and S. Amjadi. "Preparation of nano-sized Pb²⁺ imprinted polymer and its application as the chemical interface of an electrochemical sensor for toxic lead determination in different real samples." *J. Hazard. Mater.* 190(1–3) (2011): 451–459.

116. Z. Wang, Y. Qin, C. Wang, L. Sun, X. Lu, X. Lu. "Preparation of electrochemical sensor for lead(II) based on molecularly imprinted film." *Appl. Surf. Sci.* 258(6) (2012): 2017–2021.

117. T. Alizadeh, M.R. Ganjali, P. Nourozi, M. Zare, M. Hoseini. "A carbon paste electrode impregnated with Cd²⁺ imprinted polymer as a new and high selective electrochemical sensor for determination of ultra-trace Cd²⁺ in water samples." *J. Electroanal. Chem.* 657(1–2) (2011): 98–106.

118. T. Alizadeh, M. Zare, M.R. Ganjali, P. Norouzi, B. Tavana. "A new molecularly imprinted polymer (MIP)-based electrochemical sensor for monitoring 2,4,6-trinitrotoluene (TNT) in natural waters and soil samples," *Biosens. Bioelectron.* 25 (2010): 1166–1172.

119. D. Udomsap, C. Branger, G. Culioli, P. Dollet, H. Brisset. "A versatile electrochemical sensing receptor based on a molecularly imprinted polymer." *Chem. Commun.* 50(56) (2014): 7488–7491.

120. M. Riskin, R. Tel-Vered, T. Bourenko, E. Granot, I. Willner. "Imprinting of molecular recognition sites through electropolymerization of functionalized Au nanoparticles: development of an electrochemical TNT sensor based on pi-donor-acceptor interactions." *J. Am. Chem. Soc.* 130(30) (2008): 9726–9733.

121. F. Wang, L. Zhu, J. Zhang. "Electrochemical sensor for levofloxacin-based on molecularly imprinted polypyrrole–graphene–gold nanoparticles modified electrode." *Sensors Actuators B. Chem.* 192 (2014): 642–647.

122. H. da Silve, J.G. Pacheco, J. McS. Magalhães, S. Viswanathan, C. Delerue-matos. "MIP-graphene-modified glassy carbon electrode for the determination of trimethoprim." *Biosens. Bioelectron.* 52 (2014): 56–61.

123. P.E. Hande, A.B. Samui, P.S. Kulkarni. "An efficient method for determination of the diphenylamine (stabilizer) in propellants by molecularly imprinted polymer based carbon paste electrochemical sensor." *Propellants Explos. Pyrotech.* 42(1) (2017): 376–380.

13

Vibration Damping by Polymers

Bikash Chandra Chakraborty
Naval Materials Research Laboratory, Ambernath, Maharashtra, India

Praveen Srinivasan
Naval Materials Research Laboratory, Ambernath, Maharashtra, India

CONTENTS

ABBREVIATIONS

CF	carbon fiber
CL	constraining layer
CLD	constrained layer damping
dB	decibel
DMA	dynamic mechanical analysis/analyzer
DSA	dynamic signal analyzer
FFT	Fast Fourier Transform
FLD	free layer damping
G	graphite

DOI: 10.1201/9781003037880-13

Symbol	Description	Unit
w	Mass per unit length	kg/m
a_{13}	Ratio of Young's modulus of constraining layer and that of the vibrating beam	dimensionless
δ	Phase angle	rad
h_0	Viscoelastic loss factor of VEM	dimensionless
η_s	System loss factor	dimensionless
ψ_{23}	Shear parameter	dimensionless
t	Relaxation time	s
t_0	A constant in Arrhenius expression for relaxation time	s
ω	Angular frequency	rad/s
ω_d	Damped natural frequency	rad/s
ω_n	Undamped natural frequency	rad/s
ξ	Damping factor	dimensionless

GPa — gigapascal
HAF — high abrasion furnace (carbon black)
Hz — Hertz
IPN — interpenetrating polymer network
KF — Kevlar fiber
MCLD — magnetic constrained layer damping
MPa — mega pascal
MS — mild steel
NBR — acrylonitrile butadiene rubber
NC — nanoclay
phr — parts per hundred part of rubber (by weight)
PBuMA — poly(butyl methacrylate)
PEMA — poly(ethyl methacrylate)
PLZT — lead lanthanum zirconatetitanate
PMMA — poly(methyl methacrylate)
PMN-PT — lead magnesium niobate-lead titanate
PU — polyurethane
PVC — poly(vinyl chloride)
PZT — lead zirconate titanate
RKU — Ross-Kerwin-Unger
R-L-C — resistance -inductance-capacitance
SBR — styrene-butadiene rubber
SONAR — sound navigation and ranging
FLD — unconstrained layer damping
VEM — viscoelastic material
WLF — Williams, Landel & Ferry

SYMBOLS WITH UNITS

Symbol	Description	Unit
a_T	Shift factor	dimensionless
b	Width of a beam	m
C_1	Constant in WLF equation for shift factor	dimensionless
C_2	Constant in WLF equation for shift factor	K
E	Young's modulus	N/m²
E_a	Activation energy of relaxation	J/mol
E′	Dynamic or storage modulus	N/m²
E″	Loss modulus	N/m²
E_1	Dynamic modulus in the rubbery region	N/m²
E_2	Dynamic modulus in the glassy region	N/m²
f_n	Natural frequency	Hz
Δf	-3 dB bandwidth	Hz
G_2	Dynamic shear modulus of VEM	N/m²
g	Acceleration due to gravity	m/s²
h	Thickness of beam	m
I	Moment of inertia	m⁴
L	Length of beam	m
m	Mass	kg
R	Universal gas constant	J/mol. K
T	Temperature	°C, K
T_g	Glass transition temperature	°C, K
T_{ref}	Reference temperature	°C, K
t	(1) Time (2) Thickness of VEM layer	(1) S (2) m
t_c	Thickness of constraining layer	m
V	Intensity of vibration (e.g. acceleration)	Volts

13.1 Introduction

Vibration is defined as the oscillation of an object around its equilibrium position. The oscillation can be periodically symmetrical, known as harmonic, or can be nonsymmetrical, known as nonharmonic motion. The oscillatory motion of an object results in a mechanical wave which propagates in the media coupled to the object. This mechanical wave is called sound. A typical example is a simple tuning fork, which produces sound on vibration. Audible sound has a frequency range of 20 Hz to 20 kHz. Below the lower limit, the wave is subsonic and above the upper limit, the wave is supersonic or ultrasonic. Sound and vibration may be required in many situations, but in many other cases, sound and vibration are to be avoided. For example, sound is required for communication, music, etc., but for automobile engines, aircraft, construction equipment, the sound produced is a noise, which is unwanted and has to be reduced as far as possible. Vibration is also detrimental to machines, increasing the chances of malfunctioning and fatigue failure. Major noise radiators are vibrating solid surfaces, commonly industrial machinery, construction equipment, domestic equipment, vehicle engines, and metallic structures. The study of sound and vibration is also very important for naval applications. An underwater submerged object like a submarine can be detected only by acoustic waves. Such an acoustic detection system is known as Sound Navigation and Ranging (SONAR). The vibration of machinery in underwater vessels can be transmitted through the hull to seawater as radiated sound. This can be detected by passive SONAR. Therefore, vibration damping of hull structures and machinery vibration is essential in order to achieve acoustic stealth capability.

Polymers are most commonly used as passive vibration damping materials due to their ability to absorb mechanical waves and dissipate them as heat. The ability to absorb vibration by a polymer depends on the thermodynamic condition, the frequency band, and the method of damping treatment. The property by which this phenomenon takes place in a polymer is called its dynamic viscoelasticity. It is therefore important to discuss this basic property of the polymer before describing the various damping methodologies.

13.2 General Vibration Control Strategies

Vibration and noise control measures should be developed as effectively as possible. A single source may transmit vibration via several parallel paths, all of which must be controlled by appropriate damping treatment. The most effective form of vibration control is implemented at the source. In most cases, a reduction of the surface vibration will produce proportional changes in radiated sound. There are four principal alternative passive control strategies:

1. *Vibration isolation*: The mechanism to reduce sound depends on the source where the sound comes from. If it is generated within a room, then the sound has to be absorbed. If it originates from outside, then to reduce the sound intensity, it is necessary to insulate the space, and if it is transmitted through the structure, then the structure needs to be isolated from the source of vibration. Vibration isolation involves the insertion of resilient (flexible) elements between the structure which is directly subjected to the source of vibration and other structures to which it is connected.

2. *Damping*: This involves the application of one of the following methods:

   (a) A viscoelastic lossy material as an add-on to the vibrating structure to convert a large proportion of the vibrational energy into heat.

   (b) An active damping system wherein a dynamic opposing force is fed to the vibrating element which is similar in frequency band and intensity but 180 degrees out of phase for cancellation. The system can be designed as feedback or feed-forward.

   (c) A combination of the two methods above.

   (d) Smart dampers such as electro-rheological damper, magneto-rheological damper, dilatant fluid (shear thickening) damper.

   (e) Piezoelectric layer for the conversion of vibration to electrical energy which is fed to an electrical load, consisting of either a resistor alone or a combination of resistor-inductor, or resistor-capacitor-inductor.

3. *Impedance mismatching*: This involves the design or modification of the structures where they are attached to the directly excited structure to reduce the transfer of vibrational energy from one to the other.

4. *Dynamic neutralization*: This involves the connection of ancillary mechanical resonators either to the directly excited structure or to connected structures.

Of all the systems above, passive damping is the most widely used for its simplicity in mathematical predictions, adoption to a large structure without sacrificing damping effectiveness, simplicity in applying to curved contours, ease of repair and maintenance, and cost-effectiveness. Moreover, passive damping can be very broad in the frequency range which other systems may not provide. For example, active damping is very effective at low frequencies, typically below 1000 Hz, and smart dampers are good for shock mitigation.

This chapter focuses on damping vibrations by using viscoelastic materials as an added damping layer to the vibrating substrate.

13.3 Viscoelasticity

The classical theory of elasticity deals with the mechanical properties of elastic solids for which, following Hooke's law, the stress is directly proportional to the strain within a limit, known as the elastic limit, and in this limit, the ratio of stress to strain in the longitudinal direction of stretching is termed Young's modulus. This is a material property and does not depend on the values of implied force or deformation within the elastic limit. The elastic modulus of a body is, however, dependent on the thermodynamic state of the body. This definition of elastic property implies that if the stress is constant at a value, the strain will also be constant. On withdrawal of the external force, the deformation instantly becomes nil. Therefore, in a cyclic deformation, as in vibration, there will not be any energy loss in an elastic body.

Similar to the extensional stretching, there can be a deformation of a body in other modes, such as compression, shear, bending (flexure), and also by volume (bulk). All these moduli are material properties. The elongational, shear, and bulk moduli are related to each other by Poisson's ratio. Modulus of the body indicates its ability to resist deformation. Therefore, it is related to the "stiffness" (similar to spring constant), defined as the force required to cause unit deflection in a body. In the formulation of a constitutive equation, stiffness or spring constant is often used instead of Elastic Modulus. Hooke's law of elasticity is expressed as:

$$\sigma = E\varepsilon \qquad (13.1)$$

where σ is the stress of the body, ε is the strain and E is the Young's modulus.

Young's modulus, shear modulus, and bulk modulus are related to each other by Poisson's Ratio as follows:

$$\text{Shear Modulus: } G = \frac{E}{2(1+v)} \qquad (13.2)$$

$$\text{Bulk Modulus: } K = \frac{E}{3(1-2v)} \qquad (13.3)$$

$$\text{Young's Modulus: } E = \frac{9KG}{3K+G} \qquad (13.4)$$

$$\text{Poisson's Ratio: } v = \frac{\varepsilon_T}{\varepsilon_L} \qquad (13.5)$$

where ε_L is the elongational strain and ε_T is the transverse strain.

On the other hand, fluids cannot resist spontaneous deformation under external force. Newton's law of viscosity states that the shear stress is directly proportional to the rate of strain and the proportionality constant is termed the coefficient of viscosity. It implies that, under constant stress, the body would undergo indefinite deformation, which is termed as flow. Therefore, the total input energy of mechanical force will be converted to flow and cannot be recovered on the withdrawal of external force. Newton's law can be expressed as

$$\text{Shear stress}: \sigma = \eta \frac{d\gamma}{dt} \qquad (13.6)$$

where η is the coefficient of viscosity and γ is the shear strain of the fluid.

Not all liquids are Newtonian and a large variety of liquids, such as polymeric fluids, are nonlinear in the relationship between stress and strain rate. Studies on polymer melts and solutions show that they exhibit different behavior, such as pseudoplastic, dilatant, thixotropic, rheopectic, viscoelastic, viscoplastic, etc. Different polymer flow behavior is modeled as the Power Law, the Bingham Model, Casson's Model, the Carreau Model, the Cross Model, the Ellis Model, and so on.[1–4] For non-Newtonian fluids, the viscosity changes with shear rate and even in some cases, with time. However, in the discussion on dynamic linear viscoelasticity, we shall restrict ourselves to Newtonian viscosity while quantifying the damping term.

These categories of elastic and viscous bodies discussed above are idealized. The behavior of many solids can be described by Hooke's law for small strains and that of many liquids can be defined by Newton's law for small strain rate. Under other conditions, deviations are observed. Even if both strain and rate of strain are infinitesimal, a system may exhibit behavior which combines liquid-like and solid-like characteristics. For example, a body which is not quite solid does not maintain a constant deformation under constant stress, but goes on slowly deforming with time, which is a phenomenon known as creep. When such a body is constrained at constant deformation, the stress required to hold it decreases gradually, which is known as stress relaxation. On the other hand, a body which is not quite liquid may, while flowing under constant stress, store some of the energy input, instead of dissipating it all as heat, and it may recover part of the deformation when stress is removed (elastic recoil). When such a body is subjected to sinusoidal stress, the strain is neither exactly in phase with the stress (as it would be for a perfectly elastic solid), nor 90° out of phase (as it would be for a perfectly viscous liquid), but is somewhere in between. A part of the input energy is stored as elastic energy and recovered in each cycle, while a part of the input energy is dissipated due to viscous friction as heat. A material that exhibits such characteristics is called viscoelastic material and the phenomenon is termed viscoelasticity. For polymers, the stress-strain behavior is nonlinear because of the viscous dissipation of strain energy. The nonlinearity is very pronounced near and above the glass transition temperature (T_g). This implies that since common elastomers have T_g below the ambient temperature, they have predominantly nonlinear behavior at the ambient temperature. However, linear behavior can be observed for a low range of strain values, even for elastomers. If in a given experiment, the ratio of stress to strain is a function of time (or frequency alone) and not a function of strain magnitude, then it is linear viscoelastic behavior. In other words, a linear viscoelastic region is defined as the region where the Young's modulus of a polymer is independent of strain value. Viscoelasticity is studied using three types of stress-strain behavior:

1. *Stress relaxation*: a time-dependent reduction of stress of a viscoelastic body at constant strain.
2. *Creep*: a time-dependent increase in strain at constant stress (load) applied to a viscoelastic body.
3. *Complex mechanical property*: a frequency-dependent phenomenon under oscillatory load (dynamic load), say, shock and vibration.

Since viscoelasticity is a combination of elastic deformation and viscous flow, there are many mathematical models to predict and describe the behavior of real materials, such as polymers, earth soils, crude oils, processed food, proteins, blood, etc. The most common models are: (1) Maxwell; (2) Kelvin-Voigt; (3) three-parameter models, such as Zener, Anti-Zener, etc.; and (4) four-parameter models, such as the Burger Model.[5,6] Common elastomers are mostly described by the Zener and Burger models which are various combinations of Maxwell and Voigt elements in a series-parallel arrangement of spring, representing the elastic part, and dashpot, representing the viscous part. As an example, soft gaskets can be effectively modeled by four parameters.[7] Earth soils are studied widely using four-parameter models with a fairly good match with practical data.[8–13]

13.3.1 Dynamic Viscoelasticity

For a viscoelastic material, the stress and strain are expressed as a simple sinusoidal form or complex numbers to represent a time-dependent function, hence the modulus is represented by a complex quantity. The real part of this complex term (storage modulus, E') relates to the elastic behavior of the material and defines the stiffness. Imaginary component (loss modulus, E'') relates to the material's viscous behavior and defines the energy-dissipative characteristics of the material.

If applied stress varies with time in a sinusoidal manner, the stress (σ) may be written as:

$$\sigma = \sigma_0 \cos \omega t \qquad (13.7)$$

where ω is the angular frequency in radians/sec and $\omega = 2\pi f$, where f is the frequency in Hz.

In Hookean solids, with no energy dissipated, the strain (ε) is given by:

$$\varepsilon = \varepsilon_0 \cos \omega t \qquad (13.8)$$

The quantities with subscript zero represent the amplitude values.

For real materials which exhibit damping, the stress and strain are not in phase, the strain lags behind the stress by the phase angle δ.

$$\text{Stress}: \sigma(t) = \sigma_0 \cos \omega t \qquad (13.9)$$

$$\text{Strain}: \varepsilon(t) = \varepsilon_0 \cos(\omega t - \delta) \qquad (13.10)$$

These can be expressed in complex form as:

$$\sigma(t) = \sigma_0 e^{i\omega t} \qquad (13.11)$$

$$\varepsilon(t) = \varepsilon_0 e^{i(\omega t - \delta)} \qquad (13.12)$$

Figure 13.1 shows the sinusoidal dynamic stress and strain with a phase difference as an example.

The phase angle also defines an in-phase and out-of-phase component of the stresses σ' and σ'' respectively. The relationships between the in-phase and out-of-phase component and delta are given by

$$\sigma' = \sigma_0 \cos \delta \qquad (13.13)$$

$$\sigma'' = \sigma_0 \sin \delta \qquad (13.14)$$

$$\left.\begin{aligned} E' &= \frac{\sigma'}{\varepsilon_0} = E * \cos \delta \\ E'' &= \frac{\sigma''}{\varepsilon_0} = E * \sin \delta \end{aligned}\right\} \qquad (13.15)$$

In complex terms:

$$E* = E' + iE'' \qquad (13.16)$$

$$E* = \sqrt{E'^2 + E''^2} \qquad (13.17)$$

and

$$\eta = \tan \delta = \frac{E''}{E'} \qquad (13.18)$$

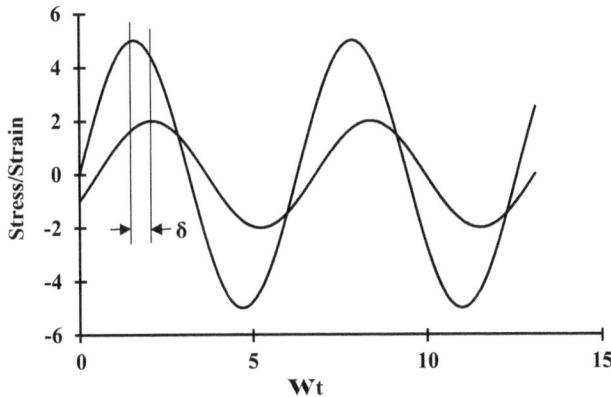

6
4
2
0
-2
-4
-6

0   5   10   15

Wt

Stress/Strain

δ

FIGURE 13.1 Dynamic stress-strain with a phase difference of δ.

η is called the loss factor, E' is termed the storage modulus or dynamic Young's modulus and E'' is termed the loss modulus. Similar expressions are valid for shear, compression, bending modes also. It may be noted that the loss factor value is the same for any mode. The loss factor accounts for the fraction of strain energy that a viscoelastic molecule will use up as loss due to internal friction (viscous dissipation).

13.3.2 Energy Loss in Dynamic Loading

As stated above, the mechanical energy of the straining process will be partly lost in a viscoelastic body due to the viscous component of the stress-strain relationship. In each cycle of straining, the loss will be constant so long as the cyclic frequency and the temperature are kept the same. The stress-strain relationship, therefore, would have hysteresis loss for a cycle of straining. This can be appreciated by a simple mathematical approach as described below.

Let the stress and strain be designated as given in Equations (13.9) and (13.10) above:

$$\sigma(t) = \sigma_0 \cos \omega t \qquad \varepsilon(t) = \varepsilon_0 \cos(\omega t - \delta)$$

$$\left(\frac{\sigma(t)}{\sigma_0}\right)^2 = \cos^2 \omega t \qquad \left(\frac{\varepsilon(t)}{\varepsilon_0}\right)^2 = \cos^2(\omega t - \delta)$$

$$\therefore \left(\frac{\sigma(t)}{\sigma_0}\right)^2 + \left(\frac{\varepsilon(t)}{\varepsilon_0}\right)^2 = \cos^2 \omega t + \left(\cos \omega t \cos \delta + \sin \omega t \sin \delta\right)^2$$

Eliminating "ωt" from the right-hand side of the above equation, we get

$$\left(\frac{\sigma(t)}{\sigma_0}\right)^2 + \left(\frac{\varepsilon(t)}{\varepsilon_0}\right)^2 = \sin^2 \delta + 2\left(\frac{\sigma}{\sigma_0}\right)\left(\frac{\varepsilon}{\varepsilon_0}\right)\cos \delta \quad (13.19)$$

Equation (13.19) represents an ellipse and the area under the ellipse represents the loss of strain energy for one cycle of loading per unit volume of the sample. Figure 13.2 shows a typical cyclic stress-strain relationship taking arbitrary values as an example.

The energy loss per cycle is the area under the loop and can be calculated by integration of infinitesimal strain energy over one cycle time $T\ (= 2\pi/\omega)$ as follows:

$$E_L = \int_0^T \sigma(t) \frac{d\varepsilon}{dt} dt$$

$$= -\int_0^{2\pi/\omega} (\sigma_0 \cos \omega t) \omega \varepsilon_0 \sin(\omega t - \delta) dt \qquad (\because T = \frac{2\pi}{\omega})$$

Upon integration, the final equation is:

$$E_L = \pi \sigma_0 \varepsilon_0 \sin \delta$$

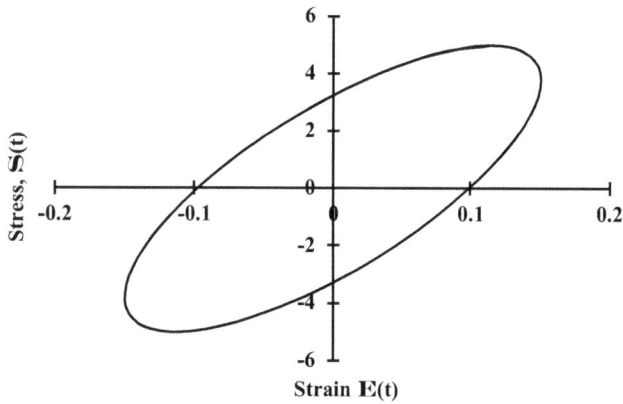

FIGURE 13.2   Dynamic stress-strain curve of a viscoelastic body.

Since

$$E'' = \frac{\sigma_0}{\varepsilon_0} \sin \delta,$$

Therefore

$$E_L = \pi E'' \varepsilon_0^2 \tag{13.20}$$

The energy loss according to Equation (13.20) is per unit volume per cycle. In terms of energy loss per unit volume per unit time, it is expressed as:

$$E_{loss} = \frac{1}{2} \omega E'' \varepsilon_0^2 \tag{13.21}$$

where angular frequency:

$$\omega = 2\pi f$$

13.3.3 Elastic Energy in Dynamic Loading

The energy stored or work done on the polymer per unit volume of the polymer sample is calculated for the half-cycle, where for the first quarter of the cycle, the energy is being stored as elastic energy and in the next quarter, it is fully released. Therefore, the energy per half-cycle is given by:

$$E_s = \int_0^{\pi/2\omega} \sigma(t) \frac{d\varepsilon}{dt} dt$$

$$= \frac{\sigma_0 \varepsilon_0}{2} \cos \delta$$

$$\therefore E_s = \frac{\varepsilon_0^2 E'}{2}$$

Therefore, work done for a full cycle is double the above, or

$$E_{stored} = \varepsilon_0^2 E' \tag{13.22}$$

The above expression is also the area under a stress-strain curve in unidirectional deformation. The unit of E_{stored} is *Jules/m³/cycle* when the modulus is in *Pa* (N/m²), dimensions are in m, and the frequency is in cycles per second or *Hz*.

13.3.4 Temperature Dependence of Dynamic Viscoelasticity

The inherent mechanical strength of any material is a direct function of temperature and polymers are no exception. However, large size and complex intermolecular and intramolecular interaction in polymers result in very interesting and useful properties in practical applications. For polymers, the large molecules are supposed to be made up of various sizes of segments ranging from a length of 6 carbon atoms to about 50 carbon atoms. Therefore, the timescales of their response to dynamic forces are different, which means that the relaxation of a molecule as such is a combination of various relaxation times of segments with different sizes, thus resulting in a combined or integrated spectrum of relaxation. Even according to the Boltzmann Superposition Principle, the time history of relaxation in an experimental event is different for different sizes of segments, thus the relaxation phenomenon takes the integration of relaxation of all segments over a definite period with their respective time history.

When a polymer chain is not undergoing any net displacement or flow as a whole, even then the segments can undergo reptation type of motion. This segmental mobility freezes below T_g, at which temperature the polymer undergoes the transition from a soft rubbery state to the stiff glassy state upon cooling. The term glass transition originated from glass which is an amorphous supercooled liquid and becomes soft at a particular temperature. The glass transition phenomenon is exhibited by only the amorphous phase of a polymer.

The mobility of the smallest segment starts in the vicinity of T_g. Thus, at a temperature below the freezing of the smallest segment, the polymer would have maximum stiffness and hence maximum elastic modulus and minimum loss in cyclic deformation. A polymer will be more like an elastic solid. Since the segmental motion is enhanced due to thermal agitation, the segments will start moving as the temperature is increased beyond the freezing temperature of the smallest segment. This segmental motion results in loss of energy due to internal friction (viscous dissipation), with a time-lag between the applied force (stress) and the deformation (strain). Consequently, the elastic modulus will decrease with a growing loss of energy. The loss gradually grows as more and more segments join the coordinated movement as the temperature keeps on increasing. At T_g, the maximum number of segments participate in the coordinated oscillatory movement and the loss of energy would be maximum, which is indicated by a peak of the loss modulus (E''). At this point, the angular frequency (ω) of the dynamic force exactly coincides with the frequency of the relaxational movement of most segments and hence the relaxation time (τ) is just the inverse of the angular frequency of the applied dynamic force:

$$\omega \tau = 1$$

The above discussions indicate that the relaxation of segments of a polymer molecule is directly responsible for its dynamic mechanical properties and the relaxation is dependent on a change in temperature. The relaxation process of a segment is decided by its size and the thermal condition. The temperature dependence of the relaxation time (τ) can be described by the Arrhenius equation:

$$\tau = \tau_0 \exp\left(\frac{E_a}{RT}\right) \qquad (13.23)$$

where E_a is the activation energy of relaxation, R is the universal Gas Constant, τ_0 is a constant, and T is the temperature in Kelvin.

A different relationship of the relaxation time with temperature was given by Williams, Landel, and Ferry (WLF), assuming that the free volume of a polymer does not change below the glass transition temperature.[14] Therefore, the relationship given by the WLF equation is valid only above the glass transition temperature. The WLF equation can be written as:

$$\log(a_T) = \log\left(\frac{\tau}{\tau_g}\right) = \frac{C_{1,g}(T - T_g)}{C_{2,g} + (T - T_g)} \qquad (13.24)$$

where $C_{1,g} = -17.44$ and $C_{2,g} = 51.6$ when reference temperature $T_{ref} = T_g$ and valid for $T = T_g + 50$ and $C_{1,g} = -8.86$ and $C_{2,g} = 101.6$ when $T_{ref} = T_g + 50$ and valid for $T = T_g + 100$.

The above expressions also give an important relationship for dynamic viscoelasticity of polymers. Both Equations (13.23) and (13.24) relate the relaxation time to the temperature. Relaxation time at a particular temperature is the inverse of the angular frequency of the dynamic stress. Hence, these expressions relate time with temperature for the dynamic viscoelastic properties of a polymer. More descriptively, it can be said that if a dynamic property is known at a temperature and frequency, then it is possible to find at which frequency the material will have the same dynamic property at a reference temperature. The ratio of the relaxation times is the shift factor (a_T), which is used to shift the frequency from the temperature dependence data of a dynamic property, such as elastic modulus, loss modulus, dynamic viscosity or loss factor.

Temperature dependence of a typical viscoelastic material poly(butadiene) rubber vulcanizate with carbon black as reinforcing filler is shown in Figure 13.3. The spectra clearly show the glassy region below -85°C, glass transition at -85°C, a rubbery plateau from -60°C to -40°C, and flow region beyond -40°C.

In the glassy region, the polymer chains are frozen like supercooled liquid, possessing glass-like behavior. Stiffness, indicated by storage modulus E, is at its highest for the material in this region and damping is typically low. The transition region is so named because the material is undergoing the transition from the glassy to the rubbery region. It is in this area that the viscoelastic material goes through the most rapid change in stiffness and possesses its highest level of damping performance. In this region, the long molecular chains of the polymer are in a semi-rigid and semi-flow state and can rub against adjacent chains. These frictional effects result in the mechanical damping characteristic of viscoelastic material. In the rubbery region, the material reaches a lower plateau in stiffness. Damping is at a lower but reasonable level. In the terminal region, the complete chains have relaxed and therefore the storage modulus drops sharply and material transforms from rubbery to a viscoelastic liquid.

13.3.5 Frequency Dependence of Dynamic Viscoelasticity

The effect of changing the frequency of a dynamic load on viscoelasticity can be explained by the phenomenological relaxation theory, which states that the relaxation of chain segments takes place at a period depending on its thermodynamic state. Obviously, at a fixed temperature, the smallest segment would have the lowest relaxation time among all the segments and its movement would match the highest frequency of the external loading. Similarly, the longest segment would have very slow

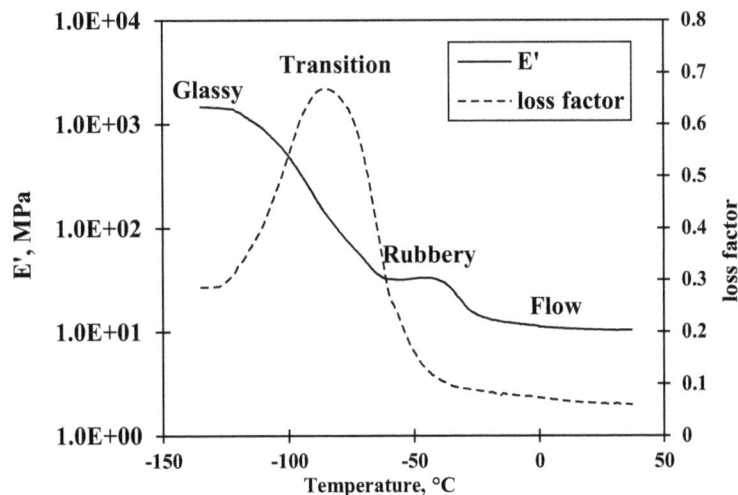

FIGURE 13.3 Dynamic viscoelasticity of poly(butadiene) rubber vulcanizate showing various regions of viscoelastic behavior.

relaxation with a large period that would match the lowest frequency of loading. Therefore, on increasing the frequency of loading from a very low value, the largest segments would start moving and since the other segments will relax very early, much before the time period of loading, the modulus will be the result of the contribution of only the larger segments. Hence, the modulus and loss both will be low. When more and more segments of different sizes join the coordinated movement with increasing frequency, the modulus would grow and so also the viscoelastic losses due to internal friction of chain segments. At a transitional point, most segments would participate in coordinated movement and hence the modulus will sharply increase and also the loss would attain maxima. The frequency at this point is termed the α-relaxation frequency of the polymer. Beyond this frequency, the segments would not match the short period of oscillation of the dynamic force and hence would behave as stiff elements, so the modulus would attain an upper plateau. As the movements of the segments would almost "freeze," the loss would be negligible. Therefore, on increasing the frequency, the modulus grows, which is a reverse phenomenon of increasing the temperature.

Considering the dynamic loading of a viscoelastic body, several simple mathematical models can be derived for the frequency dependence of dynamic viscoelasticity, provided the material is assumed to follow linear viscoelastic behavior. Many mathematical models have been developed using various spring-dashpot combinations, such as the two-parameter models of Maxwell, Kelvin-Voigt, three-parameter models like Zener, anti-Zener, four-parameter models such as Burger, etc. The most commonly applied model for dynamic viscoelasticity is the Zener model. However, four-parameter models were also used for soil movements in the very early years for prediction of oil drilling, rig construction, etc. The arrangement of spring and dashpot in this three-parameter model is shown in Figure 13.4 and the expressions for the *E'*, *E"* and *tan δ* with respect to frequency are also given. The model represents a viscoelastic body with Maxwell's arm and a spring in a parallel arrangement, which is a modification of Maxwell's model. The spring in the parallel arm represents residual stress after a long period of relaxation, which is observed for real viscoelastic materials.

The dynamic stress and strain are represented in Equations (13.11) and (13.12) above. The expressions for complex dynamic properties are given by:

$$E^* = E_1 + \frac{\omega^2 \tau^2 E_2}{1+\omega^2 \tau^2} + \frac{i\omega\tau E_2}{1+\omega^2\tau^2} \quad (13.25)$$

$$\therefore E' = E_1 + \frac{\omega^2 \tau^2 E_2}{1+\omega^2\tau^2} \quad (13.26)$$

$$E'' = \frac{\omega\tau E_2}{1+\omega^2\tau^2} \quad (13.27)$$

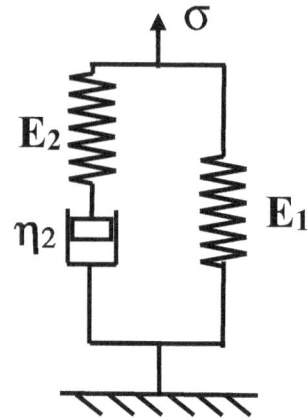

FIGURE 13.4 Zener model of three parameters representing a viscoelastic solid body.

$$\tan\delta = \frac{\omega\tau}{(E_1/E_2)(1+\omega^2\tau^2)+\omega^2\tau^2} \quad (13.28)$$

where E_1 is the elastic modulus of the arm with only spring in Figure 13.4, E_2 and η_2 are the elastic moduli of the spring and viscosity of the dashpot respectively in Maxwell's arm. Here, τ is the relaxation time, given by:

$$\tau = \frac{\eta_2}{E_2} \quad (13.29)$$

It can be seen from Equation (13.27) that the E''_{max} occurs at the condition of $\omega\tau = 1$, which was discussed above in explaining the relaxation phenomenon. Figure 13.5 shows the dependence of dynamic (storage) modulus of a Zener viscoelastic body according to Equation (13.26) and Figure 13.6 shows the loss modulus for the same body according to Equation (13.27).

13.3.6 Generalized Dynamic Viscoelastic Model

A general expression of stress and strain in an *n*-parameter model can be given as:

$$a_0\sigma + a_1\dot\sigma + a_2\ddot\sigma + ... + a_n\frac{d^{n-1}}{dt^{n-1}}\sigma = b_0\varepsilon + b_1\dot\varepsilon + b_2\ddot\varepsilon + ... + \frac{d^{n-1}}{dt^{n-1}}\varepsilon$$

where

$$\sigma = \sigma_0 e^{i\omega t} \quad \varepsilon = \varepsilon_0 e^{i(\omega t - \delta)}$$

$$\therefore \frac{\sigma}{\varepsilon} = \frac{a_0 + (i\omega)a_1 + (i\omega)^2 a_2 + ... + (i\omega)^{n-1}a_n}{b_0 + (i\omega)b_1 + (i\omega)^2 b_2 + ... + (i\omega)^{n-1}b_n} \quad (13.30)$$

The left-hand term is the complex modulus E^* as described earlier. The coefficients of the expression are to be determined by experimentation and hence it is progressively difficult to

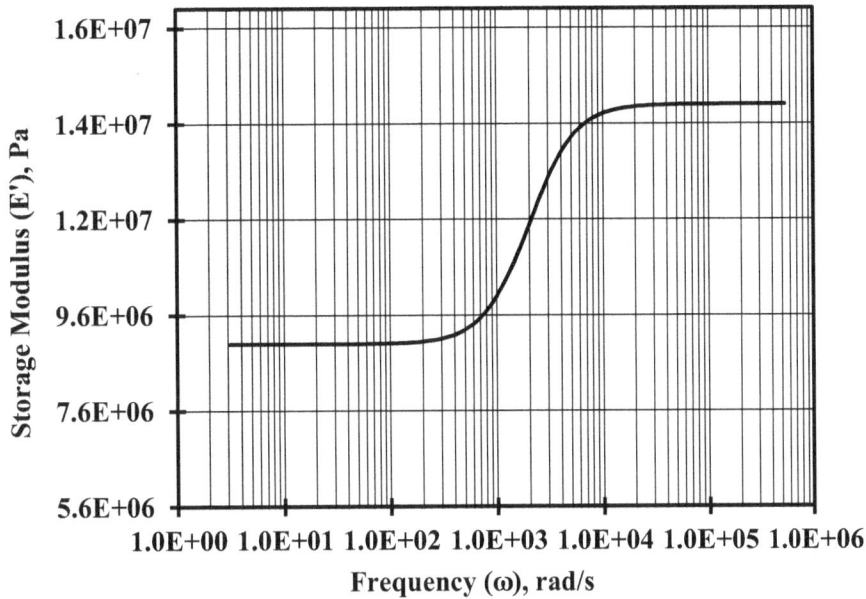

FIGURE 13.5 Storage modulus of a Zener viscoelastic body.

Note: Data generated for example only.

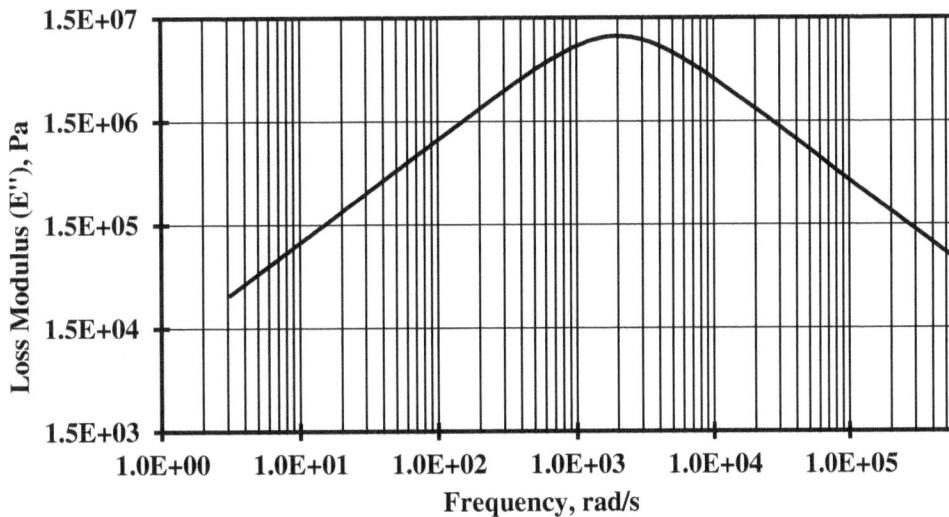

FIGURE 13.6 Loss modulus for a Zener viscoelastic body.

Note: Data generated for example only.

handle higher-order terms beyond 2 or 3. Also, the coefficients in higher-order terms must be negligible since at a sufficiently high frequency, for a real viscoelastic solid body, the complex modulus attains a constant value. This can be seen from the plot of complex modulus of nitrile rubber against frequency as shown in Figure 13.7, multiplexed at a reference temperature of 293K (20°C).

13.3.7 Combined Effect of Temperature and Frequency

As the effect of temperature on dynamic viscoelastic properties is reverse that of frequency, it is interesting to observe the shift of the α-relaxation transition (loss modulus peak) in temperature scale as the frequency is changed. Figure 13.8 shows such an effect for a cured epoxy resin measured from 60° to 100°C for the fixed frequencies: 0.2, 0.5, 1.0, 2.0, and 5.0 Hz. It is seen that the relaxation peak shifts to a higher temperature as the frequency is increased, and, hence, the relaxation time (inverse of relaxation frequency) is decreasing with increasing temperature as listed in Table 13.1. Besides, the nature of the curves also indicates that the material will be effective in damping at a higher temperature for higher frequency.

The activation energy of relaxation can be calculated using the Arrhenius Equation (13.23) from the plot of $ln(\tau)$ against $1/T$ as

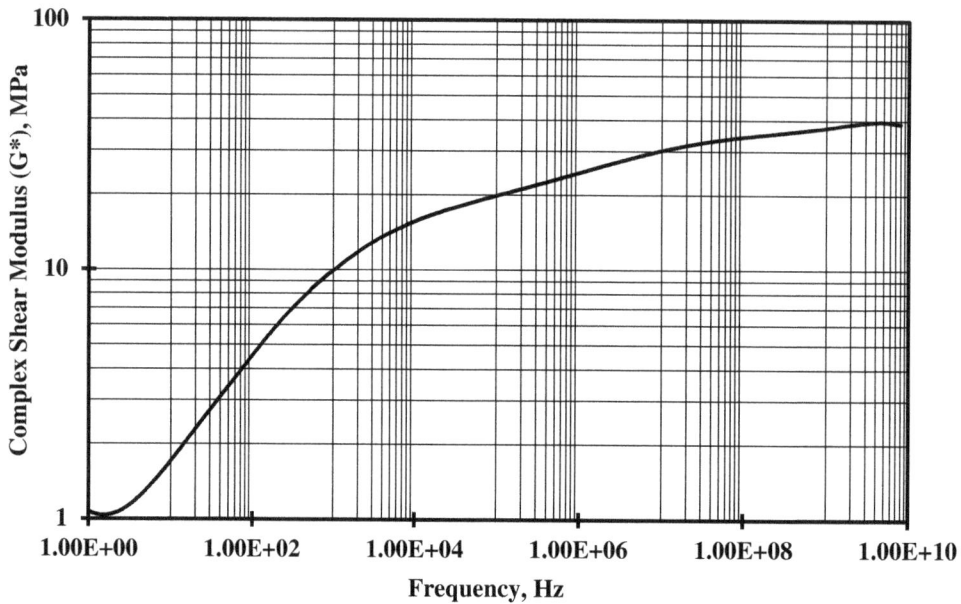

FIGURE 13.7 Frequency dependence of complex shear modulus of nitrile rubber at a reference temperature of 293 K.

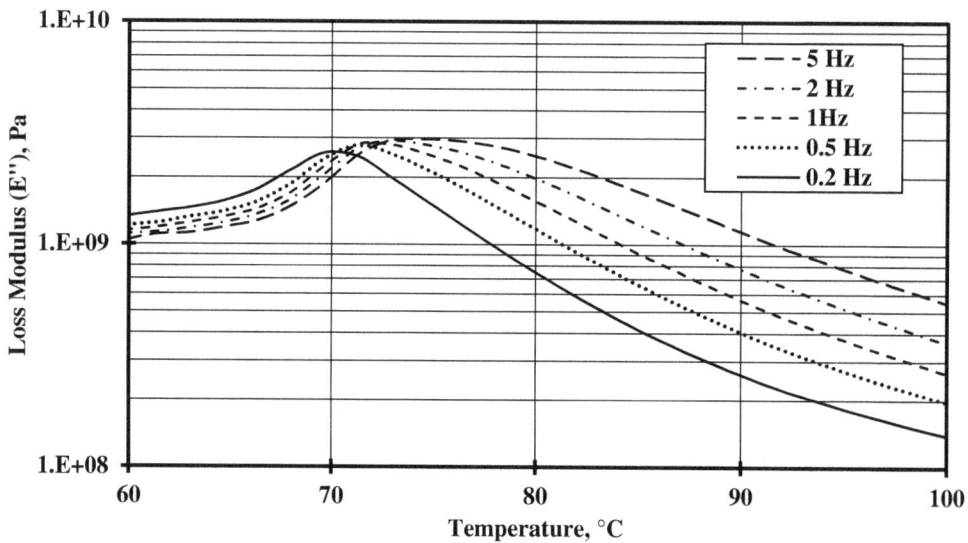

FIGURE 13.8 Loss modulus of epoxy resin at various fixed frequencies.

TABLE 13.1

Relaxation Time at Various Temperatures for An Epoxy Sample in Figure 13.8

f, Hz	T, K	τ, sec
0.20	342.60	0.80
0.50	344.30	0.32
1.00	345.00	0.16
2.00	346.10	0.08
5.00	347.69	0.03

shown in Figure 13.9. From the figure, the Activation Energy of α-relaxation of epoxy is about 640 kJ/mol. which implies that the relaxation process is very slow. This is reasonable for a cross-linked polymer having glass transition much above ambient temperature.

13.3.8 Time-Temperature Superposition

As described above, Equation (13.24) can be used to transform a temperature-dependent dynamic property to a frequency scale master curve if a set of data is made available in the temperature scale for various fixed frequencies as parameters. It may be observed that the effect of temperature is just opposite to that of frequency. To explain, it can be seen from the temperature scale data as in Figure 13.4, the storage modulus decreases with increasing temperature, but it increases with frequency as seen in Figure 13.6. The shift factors in Equation (13.24) for all temperatures above T_g with respect to a reference temperature (T_g +50) can be calculated to construct a master curve on a frequency scale. To construct a master curve, the dynamic mechanical properties of a polymer are evaluated on a temperature scale for various fixed frequencies and shift factors (a_T)

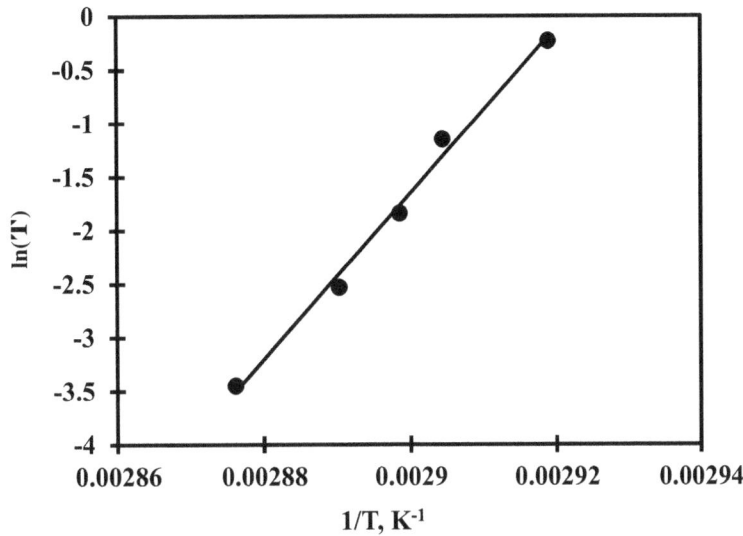

FIGURE 13.9 Arrhenius plot for α-relaxation of cured epoxy resin.

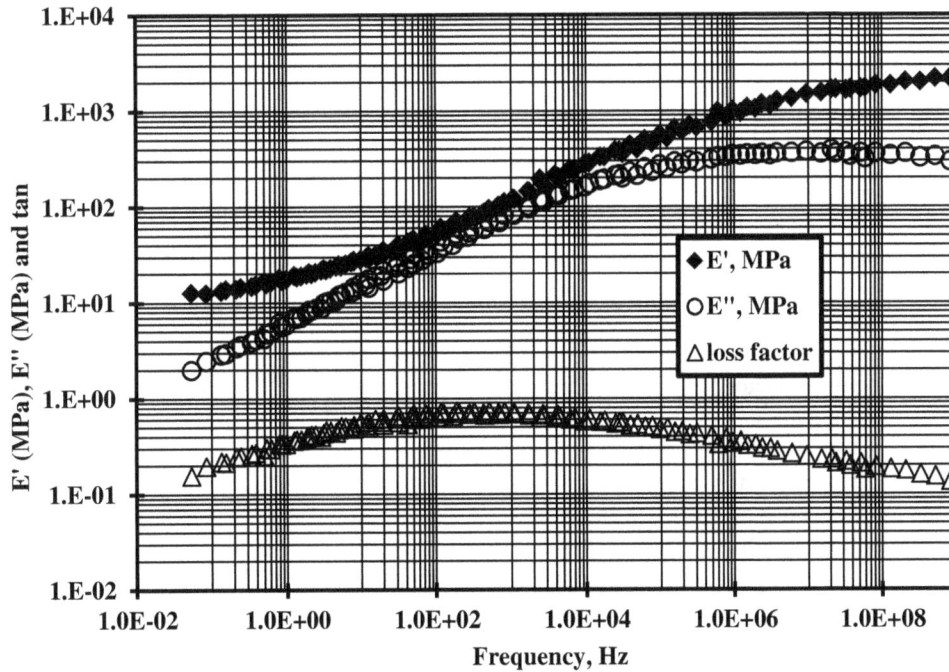

FIGURE 13.10 A typical master curve of dynamic properties of a rubber vulcanizate at 303K.

Note: T_g = 263 K.

are calculated at each temperature to shift the properties on the frequency scale using the appropriate C_1 and C_2 values. For most rubbers, the T_g value is below -10°C and the reference temperature is about 20–30°C. Hence the values C_1 = -8.86 and C_2 = 101.6 are mostly used for rubbers. In this case, T_{ref} = 303 K (30°C) taken for calculation of shift factors. Figure 13.10 shows a typical master curve on a frequency scale at a reference temperature of 30°C (303 K) generated from a temperature sweep of dynamic mechanical property of an elastomer (T_g = 263 K)

with fixed frequencies of 0.50, 0.80, 1.25, 2.0, 3.0, 5.0, 12.0, 15.0, 20.0, 30.0, and 50.0 Hz. Subsequently, the isothermal data (taken for various frequencies at each temperature) are plotted against the frequency scale. To shift the frequencies, the shift factor (taking an antilog value) of a particular temperature is multiplied by the test frequencies. Hence, a set of new frequencies is obtained, corresponding to test frequencies. This process gives the frequencies at which the dynamic property has the same values (measured at test frequencies) as that at the reference temperature. For example, at 10°C, the E' value at 0.5 Hz was 77 MPa. Correspondingly, the shifted frequency at the reference temperature (30°C) is 278 Hz. Hence, at 30°C, the E' value is 77 MPa at 278 Hz. This process is carried out for all isothermal data and finally, the master curves in frequency scale are constructed as in Figure 13.10.

The shifting of data can be done graphically also.[15] In this method, the isotherms (say, E') are plotted vs frequency, taking the temperature as parameters. The set of data at the reference temperature is kept fixed, and the remaining data are shifted to align on the reference line. The lower temperature data are shifted to the right side, i.e., the higher frequency and the higher temperature data are shifted to the left side, i.e., the lower frequency. The master curve thus obtained in the frequency scale represents the frequency-dependent property at a reference temperature chosen (for Figure 13.10, T_{ref} = 303 K).

Measurement of dynamic properties is done by a dynamic mechanical analyzer (DMA) instrument which can measure time-dependent force and deformation in various modes, such as tension, compression, bending, shear, etc. in pure sinusoidal wave form. Small samples are used in the instrument of limited force capacity (static + dynamic), typically about 10–15 N, although some instruments can go up to 500 N. Measurements are accurate for low frequencies not exceeding 50 Hz since a greater number of data can be taken for a given time of observation when the frequency is kept low. Some instruments can work up to 500 Hz, but multiplexing is not advisable when taking data at such high frequency. Extrapolation to a higher frequency using a shift factor would be more accurate with low-frequency data measured at a wide temperature range. The rate of heating is to be controlled at 2–3°C per minute to allow sufficient time for the polymer to attain the desired temperature. Soaking time at each temperature should be 1–2 minutes since polymers are generally a poor conductor of heat. The accuracy of the sample geometry would be critical to obtain correct results. Also, it is necessary to restrict the dimensions of the samples, depending on the mode of deformation, so that about 50 microns of deformation is obtained even when the material is stiff (well below its T_g). On the other hand, large deformations at the rubbery region should be avoided so that linearity in the viscoelastic property is maintained. This is necessary since the calculations of complex properties are based on the assumptions of the linear viscoelastic behavior of the materials. A scan of the strain vs storage modulus must be done before the measurement of dynamic properties, to find out the limit of strain up to which the modulus is constant as this signifies the linear region of viscoelasticity.

13.4 Polymer Materials for Vibration Damping

Every polymeric material exhibits some damping of vibration but to a different extent. The ability to absorb vibration and sound by a polymer is due to its viscoelastic loss property at the conditions of application, i.e., temperature and frequency. The viscoelastic dissipation of mechanical waves is dependent

on the modulus of the material and loss factor. System configuration and engineering design are the next important aspects. There have been, in the past, very comprehensive accounts of the dependence of chemical groups and their relative positions in a polymer chain on the damping ability of homopolymers, copolymers, and IPNs in relation to the area under the loss modulus spectrum in the temperature scale.[16–18] These findings are useful when selecting appropriate chemical groups and the corresponding polymer systems with dynamic properties favorable to damping at the application temperature and frequency band.

Mostly elastomeric materials and tailor-made blends and IPNs have been developed by various researchers, including the present authors, for vibration damping and sound attenuation.[19–34] Different fillers, such as fumed silica, carbon black, etc. as reinforcing agents[35,36] and plate types, such as graphite, mica, aluminum flakes, etc. are added to the virgin polymeric compositions to augment structural damping by enhancing the modulus and width of the loss peak. Recently, carbon nanotubes are being used to enhance the damping significantly.[32,33,37] Primary mounts of machinery require a balance between the loss factor and stiffness to reduce the intensity at resonance and provide shock mitigation. Therefore, tailoring of polymeric composition for mounts is different from surface damping such as for plates, beams, shells, frames, and contours. In the case of damping of the vibration of a plate or shell, for simplicity, the vibration damping material is brought into direct contact with a surface of the object to form a coating of single or multiple layers at a suitable thickness. The viscoelastic layer can be free on top, causing flexural damping or can be covered by a stiff material for shear damping. The damping treatment without a top constraining layer is termed free layer damping (FLD). The damping is caused by the flex extensional strain of the viscoelastic polymer. When a stiff constraining layer is added on top, the configuration is termed constrained layer damping (CLD). The damping takes place by shear deformation of the viscoelastic core in between the two stiff substrates.

A vibration damping material for use in the extensional type damping (FLD) is required to have high loss modulus, whereas a vibration damping material for use in the constrained type damping (CLD) is required to have high loss factor. In either case, it is also required that the loss properties be sufficiently high over a wide temperature and frequency range. Regarding the polymeric materials conventionally used for vibration damping, normally the glass transition region of each polymeric material is used since the loss modulus or the loss factor is largest in that area. To tailor material for the broad glass transition region, the following methods can be adopted: (1) blending of two or more kinds of polymers which are miscible with each other; (2) copolymerization; (3) synthesis of an interpenetrating polymer network of immiscible or partially miscible polymeric materials; and (4) use of nanocomposites. In the case of vibration damping materials for use in extensional type damping, often an inorganic powdery, granular, flaky, or fibrous filler is added to an organic polymeric matrix to obtain a composite material that exhibits a relatively large loss modulus.

However, the conventional vibration damping materials are still unsatisfactory or disadvantageous at certain points. As for the conventional damping material that uses only an organic polymer, whether a single kind of polymer or a combination, the loss modulus or the loss factor sharply decreases above the glass transition temperature or a low-frequency region. As for the conventional damping materials containing inorganic filler to use in the extensional type vibration damping, the actual effect of the filler is not as high as expected. Besides, an increase in the storage modulus following the addition of inorganic filler is accompanied by a decrease in the dynamic loss factor, so that the damping materials of the composite type are of little use for vibration damping of the constrained type. Short fibers and the recent development of nanomaterials such as CNT, nanoclays, graphene, etc. have proven to be more effective in enhancing damping properties, both in the case of CLD and FLD arrangements.

13.4.1 Rubbers

Rubber has many useful applications in engineering due to its special properties. Many important products that meet high requirements are made of rubber, such as tires, seals, gaskets, shock and vibration mounts, life-boats, conveyor belts, household goods, floor mats, shoes, bridge bearings, etc. To improve the properties, rubber is often reinforced by particulate fillers, such as carbon black and silica. For the filler, it is important that the particles are very small and possess a high specific surface area. For vibration and sound absorption applications, the elastomer provides both the spring and damper element. Dynamic stiffness and damping properties can be adjusted by modification or changing the viscoelastic material.

Table 13.2 lists the approximate loss factor maxima and the α-relaxation temperature at 10 Hz for a few common rubbers vulcanized using carbon black and also mineral fillers. Measurements are done by using the Dynamic Mechanical Analyzer in double cantilever mode (flexural mode). From Table 13.2, the high loss values of BIIR indicate its good damping capability, while natural rubber has the lowest loss factor among all, indicating its excellent resilient and lowest heat build-up property.

13.4.2 Polymer Blends

Since the maximum loss of mechanical energy takes place at about the glass transition temperature, none of these compositions above

can be used for the full potential of damping at ambient temperature and the desired range of frequency. A rather more realistic approach can be to use a judicious blending of elastomers and other polymers which could make a huge difference in the transition temperature, and at the same time are approximately miscible or compatible. Nitrile rubber-PVC is one such blend, which is miscible through the 20/80 to 80/20 ratio. A typical dynamic mechanical property on the frequency scale of an elastomeric composition based on nitrile rubber and poly(vinyl chloride) blend cured with sulfur and filled with carbon black is shown in Figure 13.11 (a) for the storage modulus (E') and Figure 13.11 (b) for the loss factor.

The glass transition by DMA of nitrile rubber c-black filled vulcanizate is about -22°C and that of PVC is about 84°C. Nitrile-PVC is a compatible blend and hence the glass transition of the blend will be in between these two individual temperatures. The DMA spectra show a single relaxation peak at about 485 Hz indicating compatibility of the blend constituents with the α-relaxation time of 3.28×10^{-3} seconds (calculated as $1/2\pi f$) at 20°C. The relaxation time is in the range of typical elastomeric material. However, unlike most elastomers, the relaxation peak appears at a temperature very near to the ambient and the peak is broader than c-black-filled NBR vulcanizate.

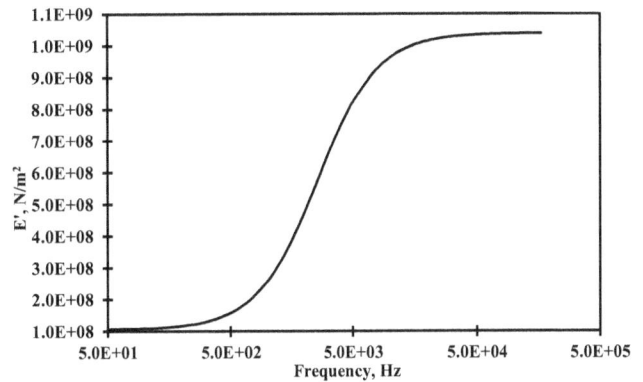

FIGURE 13.11(a) Frequency dependence of storage modulus of NBR-PVC composition.

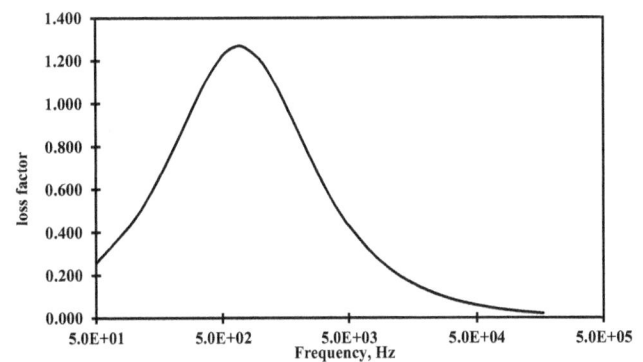

FIGURE 13.11(b) Frequency dependency of loss factor of NBR-PVC composition.

TABLE 13.2
Loss Factors of Common Elastomer Formulations

Rubber	Loss Factor Maxima	α-Relaxation Temp, K @ 10 Hz
NR + C-Black	0.12	247
PBR + C-Black	0.67	187
SBR +C-Black + clay	0.76	240
BIIR + TiO$_2$	0.92	248
EPDM + C-Black	0.32	249
PUR + C-Black + clay	0.85	246

13.4.3 IPNs

The next solution, more difficult but widely used in critical cases, is the formation of interpenetrating network polymers (IPNs), where two seemingly immiscible polymers with wide differences in relaxation temperatures, can be made as an intermingled network but with a separate cross-link structure. The intimate interaction of one component with the other makes the relaxation quite broad on both the temperature and frequency scales. Unlike immiscible blends, the IPNs do not show any split in relaxation peak but a continuous, broad hump.

Not only the relaxation is broadened, but also the overall dynamic mechanical properties, impact energy, durability, resistance to environmental stimuli, barrier properties, thermal stability, etc. are greatly improved. A large number of research papers and literature are available for many IPNs developed for various applications, especially damping of vibration and sound. A typical series of IPNs based on carboxylated nitrile rubber (XNBR) and poly (alkyl methacrylates) was developed by us[28] and evaluated for dynamic mechanical properties. The temperature scan of the loss factor shown in Figure 13.12 (a) indicates wide peaks for all IPNs XNBR-PMMA, XNBR-PEMA, and

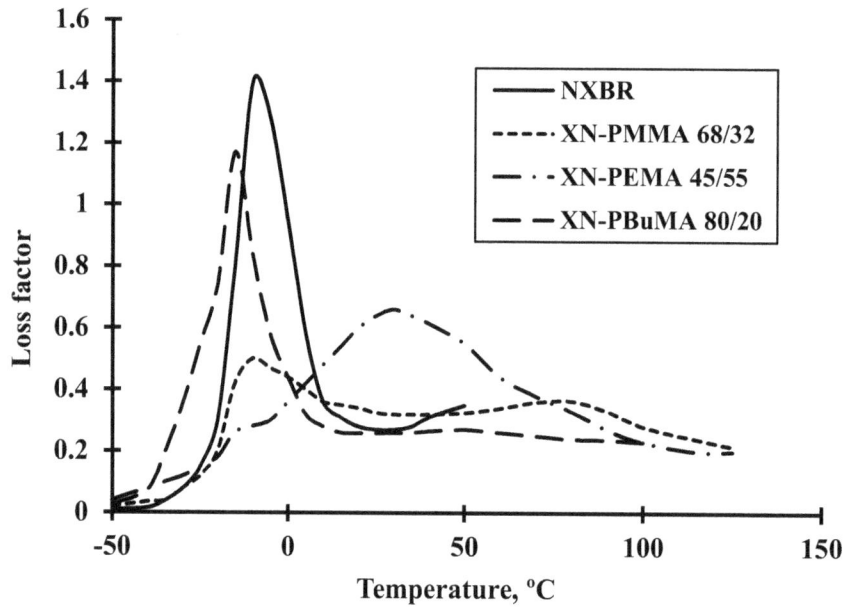

FIGURE 13.12(a)   Viscoelastic loss factor of XNBR-Poly(alkyl methacrylates) in the temperature scale.

Source: [28]. With permission from John Wiley & Sons, Copyright @2004 John Wiley & Sons.

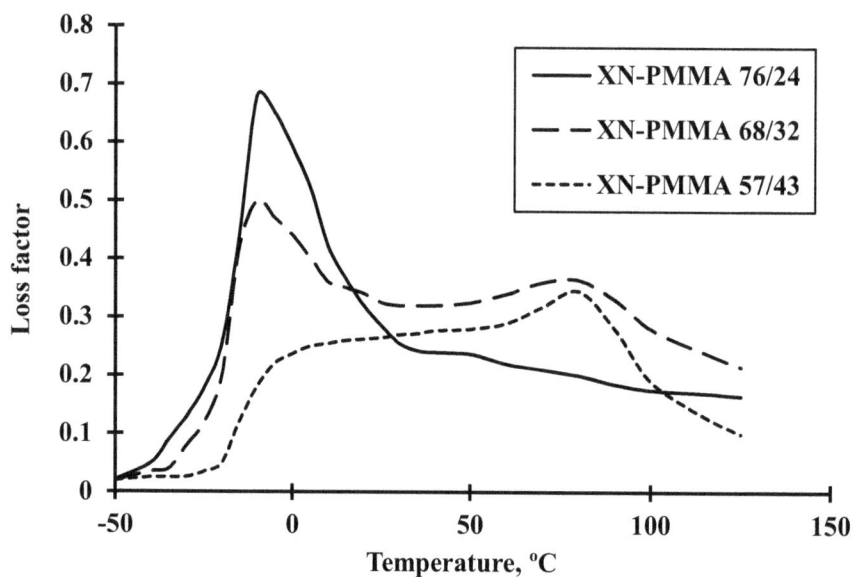

FIGURE 13.12(b)   Viscoelastic loss factor of XNBR-PMMA IPNs in the temperature scale.

Source: [28]. With permission from John Wiley & Sons, Copyright @2004 John Wiley & Sons.

XNBR-PBuMA ranging from -25°C to about 75°C compared to a narrow band of -25° to 12°C for XNBR. PMMA has a T_g of about 100°C compared to XNBR at -30°C, the IPN of these two polymers showed the broadest peak followed by XNBR-PEMA and XNBR-PBuMA. Similarly, such a broadening of the loss peak was observed with different proportions of XNBR and PMMA in XNBR-PMMA-based IPNs. Figure 13.12 (b) shows the loss factor peaks of three IPNs based on different combinations of XNBR and PMMA. It may be noted that below 30% PMMA content the breadth of the peak is less, although the peak height at 0°C for XNBR/PMMA = 76:24 is best. Although the numerical value of the loss factor alone cannot be considered for damping performance, it needs to be as high as possible in the frequency and temperature range of the application.

13.5 Elastomer-Filler Interaction

A characteristic impedance between the matrix material and embedded materials can enhance the attenuation of acoustic/vibrational waves in a structure. One means of enhancing the vibration damping of a material is by embedding high characteristic impedance particles, low characteristic impedance particles, or both high and low characteristic impedance particles within the matrix of the material. When particles with mismatched characteristic acoustic impedances are embedded within the matrix of a material that can support shearing loads, propagating acoustic energy that encounters the particles is partially reflected in random directions. That is, the propagating energy is diffused. As the propagation vectors and modes of vibration energy are effectively randomized, the probability of localized energy absorption and damping is increased.[38]

The combination of a polymeric matrix material with nanofillers and/or fillers with anisotropy (graphite flake, short fiber filler) can variously be chosen after consideration of several factors such as the vibration damping performance and other properties required in application environments.[32,33,35–48]

13.5.1 Effect of Filler Type on Tan δ_{max} Temperature

It was mentioned in Section 13.5 that the vibration damping behavior of a polymer depends on frequency and temperature. Praveen et al. studied the effect of filler geometry on the viscoelastic damping of SBR composites.[43] Figure 13.13 and Figure 13.14 show the variation of tan δ and storage modulus respectively with temperature for PU with 10 phr of individual filler.

Graphite, being a semi-reinforcing filler, has little interaction with the matrix to affect the relaxation mechanism of the matrix. As a result, the magnitude of the peak loss factor and the rubbery storage modulus remain unaffected by the graphite filler. Whereas in the case of short fiber composites and nanocomposites, the magnitude of the peak loss factor showed a significant reduction and an increase in the rubbery storage modulus, indicating better filler-matrix interaction. Figure 13.15 shows the loss factor vs frequency master curve for PU composites at a reference temperature of 25°C. The magnitude of the loss factor showed a similar trend to that observed in the temperature sweep results. However, the loss factor vs frequency graph for nanoclay-filled PU showed a significant shift toward higher frequencies. The phenomenon shows that the presence of nanoclay (platelet filler) allows for a faster relaxation of polymer molecules, leading to a shift in α-relaxation toward higher frequencies. This effect is observed in the case of thin films and structures having considerable filler-filler interaction. Quantitative equivalence between polymer nanocomposites and thin polymer films as regards the polymer glass transition has been studied by Praveen et al.[45] and Bansal et al.[41,42] The changes in the glass transition temperature with converging interparticle spacing for two filler surface treatments are quantitatively equivalent to the corresponding thin-film data with a non-wetting and a wetting polymer-particle interface.

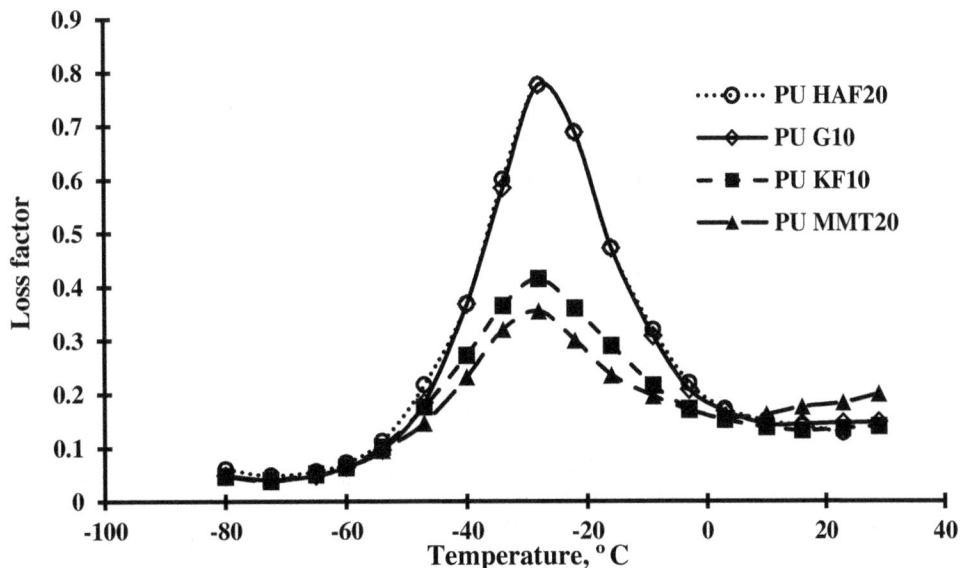

FIGURE 13.13 Effect of filler type on tan δ_{max} in temperature scale for PU.

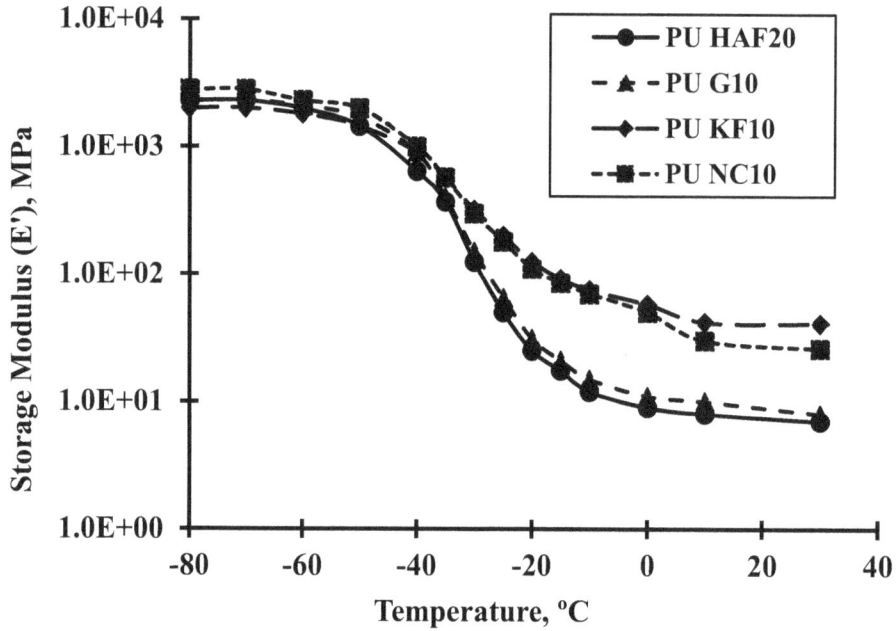

FIGURE 13.14 Effect of filler type on the shift of storage modulus for PU.

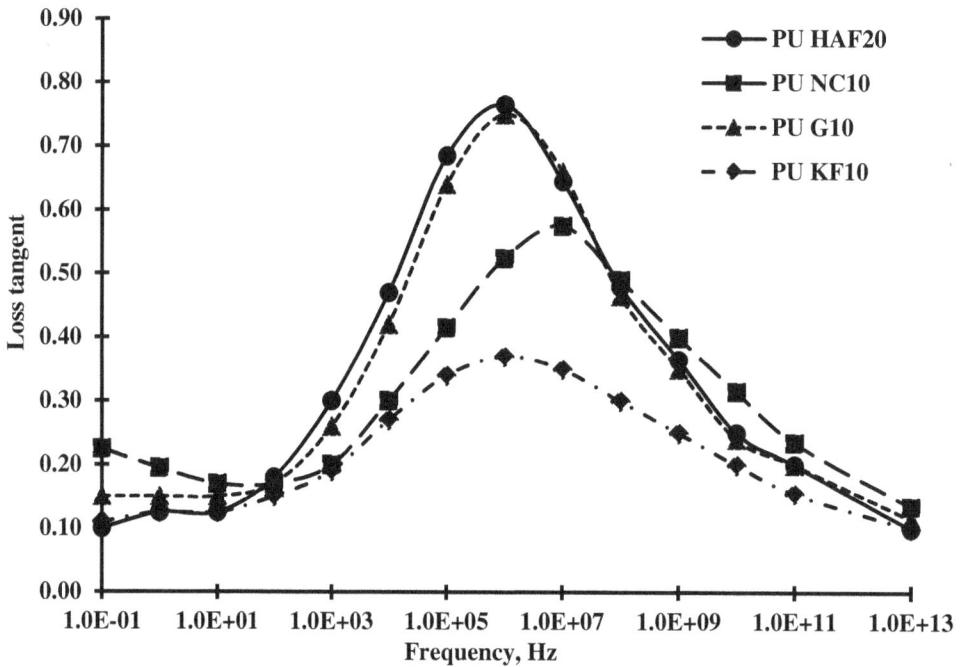

FIGURE 13.15 Loss tangent vs frequency master curve for PU composites at 25°C reference temperature.

13.6 Hysteresis

Historically, tan δ has been most frequently used as a relevant dynamic property of the material for hysteretic energy-loss processes. However, the dynamic property that gives the best correlation depends on the type of deformation applied to the material.[43] Theoretically, the energy loss of the material under cyclic deformation is described in Equation (13.20) as Jules/m³/Hz and by Equation (13.21) as Jules/m³/sec. For equal strain, E_{loss}

would be proportional to loss modulus and for constant stress deformation, it would be proportional to the loss compliance.[49]

Table 13.3 shows the compression hysteresis data of a black filled PU rubber composite with different fillers measured at constant strain deformation. The nanoclay filled system showed higher hysteresis followed by aramid short fiber and graphite. The morphology of PU nanocomposites was studied extensively by Praveen et al.[46] The TEM micrographs of PU nanocomposites

TABLE 13.3

Compression Hysteresis of PU Composites with Various Fillers

Type of Filler	Filler Loading (phr)	Storage Modulus (MPa)	tan δ Half Peak Width, °C	tan δ at 30°C	tan δ_{max}	Energy Loss J/m³ x10²	Hysteresis (%)
HAF 20	20	6.68	28	0.12	0.77	62.9	28.0
Graphite	10	7.8	28	0.15	0.77	91.8	30.73
	20	15.6	28	0.16	0.73	195.9	31.07
Aramid short fiber	5	25.4	31	0.16	0.53	319.0	31.71
	10	42.25	36	0.13	0.42	413.0	33.00
Nanoclay with black	5	12.61	28	0.16	0.62	158.3.0	41.40
	10	26.1	31	0.18	0.49	368.0	44.40
	20	64.73	36	0.20	0.37	1070.0	54.80

revealed a high level of intercalation. The clay layers appear to form long continuous structures in between the spherical carbon black particles.

The platelet nature of the nanoclay filler enables the formation of layered silicate polymer sandwich structures. The sandwich structures in a rubber matrix impart shear deformation on the intercalated polymer molecules, thus enhancing the energy dissipation under compressive deformation. Praveen et al.[45] reported compression hysteresis of SBR composites at constant deformation. Unlike PU composites, in the case of the SBR matrix, maximum energy dissipation was observed in aramid short fiber filled SBR. High compression hysteresis in short fiber filled SBR was attributed to a combination of (1) good adhesion at low strain level which increases the shear stress transfer between the fiber and matrix, thus enhancing the energy dissipation; and (2) weak adhesion at higher strains which induces mechanical friction at the interface. The studies show the influence of the matrix type on the energy dissipation characteristics of the filled polymer system.

13.7 Methods of Vibration Damping

Viscoelastic materials have found an application in vibration damping in a myriad of configurations. Two general treatments that are widely used are free layer (FLD) or unconstrained layer damping (UCLD) and constrained layer damping (CLD). These are sometimes referred to as "extensional" and "shear" damping respectively.

13.7.1 Free Layer Damping (FLD)

In this arrangement, the free viscoelastic layer is bonded to the substrate to be damped. As the plate undergoes bending, the viscoelastic material (VEM) layer is deformed, principally, as flex-extensional mode in planes parallel to the substrate surface. A typical FLD configuration is shown in Figure 13.16.

The strain in the layer is proportional to the local curvature of the plate so that the energy is proportional to the local curvature of the plate. Therefore, for full coverage by the VEM layer, the damping performance is independent of the mode of

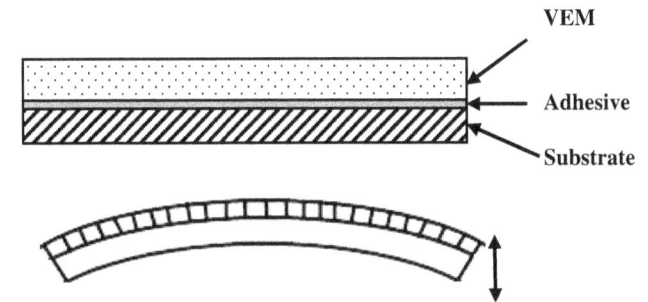

FIGURE 13.16 Typical free layer damping system.

the vibration. This fact simplifies the specification of treatment for wide frequency coverage (as there is no inherent frequency or mode shape dependence). The system loss factor of an FLD increases with:

1. the thickness ratio of the VEM to the substrate;
2. the ratio of Young's modulus of the VEM layer to that of the substrate;
3. the loss factor of the VEM layer.

13.7.1.1 Mathematical Models for Prediction

A large number of articles have been published by various researchers on mathematical modeling of FLD performance to predict the extent of attenuation by passive damping materials such as VEM on beams, plates, rings, vehicle bodies, and also for cylindrical shells.[50–60] The composite loss factor, which is also termed the system loss factor, is taken as a measure of damping in all such analysis.

There seems to be no direct relationship between the system loss factor and measured damping in terms of attenuation of vibration intensity in the decibel scale. Most of the models predict almost identical behavior in damping in the frequency scale. The most commonly used equation for an FLD treated beam is the Ross-Kerwin-Unger (RKU) equation[51] (Equation 13.31):

$$\eta_s = \eta_2 \frac{e_2 h_2 (3 + 6h_2 + 4h_2^2 + 2e_2 h_2^3 + e_2^2 h_2^4)}{(1 + e_2 h_2)(1 + 4e_2 h_2 + 6e_2 h_2^2 + 4e_2 h_2^3 + e_2^2 h_2^4)} \qquad (13.31)$$

where

η_2 = VEM loss factor,
h_2 = thickness ratio VEM: Substrate
e_2 = modulus ratio VEM: Substrate.

Oberst and Frankenfeld[50] derived a simple expression for the system loss factor from the ratio of complex stiffness of the composite beam to the stiffness of the base substrate as given in Equation (13.32) and Equation (13.33):

$$\frac{(EI)^*}{E_1 I_1} = 1 + e*h^3 + 3(1+h)^2 \frac{e*h}{1+e*h} \quad (13.32)$$

and

$$\eta_s = \frac{\mathrm{Im}(EI)^*}{\mathrm{Re}l(EI)^*} \quad (13.33)$$

where E_1 is Young's modulus and I_1 moment of inertia of the base respectively, $e*$ is the ratio of Young's modulus of VEM to base, h is the thickness ratio (VEM: Base), $(EI)^*$ is the complex stiffness of the composite beam. The complex stiffness has real and imaginary parts, hence the ratio of the imaginary to real component gives the system loss factor (η_s), as shown in Equation (13.33).

There are several such expressions with one side coverage by Nashif et al.[53] and Nakra[54] and both side coverages by Torvik.[57] However, for a beam vibration, the expressions by most researchers predict almost the same system loss factor which can be seen by a numerical calculation for a chosen system. There are studies on the prediction of damping using VEM as FLD which are based on the Finite Element Method. A few are listed in the references.[61–66] Most researchers have used the modal strain energy ratio of the composite plate/beam to that of the base plate/beam to find out the system loss factor.

In all these predictive models, the thickness ratio and the modulus ratio are deciding factors for the extent of damping using the same VEM. Since almost all the VEMs have quite a low elastic modulus compared to metals, the VEM modulus should be as high as possible without compromising the loss factor to achieve a reasonable value of system loss factor.

Figure 13.17 shows the combined effect of thickness ratio and modulus ratio on the composite loss factor (as the ratio with VEM loss factor) in an FLD treated beam, calculated as an example. It can be observed that with the storage modulus of 100 MPa of a VEM coated on a steel beam having Young's modulus of 200 GPa, the modulus ratio (e) is 0.0005 and with a thickness ratio (h) of 2, the system loss factor would be about 0.1 if the VEM loss factor is 1.0. Also, the damping increases asymptotically as the thickness ratio increases. Since the storage modulus of VEM improves with frequency, the system loss factor would also improve for the same thickness ratio. If the VEM is designed in such a way that the storage modulus and loss factor both are high in the low-frequency range, the FLD treatment would provide considerable damping.

Further, it can be seen that the system loss factor attains a constant value about the thickness ratio of 2. This implies that it is best to use a thickness maximum up to twice that of the substrate to arrive at a reasonable compromise between the weight penalty and damping. For substrates which are lower in elastic modulus, such as aluminum, the thickness can be less even with a softer VEM to obtain reasonable damping compared to steel substrates.

13.7.1.2 Example of an FLD Treatment

A steel plate of 6 mm thickness has to be treated with an NBR-PVC blend with dynamic mechanical properties as shown in Figures 13.11 (a) and 13.11 (b). The damping is to be predicted

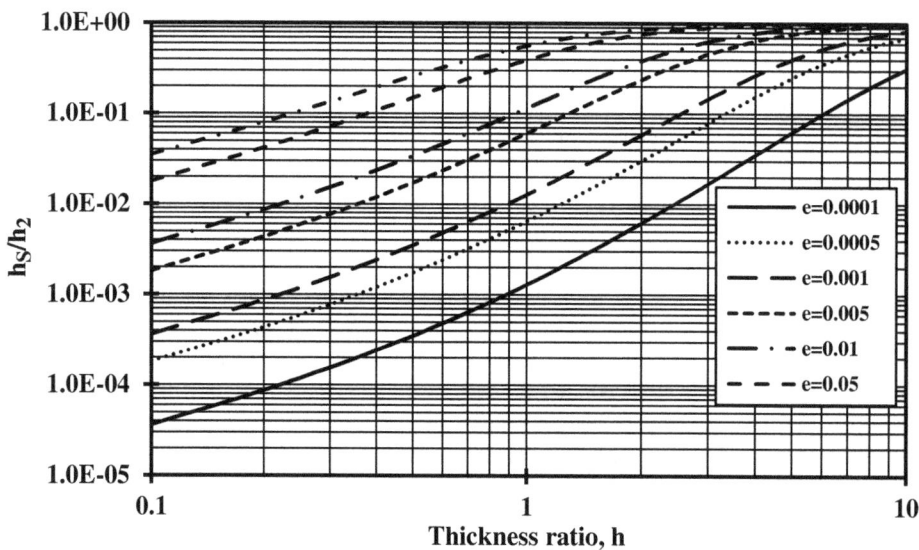

FIGURE 13.17 Dependence of system loss factor of an FLD on thickness ratio and modulus ratio.

Note: Data calculated using Equations (13.32) and (13.33).

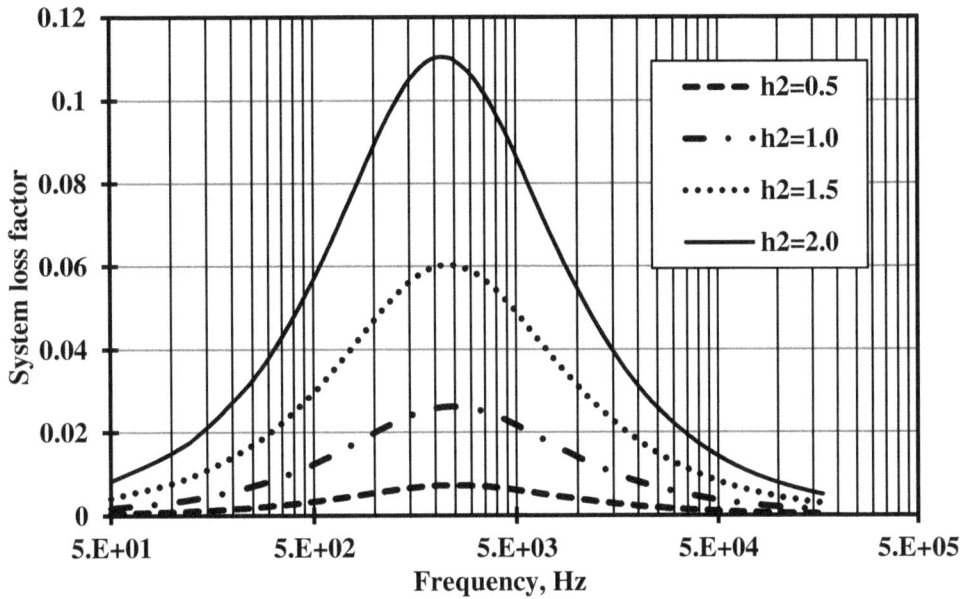

FIGURE 13.18 System loss factor of an FLD in Example 13.7.1.2 at various thickness ratios.

by a mathematical tool as described by either Equations (13.31) or (13.32). The effect of the thickness ratio on the system loss factor is to be found for an optimum coating with a maximum allowable weight penalty of 50%.

Solution: the data on dynamic modulus and loss factor in the frequency scale were extracted from Figures 13.12 (a) and 13.12 (b) respectively. Using Equation (13.31), the predictive system loss factors with respect to frequency for four cases of $h_2 = 0.5, 1.0, 1.5$, and 2.0 were calculated. The results are shown in Figure 13.18.

It can be seen that the system loss factor is sharply increased with the increase in thickness ratio and for each ratio, the peak damping is at the same frequency (approximately at 900 Hz), obviously due to the highest loss modulus of the VEM at that frequency. There is a drastic difference in the damping for the thickness ratios below and at 2. The peak damping at a ratio of 2 is almost double than that at 1.5. The system loss factor is reasonably good, about 0.08 and above for a thickness ratio of 2, at a frequency band of 600–5000 Hz, which covers the mid-frequency range. Considering the density of NBR-PVC material as 1200 kg/m³ and that of steel as 7846 kg/m³, the weight penalty for a thickness ratio of 2 is about 30.5%, which is less than the 50% restriction as stated in the example in Section 13.7.1.2.

13.7.2 Constrained Layer Damping (CLD)

The simplest constrained layer configuration is a three-layer laminate comprising the base layer to be damped, the VEM layer, and a stiff constraining layer, as shown in Figure 13.19. The principal deformation of the VEM occurs as shear, as the composite undergoes transverse (bending) motion. In bending vibration, this three-layer system exhibits an inherent dispersion from coupled bending for large wavelengths (low frequencies) to uncoupled bending for small wavelengths (high frequencies). In these low and high frequencies regimes, the elastic energy of deformation

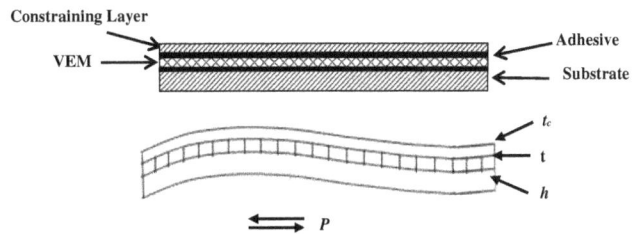

FIGURE 13.19 Constrained layer damping arrangement.

lies almost entirely in the bending and extension of the base and constraining layers. Thus, in these limits of very low and very high frequency, the VEM cannot bring about significant damping of the system. In coupled bending, the wavelength is long enough so that the shear stiffness of the VEM can cause the base and constraining members to stretch and compress rather than allowing significant shear strain in the VEM. At sufficiently large wavelengths, the composite bends as though it were perfectly bonded together by a central layer with infinite shear stiffness.

In contrast, for short wavelengths, the shear stiffness of a half wavelength of the VEM is small, relative to the extensional stiffness of the base and constraining layers. In the limits of smaller wavelengths, the laminate behaves as if the two outer layers were joined by a central layer of vanishing stiffness.

In the mid-frequency region, where the elastic energy in the VEM can represent a useful fraction of the total, the system loss factor rises and passes through a maximum. For a given amplitude of bending deflection and curvature, the strain amplitude at any point in an unconstrained layer can be prescribed. The cyclic energy dissipated per unit volume is given by $\pi \varepsilon_0^2 E''$, so the unconstrained layer of damping materials which dissipates most energy is that which has the highest loss modulus E''. The same cannot be said of the material for a constrained layer. In this case, there is an optimum value of its storage shear modulus (G'), required for most energy to be dissipated and which depends on the elastic properties and thickness of each layer in the configuration. If G' is too high, very little shear strain can occur and very little energy is dissipated. Between these extremes there exists an optimum modulus which maximizes the energy dissipation. The storage moduli of the materials normally used in thin constrained layers are much lower than those of materials used in unconstrained layers, but their loss factors are usually considerably higher.

13.7.2.1 Types of CLD Arrangement

A lot of variations are possible when constructing a CLD using various constraining layers and intensifier layers, such as the use of magnets to enhance transverse deformation, the use of a piezo-electric constraining layer with a passive R or R-L or R-L-C circuit connection so that the vibration of the piezo layer produces an electrical charge and is dissipated by the loads. In another arrangement, the VEM layer is loaded with permanent magnetic powders, and the requirement of the adhesive is eliminated for steel substrates. Active-passive combinations of multilayer CLDs are also used, wherein the cyclic deformation is taken to feed a reverse oscillatory input to neutralize the vibration, either by a feedback or feed-forward system. The VEM layer in between the piezoelectric layers is required to cater for high frequencies, typically above 400 Hz, since active controls are not suitable beyond that range. Interestingly, despite all the possible technologies to enhance CLD performance, thin layer CLDs made from highly damping polymers are best in terms of simplicity in application, maintenance, and cost.

13.7.2.2 Mathematical Models for Prediction

A large number of works have been published on predictive models of CLD performance using viscoelastic materials. The first comprehensive work was done by Ross et al.[51] and Kerwin,[67] way back in 1959. The Ross-Kerwin-Unger (RKU) equation is mostly used to predict the CLD performance of flat plates. Nakra[68–70] and Asnani et al.[71] have given simple expressions for full CLD on beams and plates. Rao[72] derived some more precise equations for other conditions which can be solved by numerical methods. Partial coverage as patches on beams, plates, and cylindrical shells has been studied and system loss factors for different locations at the various extent of coverage are compared to arrive at optimal solutions.[73–80] Presently finite element methods based on modal strain energy calculations are used for complicated structures.

For the CLD arrangement described as Figure 3.19, the RKU equation gives an expression for the complex rigidity of the composite beam (beam with CLD treatment) as follows:

$$(EI)^* = \frac{E_1 h^3}{12} + \frac{E_2^* t^3}{12} + \frac{E_3 t_c^3}{12} - \frac{E_2^* t^2}{12}\left(\frac{H_{31}-D}{1+g^*}\right) + E_1 h D^2$$

$$+ E_2^* t (H_{21}-D)^2 + E_3 t_c (H_{31}-D)^2 - \left\{0.5 E_2^* t (H_{21}-D)\right.$$

$$+ E_3 t_c (H_{31}-D)\bigg\}\left(\frac{H_{31}-D}{1+g^*}\right) \qquad (13.34)$$

where E_1 and E_3 are Young's moduli of substrate and CL, E_2^* is the complex tensile modulus of VEM, h, t and t_c are thicknesses of the substrate, VEM layer and CL respectively. D, H_{21}, H_{31}, and g^* are defined as:

$$g^* = \frac{G_2^* \lambda^2}{E_3 t t_c \pi^2} \qquad (13.35)$$

where

$$D = \frac{E_2^* t (H_{21}-0.5H_{31}) + g^* (E_2^* t H_{21} + E_3 t_c H_{31})}{E_1 h + 0.5 E_2^* t + g^* (E_1 h + E_2^* t + E_3 t_c)}$$

and

$$H_{31} = t + 0.5(h + t_c) \qquad (13.36)$$

The quantity λ is the semi-wavelength of the sinusoidal waveform of the beam under dynamic flexural deformation (vibration). The parameter g* is the shear parameter, which is a complex number. It can also be expressed by

$$g^* = \frac{G_2^* L^2}{E_3 t t_c \xi_n^2 \sqrt{C_n}} \qquad (13.37)$$

where ξ_n is expressed as:

$$\xi_n^4 = \frac{\rho_1 b h \omega_n^2 L^4}{E_1 I_1} \qquad (13.38)$$

where ω_n is the nth modal frequency of the beam and C_n is a correction factor as suggested by Rao[72] for different boundary conditions. However, the value of C_n is 1 or nearly 1 for most boundary conditions such as pinned-pinned, clamped-free, clamped-clamped, etc.

The ratio of the imaginary part of the rigidity to the real part from Equation (13.34) will give the system loss factor η_s, which is a measure of damping.

Nakra[68–70] and Asnani et al.[71] derived mathematical expressions for both natural frequencies and the system loss factor of a sandwich beam and plate vibrations. Nakra suggested an expression for the system loss factor for a simply supported beam with a constrained layer damping arrangement. The expressions are relatively simpler to calculate and are in good agreement with experimental results. The expressions for shear parameter and the system loss factor are given as Equations (13.39) and (13.40):

$$\text{shear parameter: } \psi_{23} = \frac{G_2}{E_3 h^2 \left(\frac{n\pi}{L}\right)^2} \qquad (13.39)$$

$$\text{system loss factor: } \eta_s = \frac{H \eta_0}{(1 + \alpha_{13} H_{13}^3)\left(M^2 + \eta_0^2\right) + HM}$$

$$(13.40)$$

where n = modal number, L = length of sandwich beam, G_2 and η_0 = storage shear modulus and loss factor of VEM, α_{13} = E_1/E_3, $H_{13}=t_c/h$, $H_{23} = t/h$ and E_i = Young's modulus of ith layer.

Here, the substrate is layer 3 with thickness = h, Young's modulus = E_3, VEM is layer 2 with thickness = t, dynamic shear modulus = G_2, and CL is layer 1, with thickness = t_c. Young's modulus = E_1. The parameters H and M in Equation (13.40) are given by:

$$H = \frac{3\psi_{23}}{H_{23}}\left(1+H_{13}+2H_{23}\right)^2\left(1+\eta_0^2\right) \quad (13.41)$$

$$M = 1 + \frac{\psi_{23}}{H_{23}}\left(\frac{1}{\alpha_{13}H_{13}}+1\right)\left(1+\eta_0^2\right) \quad (13.42)$$

The system loss factor (η_s) is maximum at an optimum value of the shear parameter:

$$\psi_{23,opt} = \sqrt{\frac{A_1\left(1+\eta_0^2\right)}{A_1C_1^2+B_1C_1}} \quad (13.43)$$

where

$$A_1 = 1+\alpha_{13}H_{13}^3 \quad B_1 = \frac{3}{H_{23}}\left(1+H_{13}+2H_{23}\right)^2\left(1+\eta_0^2\right)$$

and

$$C_1 = \frac{1}{H_{23}}\left(\frac{1}{\alpha_{13}H_{13}}+1\right)\left(1+\eta_0^2\right)$$

A typical CLD performance by the same NBR-PVC composition is calculated as an example in Section 13.7.2.3. Dynamic viscoelastic properties are taken from Figures 13.12 (a) and 13.12 (b).

13.7.2.3 Example

Let us consider a typical example of a clamped-free bare beam of mild steel (MS) to be treated with a VEM and an MS constraining layer (CL) with the following parameters:

L = 180 mm, h = 6 mm, t = 2 mm, t_c = 2 mm

$H_{13} = t_c/h = 0.33$, $H_{23} = t/h = 0.33$

$\alpha_{13} = E_1/E_3 = 200/200 = 1.0$

$E_1 = 200$ GPa = 200×10^3 N/mm^2

$E_3 = 200$ GPa = 200×10^3 N/mm^2

G' (G_2) and η_0 = as shown in Figure 13.12 (a) and 13.12 (b).

The CLD performance was predicted using Equations (13.39–13.42) for natural frequency modes 1, 2, and 3. The results are shown in Figure 13.20. The quantitative value of the damping is progressively lower as predicted here from mode 1 to mode 3. Also, it can be seen from Figure 13.20 that there are maxima in the damping performance (for any mode) in the CLD

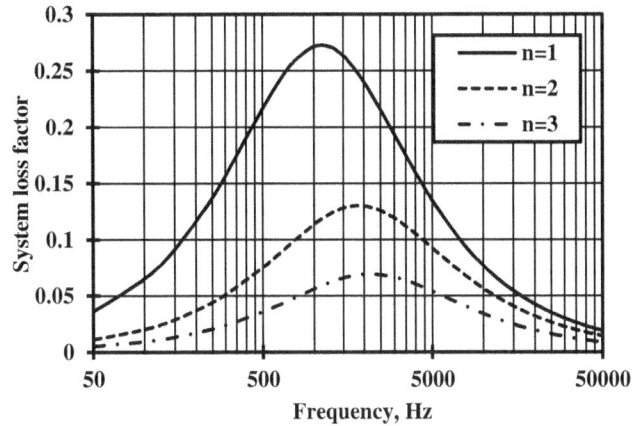

FIGURE 13.20 CLD performance of the VEM coating on MS beam for three modes using Equations (13.39–13.42) for the example in Section 13.7.2.3.

arrangement. The shear parameters for these three modes are shown in Figure 13.21. The shear parameter is lower for higher modes, in fact, it is inversely proportional to n^2 but directly proportional to G_2 (shear modulus of VEM). The system loss factor is dependent on the shear parameter and hence the mode-1 values of loss factor are the highest.

In a practical application, higher frequency intensities are always less and hence the somewhat lower system loss factor is acceptable. However, high strain rate fatigue for especially thin panels of automobiles, aircraft frames, etc. require sufficient damping at high frequency.

The damping in the first mode was also predicted for a thickness ratio of VEM to the substrate (H_{23}) = 0.5 but with the same ratio of constraining layer to the substrate (H_{13} = 0.33). Figure 13.22 shows the system loss factor for H_{23} = 0.33 compared to H_{23} = 0.50, predicted by the same equations as above. As expected, the damping was higher due to a higher thickness of the VEM. However, there is not much improvement in the loss factor when the weight penalty is considered critical.

13.7.3 Smart Damping by Shunt: CLD

Vibration damping can be enhanced for a CLD arrangement by augmenting the second type of attenuation method. A piezoelectric layer is applied on the VEM layer as a constraining layer and as the vibration takes place, the electrical charge generated in the piezo layer is conducted to a passive shunt circuit and is dissipated as heat.[81–84] The shunt can be only a resistor, or a resistor-inductor or a resistor-inductor in parallel to a capacitor. The best performance is obtained by tuning R-L-C circuit parameters according to the vibration intensity and frequency band. The advantage of this arrangement is that it does not depend on the complicated active signal drive and hence can be used for the wider frequency range. The only disadvantage is the brittleness of the piezoelectric ceramics in such a layer application. Hence, for a larger system, it is not suitable. A series of good piezoelectric ceramics, such as PZT, PLZT, PMN-PT, and PZT-Polymer 1–3 composites are now commercially available in thin sheet form which can be used in

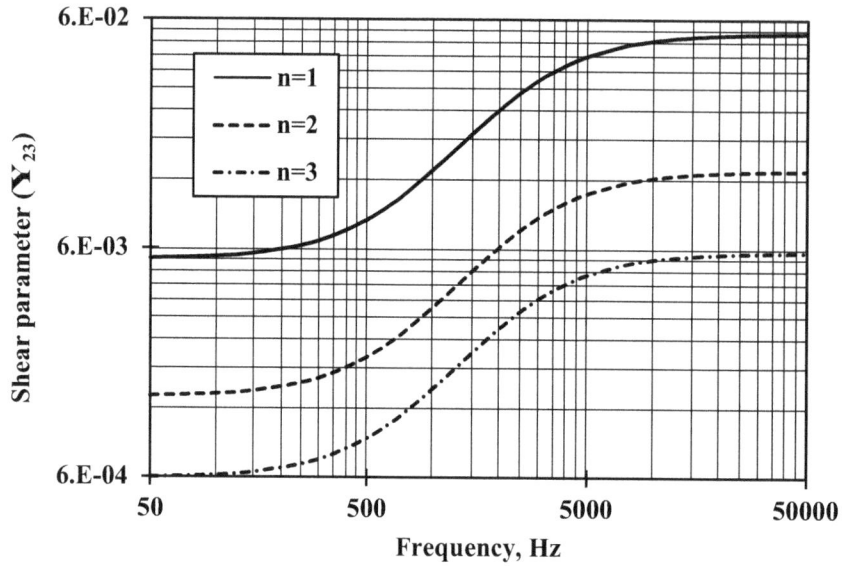

FIGURE 13.21 Shear parameter (ψ_{23}) for three modes of the CLD of the example in Section 13.7.2.3.

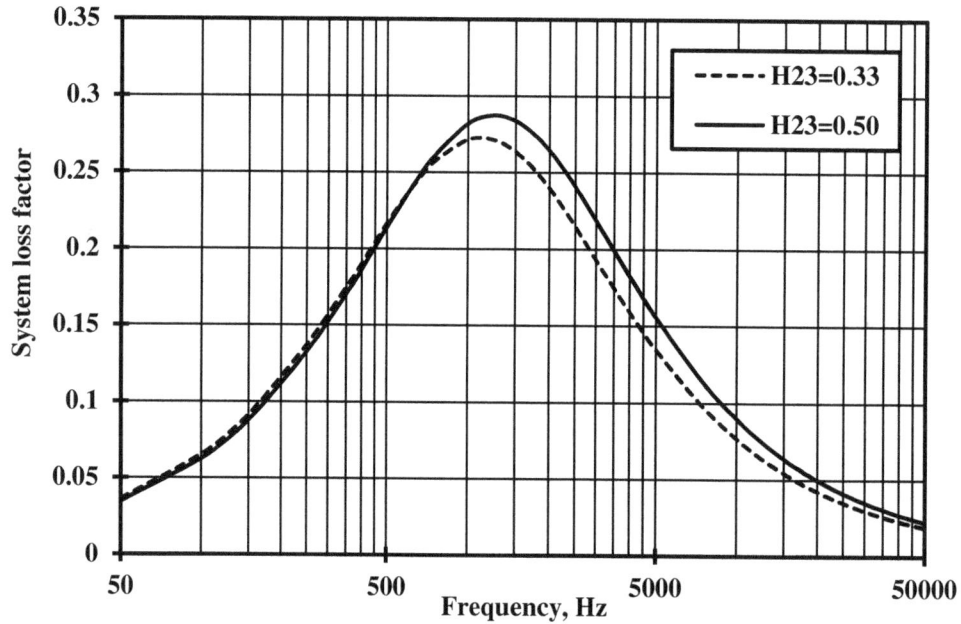

FIGURE 13.22 System loss factor for two thicknesses of VEM in the example in Section 13.7.2.3.

such an application. A more durable and tough material could be a functionally graded material of piezo layer as reported by Nazargah et al.[84]

13.7.4 Smart CLD with Magnetic Constraining Layer

A general viscoelastic damping system as CLD can be augmented by magnetic forces of attraction or repulsion without using any active source.[85–87] One possible configuration of the CLD arrangement could be the use of a viscoelastic damping layer on the substrate and magnets on both sides of the beam and at the base structure. However, the beam and the base structure must not be of a magnetic material. The magnetic attractive force increases

the shear straining of the viscoelastic layer, thus increasing the damping in the CLD system. The gap between the magnets of the beam and the base is adjusted to optimize the damping. The magnets can be arranged in repulsive mode also (N-N and S-S), but the repulsive force is always less than the attractive force, which makes the latter arrangement more preferable. Baz and Poh[85] evaluated the performances of both fully covered and segmented MCLD experimentally, using a cantilevered aluminum beam. In a different arrangement, the constraining layer itself can be a ferromagnetic material and magnets can be on the base only. The authors reported quite a good damping factor at the very low frequency band which is not possible using only the VEM layer.

13.8 Vibration Damping Characterization

Measurement of the vibration sensitivity of a substrate, either damped or undamped, can be done by free vibration or by forced vibration experiments. In the free vibration experiment, a simple instrumented hammer can be used to strike the object precisely and allow the object to vibrate freely thereafter. The hammer, fitted with a transducer, measures the force of impact. An accelerometer attached to the beam or plate picks up the dynamic signal in the time domain. Otherwise, a non-contact laser sensor can be used to measure the dynamic displacements. A dynamic signal analyzer (DSA) is used with a computer to convert the time domain signal to a frequency domain by Fast Fourier Transform (FFT). The result of the undamped and damped object is compared to find out the extent of damping in terms of system loss factor at various modes of resonance.

For a forced vibration experiment, an electrodynamic shaker is used to vibrate the object with a frequency sweep from 10 Hz to a maximum 10000 Hz in a programmable schedule as required. The dynamic signal with respect to frequency directly gives the amplitude either as intensity (voltage) ratio or power ratio with a reference acceleration in frequency scale. Comparison of the amplitudes of the undamped and damped signal gives a reduction in vibration intensity or power. The system loss factor can also be calculated from the modal peaks of the vibration spectra.

13.8.1 Structural Damping

The particle displacement in a flexural wave is normal on the surface of the plate. Strong radiation of sound can occur from flexural waves in plates and beams which are incorporated in the structures of vehicles, machines, and buildings. The function of damping is to control the free response of a system. Such responses are those in which the system's kinetic and potential energies are in balance. Free responses include: (1) resonance at a frequency equal to the natural frequency of vibration of the body; and (2) propagation of free waves in space. In the familiar case of resonance in frequency, the response to a given oscillating force in the vicinity of resonance or natural frequency is peaked, showing a maximum at the resonance frequency and a finite bandwidth of high response. Both features are controlled by the system loss factor. Therefore, a useful measure of the degree of damping is the loss factor. A true force-free response of a resonant system (once excited) would involve the exponential decay of a vibration amplitude with time. The decay is controlled by the system loss factor (η) as follows:

$$\text{Temporal decay rate}: \Delta_t = 27.3 \eta f_n \left(\text{dB} / \sec\right) \quad (13.44)$$

An intensity level L_v would be expressed in the customary logarithmic decibel (dB) measure for a response quantity "V," such as displacement, velocity, acceleration, etc.

$$L_v = 10 \operatorname{Log} [V/V_{ref}]^2 (\text{dB}) \quad (13.45)$$

The vibration response curve is used to determine several dynamic properties of a system. In an experiment recording the vibration spectra of a machine with respect to frequency, there will be a band of the frequency with peaks as different modes of resonances. From the nature of the peaks, the bandwidth, quality and damping factor can be determined.

Let us examine the first resonance peak of a forced vibration as depicted in Figure 13.23. From Figure 13.23, let us take the value U_{max}, which is the peak value, and we draw a horizontal line along a value of $U_{max}/\sqrt{2}$. We note the values of frequencies ω_1 and ω_2 from the intersections of the line with the response curve. This is a half-power line, as explained below. It is customary to relate the half-power bandwidth with the quality of the resonance peak. The vibration damping loss factors (η) of the specimens are determined by the half-power bandwidth

$$\Delta\omega_h = \omega_2 - \omega_1 \quad (13.46)$$

Figure 13.23 presents a measuring method for the loss factors using the calculated sensitivity ratio at the resonance frequencies of the specimens. The loss factor is defined as the inverse quality factor and expressed mathematically as:

$$\eta = \frac{\Delta\omega_h}{\omega_n} \quad (13.47)$$

where $\Delta\omega_h$ is the bandwidth at the half-power points; ω_2 and ω_1 are the right-side and left-side frequencies (Hz) of the half-power band (3 dB), respectively; and ω_n is the resonant frequency for the nth mode. On a decibel scale, this corresponds to -3 dB level from the peak value. At these points of frequency, the squared amplitude drops to half its peak value. For that reason, this damping estimation is also referred to as the "minus 3 dB method."

Impulse testing the dynamic behavior of mechanical structures involves striking the test object with a force-instrumented hammer and measuring the resultant motion with an accelerometer.

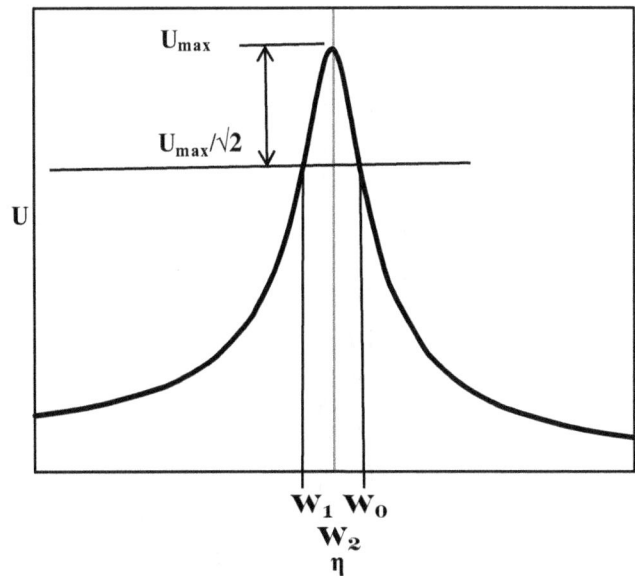

FIGURE 13.23 The vibration damping loss factors (η) by the half-power bandwidth method.

Functional transfer testing of mechanical structures depends on the ratio of the sensitivities of the force sensor to the accelerometer. This ratio can be determined with the hammer to impact a known mass instrumented with the accelerometer. The sensitivity ratio is determined from Newton's law of motion:

$$F = ma \tag{13.48}$$

$$F = \frac{V_h}{S_h} = \frac{Signal\ from\ hammer\ (V)}{Sensitivity\ of\ hammer\ (mV/kg)} \tag{13.49}$$

and

$$ma = m\frac{V_a}{S_a} = \frac{Signal\ from\ accelerometer\ (V)}{Sensitivity\ of\ accelerometer\ (mV/g)} \tag{13.50}$$

where F is the force of impact (N), m is the total mass of the pendulum (kg), and a is the acceleration due to impact (m/s^2). Therefore:

$$\frac{V_h}{S_h} = m\frac{V_a}{S_a} \tag{13.51}$$

$$\frac{V_h}{V_a} = m\frac{S_h\ (mV/kg)}{S_a\ (mV/g)} \tag{13.52}$$

$$FRF = \frac{V_a}{V_h} = \left(\frac{1}{m}\right)\left(\frac{S_a}{S_h}\right)\ kg/g \tag{13.53}$$

where FRF is the Frequency Response Function.

Figure 13.24 and Figure 13.25 show the FRF of black filled SBR filled with graphite, aramid short fiber, and nanoclay. The vibration damping loss factors (η) of the specimens were determined by the half-power bandwidth method as described earlier. The damping loss factors SBR filled with different fillers are given in Table 13.4.

In the case of SBR, graphite, and NC (low loading), this results in higher loss factor at low frequencies in the FLD configuration. Aramid was found to be least effective. In the CLD configuration, aramid was found to be most effective in all range of frequencies, especially at higher loadings.

The NC system showed similar results to that of aramid at lower frequencies. The SBR+5% clay nanocomposite has given the best damping in both FLD and CLD arrangements compared to other fibrous or particulate reinforcing fillers. The morphology of black filled SBR nanocomposites has been investigated by Praveen et al.[44] Extensive intercalation and to some extent exfoliation of the nanoclay could have resulted in numerous micro-CLDs in the SBR matrix. Slippage of the rubber chain segments

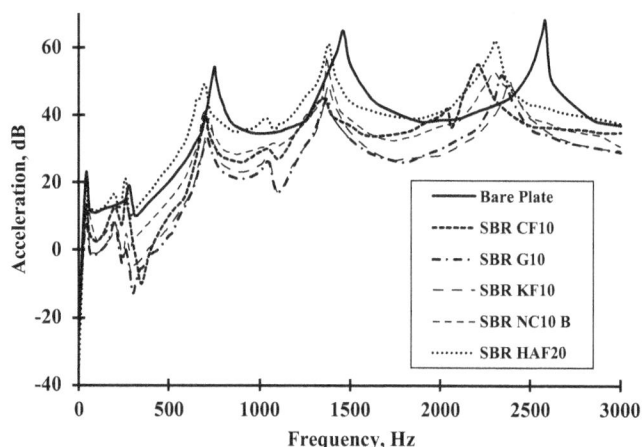

FIGURE 13.24 FLD results for SBR composites – effect of various reinforcing fillers.

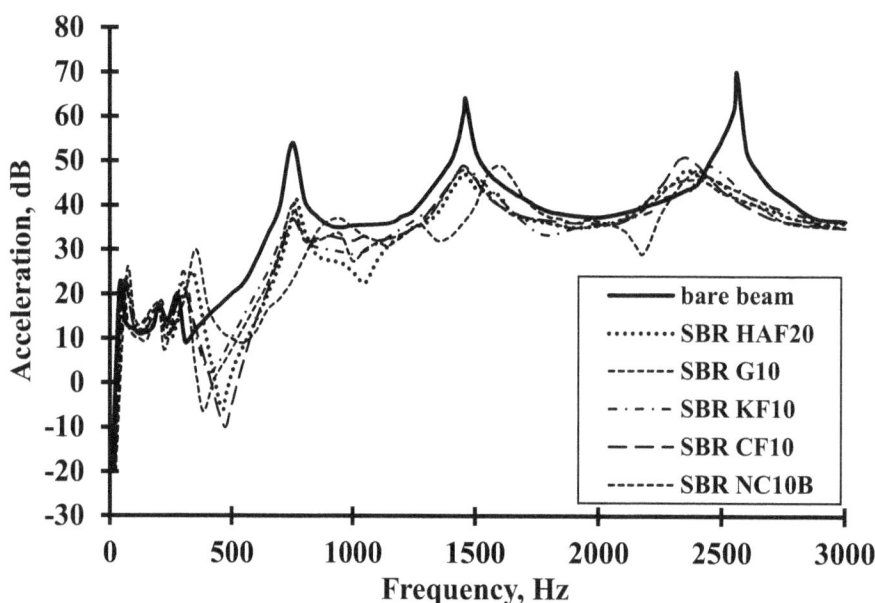

FIGURE 13.25 CLD performance of SBR composites with 10 phr filler.

TABLE 13.4

System Loss Factor (η) for FLD and CLD Using SBR Composites

Sample Designation	Phr of Fillers	Loss Factor (η) = Δf/f$_n$ (SBR)					
		FLD			CLD		
		2.2k Hz	1.5k Hz	710 Hz	2.2k Hz	1.5k Hz	710 Hz
Bare Panel	-	0.007	0.017	0.027	0.007	0.017	0.027
HAF	20	0.02	0.022	0.049	0.089	0.07	0.038
Graphite	5	0.013	0.02	0.08	0.046	0.03	0.035
(SBR-G)	10	0.022	0.015	0.05	0.056	0.072	0.042
	20	0.04	0.057	0.05	0.064	0.055	0.048
Aramid	5	0.034	0.028	0.027	0.035	0.084	0.055
(SBR KF)	10	0.031	0.014	0.035	0.053	0.066	0.068
	20	0.032	0.017	0.038	0.053	0.077	0.05
Carbon fiber	5	0.02	0.018	0.074	0.047	0.062	0.043
(SBR-CF)	10	0.025	0.026	0.031	0.042	0.034	0.053
	20	0.02	0.037	0.044	0.038	0.059	0.07
Nanoclay	5	0.037	0.015	0.092	0.083	0.05	0.031
(SBR NC)	10	0.036	0.014	0.061	0.07	0.06	0.17
	20	0.017	0.018	0.06	0.052	0.042	0.13

in relaxational oscillation could be one more reason for such a drastic difference in the loss factor compared to other fillers, particularly at lower frequencies.

In the FLD configuration, all the rubber-filler systems studied showed low values of loss factor. The inclusion of functional fillers did not improve the damping. This is because, in FLD, a high loss factor requires a high enough storage modulus since damping depends on the ratio of elastic modulus of the VEM to that of the metal substrate as seen in Figure 13.15. The rubber composites studied here had low modulus at ambient temperature and the ratio of moduli is below 10^{-3} and therefore, the system loss factor would be quite low for the FLD arrangement.

REFERENCES

1. A.M.A. de Waele. "Ostwald Called It the de Waele-Ostwald Equation." *Kolloid Zeitschrift* 47(2) (1929): 176–187.
2. M.M. Cross. "Rheology of non-Newtonian Fluids: A New Flow Equation for Pseudoplastic Systems." *Colloid Sci.* 20 (1965): 417–437.
3. R.P. Chhabra, C. Tiu. and P.H.T. Uhlherr. "Creeping Motion of Spheres Through Ellis Model Fluids." *Rheologica Acta.* 20(4) (1981): 346–351.
4. Chhabra R.P. and J.F. Richardson. *Non-Newtonian Fluid Behaviour in Non-Newtonian Flow in the Process Industries: Fundamentals and Engineering Applications.* Oxford: Elsevier, 1999, pp. 1–36.
5. J.D. Ferry. *Viscoelastic Properties of Polymers.* New York: John Wiley & Sons Inc., 1961.
6. G.V. Vinogradov and A.Y. Malkin. *Rheology of Polymers.* Moscow: Mir Publishers, 1980.
7. A.A. Alkelani, B.A. Housari, and S.A. Nassar. "A Proposed Model for Creep Relaxation of Soft Gaskets in Bolted Joints at Room Temperature." *Journal of Pressure Vessel Technology* 130 (2008): 011211-1.
8. A. Dey and P.K. Basudhar. "Applicability of Burger Model in Predicting the Response of Viscoelastic Soil Beds." In *Geo-Florida 2010: Advances in Analysis, Modeling and Design).* Florida: Geotechnical Special Publication, 199, 2010, pp. 2611–2620.
9. A. Dey and P.K. Basudhar. "Parameter Estimation of Four-Parameter Viscoelastic Burger Model by Inverse Analysis: Case Studies of Four Oil-Refineries. " *Interaction and Multiscale Mechanics* 5(3) (2012): 211–228.
10. A.M. Freudenthal and H.G. Lorsch. "The Infinite Elastic Beam on a Linear Viscoelastic Foundation." *J. Eng. Mech. Div. Proc. ASCE* 83(1) (1957): 1158-1–1158-22.
11. A.M. Freudenthal and W.R. Spillers. "Solutions for the Infinite Layer and the Half-Space for Quasi-Static Consolidating Elastic and Viscoelastic Media." *J. Appl. Phys.* 33(9) (1962): 2661–2668.
12. B.C. Hoskin and E.H. Lee. "Flexible Surfaces on Viscoelastic Subgrades." *J. Eng. Mech. Div. Proc. ASCE* 85(4) (1959): 11–30.
13. S. Murayama and T. Shibata. "Rheological Properties of Clays." *Proc. 5th ICSMFE,* 1 (1961): 269–273.
14. M.L. Williams, R.F. Landel and J.D. Ferry. "The Temperature Dependence of Relaxation Mechanisms in Amorphous Polymers and Other Glass-Forming Liquids." *J. Am. Chem. Soc.* 77 (1955): 370.
15. TA Instruments. "Thermal Analysis Application Brief: Application of Time-Temperature Superposition Principles to DMA." Number TA-144. Available at: www.tainst.com.
16. M.C.O. Chang, D.A. Thomas and L.H. Sperling. "Characterization of the Area Under Loss Modulus and Tan δ–Temperature Curves: Acrylic Polymers and Their Sequential Interpenetrating Polymer Networks." *J. Appl. Polym. Sci.* 34(1) (1987): 409–422.
17. M.C.O. Chang, D.A. Thomas and L.H. Sperling. "Group Contribution Analysis of the Damping Behavior of Homopolymers, Statistical Copolymers, and Interpenetrating Polymer Networks Based on Acrylic, Vinyl, and Styrenic Mers." *J. Polym. Sci; Part B: Polymer Physics* 26(8) (1988): 1627–1640.
18. T. Ogawa and T. Yamada. "A Numerical Prediction of Peak Area in Loss Factor for Polymers." *J. Appl. Polym. Sci.* 53 (1994): 1663–1668.

19. M. Patri, A.B. Samui and P.C. Deb. "Studies on Sequential Interpenetrating Polymer Network (IPN) Based on Nitrile Rubber and Poly(vinyl acetate)." *J. Appl. Polym. Sci.* 48 (1993): 1709–1716.

20. M. Patri, A.B. Samui, B.C. Chakraborty and P.C. Deb. "Studies on IPNs Based on Nitrile Rubber and Polyalkyl Methacrylates." *J. Appl. Polym. Sci.* 65 (1997): 549–554.

21. M. Patri, C. Reddy, V.C. Narasimhan and A.B. Samui. "Sequential Interpenetrating Polymer Network Based on Styrene Butadiene Rubber and Polyalkyl Methacrylates." *J. Appl. Polym. Sci.* 103 (2007): 1120–1126.

22. A. Mathew and P.C. Deb. "Studies on Semi-Interpenetrating Polymer Networks Based on Poly(vinyl chloride-co-vinyl acetate) and Poly(alkyl methacrylates)." *J. Appl. Polym. Sci.* 53 (1994): 1103–1106.

23. A. Mathew, B.C. Chakraborty and P.C. Deb. "Studies on Interpenetrating Polymer Networks Based on Nitrile Rubber-Poly (vinyl chloride) Blends and Poly (alkyl methacrylates)." *J. Appl. Polym. Sci.* 53 (1994): 1107–1114.

24. B. Das, D. Chakraborty, A.K. Hajra and S. Sinha. "Epoxy/ Poly (methyl methacrylate) Interpenetrating Polymer Networks-Morphology, Mechanical and Thermal Properties." *J. Appl. Polym. Sci.* 53 (1994): 1491–1496.

25. A.B. Samui, U.G. Suryavanshi, M. Patri, B.C. Chakraborty and P.C. Deb. "Sequential Interpenetrating Polymer Network Based on Nitrile–Phenolic Blend and Poly(alkyl methacrylate)." *J. Appl. Polym. Sci.* 68 (1998): 255–262.

26. A.B. Samui, V.G. Dalvi, L. Chandrasekhar, M. Patri and B.C. Chakraborty. "Interpenetrating Polymer Networks based on Nitrile Rubber and Metal Methacrylates." *J. Appl. Polym. Sci.* 99 (2006): 2542–2548.

27. N.R. Manoj, L. Chandrasekhar, M. Patri, B.C. Chakraborty and P.C. Deb. "Vibration Damping Materials Based on Interpenetrating Polymer Networks of XNBR and PMMA." *Poly. Adv. Technol.* 13 (2002): 1–5.

28. N.R. Manoj, D. Ratna, V.G. Dalvi, L. Chandrasekhar, M. Patri, B.C. Chakraborty and P.C. Deb. "Interpenetrating Polymer Networks based on Carboxylated Nitrile Rubber and Poly(Alkyl Methacrylate)s." *Polym. Eng. Sci.* 42(8) (2002): 1748–1755.

29. L.H. Sperling. "Interpenetrating Polymer Networks." In *Encyclopedia of Polymer Science and Technology.* New York: John Wiley & Sons, 2004.

30. D. Ratna, N.R. Manoj, L. Chandrasekhar and B.C. Chakraborty "Novel Epoxy Compositions for Vibration Damping Applications." *Polym. Adv. Technol.* 15 (2004): 583–586.

31. N.N. Najib, Z.M. Ariff, A.A. Bakar and C.S. Sipaut. "Correlation Between the Acoustic and Dynamic Mechanical Properties of Natural Rubber Foam: Effect of Foaming Temperature." *Materials and Design* 32 (2011): 505–511.

32. K. Sasikumar, N.R. Manoj, T. Mukundan and D. Khastgir. "Design of XNBR Nanocomposites for Underwater Acoustic Sensor Applications: Effect of MWNT on Dynamic Mechanical Properties and Morphology." *J. Appl. Polym. Sci.* 131(18) (2014): 40752.

33. K. Sasikumar, N.R. Manoj, T. Mukundan and D. Khastgir. "Hysteretic Damping in XNBR/MWNT Nanocomposites at Low and High Compressive Strains." *Composites: Part B.* 92 (2016): 74–83.

34. C. Jelena. "Interpenetrating Polymer Network Composites Containing Polyurethanes Designed for Vibration Damping." *Polimery.* 61(3) (2016): 159–165.

35. G.K. Mandal, D.K. Tripathy and S.K. De. "Effect of Silica Filler on Dynamic Mechanical Properties of Ionic Elastomer Based on Carboxylated Nitrile Rubber." *J. Appl. Polym. Sci.* 55(8) (1995): 1185.

36. K.K. Kar and A.K. Bhowmick. "Effect of Holding Time on High Strain Hysteresis Loss of Carbon Black Filled Rubber Vulcanizates." *Polym. Eng. Sci.* 38(12) (1998): 1927–1945.

37. K. Sasikumar, N.R. Manoj, R. Ramesh and T. Mukundan. "Carbon Nanotube-Polyurethane Nanocomposites for Structural Vibration Damping." *Int. J. Nanotechnol.* 9 (2012): 1061–1071.

38. W.B. Cushman and G.B. Thomas. "Acoustic Attenuation and Vibration Damping Materials." US Patent No. 5400296 A-1995 (1995).

39. M.A. Mendelsohn, F.W. Navish Jr. and K. Dongsik. "Characteristics of a Series of Energy-Absorbing Polyurethane Elastomers." *Rubber Chem. Technol.* 58(5) (1985): 997–1013.

40. G.H. Kwak, K. Inoue, Y. Tominaga, S. Asai and M. Sumita. "Characterization of the Vibrational Damping Loss Factor and Viscoelastic Properties of Ethylene–Propylene Rubbers Reinforced with Microscale Fillers." *J. Appl. Polym. Sci.* 82 (2005): 3058–3066.

41. A. Bansal, H. Yang, C. Li, B.C. Benicewicz, S.K. Kumar and L.S. Schadler. "Quantitative Equivalence Between Polymer Nanocomposites and Thin Polymer Films." *Nat. Mater.* 4 (2005): 693–698.

42. A. Bansal, H. Yang, C. Li, B.C. Benicewicz, S.K. Kumar and L.S. Schadler. "Controlling the Thermomechanical Properties of Polymer Nanocomposites by Tailoring the Polymer–Particle Interface." *J Polym Sci B: Polym Phys.* 44 (2006): 2944–2950.

43. S. Praveen, B.C. Chakraborty, S. Jayendran, R.D. Raut and S. Chattopadhyay. "Effect of Filler Geometry on Viscoelastic Damping of Graphite/Aramid and Carbon Short Fiber-Filled SBR Composites: A New Insight." *J. Appl. Polym. Sci.* 111 (2009): 264–272.

44. S. Praveen, P.K. Chattopadhyay, P. Albert V.G. Dalvi, B.C. Chakraborty and S. Chattopadhyay. "Synergistic Effect of Carbon Black and Nanoclay Fillers in Styrene Butadiene Rubber Matrix: Development of Dual Structure." *Composites: Part A* 40 (2009): 309–316.

45. S. Praveen, P.K. Chattopadhyay, S. Jayendran, B.C. Chakraborty and S. Chattopadhyay. "Effect of Nanoclay on the Mechanical and Damping Properties of Aramid Short Fibre-Filled Styrene Butadiene Rubber Composites." *Polym Intl.* 59(2) (2010): 187–197.

46. S. Praveen, P.K. Chattopadhyay, S. Jayendran, B.C. Chakraborty and S. Chattopadhyay. "Effect of Rubber Matrix Type on the Morphology and Reinforcement Effects in Carbon Black-Nanoclay Hybrid Composites: A Comparative Assessment." *Polym. Compos.* 31(1) (2010): 97–104.

47. M.B. Khan. "Intelligent Viscoelastic Polyurethane Intrinsic Nanocomposites." *Metallurgical and Materials Transactions A.* 41(4) (2010): 876–880.

48. A.R Mackintosh, R.A. Pethrick and W.M. Banks. "Dynamic Characteristics and Processing of Fillers in Polyurethane Elastomers for Vibration Damping Applications." *Journal of Materials: Design and Applications: Proceedings of the Institution of Mechanical Engineers, Part L.* 225(3) (2011): 113–122.

49. S. Futamura. "Deformation Index; Concept for Hysteretic Energy-Loss Process." *Rubber Chem. Technol.* 64 (1990): 57–64.

50. H. Oberst and K. Frankenfeld. "On the Damping of Bending Vibrations on Thin Sheet Metal by Bonded Coatings." *Acustica* 2 (1952): 181–194.

51. D. Ross, E. Ungar and E.M. Kerwin Jr. "Damping of Plate Flexural Vibrations by Means of Viscoelastic Laminae." In J.E. Ruzicka (Ed.), *Section 3: Structural Damping.* New York: American Society of Mechanical Engineers, 1959, pp. 49–87.

52. Y.P. Lu. "Forced Vibrations of Damped Cylindrical Shells Filled with Pressurized Liquid." *AIAA Journal,* 15(9) (1977): 1242–1249.

53. A.D. Nashif. D.I.G. Jones and J.P. Henderson. *Vibration Damping.* New York: John Wiley & Sons, 1985.

54. B.C. Nakra. "Vibration Damping." *PINSA.* 67A(4&5) (2001): 461–478.

55. M.D. Rao. "Recent Applications of Viscoelastic Damping for Noise Control in Automobiles and Commercial Airplanes." *J. Sound Vib.* 262 (2003): 457–474.

56. T. Yamaguchi, Y. Kurosawa, S. Matsumura and A. Nomura. "Finite Element Analysis for Vibration Properties of Panels in Car Bodies Having Viscoelastic Damping Layer." *Transactions: Japan Society of Mechanical Engineers.* Series C 69 (2003): 297–303.

57. P.J. Torvik. "Analysis of Free-layer Damping Coatings." *Key Engineering Materials.* 333 (2007): 195–214.

58. M. Danti, D. Vigè and G.V. Nierop. "Modal Methodology for the Simulation and Optimization of the Free-Layer Damping Treatment of a Car Body." *Journal of Vibration and Acoustics* 132 (2010): 021001-1.

59. R.U.H. Syed, M.I. Sabir, J. Wei and D.Y. Shi. "Effect of Viscoelastic Material Thickness of Damping Treatment, Behavior on Gearbox." *Research Journal of Applied Sciences, Engineering and Technology.* 4(17) (2012): 3130–3136.

60. S.Y. Kim, C.K. Mechefske and I.Y. Kim. "Optimal Damping Layout in a Shell Structure Using Topology Optimization." *J. Sound Vib.* 332(12) (2013): 2873–2883.

61. F. Cortés and M.J. Elejabarrieta. "Structural Vibration of Flexural Beams with Thick Unconstrained Layer Damping." *Intl. J. Solids and Structures* 45 (2008): 5805–5813.

62. F. Cura`, A. Mura and F. Scarpa. "Modal Strain Energy Based Methods for the Analysis of Complex Patterned Free Layer Damped Plates." *J Vib. Control* 18(9) (2011): 1291–1302.

63. J.S. Moita, A.L. Araújo, C.M.M. Soares and C.A.M. Soares. "Finite Element Model for Damping Optimization of Viscoelastic Sandwich Structures." *Adv. Eng. Softw.* 66 (2013): 34–39.

64. X. Zhang and Z. Kang. "Topology Optimization of Damping Layers for Minimizing Sound Radiation of Shell Structures." *J. Sound Vib.* 332 (2013): 2500–2519.

65. K. Kishore Kumar, Y. Krishna and P. Bangarubabu "Damping in Beams Using Viscoelastic Layers." *P. I. Mech. Eng. L-J. Mat.* 229(2) (2015): 117–125.

66. T. Yamamoto, T. Yamada, K. Izui and S. Nishiwaki. "Topology Optimization of Free-Layer Damping Material on a Thin Panel for Maximizing Modal Loss Factors Expressed by Only Real Eigenvalues." *J. Sound Vib.* 358(8) (2015): 84–96.

67. E.M. Kerwin Jr. "Damping of Flexural Waves by a Constrained Viscoelastic Layer." *J. Acoust. Soc. Am.* 31(7) (1959): 952–962.

68. B.C. Nakra. "Vibration Control with Viscoelastic Materials I." *Shock Vib. Digest* 8 (1976): 3–12.

69. B.C. Nakra. "Vibration Control with Viscoelastic Materials II." *Shock Vib. Digest* 13 (1981): 17–20.

70. B.C. Nakra. "Vibration Control with Viscoelastic Materials III." *Shock Vib. Digest* 16 (1984): 17–22.

71. N.T. Asnani and B.C. Nakra. "Vibration Damping Characteristics of Multilayered Beams with Constrained Viscoelastic Layers." *J. Eng. Ind,* 98(3) (1976): 895–901.

72. D.K. Rao. "Frequency and Loss Factors of Sandwich Beams under Various Boundary Conditions." *J. Mech. Eng. Sci.* 20(5) (1978): 271–282.

73. L.J. Pulgarno. "Effectiveness of Partial Coverage for Constrained Layer Damping." Paper presented at 64th Meeting, Acoustical Society of America, Washington, DC, 1962.

74. D.S. Nokes and F.C. Nelson. "Constrained Layer Damping with Partial Coverage." *Shock Vib. Bull.* 38(3) (1968): 5–10.

75. S. Markus. "Damping Mechanism of Beams Partially Covered by Constrained Viscoelastic Layer." *ACTA Technica CSAV* 2 (1974): 179–194.

76. K. Lall, N.T. Asnani and B.C. Nakra. "Vibration and Damping Analysis of Rectangular Plate with Partially Covered Constrained Viscoelastic Layer." *J Vib Acoust* 109 (1987): 241–247.

77. Y.P. Lu, A.J. Roscoe and B.E. Douglas. "Analysis of the Response of Damped Cylindrical Shells Carrying Discontinuously Constrained Beam Elements." *J. Sound Vib.* 150(3) (1991): 395–403.

78. S.W. Kung and R. Singh. "Development of Approximate Methods for Analysis of Patch Damping Design Concepts." *J. Sound Vib.* 219(5) (1999): 785–812.

79. L.H. Chen and S.C. Huang. "Vibration Attenuation of a Cylindrical Shell with Constrained Layer Damping Strips Treatment." *Comput. Struct.* 79(14) (2001): 1355–1362.

80. A. Kumar and S. Panda. "Optimal Damping in Circular Cylindrical Sandwich Shells with a Three-Layered Viscoelastic Composite Core." *J. Vib. Acoust.* 139 (2017): 1–12.

81. N.W. Hagood and A.V. Flotow. "Damping of Structural Vibrations with Piezoelectric Materials and Passive Electrical Networks." *J. Sound Vib.* 146(2) (1991): 243–268.

82. T. Delpero, A.E. Bergamini and P. Ermanni. "Identification of Electromechanical Parameters in Piezoelectric Shunt Damping and Loss Factor Predictions." *J. Intel. Mater. Syst. Struct.* 24(3) (2012): 287–298.

83. O. Thomas, J. Ducarne and J-F. Deu. "Performance of Piezoelectric Shunts for Vibration Reduction." *Smart Mater. Struct.* 21 (2012): 15008.

84. M. Lezgy-Nazargah, S.M. Divandar, P. Vidal and O. Polit. "Assessment of FGPM Shunt Damping for Vibration Reduction of Laminated Composite Beams." *J. Sound Vib.* 389 (2017): 101–118.

85. A. Baz and S. Poh. "Performance Characteristics of the Magnetic Constrained Layer Damping." *Shock and Vibration,* 7(2) (2000): 81–90.

86. M. Ruzzene, J. Oh and A. Baz. "Finite Element Modeling of Magnetic Constrained Layer Damping." *J. Sound Vib.* 23(4) (2000): 657–682.

87. H. Zheng, M. Li and Z. He. "Active and Passive Magnetic Constrained Damping Treatment." *Int. J. Solids & Structures.* 40 (2003): 6767–6779.

14

Solid Polymer Electrolyte Membranes

Swati S. Rao
Naval Materials Research Laboratory, Ambernath, India

Manoranjan Patri
Naval Materials Research Laboratory, Ambernath, India

CONTENTS

Abbreviations

SPE	solid polymer electrolytes
PEMFC	polymer electrolyte membrane fuel cell
MEA	membrane electrode assembly
BAM3G	Ballard Power Systems developed membranes
IEC	ion exchange capacity
PVDF	poly(vinylidene fluoride)
PFA	poly(tetrafluoroethylene-*co*-perfluoropropyl vinyl ether)
FEP	poly(tetrafluoroethylene-*co*-hexafluoropropylene)
ETFE	poly(ethylene-*alt*-tetrafluoroethylene)
PEEK	poly (ether ketone)
SPEEK	sulphonated poly(ether ketone)
DMFC	direct methanol fuel cell
PBI	polybenzimidazole
ABPBI	poly(2,5-bezimidazole)
POSS	polyhedral oligomericsilsesquioxanes
PILs	protic ionic liquids
AILs	aprotic ionic liquids
SPI	sulphated polyimides
OPBI	Poly(oxyphenylene benzimidazole)

DOI: 10.1201/9781003037880-14

14.1 Introduction

With the demand for energy increasing, accompanied by the depletion of fossil fuels, alternate sources of energy are gaining popularity. These energy sources, such as primary or secondary batteries, fuel cells, flow batteries or hybrid devices, require stable electrolytes to compete with the conventional type of energy sources in terms of durability and performance. Although many materials have proven their worth in these devices, they have certain limitations, such as leakage, corrosion of electrode materials or drying out, if it is a liquid electrolyte. Solid polymer electrolytes overcome these limitations and offer a suitable candidate for such applications.

Solid polymer electrolytes (SPE) are materials, which, in the form of membranes or gels, can conduct ions generated in an electrochemical reaction. These have the advantage of being flexible in their usage in terms of weight and performance. They perform a dual role in many applications wherein they behave as an electrolyte as well as a separator between the anode and the cathode.

The fact that solids such as silver sulfide could conduct silver ions was explained by Faraday way back in 1834. The pioneering work carried out by Wright (Fenton et al. 1973) on poly-ethylene oxide complexed with alkali metal salt led the way to research on a variety of polymer materials. Since then, polymer electrolytes with different ion transport capability, such as Li^+, Na^+, Mg^{2+}, H^+, etc. have been reported for a number of electrochemical applications (Agarwal and Pandey 2008; Phair and Badwal 2006).

Among the various types of SPEs studied, ion exchange membranes play a major role. Ion exchange membranes are basically comprised of three parts: a polymer backbone, an ion exchange group with either a positive or negative charge capable of undergoing an oxidation-reduction reaction, and water. The backbone gives mechanical strength; the ionic group provides electrical property, and the transport property is derived from water. Electrical neutrality is maintained in ion exchange membranes by counter-ions or gegen ions that are oppositely charged. These membranes swell in water as the ions are hydrated and frequently contain sufficient water in their structures to form a second continuous phase. This water-swollen membrane has more space between the adjacent polymer chains to adjust their positions to allow ions to pass. These membranes have been successfully employed for various electrochemical applications, ranging from electrodialysis to fuel cells. These membranes are proton-conducting and the one most effectively used is Nafion® from DuPont.

This chapter will deal with the application of solid polymer electrolyte membranes exclusively for fuel cells. Fuel cells are a class of electrochemical devices which convert the chemical energy of the reaction between a fuel, such as hydrogen, and an oxidant, such as oxygen or air, into electrical energy. These devices are environmentally-friendly since the by-product obtained is water. As can be seen from Figure 14.1, the membrane is the heart of a polymer electrolyte membrane fuel cell (PEMFC). It is sandwiched between two electrodes, the anode and cathode, to form the membrane electrode assembly (MEA), which is the basic unit where fuel cell reactions take place and electricity is generated. The membrane is responsible for the transport of protons generated at the anode toward the cathode.

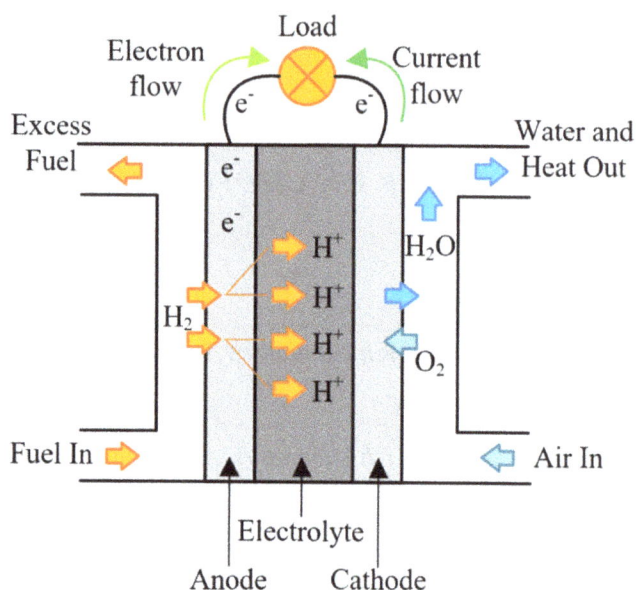

FIGURE 14.1 PEM fuel cell.

The functional requirements of membranes are high proton conductivity, good chemical and thermal stability, being impermeable to gases, and adequate mechanical stability. Every property of the membrane material influences its ultimate performance in a fuel cell since it is subjected to stringent operating conditions.

14.1.1 Ion Exchange Capacity and Water Uptake

The ion exchange capacity (IEC) defines the number of ion-exchangeable groups present per equivalent weight of the polymer and is a direct measure of the functionality of the membrane. It is obvious that the higher the number of ionic groups, the higher will be the conductivity. Also, the water uptake, another important parameter, is related to the number of ionic sites as this is the measure of the amount of water a membrane absorbs when exposed to bulk water or its vapor, as the case may be. It is desirable to have a higher number of ionic groups for higher conductivity, however, this leads to extensive swelling of the membrane. This results in poor mechanical properties and thus reduced performance. Thus, it is necessary to maintain a balance between the water uptake and the conductivity by varying the IEC to derive the best performance from the membrane material.

14.1.2 Proton Conductivity

As mentioned in Section 14.1.1, polymer electrolyte membranes contain ionizable acid groups which dissociate into immobile anionic and mobile cationic moieties. The cations, particularly the protons in the case of proton exchange membranes used for fuel cell applications, are the mobile species which are responsible for the conduction mechanism and the overall functioning of the membrane as an electrolyte. Transport of protons in a membrane takes place via two mechanisms, namely, the Grotthuss mechanism and the vehicular mechanism. The Grotthuss mechanism, commonly known as the hopping mechanism, involves the movement of the proton through a hydrogen-bonded framework. In the presence of excess water, the proton transfer mechanism is similar to that of bulk water. Here the proton is transferred through the formation and cleavage of covalent bonds. In the vehicular mechanism, the transfer of the proton takes place by diffusion of the proton. Here the water molecules act as vehicles to transport the proton species from the anode to the cathode by the formation of H_3O+. The mechanism through which proton transfer takes place depends on the humidity level and temperature of operation of the fuel cell.

It is interesting to note that the polymer morphology plays a significant role in the ion-conducting properties of the membranes. Nafion® is known to possess very high ionic conductivity due to the phase separation of the hydrophilic and hydrophobic regions of the polymer backbone and ionic groups. Ionic clusters are formed within the polymer membrane, when swollen in water, which are interconnected and function as proton-conducting pathways or channels. These channels swell or shrink depending on the amount of "free" water present within the clusters. Hydrocarbon-based membranes exhibit lower conductivities when compared to Nafion®, due to poor water channel formation because of weak phase separation between hydrophobic and hydrophilic moieties. Further, the dependence of conductivity on the humidity lowers the operational temperature and increases the system complexity. The introduction of non-volatile liquids, such as ionic liquids or phosphoric acid as proton conductors, promotes the transport of protons by the Grotthuss mechanism. This enables the fuel cell to be operated at higher temperatures and under dry conditions. Also, in order to compensate for the loss in conductivity due to structural limitations, inorganic nano-materials have been introduced into the membrane matrix. These materials, if ionic in nature, can create proton-conducting pathways for improved performance.

14.1.3 Thermal and Mechanical Stability

Thermal stability is another important property which influences the operating temperature and ultimately the performance of the fuel cell. It depends on the chemical and physical nature of the membrane material. The detailed study of this property helps to design membranes with better thermal stability. Nafion® membranes have high thermal stability and desulphonation occurs at ~280°C. Most of the aromatic polymers possess excellent thermal stability, however, desulphonation occurs at lower temperatures of 230°C. Also, thermal stability in a wide range starting from below ambient to 200°C is essential in terms of durability. The membrane should be able to withstand temperature changes from start-up to shut down. Temperature variation affects the molecular conformation of the polymer which can reduce its mechanical integrity, resulting in poor or reduced performance. The temperature cycling alone has a low effect. However, temperature cycling, when coupled with humidity changes, has a detrimental effect on membrane durability (Tang and Pan 2008).

Another essential property for qualifying membranes is the mechanical stability or dimensional integrity of the membrane under fuel cell conditions. In the fuel cell environment, the membrane is subjected to compressive load since it is sandwiched between two electrodes to form the MEA. A membrane has to possess reasonably good mechanical strength to remain a free-standing film on undergoing hot-pressing during fabrication of

the MEA and when swollen in water. The process of hot pressing imposes local stress on the membrane, leading to mechanical failure in the form of thinning or pin-hole formation. This property is strongly influenced by the presence of water. During fuel cell operation, changes in humidity lead to anisotropic swelling of the membrane in the constrained membrane electrode assembly. This causes cross-over of gases, resulting in degradation of the polymer matrix due to peroxide formation.

Mitigation of thermal and mechanical degradation can be carried out by the introduction of inorganic fillers or reinforcing materials. Reinforcement using polytetrafluoroethylene (PTFE) is known to reduce the mechanical stress upon variation in temperature and humidity. Also, the addition of inorganic fillers such as nano-silica or layered inorganic proton conductors can have a positive influence on both the temperature as well as the mechanical stability. Cross-linking of hydrocarbon polymers has also proved to be effective in improving the durability of membranes in the long term.

Ion exchange membranes for fuel cell applications can be made from different polymers, such as polystyrene, polysulphone, polyethylene, and fluorinated polymers. Although the ion exchange membranes are similar in chemical structure to ion exchange resins, they differ in the mechanical requirements of the process. Membranes, synthesized from these resins, are mechanically weak since cation resins are brittle while anion resins are soft. They are dimensionally unstable and changes in electrolyte concentration or in temperature may cause major changes in the water uptake and, hence, in the volume of the resin. Thus, it is generally not possible to use sheets of materials that have been prepared in the same way as a bead resin.

14.2 Types of Membrane

14.2.1 Perfluorinated Sulphonic Acid Membranes

This polymer has a Teflon-like backbone with a sulphonated side chain attached by an ether linkage. These polymers are superior to any other ion exchange membranes. The membranes, prepared by using this polymer, exhibit high chemical and thermal stability. The chemical stability in various conditions of these polymers has been confirmed by several studies (Appleby and Flouks 1989). The first commercially available membrane material was Nafion® 120 and was followed by Nafion® 117. In 1988, the Dow Chemical Company developed its own perfluorinated membrane. This membrane was marketed under the name Dow XUS. Dow XUS and Nafion® have similar structures, but XUS had low equivalent weight and more pendant sulphonic acid groups per – $CF2$ units (Srinivasan et al. 1991). The structures of Nafion® and Dow XUS are shown in Figure 14.2.

A short side-chain perfluorinated ionomer (the Dow membrane) with no fluoroether group in the pendant side-chain comprising only two CF2 groups was developed by the Dow Chemical Company (Ezzell et al. 1980).

All the above membranes maintain very long service life under electrosynthesis, fuel cells, and electrolysis (Nuttle and Brown 1971; Appleby and Yeager 1986; Riedinger and Faul 1988). But the major drawback of these membranes, besides their high cost (Steck

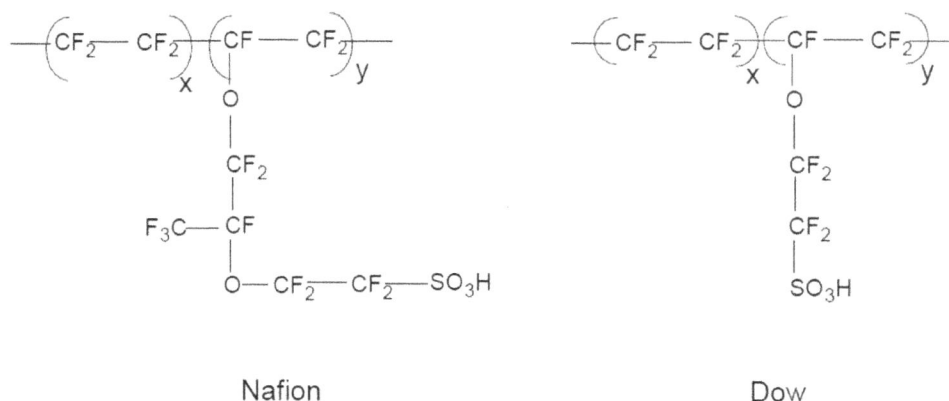

Nafion　　　　　　　　　　　　Dow

FIGURE 14.2 Structure of perfluorinatedsulphonic acid membranes.

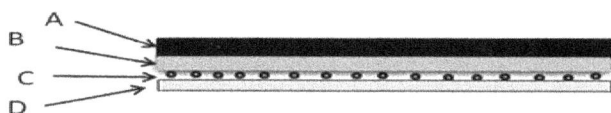

FIGURE 14.3 Schematic diagram of the composite self-humidifying membrane:(A) Nafion® layer; (B) Pt self-humidifying layer; (C) porous PTFE membrane; (D) Nafion layer.

et al. 1995), is that these membranes cannot be used at temperatures higher than 100°C and RH less than 30%. This is because the conductivity mechanism in these membranes is dependent on the amount of water present within the membrane matrix. Also, they are not suitable for use in direct methanol fuel cells, due to high methanol cross-over. To eliminate this drawback, composites of these membrane materials with reinforcing materials, solid proton conductors or hygroscopic oxides have been made. Nafion®/PTFE composite membranes are made using porous PTFE as a substrate as shown in Figure 14.3 (Liu et al. 2003a; Liu et al. 2003b). The composite membranes exhibit higher performance compared to pure Nafion® under both humidified and no-humidified conditions.

Composite membranes such as Nafion®/zirconium phosphate and Nafion®/SiO₂ have been developed (Yang et al. 2004; Pereira et al. 2008). The composite membranes show better fuel cell stability compared to pristine Nafion® membranes. It is observed that the sensitivity to change in relative humidity was less for composite membranes due to better interfacial compatibility and bonding between the silica particles and the Nafion® matrix.

14.2.2 PEM Based on Styrene and Styrene Derivatives

The PEM used in the Gemini program was made of cross-linked polystyrene sulphonic acid (Prater 1990). The moderate performance of this membrane led to the development of membranes based on modified styrene. The grafting of poly (sodium styrene sulphonate) to polystyrene is done by using free radical polymerization (Ding et al. 2002). The conductivities of these grafted copolymers reach up to ~ 0.24 S/cm. The DAIS-Analytic Corporation (Wnek et al. 1995) has developed sulphonated derivatives of membranes based on tri-block copolymers of styrene-co-ethylene-co-butylene. Styrene-based PEM can also

be made by a radiation grafting technique. Here the monomer is grafted onto any fluoropolymer base of choice such as poly-tetrafluoroethylene (PTFE), poly(vinylidene fluoride) (PVDF), poly(tetrafluoroethylene-*co*-perfluoropropyl vinyl ether) (PFA), poly(tetrafluoroethylene-*co*-hexafluoropropylene) (FEP), and poly(ethylene-*alt*-tetrafluoroethylene) (ETFE). Functional groups can be introduced by post sulphonation or carboxylation. This method is suitable for producing tailor-made membranes with the desired properties and targeted applications. Synthesis of a copolymer of styrene and acrylic acid membranes by radiation grafting onto FEP has also been carried out (Phadnis et al. 2003). The membranes with tailor-made properties could be synthesized by varying the reaction parameters and monomer concentration. However, these membranes are inferior to the perfluorinated membranes in terms of their chemical/oxidative stability. The grafted chains undergo degradation due to the presence of tertiary hydrogen atoms on the alpha carbon atom in the polystyrene structure. The decay is caused due to the chemical attack of the peroxide radicals generated during the fuel cell reaction, which leads to a loss in functionality of the membrane. Attempts to replace the hydrogen by methyl or fluorinated groups have been fruitful, as can be seen from Section 14.2.3.

14.2.3 Membranes Based on Partially Fluorinated Styrene

Poly α-β-β-trifluoro styrene, synthesized by Prober (Prober 1953) can be sulphonated (Hodgon et al. 1967; Hodgon 1968). Poly α-β-β-trifluoro styrene sulphonic acids with different ion exchange capacities can be synthesized by carefully varying the reaction parameters (Figure 14.4). These polymers are chemically stable compared to the non-fluorinated analogs. Ballard Power Systems (Wei et al. 1995) developed membranes (BAM3G) based on these partially fluorinated polymers. To further improve the mechanical and dimensional stability, a macrocomposite BAM3G membrane was developed (Steak and Stone 1999).

14.2.4 Phenol-Formaldehyde Membranes

Over 60 years ago, G.E. Laboratories proposed the use of phenol-formaldehyde as PEM candidates. Linear and cross-linked

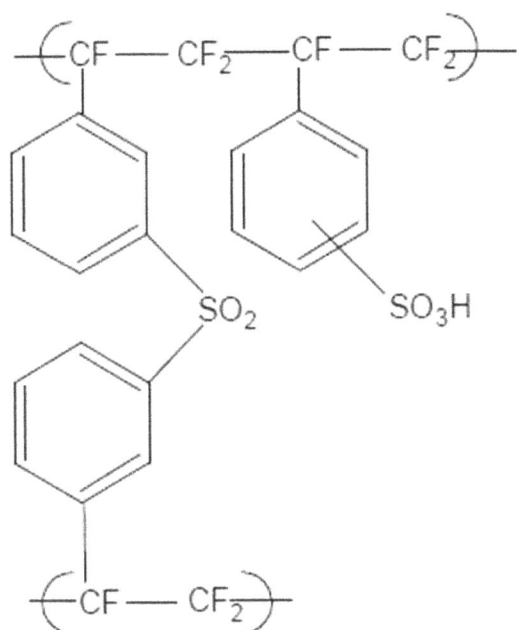

FIGURE 14.4 Structure of cross- linked sulphonated poly-α-β-β-trifluorostyrene.

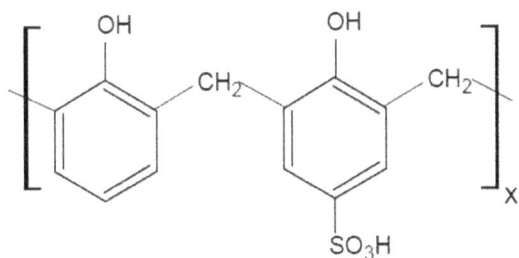

FIGURE 14.5 Structure of sulphonated phenol-formaldehyde.

phenol-formaldehyde polymers are sulphonated (Figure 14.5) (Penner 1986). However, poor performance compared to polystyrene sulphonic acid membranes was observed due to degradation caused by acid-catalyzed hydrolysis of the membranes.

14.2.5 Membranes Based on Non-Fluorinated Ionomers

The ionomer-based membranes are made from materials like polyimides (Mittal 1984), polybenzimidazoles (Linkous 1993), polyethersulphones (Johnson et al. 1984), polyetherketones (Souczk-Guth et al. 1999), and poly(phenylquinoxalines) (Savadogo 1998). All these materials have high thermal and chemical stability.

To impart proton conductivity, these materials are sulphonated to various degrees of sulphonation depending on the desired property. The sulphonation of these materials has been carried out in any of the following ways:

1. direct sulphonation of the polymer backbone;
2. chemical grafting of sulphonated monomers onto the polymer;

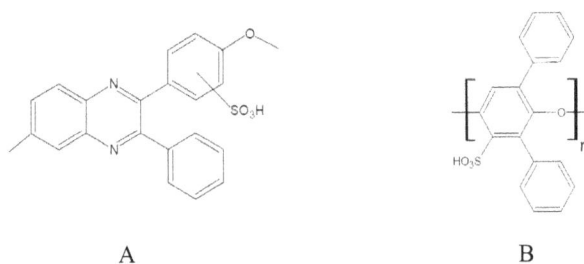

A   B

FIGURE 14.6 Structure of (A) Sulphonatedpoly(phenylquinoxaline) and (B) sulphonated *poly (2,6-diphenyl-4-phenylene oxide)* based on poly (phenylquinoxaline).

3. radiation grafting of monomers to the polymer and subsequent sulphonation;
4. polymerization of monomers which bear sulphonic acid groups.

14.2.6 Membranes Based on Sulphonated Poly (Phenylquinoxaline), Poly (Substituted Phenylene Oxide), and Polyphenylenes

Ballard Advanced Materials developed membranes called BAM1G, presented in Figure 14.6.

The fuel cell performances of these membranes are comparable to Nafion® but show limited durability in direct hydrogen fuel cells. To overcome this drawback, membranes based on sulphonated poly (2,6-diphenyl-4-phenylene oxide) and sulphonated poly (arylethersulphone) have been developed. Synthesis of sulphonated poly(phenylene oxide) is done by electropolymerization of substituted phenols (Kamloth and Josowicz 1992). An appropriate quantity of 2-allyl phenol is added to provide functional groups for thermal cross-linking. Sulphonated polyphenylenes are the most desirable materials for PEM applications since they do not contain any chemically degradable groups. However, they suffer from poor solubility and are not flexible to make mechanically stable membranes. To attain the desirable membrane properties of polyphenylene-based PEMs, a number of possible structures can be designed. Synthesis of a copolymer with segments of the perfluorocyclobutyl moiety and polyphenylene is performed, which, on sulphonation, results in a proton exchange membrane with improved properties (Fuller et al. 2009). An ionomer consisting of segments of sulphonated polyphenylenes and a combination of meta and para polyphenylenes is also synthesized (Miyaki et al. 2017). These membranes exhibit proton conductivity of 0.22S/cm and have excellent mechano-chemical stability in both dry and humid conditions. The block polymer of polyphenylene and polyethersulphone is easily soluble in organic solvents and has proton conductivity comparable to Nafion® 211, besides being thermally and mechanically stable (Jang et al. 2017). Conjugated polyphenylenes containing tetraphenylethylenes show good chemical stability and have proton conductivities comparable to Nafion® (Lim et al. 2014). Thus, it is apparent that the membrane morphology plays an important role in determining the ultimate properties of membranes.

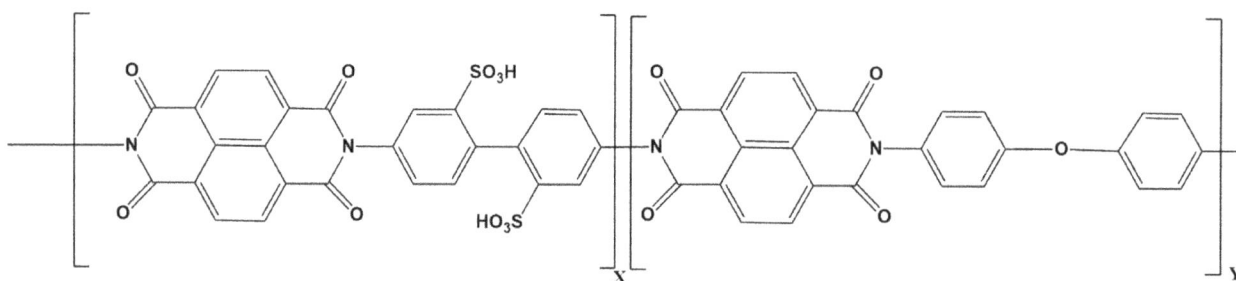

FIGURE 14.7   Structure of sulphonated naphthalenic polyimide.

14.2.7 Membranes Based on Polyimide

Polyimides are regarded as high-performance materials possessing high thermal stability and excellent mechanical properties. Synthesis of proton exchange materials can be done by the incorporation of sulphonic groups in polyimide (Bernard and Sherrington 1999; Gunduz and McGrath 2000). The synthesis of sulphonated polyimides from commercially available monomers is done by copolymerizing 4,4'-diamino-2,2'-biphenyl sulphonic acid with 4,4'-oxydianiline and ether of two dianhydrides, 4,4'-oxydiphthalic anhydride or 1,4,5,8-naphthalene tetracarboxylic anhydride (Gebel et al. 1993; Vallejo et al. 1999; Genies et al. 2001) (Figure 14.7). However, the sulphonated phthalic polyimides failed before 100 hours of testing of the fuel cell while the naphthalic polyimides are stable over 3000 hours.

Synthesis of sulphonated polyimides can be done by using sulphonated monomers and comonomers. The proton conductivities are quite satisfactory at about 0.023 S/cm. Direct sulphonation of polyimide copolymers can lead to proton conductivity up to 0.1 S/cm in some cases (Hong et al. 2002). Hydrolytic solubility of the above-mentioned membranes at 80°C remains a major issue. Randomly sulphonated polyimides with controlled degrees of sulphonation possess higher proton conductivity than Nafion® at temperatures above 140°C (Ye et al. 2006). Sulphonated fluorinated poly(ether imide) copolymers can be synthesized through a one-step high-temperature polycondensation reaction (Mistri et al. 2012). Introduction of an ether linkage leads to an improvement in solubility, and the thermal and mechanical stability of these membranes. Membranes can be made from polyimide-based membranes containing polysiloxane and crown ether moieties (Ivanov et al. 2018). High thermal stability and good solubility are possible with the membranes but have low mechanical properties. These membranes can be designed for low to medium temperature fuel cell applications. Blend membranes of sulphonated polyimide and polybenzimidazole have higher conductivities than Nafion® at 120°C due to the acid-base interaction of sulphonic acid groups on polyimide and benzimidazole rings in polybenzimidazole (Figure 14.8) creating proton-conducting pathways through the membrane matrix (Suzuki et al. 2012). Membranes exhibit stability even after 1000 hours of fuel cell operation.

14.2.8 Membranes Based on Polysulphone

Polysulphone from Udel and polyethersulphone from Victrex are the prominent members (Nolte et al. 1993). Sulphonation

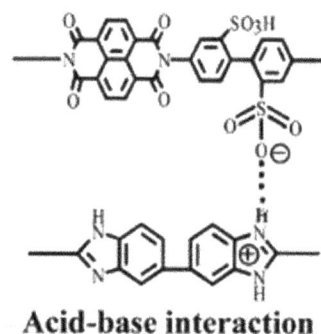

FIGURE 14.8   Scheme of acid-base interaction in sulphonated polyimide and polybenzimidazole membrane.

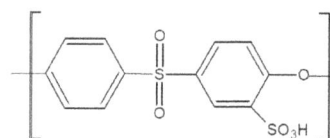

FIGURE 14.9   Structure of sulphonatedxpoly(arylethersulphone).

of these polymers leads to different types of sulphonated membranes (Figure 14.9). One of the methods is the introduction of a sulphonic acid group at the ortho position to the ether bridge of the bisphenol-A portion and the other method is for the diarylsulphone (Kerres et al. 1996). The former sulphonated polymer leads to water solubility for low levels of sulphonation (30%) preventing its application as PEM. But the latter sulphonated polymer leads to materials, which become water-soluble only when the sulphonation level exceeds 95%. Proton conductivity comparable to Nafion® can only be achieved after 90% sulphonation. But this results in an increase in water uptake of more than 70% and as the temperature increases, the water uptake increases further, thereby the membrane disintegrates, leading to a loss in mechanical property. The membranes were cross-linked during casting by polymerizing the sulphonic acid with suitable diamines to overcome the increase in water uptake but this resulted in a decrease in proton conductivity. Another method of obtaining sulphonated polysulphone was adopted by Naim et al. (Naim et al. 2004). Membranes with different degrees of sulphonation were prepared by varying the molar ratio of sulphonating agent, trimethylsilylchlorosulphonate in

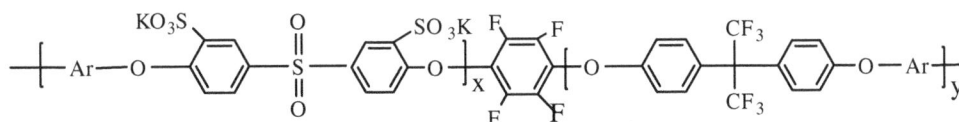

FIGURE 14.10 Structure of hydrophilic-hydrophobic multiblock copolymer.

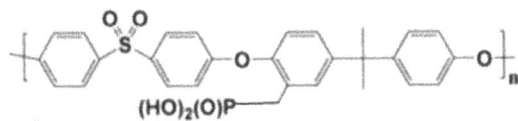

FIGURE 14.11 Structure of polyphosphonated polysulfone.

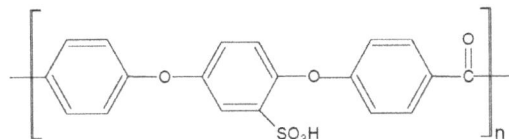

FIGURE 14.12 Structure of sulphonatedpoly(ether ether ketone).

the polymer. They could achieve a sulphonation degree of up to 73%, however, the membranes did not possess conductivity comparable to Nafion®.

Hydrophilic-hydrophobic multiblock copolymers with higher block lengths show better performance with regard to proton conductivity, presumably due to their more distinct nanophase separation and better connectivity among the ionic domains (Yu et al. 2007; Lee et al. 2010) (Figure 14.10).

Proton conductive membranes made from poly(arylene ether sulphone) bearing perfluoroalkylsulphonic acids, derived by polymer post-modification, show higher water uptake, swelling, and proton conductivity compared to Nafion® membranes (Zeng et al. 2017).

Polymers bearing phosphonic acid groups are of interest due to their high thermal stability in comparison to their sulphonic acid counterparts. Chemical modification of polymer backbones can be done to introduce phosphonic acid linkages (Stone et al. 2000; Lafitte and Jannasch 2007). The synthesis of phosphonated polysulphone by the post-phosphonation process achieved a phosphorylation degree of 150% (Abu-Thabita et al. 2010). Polysulphone is first chloromethylated and then phosphonated to get the phosphonated membrane (Figure 14.11). A conductivity value of 12 mS/cm at 100°C under humid conditions can be achieved together with the thermal stability of 252°C and low methanol permeability, which makes these membranes attractive for direct methanol fuel cell applications. The membranes comprising phosphonic acid functionalized polysulphone and triazolyl functionalized polysulphone can be made (Yue et al. 2017). A three-fold improvement in properties over individual acid or base component can be achieved due to the synergetic proton conductive effect.

14.2.9 Membranes Based on Poly (Aryl Ether Ketone)

These are polymers that consist of sequences of ether and carbonyl linkages between phenyl rings. Poly (ether ether ketone) (PEEK) is the most promising alternative membrane material due to its high thermal and mechanical stability and easy film-forming property. It is available commercially under the name Victrex PEEK. The functionality required for proton conductivity is introduced through the sulphonation of PEEK in concentrated sulphuric acid (Figure 14.12). Like sulphonated poly

(arylethersulphone), the complete sulphonation of PEEK leads to a water-soluble product. A sulphonation level of around 60% is a good compromise between the conductivity and mechanical properties of the membranes (Parnian et al. 2017). The SPEEK (sulphonated poly(ether ether ketone)) samples show progressive deterioration of thermal stability, an increase in glass transition temperature, a decrease in the elastic modulus, and a decrease in the chemical stability with increasing the DS (degree of sulphonation). The proton conductivity of SPEEK membranes increases with DS, and for each membrane increases with increasing temperature and humidity.

Sulphonated PEEK can be chemically cross-linked to reduce membrane swelling and increase mechanical strength (Helmer-Metzmann et al. 1995; Yen et al. 1998). The blending of PEEK with polyetherimides or polyethersulphones, aimed at improving the mechanical stability (low swelling) of the membrane, does not exhibit good performance in a fuel cell environment (Swier et al. 2005; 2006).

The SPEEK membranes with pendant sulphonic acid phenyl groups are derived under mild sulphonation condition compared to that used with sulphonation of PEEK (Liu et al. 2006). The membrane exhibits good dimensional stability in hot water unlike sulphonated SPEEK, which tends to swell or dissolve. Further, the other properties are also suitable for fuel cell applications.

Mechanically stable films of cross-linked PEEK followed by grafting of styrene and sulphonation can be produced by using radiation grafting technique (Chen et al. 2010). These membranes exhibit higher cell performances than Nafion® in DMFC at high temperatures. The acid-base blending of SPEEK with non-conjugated diamine provides a two-fold increase in the thermo-mechanical properties and oxidative stability in the membranes compared to sulphonated SPEEK (Hande et al. 2016) (Figure 14.13). The ionic cross-link creates a constrained amorphous region of reduced mobility.

The electrolyte can be designed by blending SPEEK with the sulphonated-fluorinated block copolymer (Kim et al. 2017). This strategy has the advantage of increasing proton conductivity with the increase in weight percentage of SPEEK in the blend while retaining the mechanical properties almost unchanged. The formation of good phase-separated morphology helps in improving the properties.

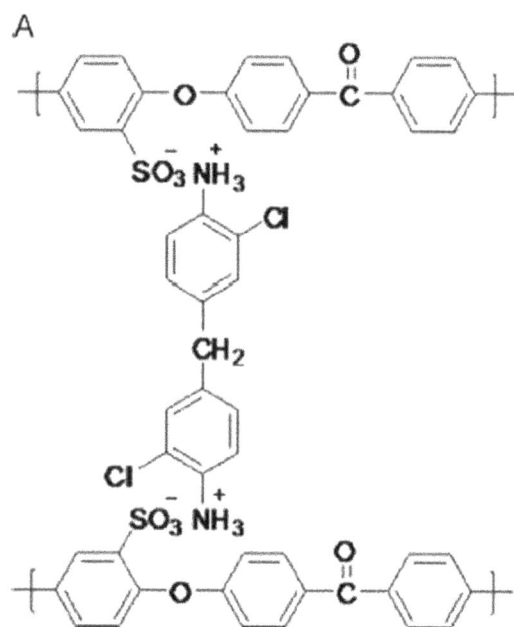

FIGURE 14.13 Schematic of acid-base interaction between SPEEK and MOCA.

14.2.10 Membranes Based on Polybenzimidazole

As discussed in the preceding sections, membranes derived from hydrocarbon-based polymers have been found to be suitable for fuel cell operation as an alternative to Nafion®. However, most of these membranes still depend on humid operational conditions to be useful under real fuel cell conditions. This limits their usage to temperatures of 120°C. First, when a polymer electrolyte fuel cell is operated at temperatures below 120°C, the requirement of pure hydrogen gas becomes stringent. Hydrogen gas with a minimum 99.999% purity is required, otherwise slight traces of carbon monoxide will poison the platinum catalyst and lead to a drop in performance. Second, water management is a major concern. Thus, more complexity is added to the system as a whole in terms of commercialization. When the fuel cell is operated at 150°C and above, the tolerance of the catalyst to carbon monoxide increases to 10,000 ppm. Polybenzimidazole (PBI) membranes have been found to be the most suitable for application in fuel cells operating at a higher temperature (Mader et al. 2008). These membranes offer the advantage of high thermal (Samms et al. 1996) and mechanical stability (Appleby 1996; He et al. 2006) and high conductivity (He et al. 2003; Ma et al. 2004) under dry conditions. PBI Performance Innovations Inc. is the main global manufacturer of PBI products, including powder, solution (with DMAc as the solvent), and PBI fibers in various forms. Those prominent among different types of PBI polymers are poly [2,2'-(meta-phenylene),5,5'-(bibenzimidazole)] (PBI) (Li et al. 2004) and poly(2,5-benzimidazole) (ABPBI) (Asensio and Gomez-Romero 2005) (Figure 14.14). ABPBI polymer has a simpler structure compared to PBI. It is synthesized from a single monomer and hence is less expensive. Also, since it has a higher number of basic sites, it tends to absorb more acid compared to PBI at the same acid bath concentration.

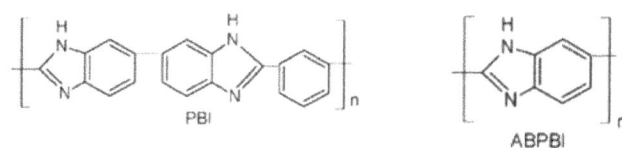

FIGURE 14.14 Structure of PBI and ABPBI.

PBI is doped with strong acids such as sulphuric and phosphoric acid to make it proton-conducting. The proton conductivity follows the order $H_2SO_4>H_3PO_4>HNO_3>HClO_4>HCl$ (Xing and Savadogo 1999). Most frequently, phosphoric acid is used as the dopant since it is a good conductor at higher temperatures. Doping in phosphoric acid is carried out by immersing the membranes in different concentrations of acid baths either at room temperature or at an elevated temperature. Doping takes place at the basic sites of the benzimidazole ring. The level of doping is determined by the weight gain of the polymer. The proton conductivity depends on the number of moles of acid present per repeat unit of the polymer. It has been found that bound and unbound acid molecules are found in doped membranes. Proton transport by the Grotthuss mechanism takes place between the bound acid and the acid anions while the unbound or free acid contributes to conductivity by the vehicular mechanism. Also, hydrogen bonding between the free acid molecules is essential for efficient proton transport. Therefore, an acid doping level of 5–6 moles per repeat unit is required for fuel cell applications. This gives an advantage in terms of water management in fuel cells since proton conduction does not require a humidified environment.

As has been seen above, the conductivity of PBI increases with an increase in acid content or the doping level of the acid. However, the mechanical strength is lowered at higher doping levels. To overcome this issue, several different approaches have been investigated which bring about a compromise between these two critical parameters for the membranes' performance. Synthetically modified PBI structures and acid-base blends with partially fluorinated sulphonated aromatic polyether can be used to minimize the problem (Li et al. 2010). Membranes obtained after doping show reduced swelling which leads to an enhancement in mechanical strength, proton conductivity, and fuel cell performance. Also, by modifying the structure of the polymer backbone, physicochemical properties can be tailored to achieve the desired results (Quartarone and Mustarelli 2012). A series of polymers are made by altering the number of nitrogen atoms in the polymer backbone by using different monomers. It was inferred that increasing the polymer molecular weight and basicity had a positive influence on the acid doping level, retention of acid and proton conductivity, along with good dimensional stability.

Other methods of enhancing the properties of PBI membranes include cross-linking, blending with other polymers or the addition of inorganic materials to form composites. Ionic cross-linking with acidic polymers such as SPEEK or sulphonated polysulphones can also be adopted. The acid-base polymer blends form hydrogen bonds and ionic cross-linking within the polymer matrix, which leads to a marked improvement in the thermal stability of the membranes. In DMFC, a comparable performance of these membranes with Nafion® is observed. PBI blends with partially fluorinated arylene polyethers exhibit high mechanical strength even at high doping levels of 11–12 moles of acid per repeat unit with proton conductivity values of 0.12S/cm at 175°C (Li et al. 2008). The selection of partially fluorinated polyether and its sulphonation and blending with PBI lead to a remarkable improvement in the membrane properties.

Polybenzimidazole can also be covalently cross-linked thermally or through an amide-type of linkage. Several types of covalent cross-linkers such as terephthaldehyde, divinylsulphone, *p*-xylene dichloride, and many more are possible (Sansone 1987; Wang et al. 2006; Xu et al. 2007); p-xylene-dibromide as a covalent cross-linker is also possible (Li et al. 2007). The cross-linking takes place through the amide-type of linkage which improves the oxidative stability of membranes.

14.3 Composite Membranes

Another method of making PEMs is by incorporating inorganic compounds in the polymer matrix. These are called inorganic-organic composite or hybrid membranes. Composite membranes are promising candidates since they combine the properties of both the polymer matrix as well as the organic or inorganic component. In addition to enhancing the mechanical and conducting properties, an inorganic component may assist in improving the thermal stability, water absorption, reactant cross-over resistance, and other properties of the polymer membranes. Various types of inorganic proton conductors are possible with different types of polymer matrix (Nagaral et al. 2010). Hydrophilic oxides such as silica are known to enhance the water uptake and water retention capacity of the membranes, thereby improving proton conductivity. Hybrid membrane based on polyvinyl alcohol and nano-porous silica can be made (Beydaghi et al. 2011). These water-insoluble hybrid materials are confirmed as potential candidates for PEMFC applications. A hybrid nano-composite of OPBI and a modified nano-silica exhibits increased proton conductivity with increasing nano-silica content in the polymer along with an improvement in the thermal and mechanical properties of the membranes (Ghosh et al. 2011). Similarly, Nafion® and SPEEK-based composite membranes can be designed (Pereira et al. 2008; Du et al. 2012; Gashoul et al. 2017). Organoclay can be used as a pseudo-cross-linker in SPEEK (Hande et al. 2011). Incorporation of organoclay into cross-linked SPEEK imparts an additional degree of cross-linking to the membranes. As a result, the cross-linked SPEEK/clay nanocomposite membranes acquire higher tensile strength, modulus, and lower elongation at break compared to pristine and neat cross-linked SPEEK. Further, the oxidative stability of SPEEK improves remarkably with cross-linking and clay addition. These improvements in the thermo-mechanical, barrier and oxidative stability of the membranes occur without significantly affecting the proton conductivity.

POSS (polyhedral oligomericsilsesquioxanes) is another material that has multi-functionality and is used as an efficient building block for hybrid materials. Different types of POSS materials are used to tailor the properties of membranes for PEM applications. Polysulphone/POSS nano-fiber composite membranes are made by using the process of electrospinning with the incorporation of polyurethane in the nano-voids of the fiber (Choi et al. 2010). The membranes exhibit proton conductivity of the order of 2.5 times higher than Nafion® and 4.3 times higher than sulphonated polysulphone membranes. The formation of the nanofiber network in the inert polymer matrix also helps in controlling the membrane swelling. POSS materials are also used in PBI and ABPBI polymers to enhance the proton conductivity and reduce acid leaching during fuel cell operation in the long term (Aili et al. 2014; Lui et al. 2016). Comb-shaped grafting of multi-sulphonated polyhedral oligosilsesquioxane grafted onto poly(arylene ether sulphone) is another interesting route to synthesize membranes with improved physico-chemical properties (Wu et al. 2017). The introduction of POSS into the polymer matrix leads to significant phase separation and well-defined interconnected hydrophilic networks. This results in enhanced properties of the membranes for use in fuel cells.

Composite membranes have been reviewed extensively by Asier Goni-Urtiaga et al. Among the solid acids as electrolyte materials, hetero-poly acid has received more attention due to its simple preparation and strong acidity (Goni-Urtiaga et al. 2012). However, they are soluble in water and also are a good catalyst for the formation of free radicals from H_2O_2 (Huang et al. 2008). To avoid this, a multifunctional catalyst -$CsxH3_xPW12O40$/CeO_2 is incorporated into the ABPBI polymer to form a composite membrane, which improves the conductivity and oxidative stability of the membrane (Qian et al. 2013). Also, the fuel cell performance is better than pristine polymer. A marked improvement in the ABPBI membranes properties is possible with the introduction of ZrP in the polymer (Rao et al. 2017). Fuel cell tests show a steady performance for more than 800 hours.

Ionic liquids are another interesting class of materials that have garnered attention due to their high intrinsic conductivity and high stability. They are non-volatile due to the strong electrostatic force between them and also they possess low flammability. Ionic liquids are molten salts at room temperature and are composed of organic cations and various anions such as I^-, BF_4^+,Pf^{6-}, etc. With such interesting properties, the ionic liquids can be used to design fuel cell membranes for high-temperature applications (Diaz et al. 2014). Ionic liquids can be classified as aprotic and protic ionic liquids. Protic ionic liquids (PILs) are formed by proton transfer from a Brönsted acid, AH, to a Brönsted base and have a mobile proton located on the cation. Hence these have been studied by a number of research groups for their applicability in fuel cells. Aprotic ionic liquids (AILs) contain substituents other than a proton (typically an alkyl group) at the site occupied by the labile proton in an analogous protic ionic liquid. These have been used more in the case of studies on lithium-ion batteries. Hybrid membranes, prepared from a mixture of polymerizable oils, protic ionic liquids, and nano-silica particles, exhibit a conductivity value around 1×10^{-2} S.cm^{-1} at 160° C under anhydrous conditions (Lin et al. 2010). However, retention of PIL in the polymer network remains a challenge, which can be overcome by design and the synthesis of polymer networks with the interacting groups that can hold the PIL. Sulphonated polyimides (SPI) can be doped with PIL to make composite membranes (Chen et al. 2014; 2015). An improvement in the proton conductivity up to 1000 times more than pristine membranes is achieved by making composite membranes of SPI with PIL. Use of an aprotic liquid in making a composite membrane with cross-linked SPEEK results in high thermal stability and improvement in the retention of ionic liquid using cross-linked SPEEK (Malik et al. 2015).

14.4 Characterization Techniques for Polymer Electrolyte Membranes

A PEM fuel cell comprises several components which are individually responsible for the overall functioning of the fuel cell.

14.4.1 Current-Voltage Characteristics

The performance of the fuel cell is determined by the polarization curve which represents the current-voltage characteristics of the membrane electrode assembly. A typical polarization curve is represented in Figure 14.15. It has a characteristic shape which is the result of four over-potentials or irreversibilities.

The first over-potential loss is due to the mixed potential of the oxygen electrode and low rate of hydrogen cross-over which shows a deviation from the theoretical voltage of the fuel cell. The second loss is due to the activation polarization where the voltage loss is a result of overcoming the energy barrier during electrode reactions. This takes place predominantly on the oxygen side of the fuel cell and is associated with the electrode kinetics. The third loss is from the ohmic over-potential. This occurs due to resistance to the ionic current in the membrane and electronic current in the electrode. The contribution of the loss due to the electronic resistance is negligible in comparison to the resistance offered by the membrane. Hence this performance loss is noteworthy when designing materials for membranes. The final loss is due to mass or concentration polarization. This occurs at high current densities and is due to the concentration loss of either the fuel or oxidant at the respective electrodes.

From the foregoing discussion, it is evident that the role of the membrane is significant in the performance of the fuel cell. Thus, qualifying the materials before introduction into a fuel cell is necessary to ascertain its feasibility of application. The

characterization is done on free-standing membranes and on membrane electrode assembly. The functional requirements of the membranes were covered in Section 14.3. Here the techniques used to determine those properties will be discussed.

The membrane is characterized in terms of its physico-chemical properties, such as ion exchange capacity, water uptake, and proton conductivity. Stability of the membranes is studied by evaluating the thermal and mechanical properties. Performance characteristics under fuel cell conditions are determined in a unit fuel cell.

The ion exchange capacity of the membranes is measured using the traditional titration method. A typical measurement involves soaking the membrane in a solution of a known concentration of sodium hydroxide for a fixed time period. Change in normality is then determined after titration against standard HCl. An average of three readings is taken and the IEC is calculated from the formula:

$$IEC\left(meq\,/\,g\right) = \frac{\left(N1 - N2\right)}{W} \times V$$

where N1 is the initial normality of NaOH, N2 is the final normality of NaOH, V is the volume of NaOH taken in ml, W is the weight of membrane in gm.

Water uptake is measured by the gravimetric method. Typically, the membranes (thoroughly dried) measuring 2" x 2" are taken in a round bottom flask containing distilled water. The membranes are refluxed for 24 hours and left overnight at room temperature in the water to equilibrate.

The membranes are taken out, patted dry on absorbent paper to remove any water adhering to the surface. The weight of these membranes was noted down. The membranes are then dried in a vacuum oven at 80°C for 24 hours and the dry weight is noted down. The water uptake is calculated from the equation:

$$Water\ uptake = \frac{W_w - W_d}{W_w} \times 100$$

where W_w is the weight of the wet membrane and W_d is the weight of the dried membrane.

The ionic conductivity can be measured by using alternative or direct current methods. The former is more suitable with a fixed frequency or applying transient techniques. The ionic conductivity depends not only on the synthesis parameters but also on the synthesis procedure.

The resistance of the membranes is measured using 0.1 and 0.5 M $H2SO_4$ by applying a DC pulse current and measuring the potential difference between the two Calomel reference electrodes. The conductivity set-up consists of a membrane holder, in which the membrane to be measured is fixed. The membrane separates the cell into two compartments. Two platinum electrodes are inserted into these compartments. The electrolyte is continuously flushed through the compartment under gravity flow. The potential difference is measured in both modes by switching the polarity of the electrodes at different current densities.

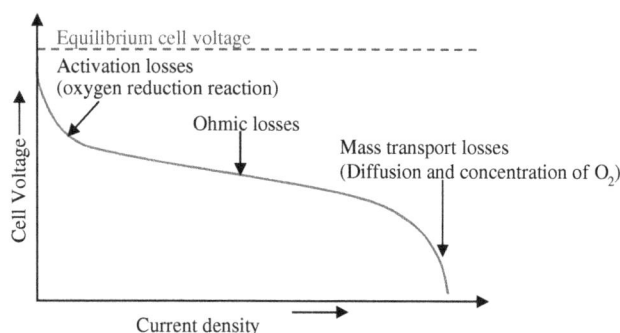

FIGURE 14.15 Typical polarization curve for fuel cell.

In the AC method, an AC signal in the frequency range 1–10 mHz is applied and the impedance data is recorded. The resistivity is calculated from the equation:

$$\rho = \frac{RA}{L}$$

where ρ is the resistivity, R is the value of resistance obtained from the Nyquist plot, A is the area of the electrode, and L is the path length.

The measurements can be carried out in two ways, either by the two-probe method or the four-probe method. In the four-probe method, four probes are placed equidistant from each other. The two in the center supply current while the two at the edges measure the potential difference. In this case, the surface conductivity can be measured. In-plane measurements can be made by placing the current probes on the opposite sides of the voltage probes, however, this does not give accurate values. The best method to determine in-plane conductivity is by the two-probe technique. Here, the membrane is sandwiched between two electrodes of known dimensions and the current-generating probes also function as voltage-measuring probes. The other advantage of this configuration is that the test cell is easy to assemble and measurements can be done at different humidity levels since the membrane is exposed to the water vapor through the open window. The conductivity values obtained depend on various factors, such as the cell configuration, sample dimensions, contact resistance of the probes, temperature, and humidity levels during measurements.

In-situ conductivity measurements can also be carried out by the electrochemical impedance spectroscopy technique where the measurements are done on the membrane electrode assembly. Here, the original anode behaves as the reference electrode and the cathode is connected to the working electrode of the potentiostat.

This method gives the in-plane conductivity of the membrane. Two methods, namely, the current interrupt and high-frequency resistance method, can be used to determine the conductivity. For more details, see (Gomadam and Weidnern 2005).

14.4.2 Thermal and Mechanical Properties

Thermal stability of the membranes is determined through a thermogravimetric analyzer. Typically, the membranes are dried till constant weight before the test. They are then subjected to heating from ambient temperature to 800°C in an air/nitrogen atmosphere. The initial decomposition temperature is measured from the TGA plot. It was obtained from the first point of inflection in the primary thermogram. This gives an approximate estimate of the degradation temperature of the membrane since the temperature may differ under actual fuel cell conditions due to the influence of other electrochemical and physical factors. Another important parameter that is taken into consideration while designing membrane materials is the glass transition temperature. This is important in terms of the procedure used for making the membrane electrode assembly, wherein the membrane is usually hot-pressed along with the electrodes. The thermal stability, as well as the mechanical stability, is crucial under these circumstances to achieve defect-free MEAs. The values of the glass transition temperature are recorded from the DSC thermograms or in some cases from the plot obtained during dynamic mechanical analysis.

Mechanical stability is generally determined by the tensile behavior of the membranes. The test samples are cut into strips of length 10 cm and width 1 cm using a standard die. This test is carried out for both dry and wet samples. The samples are then clamped between a fixed claw and a movable claw at a pre-fixed distance. They are then stretched in the vertical direction at a

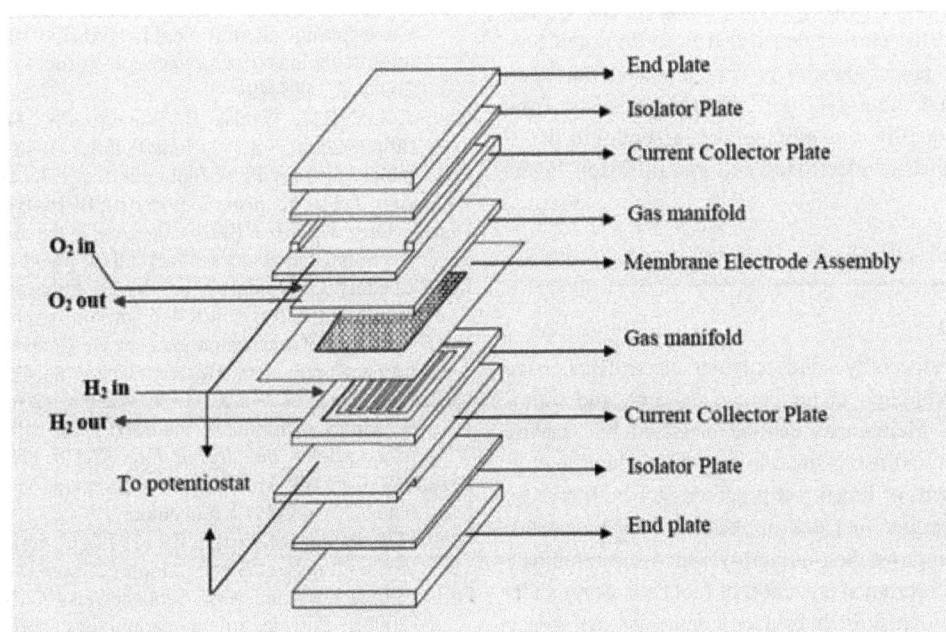

FIGURE 14.16 Schematic representation of a unit fuel cell.

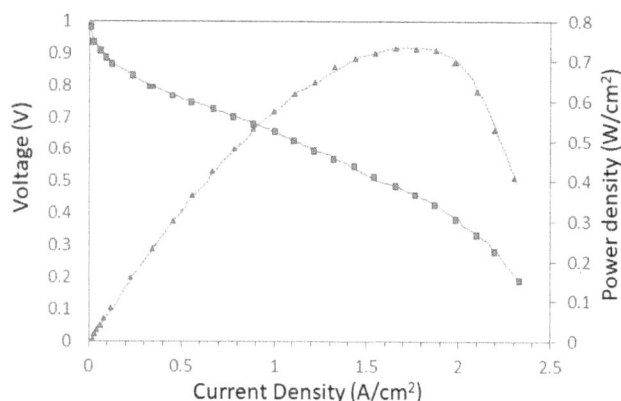

FIGURE 14.17 Power density representation of fuel cell performance.

known cross-head speed and load. The tensile strength and elongation at break are calculated from the stress-strain curve.

14.4.3 Fuel Cell Tests

The performance of the membrane under actual fuel cell conditions is determined in a unit cell. This consists of a single MEA with an active area of 10–25 sq cm, graphite flow field plates, gaskets, current collector plates, and end plates as shown in Figure 14.16. This MEA is tested at different operating conditions such as temperature, pressure, flow rate, and humidity levels. In most cases, an automated test station is used to carry out the evaluation. The current-voltage curve obtained indicates the performance of the MEA under different conditions (Figure 14.17).

Performance of the MEA is determined and reported in terms of current density and power density. Current density is proportional to the rate of electrochemical reaction taking place in the fuel cell. This is the value obtained by dividing the current by the active area. Power density is the product of the voltage of the fuel cell and the current density. It is to be noted here that the maximum power density is dependent on the operating conditions and other fuel cell components apart from the membrane. To qualify a membrane, it is essential to determine the performance of electrodes and gas diffusion layers independently.

14.5 Conclusion

Solid electrolytes, especially solid polymer electrolytes, have been a major breakthrough in membrane research and ultimately in fuel cells. Membranes can be designed by varying different parameters and using suitable materials which lead to improvements in terms of functional properties for both low- as well as high-temperature fuel cell applications. To be useful, the materials have to prove their durability which still remains a major obstacle in the commercialization of fuel cells using alternate membranes. Progress in hybrid or composite membrane may pave the way for cheaper and highly efficient membranes for fuel cells.

REFERENCES

Abu-Thabita NY, Ali SA, Zaidi SMJ. (2010). New highly phosphonated polysulphone membranes for PEM fuel cells. *J. Membr. Sci.* 360(1–2): 26–33.

Agrawal RC, Pandey GP. (2008). Solid polymer electrolytes: Materials designing and all-solid-state battery applications: an overview. *J. Phys. D, Appl. Phys.* 41(22): 223001.

Aili D, Allward T, Alfaro SM, Hartmann-Thompson C, Steenberg T, Hjuler HA, Li Q, Jensen JO, Stark EJ. (2014). Polybenzimidazole and sulphonated polyhedral oligosilsesquioxane composite membranes for high-temperature polymer electrolyte membrane fuel cells. *Electrochim. Acta.* 140: 182–190.

Appleby AJ. (1996). Issues in fuel cell commercialization. *J. Pow. Sources* 58(2): 153–176.

Appleby AJ, Foulks FR. (1989). *Fuel Cell Handbook.* New York: Van Nostrand Reinhold.

Appleby AJ, Yeager EB. (1986). Solid polymer electrolyte fuel cells (SPEFCs). *Energy* 11(1–2): 137–152.

Asensio JA, Gomez-Romero P, (2005). Recent developments on proton conducting poly2,5-benzimidazole)(ABPBI) membranes for high-temperature polymer electrolyte membrane fuel cells. *Fuel Cells* 5(3): 336–343.

Bernard RJ, Sherrington DC. (1999). Paper presented at 5th European Tech. Symp. Polyimides and High-Performance, Functional Polymers Proceedings, Montpellier, France.

Beydaghi H, Javanbakht M, Amoli HS, Badiei A, Khaniani Y, Ganjali MR, Norouzi P, Abdouss M. (2011). Synthesis and characterization of new proton conducting hybrid membranes for PEM fuel cells based on poly(vinyl alcohol) and nanoporous silica containing phenyl sulphonic acid. *Intl. J. Hyd. Eng.* 36(20): 13310–13316.

Chen B-Q, Wong J-M, Wu T-Y, Chen L-C, Shih C. (2014). Improving the conductivity of sulphonated polyimides as proton exchange membranes by doping of a protic ionic liquid. *Polymers* 6(11): 2720–2736.

Chen B-Q, Wu T-Y, Wong J-M, Chang Y-M, Lee H-F, Huang W-Y, Chen A-F. (2015). Highly sulphonated diamine synthesized polyimides and protic ionic liquid composite membranes improve PEM conductivity. *Polymers* 7(6): 1046–1065.

Chen J, Li D, Koshikawa H, Asano M, Maekawa Y. (2010). Crosslinking and grafting of polyetheretherketone film by radiation techniques for application in fuel cells. *J. Membr. Sci.* 362(1–2): 488–494.

Choi J, Lee KM, Wycisk R, Pintauro PN, Mather PT. (2010). Sulphonated polysulfone/POSS nanofiber composite membranes for PEM fuel cells fuel cells and energy conversion. *J. Electrochem. Soc.* 157(6): B914–B919.

Díaz M, Ortiz A, Ortiz I. (2014). Progress in the use of ionic liquids as electrolyte membranes in fuel cells. *J. Membr. Sci.* 469: 379–396.

Ding J, Chuy C, Holdcroft S. (2002). Enhanced conductivity in morphologically controlled proton exchange membranes: Synthesis of macromonomers by SFRP and their incorporation into graft polymers. *Macromolecules* 35(4): 1348–1355.

Du L, Yan X, He G, Wu X, Hu Z, Wang Y. (2012). SPEEK proton exchange membranes modified with silica sulphuric acid nanoparticles. *Intl. J. Hyd. Eng.* 37(16): 11853–11861.

Ezzell BR, Carl WP, Mod WA. 1980. Preparation of vinyl ethers. US Patent No. 4358412, November 9.

Fenton DE, Parker JM, Wright PV. (1973). Complexes of alkali metal ions with poly(ethylene oxide). *Polymer* 14(11): 589.

Fuller TJ, MacKinnon SM, Schoeneweiss CMR, Gittleman CS. (2008). Polyelectrolyte membranes comprised of blends of PFSA and sulphonated PFCB polymers. US Patent No. 20090278083A1.

Gashoul F, Parnian MJ, Rowshanzamir S. (2017). A new study on improving the physicochemical and electrochemical properties of SPEEK nanocomposite membranes for medium temperature proton exchange membrane fuel cells using different loading of zirconium oxide nanoparticles. *Intl. J. Hyd. Eng.* 42(1): 590–602.

Gebel G, Aldebert P, Pineri M. (1993). Swelling study of perfluorosulphonate ionomer membranes. *Polymer* 34(2): 333–339.

Genies CC, Mercier RR, Sillion BB, Cornet N, Gebel G, Pineri M. (2001). Soluble sulphonated naphthalenic polyimides as materials for proton exchange membranes. *Polymer* 42(2): 359–373.

Ghosh S, Maity S, Jana T. (2011). Polybenzimidazole/silica nanocomposites: Organic-inorganic hybrid membranes for PEM fuel cell. *J. Mater. Chem.* 21: 14897–14906.

Gomadam PM, Weidnern JW. (2005). Analysis of electrochemical impedance spectroscopy in proton exchange membrane fuel cells. *Int. J. Energy Res.* 29: 1133–1151.

Goni-Urtiaga A, Presvytes D, Scott K. (2012). Solid acids as electrolyte materials for proton exchange membrane (PEM) electrolysis: Review. *Intl. J. Hyd. Eng.* 37(4): 3358–3372.

Gunduz N, McGrath JE. (2000). Wholly aromatic five- and six-membered ring polyimides containing pendant sulphonic acid functional groups. *Polym. Prep.* 41(2): 1565–1566.

Hande VR, Rath SK, Rao S, Patri M. (2011).Cross-linked sulphonated poly (ether ether ketone) (SPEEK)/reactive organoclay nanocomposite proton exchange membranes (PEM). *J. Membr. Sci.* 372(1–2): 40–48.

Hande VR, Rath SK, Rao S, Praveen S, Sasane S, Patri M. (2016). Effect of constrained amorphous region on properties of acid-base polyelectrolyte membranes based on sulphonated poly(ether ether ketone) and a nonconjugated diamine. *J. Membr. Sci.* 499: 1–11.

He R, Li Q, Bach A, Jensen JO, Bjerrum NJ. (2006). Physicochemical properties of phosphoric acid doped polybenzimidazole membranes for fuel cells. *J. Membr. Sci.* 277(1–2): 38–45.

He R, Li Q, Xiao G, Bjerrum NJ. (2003). Proton conductivity of phosphoric acid doped polybenzimidazole and its composites with inorganic proton conductors. *J. Membr. Sci.* 226(1–2): 169–184.

Helmer-Metzmann F, Osan F, Schneller A, Ritter H, Ledjeff K, Nolte R, Thorwirth R, (1995). Polymer electrolyte membrane, and process for the production thereof. US Patent No. 5438082.

Hodgdon RB. (1968). Polyelectrolytes prepared from perfluoroalkylaryl macromolecules. *J. Polym. Sci.* 6(1): 171–191.

Hodgdon RB, Enos JF, Aiken EJ (1967). Sulphonated polymers of a α, β, β-. triflorostyrene with applications to structure and cells. US Patent No. 3341366 A.

Hong YT, Einsla B, Kim YS, McGrath JE. (2002). Synthesis and characterization of sulphonated polyimides based on six-member ring as proton exchange membrane. *Polym. Prepr.* 43(1): 666–667.

Huang QZ, Zhuo LH, Guo YC. (2008). Heterogeneous degradation of chitosan with H2O2 catalyzed by phosphotungstate. *Carbohydrate Polym.* 72(3): 500–505.

Ivanov VS, Yegorov AS, Rantovichallakhverdov G, Men'shikov VV. (2018). Synthesis and investigation of polyimide-based proton-exchange membranes containing polysiloxane and crown ether moiety. *Orient. J. Chem.* 34(1): 255–264.

Jang H, Ryu T, Sutradhar SC, Ahmed F, Choi K, Yang H, Yoon S, Kim W. (2017). Studies of sulphonated poly(phenylene)-block-poly(ethersulfone) for proton exchange membrane fuel cell. *Intl J. Hyd. Eng.* 42(17): 12768–12776.

Johnson BC, Yilgor I, Tran C, Iqbal M, Wightman JP, Lloyd DR, McGrath JE. (1984). Synthesis and characterization of sulphonated poly(arylene ether sulfones). *J. Polym. Sci: Polym. Chem.* 22(3): 721–737.

Kamloth KP, Josowicz M. (1992). Electrochemical preparation of semipermeable polymer membranes on carbon fiber microelectrodes for selective amperometric detection of cations. *Ber. Bunsen. Phys. Chem.* 96: 1004–1017.

Kerres J, Cui W, Reichle S. (1996). New sulphonated engineering polymers via the metalation route. I. Sulphonatedpoly(ethersulfone) PSU Udel® via metalation-sulphonation-oxidation. *J. Polym. Sci. Part A: Polym. Chem.* 34: 2421–2438.

Kim AR, Vinothkannan M, Yoo DJ. (2017). Sulphonated-fluorinated copolymer blending membranes containing SPEEK for use as the electrolyte in polymer electrolyte fuel cells (PEFC)". *Intl. J. Hyd. Eng.* 42(7): 4349–4365.

Lafitte B, Jannasch P. (2007). On the prospects for phosphonated polymers as proton exchange fuel cell membranes. *Adv. Fuel Cells* 1: 119–185.

Lee HS, Roy A, Lane O, Lee M, McGrath JE. (2010). Synthesis and characterization of multiblock copolymers based on hydrophilic disulphonatedpoly(arylene ether sulfone) and hydrophobic partially fluorinated poly(arylene ether ketone) for fuel cell applications. *J. Polym. Sci. Part A-Polym. Chem.* 48(1): 214–222.

Li Q, He R, Jensen JQ, Bjerrum NJ. (2004). PBI-based polymer membranes for high-temperature fuel cells—Preparation, characterization, and fuel cell demonstration. *Fuel Cells* 4(3): 147–159.

Li Q, Jensen JO, Pan C, Bandur V, Nilsson MS, Schönberger F, Chromik A, Hein M, Haring T, Kerres J, Bjerrum NJ. (2008). Partially fluorinated arylenepolyethers and their ternary blends with PBI and H3PO4. Part II. Characterization and Fuel Cell Tests of the Ternary Membranes. *Fuel Cells* 08 (3–4): 188–199.

Li Q, Pan C, Jensen JO, Noye P, Bjerrum NJ. (2007). Cross-linked polybenzimidazole membranes for fuel cells. *Chem. Mater.* 19(3): 350–352.

Li QF, Rudbeck HC, Chromik A, Jensen JO, Pan C, Steenberg T, Calverley M, Bjerrum NJ, Kerres J. (2010). Properties, degradation and high-temperature fuel cell test of different types of PBI and PBI blend membranes. *J. Membr. Sci.* 347(1–2): 260–270.

Lim Y, Lee D, Choi S, Lee S, Jang H, Lee S, Hong T, Kim W. (2014). Synthesis and characterization of sulphonated polyphenylene containing DCTPE for PEMFC potential application. *Intl. J. Hyd. Eng.* 39(36): 21531–21537.

Lin B, Cheng S, Qiu L, Yan F, Shang S, Lu J. (2010). Protic ionic liquid-based hybrid proton-conducting membranes for anhydrous proton exchange membrane application. *Chem. Mater.* 22(5): 1807–1813.

Linkous CA. (1993). Development of solid polymer electrolytes for water electrolysis at intermediate temperatures. *Int. J. Hydrogen Energy* 18(8): 641–646.

Liu B, Robertson GP, Kim D-S, Guiver MD, Hu W, Jiang Z. (2006). Aromatic poly(ether ketone)s with pendant sulphonic acid phenyl groups prepared by a mild sulphonation method for proton exchange membranes. *Macromolecules* 40(6): 1934–1944.

Liu F, Yi B, Xing D, Yu J, Fu Y. (2003b). Development of novel self-humidifying composite membranes for fuel cells. *J. Pow. Sources* 124(1): 81–89.

Liu F, Yi B, Xing D, Yu J, Zhang H. (2003a). Nafion/PTFE composite membranes for fuel cell applications. *J. Membr. Sci.* 212(1–2): 213–223.

Liu Q, Sun Q, Ni N, Luo F, Zhang R, Hu S, Bao X, Zhang F, Zhao F, Li X. (2016). Novel octopus-shaped organic-inorganic composite membranes for PEMFCs. *Intl. J. Hyd. Eng.* 41(36): 16160–16166.

Ma YL, Wainright JS, Litt MH, Savinell RF. (2004). Conductivity of PBI membranes for high-temperature polymer electrolyte fuel cells. *J. Electrochem. Soc.* 151(1): A8–A16.

Mader J, Xiao L, Schmidt TJ, Benicewicz BC. (2008). Polybenzimidazole/ acid complexes as high-temperature membranes. In G.G. Scherer (Ed.), *Fuel Cells II, Advances in Polymer Science*. Berlin: Springer-Verlag.

Malik RS, Verma P, Choudhary V. (2015). A study of new anhydrous, conducting membranes based on composites of aprotic ionic liquid and cross-linked SPEEK for fuel cell application. *Electrochim. Acta* 152: 352–359.

Mistri EA, Mohanty AK, Banerjee S. (2012). Synthesis and characterization of new fluorinated poly(ether imide) copolymers with controlled degree of sulphonation for proton exchange membranes. *J. Membr. Sci.* 411–412: 117–129.

Mittal KI. (Ed.) (1984). *Polyimides: Synthesis, Characterisation, and Applications*. Vol. 2. New York: Plenum.

Miyake J, Taki R, Mochizuki T, Shimizu R, Akiyama R, Uchida M, Miyatake K. (2017). Design of flexible polyphenylene proton-conducting membrane for next-generation fuel cells. *Sci. Adv.* 3(10): EAA0476.

Nagarale RK, Shina W, Singh PK. (2010). Progress in ionic organic-inorganic composite membranes for fuel cell applications *Polym. Chem.* 1: 388–408.

Naim R, Ismail AF, Saidi H, Saion E. (2004). Development of sulphonated polysulfone membranes as a material for proton exchange membrane (PEM). Available at: http://eprints.utm.my/id/eprint/1037/

Nolte R, Ledjeff K, Bauer M, Mulhaupt R. (1993). Partially sulphonatedpoly(arylene ether sulfone): A versatile proton conducting membrane material for modern energy conversion technologies. *J. Membr. Sci.* 83(2): 211–220.

Nuttal LJ, Brown WA. (1971). Solid polymer electrolyte water electrolysis system. Paper presented at SAE/ASME/A Conference, Los Angeles.

Parnian MJ, Rowshanzamir S, Gashoul F. (2017). Comprehensive investigation of physicochemical and electrochemical properties of sulphonated poly (ether ketone) membranes with different degrees of sulphonation for proton exchange membrane fuel cell applications. *Energy* 125: 614–628.

Penner SS. (1986). Solid polymer electrolyte fuel cells (SPEFCS). In *Assessment of Research Needs for Advanced Fuel Cells*. Oxford: Elsevier, pp. 137–152.

Pereira F, Vallé K, Belleville P, Morin A, Lambert S, Sanchez C. (2008). Advanced mesostructured hybrid silica–Nafion membranes for high-performance PEM *Fuel Cell. Chem. Mater.* 20(5): 1710–1718.

Phadnis S, Patri M, Hande V, Deb PC. (2003). Proton exchange membranes by grafting of Styrene-Acrylic acid onto FEP by pre-irradiation technique: 1. Effect of synthesis conditions. *J. Appl. Polym. Sci.* 90(9): 2577.

Phair JW, Badwal S. (2006). Review of proton conductors for hydrogen separation. *Ionics* 12(2): 103–115.

Prater K. (1990). The renaissance of the solid polymer fuel cell. *J. Pow. Sources* 29(1–2): 239–250.

Prober M. (1953). The synthesis and polymerization of some fluorinated styrenes. *J. Am. Chem. Soc.* 75(4): 968–973.

Qian W, Shang Y, Wang S, Xie X, Mao Z. (2013). Phosphoric acid doped composite membranes from poly (2,5-benzimidazole) (ABPBI) and CsxH3–xPW12O40/CeO2 for the high-temperature PEMFC. *Intl. J. Hyd. Eng.* 38(25): 11053–11059.

Quartarone E, Mustarelli P. (2012). Polymer fuel cells based on polybenzimidazole/H3PO4. *Energy Environ. Sci.* 5(4): 6436–6444.

Rao S, Hande VR, Sawant SM, Patri M. (2017). Indian patent application no. 201711038293.

Riedinger H, Faul W. (1988). The focusing of membrane R&D on areas of commercial potential. *J. Membr. Sci.* 36: 5–18.

Samms SR, Wasmus S, Savinell RF. (1996). Thermal stability of proton conducting acid doped polybenzimidazole in simulated fuel cell environments. *J. Electrochem. Soc.* 143(4): 1225–1232.

Sansone MJ (1897). Crosslinking of polybenzimidazole polymer with divinylsulfone. US Patent No. 4666996996.

Savadogo O. (1998). Emerging membranes for electrochemical systems. I. Solid polymer membranes for fuel cell systems. *J. New Mater. Electrochem. Syst.* 1: 47–66.

Soczka-Guth T, Baurmeister J, Frank G, Knauf R.. (1999). Method for producing a membrane used to operate a fuel cell and electrolyzers. US Patent No. 635514999/29763.

Srinivasan S, Velev OA, Parthasarathy A, Manko DJ, Appleby AJ. (1991). High energy efficiency and high power density proton exchange membrane fuel cells — electrode kinetics and mass transport. *J. Pow. Sources* 36(3): 299–320.

Steck AE, Stone C. (1999). α, β, β-trifluorostyrene-based composite membranes. US Patent No. 5985942.

Steck O, Savadago PR, Roberge TNV. (1995). Paper presented at International Symposium on New Materials for Fuel Cell Systems, Montreal, July.

Stone C, Daynard TS, Steck AE. (2000). Phosphonic acid functionalized proton exchange membranes for PEM fuel cells. *J. New Mater. Electrochem. Syst.* 3: 43–50.

Suzuki K, Iizuka Y, Tanaka M, Kawakami H. (2012). Phosphoric acid-doped sulphonated polyimide and polybenzimidazole blend membranes: High proton transport at wide temperatures under low humidity conditions due to new proton transport pathways. *J. Mater. Chem.* 22(45): 23767–23772.

Swier S, Ramani V, Fenton JM, Kunz HR, Shaw MT, Weiss RA. (2005). Polymer blends based on sulphonatedpoly(ether ketone ketone) and poly(ether sulfone) as proton exchange membranes for fuel cells. *J. Membr. Sci.* 256(1–2): 122–133.

Swier S, Shaw MT, Weiss RA (2006). Morphology control of sulphonatedpoly(ether ketone ketone) poly(ether imide) blends and their use in proton-exchange membranes. *J. Membr. Sci.* 270(1–2): 22–31.

Tang HL, Pan M. (2008). Synthesis and characterization of a self-assembled Nafion/silica nanocomposite membrane for polymer electrolyte membrane fuel cells. *J. Phys. Chem. C* 112(30): 11556–11568.

Vallejo E, Pourcelly G, Gavach C, Mercier R, Pineri M. (1999). Sulphonated polyimides as proton conductor exchange membranes. Physicochemical properties and separation H+/Mz+ by electrodialysis comparison with a perfluorosulphonic membrane. *J. Membr. Sci.* 160(1): 127–137.

Wang KY, Xiao Y, Chung T-S. (2006). Chemically modified polybenzimidazole nanofiltration membrane for the separation of electrolytes and cephalexin. *Chem. Eng. Sci.* 61(17): 5807–5817.

Wei J, Stone C, Steck AE. (1995). Trifluorostyrene and substituted trifluoro-styrene copolymeric compositions and ion-exchange membranes formed therefrom. US Patent No. 5422411.

Wnek GE, Rider JN, Serpico JM, Einset AG, Ehrenberg SG, Raboin. (1995). Paper presented at 1st International Symp. on Proton Conducting Membrane Fuel Cell, Electrochem. Soc. Proc. 23, 247-251.

Wu Z, Tang Y, Sun D, Zhang S, Xu Y, Wei H, Gong C. (2017). Multi-sulphonated polyhedral oligosilsesquioxane (POSS) grafted poly(arylene ether sulphone)s for proton conductive membranes. *Polymer* 123: 21–29.

Xing B, Savadogo O. (1999). The effect of acid doping on the conductivity of polybenzimidazole (PBI). *J. New Mater. Electrochem. Syst.* 2(2): 95–101.

Xu H, Chen K, Guo X, Fang J, Yin J. (2007). Synthesis of hyperbranched polybenzimidazoles and their membrane formation. *J. Membr. Sci.* 288 (1–20): 255–260.

Yang C, Srinivasan S, Bocarsly AB, Tulyani S, Benziger JB. (2004). A comparison of physical properties and fuel cell performance of Nafion and zirconium phosphate/Nafion composite membranes. *J. Membr. Sci.* 237(1–2): 145–161.

Ye X, He Bai H, Ho W. (2006). Synthesis and characterization of new sulphonated polyimides as proton-exchange membranes for fuel cells. *J. Membr. Sci.* 279(1–2). 570–577.

Yen S-PS, Narayanan SR, Halpert G, Graham E, Yavrouian E. (1998) Polymer material for electrolytic membranes in fuel cells. US Patent No. 5795496.

Yu X, Roy A, Dunn S, Yang J, McGrath JE. (2007). Synthesis and characterization of sulphonated-fluorinated, hydrophilic-hydrophobic multiblock copolymers for proton exchange membranes. *Macromol Symp.* 245: 439–449.

Yue B, Zeng G, Zhang Y, Tao S, Zhang X, Yan L. (2017). Improved performance of acid-base composite of phosphonic acid functionalized polysulfone and triazolyl functionalized polysulfone for PEM fuel cells. *Solid State Ionics* 300: 10–17.

Zeng Y, Gu L, Zhang L, Cheng Z, Zhu X. (2017). Synthesis of highly proton-conductive poly(arylene ether sulfone) bearing perfluoroalkyl sulphonic acids via polymer post-modification. *Polymer* 123: 345–354.

Index

Note: Figures are indicated in *italic* font, tables in **bold**.

For Product Safety Concerns and Information please contact our EU
representative GPSR@taylorandfrancis.com
Taylor & Francis Verlag GmbH, Kaufingerstraße 24, 80331 München, Germany